W0080203

Springer Series in Optical Sciences Volume 8
Edited by David L. MacAdam

Springer Series in Optical Sciences

Edited by David L. MacAdam

Volume 1 **Solid-State Laser Engineering**
By W. Koechner

Volume 2 **Table of Laser Lines in Gases and Vapors** 2nd Edition
By R. Beck, W. Englisch, and K. Gürs

Volume 3 **Tunable Lasers and Applications**
Editors: A. Mooradian, T. Jaeger, and P. Stokseth

Volume 4 **Nonlinear Laser Spectroscopy**
By V. S. Letokhov and V. P. Chebotayev

Volume 5 **Optics and Lasers** An Engineering Physics Approach
By M. Young

Volume 6 **Photoelectron Statistics** With Applications to Spectroscopy and Optical Communication
By B. Saleh

Volume 7 **Laser Spectroscopy III**
Editors: J. L. Hall and J. L. Carlsten

Volume 8 **Frontiers of Visual Sciences**
Editors: S. J. Cool and E. L. Smith III

Volume 9 **High-Power Lasers and Applications**
Editors: K.-L. Kompa and H. Walther

Volume 10 **Detection of Optical and Infrared Radiation**
By R. H. Kingston

Volume 11 **Matrix Theory of Photoelasticity**
By P. S. Theocaris and E. E. Gdoutos

Volume 12 **The Monte Carlo Method in Atmospheric Optics**
By G. I. Marchuk, G. A. Mikhailov, M. A. Nazaraliev, R. A. Darbinjan, B. A. Kargin, and B. S. Elepov

Frontiers in Visual Science

Proceedings of the University of Houston College of Optometry Dedication Symposium, Houston, Texas, U.S.A., March, 1977.

Editors

Steven J. Cool and Earl L. Smith, III

With 533 Figures

Springer Science+Business Media, LLC

Dr. STEVEN J. COOL
College of Optometry
University of Houston
Houston, Texas 77004
USA

Dr. EARL L. SMITH, III
College of Optometry
University of Houston
Houston, Texas 77004
USA

Dr. DAVID L. MACADAM
68 Hammond Street
Rochester, N.Y. 14615
USA

Acknowledgements

The dedication symposium for the new College of Optometry building at the University of Houston, the proceedings of which are presented here, was made possible, in part, by financial support from the following, and we wish to acknowledge with grateful thanks their supportive efforts:

American Optical Co.	House of Vision
American Optometric Association	Omega Optical
Bausch & Lomb Optical Co., Soflens Division	Styl-Rite Optics

Library of Congress Cataloging in Publication Data

Main entry under title:

Frontiers in visual science.

(Springer series in optical sciences; 8)
Bibliography: p.
Includes index.
1. Vision disorders—Congresses. 2. Vision—Congresses. 3. Optometry—Congresses. I. Cool, Steven J. II. Smith, Earl L
RE91.F7 617.7 78-24191

Originally published by Springer-Verlag New York, Inc. in 1978
Softcover reprint of the hardcover 1st edition 1978

9 8 7 6 5 4 3 2 1

ISBN 978-3-662-15815-9 ISBN 978-3-540-35397-3 (eBook)
DOI 10.1007/978-3-540-35397-3

Dedication

This volume is dedicated to Chester H. Pheiffer, O.D., Ph.D., whose seventeen years of service as Dean of the University of Houston College of Optometry raised the College to the position of national and international prominence that it now enjoys. Without Dean Pheiffer's commitment and motivation, the new College building, whose dedication the symposium papers presented here celebrated, would never have become a reality.

Foreword

The papers included in this volume were presented as a part of the dedication of a new clinical/teaching/research facility for the University of Houston College of Optometry, March 27-31, 1977. These papers were intended to cover the "state of the art" knowledge in all areas of visual system investigation. While we may not have quite reached our goal of covering all areas, the papers presented here cover a broad cross-section of investigations in vision. However, without doubt, the intention of "state of the art" coverage was achieved in all areas discussed. From the beginning, with the presentation of Nobel Laureate, Ragnar Granit, to the end, with consideration of Vision Health Care Delivery Systems, each speaker was thorough in treatment of his/her subject. From studies of the cornea and of contact lenses, through examination of crystalline lens function, ocular pathologies and retinal function, the eye is very thoroughly considered. Much of this volume covers material dealing with the process of vision after coding of information in the eye. Psychophysical studies of vision compare and contrast with neurophysiological studies of visual function; and a very thorough section on the development of visual system function should prove valuable to a wide cross-section of teachers, researchers, and clinicians. All-in-all, the contents of this volume represent a vast array of knowledge about the visual system, and this should be a valuable teaching/research resource for many years.

The new College of Optometry facility, for which the contents of this volume were a dedicatory symposium, is the end result of the dedicated and commited efforts of many people. In addition to Dr. Chester H. Pheiffer, Dean of the College, one has to acknolwedge the tremendous efforts of the College's Building Committee Members: Drs. Troy Fannin, James Koetting, J. Floyd Williams and Nelson Reber. Their gargantuan efforts "above and beyond the call of duty" made the new facility what it is. And last, but by no means least, Dr. Donald G. Pitts, Associate Dean of the College and Chairman of the Building Committee, put more of himself into the design and construction of this facility than anyone could reasonably be expected to do. The countless nights and weekends that he spent working over plans and worrying over details can never be adequately compensated for. It is because of Dr. Pitts' sacrificial efforts, more than anything else, that the College facility is the magnificent clinical/research/teaching edifice that it is. One is compelled to say: "Thank you, Don. We don't really know how you did it; but we're damn glad you did!" Because of Dr. Pitts efforts and those of everyone else in the College, the new building became a reality and the dedication symposium, of which this is the proceedings volume, was possible.

Organizing a symposium such as this is an unbelievably involved and complex process. Without the help of many, many people, it could never have happened. The entire faculty and staff of the College were involved in many ways too numerous to enumerate here. Especially, however, we should like to thank: Douglas Miller, Gerry Smith and David Castano for their invaluable technical assistance; Felix Barker, Roger Boltz, Mel Kalich, Ruth Manny, Gregg Maguire, and Ralph Parkansky for their help with the closed-circuit T.V. coverage; Debbie Carlisle and Myrta Kennon for all of their efforts in handling uncountable organizational details and minutia; Debbie Carlisle, Myrta Kennon, Tijuana Rhodes, and Poette Wadley for their long hours of typing and proof-reading programs, papers, and, finally, the entire manuscript for this volume; and Ann Ewart, Assistant Controller of the University of Houston, who bent over backward to try to facilitate the financial aspects of implementation of the symposium. And to the many who worked on the symposium but whose names we have forgotten to mention, our apologies for omitting you by name and our thanks for your efforts.

Steven J. Cool
Earl L. Smith, III
August 1, 1978

Contents

I. Keynote Presentation

The Significance of Antidromic Potentiation and Induced Activity in the Retina

Ragnar Granit
The Nobel Institute for Neurophysiology, Karolinska Institutet
S-104 01 Stockholm, Sweden
and
The Kerckhoff Institute of the Max-Planck Gesellschaft
D-6350 Bad Nauheim, W. Germany

Our work on the role of the centrifugal gamma fibres to the muscle spindles in the late forties and early fifties made me consider centrifugal fibres to the mammalian eye. RAMON Y CAJAL (1) has reported the existence of such fibres in the retina of the dog and I did not think it very likely that he could have been mistaken. Others have since shown less faith in the Old Master, but in the laboratory of another highly competent histologist (POWELL) centrifugal fibres have again been found in the optic nerve of the cat. I quote: "These electron microscopic observations of the retina following lesions of the central visual pathway may be accepted as valid evidence for the presence of centrifugal fibres to the retina in the mammal" (BROOKE, DOWNER and POWELL (2)).

I mention this to explain why my stereotaxic attack on this problem made me in the first instance go for the superior colliculus in the cat, rather than for the lateral geniculate body or the optic tract itself (GRANIT (3)).

At the time it was held that merely slow fibres pass to the colliculus and so it came as a surprise that it proved very easy to activate the large ganglion cells by antidromic shocks to that region. When the pick-up electrode was shifted from the point of entry of the optic tract toward the periphery, the latent period increased from about 1.2 ms at the lamina cribrosa to between 4 and 6 ms further out. I concluded that, inasmuch as timing has informative relevance, the retina is admirably organized for translating surface coordinates into time coordinates, provided that the eye moves. These relations were then systematically explored by DODT (8). I was thinking of the familiar Pulfrich effect.

I then tried tetanizing antidromically for some 10 or 20 s and, to my great surprise, as soon as stimulation was stopped, the isolated ganglion cell started firing at a rate greatly exceeding its previous spontaneous activity. Orthodromic potentiation was well known at the time but nobody had yet succeeded in obtaining an antidromic potentiation. The prevailing notion was that the antidromic spike entering its axonal ganglion probably extended its depolarizing action into the dendrites but not any further. In motoneurons the most striking effect known today is that of DECIMA and GOLDBERG (5,6): if an antidromic ventral-root shock is suitably timed relative to a conditioning, adjacent dorsal root potential, the shock fires a dorsal root spike.

In my experiments all cells did not produce a post-tetanic potentiation but all those that did had to be driven by the antidromic shocks. This suggested that the post-tetanic discharge hardly could be a centrifugal phenomenon, even though a centrifugal contribution could not be excluded. Driving of the ganglion spike was a too obvious conditio sine qua non. Frequency of tetanization and its duration were decisive in determining the duration and firing rate of the post-tetanic discharge. This sometimes reached values as high as 200-300 per s maintained for minutes. As I remember, most of my preparations were BREMER"S encephale isole, some were on pentobarbitone anaesthesia. The facilitation sometimes lasted for a couple of minutes and active cells often began firing already during tetanization, rate of stimulation permitting. The effect could be obtained also from the geniculate body and the optic tract but from these structures it was commonly complicated by inhibitory phenomena.

An idea of the degree of post-tetanic facilitation could be gained by translating firing rates into light intensity. Thus, for instance, a test flash of 5 l.c. was raised in effectiveness in the post-tetanic state so as to correspond to one of 600 l.c. DODT (7), later experimenting with rabbits, found the flicker-fusion frequency of a given test light raised from 16.5 before to 35 flashes per second after antidromic tetanization.

Decisive for my conclusion that the antidromic spikes actually entered the retina were a number of experiments on interference between tetanization and light stimulation. Thus, for instance when the antidromic spike failed to enter the ganglion cell during an inhibitory phase of the light test, it could be made to do so by merely increasing stimulus strength of the shock thereby bringing in other collicular terminals.

Recent experiments by others have since pushed the study of antidromic potentiation one step further and so I shall not review my old work in greater detail. Sixteen years later the problem was taken up by FUKADA (8) who confirmed my findings and connected them with ENROTH-CUGELL and ROBSON'S (9) important subdivision of the retinal ganglions into X- and Y-cells. FUKADA showed that the potentiation only occurred in the transiently responding Y-cells which he called Type I and not in the tonic Type II or X-cells which to a stationary stimulus, focused on the centre of the receptive field, respond with sustained discharge. The significant point here is that the post-tetanic effect was confined to an identified cell, even though at that stage the X-Y identification was tentative.

The next step (FUKADA and SAITO (10)) was to demonstrate that a similar long lasting after-discharge followed a flickering stimulus to the Type I-cell receptive-field organization. FUKADA proposed the term "induced activity" which from now on I, too, intend to make use of. SAITO and FUKADA (11) confirmed the capacity of flickering stimulation to elicit induced activity and studied the responses of Type I and Type II cells to intermittent light. Even when they combined antidromic and flickering stimulation, only the Type I cell proved capable of generating induced activity.

These findings were confirmed by CLELAND and LEVICK (12) who found induced activity only with their transient class of cell, apparently FUKADA'S Type I. From latency studies of the ganglion spike, combining light and optic tract stimulation, they arrived at the conclusion that the induced discharge "is associated with the appearance of an active spike-generating focus located somewhere along the axis of that cell".

The optic tract loses its myelin sheath at the <u>lamina cribrosa</u> and so the axons in their intraretinal course may acquire the complex properties of dorsal root C fibres. These are known from GASSER'S (13,14) studies of their spikes and after-potentials. This analogy may or may not be valid, but if it be, then one would expect the effect of a long-lasting tetanus to emerge as GASSER'S (14) P_2 or second positive after-potential which is of very long duration. During this hyperolarization negative after-potentials of spikes are increased. It is not easily understood how a positive after-potential could be conducive to facilitation of the ganglion. If on the other hand after loss of its myelin sheath a fibre retains the original properties of A fibres, a tetanus would probably be followed by a brief hyperpolarization, P_1, rapidly changing into a depolarization, also of relatively brief duration compared with the final, long-lasting, positive P_2, the only event of a duration long enough to approach that of an induced discharge. However, the polarity of P_2 is of the wrong sign.

In the intraretinal optic tract fibres of the monkey OGDEN and MILLER (15) noted an "intense negative post-tetanic overshoot". There was little in the way of positive after-potential. While this transient effect may aid the ganglion cell in forwarding OGDEN'S P-wave into the internal plexiform layer--I shall come to it below--no correlation is thereby established with the long-lasting induced discharges, so far not at all studied antidromically in the monkey retina. The post-tetanic negativity of OGDEN and MILLER is too brief to explain the long-duration of the induced activity. These problems clearly require more experimentation, an attack with microelectrodes on the internal plexiform layer.

At the moment we had better hold onto the two most significant new observations in this field: (i) that the induced activity can be elicited both by flicker from the orthodromic end as well as by repetitive antidromic stimulation (ii) that both routes of activation presuppose a specific set of large ganglions, apparently those representing the final common path of the synaptic organization that are driving the Y-cells of ENROTH-CUGELL and ROBSON. An explanation based on a purely extraretinal axonic focus, not yet demonstrated, suffers from the weakness of not being able to acount for the orthodromic effect of flicker and for the restriction of induced activity to merely one type of the approximately 200,000 fibres counted by HUGHES and WÄSSLE (16) in the cat's optic tract.

Some of the steadily multiplying studies which now are devoted to cat ganglion cells seem to be of particular interest in the present connection. BOYCOTT and WÄSSLE (17) described three main types of ganglions: large alpha cells with dendritic networks spreading laterally up to 1000μm, smaller beta cells with a field diameter of 25-300μm, and still smaller gamma cells with a dendritic field between 180 and 300μm. The identification proposed was: Y-alpha, X-beta, and W-gamma. The identification was based on the size of the perikaryon in combination with that of the dendritic network. For all cells the latter expands in size towards the periphery. I shall only be concerned with the Y-alpha type. It is generally accepted that the larger the perikaryon, the greater also the axonal diameter and hence the conduction velocity. The ganglion cells responding with induced activity are found among the large ones that are supported by extensive dendritic networks.

This identification was fully supported by HOFFMANN (18) and by CLELAND, LEVICK and WÄSSLE (19) who added the further specification that the Y-cells are the brisk-transient units of CLELAND, DUBIN and LEVICK (20) and CLELAND and LEVICK (21).

According to a suggestion by OGDEN (22) the antidromic spikes in the optic tract may enter the inner plexiform layer by mediation of the tight junctions discovered by DOWLING and BOYCOTT (23), which for good reasons were held to be electrical in nature. SAITO and FUKADA similarly assumed these junctions to give access to the internal plexiform layer. The tight junctions are axosomatic ones between bipolar terminals and somata of ganglion cells. DOWLING and BOYCOTT did not find them in all cone portions of the primate retina and suggested that they were characteristic of rod bipolars.

When OGDEN endowed them with the role of gate openers to the internal plexiform layer, this was done in order to explain the positive P-wave that he and BROWN (24) had found in that layer in response to antidromic shocks. OGDEN did not find any P-waves in the cat retina. In similar work GOURAS (25) recorded a graded potential at the internal surface of the monkey retina. This potential became positive in the internal plexiform layer, had a shorter latency in the periphery, longer in the centre where it also was larger and more drawn-out. The views of these two authors on the nature of the P-wave differ, but a more serious difference from the present point of view is that tight junctions were not found in the centre where the positive wave of GOURAS had its maximum size.

As such it is of course a plausible notion that an antidromic spike--whatever it does afterwards--is gated into the internal plexiform layer by such apposition contacts. Again, however, we must conclude that there is room for more studies of the microphysiology of this region.

Returning to the question of why the Y-cells or, probably, only certain Y-cells generate induced activity, it stands to reason that a large dendritic network, better than a small one, by chance alone is bound to provide more targets for the output of those amacrines that are charged with the task of maintaining lateral spread of excitation. Additionally, it is pointed out by DOWLING and BOYCOTT (23) that "no ganglion cell dendritic spread is large enough to account for these effects (meaning the long-distance McIlwain effect); the only direct pathways in the retina for the peripheral effects are via the amacrine-amacrine synapses" (p. 107).

However, since the explanation of specificity in producing induced activity somehow implicates Y-cells, it is not permissible to neglect their curious property of non-linear summation within the receptive field. To ENROTH-CUGELL and ROBSON (9) this was an essential criterion in their definition of Y-cells. I call this property curious, because my own experience with motoneurons, in both intra- and extracellular work, is that, within a large range of firing rates highly complex reflexes add in a strictly linear fashion (summary, GRANIT, 26). Similarly ENROTH-CUGELL and ROBSON found linear summation in the X-cells.

I therefore suggest that the non-linear summation of the Y-cells is a consequence of positive feedback within the amacrine circuits that support their activity. This, at the same time, would explain their proneness to excessive activity, noted also by ENROTH-CUGELL and ROBSON (9) when they stated that "the mean discharge of the Y-cells (unlike that of X-cells) was greatly increased when grating patterns drifted across their receptive fields". The nowadays commonly used Y-cell criteria, high conduction velocity, transient response, and a more peripheral location, need not in every case tally with that of non-linearity. In present day usage of the X-

Y nomenclature the original Y definition is mostly neglected. If we had had experiments correlating induced activity with non-linearity of summation, we would now be better off in discussing the nature of Y-specificity.

Of relevance for this line of thinking are some data by DUBIN (27) dealing with the serial synapses of KIDD (20). DUBIN found them to be characteristic of amacrines and published a table showing among other correlations the percentages of amacrine synapses in serial configuration in different animals. These are some of his figures: human parafovea 1.9; monkey fovea 2.5, parafovea 7.1, periphery 5.1; cat (2 animals) 8.2 and 7.5 respectively; rabbit (2 animals) 10.0 and 15.2 respectively. With these challenging figures we again come up against questions of correlation which only can be answered by appropriately designed experiments.

One task of those Y-cells which respond as if they were actuated by positive feedback could be to facilitate the recording of movement in the peripheral visual field and conduct the message at maximum speed to the cortex. This notion presupposes that the X-Y differentiation be maintained up to the central visual stations. For the geniculate body this has been found to be the case, in the cat (HOFFMANN, STONE, and SHERMAN (20); FUKUDA and STONE (30) and in the monkey (DREHER, FUKUDA and RODIECK (31). In this animal the Y-cell projections are found in the magnocellular layer, the X-cells in the parvocellular layer.

In addition to serving as transient fast detectors of movement, the Y-cells also contain information on luminosity. In now forgotten papers and in reviews (GRANIT (32) being the latest) the evidence was summarized that long ago led me to the conclusion that the dominators also in the cat are composite curves carrying a message of luminosity and not one of color. It was shown that the same, large ganglion cell could serve as both scotopic and photopic dominator, this being true also for the retina of the cat. The destination of a message that has this character could hardly be a specifically color-sensitive mechanism in the cortex. I had, in fact, postulated that in all animals the dominator originated a luminance channel.

In now proceeding to discuss some results of primate physiology in terms of Y- and X-cells, I am fully aware of gaps in our knowledge that have to be bridged by hopotheses. I am defending myself with an enlightening quotation referable to PEYTON ROUS: "Yet since what one thinks determines what one does in cancer research, as in all else, it is as well to think something" (from obituary by DULBECCO (33)). And, to begin with, I think that the MCILWAIN effect (34) and its younger descendant, the "shift-effect" of FISCHER and KRÜGER (35), may well be exponents of the particular Y-cells that give induced discharges. It was pointed out by WERBLIN and COPENHAGEN (36) that the MCILWAIN effect is restricted to the Y-cells.

The relation between spectral sensitivity, conduction velocity, and phasic versus tonic properties has been studied in the monkey by GOURAS (37,25), later continued in work with De MONASTERIO and TOLHURST (38, 39, 40). Antidromic stimulation differentiated two main groups of fibres, large ones responding phasically with a conduction velocity of 3.8 m/sec and small tonically responding fibres conducting at 1.8 m/sec. The small ones were found everywhere but had their greatest density in the center. They had opponent color properties and thus the two opposing regions of the receptive field had different color sensitivities, e. g., one red, the other green. The large phasic cells represented the same spectral sensitivity in both center and periphery of the receptive field and so the antagonism between center and surround did not differentiate wavelength. In this lot would be found the Y-cells with dominator properties or, in other works, the fast luminosity instrument of vision. I shall come to some other papers, psychophysical or based on evoked potentials that separate luminance from color channels, but let me, to begin with, consider the flicker phenomenon.

My first question is so obvious that I do not think it has ever been raised in the present era of sophisticated search for detectors: why is it that heterochromatic photometry is possible by the flicker method? My answer is that this is a fairly selective response of fast Y-cells of the dominator type specializing on transients. There would be more of them in the periphery (ENROTH-CUGELL & ROBSON (9)), hence more facilitation by interaction in the peripheral retina (GRANIT (41)).

In 1929-30, when I was keen on proving that psychophysics could be translated into the kind of neurology that Sherrington's laboratory had pushed into the foreground of research, I used flicker fusion as an index of excitability. Comparisons were made between center and periphery at 10°, area and intensity of the stimulus being varied. For the photopic fusion frequency as a function of intensity, one had the approximation known as the Ferry-Porter rule

$$f = a \text{ Log } I + b.$$

It was known at the time that stimulus area played a role for the fusion frequency but until our work (GRANIT and HARPER (42)) there had been no systematic analysis of it. A similar relationship was found to hold for area,

$$f = c \text{ Log } A + d.$$

By combining these two rules into one, the equation may be formulated as

$$f = \alpha \text{ Log } I \text{ Log } A + \beta \text{ Log } I + \gamma \text{ Log } A + \delta$$

Tabulating the values of these constants for center and periphery, they came out as in

Table 1

Center	0.90	4.76	1.79	15.40
Periphery	1.68	4.87	4.28	14.03

These values show that the constants α and γ which enter the equation in terms containing Log Area are the ones that undergo a significant increase from center towards periphery. The potent peripheral spatial summation could also be demonstrated with stimuli separated by a portion of the illuminated background. For later contributions to this problem, see BROWN (43).

The assumption that in the peripheral retina there are more Y-cells, of the kind that interact by mutual facilitation implicates a cellular substrate whose existence in 1930 merely could be adumbrated. It was not at the time possible to think in terms of a cellular differentiation that today has become the goal of a steadily increasing number of publications dealing with retinal ganglion cells.

For the cat a study by SAITO and FUKADA (11) differentiates between flicker in Type I and Type II ganglion cells. In the Type I cells the number of spikes per flash increased toward a maximum and then fell off, as repetition rate of stimulation was increased. The TYPE II cells followed rate of stimulation over a wide range with low and constant average spike frequencies. As stated above, only the Type I cells were capable of generating induced activity.

The psychophysical study of flicker and flicker fusion is a highly formalized field, accessible to quantification from several points of view, e.g. waveform, stimulus intensity, adaptive changes etc. But today, when our interest is centered on cellular identifications, other properties of the perception of flicker should in the first instance attract our attention. One of them is the peculiarly unpleasant sensation of violent flicker that at a certain rate of intermittent stimulation below the fusion point is such a striking experience. If the Y-cells of man, like those of the cat, possess an optimum of spike frequency at a certain rate of intermittent stimulation (SAITO and FUKADA (11)), it may well be that self-excitation of their amacrine loops also is at an optimum at those same stimulus rates.

On the assumption that intermittent stimulation at certain rates is particularly prone to stir up self-excitation, it would, of course, be interesting to study the visual system immediately after some 10-20 seconds of flicker. From the work by myself, DODT, FUKADA and others, reviewed above, one would expect characteristic facilitatory after-effects to occur. This work, to be sure, was restricted to the cat but one would like to have psychophysical experiments in man to fill out the picture. A great deal more could also be done from the point of view of flicker with the cat retina and optic nerve.

When I in 1945 gave up experimental work on color reception I tried to collect what psychophysical evidence there was in favor of some measure of separation of color and luminance channels (GRANIT (44)) but today this task would be a great deal easier. It is no exaggeration to state that the electrophysiological evidence in favor of spectral information being carried by broad-band dominators and narror-band curves of the type I used to call modulators now has become so convincing that, if psychophysicists fail to find either or both of these channels, one would be entitled to put down their failure to inadequate methods.

KING-SMITH and CARDEN (45) set out to test the idea that visual detection can be based on either channel, depending on which one in a given situation has the lower threshold. They had a white background illumination of 1000 td on which was presented a low test flash, colored, or a white of the same intensity relative to threshold. When the test flash durations were 200 ms all stimuli except yellow were mediated by the chromatic system. But when time of exposure was cut down to 10 ms the chromatic peaks disappeared and what remained was a broad-band curve with maximum at 555 nm. Thus the opponent color system needed a longer integration time than the luminance mechanism. Flicker was found to give the same effect as shortening of time of exposure.

It is interesting to note in ZRENNER'S (46) experiments, in which evoked potentials and psychophysical measurements were compared, that against a white background of 30,000 td and a 10 ms exposure of the test stimulus the chromatic effect was strong in the evoked potentials but barely visible in the psychophysical sensory-threshold measurements. A much longer exposure time was needed for demonstration of color specificity by the psychophysical approach. In monkeys PADMOS and NORREN (27) using evoked potentials in otherwise very similar experiemnts found intermittent stimulation merely to trace the well-known heterocrhomatic flicker curve while single exposures gave the three spectral peaks studied in several papers by SPERLING and his co-workers (SPERLING et al. (48)). These are at 450, 530-540, and 610 nm with large dips at 480-490 and 570-590 nm. The technique of SPERLING et. al., was behavioral but also psychophysical inasmuch as it made use of trained monkeys rewarded for correct responses.

It is not my intention to discuss color mechanisms. The interest here is focused on the broad-band dominance in the spectral flicker curve by comparison with the prominent peaks and dips in the curves based on single stimuli. The peaks are too narrow and too far removed from the maxima of the three retinal photopigments to represent simple projections of the latter. The nature of the interactions involved has been analyzed by SPERLING and HARWERTH (49).

We need not fall back on psychological interpretations in making our comparison between the two curves. The results of PADMOS and NORREN (47) and those of ZRENNER (46) show that the difference between "flicker curves" and "color curves" also holds for the recording of evoked potentials from monkey and from man who on the evidence of SIDLEY and SPERLING (50) has the same receiving apparatus as the monkey. Rather interesting is the fact that the psychophysically determined color peaks and dips do not come out at short exposures while they do so in the records of evoked potentials. This is not the first experiment in which conscious awareness is shown to be time-consuming. LIBET (51), stimulating the somatosensory area in patients, found each repetitive shock to produce an evoked potential but mobilization of conscious awareness required maintained, iterative stimulation for about half a second.

The gist of my argument should now be clear enough; a luminance channel takes its retinal origin in Y-cells with a dominator distribution of spectral sensitivity. Intermittent stimulation, as employed also in heterochromatic photometry with fusion frequency as an index of brightness, favors this channel of information. The chromaticity channel is likely to be based on the more slowly conducting of X-cells but at the moment it is not possible to conclude that the two channels are wholly independent and incapable of interaction. The degree of their segregation and conditions for their interaction must be established by further experimentation. I have referred above to the work on these lines commenced by GOURAS and his colleagues.

The evidence in favor of self-excitation in Y-cells should not be construed to imply that all Y-cells necessarily have this capacity developed to the degree found in those which in the cat are capable of induced activity. As pointed out above, this antidromic potentiation has not yet been studied in primates. But I have drawn attention to the similarity of the induced effects by flicker and by antidromic stimulation (FUKADA) because it suggests means of approaching the related problems of luminance specificity, Y-cells, their self-excitation, distance effects of the McIlwain type and antidromic potentiation. Intermittent light may well be a good substitute for antidromic stimulation.

I understood from the invitation that I was supposed to speak about my own work in vision, a real challenge considering that it lies so far back. I did not see the need for a fresh summary of it, even though I often find it misinterpreted. Thus, inasmuch as my own work appears in this presentation of recent developments, it serves merely as a kind of accompaniment in the background for a set of ideas that have acquired their form in the last decade.

Summary

"The significance of antidromic potentiation and induced activity in the retina", being the title of this lecture, is held to be that the two effects really are identical and that both are the outward sign of the existence of a specific organization in the retina. For this reason they can serve as a valuable criterion for identifying activity in this organization.

This activity is assumed to be in the nature of a self-excitation by positive feedback in the amacrine circuits of certain Y-cells. Relevant literature has been reviewed.

These Y-cells whose spectral response curve is of the dominator type play a prominent role in stimulation by intermittent light and in the perception of luminosity.

A number of properties of intermittent stimulation are mentioned and held to motivate a renewal of the attention of visual experimenters of "flicker" and its after-effects.

References

1. S. Ramon y Cajal, Die Retina der Wirbeltiere, (Bergmann, Wiesbaden, 1894).
2. R. N. L. Brooke, J. de C. Downer and T. P. S. Powell, Nature (Lond) 207, 1365 (1965).
3. R. Granit, J. Neurophysiol. 18, 388 (1955).
4. E. Dodt, Experientia 12, 34 (1956).
5. E. E. Decima and L. J. Goldberg, J. Physiol. (Lond.) 207, 103 (1970).
6. E. E. Decima and L. J. Goldberg, Brain Res. 57, 1 (1973).
7. E. Dodt, Documenta Ophth. 18, 259 (1964).
8. Y. Fukada, Vis. Res. 11, 209 (1971).
9. C. Enroth-Cugell and J. G. Robson, J. Physiol. (Lond.) 187, 517 (1966).
10. Y. Fukada and H. Saito, Vis. Res. 11, 227 (1971).
11. H. Saito and Y. Fukada, Vis. Res. 13, 263 (1973).
12. B. G. Cleland and W. R. Levick, J. Physiol. (Lond.) 244, 60P (1978).
13. H. S. Gasser, J. Gen. Pyysiol. 33, 651 (1950).
14. H. S. Gasser, J. Gen. Physiol. 41, 613 (1958).
15. T. E. Ogden and R. F. Miller, Vis. Res. 6, 485 (1966).
16. A. Hughes and H. Wässle, J. Comp. Neurol. 169, 171 (1976).
17. B. B. Boycott and H. Wässle, J. Physiol. (Lond.) 240, 397 (1974).
18. K. P. Hoffmann, J. Neurophysiol. 36, 409 (1973).
19. B. G. Cleland, W. R. Levick and H. Wässle, J. Physiol. (Lond.) 248, 151 (1975).
20. B. G. Cleland, M. W. Dubin and W. R. Levick, J. Physiol. (Lond.) 217, 473 (1971).
21. B. G. Cleland and W. R. Levick, J. Physiol. (Lond.) 240, 421 (1974).
22. T. E. Ogden, in Structure and Function of Inhibitory Neuronal Mechanisms, ed. by C. von Euler, S. Skoglund and U. Söderberg, (Pergamon Press, Oxford, 1968), p. 89.
23. J. E. Dowling and B. B. Boycott, Proc. Roy. Soc. B. 166, 80 (1966).
24. T. E. Ogden and K. T. Brown, J. Neurophysiol. 27, 682 (1964).
25. P. Gouras, J. Physiol. (Lond.) 204, 407 (1969).
26. R. Granit, The Basis of Motor Control, (Academic Press, London, 1970).
27. M. W. Dubin, J. Comp. Neurol. 140, 479 (1970).
28. M. Kidd, J. Anat. Lond. 96, 179 (1962).
29. K. P. Hoffmann, J. Stone and S. M. Sherman, J. Neurophysiol. 35, 518 (1972).
30. Y. Fukada and J. Stone, J. Neurophysiol. 37, 749 (1974).
31. B. Dreher, Y. Fukada and R. W. Rodieck, J. Physiol. (Lond.) 258, 433 (1976).
32. R. Granit, in The Eye, vol. 2, ed. by H. Dawson, (Academic Press, New York, 1962), p. 537.
33. R. Dulbecco, Nat. Acad. Sci. Biographical Memoirs 48, 275 (1976).
34. J. T. McIlwain, J. Neurophysiol. 27, 1154 (1964).
35. B. Fischer and J. Krüger, Exp. Brain Res. 21, 225-227 (1974).
36. F. S. Werblin and D. R. Copenhagen, J. Gen. Physiol. 63, 88 (1974).

37. P. Gouras, J. Physiol. (Lond.) 199, 533 (1968).
38. F. M. De Monasterio and P. Gouras, J. Physiol. (Lond.) 251, 167 (1975).
39. F. M. De Monasterio, P. Gouras and D. J. Tolhurst, J. Physiol. (Lond.) 251, 197 (1975).
40. F. M. De Monasterio, P. Gouras and D. J. Tolhurst, Vis. Res. 16, 674 (1976).
41. R. Granit, Amer. J. Physiol. 94, 49 (1930).
42. R. Granit and P. Harper, Amer. J. Physiol. 95, 211 (1930).
43. J. L. Brown, in Vision and Visual Perception, ed. by C. H. Graham, (Wiley, New York, 1965), p. 251.
44. R. Granit, in Sensory Mechanisms of the Retina, (Oxford Univ. Press. 1947; Hafner, New York, 1963).
45. P. E. King-Smith and D. Carden, J. Opt. Soc. Amer. 66, 709 (1976).
46. E. Zrenner, Documenta Ophthl. Proc. Ser. 13, 21 (1977).
47. P. Padmos and D. V. Norren, Vis. Res. 15, 1103 (1975).
48. H. G. Sperling, N. A. Sidley, W. S. Dockens and C. L. Joliffe, J. Opt. Soc. Amer. 58, 263 (1968).
49. H. G. Sperling and R. S. Harwerth, Science 182, 180 (1971).
50. N. A. Sidley and H. G. Sperling, J. Opt. Soc. Am. 57, 816 (1967).
51. B. Libet, in In Brain and Conscious Experience, ed. by J. C. Eccles, (Springer-Verlag, New York, 1966), p. 165.

II. Ocular Physiology and Pathology

The Sensitivity of the Cornea in Normal Eyes

Michel Millodot
University of Wales Institute
of Science and Technology
Cardiff CF1 3NU, G.B.

Introduction

The human cornea is probably endowed with the greatest density of nerve fibres of any tissue in the body. For this reason it is assumed to be the most sensitive structure (1) a characteristic which is, of course, essential to elicit the palpebral reflex which shuts the eyelids and therefore protects the eye. The cornea is innervated by the long and short ciliary nerves which are branches of the ophthalmic division of the fifth cranial nerve. These nerves lose their myelin sheath as they enter the cornea from the limbus so as not to interfere with the transparency. Only free nerve endings and supposedly some Krause end bulbs have been observed histologically (2). In this paper we shall review some of the factors which influence corneal sensitivity in normal (that is, not pathological) eyes. This information is essential as the basis from which to differentiate what is normal from what is pathological.

Sensibilities of the Cornea

It seems to be accepted nowadays that the human cornea can feel touch as well as pain sensation. Evidence supporting this fact was provided by the experiment of GRANT, *et al.* (3) who cut the descending root fo the fifth cranial nerve in patients suffering *from* trigeminal neuralgia and established that although the pain threshold was abolished, touch sensation was still present. Indeed, most of the stimulations made with an aesthesiometer evoke a tactile sensation. For that reason this threshold was named Corneal Touch Threshold (CTT). The sensibilities of the cornea to heat and cold are still debated. NAFE and WAGONER (4) and KENSHALO (5) feel that the cornea could not distinguish between hot and cold whereas LELE and WEDDELL (6) support the opposite view. In any case these sensations are likely to be minimal and quite different than in other parts of the body due to the type of nerve cells present in the cornea.

Methods

Corneal touch threshold was first assessed by VON FREY (1) with horse hairs of different lengths attached with wax to the tips of glass rods. Since that time various instruments have been devised (see reviews in COCHET and BONNET (7) and MILLODOT (8)). However, at present the most appropriate clinical instrument is the Cochet-Bonnet aesthesiometer (which is based on the instrument devised by BOBERG-ANS (9)) provided various precautions are taken which are described by MILLODOT (10,11).

Apprehension

The assessment of corneal sensitivity creates a certain amount of apprehension on the part of the subject, especially when testing the central part of the cornea since the subject sees the instrument approaching his or her eye. In this latter instance the subject tends to blink a great deal more and objective determination based on the palpebral reflexes correlate less well in the centre than in the periphery of the cornea where the correlation is almost perfect (10). This information is very valuable when testing infants or animals.

The effect of apprehension created by seeing the aesthesiometer moving towards the centre of the cornea was investigated by BONNET and MILLODOT (12). They determined the CTT in the centre and near the limbus in two conditions; (a) in visible light, (b) in a completely dark room. In the latter condition the subject's eye was illuminated in infrared light and a converter was used so that the experimenter could perform the measurements. The results are shown in Fig. 1. CTT was significantly higher in the centre of the cornea in the dark than in visible light, but the same in the periphery of the cornea. The difference between the centre and the periphery is still significant proving that this can be attributed to some inherent biological differences. In visible light the patient sees the instrument faintly in his peripheral visual field but this is not arousing any apprehension relative to what occurs when making the central measurements.

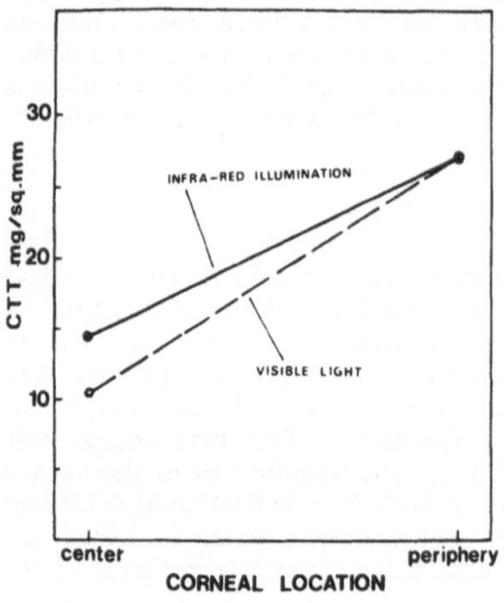

Fig. 1 CTT measured in the centre and the periphery of the cornea in two different conditions.

Scaling of Corneal Sensitivity

Further evidence of the types of sensations felt by the cornea was produced in an experiment in which judgements of corneal sensitivity were made as a function of increased pressure exerted on the cornea (13). Using the technique of magnitude estimation with a free modulus (14) corneal sensitivity was scaled for eight different pressures, with sixteen subjects. The results are shown in Fig. 2. It was found that corneal sensitivity varied as a power function of the pressure applied on the cornea according to the following equation

$$R = 6.17S^{1.01}$$

where R is the magnitude estimation, S the pressure, and the exponent is 1.01.

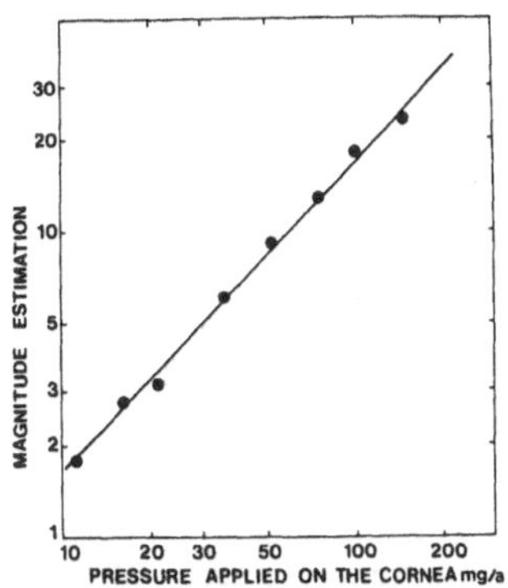

Fig. 2 Relationship between the pressure applied on the cornea the subject's estimation of magnitude (modified from MILLODOT (13)).

All sensations reported in this experiment were touch except, however, for the highest pressure (about 10 times above CTT) which evoked a sensation of pain in almost all subjects.

Corneal Eccentricity

The anatomical distribution of nerve fibers throughout the cornea shows a greater density of nerve fibers in the centre of the cornea than in the periphery (2). Hence one is led to assume that corneal sensitivity should be greater in the centre than in the periphery. This is indeed a common clinical observation and it has been reported by several authors (7,9,15). The results of these last authors are shown in Fig. 3 and are in very good accord with earlier reports. It can be seen from the data that corneal sensitivity (CTT^{-1}) is greatest in the centre and diminishes significantly towards the periphery. There is very little difference between the nasal and temporal sectors of the cornea. The upper and lower regions are less sensitive thanother parts of the cornea but the upper peripheral point is peculiarly less sensitive. Although the reason why this is so, is not yet confirmed, it is likely to be related to the diminution of oxygen of this corneal area due to the fact that it is usually covered by the upper eyelid.

Age

It has been noted that, in general, older people seem less sensitive than young people. On the other hand, ZOBEL (16) found that corneal sensitivity increased slowly up to the age of 45 to 50 and decreased sharply thereafter. Thus more evidence was needed

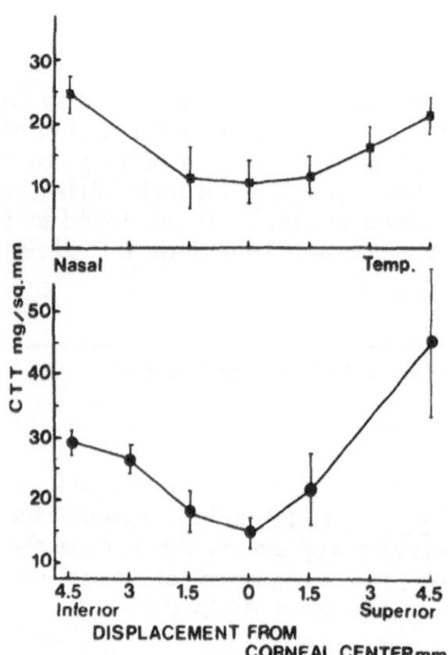

Fig. 3 CTT as a function of corneal eccentricity (modified from MILLODOT and LARSON (15)).

to elucidate this question and this was provided by MILLODOT (17). He measured corneal sensitivity in 205 people of different ages and his data are shown in Fig. 4. It is clear that corneal sensitivity remains practically the same from 7 to about 40 years of age and becomes significantly lower in the fifth decade of life. These data are in very good accord with those obtained by BOBERG-ANS (18). It is felt that ZOBEL'S results can be accounted for by the fact that he used VON FREY'S hairs which are rather inadequate for precise and reliable measurements. More data, though, in the first seven years of life are still needed which would be useful in advising about contact lenses in young children.

Eye Colour

It has sometimes been noted clinically, especially by contact lens practitioners, that

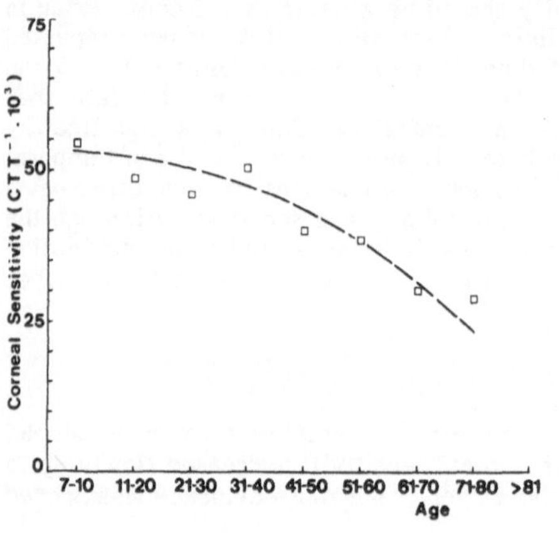

Fig. 4 Relationship between CTT and age (modified from MILLODOT (17)).

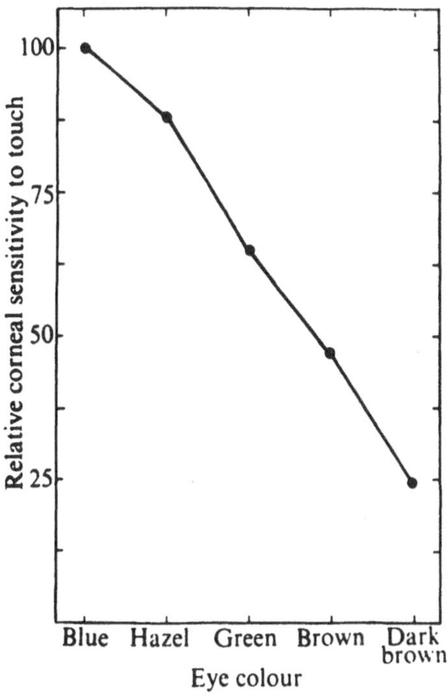

Fig. 5 Relative corneal sensitivity as a function of eye color. The sensitivity is the reciprocal of CTT and the group with the greatest sensitivity has been assigned 100% (from MILLODOT (19)).

people with dark brown eyes seem less sensitive than blue eyed people. In order to test this observation an experiment (19) was carried out to compare corneal sensitivity among people with different eye colours from blue to brown as well as a group of Chinese, Indians and Negroes with dark brown irises. The results of this investigation are shown in Fig. 5. It was found that brown eyed people, on average, had significantly less sensitive corneas than blue eyed people and people with dark brown eyes were less sensitive, on average, than any of the other groups. The reason for this difference is, as yet, unknown and it may be that it is not due to inherent differences in the eye since people with two different coloured irises (a condition called heterochromia) say one blue eye and one brown eye, have the same corneal sensitivity in both eyes (20) although this is not an unequivocal proof.

Physiological Variations

Physiological changes such as menstruation and pregnancy are known to produce an increase in the retention of water in the body (21) and as clinicians have noticed that women wearing contact lenses were apt to experience some discomfort in these circumstances, more knowledge about the sensitivity of the cornea was warranted.

In an initial study, MILLODOT and LAMONT (22) monitored corneal sensitivity in females over a 28 day period. They found that corneal sensitivity was reduced during the premenstruum and about the time of menstruation. It remained about the same throughout the cycle in females taking contraceptive pills. Recently MILLODOT (23) measured corneal sensitivity in pregnant women at different times during and after pregnancy. The results indicated that CTT becomes significantly higher after 31 weeks of pregnancy as is illustrated in Fig. 6. Both studies lead to the conclusion that generalised edema is accompanied by the diminution of corneal sensitivity.

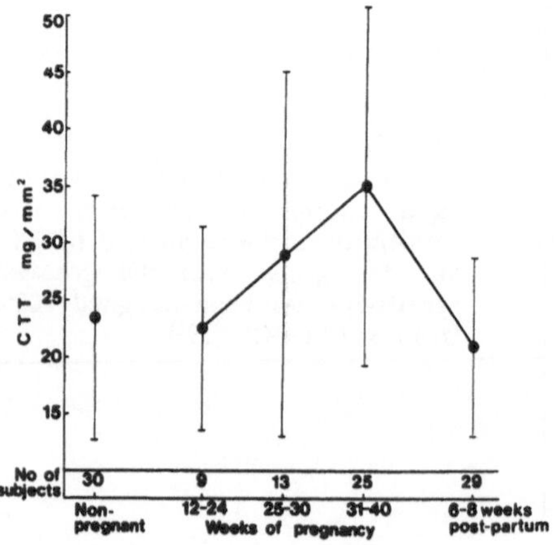

Fig. 6 CTT measured in a control group of non-pregnant women and in pregnant women during pregnancy and after delivery (from MILLODOT (23)).

Effect of Contact Lenses

Contact lenses represent a marked challenge to the integrity of the cornea. Thus it was logical to assess corneal sensitivity after wearing contact lenses. Various clinical observations had already pointed out the effect of contact lenses on corneal sensitivity (e.g., 7,9,24-27). However more systematic data were still needed. And even prior to these measurements was the need for a control study of the possible variation of corneal sensitivity throughout the day in people who do not wear contact lenses. It was indeed shown that corneal sensitivity does not remain constant but exhibits a diurnal variation, being less sensitive in the morning and more sensitive in the evening (28). This result is in good accord with the variations in corneal thickness throughout the day which is greatest in the morning and least in the evening (29).

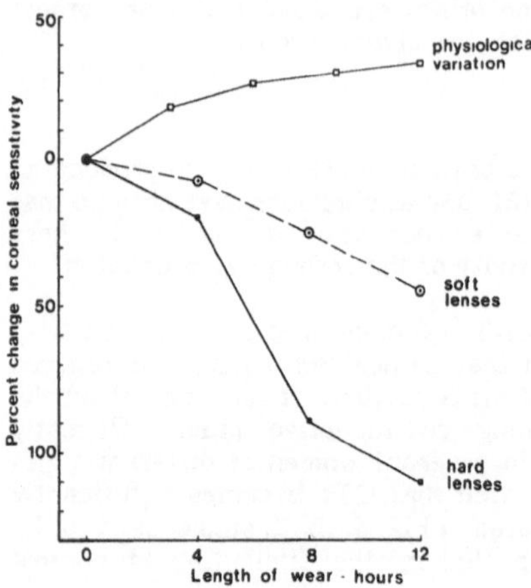

Fig. 7 Mean change in corneal sensitivity as a function of the length of wear of hard and soft contact lenses, compared to the diurnal variation (from MILLODOT (11)).

In a series of experiments (11,30,31) on the effect of hard and soft contact lenses it was demonstrated that all subjects displayed some loss of corneal sensitivity which increased as a function of the number of hours of wear. This effect was twice as large with hard lenses than with soft lenses (see Fig. 7). However, some subjects wearing very high water content soft lenses did not exhibit any loss of corneal sensitivity over the 12 hour period of the experiment. An important finding was the fact that after the lenses were removed, corneal sensitivity recovered almost (but not completely) within one hour. This information has important implications in patient care, since overwear of hard lenses especially, leads to a large reduction of corneal sensitivity with the possible risk of infections by the unaware patient.

Conclusion

The results described in this paper give rise to some comment. Corneal sensitivity appears to be related to localized corneal edema. Specifically there is a reduction of corneal sensitivity with corneal hydration as, for example, after the wear of contact lenses and after waking in the morning, but further research is needed on this relationship.

There is also a reduction of corneal sensitivity in cases of generalized edema such as during menstruation and pregnancy although corneal edema may be minimal in these conditions. Moreover corneal sensitivity may also be related to the melanin content of different people. Thus we see that corneal sensitivity varies with different conditions and the common denominator, if any, has yet to be elucidated. Nonetheless it is believed that monitoring of corneal sensitivity (or rather CTT) may be of great clinical value.

References

1. M. von Frey, Abdruck a.d. Berichten d. math. physik. Klasse d. kgl. saches. Gesellschaft d. Wissenchaften. Leipzig 46, 185 (1894).
2. E. Wolf, The anatomy of the eye and orbit, (H. K. Lewis, London, 1958).
3. F. C. Grant, R. A. Groff and F. H. Lewey, Arch. Neurol. and Psychiat. 43, 498 (1940).
4. J. P. Nafe and K. S. Wagoner, Amer. J. Psychol. 49, 631 (1937).
5. D. R. Kenshalo, J. Applied Physiol. 15, 987 (1960).
6. P. P. Lele and G. Wedell. Brain 79, 119 (1956).
7. P. Cochet and R. bonnet, Clin. Ophthalmol. 4, 1 (1960).
8. M. Millodot, Atti Fond. G. Ronchi 29, 889 (1974).
9. J. Boberg-Ans, Brit. J. Ophthal. 39, 705 (1955).
10. M. Millodot, Acta Ophthal. Kbh. 51, 325 (1973).
11. M. Millodot, Acta Ophthal. Kbh. 54, 721 (1976).
12. R. Bonnet and M. Millodot, Clin. Ophthalmol. 6, 74 (1965).
13. M. Millodot, Psychon. Sci. 12, 401 (1968).
14. S. S. Stevens, In Sensory Communication, edited by W. A. Rosenblith, (MIT. Press, Cambridge, 1961).
15. M. Millodot and W. L. Larson, Amer. J. Optom. 46, 261 (1969).
16. H. Zobel, Graefes Arch. Ophthalmol. 139, 668 (1938).
17. M. Millodot, Invest. Ophthalmol. 16, 240 (1977).
18. J. Boberg-Ans, Acta Ophthal. Kbh. 34, 149 (1956).
19. M. Millodot, Nature 255, 151 (1975).
20. M. Millodot, Invest. Ophthalmol. 15, 861 (1976).
21. F. E. Hytten and I. Leitch. The physiology of human pregnancy, (Blackwell Publications, Oxford, 1971).
22. M. Millodot and A. Lamont, Brit. J. Ophthal. 58, 752 (1974).
23. M. Millodot, Brit. J. Ophthal. 61, 646 (1977).

24. H. Hamano, Contacto 4, 41 (1960).
25. K. E. Schirmer, Brit. J. OPhthal. 47, 493 (1963).
26. J. M. Dixon, Amer. J. Ophthal. 58, 424 (1964).
27. J. Morganroth and L. Richman, J. Ped. Ophthal. 6, 203 (1966).
28. M. Millodot, Brit. J. Ophthal. 56, 844 (1972).
29. D. R. Gerstman, J. Micros. 96, 385 (1972).
30. M. Millodot, Acta Ophthal. Kbh. 52, 603 (1974).
31. M. Millodot, Acta Ophthal. Kbh. 53, 576 (1975).

Discussion

Q. Dr. Kerns: What is the mechanism behind reduced corneal sensitivity?

A. I do not know the answer. I assume it could be a difference in the concentration of the various ions Na^+, K^+, Cl^- and so on which affects the triggering of nervous impulses. Whenever you have a loss of corneal sensitivity you have edema and that would affect the ion concentration. That is one possibility. Ach is shown in large quantities across the cornea, that could also be a possibility.

Q. Dr. Kerns: Patients frequently report an itching just after they remove their contact lenses. Is that related to the recovery of corneal sensitivity?

A. I do not think so. I think recovery takes some time. This is why when you have a patient that is over wearing his or her lenses it takes several hours before he or she feels the irritation induced by over-wearing the lenses. The itching though, may be related to the eyelids or a change in the composition of tears.

Q. Dr. Kerns: What are the long range effects of contact lens wear on corneal sensitivity?

A. My work is not completed but I am not finding any significant difference for patients wearing hard contact lenses for one to two years. But there is a significant loss of corneal sensitivity after five to seven years of contact lens wear, and this appears to get worse with the years of wear. Recovery appears to take between one and four months.

The Corneal Environment: Osmotic Responses

Richard M. Hill
Jack E. Terry
The Eye Physiology Laboratory
College of Optometry
The Ohio State University

Perfect isotonicity (i.e., corresponding to a 0.90% NaCl solution) is now known to rank amongst the rarest of open eye tear conditions (1). Based on a series of 420 samples from six young, healthy subjects, an average value of 0.97% (310 mOs/Kg) was found for the waking hours. The most common range of scatter observed was from about 0.93 to just over 1.00% in the course of a day. In contrast, the average value associated with periods of prolonged lid closure was 0.89% (285 mOs/Kg), or just mildly hypotonic.

The marked hypertonic shift associated with the open eye condition above, might be taken as an indicator of the potency of evaporation as a tear influencing factor.

But there is, of course, at least one other factor which can produce an even more rapid and potentially as severe a shift - always in the hypotonic direction: the reflex response to irritation. The objective of this study was to osmotically quantify that response in relation to a very specific stimulus - the introduction of a rigid contact lens to the new wearer's eye.

Methods

To quantitatively explore the impact of reflex stimulations on the tonicity of the tears, each of the six subjects in the original study was fitted with rigid (polymethylmethacrylate) contact lenses using best fit principles. Each began with two hours of wear on the first day, with an additional wearing hour being added on each successive day. Adaptation was completed by all within 12 days without undue difficulty.

Each tear sample was passively collected from the lower cul de sac of the best managed eye by a capillary tube. Immediately on gathering the required tear volume of 5 microliters, the tube was sealed to prevent evaporation. Similarly, transfer of the sample into the measuring chamber was completed with less than two seconds of air exposure. Using a precision thermocouple hygrometer, the dew point depression value of each tear sample was determined and expressed in milliosmols/kilogram (mOs/Kg). Reliability of the method was such that 78 calibration readings made on a standard 290± 2 mOs/Kg reference solution in the course of the study averaged 289.9.

Results

Three references were used during each adaptation day to monitor the osmotic pressure responses of these subjects to their contact lenses: (1) just previous to lens insertion, (2) at the midpoint of the wearing period (always at 12 noon with the wearing day expanding symmetrically about that hour), and (3) just previous to lens removal.

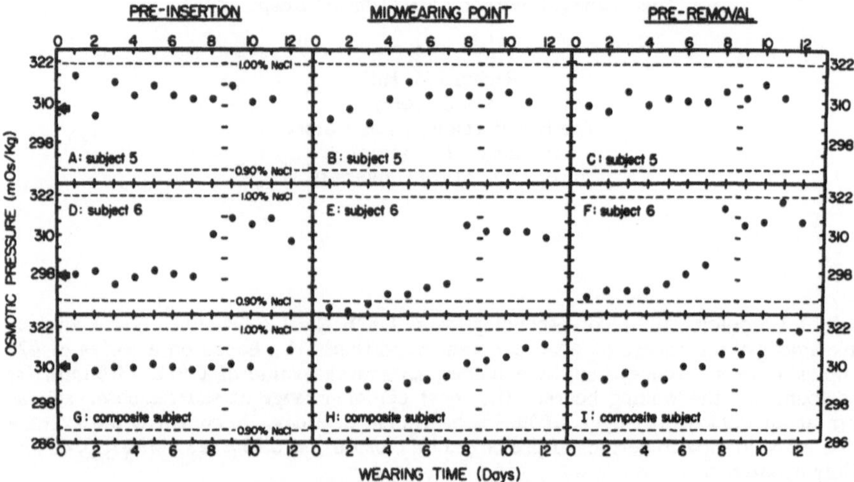

Fig. 1 Tear osmotic pressures for the most stable (Subject 5), the most disjunctive (Subject 6), and the "composite" (population averaged) case as adaptation to rigid contact lenses progressed. The arrow at the left ordinate scale of each subject sequence represents the average prefitted tear osmotic value for that case.

The daily progressions of tear osmotic pressure as adaptation proceeded are shown in Fig. 1 for two extreme individuals in the study: Subject 5 who showed the least change and Subject 6 who showed the most disjunctive and abrupt adjustment. For comparison, a "composite subject" based on the averaged data of the entire population is shown as well. the arrows at the left ordinate scales indicate the prefitting tonicity values for each case.

Discussion

In terms of their prefitted tear osmotic average of 310 mOs/Kg (0.97%), this population corresponded closely to other groups reported in the literature. For example, MASTMAN found that 70% of his subjects (unfitted) fell within the range of 298 to 315 mOs/Kg (equivalent to 0.93 to 0.98% NaCl solution) (2).

On receiving their rigid contact lenses, however, a variety of osmotic responses were observed amongst these subjects ranging from a very slight to a considerable hypotonic (relative to prefitted levels) response; but all, within 4 to 8 days, regained their original tonicity levels and then commonly exceeded them to establish new relatively hypertonic levels during their wearing periods. Of interest here too was the observation that the preinsertion samples (i.e., immediately previous to lens insertion each morning) in several of these cases, reflected a similar mild hypertonic shift. Although an increased inter-blink interval, permitting greater evaporation to occur might be an explanation in some cases, not all subjects showed a consistent change in blink frequency. Other mechanisms not yet identified may be in play as well.

As suggested by previous clinical investigators concerned with transient corneal edema (3, 4), there indeed appears to be a measurable hypotonic shift in the tear osmotic pressure of most new contact lens wearers. The basis for those events is presumably the low initial sensitivity threshold of the lid margins with associated reflex tears. As reported previously (5), however, lid thresholds can increase (i.e., sensitivity be reduced) by as much as 2 log units over the first two weeks of lens wear, having altered sufficiently within the first 4 to 8 days to correspond well with the general recovery of osmotic balance observed here.

Summary

The tear osmotic pressures of 6 subjects were monitored throughout their adaptations to rigid contact lenses. Relative to prefitting baselines established for each previous to this study, these subjects showed a range of hypotonic shifts on initially receiving their lenses. All such imbalances were rectified within 4 to 8 days of lens wear however, and a new relatively hypertonic level of tear osmotic pressure was then commonly established. Once adaptation was well advanced, this new level was frequently reflected in pre-insertion samples as well.

References

1. J. E. Terry and R. M. Hill, Arch. Ophthal. 96, 120 (1978).
2. G. J. Mastman, E. J. Baldes and J. W. Henderson, Arch. Ophthal. 65, 509 (1961).
3. S. Mishima, Survey of Ophthal. 13, 57 (1968).
4. R. B. Mandell, Arch. Ophthal. 83, 3 (1970).
5. G. E. Lowther and R. M. Hill, Am. J. Optom. 45, 587 (1968).

A Common Ocular Pathology Spheroidal Degeneration Of the Cornea and Conjunctiva

Calvin Hanna, Ph.D.
Department of Pharmacology
University of Arkansas for Medical Sciences
Little Rock, Arkansas 72201

The most common corneal pathology is the development of small, white to golden spheroids located in the stroma and composed of protein, RNA, and collagen (1). These bodies can appear spontaneously in the cornea and conjunctiva of the aged or around the site of corneal and conjunctival trauma (2). The spheroids of the aged have different corneal patterns, depending on the person's geographic location or insulting agent (3). Until an etiologic or chemical structure is known, we prefer the descriptive term of spheroid degeneration of the cornea and conjunctiva.

Clinical Characteristics and Course

The spheroids can vary greatly in appearance. They can resemble minute white droplets (Fig. 1) and were at one time thought to be trapped ointment or lipids (4) and as a result have been called degeratio corneae sphaerularis elaioides (5). What appears to be droplets at low magnification are spheroids of many shapes and sizes at higher magnification (1). This form of the degeneration which consists of isolated spheroids is quite difficult to detect with a slit lamp unless indirect illumination is used. As a result, the most common form of corneal pathology was thought to be rare until the 1970's (2). The degenerative condition of isolated spheroids in the aged progresses very slowly and is usually of no consequence to the patient. These minute bodies can also be found in the conjunctiva, and as with the cornea are located in the horizontal meridian near the limbus in normal eyes in the older age groups.

A variety of patterns and colors of spherules usually develop around sites of trauma at any age (4). These golden, yellow to white bodies may develop in various

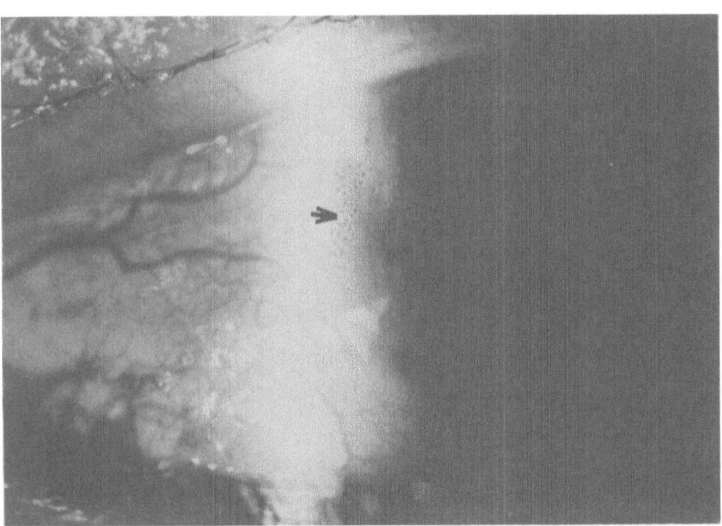

Fig. 1 Eye containing small spheroids (arrow) of the primary type of spheroid degeneration located in the peripheral cornea.

patterns including solid plaques, groups of or single spheroids depending on the type of trauma; these spheroids tend to increase in number. The ocular trauma may be associated with a penetrating ocular wound, herpes virus infection, or lattice dystrophy of the cornea. In addition, the trauma may result from ice crystals of the arctic (5) or blowing sand of the desert (7,8,9). The specialized type of trauma related to sand and ice produces band-shaped corneal spheroids arranged across the horizontal meridian and it occurs in the African desert (9) or in the eskimos of Labrador (6). There are at least twenty names which fit the description of spheroid degeneration of the cornea and conjunctiva (4,10). Many of these names relate to the geographic locations of the ocular pathology, for example, Labrador keratopathy (6) or possible underlying cause chronic actinic keratopathy (11).

We have divided the clinical picture into the three basic patterns of spheroid degeneration: (1) Corneal form - primary type, i.e., not associated with other ocular pathology; (2) Corneal form - secondary type, i.e., form only with associated corneal pathology or trauma; (3) Conjunctival form, more commonly found after the age of 60 years and frequently associated with pinguceula.

Pathology

Spheroidal degeneration has one common structural form regardless of whether it is from the cornea, conjunctiva, surround ocular trauma or part of a pinguecula. Under light microscopic examination the toludine blue, basic fuchsin and hematoxylin staining material can occur as a few minute particles or clump of large masses of material that may break through the epithelium of the cornea (11). The spheroids may appear to be rounded, may have elongated or angular shapes (Fig. 2). Under the slip lamp observation, these bodies appear to be droplets, but that general resembalance disappears at higher magnification. The intense staining of the

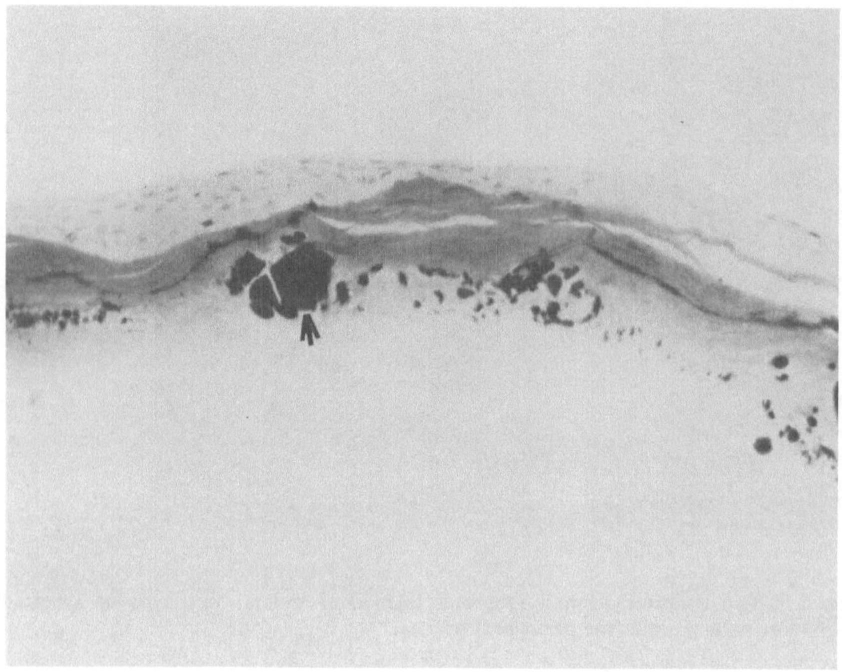

Fig. 2 Toluidine blue stained biopsy of peripheral cornea containing primary type of spheroid degeneration (arrow) at a magnification of 200 times.

spheroids with toludine blue, hematoxylin and methylene blue indicates a similarity to or are products of the cytoplasm such as nucleic acids. They also stain with basic fuchsin indicating a collagen component. The light staining with eosin, PAS alcian blue, congo red, Sudan black tend to eliminate the spheroids' being composed of mucopolysaccharides, amyloids and lipids. Electron microscopic evidence shows the spheroids to be finely granular in structure (Fig. 3) and appear to be collected on bands of collagen (Fig. 4). It is possible that the spheroids are an exudate of fibroblastic cells of the cornea and conjunctiva. With age, or after physical insult the fibroblastic cells may exude a cellular material rich in RNA which coats surrounding collagen fibrils. In time the minute accumulations of the exudate could coalesce into larger and larger concentrations that can be observed with a slit-lamp as spheroidal degeneration.

Brownstein, Rodrigues and Albert stained the spheroids with uranium and lead together with silver tetraphenyl porphrine sulfonate. (12) This stain is positive for elastin of the aorta and is absent when the aorta is pretreated with salivary elastase. Because the staining properties were weak for the spheroids and not altered by elastase as compared to elastin of the aorta they chose the term corneal elastosis

Fig. 3 Edge of spheroid (arrow) located in area of large dense body indicated by arrow in Fig. 2. Low contrast and high magnification of 100,000 times of the spheriod showing very fine granular nature of the dense spheroid.

and wrote on the "elastotic nature of hyaline corneal deposits" (12). Further the spheroids autofluoresce brilliantly under ultraviolet light illumination while most hyaline aortic elastin, the elastin of the pinguecula, autofluoresce weakly, if at all.

Differential Diagnosis

Ophthalmologists in the past have suggested that ocular ointment used immediatly postopertively may lead to ointment entrapment on healing. Entrapment of ointment in the stroma of the cornea does not occur unless the material is experimentally forced under pressure into the stroma (13,14). Once the ointment is entrapped into the cornea, the resulting droplets are not characteristic of spheroidal degenration. The droplets do not autofluoresce under ultraviolet illumination and they are easily identified without the use of a slit-lamp. The ointment cannot be trapped in the crevice of a wound for long unless a pressure dressing is used. Once the pressure dressing is removed the pseudoentrapment of ointment disappears (15). There do not appear to be other corneal conditions that resemble spheroidal

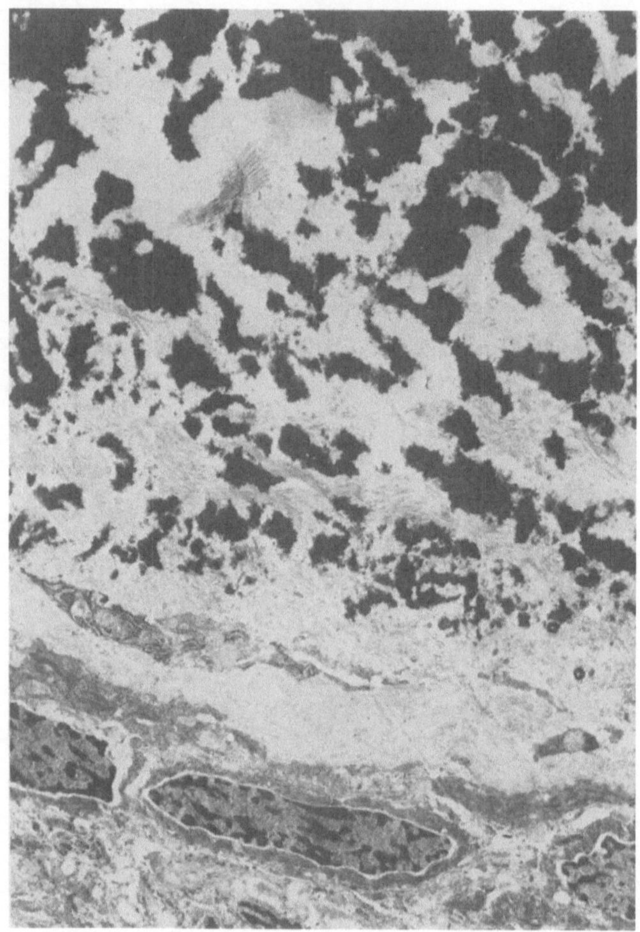

Fig. 4 What appears to be a single non-dense spheroid at low magnification turns out to be many minute dense spheroids located on collagen fibrils of the corneal stroma at 5,000 times magnification using uranium-lead staining.

degeneration of the cornea. Changes in the conjunctiva, such as cysts, lymphangiectasis, xanthomatous patches or the presumed forward migration of fat cells to the episcleral from the orbital fat pads are characteristically quite different from the spheroids of the conjunctiva (16).

References

1. C. Hanna and F. T. Fraunfelder, American J. Ophthalmology, 74, 829 (1972).
2. F. T. Fraunfelder, C. Hanna and J. M. Parker, American J. Ophthalmology, 74, 821 (1972).
3. R. S. Bartholomew, Documenta Ophthalmologica, 43, 325 (1977).
4. F. T. Fraufelder and C. Hanna, American J. Ophtahlmology, 76, 41 (1973).
5. L. Lugli, Arch. Ophthalmology, 134, 211 (1935).
6. A. Freedman, Arch. Ophthalmology, 74, 198 (1965).
7. G. Falcone, Riv. Ital. Trac., 6, 3 (1954).
8. G. B. Bietti, P. F. Guerra and P. F. Ferraris de Gaspare, Bull. Mem. Soc. Franc. Opht., 68, 101 (1955).
9. F. C. Rogers, British J. Ophthalmology, 57, 657 (1973).
10. J. Freedman, British J. Ophthalmology, 57, 688 (1973).
11. G. K. Kilintworth, American J. Pathology, 67, 327 (1972).
12. S. Brownstein, M. M. Rodrigues, B. S. Fine and E. N. Albert, American J. Ophthalmology, 75, 799 (1973).
13. C. Hanna, F. T. Fraufelder, M. Cable and R. E. Hardberger, American J. Ophthalmology, 76, 193 (1973).
14. F. T. Fraunfelder, C. Hanna, M. Cable and R. E. Hardberger, American J. Ophthalmology, 76, 475 (1973).
15. F. T. Fraunfelder, C. Hanna and A. H. Woods, Archives Ophthalmology, 93, 311 (1975).
16. F. T. Fraunfelder and C. Hanna, American J. Ophthalmology, 79, 262 (1975).

New Studies on Fluctuations of Accommodation

Yves Le Grand
Museum of Natural History
Paris, France

It is a curious thing to find a laboratory working on physiological optics in the Museum of Natural history! This Museum is about 350 years old and was founded under King Louis XIII for the cultivation of pharmacological plants, so that the poor people in Paris might cure themselves without paying money. Ten years after its foundation, there were already more than 3,000 different species of plants in this garden, which is still named the "Jardin des Plantes", which means the Botanical Garden. During the French Revolution, at the end at the 19th century, some animals came from the royal gardens of Versailles and the "Jardin des Plantes" began to be also a zoological garden. Since its foundation, this garden was also a place for scientific research: zoology, botany, geology, chemistry. Now 35 laboratories are around the "Jardin des Plantes"; all deal with Natural History, and constitute the Museum.

My laboratory is devoted to physics applied to biology, and was founded in 1836 for the first of the BECQUEREL, a genealogy of French physicists. This first BECQUEREL, ANTOINE-CESAR, was an officer in the Napoleonic army. He was very tall. During the war in Spain in 1810, he was terribly disturbed, because the spanish people suffered much: in the Prado Museum in Madrid you may see GOYA paintings showing the horrible struggle between the Spanish population and French soldiers. BECQUEREL left the army and began to study electricity, especially in animals and plants. He was followed by his son EDMOND who was only 6 ft tall and who studied phosphorescence. The grandson HENRI BECQUEREL, about 5 ft 6 in tall, who discovered radioactivity in 1896 by putting uranium salts on a photographic plate, and he shared in 1902 the first Nobel prize in Physics with PIERRE and MARIE CUIRE. HENRI BECQUEREL was a very clever man, for example he missed the Zeeman effect only because his magnet was insufficiently powerful. He died in 1908 and his son, JEAN BECQUEREL, was my teacher in the Museum; he was a specialist of optics at low temperatures, and worked in Holland in the Kamerlingh Onnes laboratory in Leiden, where liquid helium was first obtained. He retired in 1949, and as he was only 5 ft 2 in tall, the BECQUEREL genealogy stopped, happily for me who was the assistant of JEAN BECQUEREL and succeeded him. In the same way as the first BECQUEREL, I came back to physics applied to biology, especially to vision.

My Paris laboratory is rather small, there are only about 15 research workers, and not a large quantity of money, but we do our best. Four teams are working, the first devoted to color vision and animated by Dr. PARRA, a specialist of colorimetry and especially of the problem known as "MacAdam ellipses", which according to PARRA are not ellipses; another team deals with stereoscopic vision, with Miss BOURDY, who worked in U.S.A. some years ago with Profs. FRY and BLACKWELL; the third is devoted to corneal lenses (Dr. BONNET) and the last is studying visual fatigue, and it is about this subject that I will try to give you now some indications.

In this paper I shall speak about microfluctuations of accommodation in the human eye, and describe a new infrared optometer built in the Paris Institute of Optics by Dr. PIERRE DENIEUL, a young research fellow of my laboratory, with the aid of Dr. JACQUES SIMON.

To begin, I shall give a short historical retrospect about this curious phenomenon of the microfluctuations of accommodation. When a subject stares at a target at a given distance, his lens adjusts its form through the traction of zonula fibers by the ciliary muscle, so as to give a sharp image of the target on the retina. But this adjustment of the lens is not static: ARNULF et al. were the first to describe (1951) small fluctuations near the equilibrium position, these fluctuations amounting to a mean value of 0.1 diopter, in the frequency bandwidth 0-5 cycles per sec. Various techniques have been used for recording these fluctuations. For instance, ARNULF et al. used an ophthalmoscopic method to photograph the retinal image, the target being a radial grating illuminated by an electronic flash. They found that astigmatism is continuously varying, so it is necessary to use numerous photographs in order to obtain a statistical description. Recently in the Madrid Institute of Optics, ARNULF has performed successfully a 24 images per second kino film, with a laser as point source and a brightness electronic amplifier before the photographic camera which records the retinal image. The use of a laser gives a well-known artefact, the granularity of the image named "speckles", the fluctuations both in focusing and of astigmatism are nevertheless very apparent in this technique. In Cambridge, CAMPBELL (1954) used another method for continuous recording: he measured the defocusing produced by the fluctuations of accommodation, with a parallax optometer using two beams of light entering the eye by different points of the pupil. An inconvenience is the necessity of using a mydriatic in order that the pupil diameter should be sufficiently large, so the vision is somewhat artificial. In ARNULF's method, on the contrary, the pupil was natural. By recording simultaneously microfluctuations in the two eyes, CAMPBELL proved that oscillations are synchronous, so they are not due to a muscular noise, but to a servomechanism acting on both eyes from an upper nervous center: there is a feedback loop which involves the retinas, the cortical visual centers and the ciliary muscles, in order to ensure the sharpest retinal images.

The new apparatus built by PIERRE DENIEUL was designed so as to obtain the following performances: sensitivity of 0.01 diopter, range of ametropia ± 10 diopters, the entering beam must cover the whole natural pupil and the visual field must be sufficiently free to avoid instrument myopia and allow binocular vision of the fixation target illuminated in visible light, whereas the measuring beam was infrared so as not to interfere with vision (870 nm).

The theoretical principle of the apparatus is to image a black and white grating object (Fig. 1) on the retina and to analyze the ophthalmoscopic image, after a double passage through the eye, by a moving slit and a photoelectric receiver, which measures the modulation factor M, that is the ratio of maximal and minimal illumination in the image. The defocusing, $\Delta \pi$, due to fluctuations of accommodation induces a decrease, Δ M, in the modulation factor, and the measure of Δ M gives $\Delta \pi$. As already said, the measuring target is illuminated in infrared whereas the fixation target is in visible light, this fixation target is not seen on Fig. 1.

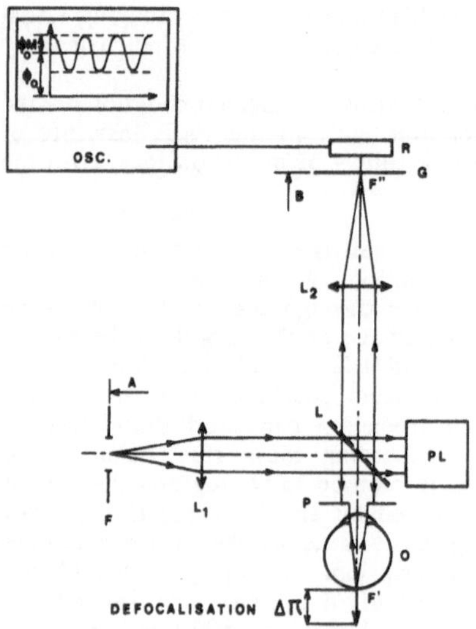

Fig. 1 Schematic diagram of infrared optometer. (See text for explanation).

Actually Dr. DENIEUL's apparatus uses a fixed source slit and an analyzing movable grating. The energy flux in the ophthalmoscopic image is measured by a photomultiplier sensitive to infrared, and an electronic circuit allows the simultaneous recording of the mean photometric level of the signal and the ratio of maximal and minimal responses, hence the Δ M modulation variation due to the fluctuations of accommodation. In order to deduce the $\Delta \pi$ defocusing value, it is necessary to record a calibration curve: the best sensitivity is obtained in the linear parts of the curve, where a small variation on defocusing induces a large variation of contrast. In order to place the functioning point on the best slope of the curve, the slit source is placed systematically out of focus, which means paradoxically that sensitivity is better when operating on out of focus images than on perfectly focused ones. Astigmatism can be tested by turning simultaneously the slit and the grating.

After a rather long period of adjustment, the optometer works now satisfactorily. Numerous difficulties have been met: blinking, pupillary movements, eye movements, eyelids, corneal reflection, and so on. A surprising result is the poor optical quality of the infrared image, the maximum of the modulation factor being 0.25-0.30 with normal pupil diameters. A partial explanation is found in the chromatic aberration of the eye: between visible light and 780 nm the hyperopia would be 0.8 diopter, in theory, and in fact Dr. DENIEUL finds 1.0-1.5 diopter: it seems that light is reflected in a plane in front of the retina, perhaps on retinal vessels as already suggested by CORNSWEET (1970).

Some preliminary results are shown in Fig. 2 for vision at far distance, that is accommodation relaxed, and a pupil of 5 mm diameter: the upper curve gives the

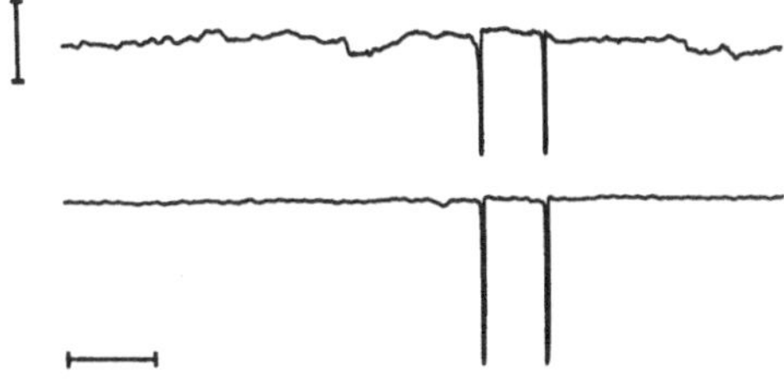

Fig. 2 Measurements of accommodative microfluctuations with accommodation relaxed. (See text for explanation).

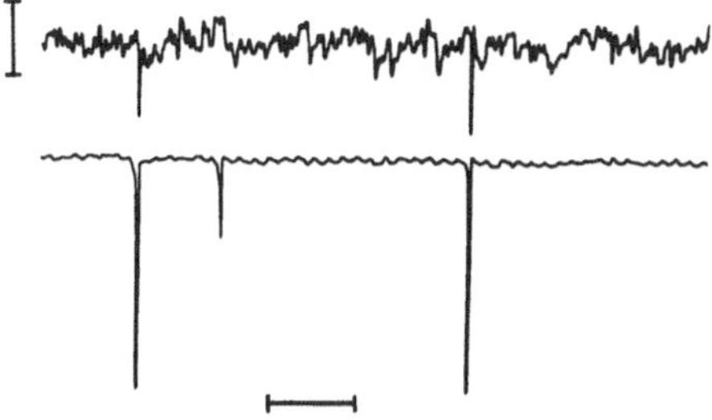

Fig. 3 Measurements of accommodative microfluctuations with a two diopter accommodative stimulus. (See text for explanation).

values, and the lower one the mean response; the scales given are 0.5 diopter in ordinates and 5 sec in abscissas; the large variations in both curves are due to blinking. Fig. 3 is relative to the same subject and same conditions, except that the accommodation is now of 2 diopters; fluctuations of about 0.2 diopters and mean frequency of about 2 cycles per sec are clearly seen.

Dr. DENIEUL's aim is now to use his optometer on a large statistical scale to measure the influence of various parameters on the microfluctuations of accommodation, and possibly to find an objective measure of visual fatigue through the amplitude of the microfluctuations when all other factors are kept constant. It will be a long and perhaps tedious work, but all those who have worked experimentally in the field of physiological optics are well aware of these difficulties.

Discussion

Q. Dr. Fry: I am interested in the correlation between the pupil fluctuation and the accommodative fluctuation. Particularly in reference to discomfort, that is to say, the discomfort may arise from the pupil, and you are getting an indirect effect.

A. Yes, Dr. DENIEUL will also now have a record of the pupillary variation. He uses a full aperture pupil. It is not an artificial one, but we will record simultaneously in his apparatus fluctuations of accommodation and fluctuations of the pupillary diameter. But it is not seen in this figure.

Q. Could I ask you a question generally? I know that several people are interested in this, and if anyone can throw any light on the present status of the correlation between the pupil and the ciliary muscle, I would like some information on it. I think Dr. STARK is geared up to do this sort of thing.

A. Dr. Stark: I do not think there is much correlation between the accommodative fluctuation and the pupillary fluctuation. Perhaps the gains between the two systems which are linked, of course, is small enough that the fluctuations do not show much correlation.

A. Dr. LeGrand: It seems that when the subject suffers more and more of visual fatigue, both tend to increase. He has fluctuations of the pupil diameter, and simultaneously increases in the fluctuation of accommodation, but I am sure this phenomenon has some link, but not a very direct one, I think.

Q. I had in mind the possibility that the accommodative mechanism may be the source of the fluctuation in the pupil.

A. I do not know what physiologists think about it, it is more complicated. There is a sure link, but it is from a physiological point of view, more complicated than the direct relation. You know that the pupil diameter is sensitive to an extraordinary great number of phenomena and when you shout very loudly you see the pupil diameter of your wife begin to make some fluctuations, although there is no difference in light.

The Etiopathogensis of Primary Glaucoma

Jess B. Eskridge, O.D., Ph.D.
School of Optometry
University of Alabama in Birmingham

Introduction

Glaucoma has been considered as an elevated intraocular pressure that is not compatible with health and function of the eye. Recent studies have added to our understanding and primary glaucoma is now considered as an ocular condition of decreased visual function, manifested by a visual field loss that is associated with an unphysiological intraocular pressure.

There are many differences that are present between normal and glaucomatous eyes, and there are several changes from the normal that occur in glaucomatous eyes. The purpose of this paper is to discuss some of the differences that exist between normal and glaucomatous eyes and to discuss some of the changes that occur in the anterior chamber and in the posterior segment of the eye in glaucoma in an effort to better understand the etiopathogensis of primary glaucoma and its significance in patient care.

Anterior Chamber Changes

It has long been realized that shallowness of the anterior chamber is associated with primary angle-closure glaucoma. The narrow angle that is associated with the shallow anterior chamber restricts access to the trabecular meshwork and increases the resistance to the outflow of the aqueous humor. BARKAN (1) fully described the relationship between the forward position of the anterior surface of the crystalline lens, the narrowing of the anterior chamber angle, and the process of primary angle-closure glaucoma. LOWE (2) compared the physical measurements of normal eyes to eyes with primary angle-closure glaucoma, and concluded that the shallow anterior chamber in primary angle-closure glaucoma was due to an anterior positioning of the crystalline lens and an increased lens thickness. TOMLINSON and LEIGHTON (3) compared the ocular dimensions of patients with primary angle-closure glaucoma to those of matched normal control subjects and found the primary angle-closure glaucoma patients to have smaller corneal heights, smaller corneal diameters, and thicker crystalline lenses positioned more anteriorly than the normal control subjects. They concluded that these differences were responsible for the decreased anterior chamber depth and the narrow angle in primary angle-closure glaucoma, and that the anterior chamber depth is a useful diagnostic sign for identifying patients with primary angle-closure glaucoma and patients who are angle-closure glaucoma suspects.

The depth of the anterior chamber has generally been considered to be of little diagnostic concern in primary open-angle glaucoma. If there is a difference between the anterior chamber depth in normal patients and primary open-angle glaucoma patients, perhaps it may be a contributing factor in primary open-angle glaucoma and therefore have etiological significance. Further, if it is a factor, the basic etiological differentiation between primary angle-closure glaucoma and primary open-angle glaucoma would be less distinct. FRANCOIS et al. (4) have shown that the anterior chamber depth is directly related to the outflow of aqueous humor from the eye in normal and pathologically affected eyes, i.e., the shallower the anterior chamber the greater the reduction in the facility of aqueous outflow. TORNQUIST and BRODEN (5) measured the anterior chamber depth in 145 primary open-angle glaucoma patients and compared the measurements to 398 normal subjects. They concluded that the anterior chamber depth was less in the open-angle glaucoma patients than it was in normals, and suggested that the shallow anterior chamber predisposes the patient to a decreased aqueous outflow. TOMLINSON and LEIGHTON (6) also compared the ocular dimensions of primary open-angle glaucoma patients with those of matched normal control subjects and found that the depth of the anterior chamber in primary open-angle glaucoma was significantly less than in the normal, and that the crystalline lens was significantly thicker in the primary open-angle glaucoma patients.

These results demonstrate that the anterior chamber depth in both primary angle-closure and open-angle glaucoma is reduced from that in the normal, that the lens thickness is increased in both primary angle-closure and open-angle glaucoma over that of the normal, and that the flow of aqueous humor out of the eye is probably reduced in open-angle as well as angle-closure glaucoma. Though the differences in angle size and chamber depth from the normal in primary open-angle glaucoma are only about half as great as those in primary angle-closure glaucoma, the differences are significant and should be considered in the etiology of primary open-angle glaucoma.

The aqueous humor leaves the anterior chamber of the eye by passing through the pores in the trabecular meshwork and then through the endothelial cell lining into Schlemm's canal. The aqueous is transferred through the endothelium into Schlemm's canal by endothelial vacuoles (7). JOHNSTONE and GRANT (8) determined that this process of aqueous outflow was pressure regulated in that the number and size of the endothelial vacuoles changed as the intraocular pressure was raised and lowered. They also found that when the intraocular pressure is reduced so that the pressure in Schlemm's canal is higher than that in the anterior chamber, the trabecular meshwork collapses and forms a blood-aqueous barrier so that blood cannot enter the anterior chamber.

BILL (9) feels that an active phagocytotic process occurs during the aqueous transfer in which the endothelial cells of the trabecular meshwork engulf and digest the debris in the aqueous outflow thereby providing a self-cleaning mechanism. If this phagocytotic process is decreased due to reduced function of the endothelial cells, the outflow resistance would increase resulting in a decreased aqueous outflow and an increased intraocular pressure. The efficiency and effectiveness of the process could also be reduced if the inflow of debris or undigestible particles was increased. He also pointed out that corticosteroids which increase outflow resistance also inhibit phagocytosis. This reduced phagocytosis could occur in both open-angle and angle-closure glaucoma.

TRIPATHI (7), with the use of an electron microscope, evaluated the trabecular meshwork and Schlemm's canal in enucleated eyes that had had primary open-angle or angle-closure glaucoma. He found a quantitative and qualitative depletion of the endothelial vacuoles in the trabecular wall of Schlemm's canal in the eyes that had had primary open-angle glaucoma as well as those that had had angle-closure glaucoma. This indicates that a resistance to aqueous outflow in the trabecular wall of Schlemm's canal could be present for both primary open-angle and angle-closure glaucoma. He postulated that the resistance could be due to a mechanical build-up of extracellular materials, or to enzymatic, nutritional, or immunological factors.

That such a resistance does occur in primary open-angle glaucoma has been shown by WOLLENSAK and MILDNER (10) who measured the pressure in Schlemm's canal and in the anterior chamber in normal patients, and patients with primary open-angle glaucomatous eyes. They found the pressure in the anterior chamber to be about 1.4 times the pressure in Schlemm's canal in the normal eyes but significantly greater than 1.4 in most of the primary open-angle glaucomatous eyes.

NESTEROV (11) also maintains that there is a decreased aqueous outflow in open-angle glaucoma, and attributes that decrease to a partial blockage of Schlemm's canal. He postulates that the increased intraocular pressure pushes the inner wall of Schlemm's canal toward the outer wall, narrowing the lumen of the canal, and decreasing the effective outflow of aqueous humor. Using this concept as a basis, CAIRNS (12) devised a surgical procedure for primary open-angle glaucoma that he called goniospasis. The procedure consists of placing a suture in the iris and then pulling it away from the anterior chamber angle. He performed this operation on several primary open-angle glaucoma patients and obtained a significant increase in aqueous humor outflow.

There are other studies that also indicate that changes do occur in the anterior chamber in primary open-angle glaucoma. SEGAWA (13) has shown that there is an increase in the extracellular materials containing acid mucopolysaccharide and glycoprotein in the endothelial lining of Schlemm's canal in primary open-angle glaucoma, and that the greater the amount, the greater the resistance to aqueous outflow. He also pointed out that the presence of corticosteroids allows an increase in the mucopolysaccharide protein complex deposits. Perhaps the increase in intraocular pressure in the presence of corticosteroids in primary open-angle glaucoma patients and glaucoma suspects is the result of an increase of this extracellular material in the endothelium due to a reduction in endothelial phagocytotic activity (BILL) which produces a decrease in the aqueous outflow. SEGAWA also pointed out that aqueous humor contains plasminogen, a chemical substance which decreases the formation of this extracellular material, and therefore probably assists in the regulation of aqueous outflow by preventing the build-up of the extracellular deposits in the aqueous outflow pathways.

Many studies have shown that primary open-angle glaucoma increases in frequency and severity with advancing age. COLEBRANDER (14) has indicated that collagen is destroyed in old age so that the trabeculae become limp and stick together resulting in an increase in resistance to aqueous outflow. This suggests that primary open-angle glaucoma could be due to a connective tissue disorder of the trabecular meshwork. WALTMAN and YARIAN (15) found that positive antinuclear antibody reactions occurred almost six times more often in patients with primary open-angle glaucoma than in the normal. Such reactions suggest the presence of a connective tissue disorder or an altered immune function. The study of BECKER et al. (16) showing the presence of plasma cells and gamma-globulin in the trabecular meshwork of eyes with primary open-angle glaucoma also suggests some trabecular tissue damage or an antigen-antibody reaction which may alter trabecular structure and function. Further research is needed to determine if glaucoma is associated with a connective tissue disorder or an altered immunological function.

One of the most significant clinical studies showing that physical changes in the anterior chamber do occur in primary open-angle glaucoma is a report of KIMURA and LEVENE (17). They performed gonioscopy on 110 patients with primary open-angle glaucoma and on an equal number of matched normal controls. They found three significant differences between the two groups. There were more iris processes at all levels of insertion, more trabecular pigment, and a more anterior insertion of the iris root in the primary open-angle glaucoma patients than in the normal controls.

These studies show that distinct cellular and morphological changes occur in the anterior chamber in open-angle as well as angle-closure glaucoma and suggest that the cause of increased intraocular pressure in primary open-angle glaucoma and primary angle-closure glaucoma is an increased resistance to aqueous humor outflow. This information would also suggest that the classification of glaucoma should be based on the angular subtense of the anterior chamber angle, which can be clinically assessed, rather than the resistance to aqueous humor outflow. Wide-angle glaucoma and narrow-angle glaucoma would therefore be more appropriate terms than open-angle and angle-closure glaucoma.

Posterior Segment Changes

Although the changes in the eye that produce an increase in the intraocular pressure occur in the anterior chamber, the changes in the eye that produce the visual impairment in glaucoma occur in the posterior segment of the eye in and around the anterior portion of the optic nerve. The changes in the posterior segment are due to the effects of an unphysiological intraocular pressure brought about by a frank increase in the intraocular pressure, to an imbalance between the intraocular pressure and the systemic blood pressure in the vessels supplying the anterior segment of the optic nerve, or both.

Though the vasogenic theory to account for the changes that occur in the posterior segment of the eye does not conveniently explain all the data, there is considerable evidence in its favor and general agreement of its involvement.

GALLOIS (18), as early as 1933, pointed out the relationship between a decrease in the systemic blood pressure and the progression of the visual field loss in glaucoma. LAUBER (19) observed that the vascular tension in the retina of tabetic patients with atrophy of the optic nerve was low, and indicated that when the difference between the intraocular pressure and the arterial pressure was diminished, there would be a decrease in the capillary circulation and a loss of oxygen to the retinal tissue resulting in diminution of function and visual field changes. He pointed out that the same mechanism applies when there is an increase in the intraocular pressure in patients with glaucoma, and indicated that patients with high systemic blood pressure can have higher intraocular pressures without developing glaucoma. HARRINGTON (20) focused attention on the clinical relationship of blood pressure and visual field loss by presenting clinical evidence that arterial insufficiency and decreased blood flow in the anterior portion of the optic nerve are the main factors in the production of visual field loss in glaucoma. He demonstrated with clinical patient records that visual field losses occurred with intraocular pressures that were formerly compatible with health and function when the systemic blood pressure was lowered by medical treatment for systemic hypertension. DRANCE et al. (21) also studied this relationship by comparing low tension glaucoma patients with ocular hypertensive patients matched for age and sex. They found a significantly higher rate of low systemic blood pressure in the low tension glaucoma patients than in the ocular hypertensive controls. HITCHINGS and SPAETH (22) studied the blood supply to the optic nerve head using fluorescein angiography in low-tension and high-tension wide-angle glaucoma patients matched for visual field loss. They found no difference in the circulation times between the two groups, suggesting that a decreased vascular supply to the optic nerve head can result from an unphysiologically high intraocular pressure or to an unphysiologically low blood pressure in the vessels supplying the optic nerve head.

HAYREH (23), in a summarizing paper, maintains that the primary blood supply to the anterior portion of the optic nerve comes from the posterior ciliary arteries by way of the peripapillary choroidal circulation, and that the central retinal artery supplies only a thin superficial layer of tissue on the optic nerve head, and possibly some tissue in the retrolaminar region. The peripapillary choroidal circulation is more vulnerable to extracellular pressure than the retinal circulation, so that with a given rise in intraocular pressure or a fall in perfusion pressure in the posterior ciliary arteries, blood flow decreases and a vascular insufficiency develops in the anterior portion of the optic nerve. If this insufficiency continues over a given period of time, an ischemic condition results with corresponding optic disc cupping and visual field loss. As the cupping of the optic disc increases, the lamina cribrosa which is linear in normal eyes, begins to bow backwards and take on a spherical concave shape. With this shape it is possible that the optic nerve fibers passing through the fenestrated connective tissue sheets of the lamina cribrosa would be subjected to a "shearing" pressure resulting in additional decreased function and visual field loss.

BLUMENTHAL et al. (24) evaluated the blood flow in the retina and the peripapillary choroidal area in normal patients with the use of fluorescein angiography at normal intraocular pressures and at elevated intraocular pressures induced with a suction ophthalmodynamometer. They found that the vessels supplying the retina would function at higher intraocular pressures than those supplying the peripapillary choroidal area. TSUKAHARA et al. (25) measured the flow of blood in the peripapillary choroidal vessels and the prelaminar disc capillaries in glaucomatous eyes with spontaneously elevated intraocular pressures and after the intraocular pressures had been reduced. They found that the filling of the peripapillary choroidal vessels and the prelaminar disc capillaries was retarded when the intraocular pressure was elevated. They concluded that the prelaminar capillaries originate from the peripapillary choroidal vessels because the filling of the prelaminar capillaries and the choroidal vessels was nearly simultaneous, and that the blood flow in these capillaries and choroidal vessels is decreased at the elevated intraocular pressure levels present in glaucomatous eyes.

Another study by KORNZWEIG et al. (26) indicates the relationship of the peripapillary choroidal circulation and glaucoma. They evaluated fifteen postmortem eyes of patients with a known long standing history of glaucoma, and found a selective atrophy of the radial peripapillary capillaries and that these capillaries had a unique anatomy. They concluded that because of their unique anatomy (length, position, and infrequent anastomosing) these capillaries are more vulnerable to damage from an increase in the intraocular pressure.

ARMALY and ARAKI (27) have shown that rapid, short-lasting elevations in intraocular pressure that would occur in angle-closure glaucoma also produce a reduction in the blood-flow rate in the optic nerve of the rhesus monkey. The reduction in the blood-flow rate was also greater when the systemic arterial pressure was reduced. They postulated that the effect was due to interference in the blood-flow in the branches of the short ciliary arteries that go through the sclera then back to the optic nerve.

ANDERSON (28) has proposed an additional mechanism to account for some of the early changes that occur in the posterior segment of the eye. His studies (29,30) have shown that the glial tissue framework supporting the nerve fibers occupies almost the entire surface of the optic nerve head and about half of the volume of the optic papilla. The capillaries feeding the optic nerve fibers are located in and supported by this glial tissue. ANDERSON proposes that the glial tissue is sensitive to pressure, and that an increase in intraocular pressure results in damage and deterioration of the glial tissue from the mechanical stress. The loss of glial tissue support of the capillaries in a given area exposes the capillaries in that specific area to an unphysiological extravascular pressure, resulting in a localized closure of these capillaries and the loss of blood supply to the adjacent nerve fibers. ANDERSON feels that the visual field loss results from a vascular insufficiency following the astroglial tissue breakdown, but that the cupping of the optic disc results from the glial tissue deterioration first and then by degeneration of the nerve fiber tissue.

This concept has several supporting features. Due to the localized closure of capillaries, the small variably located paracentral scotomas that occur in glaucoma can be easily accounted for. This theory can also account for the splinter hemorrhages that occur on the optic disc in glaucoma. BEGG, DRANCE, and SWEENEY (31), have described these hemorrhages as a sign of acute ischemic optic neuropathy in primary open-angle glaucoma. Further, since some of the changes in the glial tissue would occur before capillary closure, this could account for the changes in the cupping of the optic disc preceding the losses in the visual field. The loss of support and protection of the capillaries by the glial tissue might also account for the continuing optic disc cupping and loss of visual field by an intraocular pressure level that was once compatible with health and function.

There is another change in the posterior segment associated with an elevated intraocular pressure that could also be part of the process of visual impairment. In the normal state, there is a continuous rapid and slow flow of protein from the retinal ganglion cells through the optic nerve to the lateral geniculate nucleus. LEVY (32) has shown that even with minor elevations in the intraocular pressure there is a reduction in the slow axonal flow of protein in the optic nerve. ANDERSON and HENDRICKSON (33) found a reduction in the rapid transport of protein in the optic nerve with moderately elevated intraocular pressures. Some of the blockage occurred in the retina, but most of it occurred in the region of the lamina cribrosa. It is possible that if this blockage is present for a given period of time that axon damage could occur resulting in a visual field loss. It has not been determined whether the reduction in the flow of protein in the optic nerve fibers is due to pressure changes along, vascular insufficiency, or both.

One final area of concern in the pathogenesis of primary glaucoma is an explanation for the variation in susceptibility of eyes to the effects of pressure and further to account for the individual variations in types of visual field loss. ANDERSON (28) has suggested that the supporting glial tissue might be more resistant to pressure in some patients than others with the consequence that those patients could withstand higher intraocular pressure levels without visual impairment. The variation is areal susceptibility of glial tissue in a particular disc might also account for the variation in the location and type of visual field loss.

HAYREH (23) states that the individual visual field variations are due to the nature of the blood supply to the anterior portion of the optic nerve. He points out that the blood supply to the retrolaminar portion of the optic nerve can come from two sources, the central retinal artery and the posterior ciliary arteries. In some cases more of the blood supply comes via the central retinal artery, in other cases nearly all of it comes from the posterior ciliary arteries, and in some eyes there will be a combination. In those eyes where the major blood supply to the retrolaminar tissue comes from the central retinal artery, a greater resistance to the effects of intraocular pressure will be present, and the eye will be less susceptible to visual field losses since the retinal circulation is not as easily affected by increased intraocular pressure as the choroidal circulation. In those eyes where the major blood supply to the retrolaminar tissue comes from the posterior ciliary arteries, a reduced resistance to the effects of intraocular pressure will be present, and the eye will be more susceptible to visual field losses.

FLYNN (34) proposed that the eyes which are more susceptible to the effects of increased intraocular pressure are those in which part of the optic nerve is supplied by a blood vessel that has an anastomotic channel to a vascular bed outside of the influence of the intraocular pressure. When the intraocular pressure increases in such eyes, blood is "shunted" away from this part of the optic disc through the anastomotic channel resulting in a vascular insufficiency in that part of the optic disc and the corresponding visual impairment. Accordingly, those areas of the optic nerve that do not have the anastomosing channels would be more resistant to the effects of intraocular pressure variations. Some studies of the vasculature of the anterior portion of the optic nerve give significance to this theory. LIEBERMAN et al. (35) used a histologic approach to study the blood vessel supply in the anterior portion of the optic nerve. They question the dominant role of the peripapillary choroidal circulation in the vascular supply to the anterior area of the optic nerve. They maintain that the retrolaminar area is profusely supplied with vessels from the pial sheath around the optic nerve and small branches from the central retinal artery, that the lamina cribrosa receives its supply from short posterior cilliary arteries without passing through the peripapillary choroidal circulation, and that the prelaminar area receives its supply directly from smaller short posterior ciliary branches and branches from the choroidal circulation. They do not deny some choroidal circulation at all areas, but maintain that the major vascular supply does not come via the peripapillary choroidal circulation. They also reported significant anastomosing among the vessels and the formation of longitudinal vessels that extend from the retrolaminar area to the optic nerve head. FRANCOIS and NEETENS (36) previously discussed this axial vascularization of the optic nerve, and pointed out that the vessels in the pial sheath around the optic nerve come from the ciliary and muscular arteries. This rather diverse and anastomotic circulation system does lend great support to FLYNN'S proposal.

Recapitulation

It appears that primary glaucoma, a specific type of decreased visual function, results from an unphysiological intraocular pressure. The unphysiological intraocular pressure results from a decreased aqueous outflow or a decrease in systemic blood pressure in the vessels supplying the optic nerve head and the anterior portion of the optic nerve, or a combination of these two. The decreased aqueous outflow can be due to interference from the iris, from iris processes, from pigment or other debris in the trabecular meshwork, from blockage of Schlemm's canal, from increased pressure in the aqueous veins, or from trabecular tissue changes or degeneration. It has been customary to refer to glaucoma as angle-closure if the iris is physically impeding the outflow of aqueous humor, and as open-angle if the anterior chamber angle is relatively large and the root of the iris is physically away from the trabecular meshwork. Though there are probably patients whose decreased aqueous outflow is due to a singular cause, that is, iris interference, or iris process interference, or pigment interference, etc., the information discussed above would suggest that most causes of decreased aqueous outflow have a multifactor etiological basis. It also points out that the anterior chamber depth is reduced in open-angle glaucoma from that in the normal, that there are anatomical differences in the anterior chamber in some open-angle glaucoma patients, that there is an increase in the extracellular material in the anterior chamber angle in open-angle glaucoma, and that a decrease in the aqueous outflow can occur in open-angle as well as angle-closure glaucoma. This indicates that a thorough evaluation of the anterior chamber and angle is necessary for all patients suspected of having glaucoma regardless of type.

The pathogenesis of the optic disc cupping may be due to glial tissue deterioration caused by mechanical stress from an abnormally elevated intraocular pressure or to an imbalance between the extravascular and intravascular pressure in the vessels supplying the anterior segment of the optic nerve. The pressure imbalance may be due to an increase in the intraocular pressure, or to a decrease in the blood pressure in the involved vessels, or to a combination of these two, and results in tissue ischemia and degeneration. The pathogenesis of the visual field loss appears to be due to a vascular insufficiency resulting from a pressure imbalance.

These pathological changes are not simply a function of the angular subtense of the anterior chamber angle, but are evidence of a chronic unphysiological intraocular pressure and can and do occur in patients with narrow or wide anterior chamber angles. DOUGLAS et al. (37) and HORIE et al. (38) have also shown that the visual field and optic nerve head changes in primary chronic angle-closure glaucoma are similar to those that occur in primary open-angle glaucoma. This information strongly suggests that although there are specific differences, the basic etiopathogenesis of primary angle-closure glaucoma and primary open-angle glaucoma is similar.

References

1. O. Barkan, Am. J. of Ophth. 34, 332 (1954).
2. R. Lowe, Br. J. of Ophth. 54, 161 (1970).
3. H. Tomlinson and D. A. Leighton, Br. J. of Ophth. 57, 475 (1973).
4. J. Francois, M. Rabaey, A. Neetens and L. Evans, Arch. of Oph. 59, 683 (1958).
5. R. Tornquist and G. Broden, Acta. Ophth. 36, 309 (1958).
6. A. Tomlinson and D. A. Leighton, Br. J. of Ophth. 58, 68 (1974).
7. R. C. Tripathi, Br. J. of Ophth. 56, 157 (1972).
8. M. A. Johnstone and W. M. Grant, Am. J. Ophth. 75, 365 (1973).
9. A. Bill, Invest. Ophth. 14, 1 (1975).
10. J. Wollensak and I. Mildner, Tran. Am. Acad. Ophth. and Otol. 79, 340 (1975).
11. A. P. Nesterov, Am. J. Ophth. 70, 691 (1970).
12. J. E. Cairns, Ann. of Ophth. 8, 1417 (1976).
13. K. Segawa, Jap. J. Ophth. 19, 317 (1975).
14. M. D. Colenbrander, Ophthalmologica 162, 276 (1971).
15. S. R. Waltman and D. Yarian, Invest. Ophth. 13, 695 (1974).
16. B. Becker, E. V. Keats, and S. L. Coleman, Arch. of Ophth. 68, 643 (1962).
17. R. Kimura and R. Z. Levene, Am. J. Ophth. 80, 56 (1975).
18. J. Gallois, Bull. Soc. D'Ophth. 45, 110 (1933).
19. H. Lauber, Arch of Ophth. 16, 555 (1936).
20. D. O. Harrington, Am. J. of Ophth. 47, Part II, 177, (1959).
21. S. M. Drance, V. P. Sweeney, R. W. Morgan and F. Feldman, Canad. J. Ophth. 9, 399 (1974).
22. R. A. Hitchings and G. L. Spaeth, Br. J. of Ophth. 61, 126 (1977).

23. S. S. Hareh, Br. J. of Ophth. 58, 863 (1974).
24. M. Blumenthal, M. Best, M. A. Galin and K. A. Gitter, Am. J. Ophth. 71, 819 (1971).
25. S. Tsukahara, S. Nagataki, M. Sugaya, S. Yoshida, and Y. Komura, Jap. J. Ophth. 19, 386 (1975).
26. A. L. Kornzweig, I. Eliasoph, and M. Feldstein, Arch. Ophth. 80, 696 (1968).
27. M. F. Armaly and M. Araki, Invest. Ophth. 14, 724 (1975).
28. D. R. Anderson, in Symposium on Glaucoma, (C. V. Mosby, 1975, p. 81).
29. D. R. Anderson, Arch. Ophth. 82, 800 (1969).
30. D. R. Anderson and W. F. Hoyt, Arch. Ophth. 82, 506 (1969).
31. I. S. Begg, S. M. Drance, and V. P. Sweeney, Can. J. Ophth. 5, 321 (1970).
32. S. N. Levy, Invest. Ophth. 13, 691 (1974).
33. D. R. Anderson and A. Hendrickson, Invest. Ophth. 13, 771 (1974).
34. J. Flynn, in Symposium on Glaucoma, (C. V. Mosby, 1975).
35. M. F. Lieberman, A. E. Maumenee and W. R. Green, Am. J. of Ophth. 82, 405 (1976).
36. J. Francois and A. Neetens, Doc. Ophth. 26, 38 (1969).
37. G. R. Douglas, S. M. Drance and M. Schulzer, Arch. Ophth. 93, 409 (1975).
38. T. Horie, Y. Kitazawa, and H. Nose, Jap. J. Ophth. 19, 108 (1975).

Biogenic Monoamines and Amino Acids as Retinal Neurotransmitters

B. Ehinger
Departments of Ophthalmology and Histology
University of Lund
S-223 62 Lund, Sweden

Biogenic Amines

Dopamine was not the first substance presumed to be a neurotransmitter in the retina, but today it is probably the best proven. The first indication of this role was obtained in 1963. With the aid of the formaldehyde histofluorescence method of FALCK and HILLARP it was shown that the formaldehyde treatment induced fluorescence in certain neurons of the retina, indicating that they contain a catecholamine which was later shown to be dopamine (Table I). The dopamine containing cell bodies were found among the amacrine cells (Fig. 1) and their processes distribute in one to three sublayers in the inner plexiform layer, depending on the animal species. Figure 2 summarizes data from the literature (1-15). These cells were named junctional cells, and, as will be discussed below, they are most likely dopaminergic, i.e. their transmitter is dopamine. The number of these cells has never been accurately assessed, but they can be estimated to constitute a few per cent of the amacrine cell bodies. As will be discussed below, they have the type of synapse ascribed to amacrine cells and may thus be regarded as a subpopulation of amacrine neurons. Further, a few dopaminergic cell bodies can be found in the inner plexiform layer and among the ganglion cells. Their branches are found in the inner plexiform layer, mixing indistinguishably with the processes of the dopaminergic junctional cells. The dopaminergic neurons among the ganglion cells are few, only a few per cent of the number of junctional cells. They have been called alloganglion cells. The dopaminergic neurons in the inner plexiform layer are fewer still, and have been called eremite cells. Since the detailed morphology of the branches of these cells is unknown and also which contacts they make with other cells it is not possible to tell whether they represent special classes of cells. The simplest explanation is that they are displaced dopaminergic junctional cells. Neither the alloganglion cells nor the other dopaminergic neurons have been found to send axons to the optic nerve (16) and are therefore entirely intraretinal.

Table I.
Catecholamine Concentration in the Retina

Species	Concentration (nmoles/g wet wt.) Dopamine light	Dopamine dark	Noradrenaline light	Noradrenaline dark	Adrenaline light	Adrenaline dark	References
Rabbit	1.2		trace				Haggendal and Malmfors (1)
	0.7		trace				Haggendal and Malmfors (2)
	1.5*	1.0*					Nichols et al (19)
	1.5ℓ		trace		trace		da Prada (34)
	8.3	14.4	trace	trace	trace	trace	Drujan et al (74)
Rat	3.3 †	2.1 †					Nichols et al (9)
	2.3 §	2.0 §	trace		trace		da Prada (34)
Rat (7 days old)	0.04ℓ		trace		trace		da Prada (34)
Mouse	2.8		trace		trace		da Prada (34)
Guinea Pig	1.6✝	1.1✝					Nichols et al(9)
	1.0ℓ		trace		trace		da Prada (34)
Chicken	0.6ℓ		trace		trace		da Prada (34)
Cat	0.8ℓ		trace		trace		da Prada (34)
Hamster	1.9ℓ		trace		trace		da Prada (34)
Goldfish	4.3ℓ		trace		trace		da Prada (34)
Frog (Rana esculenta)	4.8ℓ		trace		trace		da Prada (34)
Frog (species unknown)	5.6	9.0	trace	trace	4.4	5.4	Drujan et al (74)
Toad (species unknown)	1.1	2.0	trace	trace	0.7	1.1	Drujan et al (74)
Toad (Bufo marinus)	2.7	3.3					Drujan et al (74)
Pigeon central part	0.16		0.39				Stoeckel et al (75)
Pigeon peripheral part	0.97		0.51				Stoeckel et al (75)

ℓ Assuming 1 g wet weight = 0.12 g protein
* Assuming a retinal weight of 60 mg
† Assuming a retinal weight of 10 mg
✝ Assuming a retinal weight of 20 mg

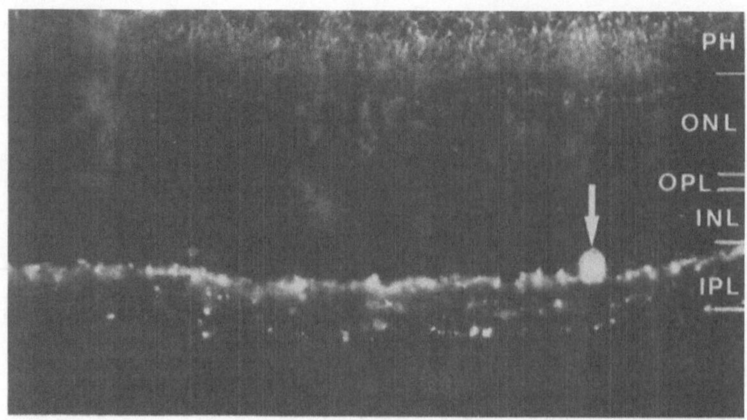

Fig. 1 Fluorescence micrograph of rabbit retina treated to show the dopaminergic neurons. The arrow points to a fluorescent dopaminergic cell body. There are also fluorescent terminals in three more or less well defined sublayers in the inner plexiform layer (IPL). INL, inner nuclear layer; OPL, outer plexiform layer; ONL, outer nuclear layer; PH, photoreceptor outer limbs. From Ehinger and Floren (28). X 350.

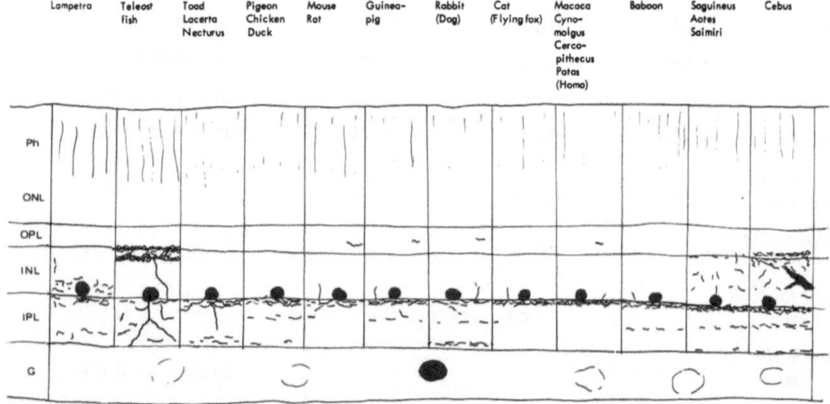

Fig. 2 Summary diagram of the distribution of dopaminergic neurons in the species so far investigated. Some species' names appear in parentheses indicating that significant information has been omitted (for clarity) and that the original publications should be consulted for these details. Eremite cells occur infrequently in all species, but have not been included in the drawing. Alloganglion cells occur in all species (except pigeon) in the ganglion cell layer (G) and this layer has therefore not been divided between the different species. Designation of layers like in Fig. 1.

In teleost fish and New World monkey there are, in addition, a set of terminals originating from the junctional cells, but reaching out to the outer plexiform layer where they surround the horizontal cells (8, 17, 18). These neurons transmit signals centrifugally within the retina, as will be further discussed below in connection with electron microscopy. Similar neurons have been observed in rabbits, dolphines, cats and New World monkeys with other techniques (19-25), but in these species they must use a transmitter other than dopamine or any related substance because they cannot be forced to accumulate dopamine or similiar amines either by uptake from the extracellular space or by inhibiting the degrading enzyme, monamine oxidase, while pushing the synthesis by applying precursors (see ref. 13, p. 152). These nuerons form a new class, called interplexiform cells, and are of special interest because they can form part of feedback control loops in the retina.

Fluorescene microscopy clearly shows that dopamine is present presynaptically in certain retinal neurons, which is a main requirement for a transmitter. It has also been shown (26) that dopamine can be released by light stimulation, that there are efficient recapturing systems for it (26-28) and that the enzymes involved in its synthesis and metabolism are present in the retina (26, 29-32, see also ref. 33). A light-induced increase in the rate of synthesis of dopamine has been reported in rat retina (34). Further, dopamine has been shown to inhibit the firing rate of most retinal ganglion cells (35-37). Finally, in intracellular recordings from horizontal cells in the goldfish retina, HEDDEN and DOWLING (38) showed that dopamine acts to isolate the center response from the influence of the surround. From the above experiments we may conclude that dopamine is very likely a retinal neurotransmitter in a variety of amacrine cells (the junctional cells) and, in some species, in dopaminergic interplexiform cells, as will be further discussed below. However, we do not know the detailed postsynaptic effects of dopamine (or the dopaminergic neurons) on more than a few cell types.

One way of getting more detailed information on the function of the dopaminergic neurons is to analyze which cells they contact. This has to be done in the electron microscope, where dopaminergic neurons can be visualized by making them accumulate either 6-hydroxydopamine or 5,6-dihydroxytryptamine. The former substance will not form fluorophores with formaldehyde, and its uptake can therefore not be readily monitored. Moreover, the labelling becomes rather discrete and difficult to observe, so that the work becomes rather time consuming. 5,6-dihydroxytryptamine gives more clearcut labelling (Fig. 3) and its uptake can be checked with the FALCK and HILLARP method. When that was done it was noted that not only do the dopaminergic junctional cells accumulate 5,6-dihydroxytryptamine, but also a second set of neurons. These have now been studied in more detail (28, 39, 40). They form a special set of amacrines (here tentatively called the I-neurons) with their processes slightly differently distributed than the dopaminergic junctional cells (Fig. 4). A method was worked out to remove these neurons with the aid of the neurotoxin, 5,7-dihydroxytryptamine (28) to make it possible to study the dopaminergic junctional cells in the electron microscope without interference from the I-neurons. When this was done (41) it was noticed that in rabbits the dopaminergic junctional cells made synapses of the conventional type onto amacrines. The labelled neurons are characterized by their small (400-600 A) dense cored synaptic vesicles, by increased density of their cytoplasm and mitochondrial and cellular membrane. There is normally also some mitochondrial swelling (Fig. 3). No contacts were found onto bipolars or ganglion cells. The input is from amacrines, onto the intervaricose parts of the processes. Thus, in rabbits the dopaminergic junctional cells form a set of interneurons, connecting different amacrines only.

In goldfish, 5,6-dihydroxytryptamine will in the outer plexiform layer label terminals making synapses onto either horizontal or bipolar cells (18). No input has been found in this layer. Instead, input is seen only in the inner plexiform layer, coming from amacrines. The labelled cells must thus be transmitting information centrifugally within the retina from amacrines in the inner plexiform layer to horizontal and bipolar cells in the outer plexiform layer (Fig. 5). This special class of

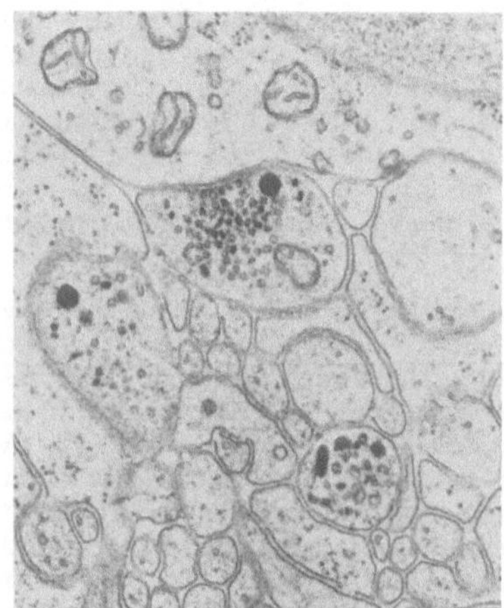

Fig. 3 Electron micrograph from rabbit retina from which the indoleamine accumulating terminals had been removed and in which the dopaminergic terminals were labelled with 5,6-dihydroxytryptamine. Three labelled processes can be seen with both small and large dense-cored synaptic vesicles. One of the processes is making a synapse with an amacrine cell body. X 40 000.

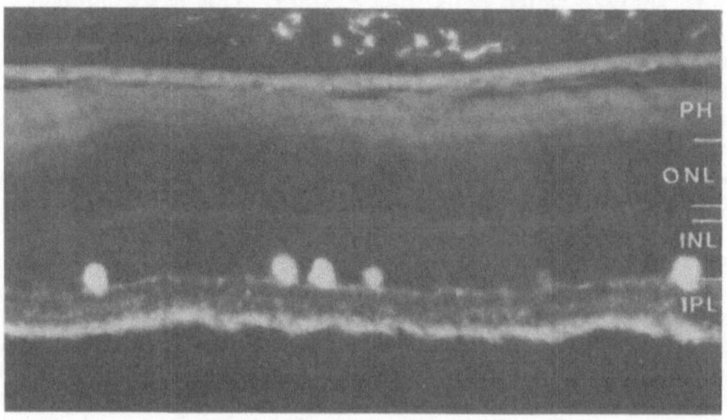

Fig. 4 Fluorescence micrograph four hours after the injection of an indoleamine (5,7-dihydroxytryptamine) for the demonstration of the indoleamine accumulating neurons. There are several fluorescent perikarya in the innermost part of the inner nuclear layer (INL) and three sublayers of fluorescent terminals in the inner plexiform layer. The innermost sublayer is best developed. Compare with the distribution of dopaminergic neurons (Fig. 1). Designation of layers as in Fig. 1. From Ehinger and Floren (39). X 330.

cells has been named the interplexiform cells. They can be expected to form part of regulatory feedback loops within the retina. It is as of yet a matter of speculation which function they regulate, but it is of importance to recognize the possible existence of such loops and to note that these neurons occur in substantial numbers, making it likely that they have a significant role in the function of the retina. It

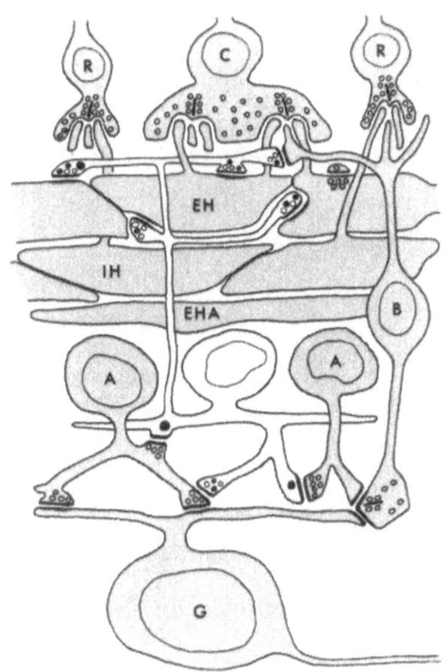

Fig. 5 A schematic diagram of the synaptic connections of the interplexiform cells of the goldfish retina. The input to these neurons is in the inner plexiform layer from amacrine cells (A). The interplexiform cell processes make synapses onto amacrine cell processes in the inner plexiform layer but never do they contact the ganglion cells (G) or their dendrites. In the outer plexiform layer, the processes of the interplexiform cells surround the external horizontal cells (EH). They make synapses on the external horizontal cells perikarya and onto bipolar cell dendrites. The interplexiform cell processes have never been observed as postsynaptic elements in the outer plexiform layer at either rod (R) or cone (C) receptor terminals or at the occasional external horizontal synapse. Also no synapses are seen between interplexiform cell processes and elements of the intermediate and internal horizontal cell layers. IH, intermediate (rod) horizontal cell; EHA, external horizontal cell axon process; B, bipolar cell. From Dowling and Ehinger (41).

should further be observed that they are dopaminergic only in teleost fish and some New World monkeys. As discussed above, interplexiform cells occur also in other species (cat, rabbit, dolphin, New World monkey) but in these species they must use a transmitter other than dopamine or any related substance.

Glycine and GABA

Glycine and GABA are strongly suspected to be neurotransmitters in the CNS, where they have powerful inhibitory effects. They occur in the retina in amounts similar to those in the brain (Table II) and inhibit the firing of ganglion cells (ref. 33 or 42). Both are taken up by the retina by active, high affinity (Km less than 5×10^{-5} M) mechanisms in all species so far studied (43-48) and such an uptake mechanism is often believed to be associated with a transmitter function for the substance. A number of investigations have shown that both glycine and GABA depress the b wave of the e.r.g. (49-52). In one study GABA paradoxically increased the b wave (52). The explanation for this is not clear. Both strychnine and picrotoxin, which are inhibitors of the glycine and GABA receptors respectively, have also been shown to affect the e.r.g. However, the e.r.g. is a complex response without well defined origin, and these results are therefore hard to evaluate. More direct proof for the transmitter role of glycine and GABA was obtained by WYATT and DAW (53). In rabbits, directionally sensitive ganglion cells were found to lose this sensitivity when picrotoxin was injected. Strychnine had a different effect. Ganglion cells operating as local edge detectors lost this function when strychnine was injected. Both these functions are believed to be mediated by different sets of amacrines in the inner plexiform layer, and the results thus support the presumption that glycine and GABA are neurotransmitters in subclasses of amacrine cells. Further, KIRBY and ENROTH-

Table II.
The Concentration of Glycine and γ-Aminobutyric Acid in Retina.

Species	Concentration µmoles/gm wet wt. Glycine light	dark	Gaba light	dark	References
Rat	2.30	2.52	1.92	1.93	Starr (76)
	1.03		1.55		Pasantes-Morales et al (77)
		5.34			*Macaione (78)
	2.52		4.44		*Macaione et al (79)
	2.21		3.20		Voaden (55)
Rat (age 21 days)	5.89		3.72		Lund Karlsen and Fonnum (80)
Rat	2.2		2.8		Voaden et al (81)
Rat (4 days old)			0.8		*Macaione et al (82)
Rat (21 days old)			5.3		Macaione et al (82)
Rat (90 days old)			5.8		
Mouse	2.03	2.09	2.00	1.63	*Cohen et al (83)
Mouse	3.1		2.0		Orr et al (66)
Rabbit			0.51		*Kuriyama et al (84)
			1.21		Davis et al (85)
	3.21		3.54		Voaden (55)
Rabbit	2.1		2.7		Voaden et al (81)
Rabbit (11 days old)			0.7		Davis et al (85)
Cat	2.0		2.0		Voaden et al (81)
Ox			5.1		*Macaione (78)
Chicken	2.01	2.00	3.69	3.43	Starr (76)
Pigeon	4.00		3.81		Voaden (55)
Pigeon red spot	4.0		5.3		Voaden et al (81)
Pigeon periphery	5.6		5.2		Voaden et al (81)
Frog	1.62	1.68	2.70	2.25	Starr (76)
			3.48	1.39	*Graham et al (86)
	1.17		1.12		Kennedy and Voaden (62)
Frog	(4.4umol/ retina)		(2.8umol/ retina)		Yates and Keen (63)
Goldfish	0.63	0.70	1.96	1.42	Starr (76)
			3.15	1.65	Lam (87)
Brain	0.55 -	1.45	0.83 -	2.27	McIlwain and Bachelard (88)

*Recalculations based on 1 gm wet wt. = 0.12 gm protein
= 0.20 gm dry wt.

CUGELL (54) found that the influence of the surround was substantially reduced by the GABA receptor blockers, bicuculline and picrotoxin, in the Y class ganglion cells in cats. The X class of ganglion cells was virtually unaffected. Again, this argues for GABA as the transmitter in certain amacrines.

There is no method for visualizing endogenous glycine or GABA directly in the microscope, but the neuronal uptake can be used for labelling them with either tritiated glycine or GABA which can then be localized autoradiographically (see ref. 55). Glycine will in rabbits label a set of cell bodies among the amacrines, as will GABA when injected intravitreally. There will also be radioactivity in the inner plexiform layer; in the case for glycine diffusely through the layer and with GABA with a tendency to form a sublayer in the innermost part of the inner plexiform layer. Both have been shown to be retained in the unchanged form to a significant extent indicating the presence of protecting stores (56,57).

Glycine is not taken up by glia to any significant extent, but GABA is. However, the turnover of (^{3}H) -GABA is in rabbits higher than in neurons, and the glial radioactivity therefore disappears long before the (^{3}H)-GABA in neurons (89). This enables us to use retinas loaded with either (^{3}H)-glycine or (^{3}H)-GABA for different types of efflux experiments and to state the origin of the released radioactivity with a reasonable degree of confidence.

When the retina labelled with (^{3}H)-glycine is set up in a superfusion system (58,59) radioactivity is released as seen in Fig. 6. A flashing light will cause a significant increase in the rate of radioactivity release. A steady light will not

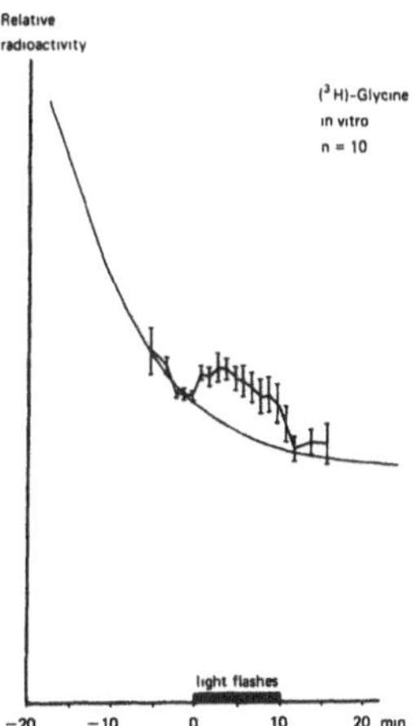

Fig. 6 Efflux of radioactivity from rabbit retinas preloaded with (^{3}H)-glycine and stimulated with light flashes (2 per sec.). The continuous line is the spontaneous efflux in the dark. The light flashes cause a significant increase ($p < 0.001$) in the release rate. The vertical bars indicate the standard error of the mean. 10 experiments. From Ehinger and Lindberg-Bauer (59).

release any radioactivity, which is in good accordance with the notion that amacrines respond only to transient stimuli. Similar results have been obtained in retinas labelled with (^{3}H)-GABA (60,61). If in this case experiments are run at comparatively short times after the injection of (^{3}H)-GABA when radioactivity is mainly present in glia, no light-induced changes can be seen. Neither is any such release seen in retinas labelled with (^{3}H)-valine, which is mainly taken up by glia (59). This indicates that despite the fact that glial cells display pronounced changes in membrane potentials in response to light driven activity in the retina, there is no light inducable release of what has been taken up.

The experiments have thus shown that it is possible to obtain light driven release of radioactivity from neurons that have accumulated either (^{3}H)-glycine or (^{3}H)GABA, which is a requirement for accepting either substance as a retinal neurotransmitter.

An important observation on the release of (^{3}H)-GABA is that both pentobarbital and to a certain extent also amino-oxyacetic acid inhibit it (60,61). We constantly failed to obtain GABA release until we ceased using pentobarbitone to kill the rabbits and lowered AOAA to less than 10^{-5}M in the superfusion medium.

Taurine

Taurine is present in the photoreceptors in high concentrations (62-66). (^{3}H)-taurine is in rabbits largely taken up by glia, but, in incubation experiments, also in the photoreceptors (Fig. 7). If taurine is a retinal neurotransmitter, one would thus expect it to be so in photoreceptors.

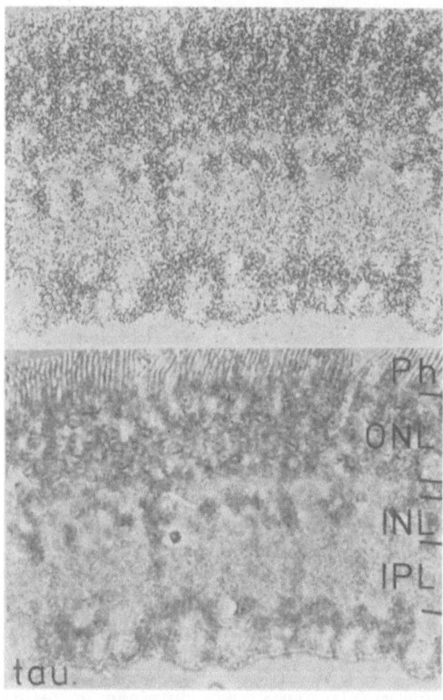

Fig. 7 Autoradiogram of rabbit retina incubated 15 min at 37°C in 1,3x10^{-6}M (^{3}H)-taurine. Above, focus on the grains; below, focus on the section. There is radioactivity in the photoreceptors and in structures with the distribution of glial cells in the inner layers. X 480.

PASANTES-MORALES, URBAN, KLETHI and MANDEL (67) reported that (^{35}S)-taurine could be released from chicken retina by a flash from a xenon tube placed a few millimeters from the retina. A single flash was sufficient, and resulted in a prolonged release. This would not fit with taurine as a photoreceptor transmitter because light is thought to diminish the transmitter release in photoreceptors (see ref. 68). We have used the same equipment used for showing glycine and GABA release to try to find any light induced changes in taurine release in rabbits, but have been able to detect neither any increase nor any decrease in the release of radioactivity in a total of 18 experiments. The superfusion speed was 1 ml/min, and the superfusate was collected in 1 ml fractions. We have not placed the flash tube close to the retina like PASANTES-MORALES did because of the risks of getting non-specific effects by the electric fields surrounding the tube or by its heat.

Conclusion

Apart from the three substances discussed here in some detail there are now a few others suspected to be transmitters in the retina. Curiously enough, they are almost all presumed to act in different varieties of amacrines (Fig. 8). Except for dopamine we have very little information about the spread and contacts of the different varieties. It is particularly remarkable that we have no indication as to the possible transmitter of the ganglion cells, whose axons exclusively form a thick and readily accessible bundle, the optic nerve. It is quite likely that for many or most of the retinal neurons we shall have to search for their transmitters among classes of substances hitherto not recognized as putative transmitters. The many transmitters found already among amacrines suggest that the final list is not going to be short.

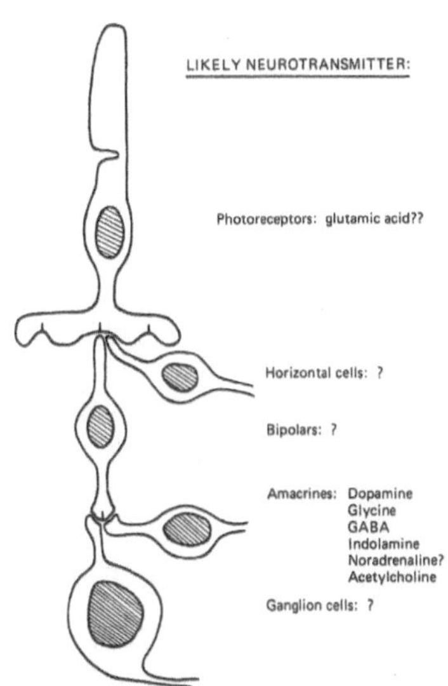

Fig. 8 Most synapses in the retina are of the chemical type, but the transmitters involved are by and large only known for subclasses of amacrines. Glutamic acid (or possibly, aspartic acid) may be transmitter in photoreceptors because horizontal cells respond strongly to these substances (see ref. 68). The transmitters of horizontal cells and bipolars are unknown although GABA reuptake has been observed in horizontal cells of a number of cold-blooded vertebrates (45, 69). GABA synthesis was also demonstrated in isolated horizontal cells of goldfish (70). In amacrines, dopamine, glycine, GABA and an indoleamine are all likely neurotransmitters, as discussed in the text. There is also good evidence that acetylcholine is the transmitter of some amacrines (see ref. 71 and 72). The evidence for noradrenaline is rather indirect (15). The transmitterof the ganglion cells is unknown although Henke et al (73), obtained evidence suggesting glutamate for this role in pigeons.

References

1. J. Haggendal and T. Malmfors, Acta Physiol. Scand. 59, 295 (1963).
2. J. Haggendal and T. Malmfors, Acta Physiol. Scand. 64, 58 (1965).
3. T. Malmfors, Acta Physiol. Scand. 58, 99 (1963).
4. B. Ehinger, Z. Zellforsch 71, 146 (1966).
5. B. Ehinger, Invest. Ophthalmol. 5, 42 (1966).
6. B. Ehinger, Z. Zellforsch 82, 577 (1967).
7. A. M. Laties and D. Jacobowitz, Anat. Rec. 156, 383 (1966).
8. B. Ehinger, B. Falck and A. M. Laties, Z. Zellforsch 97, 285 (1969).
9. E. Scheie and A. M. Laties, Herpetologica 27, 77 (1971).
10. W. Nichols, D. Jacobowitz and M. Hottenstein, Invest. Ophthalmol. 6, 642 (1967).
11. B. Ehinger and B. Falck, Z. Zellforsch 100, 364 (1969).
12. B. Ehinger, Z. Zellforsch 139, 171 (1973).
13. B. Ehinger, In Transmitters in the visual process. ed. by S. L. Bonting,(Pergamon Press, 1976).
14. Y. Sano, H. Yoshikawa and M. Konishi, Arch. Histol. Jap. 30, 75 (1968).
15. B. Ehinger, K. Holmberg and P. Ohman, Acta Zool. 58, 117 (1977).
16. B. Ehinger and B. Falck, Albrecht v. Graefes Arch. Klin, Exp. Ophthal. 178, 295 (1969).
17. J. E. Dowling and B. Ehinger, Science 188, 270 (1975).
18. J. E. Dowling and B. Ehinger, Proc. Roy. Soc. (London) Ser. B. In press (1978).
19. B. B. Boycott, J. E. Dowling, S. K. Fisher, H. Kolb and A. M. Laties, Proc. Roy. Soc. Lond. Ser. B. 191, 353 (1975).
20. A. Gallego, Vision Res. Suppl. 3, 33 (1971).
21. A. Gallega, Arch. Soc. Esp. Oftalmol. 31, 299 (1971).
22. W. W. Dawson and J. M. Perez, Science 181, 747 (1973).
23. H. Kolb, J. Neurocytol. 6, 131 (1977).
24. H. Kolb and R. W. West, J. Neurocytol. 6, 155 (1977).
25. C. W. Oyster and E. S. Takahashi, Proc. Roy. Soc. Ser. B. 197, 477 (1977).
26. S. G. Kramer, Invest. Ophthalm. 10, 438 (1971).
27. S. G. Kramer, A. M. Potts and Y. Mangnall, Invest. Ophthalmol. 10, 617 (1971).
28. B. Ehinger and I. Floren, Exp. Eye Res. In press (1977).
29. K. Mizuno, Eye, Ear, Nose and Throat 39, 493 (1960).
30. O. Eranko, M. Niemi and E. Merenmies, In The structure of the eye ed. by G. K. Smelsert (Acad. Press, New York, 1961).
31. A. Mustakallio, Acta Ophthalmol. Suppl. 93, 1 (1967).
32. S. G. Kramer, In Transmitters in the visual process, ed. by Bonting (Pergamon Press, New York, 1976).
33. L. T. Graham, Jr. In The eye ed. by H. Davson and L. T. Graham, Jr. (Academic Press, New York, 1974).
34. M. da Pranda, In Nostriated dopaminergic neurons ed. by E. Costa and G. L. Gessa (Raven Press, New York, 1977).
35. M. Straschill and J. Perwein, Arch. Europ. J. Physiol. 312, 45 (1969).
36. M. Straschill and J. Perwein, In Proceedings of the Golgi Centennial Symposium ed. by M. Santini (Raven Press, New York, 1975).
37. A. Ames and D. A. Pollen, J. Neurophysiol. 32, 424 (1969).
38. W. L. Hedden and J. E. Dowling, Proc. Roy. Soc. (London) Ser. B. In Press (1977).
39. B. Ehinger and I. Floren, Cell and Tissue Res. 175, 37 (1976).
40. J. Floren and C. Hanson, In preparation (1978).
41. J. E. Dowling and B. Ehinger, J. Comp. Neurol. 180, 203 (1978).
42. A. van Harreveld, Progress in Neurobiology 8, 1 (1976).
43. A. Bruun and B. Ehinger, Invest. Ophthalmol. 11, 191 (1972).
44. M. J. Neal, D. G. Peacock and R. D. White, Brit. J. Pharmacol. 47, 656 (1973).

45. M. J. Voaden, J. Marshall and N. Murani, Brain Res. 67, 115 (1974).
46. M. S. Starr and M. J. Voaden, Vision Res. 12, 549 (1972).
47. M. Goodchild and M. J. Neal, Brit. J. Pharmacol. 47, 529 (1973).
48. M. J. Neal, In Transport phenomena in the nervous system; Physiological and pathological aspects. ed. by A. Lajtha (Plenum Press, London, New York, 1975).
49. N. W. Scholes and E. Roberts, Biochem. Pharmacol. 13, 1319 (1964).
50. S. Z. Kramer, P. A. Sherman and J. Seifter, Int. J. Neuropharmacol. 6, 463 (1967).
51. H. Pasantes-Morales, N. Bonaventure, N. Wioland and P. Mandel, Neurosci. 5, 235 (1973).
52. M. S. Starr, Exp. Eye Res. 21, 79 (1975).
53. H. J. Wyatt and N. W. Daw, Science 191, 204 (1976).
54. A. W. Kirby and C. Enroth-Cugell, J. Gen. Physiol. 68, 465 (1976).
55. M. Voaden, In Transmitters in the visual process, ed. by S. L. Bonting (Pergamon Press, 1976).
56. B. Ehinger and B. Falck, Brain Res. 33, 157 (1971).
57. B. Ehinger, Brain Res. 46, 297 (1972).
58. B. Ehinger and B. Lindberg, Nature 251, 727 (1974).
59. B. Ehinger and B. Lindberg-Bauer, Brain Res. 113, 535 (1976).
60. B. Bauer and B. Ehinger, Experientia 33, 470 (1977).
61. B. Bauer, Acta Ophthalm. 56, 270 (1978).
62. A. J. Kennedy and M. J. Voaden, Biochem. Soc. Transactions 2, 1256 (1974).
63. R. A. Yates and P. Keen, Brain Res. 107, 117 (1976).
64. M. J. Voaden, N. Lake, J. Marshall and B. Nathwani, Exp. Eye Res. In press (1977).
65. T. H. Orr, A. I. Cohen and J. A. Carter, Exp. Eye Res. 23, 377 (1976).
66. H. T. Orr, A. I. Cohen and O. H. Lowry, J. Neurochem. 26, 609 (1976).
67. H. Pasantes-Morales, P. F. Urban, J. Klethi and P. Mandel, Brain Res. 51, 375 (1973).
68. H. Ripps, M. Shakib and E. D. MacDonald, J. Cell Biol. 70, 86 (1976).
69. D. Lam and L. Steinman, Proc. Nat. Acad. Sci. 68, 2777 (1971).
70. D. Lam, Nature 254, 345 (1975).
71. R.H. Masland and A. Ames, J. Neurophysiol. 39, 1220 (1976)
72. R. H. Masland and C. J. Livingstone, 39, 1210 (1976).
73. H. Henke, T. M. Schenker and M. Cuenod, J. Neurochem. 26, 131 (1976).
74. B. D. Drujan, J. M. Diaz Borges and N. Alvarex, Life Sci. 4, 473 (1965).
75. M. E. Stoeckel, G. Roussel, J. Zwiller, B. Madarasz and A. Porte, Cell and Tissue Res. 173, 335 (1976).
76. M. S. Starr, Brain Res. 59, 331 (1973).
77. H. Pasantes-Morales, J. Klethi, M. Ledig and P. Mandel, Brain Res. 41, 494 (1972).
78. S. Macaione, J. Neurochem. 19, 397 (1972).
79. S. Macaione, P. Ruggeri, F. de Luca and G. Tucci, J. Neurochem. 22, 887 (1974).
80. R. Lund Karlsen and F. Fonnum, J. Neurochem. 27, 1437 (1976).
81. M. J. Voaden, N. Lake, J. Marshall and B. Morjaria, Exp. Eye Res. 25, 279 (1977).
82. S. Macaione, R. Campisi and A. Albanese, Bollettino della Societa Italiana de Biologia Sperimentale 46, 785 (1970).
83. A. I. Cohen, M. McDaniel and H. Orr, Invest. Ophthalmol. 12, 686 (1973).
84. K. Kuryama, B. Sisken, B. Haber and E. Roberts, Brain Res. 9, 165 (1968).
85. J. M. Davis, W. A. Himwich and H. C. Agrawal, Dev. Psychobiol. 2, 34 (1969).
86. L. T. Graham, Jr., C. F. Baxter and R. N. Lolley, Brain Res. 20, 379 (1970).
87. D. Lam, J. Cell. Biol. 54, 225 (1972).
88. H. McIlwain and H. S. Bachelard, Biochemistry of the Central Nervous System (Churchill Livingston, London, 1971).
89. B. Ehinger, Exp. Eye Res. In press (1977).

Lipid-Rhodopsin Interactions in Photoreceptor Membranes

Robert E. Anderson
Cullen Eye Institute
Baylor College of Medicine
6501 Fannin
Houston, Texas 77030

Introduction

Lipids are present in all living cells and serve a variety of functions. Today, I will discuss the chemistry of the lipids of photoreceptor membranes and some of their interactions with rhodopsin, the rod visual pigment.

An early model of the arrangement of lipids in biological membranes came from DANIELLI and DAVSON (1) who envisioned the phospholipids existing in a bilayer with the hydrophobic fatty acid chains opposing each other and the proteins on the outer surface of the bilayer. This model explained many of the observed lipid properties of membranes, but the placement of protein on the surface only was inconsistent with experimental findings. Some membrane proteins were dissassociable from the phospholipids by salt solutions, while others were not, which suggested that these proteins were bound to the phospholipids through non-ionic bonds. Transport of non-lipid compounds across the lipid bilayer was difficult to understand if the proteins were only on the surface, and the presence of globular proteins in membranes was not consistent with the hypothesis that they were only surface molecules. These and other considerations led SINGER and NICOLSON (2) to propose the fluid mosaic model which differs from earlier models primarily in the placement of the protein. The lipid bilayer still persists but some of the proteins are intrinsic membrane components with their hydrophobic regions imbedded in the hydrophobic core of the bilayer. This arrangement allows for movement of protein within the bilayer and predicts a dynamic role for lipid in the control of membrane structure and function.

Lipid Chemistry or Photoreceptor Membranes

The predominant lipids of photoreceptor membranes are the glycerolipids, schematically depicted in Fig. 1. R and R^1 represent fatty acids esterified to positions -1 and -2, respectively. Saturated fatty acids preferentially occupy position -1 and polyunsaturates occupy position-2 (3). The functional group on position-3, depicted as X in the figure, determines whether the glycerolipid is a neutral lipid or a phospholipid. When X is phosphorus, the lipid is the phospholipid phosphatidic acid.

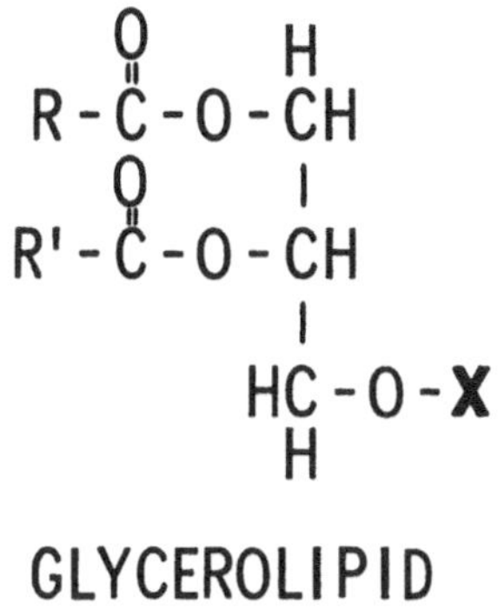

$$
\begin{array}{l}
\quad\ \ O \qquad H \\
R-\overset{\|}{C}-O-CH \\
\quad\ \ O \qquad\ | \\
R'-\overset{\|}{C}-O-CH \\
\qquad\qquad\quad | \\
\qquad\qquad HC-O-\mathbf{X} \\
\qquad\qquad\ \ H
\end{array}
$$

GLYCEROLIPID

Fig. 1 Schematic of glycerolipids.

Addition of a choline, serine, ethanolamine, or inositol molecule to the phosphorus of phosphatidic acid results in the common glycerophospholipids of photoreceptor membranes: phosphatidyl choline, phosphatidyl serine, phosphatidyl ethanolamine, and phosphatidyl inositol. Phosphatidyl choline and phosphatidyl ethanolamine are the two major phospholipid classes in photoreceptor membranes. Occasionally, small amounts of lysophospholipids, which contain only one fatty acid, are found. If X is hydrogen, the compound is the neutral lipid 1, 2-diglyceride; if X is a fatty acid, the lipid is a triglyceride. Glycerolipids make up 84.8% of the total lipid of frog photoreceptor membranes, 1.6% of which is neutral lipid and 83.2% phospholipid (Table I, (4, 5)).

Table I

Lipid class composition of frog photoreceptor membranes (4, 5)

Lipid Class	Mole Percentages
Glycerolipids (GL)	
Phospholipids	
Phosphatidyl choline	38.1
Phosphatidyl ethanolamine	29.1
Phosphatidyl serine	10.8
Phosphatidyl inositol	1.9
Lysophosphatidyl choline	0.6
Lysophosphatidyl ethanolamine	0.4
Unidentified	2.3
Neutral Lipids	
1, 2-diglycerides	1.6
Sphingolipids (SL)	
Sphingomyelin	1.6
Free Fatty Acids (FFA)	5.6
Cholesterol (CHOL)	8.0
Total Neutral Lipids (GL + FFA + CHOL)	15.2
Total Phospholipids (GL + SL)	84.8

Sphingmyelin, a minor component of photoreceptor membranes, is a member of the sphingolipid family, depicted in Fig. 2. This lipid class differs from the

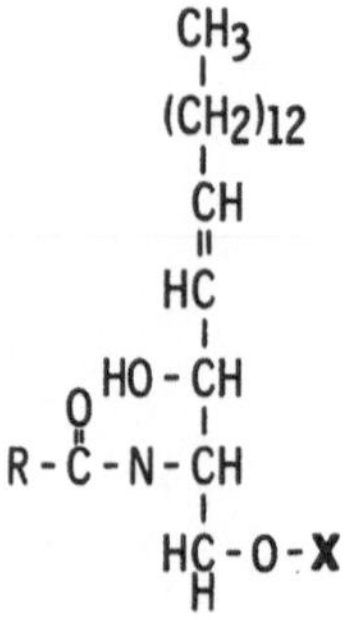

```
          CH3
           |
         (CH2)12
           |
           CH
           ||
           HC
           |
     O HO - CH
     ||     |
R - C - N - CH
            |
          HC - O - X
           H
```

SPHINGOLIPID

Fig. 2 Schematic of sphingolipids.

glycerolipids in several respects: the backbone of the molecule is a long chain sphingosine base, the free hydroxyl is never esterified, and the R-group is usually a long chain saturated or monounsaturated fatty acid linked to sphingosine through an amide linkage. If X is hydrogen, the compound is a ceramide; if X is phosphoryl choline, the compound is sphingomyelin; when X is carbohydrate, the compound is a cerebroside; and, if sialic acid is attached to a cerebroside, the product is a ganglioside. There have been reports that cerebrosides (6) and gangliosides (7, 8) are components of photoreceptor membranes and whole retina, although in my laboratory we have not found these lipids in photoreceptor membranes. Our feeling is that they are probably contamination from the neural retina during the isolation of photoreceptor membranes.

The R groups on the glycerolipids and sphingolipids represent fatty acids, four families of which are depicted schematically in Fig. 3. The short-hand nomenclature used for fatty acids is as follows: The number before the colon is the number of carbons in the chain; the number after the colon is the number of double bonds; and the number following the omega is the position of the first double bond from the methyl terminus of the molecule. Stearic acid is representative of the class of saturated fatty acids which, along with oleic acid (ω 9 family), can be synthesized by vertebrates. The ω 3 and ω 6 families cannot be synthesized and must be obtained in the diet. Elongation and desaturation of the three unsaturated fatty acids lead to the formation of long chain polyunsatured fatty acids by the pathways shown in Fig. 4. Since vertebrates cannot introduce double bonds between existing double bonds and the methyl terminus, the fatty acid family remains the same regardless of the number of added carbon atoms and/or double bonds.

The polyunsaturates commonly found in the retina are boxed in Fig. 4. Although 22:6 ω 3 is by far the major component in vertebrate photoreceptors (9), 20:5ω 3 is the major polyunsaturate of the rhabdomeric retinas of moth (10) and *Limulus* (11); the squid contains large amounts of 20:4ω 6, 20:5ω 3, and 22:6ω 3(12). 20:3ω 9 is found in rats only after deprivation of essential fatty acids (13).

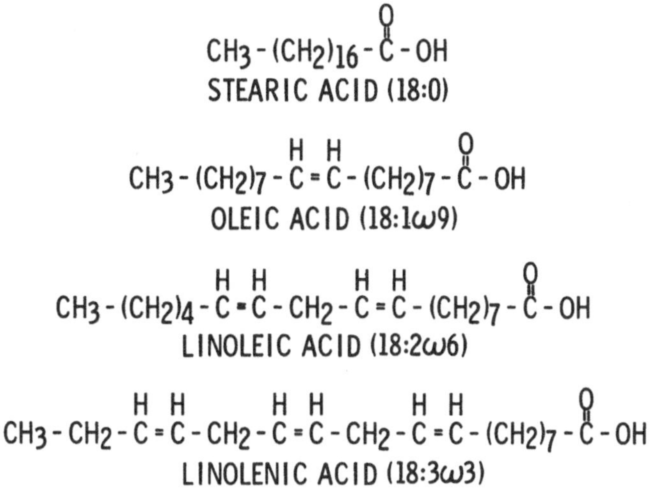

$$CH_3-(CH_2)_{16}-\overset{O}{\overset{\|}{C}}-OH$$

STEARIC ACID (18:0)

$$CH_3-(CH_2)_7-\overset{H}{C}=\overset{H}{C}-(CH_2)_7-\overset{O}{\overset{\|}{C}}-OH$$

OLEIC ACID (18:1ω9)

$$CH_3-(CH_2)_4-\overset{H}{C}=\overset{H}{C}-CH_2-\overset{H}{C}=\overset{H}{C}-(CH_2)_7-\overset{O}{\overset{\|}{C}}-OH$$

LINOLEIC ACID (18:2ω6)

$$CH_3-CH_2-\overset{H}{C}=\overset{H}{C}-CH_2-\overset{H}{C}=\overset{H}{C}-CH_2-\overset{H}{C}=\overset{H}{C}-(CH_2)_7-\overset{O}{\overset{\|}{C}}-OH$$

LINOLENIC ACID (18:3ω3)

Fig. 3 Schematic of four families of fatty acids: saturates, ω 9 monounsaturates, and ω 6 and ω 3 polyunsaturates.

ELONGATION AND DESATURATION PATHWAYS

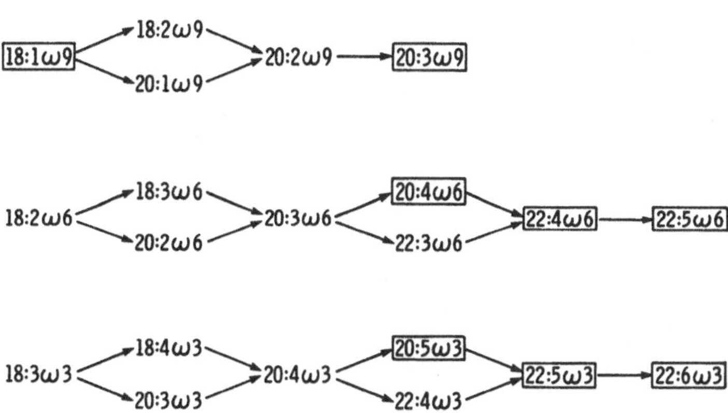

Fig. 4 Pathways of elongation and desaturation of retina fatty acids. Those acids commonly found in the retina are shown in the boxes.

Table II contains the fatty acid composition of phosphatidyl ethanolamine from several vertebrate and invertebrate species. The compositional similarity which exists between all of the vertebrates that we have examined is not reflected in the invertebrates.

The polyunsaturates of the retina have methylene interrupted cis-double bonds, depicted in the molecular models shown in Fig. 5. In general, the greater the degree

Table II

The fatty acid composition of phosphatidyl ethanolamine from several species.

Acid	Rat (30)	Frog (5)	Cattle (31)	Limulus (11)	Squid (12)
			(mole percentage)		
14:0[a]	0.2	0.2	-	-	0.3
15:0	-	0.1	-	-	-
16:0 DMA[b]	0.4	0.6	-	1.4	-
16:0	5.5	6.4	15.7	2.2	3.6
16:1	0.3	1.1	-	0.9	0.1
17:0	-	-	-	0.5	0.4
18:0 DMA	1.2	0.9	0.5	12.0	-
18:0	28.1	13.0	29.4	11.7	3.7
18:1	3.2	6.0	4.6	4.3	0.9
18:2ω6	0.6	0.8	0.8	1.8	-
18:3ω3	0.4	-	-	0.8	-
20:0	-	-	-	11.6	-
20:1	-	-	-	-	0.6
20:2	0.8	-	-	-	-
20:3ω9	-	0.5	0.2	-	0.2
20:4ω6	2.9	6.5	3.0	12.1	34.0
20:5ω3	0.1	-	-	40.5	7.5
22:4ω6	0.2	7.0	1.4	-	-
22:5ω6	1.1	2.2	4.3	-	-
22:5ω3	-	3.2	0.6	-	0.1
22:6ω3	54.8	51.1	39.5	-	47.3

[a]-Fatty acid nomenclature is given in the text.

[b]-DMA are dimethyl acetals produced by the acid methanolysis of plasmalogens.

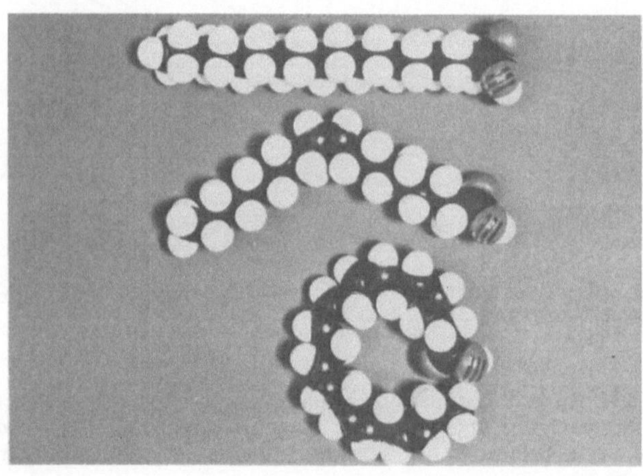

Fig. 5 Molecular models of the fatty acids 18:0, 18:1ω9, and 22:6ω3.

of cis-unsaturation, the greater the disorder in molecular packing, and the lower the melting temperature. Molecular models of 18:0, 18:1 ω 9, 22:6 ω3 in phosphatidyl ethanolamine are given in Fig. 6. In the top molecule, 18:0 and 18:1 ω 9 are the two

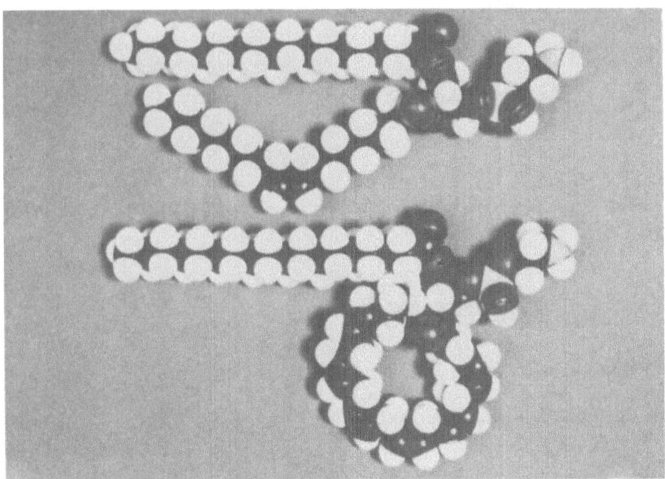

Fig. 6 Molecular models of phosphatidyl ethanolamine containing 18:0 and 18:1ω9 (top) and 18:0 and 22:6ω3 (bottom).

constituent fatty acids. Bilayers containing a predominance of these types of phosphatidyl ethanolamine molecules would have a higher melting temperature (and thus be less fluid) than those containing phosphatidyl ethanolamine molecules composed of 18:0 and 22:6 ω3. The latter is the predominant specie of phospholipids of vertebrate photoreceptor membranes, assuring a fluid bilayer.

Neutral lipids were once though to be insignificant components of photoreceptor membranes. However, recently Dr. REX WIEGAND of our department found 15.2 mole % neutral lipids in the total lipids in the photoreceptor membranes of frogs (see Table I). The major component is cholesterol, the structure of which is shown in Fig. 7. Cholesterol has the property of increasing the fluidity of membranes which contain a predominance of saturated fatty acids, but decreasing the fluidity of membranes that contain high levels of polyunsaturated fatty acids.

Location of Rhodopsin in the Lipid Bilayer

Rhodopsin is a glycolipoprotein containing the carbohydrates mannose and N-acetylglucosamine (14) and the lipid 11-cis-retinaldehyde (15) in covalent linkage. The carbohydrate moiety is exposed on the outer surface of the rod outer segment plasma membrane, and on the intradiscal surfaces of the discs (16). Most of the available evidence supports the notion that rhodopsin spans the bilayer (summarized in Ref. 18) with its chromophore oriented in the plane of the bilayer (17). The hydrophylic (carbohydrate and polar amino acids) and hydrophobic (retinaldehyde and non-polar amino acids) groups orient rhodopsin in the membrane and assure its directionality. The fluid nature of the bilayer allows rhodopsin to undergo rotational (19) and lateral (20) diffusion in the plane of the membrane, but due to the amphipathic nature of rhodopsin, it probably does not tumble through the bilayer.

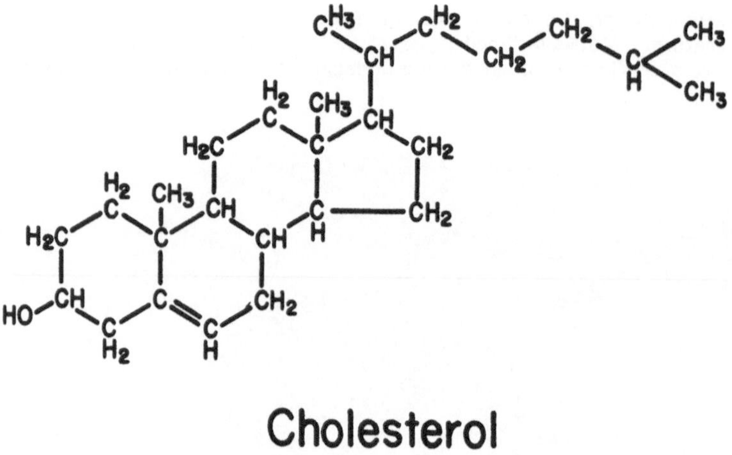

Cholesterol

Fig. 7 Schematic of cholesterol.

There is no migration of rhodopsin between individual discs or between discs and the plasma membrane.

The Function of Lipids in Photoreceptor Membranes

A large portion of rhodopsin resides in the lipid bilayer where it is bound rather strongly to lipid through non-covalent hydrophobic-hydrophobic interactions. Our unpublished data show that 30 moles of phospholipid remain tightly bound to rhodopsin in the presence of the non-ionic detergent emulphogene. Others have shown that non-denaturing solvents cannot extract all of the phospholipid from photoreceptor membranes (21). The hydrophobic domain thus generated by these strong non-covalent interactions provides a stable environment for the chromophore and protects the linkage of 11-*cis*-retinaldehyde and opsin from chemical and thermal insults. For example, rhodopsin in digitonin or emulphogene solutions does not react with sodium borohydride or hydroxylamine in the dark, whereas rhodopsin in detergents which completely dissociate the lipid reacts readily with these two reagents (22). Our unpublished data (22) and the recent report of BONTING, *et. al.* (23) show that the thermal stability of rhodopsin is less in delipidated preparations than in native or reconstituted membranes.

Following photon capture, rhodopsin undergoes a series of conformational changes within the membrane, one of which (probably the metarhodopsin I-metarhodopsin II transition) is responsible for the initiation of the ionic events involved in visual excitation. STEWART, *et. al.* (24), APPLEBURY, *et. al.* (25), and BONTING, *et. al.* (23) have shown that phospholipid participates in this transition.

ZORN and FUTTERMAN (21), SHICHI (26) and BONTING, *et. al.* (23) showed that phospholipid was necessary for the regeneration of rhodopsin. Although HONG and HUBBELL (27) found that digitonin can promote regeneration in the absence of phospholipid, regeneration of rhodopsin under the physiological conditions certainly depends on the presence of a lipid bilayer. DRATZ (28) has recently reported that the oxidation products of long chain polyunsaturated fatty acids inhibit the regeneration of rhodopsin.

Specific fatty acids appear to be important to the electrical activity of photoreceptor membranes (29). The electroretinogram of rats raised on diets that were identical except for single fatty acid substitutions had the greatest a-wave amplitude in the group that was supplemented with 18:3 ω 3, the precursor of 22:6ω 3. The groups fed no fatty acid supplement or 18:1 ω 9 had the lowest amplitude. The fatty acid composition of the photoreceptor membranes correlated with the fatty acid supplements, suggesting that specific fatty acid substitutions in protoreceptor membranes were responsible for the changes in the electrical activity.

Conclusions

Photoreceptor membranes contain rhodopsin imbedded in a fluid lipid bilayer composed predominantly of phospholipids containing large amounts of long chain polyunsaturated fatty acids. In addition to providing a stable matrix for the protein, the lipids participate in the dark reactions that follow photobleaching and in the regeneration of rhodopsin. Fatty acids of the ω3 family appear to have some as yet undefined functional role in the electrical events of visual excitation.

Acknowledgements

The collaboration of the following persons in the studies reviewed in this paper is gratefully acknowledged: R. M. Benolken, Peter A. Dudley, Rex D. Wiegand, Eston O. Plante, Maureen B. Maude, Margaret B. Jackson, and Paula A. Kelleher. This research was supported in part by grants from the Retina Research Foundation (Houston), the National Institutes of Health (EY 00871), and the National Science Foundation (#BNS-75-07-197 to R. M. Benolken).

References

1. J. F. Danielli and H. Davson, J. Cell Comp. Physiol. 5, 495 (1935).
2. S. J. Singer and G. L. Nicolson, Science 172, 720 (1972).
3. R. E. Anderson and L. Sperling, Arch. Biochem. Biophys. 144, 673 (1971).
4. R. D. Wiegand and R. E. Anderson, ARVO Abstracts (1977).
5. R. E. Anderson and M. Risk, Vis. Res. 14, 129 (1974).
6. W. T. Mason, R. S. Fager and E. W. Abrahamson, Biochem. 12, 2147 (1973).
7. S. Edel-Harth, H. Dreyfus, P. Bosch, G. Rebel, P. F. Urban and P. Mandel, FEBS Letters 35, 284 (1973).
8. H. H. Hess, P. Stoffyn and K. Sprinkle, J. Neurochem. 26, 621 (1976).
9. R. E. Anderson and M. B. Maude, Biochem. 9, 3624 (1970).
10. D. Zinkler, Verh. Deutsch. Zool. Ges. 67, 28 (1975).
11. R. M. Benolken, R. E. Anderson and M. B. Maude, Biochim. Biophys. Acta. 413, 234 (1975).
12. R. E. Anderson, R. M. Benolken, P. A. Kelleher, M. B. Maude and R. D. Wiegand, Biochim. Biophys. Acta. In Press (1978).
13. R. E. Anderson and M. B. Maude, Arch. Biochem. Biophys. 151, 270 (1972).
14. J. Heller and M. A. Lawrence, Biochem. 9, 864 (1970).
15. D. Bownds, Nature 216, 1178 (1967).
16. P. Rohlich, Nature 263, 789 (1976).
17. P. A. Liebman, Biophys. J. 2, 161 (1972).
18. S. E. Ostory, Biochim. Biophys. Acta 463, 91 (1977).
19. P. K. Brown, Nature 236, 35 (1972).
20. M. M. Poo and R. A. Cone, Exp. Eye Res. 17, 503 (1973).
21. M. Zorn and S. Futterman, J. Biol. Chem. 246, 881 (1971).
22. E. O. Plante and R. E. Anderson, Unpublished Observations.
23. S. L. Bonting, P. J. G. M. van Breugel and F. J. M. Daemen, in Function and Biosynthesis of Lipids, Vol. 83, ed. N. G. Bazan, R. R. Brenner, and N. M. Guisto (Plenum, New York, 1977) p. 175.

24. J. G. Stewart, B. N. Baker, E. O. Plante and T. P. Williams, Arch. Biochem. Biophys. 172, 246 (1976).
25. M. Applebury, D. M. Zuckerman, A. A. Lamola and T. M. Jovin, Biochem. 13, 3448 (1974).
26. H. Shichi, J. Biol. Chem., 6178 (1971).
27. K. Hong and W. L. Hubbell, Biochem 12, 4517 (1973).
28. C. C. Farnsworth and E. A. Dratz, Biochim. Biophys. Acta. 443, 556 (1976).
29. T. G. Wheeler, R. M. Benolken and R. E. Anderson, Science 188, 1312 (1975).

Lipid Function in Excitable Membranes

R. M. Benolken
Sensory Sciences Center
University of Texas
Graduate School of Biomedical Sciences
Houston, Texas 77030

Excitable membranes are molecular bilayers which regulate ionic transport rates as a function of stimulus parameters, and in this way, generate the electrical messages of the nervous system. HODGKIN and HUXLEY (1) have provided an elegant description of the excitable characteristics of the squid axon. The HODGKIN-HUXLEY equation describes how ionic transport rates are regulated selectively by the squid membrane, but it does not provide a molecular explanation for the way the lipids and proteins of the membrane accomplish this regulation (2-7). Our understanding of the molecular events in the excitable process is very incomplete at the present time.

In the following, we examine two experimental approaches to the problem of identifying lipid contributions to the excitable properties of intact membrane systems. The first approach involves molecular dissection of invertebrate neural membranes by enzymes which selectively degrade membrane phospholipids. The second exploits biosynthetic pathways to control the fatty acid composition of membrane phospholipids in vertebrate photoreceptors. The approaches are complementary, the strengths of one compensating somewhat for the weaknesses of the other.

Molecular Dissection of Excitable Membranes

Phospholipids are the principal lipid components of membrane bilayers. The major phospholipids are characterized by a 3-carbon glycerol backbone with fatty acids esterified at carbon positions 1 and 2. The phosphate group esterified at the C-3 position is linked to a base group, where the latter is most frequently choline or ethanolamine (see ROBERT E. ANDERSON, this volume). The phosphate and base group at C-3 define the polar domain of the phospholipid molecule, while the fatty acid chains at C-1 and C-2 establish the hydrophobic inner core of the bilayer. Various phospholipases are available which hydrolyze specific phospholipid linkages, and these have been employed to identify functional groups of membrane phospholipids (3). Phospholipases A_1 and A_2 cleave the fatty ester bonds at C-1 and C-2 respectively, and phospholipase C splits the phosphate group at C-3.

ABBOTT et. al., (8) applied a mixture of phospholipases A_1 and A_2 externally to squid axons with negligible effect on the electrical response properties of the cell. However when applied internally to a continuously perfused axon, the same phospholipase mixture suppressed the resting membrane potential as well as the amplitude and rate of rise of the action potential. Voltage clamp data indicated the resting potential and steady-state current were not altered substantially until after the transient inward current was suppressed. Consequently, the initial target of the phospholipase mixture appears to be the sodium conductance mechanism which increases the transport rate of sodium ions during the neural response (2). HINZEN and TAUC (9) extended this analysis to Aplysia neurons where purified phospholipase A_2 from bee venom behaved very much like the mixture of A_1 and A_2 in the squid study. Again there was a marked suppression of the transient inward current with little effect on late outward current (9). These results implicate the fatty acid linkage at the C-2 position of membrane phospholipids with the sodium gating mechanism (10) which initiates the action potential of the neural response.

Phospholipase C from Bacillus cereus behaved very differently from phospholipase A_2 in Aplysia neurons (9). Internal application of phospholipase C increased late outward current of the neural response. The late outward current is associated with an increase in potassium conductance as the neural membrane repolarizes toward the resting state after the peak of the action potential (2). Hence, the phospholipase C result appears to implicate the head group linkage with regulation of potassium transport. GOLDMAN presents an interesting discussion of ionic interactions with phospholipid head groups in his model of axonal excitability (11).

Why external application of phospholipases had little or no effect on the electrical characteristics of the membrane is not clear. One possibility for squid is that the enzymes did not penetrate the Schwann cell membranes which surround the outer envelope of the axonal membrane (8). Another possibility is that the asymmetry of phospholipid class distributions between the inner and outer layers of the bilayer (12) accounts for the observed electrical differences.

Phospholipase activity may modify membrane transport properties as a consequence of reaction products which remain membrane-bound (3, 8). For example, phospholipase A_2 removes C-2 fatty acids and also produces membrane-bound lysophospholipids (in this case, those with a single C-1 fatty acid). Lysophospholipids exhibit detergent properties, and they may solubilize membrane patches to modify transport characteristics (13). Available data do not argue compellingly for or against a membrane solubilization model, but two independent observations suggest it is an unlikely possibility in this instance. Membrane solubilization should be relatively non-selective, and a general increase in leakage current would be expected rather than the observed selective suppression of transient sodium current (8, 9). Further, as discussed in the following section, the electrical characteristics of excitable membranes can be modified by altering membrane fatty acids without measurably altering the lysophospholipid composition of the membranes.

Biosynthetic Control of Membrane Fatty Acid Composition

The outer segment of a rat photoreceptor cell contains a tightly-packed array of many hundreds of flattened disc membranes (14). Most of the membrane-bound protein in outer segment membranes is the photopigment rhodopsin. Outer segments are highly specialized for regulating ionic transport as a function of absorbed quanta; the excitation sequence probably includes quantum absorption by rhodopsin in disc membranes which releases intermediate transmitters which in turn decrease sodium conductance of the plasma membrane of the outer segment (15, 16).

Rhodopsin is assembled in the inner compartment of the rod cell and is transported in membrane-bound form to the basal region of the outer segment where it is incorporated into the plasma membrane (17, 18). The basal region of the plasma membrane is invaginated, and the invaginations seal off to form the disc membranes (14). As new disc membranes are added to the basal region of the outer segment, the oldest disc membranes at the apical tip are cast off in packets which are taken up by cells of the pigment epithelium (19). Turnover and renewal of outer segment membranes proceeds continuously (14), and the photoreceptor membranes of rat rods are renewed every 10 days.

The renewal process also involves lipid synthesis, since opsin is packaged and transported to the outer segment in membrane-bound form (17). The lipid synthesis of concern here involves the fatty acids of the membrane phospholipids, and the biosynthetic pathways for mammalian fatty acids are shown in Fig. 1. The fatty acids are identified as follows: the first number specifies the number of carbon atoms, the number after the colon indicates the number of double bonds, and the number after ω specifies the position of the first double bond from the methyl terminal.

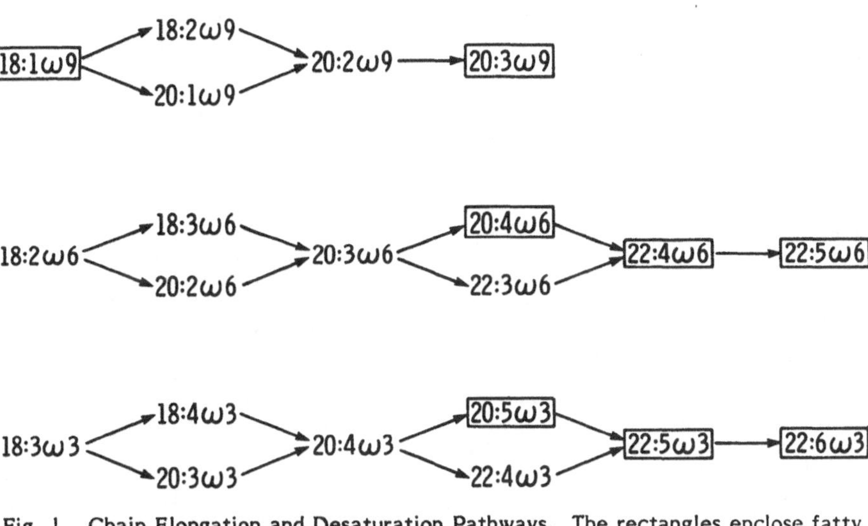

Fig. 1. Chain Elongation and Desaturation Pathways. The rectangles enclose fatty acids observed in photoreceptor membranes.

Vertebrates are able to synthesize ω 9 and saturated fatty acids de novo. However dietary ω6 precursors are required for synthesis of fatty acids of the ω 6 family, and ω3 precursors are required for synthesis of fatty acids of the ω 3 family.

As shown in Table 1, ω3 and ω6 fatty acids are both found in photoreceptor membranes, with 22:6 ω3 the major polyunsaturate of vertebrate outer segment membranes. Fatty acids are not uniformly distributed between C-1 and C-2 of membrane phospholipids. The unsaturated fatty acids are preferentially esterified at the C-2 position of vertebrate photoreceptor membranes (20,21).

Table 1 COMPOSITION OF THE MAJOR FATTY ACIDS FROM PHOSPHATIDYL ETHANOLAMINE OF PHOTORECEPTOR MEMBRANES (in relative mole percent)

Fatty Acid	Rat (22)	Cattle (23)	Frog (21)	Limulus (24)	Moth (25)	Squid (26)	Squid (27)
16:0	---	---	---	---	---	13	---
16:0	7	16	6	2	4	6	4
18:0	34	30	13	12	10	11	4
18:1ω9	4	5	13	4	21	5	1
20:4ω6	8	3	7	12	1	5	34
20:5ω3	---	---	---	41	40	6	8
22:6ω3	45	40	51	---	---	3	47

Since mammalian photoreceptors turnover and are renewed every 10 days, the percentage of ω3 and ω6 fatty acids can be controlled in photoreceptor membranes by dietary manipulation of ω3 and ω6 precursors. After 2 generations of deprivation of these precursors, the percentage of 22:6 ω3 in photoreceptor membranes of albino rats was about half the normal value, and the a-wave of the electroretinogram (ERG) was also reduced substantially relative to the b-wave (28). Additionally, we could not detect significant changes in the rhodopsin content of the retinas or in the phospholipid class composition of the photoreceptor membranes. The light response of the photoreceptor cell membranes appeared to be reduced selectively as a consequence of molecular replacement of ω3 and ω6 fatty acids.

A second series of experiments explored the relationship between the response of the photoreceptor cell and the molecular configuration of membrane fatty acids (29). A control diet, designated fat free, was supplemented with vitamins but was deficient in ω3, ω 6, and ω 9 precursors. Purified ω3, ω6, or ω9 precursors were added to the fat free diet as shown on the abscissa of Fig. 2. Groups of mature rats were maintained on the respective supplements for 40 days. Then the average values of the a-wave and b-wave were measured, and the average ERG values for the various supplements were normalized to the corresponding values for fat free controls as shown on the ordinate of Fig. 2. As expected, the ERG response of the group supplemented with the ω 9 precursors behaved like the fat free response. The a-wave of the ERG is generated primarily by the response currents of illuminated photoreceptor cells (30), and hence the electrical response changes observed with molecular replacement of membrane fatty acids can be assigned to the plasma membranes of the photoreceptor cells with considerable confidence. Whether these

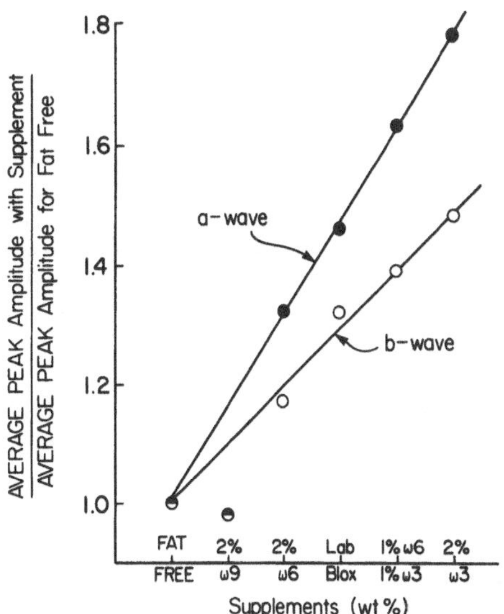

Fig. 2. Ordinate is average peak ERG amplitudes of a-wave (filled circles) and b-wave (open circles) normalized to the corresponding values for fat free. Precursor supplements are shown on the abscissa. The average peak a-wave amplitudes for 2% ω 9, 2% ω 6 and 2% ω 3 precursors were all different from each other with a t-test confidence of no less than 99% for samples of 12 to 16 eyes.

response changes can be assigned specifically to the plasma membrane of the outer segment is much less certain. Light absorption in the outer segment selectively decreases the sodium conductance of the plasma membrane, but we do not know whether fatty acid replacements modify the sodium transport properties of the plasma membrane directly or whether they modify transmitter release by the disc membranes. In either case the response currents of the plasma membrane discriminate between 3 molecular configurations of membrane fatty acids.

Discussion

The phospholipase and fatty acid replacement studies differ in their resolution of the electrical and chemical properties of excitable membranes. The phospholipase studies presently have the advantage in electrical resolution, and by an impressive margin. Spatial voltage clamps of large neural cells of squid and Aplysia permit measurement of transient and steady-state conductance values for these membranes. By contrast the a-wave of the ERG can only assign electrical changes to the average response of photoreceptor cells. The photoreceptor preparation clearly has the advantage for chemical resolution. Highly purified photoreceptor membrane fractions can be isolated from outer segments in good yields (17). By comparison the excitable membrane of the squid axon accounts for only about 0.001% of the preparation (8), and it is intimately associated with myelin which constitutes about 80% of the membrane envelope. Myelin contamination sharply alters fatty acid profiles as shown for nerve preparations with different percentages of myelination in the Gar (5).

These experimental systems differ significantly in other ways. The phospholipase experiments involve an invertebrate, 2-compartment transport system which is specialized for generating an all-or-none neural impulse, whereas the molecular replacement experiments involve a vertebrate, 3-compartment transport system which is specialized for generating a graded receptor response. Yet the general conclusions from the two approaches are in good agreement. Both implicate the fatty acid linkages of membrane phospholipids with sodium transport. Additionally, phospholipase C data from Aplysia neurons suggest the polar head groups of membrane phospholipids are involved in regulation of potassium transport.

Recently the function of membrane fatty acids has been explored in the central nervous system of the rat. LAMPTEY and WALKER (31) altered the fatty acid composition of rat brains by dietary manipulation of $\omega 3$ and $\omega 6$ precursors. Learning deficiencies on a visual task were observed for animals deprived of $\omega 3$ fatty acids, although as noted (31), these behavioral data could not differentiate between learning disabilities and impaired visual discrimination.

Acknowledgement

I am very grateful for grant BNS-75-07197 from the National Science Foundation and for collaborative association with Robert E. Anderson and Thomas G. Wheeler. The manuscript benefitted from the thoughtful criticism of Elnora S. Harcombe.

Reference

1. A. L. Hodgkin and A. F. Huxley, J. Physiol. 117, 500 (1952).
2. A. L. Hodgkin, Proc. Roy. Soc. B 148, 1 (1957).
3. J. M. Tobias, J. Comp. Physiol. 46, 183 (1955).
4. I. Tasaki and T. Takenada, PNAS (USA) 52, 804 (1964).
5. G. K. Chacko, D. E. Goldman and B. E. Pennock, Biochim. Biophys. Acta 280, 1 (1972).
6. T. Narahashi, Physiol. Rev. 54, 813 (1974).
7. E. Roux, ed., Electrical Phenomena at the Biological Membrane Level. (Elsevier, 1977).
8. N. J. Abbott, T. Deguchi, D. T. Frazier, K. Murayama, T. Narahashi, A. Ottolenghi and C. M. Wang, J. Physiol. 220, 73 (1972).
9. D. H. Hinzen, and L. Tauc, J. Physiol. 268, 21 (1977).
10. R. D. Keynes and E. Rojas, J. Physiol. 255, 157 (1976).
11. D. E. Goldman, Biophys. J. 4, 167 (1964).
12. M. S. Bretscher and M. C. Raff, Nature 258, 43 (1975).
13. G. Wald, Exper. Eye Res. 18, 333 (1974).
14. R. W. Young, Invest. Ophthal. 15, 700 (1976).
15. W. A. Hagins, Ann. Rev. Biophys. Bioeng. 1, 131 (1972).
16. M. Montal and J. I. Korenbrot, in The Enzymes of Biological Membranes. Vol. 4, ed. by A. Martonisi (Plenum Press, 1976).
17. D. S. Papermaster, C. A. Converse and J. Siu, Biochem. 14, 1343 (1975).
18. S. F. Basinger, D. Bok and M. Hall, J. Cell. Biol. 69, 29 (1976).
19. R. W. Young and D. Bok, J. Cell Biol. 42, 392 (1969).
20. R. E. Anderson and L. Sperling, Arch. Biochem. Biophys. 144, 673 (1971).
21. R. E. Anderson and M. Risk, Vis. Res. 14, 129 (1974).
22. R. E. Anderson and M. B. Maude, Arch. Biochem. Biophys. 151, 270 (1972).
23. R. E. Anderson, M. B. Maude and W. Zimmerman, Vis. Res. 15, 1087 (1975).
24. R. M. Benolken, R. E.Anderson and M. B. Maude, Biochim. Biophys. Acta, 413, 234 (1975).
25. D. Zinkler, Verh. Deutsch. Zool. Ges., 67, 28 (1975).
26. W. T. Mason, R. S. Fager and E. W. Abrahamson, Biochim. Biophys. Acta 306, 67 (1973).
27. R. E. Anderson, R. M. Benolken, P. A. Kelleher, M. B. Maude and R. D. Wiegand, Chemistry of the Photoreceptor Membranes of squid. (in preparation) (1978).
28. R. M. Benolken, R. E. Anderson and T. G. Wheeler, Science 182, 1253 (1973).

29. T. G. Wheeler, R. M. Benolken and R. E. Anderson, Science 188, 1312 (1975).
30. T. Tomita, Quart. Rev. Biophys. 3, 179 (1970).
31. M. S. Lamptey and B. L. Walker, J. Nutr. 106, 86 (1976).

Photoreceptor Shedding in the Frog Retina

Scott F. Basinger
Department of Ophthalmology
Cullen Eye Institute
Baylor College of Medicine
Houston, Texas

Introduction

The vertebrate retina can be subdivided into two distinct areas, the neural retina and the sensory retina. The sensory retina (see Fig. 1) contains two types of photoreceptor cells, the rods and cones, and is backed by a single layer of epithelial cells, the pigment epithelium. Light initially received by the photoreceptors is transformed into a neural message and then further processed and propagated by the cells of the neural retina. The sensory information then leaves the retina via the optic nerve and passes on to the brain. The maintenance of normal, healthy photoreceptors is essential to the visual process, and in recent years a great deal of research attention has been turned toward an understanding of the physiology of the photoreceptor. This report describes one area of this research, and will focus on the outer segment of the frog photoreceptor cell.

Vertebrate photoreceptor outer segments are composed of stacks of membrane lamella (discs) surrounded by a plasma membrane. In the rod outer segment, the disc membranes are isolated and free-floating except at the base, where they are continuous with the plasma membrane. On the other hand, the cone outer segment is thought to be composed of a continuous membrane with numerous continuities between the disc membrane and the surrounding plasma membrane. The membranes of the outer segment are known to contain the visual pigment molecules (rhodopsin in the rods) responsible for receiving light energy and initiating the visual process.

It is now firmly established that the outer segments of photoreceptor cells in the vertebrate retina are constantly renewed throughout the life of the animal (1). The renewal process was first observed in rod outer segments, and has recently been inferred for cone outer segments as well. Since photoreceptor outer segments remain fairly constant in length throughout adult life, the ongoing renewal process must be balanced by an equivalent disposal process. It is now known that outer segments are kept at a constant length through a process of outer segment shedding, first

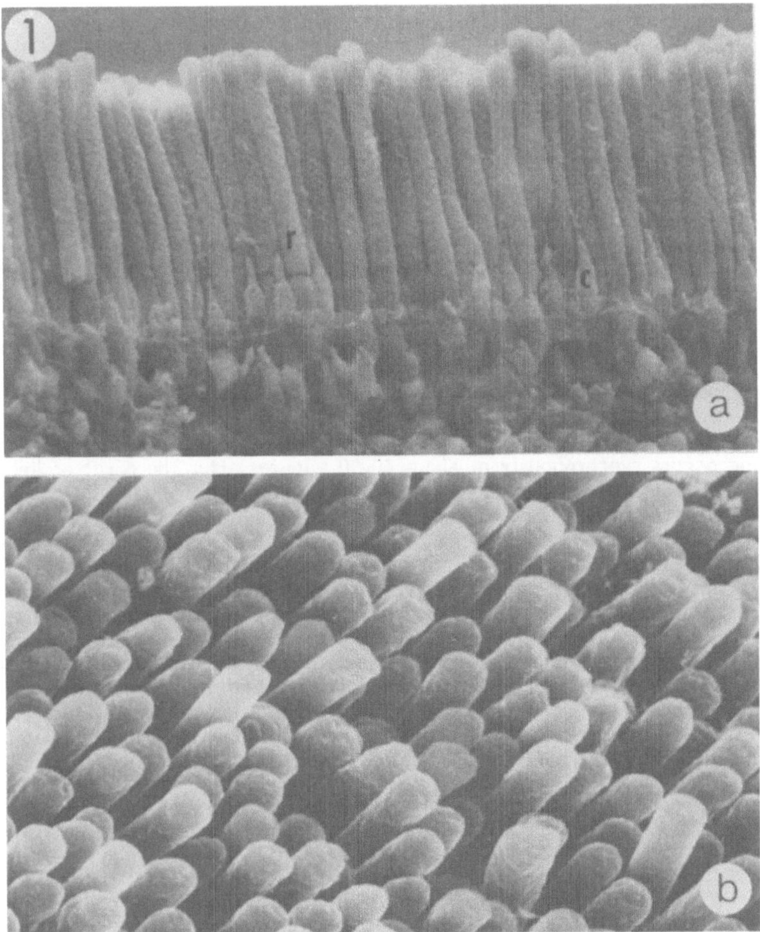

Fig. 1 Scanning electron micrographs of the frog retina.
a. Longitudinal section showing the rod (r) and cone (c) photoreceptors. 850X.
b. View of the outer segment tips, the site of outer segment shedding, normally covered by the pigment epithelium. 1500X.

discovered in rod outer segments (2) and later in cones (3). In both organelles, shedding occurs when the tip of the outer segment is pinched off, then engulfed and phagocitized by the retinal pigment epithelium. Shed outer segment tips, termed phagosomes, have been observed in the pigment epithelium of numerous vertebrate species, but the mechanism of the shedding process is poorly understood. Recently, however, LAVAIL (4) found that rat rod outer segments shed shortly after the onset of light when maintained on a diurnal lighting cycle of twelve hours light:twelve hours dark. This process was confirmed in the frog retina by BASINGER, et. al. (5) and in a number of other vertebrate species (1, 6). Recently, YOUNG (7-9) has presented evidence which suggests that cone outer segments shed in a synchronous manner shortly after the onset of the dark period.

The remainder of this report will describe the shedding process in the frog retina and describe some of the characteristics of shedding in an attempt to better understand what controls this process.

Dark and Light Regulate Rod Outer Segment Shedding

In adult frogs maintained at room temperature on a diurnal lighting cycle of 14 hours light:10 hours dark, approximately 20% of the photoreceptors shed each day. This process is pictured in Fig. 2a - 2d. Figure 2a shows a retina examined at 8 a.m., immediately prior to the onset of light. No phagosomes are visible in the pigment epithelium, although a number of small, dark staining bodies are present (arrow), probably representing the final digestive stage of phagosomes shed during the previous day. Figure 2b shows a retina examined at 10 a.m., two hours after the onset of light. A number of recently shed phagosomes are present above the tips of the outer segments (arrows) in the apical portion of the pigment epithelium. The size and shape of these phagosomes suggests that the shedding process occurred almost synchronously, and examination of earlier time points between 8 and 10 a.m. confirm that shedding occurs synchronously at about 60-90 minutes after the onset of light. Extensive examinations of longitudinal sections at this time reveal that about one out of every five outer segments (20%) has shed on a given day. The fate of the newly shed phagosomes is revealed in Fig. 2c, a retina taken at 6 p.m., ten hours after the onset of light. Extensive digestion of the phagosomes has occurred, and they now appear as dense, dark-staining bodies on the basal side of the pigment epithelium (arrows). Their uniform size and shape suggests that the digestion process, like the shedding process, occurs in a nearly synchronous manner. Figure 2d shows a retina taken at 11 p.m., one hour after the onset of the dark phase. The phagosomes are now extensively digested, and are barely recognized at the light microscope level. No new shedding has occurred, and examination of retinas taken at hourly intervals throughout the day show that shedding occurs shortly after the onset of light and at no other time throughout the complete 24-hour diurnal cycle.

To learn if rod outer segment shedding was a circadian process, or alternately one in which shedding was initiated by light stimulation, retinas from frogs maintained in the dark past the normal onset of light were examined for the presence of phagosomes. Figure 2e shows a retina taken at 10 a.m. and Fig. 2f a retina taken at 10 p.m., both from frogs prevented from receiving the normal onset of light at 8 a.m. No phagosomes are present in the pigment epithelium, suggesting the shedding process in the frog retina is not circadian, but requires light stimulation. Further evidence for this conclusion comes from the observation (not shown) that if frogs are given light stimulation at midnight, eight hours before the normal onset of light, shedding again occurs at its normal 20%. Thus, shedding is initiated by the onset of light rather than occurring at the same time each day without respect to the lighting conditions.

Since only one of every five rod outer segments sheds each day, cross-sections through the apical portion of the pigment epithelium were examined to determine if the pattern of shedding correlated in any way with this tissue layer. In the frog retina, each pigment epithelial cell covers approximately nine rod outer segments, and while some pigment epithelial cells showed as many as four phagosomes, others showed none, suggesting the pigment epithelium is not a determining factor as to which outer segments are shed on a given day.

Newly shed phagosomes in the frog retina measure approximately 10-15% of the outer segment length, and it is known that frogs at room temperature renew approximately 2-2 1/2% of their outer segment length each day. Thus, shedding a

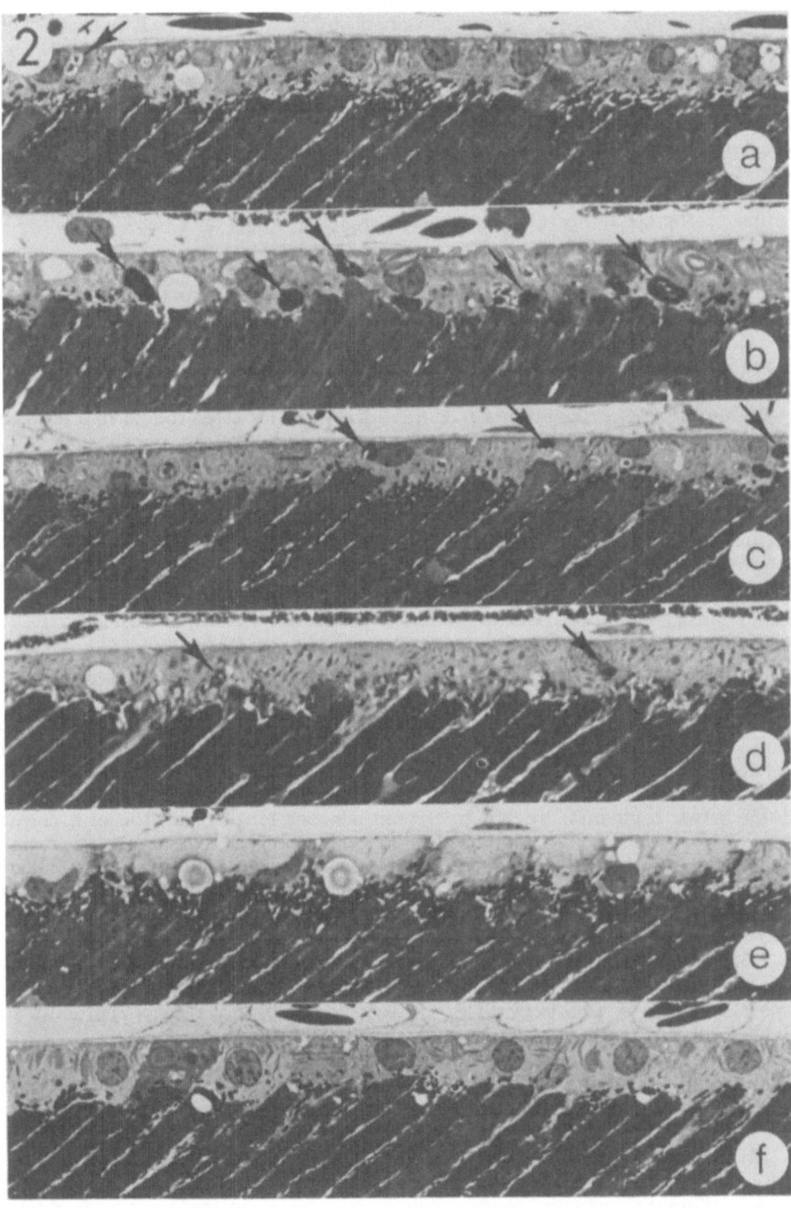

Fig. 2 See following page for legend.

Fig. 2 Light micrographs of longitudinal sections through frog rod outer segments and pigment epithelium. Frogs were maintained at 22° C on a diurnal lighting cycle of 14 hours light (8 a.m. - 10 p.m.) and 10 hours dark (10 p.m. - 8 a.m.) and retinas examined at the indicated times.

a. 8 a.m., just prior to the onset of light. No phagosomes are present in the pigment epithelium, only residual bodies (arrow) representing the final digestive stages of phagosomes shed the previous day.

b. 10 a.m., 2 hours after the onset of light. A number of newly shed phagosomes (arrows) appear in the apical portion of the pigment epithelium. Quantitation of the shedding response reveals that about one of every five (20%) of the photoreceptors have shed.

c. 6 p.m., 10 hours after the onset of light. The phagosomes shed earlier in the day are now displaced toward the base of the pigment epithelium (arrows) and have been significantly digested.

d. 11 p.m., 1 hour after the onset of darkness. By this time, the phagosomes (arrows) are not easily seen at the light microscope level, and are almost completely digested.

e. 10 a.m., dark, retina from a frog kept dark 2 hours after the normal onset of light. Shedding of phagosomes has not occurred due to the absence of light stimulation. The pigment epithelium nuclei have migrated to the apical side of the cell.

f. 10 p.m., dark, retina from a frog kept dark for the entire light period. No new phagosomes are present, again due to the lack of light stimulation. Thus, synchronous shedding occurs within the first 2 hours after the onset of light, and does not occur in the absence of light stimulation.

Fig. 3 Light micrographs of longitudinal sections through the rod outer segments and pigment epithelium of frogs kept under non-diurnal lighting conditions.

a. Retina from a frog kept in total darkness for 5 complete diurnal cycles, then examined at 10 a.m., normally near the peak time of shedding. The absence of light stimulation has almost completely prevented shedding, with only an occasional phagosome (arrow) seen in the pigment epithelium. Due to the continuation of renewal, the outer segments are ten to fifteen percent longer than those in diurnal controls.

b. Retina from a frog kept in total darkness as above, then exposed to 2 hours of light. Numerous new phagosomes are present in the pigment epithelium. At least 50% of the outer segments have shed, and some of the phagosomes are nearly twice the normal size.

c. Retina from a frog kept in constant light for 5 complete diurnal cycles, then examined at 10 a.m. Virtually no phagosomes are observed in the pigment epithelium, due to the absence of a dark period. As in dark-kept frogs, the outer segments are ten to fifteen percent longer than normal.

d. Retina from a frog kept in constant light as above, then given 2 hours of darkness followed by 90 minutes of light stimulation. Virtually all of the outer segments have shed, emphasizing the importance of a dark period in the shedding process.

e. Retina from a frog kept in total darkness for 3 months, then exposed to 2 hours of light. Extensive shedding has occurred, and the outer segments are severely degenerated. Pigment epithelial morphology is abnormal, and these cells are thicker than normal.

f. Retina from a frog kept in constant light for 3 months, then given 2 hours of darkness and 2 hours of light stimulation. Extensive shedding has occurred, and in some cases outer segments have shed as many as four phagosomes. The pigment epithelium is considerably thicker than in normal animals, and although not shown, the rod outer segments are over 50% longer than normal.

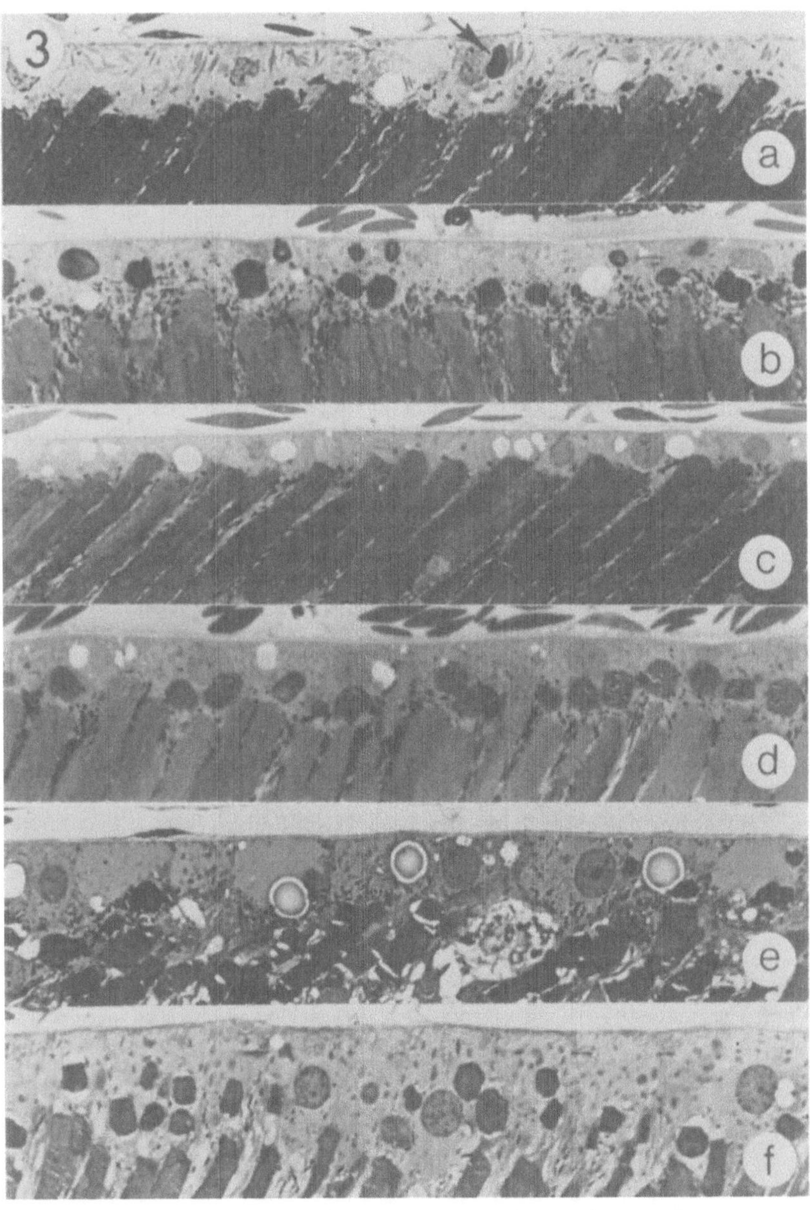

Fig. 3 See preceding page for legend.

phagosome every five days would serve to keep the outer segment at a constant length. This correlates very well with our data showing that 20% of the outer segments shed each day, and implies that the length of the outer segment is an important factor in the shedding process. It is tempting to suggest that only the longest of each five outer segments shed each day, but this speculation remains to be supported by further experimental evidence. Other evidence which suggests that shedding plays an important role in maintaining the outer segments at a constant length can be seen by examining the effect of temperature on the shedding process. In animals maintained at 10°C, where the outer segment renewal rate is greatly reduced, only about 10% of the outer segments shed each day, although the phagosomes are of normal size. Similary, in frogs maintained at 32°C, where outer segment renewal is significantly increased, about half the outer segments shed phagosomes each day, again of normal dimensions.

Effect of Prolonged Light or Dark Exposure

Frogs removed from the normal diurnal cycle and placed in either constant darkness or constant light have revealed a number of interesting features of the shedding process. Figure 3a shows the retina from a frog maintained in constant darkness for five days. The outer segments in this animal are 10 to 15% longer than in normal diurnal controls, and only an occasional phagosome is seen in the pigment epithelium. With outer segment renewal continuing in the dark and reduced shedding taking place, longer outer segments would be expected. Figure 3b shows a retina from a companion animal maintained in the dark for five days then given two hours of light stimulation. Approximately 55% of the outer segments have shed, and some appear to have shed phagosomes twice normal size. This indicates that depriving the frogs of their normal light period severely inhibits, but does not prevent, shedding from occurring, and the ongoing renewal process causes an increase in the length of the outer segments. When light stimulation is used to initiate shedding in light deprived animals, it occurs at almost three times the normal 20%.

Perhaps more interestingly, we have found that depriving frogs of the normal dark period has an even more dramatic effect upon the shedding process. Figure 3c is a retina from a frog maintained for five days in continuous light. As with the prolonged dark frogs, the outer segments are significantly longer than in diurnal controls, and virtually no phagosomes are observed in the pigment epithelium. Figure 3d shows a retina from a frog maintained in constant light for five days, then given two hours of darkness followed by one and one half hours of light stimulation. A large burst of synchronous shedding has occurred and quantitation reveals that nearly every outer segment has shed. This might be expected if all outer segments were prevented from shedding for five full days, and then stimulated to shed. This demonstrates dramatically the importance of a dark period to the shedding process, and indicates that a dark "priming" period followed by light stimulation is responsible for the normal synchronous burst of shedding which occurs in animals maintained on a diurnal light-dark cycle.

Under extreme conditions, prolonged exposure to either constant darkness or constant light followed by a change in lighting conditions necessary to initiate shedding can have a pathological effect upon the outer segments. Figure 3e and 3f show the retinas of frogs maintained in constant darkness or constant light for three months, then stimulated to shed and examined two hours later. Prolonged darkness resulted in a severe retinal degeneration (Fig. 3e), and constant light in greatly elongated outer segments (Fig. 3f). The pigment epithelium in both is abnormal. It should be stressed that this is an occasional effect seen under extreme conditions of light or dark deprivation.

Regulation of Outer Segment Shedding

One of the major goals of our research is to describe in detail the chain of events which eventually result in the shedding of outer segment tips. At the present time, it is fairly clear that at least three things are important to, and perhaps necessary for, outer segment shedding: 1) attainment of a certain length by the outer segment, 2) a period of darkness to "prime" the outer segments for shedding, and 3) a period of light stimulation to "trigger" shedding. In other words, once renewal has caused the outer segment to grow beyond some criterion length, a dark period followed by light stimulation will result in shedding. Our experiments suggest that one hour or less of darkness is sufficient for "priming" and approximately five minutes of normal room illumination (450 lux) is sufficient for light stimulation, but these data reveal little about the molecular mechanism involved in the shedding process.

We have made a number of preliminary attempts to identify the primary light receptor for the shedding process. Obvious candidates are the rods and cones of the lateral eyes, or perhaps the pineal organ, which in amphibians possess photoreceptors thought to contain rhodopsin. Of course, even if the primary light receptor were to be identified as the photoreceptors of the lateral eyes, higher centers of the animal might still be involved in the shedding process through hormone release after stimulation via the optic nerve. Thus, the pituitary and the pineal glands, and their respective neuroendocrine products, are possible candidates for a role in the shedding process. In a preliminary experiment intended to examine the action spectrum of the primary light receptor for shedding, we found that when frogs were stimulated with an equal flux of blue (400 nm) light, green (520 nm) light, or red (620 nm) light only the green was effective in initiating shedding, although melanin granule migration occurred with all three colors. This experiment only shows that the primary light receptor is a pigment which absorbs in the green part of the spectrum, and of course the outer segments in both lateral eyes and pineal fit this criteria. However, some support for intraocular control of shedding was provided by a recent observation that frogs maintained in constant light for 20 days had shed spontaneously in one eye but not the other (10).

In summary, our experiments have shown that shedding in the frog retina is not a true circadian process, but one controlled and entrained by a dark period followed by light stimulation. Light deprivation, and particularly dark deprivation, are both effective in preventing shedding, and under extreme conditions, can result in severe degeneration of the outer segments. Finally, our initial results suggest that the mechanism for shedding is not contained solely within the eye, but may involve the higher centers of the animals, possible the pineal or the pituitary.

Acknowledgement

This work was supported by grants from the Retina Research Foundation, Houston, Texas, and the USPHS, No. EY 01406. The excellent assistance of Rosemary Hoffman, Michael Matthes, and Luis Marroquin is gratefully acknowledged.

References

1. R. W. Young, Invest. Ophthal. 15, 700 (1976).
2. R. W. Young, J. Cell Biol. 33, 61 (1967).
3. D. H. Anderson and S. K. Fisher, J. Ultrastruct. Res. 55, 119 (1976).
4. M. M. LaVail, Science 194, 1071 (1976).
5. S. Basinger, R. Hoffman and M. Matthes, Science 194, 1074 (1976).
6. J. G. Hollyfield, J. C. Besharse and M. E. Rayborn, Exp. Eye Res. 23, 623 (1976).
7. R. W. Young, J. Ultrastruct. Res. 61, 172 (1977).

8. R. W. Young, Invest. Ophthal. 17, 105 (1978).
9. W. T. O'Day and R. W. Young, J. Cell Biol. 76, 593 (1978).
10. J. G. Hollyfield and S. F. Basinger, submitted for publication.

Mapping Retinal Features in a Freely Moving Eye with Precise Control of Retinal Stimulus Position

D. Max Snodderly, W. P. Leung, G. T. Timberlake, and D. P. B. Smith
Eye Research Institute of Retina Foundation
20 Staniford Street
Boston, Massachusetts 02114

Introduction

The primate retina has evolved to cope with a diverse array of problems posed by a complex natural environment. One solution employed by the retina is regional specialization. At its center the retinal fovea has tightly packed cone receptors, exquisitely refined spatial acuity, and no rods. As one moves from the fovea into the peripheral retina, visual acuity decreases rapidly (MILLODOT, (1)) along with a decrease in cone receptors and an increase in number of rods. At only 20 degrees eccentricity, the rods have reached their maximum density,and at more peripheral locations, their density declines steadily (reviewed by RODIECK (2)). Along with these variations in receptor populations, there are corresponding variations in interneuron populations (reviewed by SNODDERLY (3)).

Attempts to relate visual performance to the underlying retinal anatomy and neurophysiology would be strengthened by the ability to specify exactly where a visual stimulus falls on this extremely heterogeneous neural network. The main obstacle preventing this is the incessant movement of the eye that occurs even when a subject is instructed to fixate a point or inspect a tiny object. In the past, measurement of small eye movements has required a cumbersome attachment to the eye, such as a contact lens, that tended to degrade visual optics. This has deterred most investigators from simultaneously monitoring fixational eye position and visual performance.

Only very recently have developments in eyetracking technology made it possible to monitor eye position without attachments to the eye or interference with normal viewing conditions (CORNSWEET and CRANE (4); CRANE and STEELE (5)). This approach employs an infrared beam to form the first and the fourth Purkinje images, which are monitored remotely via servomechanisms. We have used this Double Purkinje Image eyetracker to monitor eye position in humans while simultaneously measuring visual thresholds in the vicinity of specific retinal structures. The experiments have gone through three levels of precision with increasing degrees of refinement: 1) Restricting the direction of gaze to a small "window". 2) Measuring precise eye position at the moment of stimulus presentation, and 3) Compensating for fixational eye movements so as to control the stimulus location exactly.

One important feature of our studies is the emphasis on relating an optical measure of eye position to a psychophysical measure of retinal position. This can be viewed as a test of the use of voluntary eye movements to calibrate the eyetracker. The assumption is usually made that the eye rotates about a fixed center of rotation and brings the fixation point to the center of the fovea each time. If the eyetracker has a stable linear output, the position of retinal structures should then be predictable at any indicated eye position. However, it is important to realize that 1) the eye does not have a fixed center of rotation (Alpern (6)), and 2) it is not trivial to prove that a given subject brings the fixation point to the center of his fovea on any given trial. In fact, we believe that some subjects deviate slightly from this expected behavior during long experimental sessions. The importance of these violations of the initial assumptions can only be known when accurate measurement of retinal position is possible.

Soon we plan to apply these eyetracking techniques to single-unit recording experiments in chronic monkeys. This is a direct approach to relating local neuronal mechanisms to visual performance, and it represents one of the most demanding tasks for an eyetracker. The position of a receptive field on the retina can be specified within a few minutes of arc, and, for detailed analysis, the position of the stimulus on the retina must be known with comparable accuracy (See SNODDERLY, et al. (7) for discussion and references). The human studies we report here indicate that this level of accuracy is possible with the double-Purkinje-image eyetracker in a tightly controlled experiment.

General Methods

The subject's head was restrained with a dental-impression bite board and a 2-point forehead rest mounted on a 3-way positioning device (Fig. 1). Eye position was continuously monitored with a double-Purkinje-image eyetracker (SRI International, Menlo Park, California). The subject viewed the stimulus display with the right eye through a dichroic mirror that reflected the infrared beam of the eyetracker and transmitted the visible light of the mapping stimulus. The left eye was occluded by an eyepatch.

An optical mount immediately in front of the dichroic mirror contained a beam-splitting pellicle (National Photocolor), an electroluminescent panel (T. L. Robinson Co.) to provide a uniform background adapting field, and two Kodak Wratten 55 color filters (Fig. 2). The mapping stimulus was a point of light flashed for 5 msec on the face of a Tektronix 604 display monitor with a green P31 phosphor. Since the color of light emitted by the P31 phosphor changes slightly with beam current, the Wratten filters were chosen in part to minimize color changes and match the stimulus color to that of the adapting background. Their spectral range (480-580 nm, dominant wavelength 530 nm with Illuminant C) is also convenient for delineating vessels because it is strongly absorbed by hemoglobin.

The CRT light output as a function of Z-axis voltage was calibrated with a Gamma Scientific telephotometer (model 2020-31) with a 1-min aperture. The luminance of the measuring spot was one log unit less than the peak value at ±0.7 mm (±4.2 min of arc).

The CRT output was approximately logarithmic with input voltage over nearly

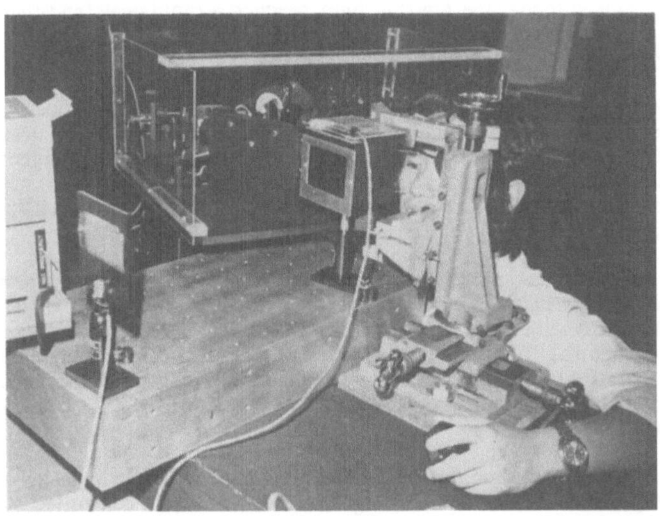

Fig. 1 Subject in position for retinal mapping. Eye position is monitored by the eyetracker overhanging the subject's right shoulder. The visual field is viewed through a dichroic mirror extending in front of the subject's right eye and a beamsplitting pellicle placed on the visual axis. A green electroluminescent panel covered by an opaque mask with 0.85 mm diam holes provides the fixation display. A cathode-ray display unit to the right of the fixation display could be positioned independently to present flashed point stimuli.

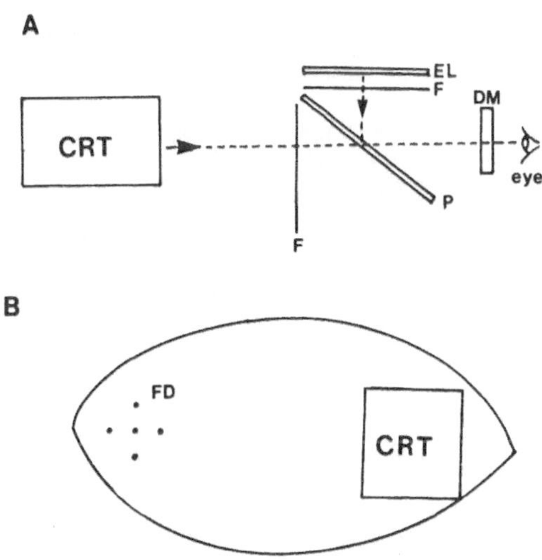

Fig. 2 A. Optical path of visual stimuli. DM, Dichroic mirror; EL, Electroluminescent panel; F, Wratten 55 filter; P, pellicle; CRT, cathode ray tube display. B. Subject's vew of visual field. The circular pellicle frame appeared elliptical in projection and both the fixation display, FD, and the CRT were included within the adaptation field.

three log units. In the pulsed mode, the emitted light reached a rapid peak, and then decayed over a period of about a second to a lower level. This made it essential to calibrate the CRT under precisely the conditions being used.

A fixation display was positioned 57.3 cm from the subject on the visual axis. It consisted of a central fixation point flanked by 4 points that were spaced 1 degree horizontally and vertically from the center (Fig. 2). For calibrations, the subject fixated each point without blinking for 5 to 10 secs. Longer fixation periods were not used because the tear film begins to break up (LEMP (8)) and the intensity of the Purkinje images is variable.

The eyetracker output was amplified, filtered for a 100 Hz bandwidth, and sampled at 200 Hz by a PDP 12 computer. Averages of eye position for the fixation periods were calculated on-line and used to establish 1) a reference zero corresponding to the central fixation point and 2) the gain of the system (volts/degrees of arc rotation). No corrections were made for tangent distortions. Analog outputs from the computer controlled the X, Y, and Z axes of the CRT display for stimulus presentation.

The subject initiated all experimental trials, including calibration, by pressing a "ready" button. Trials were cued with auditory signals from a Sonalert controlled by the computer as indicated in Fig. 3. When the subject was responding rapidly, a trial could be completed within three to four sec.

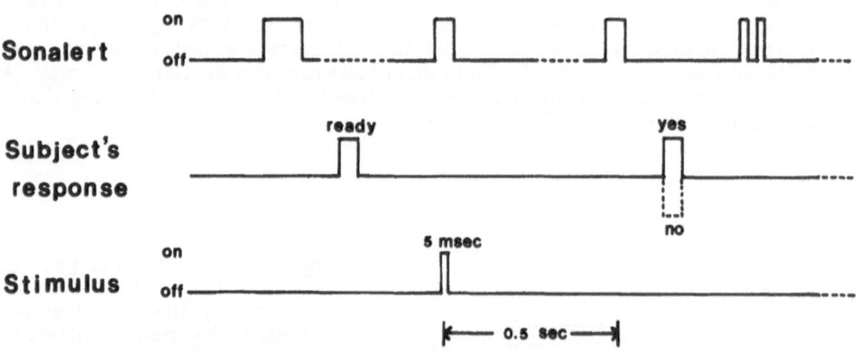

Fig. 3 Temporal patterning of a mapping trial. The computer provided auditory cues for the subject by activating a Sonalert in the manner shown on the top line. Each trial began with a 'beep' that indicated the computer was ready to accept a trial. The subject then pressed a button to initiate the trial and the stimulus was flashed for 5 msec a variable fraction of a second later, simultaneously with a second beep. The third beep signaled that the computer was ready to accept the subject's response, which consisted of closing one of two switches to indicate whether the stimulus was seen or not. A pair of closely spaced beeps informed the subject that his response had been recorded and the trial was over.

Specific Methods and Results

Level 1. Mapping retinal features with restricted deviation of gaze: the blind spot and angioscotoma

The largest and most obvious retinal feature is the optic disk. Its projection into the visual field is the blind spot, which is routinely mapped in standard clinical examinations, and is demonstrated to countless students in introductory psychology and physiology courses. With careful perimetric techniques, it can be shown that the blind spot has fingerlike extensions, the angioscotoma, that are thought to be the shadows of the retinal blood vessels (EVANS (9)). However, we have been unable to find any published account in which eye position was accurately monitored to assure that eye movements did not contaminate the results. Furthermore, only ophthalmoscopic examination of the eye has been used to verify that the angioscotoma actually conforms to the structure of the retinal vascular tree.

Our retinal mapping studies therefore began with a verification and extension of these early results. As a first level of control we restricted the range of eye position to a small "window". The mean eye position during fixation of the center of the fixation display was taken as a reference that we will call the mean fixation locus. In order for a trial to be accepted, the subject's direction of gaze had to be within ±15 min of arc of this initial reference. Usually the subject's performance was better than this, but occasionally the eye would wander outside the window and the subject would hear an auditory warning signal. A window smaller than this produced enough warnings to be frustrating to the subject.

Once the fixation criteria had been met, stimuli were flashed in a regular spatial grid in random order on the face of the CRT as previously described. The blind spot was localized by choosing a stimulus intensity that was invisible within the blind spot but easily distinguishable outside it. A low photoptic background illumination reduced the effects of scattered light so that a clear delineation of the blind spot was obtained. The subject indicated on each trial whether or not the point was seen.

First a coarse array of points 80 to 60 min of arc apart were tested, and then the edges of the blind spot were outlined by selected testing of several fine arrays of points 10 to 7.5 min of arc apart. Two separate sessions were run on two different days and a map consisting of all the points seen by the subject was constructed for each session. These two maps were juxtaposed with the aid of a fundus photograph of the same subject to form the composite map shown in Fig. 4.

Fig. 4 Blind spot and angioscotoma mapped with restricted deviation of gaze. Composite formed by juxtaposition of top half mapped in one session (9/29/76, subject GTT) and bottom half mapped the next day. Points of light were flashed in rectangular arrays throughout the entire enclosed area and all points seen by the subject are shown. Widely spaced points are 80 min arc apart in the top half and 60 min arc apart in the bottom half of the figure; closely spaced points are 10 min arc apart in the top half and 7.5 min arc apart in the lower half.

The composite map was superimposed on the fundus photograph to illustrate that the gaps formed by the unseen points correspond to the optic disc and the retinal vessels as would be predicted by the older work (Fig. 5). However, only the largest vessels were reliably shown, and the outlines of some of them disappeared rather rapidly as they diverged from the optic disk. Nevertheless, the success in matching the psychophysical map to the vessel arrangement encouraged us to improve our technique so that we could map the smaller vessels and follow them closer to the fovea.

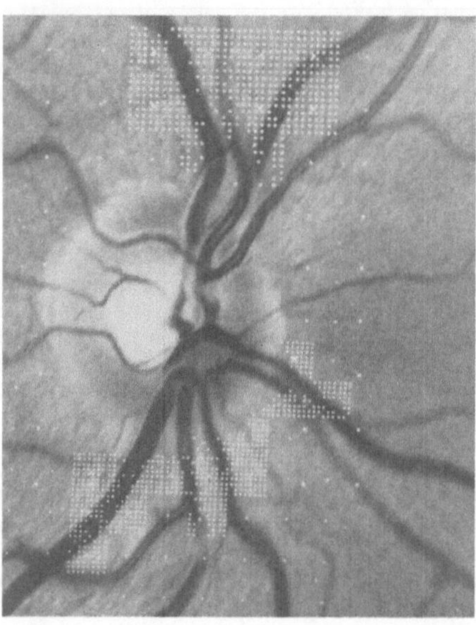

Fig. 5 Superposition of the map of Figure 4 on a fundus photograph of the same subject. Magnification selected to give the best fit.

Level 2. Recording the retinal position of the stimulus spot

The most accurate way to determine the relative spacing of the array of stimulus points on the retina is to record the exact deviation of the eye from the reference fixation locus at the instant the stimulus is flashed. Then the points seen by the subject can be placed on a map in an irregular but accurate representation of what actually happens. The paradigm illustrated in Fig. 6 makes this possible.

After the subject initiates a trial, a stimulus is flashed for 5 msec. If the stimulus occurs during a drift period the position of the eye is recorded and the trial is accepted because the retinal location can be accurately defined.

If the stimulus occurs within ± 0.2 sec of a saccade, the trial is rejected and is not displayed on the map. We applied this restriction for two reasons. First, the position of a stimulus on the retina is basically undefined during a saccade. The eyetracker has about a 2 msec time lag in its response, which produces a small position error when the eye is moving rapidly. Also, the saccade waveform frequently has large transient overshoots when the eye moves abruptly from one drift level beyond the next drift level and back. A saccade of this type is labeled "S" in the record of vertical eye position shown in Fig. 6. Part of the transient overshoot may

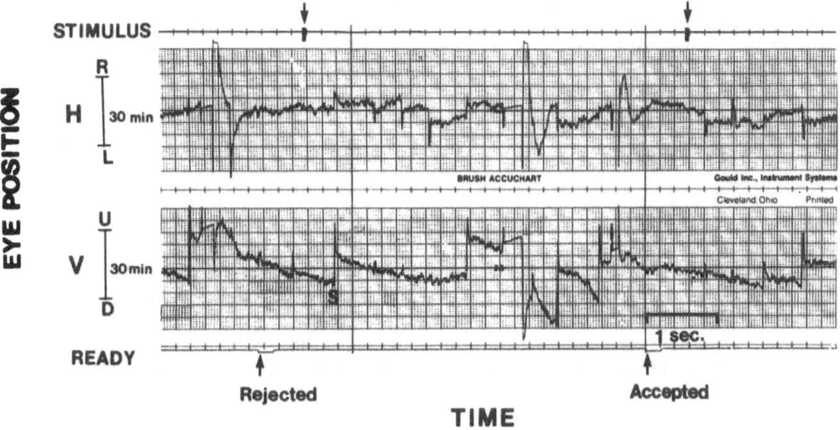

Fig. 6 Chart records of two mapping trials. Downward deflections (arrows) on the event channel of the topmost trace indicate the time of occurrence of a flashed, 5-msec stimulus spot. The next two traces are continuous records of horizontal and vertical eye position respectively. The bottom trace is a second event channel recording with a downward deflection (arrows) the time when the subject pressed a "ready" button to initiate a trial. The first trial was rejected because a small vertical saccade occurred within 200 msec of the stimulus flash. A larger vertical saccade with an overshoot typical of those recorded with the Double-Purkinje image eyetracker is labeled S.

be due to lateral displacements of the lens within the eye during the saccade (CRANE and STEELE (5)). The lens movement presumably causes small retinal image displacements but there is no way to measure the magnitude of the displacements at present; it is estimated that they are only about one-tenth the indicated magnitudes of the overshoots.

The second reason for disallowing trials when the stimulus falls near a saccade is the rise in visual thresholds before and after a saccade known as saccadic suppression. Our unpublished pilot studies indicate that small but measurable visual suppression does occur during fixational saccades under our conditions. According to earlier work the threshold should be unaffected by saccades as long as the stimulus is not within 200 msec of the saccade (See MATIN (10) for a review).

Figure 6 compares two trials, one of which was rejected because of a saccade that occurred within 200 msec of the stimulus flash. Saccades were detected on-line by the computer by continuously calculating an approximation of the average angular speed of the eye. The horizontal (H) and vertical (V) position of the eye at each instant was subtracted from the values recorded 25 msec earlier to produce the differences in eye position ΔH, ΔV. The 25 msec time epoch was chosen because it is about the duration of a small fixational saccade (DITCHBURN (10)) and it appeared to be near optimum for discriminating small saccades from instrument noise. The average angular speed of the eye during this epoch would be the Pythagorean sum $\left[(\Delta H)^2 + (\Delta V)^2\right]^{1/2}$ but we used the following numerical approximation to save computer time: Taking absolute values of H and V, the larger value was added to half the smaller one. The maximum error of this approximation is about 12%. When this number exceeded a criterion value we considered that a saccade was occurring.

For most of our experiments we used a criterion of five degrees/sec for detecting small saccades. If we define a saccade as a rapid displacement completed within 30 msec, the program detects saccades 10 min of arc or larger as judged by visual comparison with chart records. Smaller saccades can be detected at lower criterion levels, but they can not be discriminated reliably from the noise. When noise pulses begin to trigger the saccade detection routine, the subject becomes frustrated because too many trials are rejected and the experiment becomes tediously long.

Our eyetracker (designated Version IIIc by the designers) currently has a noise level of about 3 to 4 min of arc, peak-to-peak when it is adjusted to track several subjects without optimizing for each individual subject. We reduced the contribution of the high frequency noise during drift periods with no detected saccades by averaging eye position over a 0.1 sec period centered on the moment the stimulus was presented. This is preferable to simple continuous low-pass filtering which would cause the signal to decay slowly back to a correct level after all fast transients, such as large saccades.

The position accuracy that is achieved with this approach can only be estimated because it involves several statistical processes that are difficult to measure. Averaging eyetracker noise over 20 samples (0.1 sec) effectively reduces the system bandwidth by a factor of 20. If the noise has a flat power spectrum, the noise power will also be reduced by a factor of 20, and the noise amplitude will be reduced by $(20)^{1/2}$ or 4.5. This means that less than 1 min of arc error should be contributed by tracker noise. Noise reduction will be less if the power spectrum is not flat and low frequencies are especially prominent.

Drift movements will contribute to position error if there is an acceleration of the eye during the averaging period. In the worst case, a drift movement may be as fast as 30 min arc/sec and it could reverse direction at the moment the stimulus is presented. This would cause an error in the computed position of 0.75 min arc.

The largest error will be introduced by undetected saccades that are included in the average. If we assume that the saccade waveform can be approximated by a ramp lasting for 30 msec, the worst case error can be estimated. The stimulus could occur when the eye concluded a 9 min of arc saccade and then immediately reversed direction to saccade back to the original position. In this worst case the error could be as much as 6 min of arc. Smaller bi-directional saccades would produce correspondingly smaller errors and a unidirectional saccade of 9 min arc would only introduce an error of 3 min of arc.

This means that the errors should be distributed over a range of about 1 to 6 min of arc. The frequency of errors of a given magnitude depends on the frequency distribution and time of occurrence of the undetected saccades. Since it is unlikely that stimuli would fall exactly at the peak of a saccade and that many bi-directional saccades were as large as 9 min, errors as large as 6 min should have been quite infrequent.

With this improved approach, we repeated the blind spot maps on two subjects, again starting with a coarse array, and then increasing definition with smaller fine arrays of closely-spaced points. Subjects had to respond yes or no on every trial and catch trials with no stimulus were included every 6 to 10 trials, on the average. If the subject responded yes on a catch trial, the computer issued a patterned train of beeps that sounded enough like a laugh to embarrass the subject into stable and conservative performance. We wanted to bias the subject slightly toward conservative performance, in order to minimize responding to scattered light when the stimulus fell on vessels or the optic disk.

Figures 7 and 8 illustrate the effects of fixational eye movements on retinal stimulus position under our experimental conditions. The precise array of points in space is transformed into a messy scattering of points on the retina. One must be cautious in interpreting these arrays, because the random clumping of points gives the appearance of a fine structure that is illusory. In particular a blank region can be caused merely by the failure of stimuli to land on that part of the retina due to the wanderings of the eye. For this reason we have always constructed two maps, one of them showing the points seen by the subject (the positive map), and the other showing

Fig. 7 Effect of fixational eye movements on intended stimulus position. Coarse array. The outer borders and the vertical center line are formed from reference points spaced 1 degree of visual angle apart. The left panel shows the spacing of stimulus positions in a coarse array as they were presented in space. The right panel shows how fixational eye movements actually distributed these points on the retina. Subject WPL, 3/22/77.

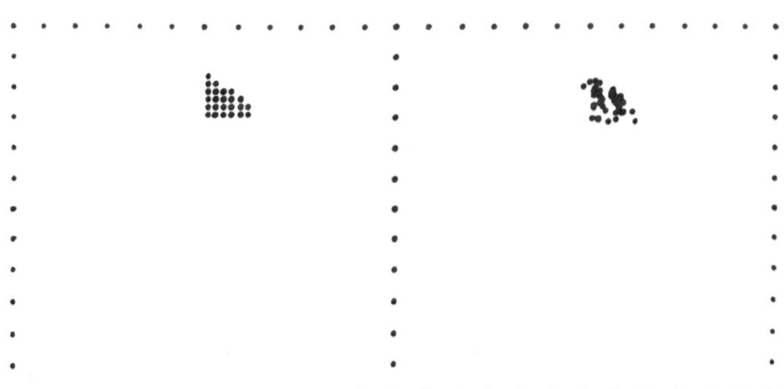

Fig. 8 Effect of fixational eye movements on intended stimulus position. Two fine arrays superimposed. Only points not seen by the subject (negative points) are shown. Same conventions as Fig. 7. Subject WPL, 3/22/77.

the points not seen by the subject (the negative map) for simultaneous comparison. If the blank regions in one map correspond to filled regions in the other, then one knows that enough points on the retina have been stimulated and the patterns within the map should be caused by retinal structures.

Our most extensive maps are shown in Fig. 9. The negative map is superimposed on a fundus photograph of the same subject in Fig. 10. It is clear that the points not seen by the subject fall primarily within the optic disk and on the retinal vessels.

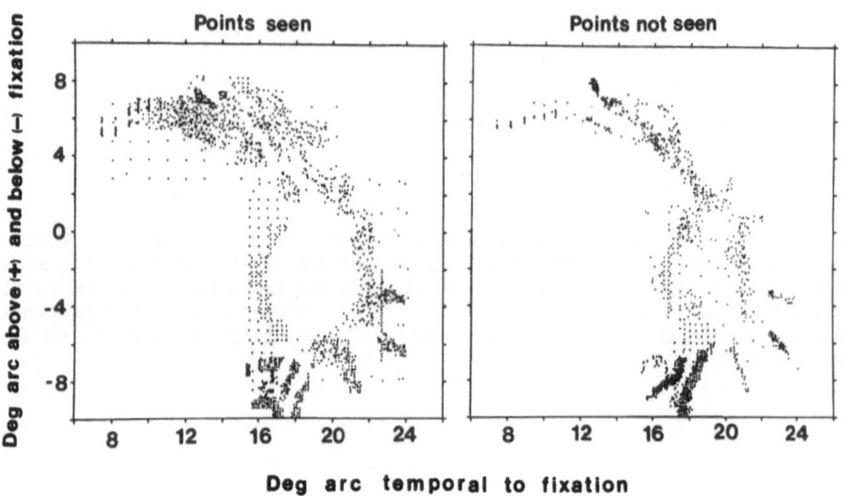

Fig. 9 Blind spot and angioscotoma of subject WPL. Composite map formed from seven sessions. The mean fixation locus is taken as the zero position and distances are indicated as horizontal and vertical degrees from that reference point. The left panel is the array of points seen by the subject (positive map); the right panel is the array of points not seen by the subject (negative map). The irregularly spaced points within the upper right regions were obtained according to Level 2 methods of recording eye position and the regular arrays of points on the left side of each panel were obtained using position compensation (Level 3).

As the points on the map accumulated we attempted to follow the vessels as close to the fovea as possible. Since detection thresholds gradually decreased when we moved from the optic disk toward the fovea, we decreased the luminance of the mapping spot until it was just visible in each new region of the retina. By doing this we were able to follow vessels in an orderly manner but still rather tediously because many of the stimuli missed the vessel when the eye made an unanticipated fixational movement.

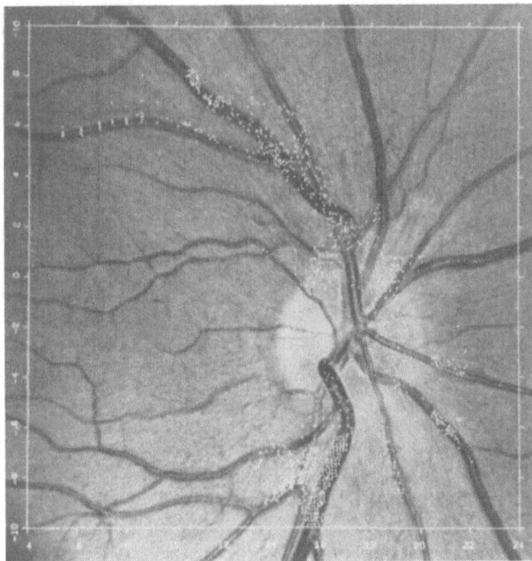

Fig. 10 Superposition of the negative map of Figure 9 on a fundus photograph of the same subject.

Level 3. Compensation for eye movements

In order to improve the efficiency of the mapping, we eventually found it most convenient to compensate for eye movements by varying the position of the stimulus. This is done by the Level 2 computer program as described above with the added feature that the program positions the stimulus at the desired location based on the direction of gaze of the subject immediately before the stimulus is flashed. We call this "position compensation" rather than image stabilization because the stimulus is only flashed briefly. As before, intervals near saccades were disallowed.

The points on the map that were determined with position compensation can be identified by the regular lines and grids that they form. Many of the points in the lower part of Fig. 9 and in the upper left are part of this group. These points are subject to a minimum position error of ± 2 min of arc because no averaging is used to reduce the noise of the eyetracker.

A relatively efficient means of locating the path of a vessel is to present the stimulus in a series of linear arrays that cross the vessel repeatedly. This was done in the upper left corner in order to follow the small horizontal artery within about 6 degrees of the fovea.

We now felt that the correspondence between the maps and the geometry of retinal vessels even within the central retina was convincing. Furthermore, the good fit over such long distances rules out the presence of large torsional movements that would be undetected by the eyetracker. This is consistent with FENDER's (12) measurements with contact lens techniques showing that torsional movements during maintained fixation are quite small.

Since the maps include data gathered during many long, separate sessions, we were aware that several types of errors were being incorporated. These included long-term electronic drifts of the eyetracker; apparent shifts in mean fixation locus associated with subjects getting on and off the bite bar; and possible differences in the calibration of the mean fixation locus from session to session. The electronic drifts were eventually identified as thermal in origin and they were reduced to less than 2 min per hour by providing additional ventilation for the servomechanism control circuitry. We then were ready to improve the precision of localizing retinal landmarks so as to gain control of the other sources of position error.

For this purpose we refined our technique to include measurement of increment threshold at each retinal point that was tested. The computer program was rewritten to incorporate a double random staircase for determining threshold (CORNSWEET (13)), and the adapting background was fixed at 0.17 millilamberts. The details of the technique and more extensive data will be presented in a later publication, but for present purposes, we only wish to demonstrate how well this allows us to localize the center of a retinal vessel.

Figure 11 shows two threshold profiles measured across the horizontal vessel near the fovea in Fig. 10 with a stimulus spacing of 10 min. A peak whose location presumably corresponded to the center of the vessel could be localized within 10 min of arc from these curves.

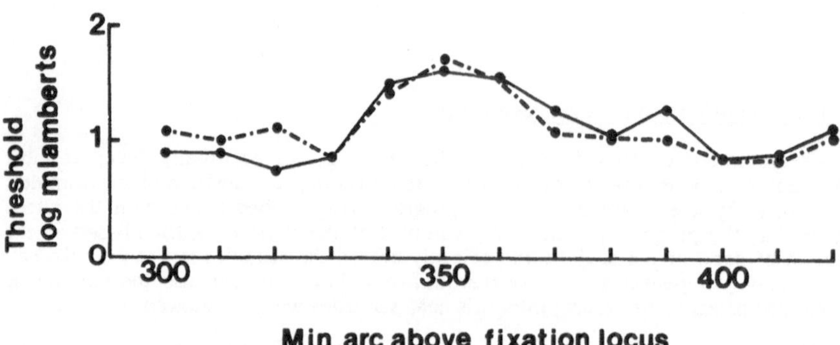

Fig. 11 Threshold profile along a vertical linear array of points intersecting the horizontal vessel in the upper left quadrant of Figure 9 at ten degrees to the right and six degrees above the fixation locus. Upward movement in the visual field corresponds to translation to the right on the horizontal axis of the graph. Two consecutive profiles were run.

In separate experiments we tested the accuracy of position compensation by determining the threshold profile across a retinal vessel before and after instructing a subject to shift gaze 1 degree from center. The location of the peak remained in the same relationship to the direction of gaze regardless of where the subject looked. This meant that position compensation was working at least as accurately as our vessel measurements would allow us to measure (i.e., 10 min arc).

In cases where the fixation shifts are small, rapid, or are diagonal, the task of giving an independent confirmation by mapping of retinal vessels is extremely difficult. It would be preferable instead to localize the position of the fovea in space by means of some sharply peaked threshold function. This is the reference point that one would like to use for many studies of visual function.

We tried to extend our scotoma mapping to the fovea by dark-adapting the subject and using an extra-foveal fixation display. We hoped to find a region of elevated threshold that would correspond to the rod-free area thought to be 1 degree or more in diameter (RODIECK (2); HOGAN, ALVARADO, and WEDDELL, (14)). To date, however, we have been unable to obtain a convincing threshold increase in the fovea under our experimental conditions. This is consistent with the small threshold increases found by STILES (15) in the dark-adapted fovea but not the larger ones reported by CROZIER and HOLWAY (16) and CRAWFORD (17). It means that we will have to devote more attention to this problem.

Summary and Conclusions

A new eyetracking technique has been combined with psychophysical threshold determinations to specify the projected locations of retinal landmarks in visual space. The retinal vessels can be followed within a few degrees of the fovea and their axis centers can be located within at least 10 min of arc.

This provides an independent validation of the eyetracking technique and verifies that it can be used to place a stimulus in a known location on the retina. It opens the way to studying visual function intensively in local retinal areas so that psychophysical data and characteristics of the underlying neural mechanisms can be compared under identical conditions.

Acknowledgments

This work was supported in part by the National Institutes of Health (Grant EY 01520), the American Heart Association (Grant 74-760), and the Massachusetts Lions Eye Research Fund. We thank Sandra Spinks and Rita Raskin for expert help in preparing the illustrations. D.M.S. also holds an appointment in the Department of Ophthalmology, Harvard Medical School, Boston, Massachusetts 02115.

References

1. M. Millodot, Br. J. Physiol. Optics 27, 24 (1972).
2. R. W. Rodieck, The Vertebrate Retina, (San Francisco, Freeman, 1973).
3. D. M. Snodderly, In Perspectives in Primate Biology, A. B. Chiarelli (Ed.), (New York, Plenum Press, 1974).
4. T. N. Cornsweet and H. D. Crane, J. Opt. Soc. Amer. 63, 921 (1973).
5. H. D. Crane and C. M. Steele, Appl. Opt. 17, 691 (1978).
6. M. Alpern, In The Eye, Vol. III, sec. ed., ed. by H. Davson, (New York, Academic Press, 1969).
7. D. M. Snodderly, H. A. Swadlow and R. B. Barlow, Exp. Brain Res. 31, 179 (1978).
8. M. A. Lemp, In The Preocular Tear Film and Dry Eye Syndromes, ed. by F. J. Holly and M. A. Lemp, Vol. 13, No. 1 of International Ophthalmology Clinics, (Boston, Little Brown, 1973).
9. J. N. Evans, An Introduction to Clinical Scotometry, (New Haven, Yale Univ. Press, 1938).
10. E. Matin, Psychol. Bull. 81, 899 (1974).
11. R. W. Ditchburn, Eye-Movements and Visual Perception, (London, Oxford Univ. Press, 1973).
12. D. H. Fender, Br. J. Ophthalmol. 39, 65 (1955).
13. T. N. Cornsweet, Am. J. Psychol. 75, 485 (1962).
14. H. J. Hogan, J. A. Alvarado and J. E. Weddell, Histology of the Human Eye, (Philadelphia, Saunders, 1971).

15. W. S. Stiles, Doc. Ophthalmol. 3, 138 (1949).
16. W. J. Crozier and A. H. Holway, J. Gen. Physiol. 22, 351 (1939).
17. M. L. J. Crawford, Brain Res. 119, 345 (1977).

Acute Effects of Alcohol and Marijuana on Vision

Anthony J. Adams, O.D., Ph.D.
University of California School of Optometry
Berkeley, California 94720

Summary

Alcohol and marijuana, both widely used socially as recreational drugs, produce changes in vision and vision performance which last for some hours after drinking or smoking. In our study, experiments were conducted double-masked using a placebo in a replicated Latin square design. In general, the oculomotor changes were more apparent than sensory changes and alcohol invariably produced larger vision deficits than marijuana for "equivalent" levels of social use of the drug.

The major oculomotor findings are 1) an increase in tonic convergence (more with alcohol), 2) the maximum velocity of smooth eye tracking is decreased by alcohol and not by marijuana, 3) optokinetic nystagmus and peripheral gaze nystagmus are affected by both drugs but more by alcohol, 4) pupil size is reduced by marijuana and unaffected by alcohol. The major visual sensory findings are 1) no change in visual acuity with either drug, 2) a marked reduction in the acuity of moving objects by alcohol and to a lesser extent by marijuana, 3) a prolonged glare recovery associated with either drug, 4) small reductions in color discrimination similar to those seen in mild protonomaly, and 5) a decrease in visual search time for alcohol but not for marijuana.

In general, combined doses of alcohol and marijuana failed to support a simple additive model for drug activity. All observed changes reached a maximum within 2 hours and lasted for up to 6 hours after drug ingestion, and most of the changes are dose related.

I. Introduction

Marijuana and alcohol are widely used as recreational or social drugs primarily for the purpose of experiencing a "high"--a state of euphoria associated with relaxation and breakdown of inhibitions. It is the undesirable effects that might occur from these drugs that has provoked societal concern for the health and general performance of users. Of the numerous short-term acute effects reported for alcohol and marijuana, many pertain to behavior or performance involving vision (e.g. automobile driving).

The studies reported here were conducted in an attempt to elucidate the extent to which vision functions are directly affected by alcohol and marijuana. All of these studies have been conducted in collaboration with MERTON C. FLOM and REESE T. JONES, (University of California, School of Optometry, Berkeley, and School of Medicine, San Francisco, respectively) and BRIAN BROWN, GUNILLA HAEGERSTROM-PORTNOY, and ARTHUR JAMPOLSKY (Smith-Kettlewell Institute of Visual Sciences, San Francisco). Our results over the past 4 years primarily address oculomotor and sensory aspects of the eye and visual system, and will be summarized within the context of these categories. In general, however, two main principles emerge: a) Alcohol produces functionally important deficits in visual performance which depend on oculomotor control, while marijuana at "equivalent" doses produces much smaller deficits; b) Visuo-sensory changes seen both with alcohol and marijuana, are quite small when compared to the oculomotor changes.

II. General Experimental Methods

All of the experiments were conducted at the Smith-Kettlewell Institute of Visual Sciences (SKIVS), Pacific Medical Center in San Francisco. A special room with living room type furnishings (e.g., soft chairs, end tables, radio, and pictures) was associated with the laboratory at SKIVS. On experimental days the subjects spent all of their time in this room except when they were actually being tested in the laboratory. Drug administration (alcohol or marijuana) occurred in this room.

More than forty male subjects participated in the experiments reported here; a number of subjects participated in more than one experiment. Each experiment usually involved 10 subjects. The subjects ranged in age from 19 to 28 years and had been screened by a psychiatrist to establish acceptability to the study. Our marijuana subjects (who must have smoked marijuana at least five times and had no "bad trips" on marijuana) are in general social drinkers who drink beer, wine, or liquor at least once a week. All of the subjects used in the alcohol studies had at least this level of alcohol experience.

Subjects were told the general nature of the study and were given a brief description of each test to be performed. Each subject was asked to eat a light (low fat) breakfast on the day of an experiment and to arrange transportation so he would not have to drive home afterwards. The subjects stayed in the laboratory after the experiment until they were essentially "down". Those who were at all high or uncomfortable at the end of the day were sent home in a taxi. Payment for serving as subject was $2.00 per hour; a bonus scheme was used for return visits.

When the subjects reported to the laboratory, they were given one or more trials on the test(s) to establish a pre-drug base-line. Glasses were worn if necessary for good distance vision; contact lenses were not worn. During an experimental day, an attending physician (psychiatrist or ophthalmologist) was either in the laboratory or was immediately available. Measurements were performed immediately after taking the drug and were repeated at regular intervals throughout the day until recovery from the drug or return to measurement baseline occurred. A light lunch was provided for the subjects at mid-day at an appropriate time between experimental trials.

The standard alcohol treatments were 1.0 and 0.5 ml/kg of 95% ethanol. The alcohol was mixed with fruit juice to bring the total volume (ml) to 3 times the subjects' body weight (Kg). This mixture, with 2 ice cubes added, was drunk through a straw in about 20 minutes from an enclosed cup. Two drops of peppermint or eucalyptus extract were placed on the lid of the cup together with 2 drops of alcohol so that the alcohol and placebo treatments looked, smelled and tasted alike. These standard alcohol treatments produced blood alcohol levels of approximatley 0.07% and 0.03% 30 minutes after finishing the drink. Blood alcohol levels were estimated by breath analysis using the intoxilyser (Omicron Systems Corporation, Palo Alto). Subjects rated their "high" on a scale of zero to 100, where zero was sober and 100 was as high as a subject had ever been.

Marijuana treatments were 0.8 gm cigarettes containing 8, 12, 15 or 22 mg of THC (Δ-9-Tetrahydrocannabinol) which were smoked for "maximum intake" in about 10 minutes. One of the investigators (R.J.) was responsible for obtaining, maintaining, preparing, assaying, and dispensing the marijuana, which was obtained through the National Institute on Drug Abuse.

All of the experiments described here were carried out double-blind, that is neither the subject nor the experimenter was told whether a drug or placebo had been given.

III. Specific Experiments

A. Oculomotor Functions

1. Tonic eye position

Vision is most highly developed at the fovea and discrimination of color and form deteriorates precipitously from the fovea to the periphery. Consequently, fine oculomotor control is necessary to place the retinal image precisely on the fovea; furthermore, binocular alignment and co-ordination calls for even greater complexity of the neural substrate. The tonic innervation to the extraocular muscles, controlled by the third, fourth and sixth oculomotor nuclei in the midbrain, determines the relative posture of the eyes when binocular fusion is disrupted. The position of the covered eye with respect to the distant target fixation by the open eye is called heterophoria. Our experiments indicate that alcohol and marijuana tend to cause the eyes to converge (become esophoric) under dissociated conditions with a real target at remote distances. The magnitude of this esophoric shift was about 4 prism diopters at approximately two hours after drinking 1.0 ml of 95% ethanol per Kg body weight; blood alcohol at this time was about 0.09 g%. Marijuana (containing 22 mg THC) also produced an esophoric shift but only of about 2 prism diopters approximately one hour after smoking (1).

2. Eye movement tracking

Ocular tracking of a smoothly moving target, a more dynamic motor function than heterophoria, involves a combination of smooth pursuit movements and abrupt saccadic movements which normally operate harmoniously through separate but carefully ochestrated and interactive neural systems. Marijuana, even at doses higher than those generally used socially, failed to alter the maximum velocity of smooth or saccadic tracking of a target pendularly oscillating at regularly increasing frequencies. By comparison, low, moderate and high alcohol doses (0.5, 1.0, and 1.75 ml/kg) impaired both components of oculomotor saccadic tracking (Fig. 1). Our data suggest that alcohol increases the central processing time required to produce smooth and saccadic tracking movements (2).

3. Pupil size and response to light

Photographic measurements of pupil size, at low and high photopic light levels were made in subjects under alcohol and marijuana intoxication (3). Alcohol had no effect

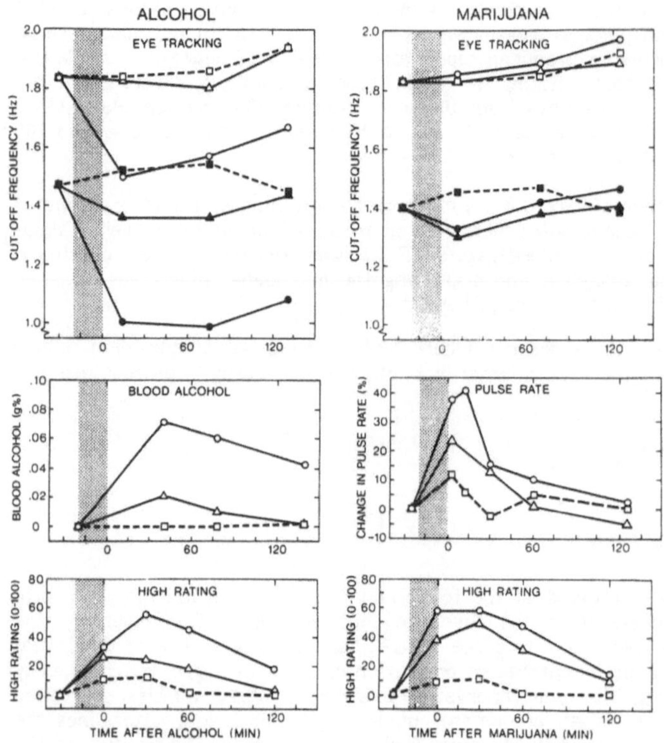

Fig. 1 In the upper panels, cutoff frequency (Hz) is shown as a function of time after ingestion of alcohol and marijuana for "saccadic tracking" (open symbols) and "smooth tracking" (solid symbols). The cutoff frequency represents the criterion endpoint of eye tracking of a sinusoidally moving target which increases in frequency from 0.5 Hz to 3.0 Hz over a period of 40 seconds. The curves were displaced slightly vertically to equate the pre-treatment values with the pre-treatment levels of the 1.0 ml/kg dose for alcohol and the 15 mg THC dose for marijuana. For alcohol, ○ : 1.0 ml/kg body weight of 95% ethanol; △ : 0.5 ml/kg body weight of 95% ethanol; □ : placebo. For marijuana, ○ : 15 mg THC; △ : 8 mg THC; □ placebo. The time course of blood alcohol levels and change in pulse rate (center panels) and the time course of the subjects' high ratings (lower panels) are also shown.

on pupil size at low photopic levels and marijuana produced a significant dose-related reduction in pupil diameter. There was no significant change produced by the combination drug treatment. Our data showing a dose-related marijuana induced pupillary constriction are consistent with our observations in earlier experiments (1) and those of HEPLER et al. (4) where a small pupillary constriction accompanies marijuana smoking. This is contrary, however, to earlier reports that pupil size increases (5) or does not change (6).

4. Reflex eye movements

Reflex eye movements in response to regularly spaced stipes moving continuously in one direction across a fairly large portion of the visual field are referred to as optokinetic nystagmus (OKN). These eye movements have been found in a qualitative way to be greatly affected by alcohol, and altered to a lesser degree by marijuana. After drinking alcohol (1 ml/Kg), the OKN pattern became noticeably irregular and the frequency and amplitude of the OKN saccades decreased. Alcohol has been shown in our other experiments to impair smooth ocular tracking, to increase saccadic latency, and to decrease saccadic velocity--all three factors being involved in OKN. Thus, after alcohol, the eyes presumably follow the stripes smoothly with a velocity less than that of the target and with less than full amplitude; there is probably a longer time required before the compensatory saccade is executed, and then it seemingly has a reduced velocity. Fewer of our subjects showed changes in OKN after smoking marijuana and then to a lesser degree than after drinking alcohol. With a very high dose of alcohol given to one of our subjects, OKN was essentially obliterated.

B. Sensory Effects

Very few alcohol induced sensory changes in vision are reported in the literature. The literature on changes with marijuana is also very limited; those few studies which examine drug induced sensory changes in vision tend to be poorly controlled. Marijuana and alcohol research in this area is characterized by the use of relatively insensitive clinical techniques which are often designed to reveal pathological change rather than small changes from the non-drug state. We have investigated visual acuity for stationary and moving targets, contrast sensitivity of the peripheral retina, recovery of the eye to glare in the central and peripheral retina, foveal color vision, and visual search.

1. Visual Acuity

After reviewing the literature up to 1940, JELLINEK and MCFARLAND (7) concluded that acuity is probably impaired after alcohol, but "it is not possible to state how great the impairment of visual acuity is, or how the degree of impairment varies with the amount of alcohol ingested". More recent experiments suggest that visual acuity is rather insensitive to alcohol; while a few studies suggest small or questionable decrements in visual acuity (8-10), others fail to show any consistent change (11-13). The literature on marijuana effects on visual acuity is fairly clear. Any effects which are present are small, and the authors of the LEDAIN report conclude, "Experimental attempts to demonstrate effects in visual acuity . . .have found no consistent change" (14).

Typically these experimenters determined Snellen acuity with conventional high contrast black-on-white letters using clinical or modified clinical techniques. However, environmental contrast levels rarely approach those of the clinical Snellen chart and consequently a test of low contrast acuity is more relevant.

We found that visual acuity for 4-position Landolt C's of high (49%) and low (12%) contrast to be unaffected by low and moderate doses of alcohol (0.5 and 1.0 ml/Kg) and marijuana (8 and 15 mg THC) during the period of six hours after taking either drug. This result suggests that neither alcohol nor marijuana taken in customary social doses affects the resolution capacity of the human eye. The relative insensitivity of static visual acuity to alcohol and marijuana is in sharp contrast to the decrements in acuity that we show to occur when the acuity targets are in motion and require coordinated eye movement behavior for their resolution.

2. Visual acuity for moving targets

Dynamic visual acuity (DVA), the resolution of detail in moving targets, is a complex task involving precise sensory and motor coordination. It is important in such practical situations as driving and flying; for example, BURG (16, 17) has shown that

DVA is significantly correlated with accident record, particularly for collisions where vehicles come from the side. In such collisions, it might be expected that detection, tracking, identification, and prediction of the path of the other vehicle would be of importance.

The components involved in DVA are (a) static visual acuity, (b) ocular pursuit of the moving target by a combination of fast (saccadic) eye movements and slower (pursuit) eye movements, and (c) interpretation of the target image which may be moving on the retina at some distance from the fovea. LUDVIGH and MILLER (18) have shown that good static acuity is a necessary, but not sufficient, condition for good DVA. It has been shown that the observed decrement in acuity, produced as the target angular velocity is increased, can be accounted for by the inability of the oculomotor system to stabilize the target image on the retina (19).

In our experiments (20), when Landolt C's were swept across the field at angular velocities between 5 and 40 degrees/second, acuity was clearly worsened after drinking alcohol and worsened to a lesser extent by marijuana (Fig. 2). We have speculated that since DVA correlates with accident record, reduction in DVA with alcohol may be an important contributing factor in alcohol-related traffic accidents.

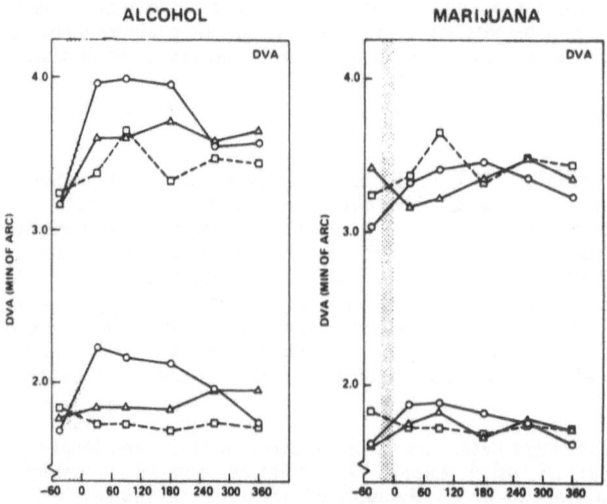

Fig. 2 Dynamic visual acuity (collapsed across target speeds and subjects) is shown as a function of time after ingestion of alcohol or marijuana, for targets of low contrast (upper set of curves) and high contrast (lower set of curves). For alcohol; 1.0 ml/kg body weight of 95% ethanol (O); 0.5 ml/kg body weight of 95% ethanol (Δ); placebo (□). For marijuana: 15 mg THC (O); 9 mg THC (Δ); placebo (□). The shaded area indicates the time of drug ingestion.

3. Glare recovery

Sudden changes in environmental light levels require the eye to readjust to achieve the same sensitivity to target contrast at the new level, a process which takes many seconds or minutes when the new environment is considerably dimmer than the previous level. During this recovery time, the eye is relatively blind to fine detail. Our experiments show that relatively low doses of alcohol significantly prolong recovery times following bright light exposure; these changes are dose-related and can be seen for several hours following alcohol ingestion.

On a given experimental day, subjects were tested before drinking and 30, 90, 180, 270 and 360 minutes after drinking (Figs. 3 and 4). At each of these test times, blood alcohol levels and subjective "high" ratings were ascertained by a second experimenter.

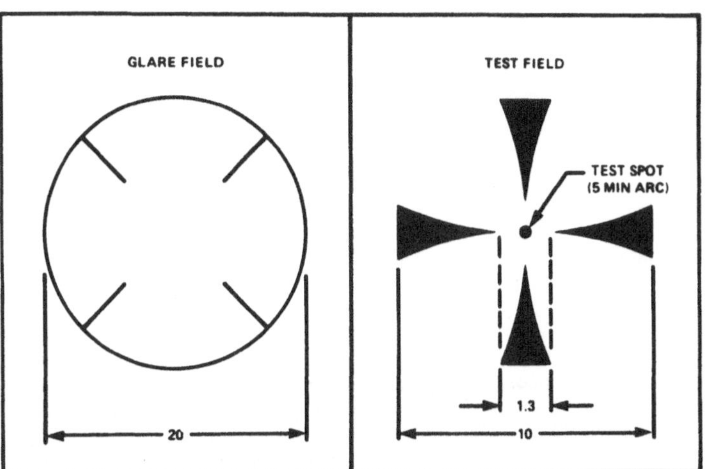

Fig. 3 In each measurement session, subjects preadapted for at least 5 min to the low photopic luminance levels of the test laboratory. Fixation was aided by four thin diagonally oriented reference lines. The subject fixated the center of a 20° circular glare field of 5.6 x 10^4 cd/m^2 with the right eye. Immediately following a 10-sec exposure to the high intensity adapting field, the subject fixated the center of the test field where a test spot (5 min of arc) was intermittently presented (125-msec flashes at 4 Hz) on a 22.6 cd/m^2 background. The contrast of the test spot was under the subject's control. When he had recovered contrast sensitivity to a fixed contrast, he turned a switch to reduce the target contrast a further fixed step below his threshold. When his contrast sensitivity had again recovered to the point of detection, he again turned the switch and the spot contrast was further stepped down. The times at which the switch was operated were recorded, and in this way the time taken to recover to each of five predetermined contrast levels (5.34, 2.00, 1.45, 0.99, and 0.64) was obtained.

In a replication experiment involving alcohol, marijuana and combined drug ingestion, we confirmed that low doses of alcohol produce significant delays in the recovery of vision following exposure to high light levels (21). Further, marijuana (8 and 15 mg THC) and a combined marijuana (15 mg THC) and alcohol (0.75 ml/Kg 95% ethanol) dose produced significant delays for at least 2 hours after drug ingestion. The combined alcohol and marijuana treatment produces little more than the effect produced by either drug alone, suggesting some antagonsim between the drugs--a suggestion which is supported by our finding of a significantly lower blood alcohol level for the alcohol dose when combined with marijuana than when taken alone.

Peripheral ocular factors are unlikely to have produced the delay. Marijuana and alcohol do not affect visual acuity even at low contrast levels (22). This essentially precludes the possibility that lens accommodation, which would blur the target, could have produced the slowing of glare recovery. A marijuana or alcohol-

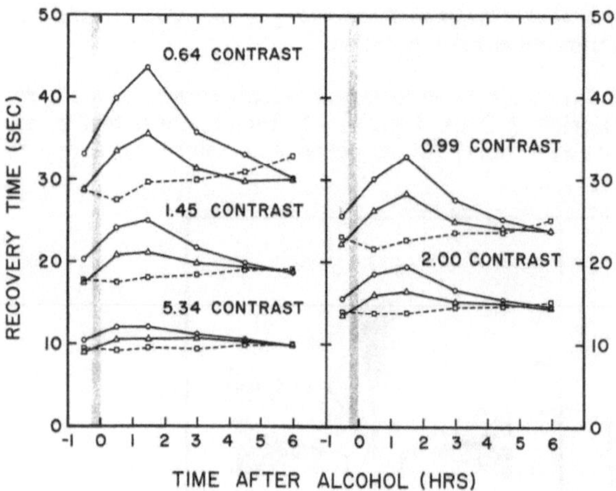

Fig. 4 Time course of glare recovery to targets of contrasts 5.34, 2.00, 1.45, 0.99, and 0.64 after ingestion of alcohol. 1.0 ml/kg body weight at 95% ethanol (O), 0.5 ml/kg body weight of 95% ethanol (△), and placebo (□). The shaded area indicates the time of alcohol ingestion. The slowing of glare recovery is statistically significant for all contrast levels 90 minutes after alcohol ingestion.

induced increase in pupil size might be anticipated to produce longer glare recovery times by increasing the effective light exposure for the glare source. Our results indicated that pupil diameter decreased slightly (7.4%) as a result of smoking marijuana and was unaltered by alcohol. In fact, pupil size seems to be irrelevant since a deliberate increase in pupil diameter from 3 mm to 8 mm failed to change the time course of glare recovery in a separate experiment on one subject (23). It is unlikely therefore that pre-retinal factors produce the increase in recovery time. The process of light and dark adaptation takes place in the retina--all the time course characteristics are seen in the electroretinogram--and it consequently seems likely that marijuana and alcohol or metabolic products of them act directly on the retina to produce the delay in glare recovery.

Regardless of the mechanism involved, the increased recovery time produced by alcohol or marijuana intoxication must be viewed as critical from a practical point of view. The detection of both large and small objects is important for safe driving. While our laboratory experiments refer only to small objects, it is reasonable to assume that recovery of contrast sensitivity for large objects is similarly affected.

The period of recovery from glare is a period of relative blindness for the individual and is thus potentially hazardous. The sky may act as an extended glare source for the automobile driver, particularly soon after sunrise and just before sunset. The sky luminance levels under these conditions may be as high as those experienced by the subjects in our experiment (24). Under certain circumstances, a driver will be forced to intermittently view very bright sky or be subjected to high luminance glare from light scattered by the windshield. Following the glare, important features of the driving environment are lost or less visible for recovery times of many seconds. Alcohol and marijuana prolong this recovery. The possible consequences of an additional 30% to 50% delay in seeing critical detail under driving conditions are obvious.

4. Color vision

The chronic use of a variety of drugs including alcohol has been reported to produce changes in color vision (25). However, there are very few reports which document short term color vision changes associated with the acute use of drugs. Short term changes in color vision pose potentially more problems than a constant defect; an unstable color world does not permit the numerous adaptations that are possible to minimize the effects of reduced color discrimination.

We have found small, yet significant, dose-related impairments in hue discrimination associated with alcohol (0.5 and 1.0 ml/Kg body weight of 95% ethanol) and marijuana (8 and 15 mg THC) intake (26). The FM 100 Hue test was used to measure hue discrimination. The impairment was predominantly located in the blue and yellow regions of the color circle, suggesting that acute doses of these drugs might produce increased confusion of blues with blue-greens or blue-purples, and yellows with yellow-green (for alcohol) and yellow-reds (for marijuana). The impairment, though small and probably clinically insignificant, lasted for at least 90 minutes after drinking alcohol and for about 60 minutes after smoking marijuana. We were unable to attribute the color impairment to central factors unrelated to color vision and concluded that the effect of the drugs was probably peripheral--possibly at the retinal level.

A large number of retinal diseases produce color vision disorders, many of them with hue discrimination losses similar to those which we found with alcohol and marijuana. KRILL and FISHMAN (27) suggest that two tests, the anomaloscope and the Farnsworth-Munsell 100-hue test (F-M 100) are particularly useful in detecting and characterizing acquired color vision defects. They find that patients with macular diseases, other than cone degeneration, are characterized by requiring more red than normal in a mixture of red/green to match a standard yellow. Further, the changes in anomaloscope setting usually occur before changes in the F-M 100 test can be detected (28). In view of the similarity of the alcohol and marijuana induced defects to the early acquired color vision defects associated with retinal disease, and further evidence that alcohol may act directly on the human retina (29, 30), we further studied the effects of alcohol and marijuana on anomaloscope settings (26). Such color mixture measurements in addition to being the first sign of retinal change in macular disease, have the advantage of being relatively immune to non-visual influences associated with drug intoxication. The red/green setting is made on a continuous scale; it is equally likely that a predominantly green or predominantly red setting would be made in the yellow match, due to carelessness or reduced motivation for example.

Our experiments demonstrate transient, small, but significant changes in the color mixture functions of the human eye associated with alcohol and marijuana intake (Fig. 5). The effects of marijuana and alcohol are additive; the combined dose produces about twice the shift in red/green mixture settings as either drug alone. In each case the shift involves a greater proportion of red in the red/green mixture which matches yellow. This mixture stimulates both the "red" and "green" photopigments and probably does not involve "blue" photopigment at all. Although the magnitude of the shift (1 to 2 units) would not be considered clinically significant (95% of the normal color vision population would be included in about ± 6 units) and would probably reflect only subtle changes in color perception, the increased red in the mixture is exactly what is found in settings made by patients with early macular disease. KRILL (28) cites this as one of the earliest measurable changes in color vision in macular disease and it is often seen before any change can be detected in visual acuity. While there is no experimental evidence to support the hypothesis that alcohol acts on the retina to produce structural changes similar to those which occur

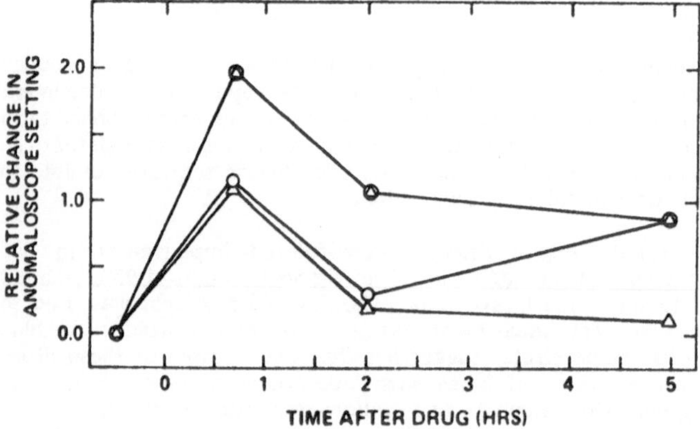

Fig. 5 Time course in change of color mixture (anomaloscope) setting after drug ingestion relative to placebo change. ⊛ : 0.75 ml/kg body weight of 95% ethanol and marijuana 14 mg THC: O : marijuana 14 mg THC: Δ : 0.75 ml/kg body weight of 95% ethanol. Numbers greater than zero represent more red in the mixture than for the placebo condition.

in early macular disease, it is nevertheless difficult to overlook the striking similarity of color vision changes, glare recovery delays (31) and stable visual acuity (22, 32) in these two situations.

There is no evidence in our study that there is any chronic change in color vision associated with the acute doses of the drug; anomaloscope settings return to pre-drug levels are unaltered. It is known, however, that chronic alcoholics have marked changes in color vision (25), and that alcohol can effect the retina directly (29).

Our results suggest that the "red-sensitive" system is preferentially affected by these drugs: Just why the "red" cone system should be more sensitive than the "green" system to alcohol and early macular disease is not clear. By contrast, diseases of the optic nerve and visual pathways usually produce a greater than normal green component in the red-green mixture ratio (33). Inasmuch as this distinction between retinal and visual pathway effects on red/green mixtures can be applied to the effects of drugs on color vision, it suggests that alcohol and marijuana may be acting at the retina rather than more centrally in the visual pathways or the cortex.

5. Visual search

Visual search, an important everyday task, typically involves the detection of a specific target in an array of other targets (or non-targets) and can include or exclude eye movements depending on the stimulus conditions, instructions, and strategies adopted. This task has been extensively investigated, in view of its applicability to civilian, industrial and military environments.

Surprisingly, we have been unable to find previous investigations of search behavior for subjects under the influence of alcohol or marijuana in spite of the fact that visual search is an important aspect of visual performance in situations where a subject may be under the influence of drugs. It might be expected that these drugs will affect visual search performance since oculomotor behavior is affected by alcohol and to a lesser extent by marijuana (2) and since marijuana has effects on short term memory (34). In view of these factors and the high face-validity of visual search tasks, such as driving, inspection, and reconnaisance, we investigated search behavior in 10 subjects after two doses of marijuana, one dose of alcohol, a combined alcohol/marijuana treatment and placebo (35).

Visual search times were measured for a single bright dot (2.8 minutes of arc diameter) in a random field of 199 dimmer dots (2.8 minutes of arc diameter). Displays were computer generated and presented to the subject on a cathode ray oscilloscope (Hewlett-Packard 1311A with type P21 (green) phosphor) at a viewing distance of 1 meter. The square display subtended 11.6° on a side. The oscilloscope screen was indirectly illuminated by a single incandescent lamp which gave the screen a luminance of 4.6 cd/m^2. All dots were placed at random in a 1000 x 1000 matrix. Each display had one bright dot in one of the 10^6 possible positions; the bright dot luminance was 41.5 cd/m^2, the dim dot luminances were 11.4 cd/m^2. The experiment was controlled by a computer program which generated the displays on-line (each display took about one second to generate), timed the subjects' responses, and checked whether or not the responses were correct.

Since our eye movement experiments revealed clear impairment from alcohol, we expected impaired visual search performance in alcohol-intoxicated subjects. However, our alcohol-intoxicated subjects showed significant search improvement and we were unable to show any performance change after marijuana. We interpret this result to mean that subjects in the alcohol and combined alcohol/marijuana conditions used an efficient search strategy, one which they would not other wise have adopted.

We have speculated that if the subjects had their saccadic search behavior substantially reduced or eliminated by the drug condition, shorter search times may be possible. The reason is that saccadic eye movements are accompanied by an approximately 100 ms period of decreased light sensitivity or suppression (26 37), 25-30 ms during the movement and 30-40 ms both immediately preceding and following the actual saccade. Our speculation is supported by a report by BELT (38) that during a driving task the number of fixations decreases by about 30% for blood alcohol levels equivalent to those used in our experiments.

The practical implications of these findings are important. Search tasks similar to ours which do not require "divided attention", may well show performance improvement with moderate doses of alcohol. However, MOSKOWITZ (39) has shown that alcohol seriously impairs performance when "divided attention" is required but does not necessarily degrade performance on single tasks. Consequently, the results attained in our experiments should be applied with caution in more complex situations since many applied tasks require that the subject attend intermittently to more than one aspect of a display/control system to achieve adequate performance.

C. Physiology of the Eye

In 1971, HEPLER *et al.* (40) reported that marijuana reduced the normal intraocular pressure (IOP) within the eye. This result led them to the experimental use of marijuana in a patient with abnormally high IOP. We confirmed the drop in IOP in normal subjects (41) in a double-blind experiment involving a marijuana placebo. Our experiments suggested that the effect of marijuana was indirect. IOP decreased only in those subjects whose marijuana experience was limited to moderate use and for whom the experimental dose produced a substantial "high" and peaceful relaxation.

IV. Comment

Drugs can affect many aspects of visual behavior which are important in everyday environments--vigilance, detection and oculo-motor functions are required. The relations between these functions are complex, and in an attempt to isolate drug effects on single tasks, we have, where possible, studied them in isolation. It is important to be aware that drugs affect all aspects of skilled performance simultaneously. The magnitude of the effects which we have shown in these isolated functions are underestimates of the decrements which can be expected when complex tasks are performed under intoxication. Such tasks require intact function throughout the hierarchy of processes involved in skilled performance.

For example, we have shown that alcohol significantly retards recovery of foveal contrast sensitivity after glare, and that alcohol reduces acuity for moving targets. Thus, in alcohol-intoxicated subjects, recognition of detail in moving targets after glare will be degraded in two ways which will interact. The subject will take longer to detect targets since his glare recovery will be prolonged. This delay in detection will mean that the error between the eye and a moving target will be increased and the already stressed oculomotor system will have even greater demands made upon it. Further, in complex tasks alcohol degrades decision making ability, producing further delays and there will also be increased latency of the normal motor response systems. In this way drug effects in the early stages of information processing by the visual system, at more basic levels of the hierarchy of processes making up skilled performance, cascade and produce greater and greater cumulative functional deficits.

Acknowledgement

This research was supported by Contract No. DADA 17-73-C-3106 from the U.S. Army Medical Research and Development Command and by National Institutes of Health Grants K02MH32904 and DA00033.

References

1. A. Jampolsky, M. C. Flom, A. J. Adams and R. T. Jones, Final Report to U.S. Army Medical Research and Development Command, Washington, D.C., Contract No. DADA17-72-C-2083 (1973).
2. M. C. Flom, A. J. Adams and B. T. Jones, Am. J. Optom. and Physiol. Optica. 53, 764 (1976).
3. B. Brown, A. J. Adams, G. Haegerstrom-Portnoy, R. T. Jones and M. C. Flom, Am. J. Ophthal. 83, 350 (1977).
4. R. S. Hepler, I. M. Frank, and J. T. Ungerleider, Am. J. Ophthal. 74, 1185 (1972).
5. W. Mayer-Gross, E. Slater and M. Roth, Clinical Psychiatry, 2nd Ed., (Cassell, London, 1960).
6. A. T. Weil, N. E. Zinberg and J. M. Nelsen, Science 162, 1234 (1968).
7. E. M. Jellinek and R. A. McFarland, Quart. J. Stud. Alc. 1, 272 (1940).
8. H. Newman and E. Fletcher, Amer. J. Med. Sci. 202, 723 (1941).
9. G. A. Brecher, A. P. Hartman and D. D. Leonard, Am. J. Ophthal. (supplement) 39, 44 (1955).
10. R. G. Mortimer, Percept. Mot. Skills 17, 399 (1963).
11. Z. W. Coslon, J. Am. Med. Assn. 155, 1525 (1940).
12. D. G. Marquis, E. L. Kelly, J. G. Miller, R. W. Gerard and A. Rapoport, Annals N.Y. Acad. Sci. 67, 701 (1957).
13. G. Verriest and E. Laplasse, Exp. Eye Res. 4 95 (1965).
14. G. Le Dain (Ed.), Cannabia. A Report of the Commission of Inquiry into the Non-medical Use of Drugs, Ottawa, Information Canada (1972).

15. Carlson, W. L., J. Safety Res. 4, 12 (1972).
16. A. Burg, Vision Test Scores and Driving Record: Additional Findings, University of California Report No. 68-27 (1968).
17. A. Burg, Visual Degradation in Relation to Specific Accident Types, Final report, Institute of Transportation and Traffic Engineering, University of California in Los Angeles (1974).
18. E. J. Ludvigh and J. W. Miller, J. Opt. Soc. Amer. 48, 799 (1958).
19. B. Brown, Vis. Res. 12, 305 (1972).
20. B. Brown, A. J. Adams, G. Haegerstrom-Portnoy, R. T. Jones and M. C. Flom, Perception and Psychophysics 18, 441 (1975).
21. A. J. Adams, B. Brown, G. Haegerstrom-Portnoy, M. C. Flom, R. T. Jones, Psychopharmacology 56, 81 (1978).
22. A. J. Adams, B. Brown, M. C. Flom, R. T. Jones and A. Jampolsky, Am. J. of Optom. and Physiol. Optics 52, 729 (1975).
23. A. J. Adams, B. Brown and M. C. Flom, Perception and Psychophysics 19, 219 (1976).
24. R. C. Hopkinson, J. Opt. Soc. Amer. 44, 455 (1954).
25. R. Cruz-Coke, Color Blindness. An Evolutionary Approach, (C. C. Thomas, Springfield, 1970).
26. A. J. Adams, B. Brown, G. Haegerstrom-Portnoy, M. C. Flom and R. T. Jones, Perception and Psychophysics 20, 119 (1976).
27. A. E. Krill and G. A. Fishman, Trans. Acad. Ophthal. Otolaryng. 75, 1095 (1971).
28. A. E. Krill, Hereditary Retinal and Choroidal Diseases, Vol. 1. (Harper and Row, Maryland, 1972).
29. H. Ikeda, Vis. Res. 3, 155 (1963).
30. J. H. Jacobson, T. Hirose, and P. E. Stokes, Ophthalmologica (Suppl.) 158, 669 (1969).
31. A. J. Adams and B. Brown, Nature 257, 481 (1975).
32. S. L. Severin, R. L. Tour and R. N. Kershaw, Arch. Ophth. 77, 2 (1967).
33. P. Grutzner, In Handbook of Sensory Physiology, VII/4: Visual Psychophysics, ed. by D. Jameson and L. N. Hervich (Springer-Verlag, New York, 1972.
34. E. L. Abel, Nature 227, 1151 (1970).
35. A. J. Adams, B. Brown, M. C. Flom, A. Jampolsky and R. T. Jones, Final Report. U.S. Army Medical Research and Development Command, Washington, D.C. Contract DAD 17-73-C-3106, (1976).
36. F. C. Volkmann, J. Opt. Soc. Amer. 52, 571 (1968).
37. P. L. Latour, Vis. Res. 2, 261 (1962).
38. B. L. Belt, Driver Eye Movements as a Function of Low Blood Alcohol Concentrations. College of Engineering, Ohio State University (1969).
39. B. Moskowitz, J. Safety Res. 5, 185 (1973).
40. R. S. Hepler and I. R. Frank, J.A.M.A. 217, 1392 (1971).
41. M. C. Flom, A. J. Adams and R. T. Jones, Invest. Ophthal. 14, 52 (1975).

III. Contact Lenses

Soft Contact Lenses: A Look into the Future

Theodore Grosvenor
Illinois College of Optometry
3241 South Michigan Ave.
Chicago, Illinois 60616

Introduction

In the year 1960, WICHTERLE and LIM published a brief review in Nature concerning their research with hydrogel materials for use as prosthetic devices (1). They stated that the demands for such a material were: 1) a structure permitting the desired water content; 2) inertness to normal biological processes; and 3) permeability for metabolites. Materials meeting these demands had to have high molecular weights; they had to contain hydrophilic groups; and it was necessary that they have enough cross-linkage to prevent absorption. Materials meeting these requirements were co-polymers of glycomonomethacrylate with several tenths percent of glycodimethacrylate. WICHTERLE and LIM mentioned almost incidentally (in the last sentence of their report) that promising results had been obtained with these materials in the manufacture of contact lenses.

During the 1960's, the soft lens developed by WICHTERLE and LIM was fitted on large numbers of patients at the Second Eye Hospital in Prague, under the supervision of DRIEFUS. As reported by MORRISON (2), and by FISHER (3), by the late 1960's Dr. DRIEFUS and his associates had fitted approximately 3,000 patients with these lenses. Six percent of the patients were reported to be wearing their lenses 24 hours per day, 36% were wearing their lenses about half the day. The only failures were those who could not handle the lenses, according to these early and perhaps overly-enthusiastic reports. The lens was made by pouring liquid monomer solution into a revolving mold, the material being polymerized in the manufacturing process. The finished lens was in the swollen state, and was then equilibrated in physiological saline solution. The resulting outside curvature of the lens was that of the mold, and the inside curvature depended upon the speed of the rotation of the mold and other factors.

Soft Lens Materials

Soft lens materials have been described in detail in a supplement published by the Encyclopedia of Polymer Science and Technology (4). The original WICHTERLE lens was made from poly (2-hydroxyethyl) methacrylate, lightly cross-linked with ethylene dimethacrylate. The Bausch and Lomb Soflens, the successor to the WICHTERLE lens, is still made by the spin-casting process and contains the same materials as did the original WICHTERLE lens.

In addition to hydroxyethyl methacrylate, many hydrogel lenses contain vinylpyrrolidone, either as a copolymer or as a graft polymer. By varying the content of vinylpyrrolidone in the lens, the water content may be increased from approximately 40% to as high as 80% or 90%. As the water content of a hydrogel material increases, its permeability to water and water-soluble substances increases and its strength tends to decrease.

It is possible to make hydrogel lenses which contain vinylpyrrolidone but not hydroxyethyl methacrylate: one such lens is apparently a copolymer of methyl methacrylate and vinylpyrrolidone. Other lenses contain neither hydroxyethyl methacrylate nor vinylpyrrolidone.

The nomenclature relative to soft lens materials is confusing, to say the least. A given lens may be identified by the manufacturer's trade name for the lens, by the chemical designation for the material for which the lens is made, by the company which manufactures the lens, or in some cases by the name of the person who developed the lens. In addition, due to the fact that hydrogel lenses are classified as drugs by the F.D.A., each hydrogel material must have a nonproprietary generic name. Generic names are assigned by the United States Adopted Names Council. The USAN Council uses the stem -filcon to identify all hydrogels used in contact lenses, with the exception of the Bausch and Lomb Soflens material, which was named polymacon. The hydrogel lenses which have received F.D.A. approval, together with the materials from which they are made and their generic names, are shown in Table 1.

Table 1

F.D.A. Approved Hydrogel Lenses*

Material	Generic Name	Trade Name (Manufacturer)
poly (2-hydroxyethyl methacrylate)	polymacon	Soflens** (Bausch and Lomb)
poly (2-hydroxyethyl methacrylate-co-vinyl pyrrolidone)	hefilcon A	Hydrocurve (Soft Lenses, Inc.) Naturvue (Milton Roy Co.)
poly (vinyl pyrrolidone-g-2-hydroxyethyl methacrylate)	vifilcon A	Sofcon*** (Warner Lambert)
poly (glycidyl methacrylate-co-vinyl pyrrolidone) and/or poly (glycidyl methacrylate-co-2-hydroxyethyl methacrylate)		Aquaflex (Union Optical) AOsoft (American Optical)

*Materials and generic names taken from Table 2, Supplement No. 1, Encyclopedia of Polymer Science and Technology.
**Spin-cast lens (all other lenses are lathe-cut).
***Approved for therapeutic use only.

Manufacturing Processes

The spin-casting process, used for the original WICHTERLE lens and for the present-day Bausch and Lomb Soflens, has already been described. During the past several years Bausch and Lomb has developed several series of lenses. All lenses for each series are made in the same mold or in molds having the same radius of curvature with the result that all lenses in a series, although having different powers, have the same front surface radius of curvature. Thus a minus one diopter lens will have a much flatter back surface curvature than a minus six diopter lens in the same series. This designation of lens curvature, by front surface rather than by back surface curvature, represents a marked departure from the convenient and practical convention used for many years in the fitting of conventional contact lenses.

All hydrogel lenses except for the Bausch and Lomb lens are manufactured by means of a lathe-cutting process. For this process, the material is already polymerized and is in the dehydrated state before manufacturing is begun. Since water would be absorbed by the material during the manufacturing process, polishing compounds must have an oil base rather than a water base. In addition, hydrogel materials swell by predictable amounts when hydrated and this swelling must be taken into account in determining the desired parameters of the finished lens.

The manufacturer of a lathe-cut soft contact lens has a wide variety of parameters available, as does the manufacturer of a lathe-cut conventional contact lens: he may make a lens of any power, of any base curve radius, of any thickness, of any optic zone size, and may make use of any desired peripheral curves. Although most manufacturers use spherical surfaces only, at least one manufacturer, Hirst Contact Lens of New Zealand, makes use of an aspheric back surface.

Fitting Procedures for Lathe-Cut Lenses

I was fortunate in having spent the years 1970 through 1975 in Canada: during this period the development of soft contact lenses proceeded rather swiftly in that country with very little interference on the part of the Federal Drug Directorate, Canada's counterpart of the F.D.A. A great many brands of lathe-cut lenses were (and still are) available, and the fitting of these lenses is a relatively simple, logical and straight-forward procedure. The back surface of the lens is typically fitted with a radius of curvature from 0.5 mm to 1 mm flatter than that of the cornea and the diameter is typically on the order of 14 mm, somewhat larger than the cord diameter of the cornea. These lenses seldom fail to center almost perfectly on the cornea, although occasionally extending further beyond the limbus temporally than nasally. A well-fitting lens typically has a blink lag of from 1 to 2 mm, and lags down by about the same amount when the patient looks upward.

If a lathe-cut soft lens fits too loosely, a lens having a steeper base curve or a larger diameter is tried, just as one would do when fitting a conventional lens. Similarly if a lens proves to fit too tightly, a lens having a flatter base curve or a smaller diameter is tried, again just as one would do when fitting a conventional lens. Optic zone diameter is varied less often with lathe-cut soft lenses than with conventional lenses, as a small optic zone usually tends to make the lens fit too loosely. One Canadian lens manufacturer routinely applies only a 0.5 mm bevel to soft lenses, with the result that a 14 mm soft lens has a 13 mm optic zone.

Soft Lenses Vs. Conventional Lenses

How do soft contact lenses compare to conventional contact lenses? When the news of the soft lens reached the Western world more than ten years ago, optometrists were immediately enthusiastic about these lenses because of their hydrophilic nature and because they were said to be permeable to oxygen and carbon dioxide. However the fact that the lenses were hydrophilic simply meant that the wearer would not have to use a wetting solution; and as early as 1964 it was found by HILL and FATT (5) that oxygen does not penetrate through the hydrogel material any more rapidly than through polymethyl methacrylate. It is now generally understood that the gas permeability of a hydrogel lens can be increased significantly either by greatly increasing the water content of the material or by making the lens extremely thin.

While manufacturers and practitioners argued about the value of the hydrophilic nature and the permeability of gel lenses, wearers of these lenses were pleased with them for an entirely different reason: the lenses were very comfortable. The superior comfort for the new wearer, when compared to conventional lenses, is due to two factors: 1) the lens material is soft and flexible; 2) the lens is large enough so that the upper and lower edges of the lens are constantly underneath the lid margins, with the result that the eyelids do not have to "mount" the edges of the lens with every blink.

The almost complete lack of eyelid sensation is not the only advantage of soft contact lenses. It is not necessary to build up or to maintain wearing time as it is with conventional lenses, with the result that soft lenses may be successfully worn on a part-time basis. This makes them ideal for use in sports, in which an additional advantage is that a well-fitting soft lens will not pop out of the wearer's eye. Another important advantage of lathe-cut soft lenses is that they almost always center on the eye beautifully, with the result that they can be used with great success for an unhappy conventional lens wearer whose chief problem is poor centration. When I was in part time optometric practice in Canada, some of my happiest soft lens wearers were former conventional lens wearers whose lenses failed to center, causing discomfort, variable vision and flare at night.

The Role of the F.D.A.

As indicated in the abstract, to look into the future of soft lenses it is necessary only to travel to Canada, Europe, Japan, Australia or New Zealand or to work with one of the many lathe-cut lenses now under F.D.A. scrutiny. There is no doubt that the restrictive policies of the F.D.A. have greatly slowed progress in soft lens development in the United States as compared to the rate of progress in other countries. High quality lathe-cut soft lenses were available from a number of manufacturers in Canada as early as 1971, and by 1973 many Canadian optometrists were fitting toric front surface lenses with a fair degree of success. This past summer I returned to New Zealand where I attended the annual conference of the New Zealand Contact Lens Society, and found that in that country and in Australia optometrists are fitting the majority of their contact lens patients with soft lenses, and are fitting toric front surface lenses routinely for patients having significant amounts of astigmatism.

Conventional "hard" contact lenses owed their development chiefly to a large number of small laboratories, many of which were owned by optometrists who constantly experimented with new lens designs. A conventional contact lens laboratory could (and still can be) set up for an expenditure of only a few thousand dollars: the manufacturing of conventional lenses was, and still is, an extremely successful "cottage industry". It is of interest that none of the traditional ophthalmic lens manufacturers was successful in breaking into the conventional contact lens market! However, due to the need to meet the very stringent F.D.A. requirements, the development of a new soft contact lens requires an expenditure of several millions of dollars and typically spans a period of several years. The result is that with a few exceptions the development of soft lenses has been concentrated in the hands of large, well-financed manufacturers.

The first contact lens to receive F.D.A. approval, the Bausch and Lomb Soflens, was approved in March, 1971 and enjoyed a monopoly position for two and one-half years until the Soft Lenses Inc. Hydrocurve lens was approved in September, 1973. These two lenses had no further competition for almost three additional years, the Milton Roy Naturvue lens being approved in May, 1976 and the Union Optical Aquaflex lens being approved in June of the same year. The Warner-Lambert Sofcon lens was approved in September, 1973, but for therapeutic use only. Thus there are presently only four "brands" of soft contact lenses available in this country for optometric use.

The fact that so few companies have been allowed to manufacture soft contact lenses has meant that each of the companies has kept so busy just trying to meet the tremendous demand for the already-approved spherical lenses that they have had little time or motivation to attempt to develop lenses for the correction of astigmatism or other special-purpose lenses. Hopefully during the next several years enough lens manufacturers will have won F.D.A. approval to establish a level of competition in the soft lens field that will be more conducive to the development of these special lenses.

In addition to the need for lenses to correct astigmatism, other problems which have plagued soft contact lenses have been microbial contamination of lenses, "coating" of lenses, allergic responses, and the possibility of induced myopia and astigmatism.

Astigmatism

Whereas conventional contact lenses are typically rigid enough to maintain their shape while on the cornea and thus eliminate keratometric astigmatism, soft lenses are sufficiently flexible that they conform to the corneal surface and therefore have no effect on astigmatism. When soft contact lenses first became available, some practitioners, as well as some manufacturers, claimed that astigmatism could be "masked" by fitting the lenses steeper, flatter, or larger than they would otherwise be fitted or by using unusually thick lenses. A number of investigators soon found that these procedures were of little, if any, value in the correction of astigmatism. In a study involving 24 subjects fitted with Griffin lenses (6) (the fore-runner to the Sofcon lens), data for 47 eyes (see Fig. 1) showed a remarkably close relationship for astigmatism measured on the front surface of the lens (while being worn) as compared to corneal astigmatism without the lens, showing that for these subjects virtually all of the corneal astigmatism was present when wearing the lenses.

In the absence of soft lenses for the correction of astigmatism, we may ask the question, "how much astigmatism can we allow to be uncorrected?" Since spherical soft lenses correct no astigmatism at all, fitting patients with spherical soft lenses is equivalent to prescribing spectacle lenses without cylinders! I have found that many patients can tolerate up to 1.00 D of uncorrected with-the-rule astigmatism; but if astigmatism is against-the-rule or oblique, the limit is 0.50 D or 0.75 D. Otherwise poor acuity or asthenopia, or both, will result.

Soft lenses have a tendency to rotate on the eyes just as hard lenses do, so the incorporation of a toric front surface requires the use of some method of lens stabilization. As with conventional lenses, stabilization can be obtained by truncation or by the provision of prism ballast, or both.

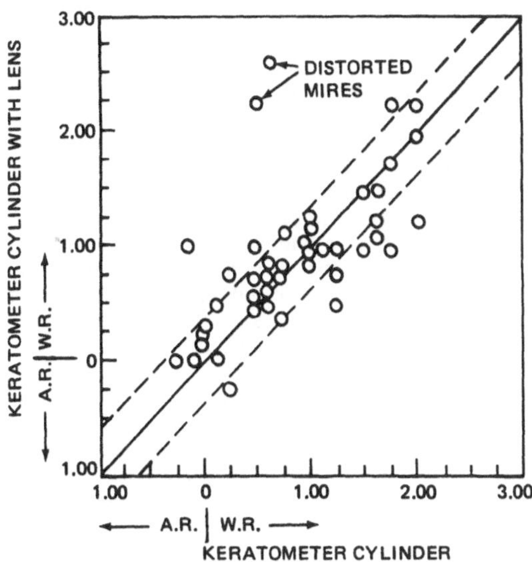

Fig. 1 See text for explanation.

In a discussion of toric contact lens fitting in Australia, HOLDEN (7) stated that in that country 80% of all new contact lenses fitted are soft lenses, and that toric soft lenses account for 25% of these. HOLDEN concluded that 1) if truncation is to be used, double truncation is necessary (a single truncation is ineffective); 2) a combination of prism ballast and a single lower truncation is effective; 3) prism ballast without truncation is effective provided the lens is thin (e.g. , 0.20 mm) and the front surface design is carefully controlled; and 4) both front and back toric surfaces can be successfully combined with these techniques. He recommended that toric soft lenses should be fitted so that they are relatively flat and mobile, so that they can be expected to respond rapidly and reliably to lens orientation forces.

BAYSHORE (8) reported on a series of 88 patients who were fitted with toric front surface soft lenses. All lenses were double truncated without prism ballast, and diameters used were 12.0 x 14.0 mm, 12.5 x 14.5 mm, and 13.0 x 15.0 mm. Of the 88 patients fitted, 26 were discontinued for various reasons, and the remaining 62 patients were classified as satisfactory. BAYSHORE's lens fitting procedures consisted of the use of a double-truncated plano trial lens set, with base curves in 0.1 mm steps. The patient was fitted with the trial lenses and wore the lenses for a period of three or four hours, using his own spectacles as an over-correction, after which the final analysis of the fitting characteristics and lens positioning was determined. The work of HOLDEN, BAYSHORE and other researchers indicates that we need have no worries about the development of toric lenses in this country, once enough manufacturers have been cleared by the F.D.A. so that the competitive urge will become manifest.

Microbial Contamination

Due to the porous nature of hydrogel materials, contact lens practitioners expressed concern about the possibility of microbial contamination of soft lenses. This concern was reinforced by well-publicized incidents of lens contamination occurring in the early 1970's, both in this country and in Canada. Until recently the only hygenic lens care systems available in this country was the boiling, or "aseptisizing" system, used in conjunction with saline or with a cleaning agent such as Softmate or Pliagel; but in Canada and in other countries two additional systems have been used. One of these systems involves soaking the lens in hydrogen peroxide and then rinsing it in sodium bicarbonate; while the other involves cleaning the lens with Preflex, storing it in Flexsol, and rinsing it in Normol. Both Flexsol and Normol are preserved with thimerosal and chlorhexidine: Flexsol contains Adsorbobase, a combination of water-soluble polymers, whereas Normol is essentially a preserved saline solution.

The efficacy of the Flexsol/Normol system was investigated in a bacterial monitoring study conducted at the University of Waterloo (9). In this study 120 clinic patients were fitted with soft contact lenses at the University of Waterloo contact lens clinic, and their tear samples and lens storage cases were cultured at regular intervals for a period of several weeks. As shown in Table 2, microbial contamination

Table 2

Microbial Contamination of Tear Samples

Sample	No. Persons Tested*	No. Samples with Growth	Bacteria		Fungi	% Growth
			Gram+	Gram-		
1	118	148	129	13	2	62.7
2	102	123	111	12	3	60.3
3	94	79	70	4	2	42.0
4	85	71	59	6	4	42.0
5	77	68	55	7	1	44.0
6	58	39	34	2	0	33.6
7	35	24	22	1	1	34.3
8	15	11	10	1	0	36.7
9	6	6	6	0	0	50.0
10	1	0	0	0	0	0

*Two samples were tested for each person (one for each eye).

of tear samples decreased during soft lens wear, indicating the possibility that residual preservative in the lenses may have been responsible for reducing the contamination. As for lens storage cases, Table 3 shows that contamination was found in nine of a total of 392 samplings, indicating that in the great majority of instances, the wearers utilized the Flexsol/Normol system in its intended manner. Those wearers whose storage cases were found to be contaminated were questioned, and it was found that without exception they had not followed instructions in hygenic care of their lenses. Whenever contamination was found, the microbiologist filled the lens storage container with fresh Flexsol and the micro-organisms were killed in a matter of minutes.

The boiling, or "asepticizing" system, used originally for the Bausch and Lomb Soflens and approved by the F.D.A. for use with this and other lenses, has been considered effective in killing all micro-organisms except for the spore-formers,

Table 3

Contamination of Lens Storage Cases

Sample	No. Patients Cases Tested	No. Patients With Contam.	% of Those Tested	Storage Wells With Contam.	% Contam.*
1	107	4	4	7	3
2	96	2	2	3	2
3	81	2	3	4	2
4	66	1	2	1	1
5	32	0	0	0	0
6	8	0	0	0	0
7	2	0	0	0	0

*Each patient's case contained two storage wells.

which are not killed at a temperature of 100° C. SNYDER, HILL and BAILEY (10) investigated the efficacy of a home-sterilization system, the Durosoft Autoclave Sterilizer, by implanting a thermister element inside the case, which provided an electrical read-out of the temperatures a lens would be subject to throughout the instrument's cycle. Averaging the results of ten such cycles, they found that the temperature inside the case reached 121° C after approximately 15 minutes and remained at or above that temperature for an average of 22 min. The authors commented that home sterilization appears to be feasible with this system, on the basis of the observed temperature performance, but that the ultimate test of the system would be to test the survival potential of all of the most resistant ocular organisms.

"Coating" of Lenses

For a period of several years after the introduction of the Bausch and Lomb Soflens, the only F.D.A. approved method of cleaning the lens, prior to boiling, was that of rubbing the lens with saline solution. The result was that protein materials originating from the pre-corneal film were deposited on the lens surfaces while the lenses were worn, and these protein materials were coagulated during the boiling process forming a "coating" which was difficult or impossible to remove. More than one practitioner found that patients who failed to follow the prescribed routine of nightly boiling were those whose lenses lasted the longest: among the patient who were faithful in boiling their lenses nightly, few if any lenses were still wearable at the end of a year. With the approval of the enzyme cleaner, Hydrocare, for use with the Bausch and Lomb lens, lens "coating" should prove to be a less severe problem and hopefully the useful life of these lenses can be significantly extended. It should be understood that the enzyme cleaner is not a prophylactic cleaner, but is intended to be used on a weekly basis to remove lens coating that has already been deposited.

When the F.D.A. approved the Hydrocurve lens, it also approved two prophylactic cleaners, Softmate and Pliagel. These cleaners are intended to be used nightly, before the lenses are boiled, in order to remove any protein materials before they have a chance to be coagulated on the lens. Softmate and Pliagel have also been approved by the F.D.A. for use the with Naturvue lens.

In Canada, where patients fitted with lathe-cut lenses are not required to boil their lenses, I found that lens coating for these patients was a rarity. Lenses worn by those patients who used the Preflex/Flexsol/Normol system, which does not require boiling, remained clear, comfortable and wearable for as long as two or three years. The announcement was recently made that Preflex, Flexsol and Normol have been approved by the F.D.A. for use with the Hydrocurve and Naturvue lenses. It should be understood however that solutions such as Flexsol and Normol cannot be used for lenses which are going to be boiled. In a study involving 8 contact lens solutions, BAILEY (11) found that all 8 solutions caused lens coating, at least to some degree, when used in conjunction with the boiling procedure. The worst offenders were Flexsol, Liquifilm, and Normol.

Allergic Reaction

A result of lens "coating" which is still not completely understood is an allergic response which tends to resemble vernal conjunctivitis. A patient who wears soft lenses successfully and has no other problems may develop follicles on the superior palpebral conjunctiva which may gradually lead to discomfort and inability to wear the lenses. The discovery that this vernal-like condition can occur as a result of soft lens wear has lead many practitioners, including myself, to evert the patient's upper lids before fitting lenses and at every follow-up examination.

REFOJO and HOLLY (12) have proposed that when a foreign body such as a contact lens comes in contact with the tear film, tear protein will be readily adsorbed to the lens surface and that this protein may become denatured and consequently may act as an antigen in provoking an immunological reaction. They have suggested that if autologous proteins were adsorbed, then denatured, and subsequently released from the contact lens surface, a constant antigenic stimulus would be present in the tear film. They tested this hypothesis by exposing hydrogel buttons to three proteins which are characteristic of those found in the pre-corneal film. The proteins used were bovine albumin, bovine gama-globulin and egg-white lysozyme. They used changes in contact angles of the surface as indicators of adsorption of the protein materials, and concluded that tear proteins readily adsorb onto the surface of hydrogel lenses, confirming their hypothesis, and that these proteins are in a somewhat denatured state if lenses are regularly boiled. They proposed that the denatured protein could desorb onto the tear film and/or come into contact with the tarsal conjunctiva where it could act as an antigen in certain predisposed patients, resulting in an allergic conjunctivitis.

For patients who have a tendency to develop this condition, the lenses should always be carefully cleaned before boiling and an enzymatic cleaner should be used regularly to remove any coating which may accumulate on the lenses.

Induced Myopia and Astigmatism

A number of investigators have found a tendency for soft lenses to cause changes in wearer's refraction. These changes are typically in the direction of increased myopia or increased with-the-rule astigmatism. In one such study, I monitored the corneal curvature and refractive error of 24 patients fitted with lathe-cut soft lenses (13). Time courses of corneal curvature and refractive changes were plotted for ten of these subjects (Figs. 2 and 3) and it was found that the most common tendency was for the occurrence of an initial corneal flattening in one or both meridians, followed by a steepening which begins at about six weeks of lens wear. For most patients, the

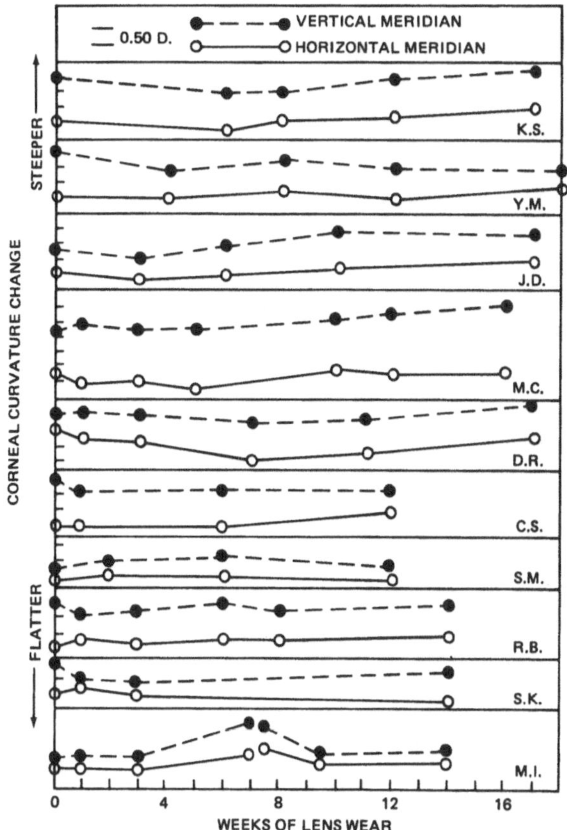

Fig. 2 See text for explanation.

corneal curvature returned to its original value after several more weeks but for a few patients the corneas steepened beyond their pre-fitting values and this steepening was accompanied with an increase in myopia. In addition, some patients showed a tendency for increased with-the-rule corneal astigmatism.

It was concluded that the induced myopia and with-the-rule astigmatism were the result of one or more of three factors: 1) lenses fitting too steeply, 2) inadequate blinking, and 3) overwearing of lenses. In all cases, re-fitting the patients with looser lenses brought about a reversal of the induced changes. Many of the patients showing these changes were poor blinkers or wore their lenses for prolonged periods (one was a college student who wore her lenses twenty hours a day while studying for exams): these patients were instructed in regard to blinking exercises and wearing schedules.

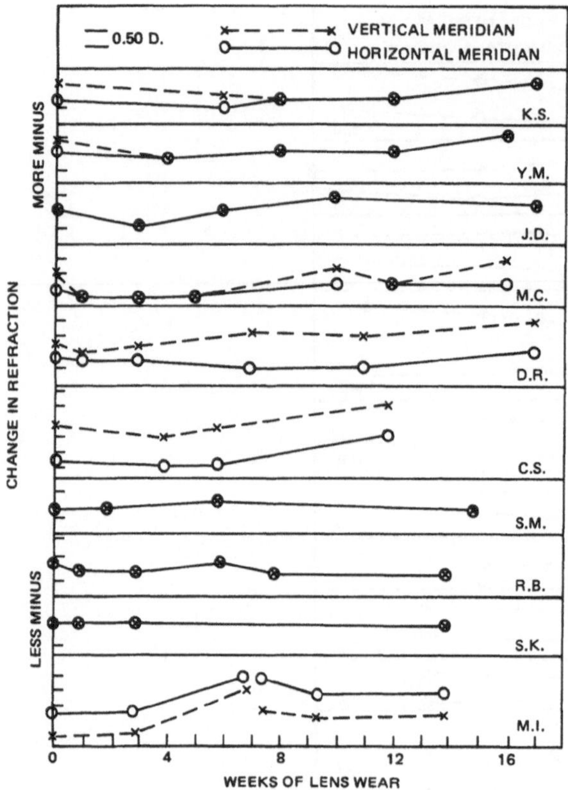

Fig. 3 See text for explanation.

Since it is known that both poor blinking habits and overwearing of lenses can result in lens dehydration with consequent steepening, it was proposed that the induced myoia was the result of a band-like pressure at or near the limbal area, causing an increase in central corneal curvature. Corneal edema, if it had occurred, would have affected the whole cornea, thus bringing about a flattening rather than a steepening of the cornea (14): such edema was probably the cause of the initial flattening occurring with most wearers. The center thickness of the lenses used in this study was in the neighborhood of 0.20 mm.

HILL has reported a number of studies in which corneal changes have occurred as a result of wearing soft contact lenses. In a recent article (15) he reported fitting Ultra-Thin Bausch and Lomb lenses, averaging about 0.05 mm in center thickness, on two groups of patients. One group consisted of 26 eyes of persons who had successfully worn conventional thickness Bausch and Lomb lenses, and the other group consisted of 28 eyes of persons who had never worn contact lenses. Hill found that for the patients who had previously worn lenses, there was a tendency for corneal curvature to flatten and myopia to decrease while wearing the ultra-thin lenses; but for the patients who had not previously worn lenses these changes did not occur.

These studies, together with the experience of a number of practitioners, indicate that changes in corneal curvature and refraction of soft lens wearers can be avoided, but that should they occur, they can usually be reversed if appropriate measures are taken soon enough.

Lenses Now Under F.D.A. Scrutiny

In his latest "contact lens update", BAILEY (16) reported that more than a dozen lenses are now in the I.N.D. (Investigative New Drug) stage and that the following soft lenses are in the N.D.A. (New Drug Application) stage:

Lens	Laboratory
Aosoft	American Optical
Gelflex	Calcon
CSI	Corneal Science Inc.
Hydralens	Ophthalmos
Softsite	Paris
Tresoft	Urocon
Sofcon	Warner-Lambert
Dura Soft	Wesley-Jessen

BAILEY also reported that at least four toric soft lenses are being investigated (Gelflex, DuraSoft, Hydrocurve and Soflens) and that Bausch and Lomb is working with a gas permeable lens and an ultra-thin (0.05 mm) lens which transmits from 5 to 6% oxygen.

At Indiana University, Dr. IRV BORISH and I have been working with a number of investigative lenses, both spherical and toric, and our most successful lens to date has been the CSI lens. This is a lathe-cut hydrogel lens made of poly (glyceral methacrylate-co-methyl methacrylate) and has the generic name, crofilcon A (4). The center thickness varies from 0.06 mm to 0.07 mm and the lens is constructed in such a way that thickness variations between various zones are minimal. As a result it is sometimes referred to as a "membrane lens". For most wearers this lens elicits little if any sensation and provides good vision for patients whose astigmatism is within the range catered for by spherical soft lenses. There is seldom if ever a problem of achieving adequate centration and movement of the lens, and we often use it as a "lens of last resort", since it regularly provides good centration and good acuity when all other hydrogel lenses fail to do so. The CSI lens is not yet available in toric form.

Extended Wear Lenses

Those practitioners and manufacturers who formerly spoke of "continuous wear" lenses are now speaking in terms of "extended wear" lenses: "continuous wear" implies that the patient wears the lenses 24 hours a day until they are ready to be replaced, but "extended" wear implies that the patient may wear the lenses overnight occasionally or even fairly often, but will remove them at regular intervals.

The main problem to be overcome in the design of an extended wear lens is that of supplying the cornea with oxygen. When well-fitting soft lenses are worn only during the waking hours the movement of the lens with each blink, although often small in amount, supplies the cornea with sufficient oxygen to meet its metabolic requirements. However, since the "pump" mechanism does not work during sleep, the cornea must receive its entire oxygen supply by transmissiton through the lens HILL and CARNEY (17) have stated that during sleep the cornea normally receives about 7% oxygen from the lid capillaries (vs. 21% in air) and we should expect an extended wear lens to supply about 10%, to allow a favorable margin of safety.

The oxygen transmission of a hydrogel lens can be increased by either (or both) of two methods: by increasing the water content of the lens or be reducing the thickness of the lens. Since the oxygen transmission of a lens is a function of the amount of water the lens contains, the "ideal" contact lens from the point of oxygen transmission would be 100% water! An example of a lens having an extremely high water content is the Permalens, which as a water content of 74%. According to practitioners who have worked with this lens, it is extremely fragile and will not last long if handled daily by the wearer (which is probably the main motivating factor for continuous wear of such a lens); and vision can vary from minute to minute with the lens.

The effect of reducing the thickness of a hydrogel lens is shown in Fig. 4, taken fromHILL and CARNEY (17) which shows that the thickness of a lens would have to be on the order of 0.03 mm in order for the oxygen transmission to reach a value of 10%. The extended wear lenses being investigated at the present time (for example those made by CSI) have thicknesses of this order.

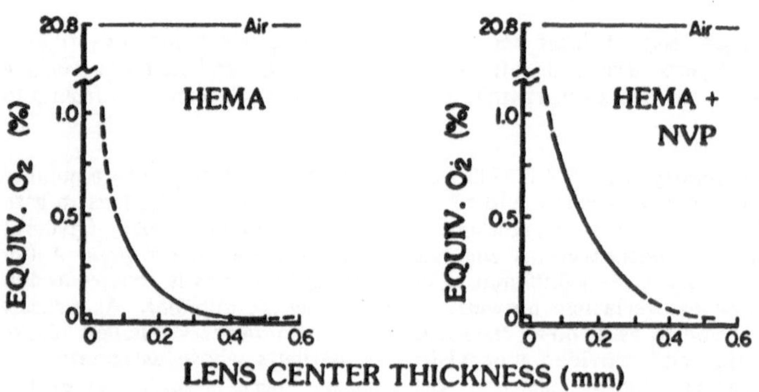

Fig. 4 See text for explanation.

According to KELLY (18) the F.D.A.'s proposed protocol for extended wear lenses is a complex one, and may seriously delay the arrival of such lenses on the market. Included in the protocol, according to KELLY, are the following provisions: (a) patients must be followed up for a minimum of one year; (b) the study must involve 500 patients, with a minimum of 25 patients per investigator; (c) pachometry must be performed on 35 patients in the study and specular microscopy on at least 100 patients; (d) the F.D.A. may make unannounced visits to clinical investigators to check on records and procedures; (e) manufacturers will be required to submit a full report of the progress of the study every three months during the course of the study; (f) extended wear lenses will have to be able to withstand patient handling; (g) manufacturers will have to prove that the parameters of the lens will not change with wear; and (h) a separate study of at least 100 aphakic patients will be required.

Summary

To look into the future of soft contact lenses it is necessary only to travel to Canada, Europe, Japan, Australia or New Zealand; or to work with one of the many lathe-cut lenses now under F.D.A. scrutiny. Lathe-cut soft lenses may be obtained in any desired base curve radius, power, diameter, optic zone size or thickness, just as lathe-cut hard lenses can. Toric front surface lenses are available for the correction of astigmatism, and some manufacturers are now experimenting with soft lenses in the form of bifocals. Still-to-be-solved problems with soft lenses include bacterial contamination, "coating" of lenses, allergic responses, and induced myopia and astigmatism.

References

1. O. Wichterle and D. Lim, Nature 185, 4706 (1960).
2. R. J. Morrison, J. Am. Optom. Assoc. 37, 211 (1966).
3. E. J. Fisher, Can. J. Optom. 29, 139 (1968).
4. Supplement No. 1, Encyclopedia of Polymer Science and Technology, John Wiley and Sons, Inc., (1976).
5. R. M. Hill and I. Fatt, Am. J. Optom. 41, 382 (1964).
6. T. Grosvenor, Am. J. Optom. 49, 407 (1972).
7. B. Holden, Int. Cont. Lens Clinic , p. 59 (Spring, 1976).
8. C. A. Bayshore, Int. Cont. Lens Clinic, p. 69 (Spring, 1975).
9. A. M. Charles, C. Murcheson and T. Grosvenor, Am. J. Optom 50, 777 (1973).
10. A. C. Snyder, R. M. Hill and N. J. Bailey, Cont. Lens Forum 2, 41 (1977).
11. N. J. Bailey, Cont. Lens Forum 1, 10 (1976).
12. M. F. Refojo and F. J. Holly, Cont. Intraoc. Lens Med. J. 3, 23 (1977).
13. T. Grosvenor, Am. J. Optom. 52, 405 (1975).
14. R. Mandell, Int. Cont. Lens Clinic 1, 32 (1974).
15. J. F. Hill, Cont. Lens Forum 2, 31 (1977).
16. N. J. Bailey, Cont. Lens Forum 2, 35 (1977).
17. R. M. Hill and L. G. Carney, Cont. Lens Forum 1, 29 (1976).
18. C. Kelly, Cont. Lens Forum 1, 22 (1976).

The State of the Art and Science of Contact Lens Fabrication and Fitting

Henry A Knoll
Soflens Division, Bausch & Lomb, Inc.
Rochester, New York

"A science teaches us to know, and an art to do, and all the more perfect sciences lead to the creation of corresponding useful arts. Astronomy is the foundation of the art of navigation, . . . chemistry is the basis of many useful arts" (WILLIAM STANLEY JEVONS, 1935-1882) (1).

As in the case of many human endeavors, the art of contact lens fabrication and fitting developed faster than the science of contact lens fabrication and fitting. In the ninety years since blown glass contact lenses were first put on eyes, the art and the science have spurred each other on, advancing slowly and cautiously at times and at other times moving rapidly on a wave of new developments.

One of the developments linked optical science with the fabrication of ground and polished glass trial sets produced by the Carl Zeiss Co. of Jena and introduced in 1930. Professor HEINE, an ophthalmologist of Kiel, Germany had worked out a system of afocal scleral contact lenses whose corneal portion cleared the cornea and whose curvatures were calculated to correct refractive errors. HEINE'S work was commented upon by Professor JAMES P. C. SOUTHALL as follows: "Contact eyeglasses seem to have an assured future and one day they will undoubtedly be used more extensively than now, not only for correcting irregular astigmatism and the more serious defects of vision but for ordinary purposes in many cases where spectacles are used at present." (2)

The mid and late thirties brought additional developments. Methods were developed to take molds of eyes from which glass and later plastic shells were formed. The use of fluorescein to judge clearance was described by OBRIG (3). Then World War II came along and progress slowed almost to a halt.

A second wave followed the end of World War II, with the full utilization of polymethyl methacrylate and the patenting of the corneal contact lens. We are now on a new wave of advancement spurred on by the improved knowledge of the oxygen requirements of the cornea and developments in the field of polymer science. The search is on for an oxygen permeable material capable of being economically made into contact lenses.

But I am getting ahead of my story. Let us first look at what might be considered the foundation sciences of the art of contact lens fabrication and fitting. Having done this I would then like to take a close look at several areas to see how things are going.

Before listing the foundation sciences I give here a definition of science taken from the Third Edition of Webster's unabridged dictionary.

> "A body of systematized knowledge comprising facts carefully gathered and general truths carefully inferred from them, often underlying a practice, usually connotating exactness and often denoting knowledge of unquestionable certainty" (4) (emphasis mine).

Exactness and unquestionable certainty are criteria which have often not been applied rigorously when evaluating new developments in the foundation sciences. I cite as an example the work being done involving tear breakup time (BUT). It is only very recently that the subject of individual variability has been examined and two studies show that the variability is so large as to cast doubt on the results reported by some of the earlier studies relating the effect of contact lens wear upon tear breakup time (5,6).

I propose that the foundation sciences of the art of fabrication and fitting contacts are as follows:

1. Anatomy
2. Physiology
3. Optics and Physiological Optics
4. Pathology
5. Material Science
6. Clinical Science
7. Psychology and Sociology

Certainly other sciences are involved in practicing the art of fabrication and fitting of contact lenses, but their consideration had best be left to those better versed in these sciences. I have reference to forensic science and statutory science. Practitioners of the art had better be aware of patent restrictions and national and local laws that pertain to the fabrication and fitting of contact lenses. This is especially so in view of the rules and regulations promulgated by the U.S. Food and Drug Administration.

Some appreciation of the state of the fundamental sciences of contact lens fabrication and fitting in 1887 can be gained from the following. DESCEMET, VON GRAFE, PURKINJE and AMICI were dead. Their works were widely published and studied. HELMHOLTZ, BOWMAN and DONDERS were in the last decade of their lives. Their works were also well known. Not bad as a start for the anatomy, physiology and optics of the eye.

We were well on the way concerning the pathology of the eye (the staining of traumatized tissues by fluorescein had been reported five years earlier).

GULLSTRAND'S description of the slit lamp in 1911 marked a truly great milestone in the advancement of the examination and understanding of the normal and abnormal cornea. GULLSTRAND contributed further to the fundamental sciences with his work on the shape of the human cornea. More on the shape of the cornea later.

It was perfectly logical in 1887 to use glass as the material to make contact lenses. Glass was serving as the material for artificial eyes. There was nothing else available until fifty years later when poly methyl methacrylate came along. PMMA was king for thirty years and is only today being challenged as the material most used in contact lenses.

Improvements in slit lamps and their wide spread utilization has greatly advanced the fudamental sciences of anatomy, physiology, pathology, and clinical science. The addition of the pachometer has added significantly to our knowledge of the corneal response to the wearing of contact lenses as well as the normal physiology of the eye.

Electron microscopy has added significantly to our knowledge of the structure of the cornea and to the surface quality of contact lenses. Most recently we have the addition of the specular microscope which enables us to view in great detail the living cells of the endothelium.

We seem to be at a point where we can speak with some certainty about the metabolic requirements of the cornea. This is particularly true as it relates to the oxygen requirement. This level of sophistication makes it feasible to predict certain necessary properties of the materials to be used in the fabrication of contact lenses. The materials must be able to transmit the oxygen needed by the cornea or else have physical characteristics which makes it possible for the lens on the eye to pump oxygen via the tears.

Two areas which differ greatly in the amount of attention they have received since the advent of contact lenses are the shape of the human cornea and the dynamics of blinking.

GULLSTRAND early on qualitatively described the shape of the human cornea as follows: "There is a central optical zone where the curvature is approximately spherical, and which extends horizontally about 4 mm, and somewhat less than this vertically, and is decentered outwards and usually also a little downwards; and that the peripheral parts are considerably flattened, decidedly more so on the nasal side than on the temporal, and usually more so upwards than downwards" (7). At another point after discussing a suggestion that a corneal section might be described by an ellipse he writes: "All that can be concluded from this is, that, in vision with a pupil of medium size, the assumption of a spherical shape for the utilized part of the cornea is provisionally the best approximation: and in cases where the eccentric parts of the cornea are predominantly effective (as in certain investigations of the lens constants) the ellipse may continue to be considered as a better approximation than any similar hypothesis" (7). Put another way, GULLSTRAND told us that the cornea was not spherical and further that it was not a surface of revolution.

Dozens of investigators have attempted to quantify GULLSTRAND'S description with varying degrees of success. In almost all cases investigators have attempted to establish the curvature of the cornea at various distances from its center, usually along a horizontal meridian. One of the exceptions to this approach has been the tack taken by an Australian physicist B.A.J. CLARK using a sophisticated photokeratoscope of his own design (8). Using a computer to reduce his data he calculates the deviation of the surface of the cornea in eight semi meridians from the

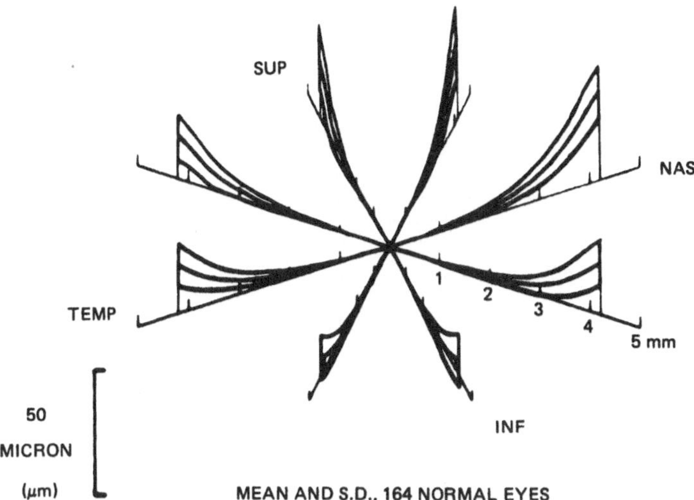

Fig. 1 Corneal asphericity (deviation from a reference sphere) of human corneas. The radial lines represent semi-meridians. Taken from Clark (8).

surface of a reference sphere. He defines corneal asphericity as the deviation, measured radially, from the reference sphere. Fig. 1 shows the summary data for 164 eyes. Shown are the means and standard deviations. Note the correspondence with GULLSTRAND's description - nearly spherical within the central 4 mm and flattening more nasally and superiorly.

One serious limitation of the system is that data cannot be obtained beyond the limits shown, i.e. approximately 4 mm from the center of the cornea. The limit is set by the numerical aperture of the lens used to produce the photokeratograph. Within the roughly 8 mm diameter great variability was found among the subjects used in this study. Individual corneal data plots are shown in Figs. 2 and 3.

In spite of the many investigations reported (see CLARK (9) for an extensive bibliography) we still do not have data describing corneal shapes out to the visible iris diameter. We can only hope that new approaches are in the wings that will allow us to characterize the full cornea on an individual basis to make possible a more scientific approach to designing and fitting contact lenses.

The second area mentioned, namely the dynamics of blinking has received very little attention. About ten years ago there appeared a short paper by LUTES (10) of Southern California in which he described some experiments involving high speed photography of blinking in subjects wearing hard contact lenses. The surprising finding was the large amplitude of the lens movements during a voluntary blink. LUTES' work has been recently repeated by a graduate student of Professor FATT'S at the University of California. Plots of lid, lens and eye movements as a function of time are shown in Fig. 4. Note that the lower lid moves very little. It would appear that most of its movement is associated with contact by the upper lid. The amplitude of the lens movement is about 1 centimeter. Finally note the eye movement and its close synchronism with the upper lid movement.

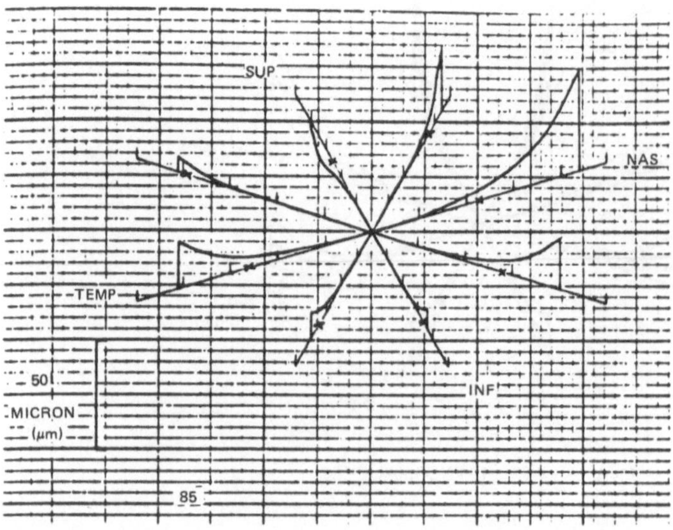

Fig. 2 Corneal asphericity (deviation from a reference sphere). Results for the right eye of a 19 year old female showing by crosses the position in each semi-meridian where the value of asphericity reaches 5 microns. Taken from Clark (8).

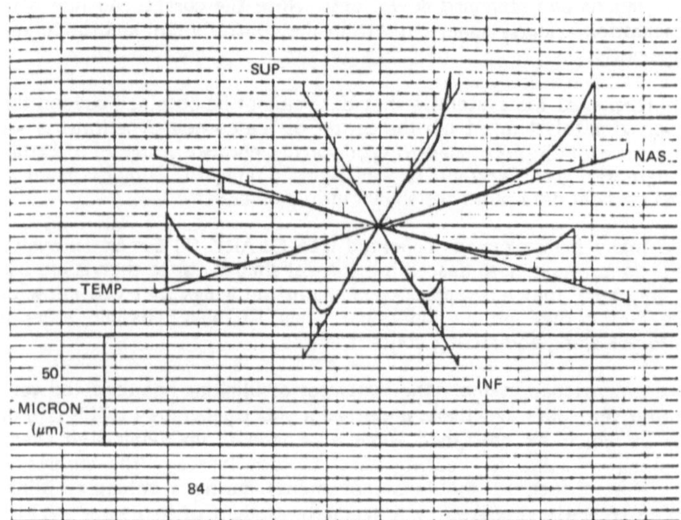

Fig. 3 Corneal asphericity (deviation from a reference sphere). Results for the right eye of a 21 year old male, showing negative asphericities. Taken from Clark (8).

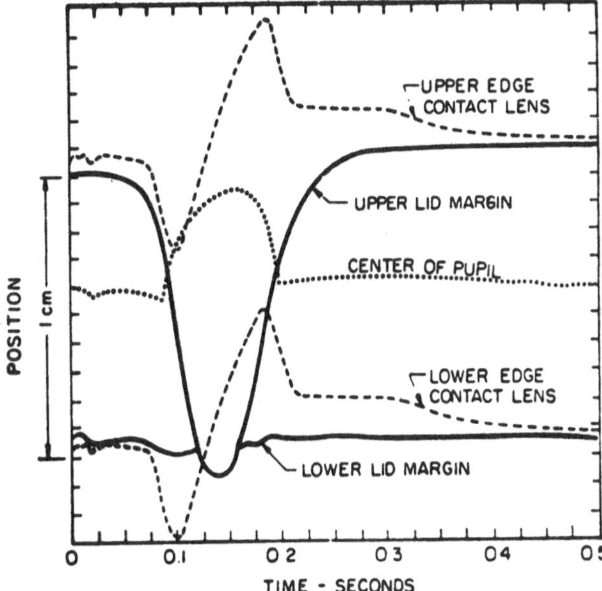

Fig. 4 Movement of lids, eye and hard contact lens during a voluntary blink. Data taken from photographic recording at 400 frames/sec. Courtesy of Professor Irving Fatt.

A number of years ago a colleague told me about his observation of the apparent vertical movement of a neon pilot light on a piece of electronic equipment during voluntary blinking. When I saw this graph it occurred to me that what my colleague may have been observing was an asynchrony between lid and eye. If one were to look at a horizontal sweep on the face of an oscilloscope and blink, perhaps it would be possible to catch the eye's displacement before the lid completely covered the pupil, or after the lid uncovered the pupil before the eye was centered again. Sure enough with a sweep speed of 20 ms/cm. two observers, in addition to myself, were able to see a small vertical upward blip with almost every voluntary blink. From the shape of the blip we know that it comes at the beginning of the blink. Then we discovered something else and that is that the direction of the blip (opposite the direction of the eye movement) changes with elevation and depression of the plane of regard. Then we had another surprise, GINSBORG and MAURICE (11) had already described this technique in the literature and had gotten results similar to ours. They had also used a vertical sweep and demonstrated that the eye movement during a blink is always toward the primary position. For example, if the eyes are turned to the right, during the blink the eye movement will be to the left. Clearly these eye movements (amplitude $\approx$ ½ mm) are independent of the upward rotation of the eye.

Contact lens fitters will readily admit that the lids play an important role in the fit of a contact lens. Yet this is one aspect of the fundamental science of contact fabrication and fitting that we know very little about. Much needs to be done.

With ninety years of contact lens fitting experience behind us a lot of experience has accummulated concerning the psychology of wearing contact lenses. On the other hand very little has been published about the perception of the average citizen of contact lenses as a means of vision correction. There have been no couplets like DOROTHY PARKER'S "Men seldom make passes at girls who wear glasses", or short stories like EDGAR ALLAN POE'S "The Spectacles" or GOETHE'S often quoted dislike of spectacles applied to contact lenses. True, contact lenses have not been around for 700 years, but the ninety years of their existence have been characterized by a communication explosion. Perhaps it's because the little disks are invisible?

References

1. Webster's New International Dictionary, Second Edition, Unabridged, G. & C. Merriam Co., Springfield, Massachusetts, 1934. Under the entry "science".
2. J. P. C. Southall, Review of Scientific Inst. 5, 54 (1934).
3. T. E. Obrig, Contact Lenses, First Ed., (The Chilton Co., Philadelphia, 1942), p. 153.
4. Webster's Third New International Dictionary, Unabridged, G. & C. Merriam, Springfield, Massachusetts, 1971. Under the entry "knowledge".
5. G. T. Vanley, I. H. Leopold and T. H. Gregg, Arch. Ophth. 95, 445 (1977).
6. M. D. Sarver, unpublished data.
7. A. Gullstrand, in v. Helmholtz's Physiological Optics, English Trans., Vol. I, pp. 314-315, 1924.
8. B. A. J. Clark, Ph.D. Thesis, University of Melbourne.
9. B. A. J. Clark, Aust. J. Optom 56, 48 and 182 (1973).
10. H. R. Lutes, Contacto 13, 13 (1969).
11. B. L. Ginsborg and D. M. Maurice, Brit. Jr., Ophth. 43, 435 (1959).

Important Diagnostic Testing of the Contact Lens Wearer

Tom F. Brungardt
Southern California College of Optometry
Fullerton, California 92631

The contact lens wearer's refractive error is neutralized by two separate systems. Analysis of each of these two systems of correcting lenses can tell much about the anatomy and physiology of the eye and the visual systems. Analysis of these two system and the results of testing any change in the correcting lenses or the visual acuity resulting will give a new dimension to diagnosis. Of most importance is the interplay between the two correcting or neutralizing systems.

Most of us are aware that a marked improvement in visual acuity occurs with the application of a hard contact lens to the keratoconic eye. This occurs because the distorted anterior corneal surface is refractively eliminated with contact lens application. Other irregularities of the corneal surface caused perhaps by trauma or disease are likewise successfully treated for vision improvement.

Unfortunately, the actual wearing of a contact lens fitted for cosmesis often produces such an irregularity of the eye that the person complains of an inability to satisfactorily wear spectacles. Such a person becomes a "contact lens cripple", dependent upon the device to correct refractively a corneal irregularity which was caused by the original cosmetic device. This condition might well mask some other and more devastating eye or visual system pathology. The analysis suggested will allow a differential diagnosis.

Interest in this analysis should be high amongst those purposefully pursuing orthokeratology.

The two neutralizing systems are: 1) the spectacle only, worn at a conventional 13 mm before the eye; and 2) a contact lens system. Both systems should correct the eye's refractive error to equal visual acuity with control of accommodation.

The contact lens system is composed of three separable lenses: the cornea, the lacrimal fluid or tear film, and the contact lens itself. Since the lacrimal lens is formed by the surfaces of the other two lenses, it is primarily dependent for its power from these surfaces. Hence, a change in corneal curvature will change the posterior surface of the lacrimal lens by an opposite and equal radius amount. A change in contact lens base curve will likewise change the anterior surface of the lacrimal lens by an equal and opposite amount of radius.

Power change of the lacrimal lens is further determined by the difference in refractive index of the contact lens, the tears, and of the cornea and by its thickness. Within fairly broad limits neither change in thickness of the lacrimal lens nor its index of refraction is an important variable to the refractive power of the system.

Corneal curvature change, expressed as vergence power (diopters), therefore is almost negated by the presence of the lacrimal lens. The actual change is in the ratio of 9 to 1; i.e., change of 9.00 D of corneal curvature will produce a 1.00 D change in total refraction in the contact lens system, or in more clinical terms, if an over-refraction change of -0.12 D is found, the cornea would have to have steepened 1.00 D . Over-refraction is that spectacle correction found necessary to obtain maximum visual acuity with accommodation controlled while a contact lens is on the cornea. Hence, whenever a change in the over-refraction occurs, corneal change is not the first suspected cause, rather a very late one.

Practically, any change in an over-refraction would be more often caused by change in the contact lens power or its base curve. Secondly, any change in eye refractive components, other than the anterior corneal surface, would cause such a change, and investigation of these tissues is next indicated. Change in the corneal curvatures would be alarmingly high for even a small change in the over-refraction and clinically could be ignored.

The eye refractive components, other than the anterior corneal curvature, would include the axial length, or segments of this length, the crystalline lens power, the corneal thickness, posterior corneal curvature, and the index of refraction of any of the eye structures or spaces. The diabetic changes his refractive error quickly without affecting corneal curvature. The aging change their refraction with cataract formation. The young become more myopic during school months and reverse the trend in summer months. Macular edema causes an increase in hyperopia. These are examples of changes in refractive error caused by other than change in the anterior corneal curvatures.

Stability of the anterior corneal curvatures is assured after the formative years (1). However, trauma or disease can change these curvatures, and wearing a contact lens is traumatic to the cornea.

Study of the two systems of refractive error determination and change in either the quantity or the quality of refraction will allow a practitioner to properly ascribe the cause of change to either the anterior cornea or to the other internal eye structures.

Consider only the visual acuities (quality) resulting from a spectacle only correction and a contact lens system. Assume that originally both correcting systems gave equal and good visual acuity. At some later date, however, the spectacle only correction allowed for a lesser acuity while the contact lens system still yielded the original acuity. The responsible tissue for these differences is the anterior corneal surface.

Secondly, change might occur in both resulting in acuities from equal and good to equal but poor. Such a change cannot be ascribed to the anterior corneal surface, but to tissues other than that surface. Such a change would dictate investigation of the integrity of the internal eye structures such as the crystalline lens and retina or to physiological factors. Indicated tests might be ophthalmoscopy, gonioscopy, tonometry, Amsler Grid and tangent screening, color testing and investigation of systemic disorders such as diabetes, multiple sclerosis, and macular edema.

Thirdly, good visual acuities in both neutralizing systems could change such that unequal loss results. The spectacle only visual acuity could fall three lines of Snellen letters while the over-refraction visual acuity, contact lens in situ, might fall only one line. Analysis of this change is first directed toward the internal eye structures, optic nerve, brain, and systemic disorders in order to account for the decrement of one line of visual acuity. The anterior corneal curvatures are responsible for the remaining two lines of visual acuity loss.

The same analysis is used to account for change in the quantity amount of refraction. In order to simplify this presentation, assume that the visual acuity resulting is unchanged. Any change in quantity amount of the spectacle only refraction is due to either or both anterior corneal change or to change in internal factors other than that surface. Change in quantity amount in the contact lens system, however, is due to the internal factors only. Hence, corneal changes are known by subtracting the changes in the two systems.

Examples of these changes are taken from a study of 25 long-term contact lens wearers and one of these is presented (2).

A patient was fitted with hard contact lenses at age 25. The spectacle correction was -0.75 -1.75 x 180. Two years later the spectacle correction was -0.25 -1.25 x 180 or a reduction of myopia of 0.50 D in each meridian without loss in visual acuity.

The contact lens power selected in trial fitting was -0.75 and the over-refraction was -0.50 D S (diopter sphere). Hence, the lacrimal lens power was +0.50 -1.75 x 180. Since the base curve is known to be 7.67 mm (44.00 D), the corneal curvatures can be calculated (3). They are, at the time of fitting, 45.25 90 and 43.50 180. The base curve was thus 0.50 D steeper than the flatter corneal curvature.

At age 27, the contact lens being worn was analyzed to be 44.25 D base curve and back vertex power of -1.00 D S. An over-refraction result of +0.50 -0.50 x 90 resulted in a visual acuity equal to the original. The corneal curvatures are now calculated to be 45.25 90 and 43.50 180. There is no change in corneal curvatures. The presently worn lens is now 0.75 D steeper, and the power of the lacrimal lens is +0.75 -1.75 x 180. The change in refractive status of less myopia is due to a change in eye structures other than the anterior corneal curvature. The actual change is -0.50 -0.50 x 180, or less myopia.

A simplified manner of calculating the change and properly ascribing any change to either anterior corneal curvature change or internal eye change is possible without ever knowing what the corneal curvatures are.

The internal eye caused changes in refraction are known by: 1) adding the power of the contact lens to the over-refraction result in each time period studied; 2) to the later period result add the change in base curve expressed as vergence power, a flatter lens takes a minus sign; 3) subtract the latter period result from the original.

In the above example, the original power before the eye was (-0.75 D S) + (-0.50

D S) = -1.25 D S. At the latter period the power before the eye was (-1.00) + (+0.50 - 0.50 x 90) = -0.50 -0.50 x 90. Since the base curve is now steeper 0.25 D, the power of the system at this latter date compared to the eye originally is -0.25 -0.50 x 90. The change in internal eye structures is thus (-1.25 D S) - (-0.25 -0.50 x 90) or -1.00 +0.50 x 90 or -0.50 -0.50 x 180.

Since subtracting the latter period spectacle from the original found spectacle gives the same result, no corneal curvature change has occurred.

Other examples can be presented to illustrate change in both the corneal curvatures and to the internal eye structures. In the study (2), from which this example was taken, visual acuity did not change in either system of neutralizing the refractive error. Additionally, the spectacle only refraction did not change for the majority of patients. However, the analysis, here presented, showed that the two identifiable causes of stability of the spectacle only refraction did change but in opposite directions.

The majority of patients became more myopic, due to the internal eye changes; but their anterior corneal curvatures flattened. The curvature flattening is obviously due to the contact lens. The contradictory increase in myopia ascribed to internal eye factors could be due to normal aging with an increase in axial length. However, certain patients' eyes changed in excess of expected aging, and new cylindrical components were introduced. These changes may be due to the crystalline lens deformation caused by unequal pressure from the contact lens on the limbal area.

Summary

The care of patients wearing contact lenses is enhanced when the causes of any change in refractive quantity and resultant visual acuity is analyzed. Change in spectacle only refraction is due to both corneal curvature change and internal eye components other than the anterior corneal surface. Change in the contact lens refraction is due only to the internal eye components. Change in the anterior corneal curvature is the difference between the change in spectacle only refraction and the contact lens refraction.

References

1. H. Knoll, Am. J. Optom. and Arch. Am. Acad. Optom. 53 (1976).
2. P. Brungardt and T. Brungardt, Paper presented to the American Academy of Optometry, December, 1975, Columbus, Ohio; paper in review.
3. T. Brungardt, J. Am. Optom. Assoc. 46, 230 (1975).

IV. Color Vision

Analysis of Human Color Vision by Exchange Thresholds

W. A. H. Rushton
Trinity College
Cambridge University

A property of vision that is particularly useful in the analysis of color vision is called "The Principle of Univariance". It was already implied in Thomas Young's formulation of trichromacy. Somewhat loosely expressed, it means that the response of a visual receptor depends upon the number of quanta it catches but not upon the wavelength of these caught quanta. Rhodopsin, for example, absorbs green light more readily than red, thus green light will excite the rhodopsin-filled rods more than will an equal energy of incident red light. But if, instead of measuring the energy of light incident, we measure the energy absorbed, then it is found that red and green lights that are equally absorbed will have identical rod-exciting effects, and will hence appear identical by rod vision.

In particular since the only difference in various lights as regards rod excitation is due to the amount absorbed and not to wavelength, the univariant result is seen as a variation in brightness and not in color. The monochrome appearance of twilight vision when only rods are active has long been recognized. A similar condition is found in the cone vision of dichromats (protanopes and deuteranopes) where every spectral light in the red-green range can be perfectly matched with yellow.

KOENIG explained this by supposing that protanopes lacked the red-sensitive cones, deuteranopes lacked the green-sensitive, and that each cone exhibited a univariant response. The alternative FICK-LEBER hypothesis, that dichromats possessed both red and green cones but confused their separate outputs, has been excluded by retinal densitometry, which is not concerned with cone outputs whether separate or confused. Dichromats were found to have only one of the two principal cone pigments. Reflectivity measurements showed that protanopes lacked the pigment of the red-sensitive cones; deuteranopes lacked the pigment of the green-sensitive cones (1). The results of this experiment are shown in Fig. 1 which plots by

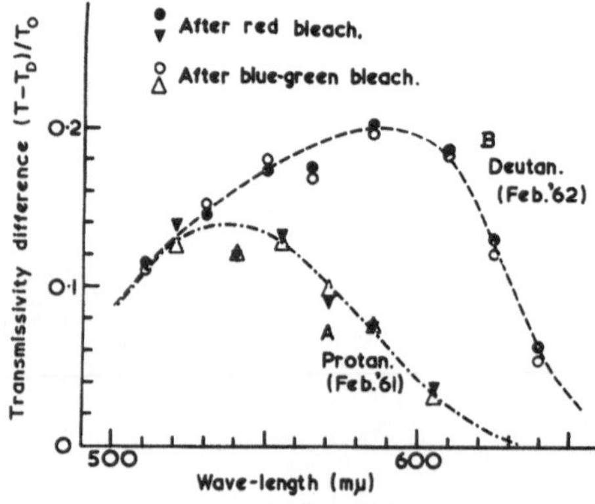

Fig. 1 Difference Spectrum of human cone pigments, measured by reflection densitometry in a protanope and a deuteranope after 50% bleaching with red light (black symbols) or with blue-green (white symbols). (From Rushton (1)).

retinal reflection the change in retinal transmissivity for light of various wavelengths when the dichromats eye has been half-bleached by red light (black symbols) or blue-green (white symbols).

It is seen that the difference spectra from the two kinds of bleaching coincide. This could not happen if FICK were right and the normal red and green cone pigments were still present in the dichromat. For, red bleaching would produce a greater change in the red part of the spectrum; green bleaching, a greater change in the green part. They could not behave identically under the two half-bleachings as is seen to be the case.

On the other hand this is exactly what would be expected from KOENIG'S hypothesis. If each dichromat has only one kind of cone, the principle of univariance requires that each cone will respond identically if the quantum catch is the same for both, which will be the case when (as here) both are bleached 50%.

Light Exchange

Suppose we have two lights I_1, I_2 that differ both in wavelength and intensity. We may excite a given class of cone by a neat exchange of one of these lights for the other, by quickly sliding P_3, the sheet polarizer of Fig. 2, between the positions where only vertical polarization is transmitted and that where only horizontal is. The condition where this exchange will not excite at all is where the quantum catch by this cone's pigment is exactly the same from the one light as from the other. The principle of univariance predicts that this exchange will have exactly the effect of a continuation of the first light without any exchange to the second. Experiment confirms this expectation. For instance, a protanope who matches I_1 with I_2 in a bipartite field will not perceive any change when I_1 is neatly exchanged for I_2 in the course of a steady presentation.

Figure 2 indicates the equipment used to produce this light exchange. Interference filters λ_1, λ_2 in the beams from the light source A control the two wavelengths, wedge R adjusts the relative intensity, polaroids P_1, P_2 polarized either vertically or horizontally, the exchange intensity being adjusted to threshold by the wedge W. The beam-splitting cubes C_1, C_2 allow the exchange lights I_1, I_2 to fall upon an unchanging background I_3 where $I_3 = \frac{1}{2}(I_1 + I_2)$, by suitable rotation of the polaroid P_4.

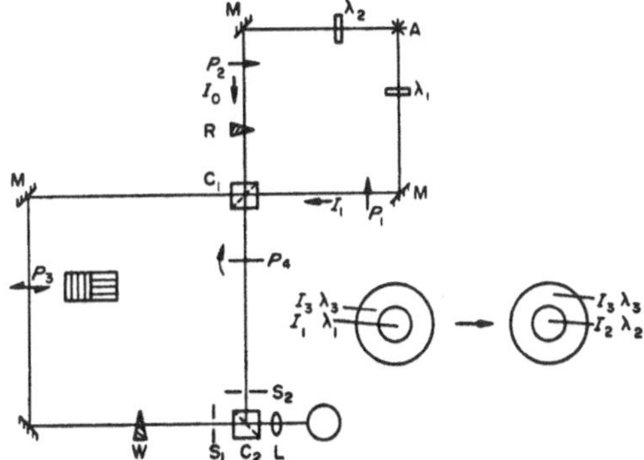

Fig. 2 Diagram of optical arrangement to deliver "light substitution" on a steady background that is the average of the two lights. A, light source, λ_1, λ_2 interference filters, P sheet polarizers. Inset, lower right shows the presentation of I_1 or I_2 superposed on steady I_3.

The measurement consists in varying the ratio of intensities I_1/I_2 by wedge R, and for each R setting finding the exchange threshold by varying the wedge W until the exchange is barely detectable. Fig. 3 shows the result when I_1, and I_2 are both lights of wavelength 540 nm, and ordinates plot w, the interposed density W, abscissae

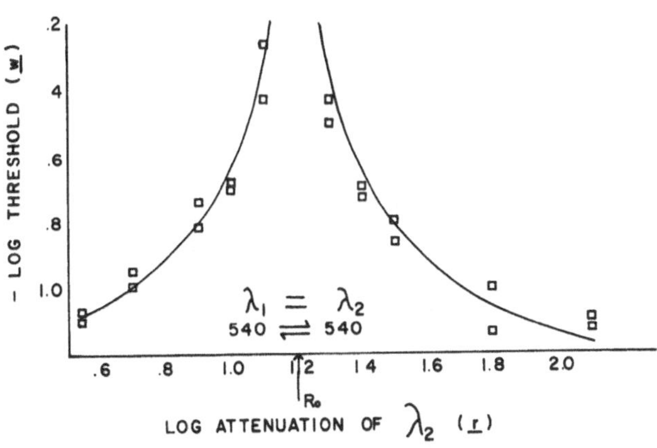

Fig. 3 Exchange thresholds when there is no colour change. Abscissae r, interposed density of wedge R in red beam. Ordinates w, interposed density of wedge W in the exchange beam. Curves are theoretical expectations from Weber's law and Univariance; circles and triangles, experimental results.

plot $\underline{r}$, the interposed density R. When $\underline{r}$ has some value r_c the intensity of I_1 is the same as that of I_2, and the exchange presents zero change. Thus the threshold at r_c must be infinite. For other $\underline{r}$-values the change is greater and a lower threshold expected. From Weber's law it is easy to calculate the expected results of Fig. 3. It is plotted as the two-branched curve, and the experimental results plotted as circles (deutan.) and triangles (protan.) fit well enough.

Conclusions

When the exchange involves no change in the wavelength of the light, the results follow a curve of fixed shape, symmetrical about a vertical axis at $\underline{r}_o$, which is the $\underline{r}$-value where the quantum catch from I_1 is equal to that from I_2.

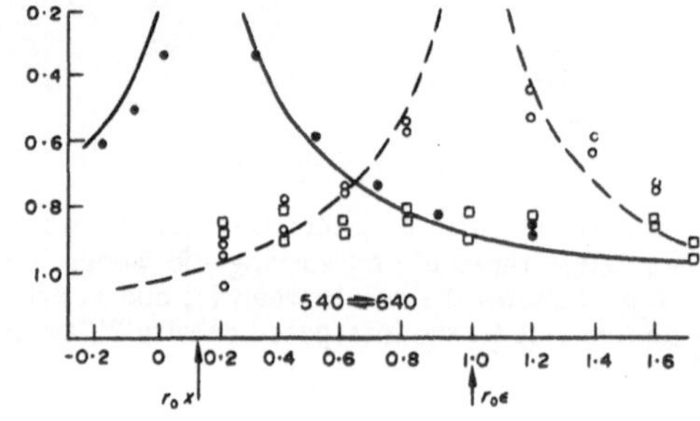

(a)

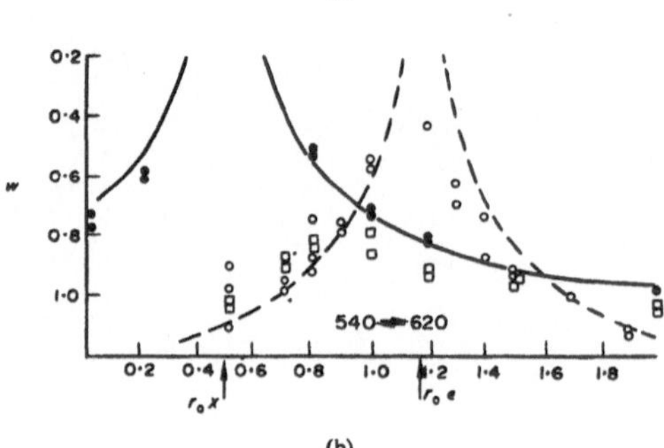

(b)

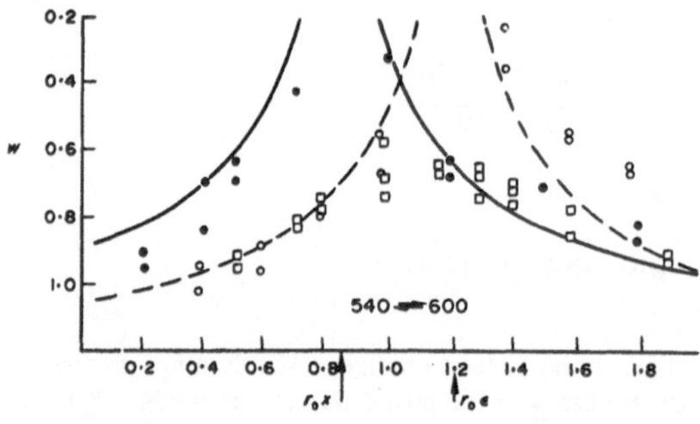

(c)

Fig. 4 Exchange thresholds plotted as in Fig. 3 when colours change between 540 nm and 640, 620 or 600. Black circles show results from protanopes, white circles from deuteranopes, squares from normal subjects. Continuous curves show expectations from chlorolabe cones, dashed curves from erythrolabe cones. Squares lie close to whichever curve lies lowest.

Exchange in Dichromats

In a dichromat who has but one class of cone active in the red-green spectral range, we may extend our analysis to exchanges where I_1 and I_2 are lights of different wavelengths. In the foregoing conclusion r_o was the r-value where the quantum catch was the same from I_1, and from I_2. For a protanope we may determine this for any two wavelengths, e.g., the r-value that gives minimum flicker when I_1 is rapidly alternated with I_2.

In Fig. 4 black circles plot results as in Fig. 3 when the wavelength is exchanged between 540 and 640, 620 or 600 nm and the subject is a protanope with only chlorolabe in his cones. The continuous branched curve is identically like that of Fig. 3, but with the axis r at the point determined by flicker. White circles give the same results with a deuteranope, and the r_o axis is also determined from his flicker results, to give the relative spectral absorption of erythrolabe in his cones. The interrupted curve is the same curve as for the protanope but laterally displaced to the new r axis. Squares in Fig. 4 plot the exchange threshold in a normal subject. They fit rather closely the lower contours of the two curves. The normal possesses both chlorolabe and erythrolabe, and his threshold is that of the protanope or that of the deuteranope which ever sees better in the particular situation.

Anomalous Trichromats

The most common type of color abnormal requires the mixture of three primary lights in order to match every spectral light. He is therefore a trichromat. He is anomalous in that the primaries must be mixed in unusual proportions. It is easy to prove that this abnormality is inconsistent with the presence of normal cone pigments with abnormal processing of the signals arising from their quantum catch.

The method of exchange thresholds allows us to eliminate the action of the normal cone pigment so the abnormal pigment is analysed alone, and yields its action spectrum.

The full analysis is too complex to be treated here. In principle the method rests upon the results of densitometry which in the protanomalous reveals the presence of chlorolabe only. Now we exchange red and green lights adjusted in intensity so that the normal cones cannot respond. But the anomalous trichromat detects the exchange, thus he must detect it by his abnormal cones. The threshold for this exchange is measured against a steady background of various wavelengths. Then, assuming the threshold for the abnormal cones is raised in proportion to the absorption by these cones from background light, we may learn the spectral absorption of the abnormal pigment.

The arguments and controls have been set out in greater detail by RUSHTON, SPITZER-POWELL and WHITE (2).

Spectral Sensitivity of Abnormal Pigments

Figure 5 shows, by the continuous line, the spectral sensitivity of the abnormal pigment in the protanomalous; Fig. 6 shows it in the deuteranomalous. The dotted line plots chlorolabe, the dashed line erythrolabe. Chlorolabe is absent in the deuteranomalous, erythrolabe absent in the protanomalous.

It is obvious that color discrimination by two cones must depend upon the ratio of the signal sizes arising from the absorption of various colored lights in the two cones. In Figs. 5 and 6 it is clear that the difference in log quantum flux at various

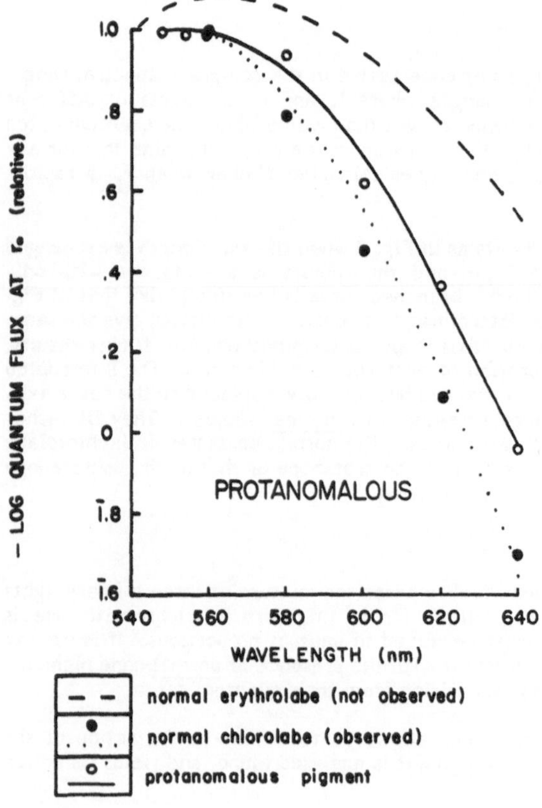

Fig. 5 Spectral sensitivity curves in anomalous trichromats measured by exchange thresholds in the protanomalous.

wavelengths is far greater between the two normal pigments than when one of these is replaced by the anomalous pigment. This accounts for the poor color discrimination of the anomalous trichromats and even more for their poor color naming.

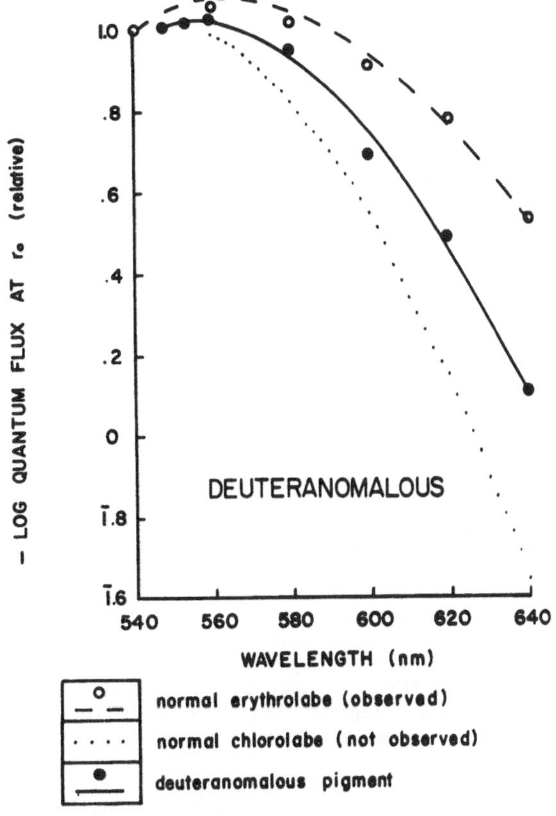

Fig. 6 Spectral sensitivity curves in anomalous trichromats measured by exchange thresholds in the deuteranomalous.

References

1. Rushton, Nature 206, 1087 (1965).
2. Rushton, Spitzer-Powell and White, Vision Res. 13, 1993 (1973).

The Distribution of Blue Receptors in Primates Eyes Revealed by Spectral Photic Damage and by Histochemical Response Experiments

H. G. Sperling
University of Texas Health Science Center
Sensory Sciences Center
6420 Lamar Fleming Ave.
Houston, Texas 77025

To understand the interactions between receptors which result in color and brightness discriminations, it would be valuable to know the spatial distribution of color receptors. It would be useful just to learn the percentage of red-, green-, and blue-sensitive cones as a function of retinal position. It would be much more useful to know the detailed classification within small regions from receptor to receptor, especially if it were true in primates, as it is in fish, that there is a regularly repeated pattern in the cone mosaic.

The following will describe a series of experiments, in which we have been engaged for about ten years, that has answered some of these questions and shows promise of answering the others. These studies involve the interplay of several disciplines, ranging from psychophysics performed on humans and monkeys to anatomical and histochemical studies performed on monkeys and fish. The work grew out of our ability to obtain psychophysical increment-thresholds on rhesus monkeys which were as detailed and quantitatively accurate as those which we had been obtaining on humans. This enabled us to obtain baseline data, then induce pathology by intense light exposure, perform subsequent psychophysical measures and then histology all on the same animal.

Figure 1 shows the typical optical system which we have used throughout our psychophysical experiments. It consists of a three path maxwellian view with a xenon arc (Osram 1600 watt) as the source for the intense spectral exposure path and for a spectral path which provides a 2° circular test light in the center of the 18° background field. The third, white background, path is from a tungsten iodide source, also aperatured to 18° and focused in the pupil of the eye. The three aperatures are in conjugate focal planes so that the viewing distance may be placed at infinity, for relaxed accommodation. The images in the pupil are smaller than the smallest natural pupil at the background intensities used. The image of the test target is largest, so that lateral movements of the head vignette the test beam, reducing its intensity, hence the probability of its detection. Since the animal works to obtain his daily fluid in the experiment, he learns to hold the beam centered in his pupil. We

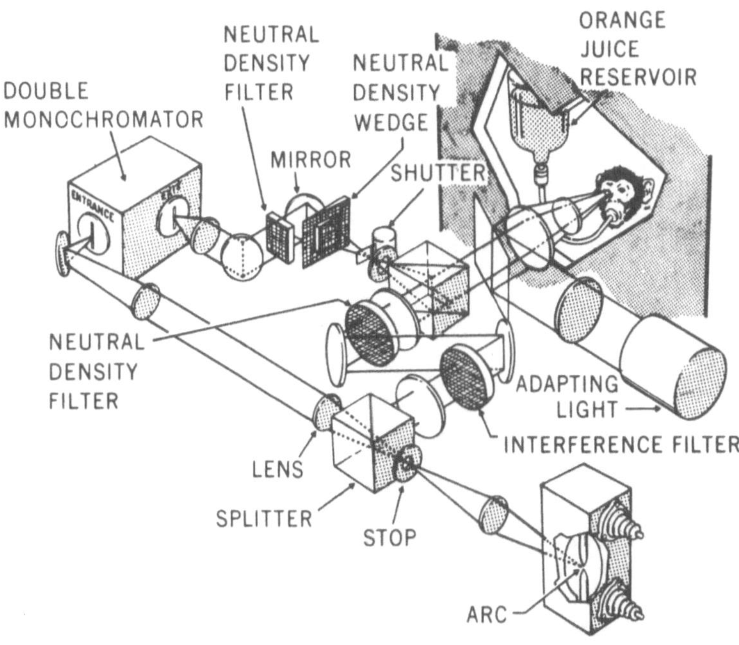

Fig. 1 Diagram of apparatus for psychophysical measurement of increment thresholds on monkeys (from Ref. (2)).

have demonstrated this by infrared television viewing. We have also shown that the animal places the test light on the fovea by making tiny laser burns on the retina of a trained monkey through the optical system during the course of threshold determinations (1).

The procedure for obtaining thresholds is described in detail by SIDLEY and SPERLING (2). Briefly, the animal is trained to both look for a test light flash in the center of the background field and to hold a lever down, when he hears a tone. If he releases the lever within 500 ms after the test light, he hears a beep each time and receives a portion of synthetic orange juice on a fraction of those times. The size of the portion and of the fraction are arranged to obtain thresholds for twenty-two wavelengths from violet to red, once in each direction in a single session, before the monkey's thrist is slaked. In a good subject, all of the animal's daily fluid needs are met in the experimental session which may last several hours. This is especially valuable in obtaining the intense light effects which will be reported.

The threshold criterion is two flashes with no response in a row in a descending intensity series. The energy value of the first is taken as the threshold. Figure 2, open circles, shows a typical increment-threshold spectral sensitivity function obtained against a 3000 Troland, 5500° K neutral background. We always plot thresholds in log reciprocal quanta as a function of equal wave number intervals. Here the thresholds are compared with the abstracted absorption spectrum of the

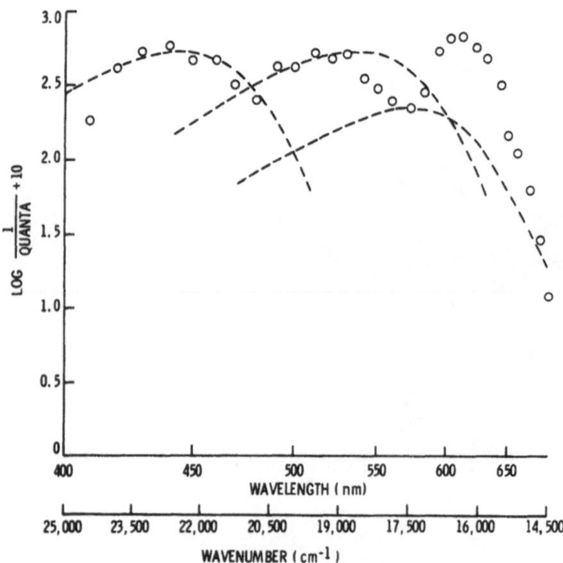

Fig. 2 Mean spectral sensitivity for one monkey compared with generalized absorption spectra of the three classes of primate cones (from Ref. (4)).

three classes of primate cones after MARKS, DOBELLE and MACNICHOL (3). A major finding in our work with monkeys and humans is that when measured against intense neutral backgrounds, we always find three peaks in the spectrum at about 440, 540 and 610 nm. The shortest wavelength peak is very well fit by the absorption spectrum of the short wavelength sensitive class of cones, corrected for pre-retinal absorptions. The middle and long wavelength peaks are too narrow to be fit by input from only one class of cones, and the 610 peak is 35-40 nm too far towards the red end of the spectrum to be fit by red-sensitive cones which peak near 570 nm. Therefore, we have proposed, and shown considerable data to support, a model of the increment-threshold sensitivity of primates (see Table 1) which assumes the three

Table I

Equations of Spectral Sensitivity (SS) Model (4), where Q is number of quanta at threshold, R, G, B the three regions of the spectrum served by the three independent channels; $\alpha 575$, $\alpha 535$, $\alpha 445$ are the absorption spectra of the cones.

$$SS_\lambda = M \quad \begin{matrix} 1/Q_R & = & K_1\ \alpha 575 - K_2\ \alpha 535 & (1) \\ 1/Q_G & = & K_3\ \alpha 535 - K_4\ \alpha 575 & (2) \\ 1/Q_B & = & K_5\ \alpha 445 & (3) \end{matrix}$$

peaks to result from three independent channels. Sensitivity is determined by the most sensitive channel, acting alone, over a spectral region. This allows us to account for the sharp dips near 480 and 580 nm and the way in which they shift with

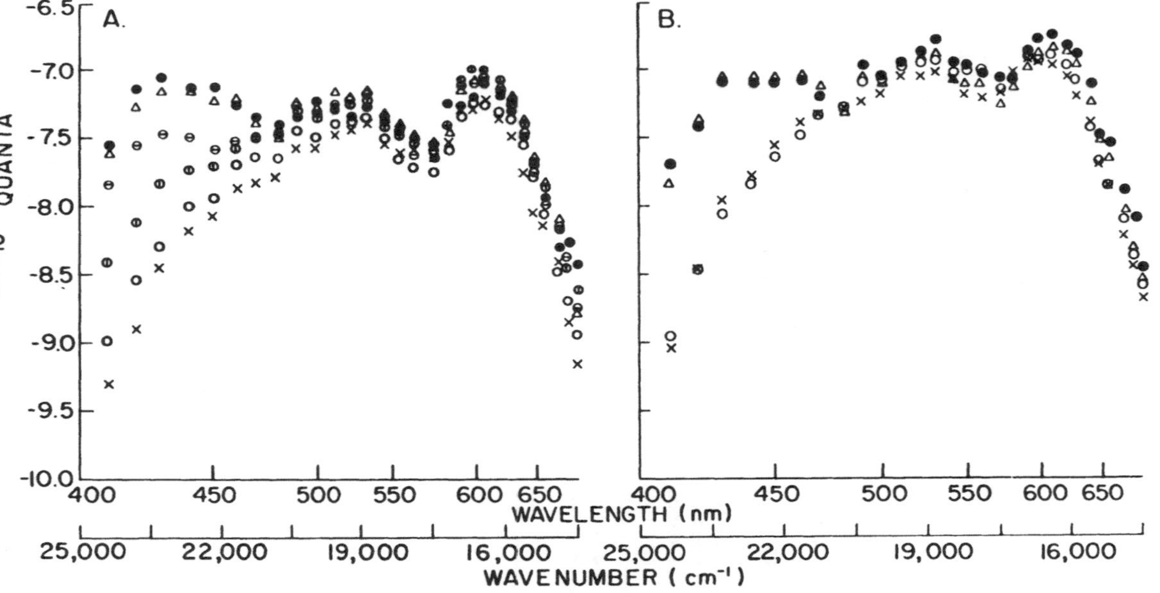

Fig. 3 Increment-threshold spectral sensitivity of two rhesus monkeys before exposure -solid circles. In Fig. 1A, 6 weeks after exposure to 4.2×10^{-4} watts/steradian at the cornea of 463 nm blue radiation for: 560 minutes - horizontally divided circles; 120 minutes - vertically divided circles; 1680 minutes - open circles. Data taken 5 months later - crosses. In Fig. 1B, the open circles were taken 2 1/2 weeks after 280 minutes exposure to 7.9×10^{-4} watts/steradian, and the crosses five months later. The open triangles are control data taken from the unexposed eye two weeks after the last exposure (from Ref. (5)).

changing adaptation. The narrowed peak at 540 nm, and the narrowed and displaced peak at 610 nm are accounted for by inhibitory interaction between the red- and green-sensitive cones in channels 1 and 2 of the model. We have found through a series of neutral and chromatic adaptation studies on monkeys and humans (4) that linear subtractive interaction between functions representing the absorptions of the red- and green-sensitive cones best models the inhibitory interactions in the red and green channels.

Following the statement and experimental testing of the model, we began studies (5, 6) of the effects of intense, prolonged spectral light exposures on these functions in rhesus. Figure 3 shows results of blue light exposures on two monkeys. These exposure intensities lie approximately midway between the upper limit of light intensities encountered in nature and the threshold for gross burn damage to the retina. They were achieved by filtering the arc lamp with a 6 nm half band-width interference filter, with peak transmission at 463 nm. The sensitivity peak at 440-445 nm, which represents the response of the blue-sensitive cones alone, was reduced in three series of seven day exposures in the monkey represented in Fig. 3A, and in one seven day series for the monkey in 3B. The exposure intensity for 3B was double that of 3A and the total exposure time was halved. The filled circles represent the pre-exposure sensitivity; the open circles post-exposure sensitivity obtained some days after each series of exposures; the crosses represent sensitivity five months later in the exposed eye; the open triangles represent the sensitivity of the unexposed eye after the entire series of exposures. It is clear from these data that repeated brief exposures for several hours each day, day after day, for a week or more reduces the sensitivity of the blue responding cones at least below that of the adjacent green responding cones, and that the green- and red-sensitive cones are unaffected by those exposures. Figure 4 shows data on the same monkeys obtained in the exposed and

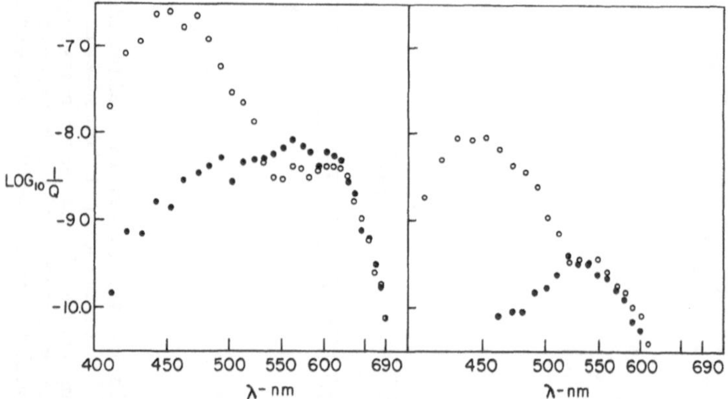

Fig. 4 Increment-threshold spectral sensitivity of blue-blinded monkeys of Fig. 3. Taken against 10,000 tds of 550 nm added to 3000 td, white, 24 months after blue exposure. Open circles are from the unexposed eye. Filled circles are from the experimental eye.

unexposed eye at the end of 24 months. These data were obtained with an intense yellow background calculated to reduce the sensitivity in the green and red regions far below that in the blue, thus revealing any recovery of the blue mechanism. It is clear that no such recovery is seen, that the blue sensitivity in the unexposed eye is quite normal, and that loss of blue response is probably total.

Figure 5 shows data obtained on one of the blue exposed monkeys at the end of 36 months. The exposed eye is represented in the right hand column; the unexposed in

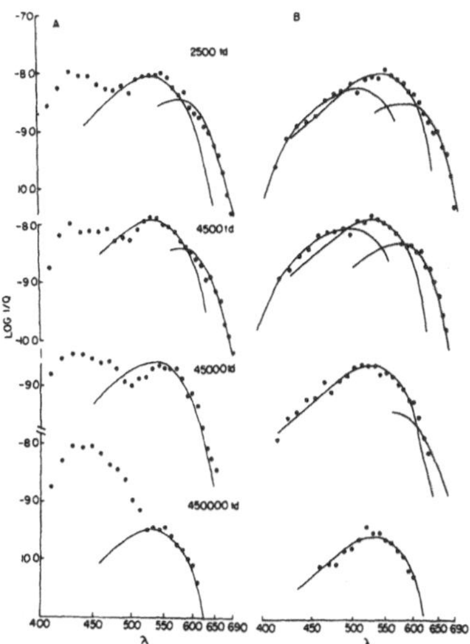

Fig. 5 Comparison of "blue-blinded" eye B with unexposed eye A of a rhesus monkey 36 months after exposure. Increment-thresholds were obtained with 2° centrally fixated spectral test lights against a 3000 td, 5500° K white background plus different amounts of 650 nm, red as labeled. The solid curves fitted to the data represent rod and cone response functions as follows: the low curve with the longest wavelength peak models the response of the red cones and is taken from monkey threshold studies against intense blue-green backgrounds; the mid-spectrum curve, modeling the green cone response, is the spectral sensitivity of protanopes by flicker photometry, taken in our laboratory. It agrees very closely with Stiles 4. The shortest wavelength curve is the absorption spectrum of rhodopsin weighted for pre-retinal absorptions (from Ref. (8)) and models rod response.

the left. These data were obtained against increasing intensities of a narrow band 650 nm red background (added to our standard 3000 td white), designed to reveal the shape of the green-sensitive cone response function. Leaving aside various results with respect to the nature of green-sensitive cone response and some evidence of blue inhibition of green-sensitive cone and of rod response, it is clear that there is no sign of blue-sensitive cone recovery at the end of three years.

Figure 6 shows data on two animals in another experiment where they were exposed to intense narrow-band green light. The full circles are pre-exposure on each of the two monkeys in Figs. 6A and 6B; the open circle data labeled 1 were taken 24 hours after the end of a series of exposures to green light. It is clear that the animals were reduced to dichromacy (deuteranopia). Their data are fit by the absorption spectrum (weighted for pre-retinal absorptions) of the blue-sensitive and red-sensitive cones, shown as the two solid curves. Then, over a period of weeks, there is gradual recovery to trichromacy. By the 18th or 30th day, (bottom curves), each monkey had

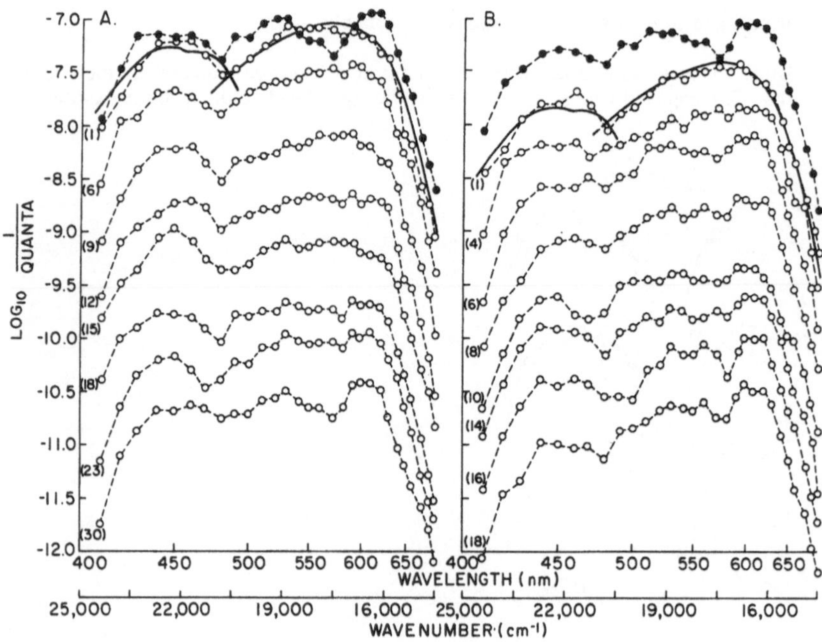

Fig. 6 Increment-threshold spectral sensitivity of two monkeys before and after exposure to green light. Solid circles are data before exposure. The open circles are data obtained at different days after exposure; the number to the left of each curve gives the number of days after exposure. The solid curves are the absorption spectra of the 575 nm and 445 nm cone photopigments. The data before exposure and 1 day after exposure are on the same scale; each successive day's measurements are displaced downward by 0.5 log unit for ease of viewing (from Ref. (5)).

recovered to very closely the same sensitivity as pre-exposure.

A further psychophysical study examined the effect of the blue exposures on the rod photoreceptors, 36 months after exposure on the same monkey as in Fig. 5. Using a newly developed behavioral technique for holding the animal's fixation away from the center of the field (7), we were able to hold the test light at 6° from the central fovea on the horizontal meridian, in the temporal field. The animal was dark-adapted for thirty minutes, a dark background was used and thresholds were obtained in both the "blue-blinded" and the unexposed eye. The results are shown for the exposed eye in Fig. 7B, the unexposed eye in 7A. The functions drawn through the data points are of rhodopsin corrected for pre-retinal absorptions (8) and part of the green-sensitive cone response function. It is clear that we did not succeed in eliminating all cone function, but that the main peak is that of rhodopsin. The sensitivity in the two eyes is nearly the same. We can conclude that if the intense light exposures had affected the rods, they have fully recovered. Since the blue sensitivity peak was completely absent in this eye, at this time, it is clear that rods do not serve the blue sensitivity

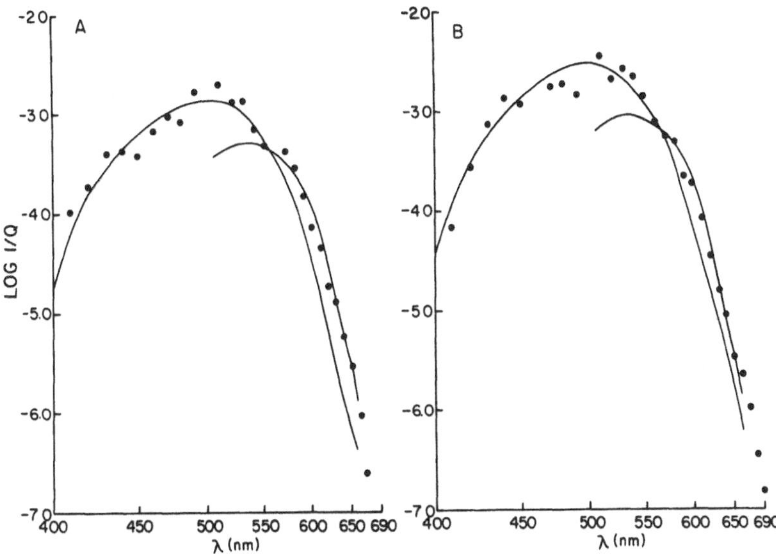

Fig. 7 Increment-thresholds at 6^{0} in peripheral retina against dark background after 30 minutes dark adaptation on monkey of Fig. 5. The solid curves model green cone response (longer wavelength peak) and rod response (shorter wavelength peak) as in Fig. 5.

function, as has been proposed (9). It is also clear that the rods are either less susceptible to these effects or recover more rapidly, despite the fact that the peak sensitivity of the rods and blue cones are close to each other in the spectrum.

In summary, these psychophysical data have shown that blue light exposure reduces or eliminates the response of the blue mechanism without affecting the green or red mechanisms. Green light exposure selectively reduces or eliminates the response of the green mechanism for days with little effect on the blue and red mechanisms. This establishes a long-term, differential effect of spectral light. From this, it is quite reasonable to hypothesize that a lesion occurs via absorption of light by the cone photopigments; the "green-blindness" is produced by the absorption of light by the green-sensitive cone photopigment present only in the green-sensitive class of cones; the "blue-blindness" by absorption in the photopigment of the blue-sensitive cones. Clearly this differential spectral effect, producing data which resemble the threshold sensitivities of congenital tritanopes and deuteranopes (10) is different from that produced by absorption of light in the non-selective pigment epithelium which accounts for thermal burns. The long-term nature of these effects suggests tissue damage to the photoreceptors themselves. To test this possibility, we produced additional blue-blinded animals by the same procedures and sacrificed them for tissue study at different itmes after exposure. We sacrificed a blue-blinded animal 15 days after exposure. In vertical one micron thick sections by light microscopy through various regions in the normal eye, the photoreceptors and pigment epithelium appear quite normal - the ellipsoids of the cones are regularly spaced and unswollen; the cone nuclei are light staining and lie just above the outer limiting membrane. Figure 8 is typical of sections of the "blue-blinded" eye. There are normal cones, swollen cones, pyknotic cones, some of which show both pyknotic nuclei and pyknotic or shrunken ellipsoids. Frequently, we see gaps in the cone

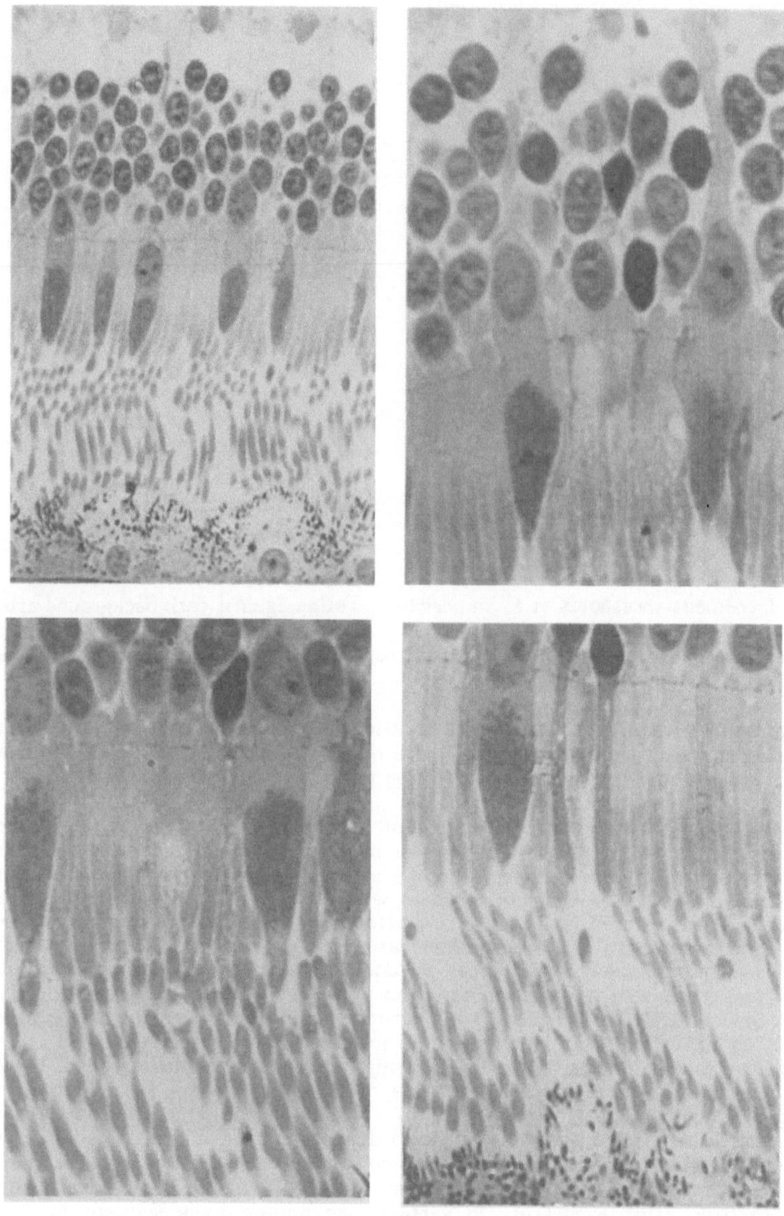

Fig. 8 Light photomicrographs of vertical 1 μm thick sections from 15 day post-exposure blue-blinded eye in parafoveal locations and shows: swollen cones, pyknotic nuclei, atrophied outer segments and gaps in the cone mosaic.

mosaic, probably indicating where cones have degenerated and been phagocytized. Table II summarizes the percentage of normal cones and cones with various degrees of damage observed at three distances from the fovea. It is clear that in a blue-blinded eye fifteen days after exposure there is a broad spectrum of damage which, in view of the normal appearance of the unexposed eye, justifies the conclusion that the effect involves selective damage to cones, but is too diverse to be useful in picking out the blue-sensitive receptors.

Table II

Percentage of 5,000 Cones Per Position

	300 Microns		500 Microns		800 Microns	
Per MM	Experimental	Control	Experimental	Control	Experimental	Control
Normal Cones	62%	92%	41%	96%	53%	88%
Swollen Cones	17%	1%	19%	0	18%	7%
Degenerate Ellipsoids	9%	0	20%	0	18%	0.5%
Pyknotic Nuclei	2.8%	0.4%	28%	0	8%	0.2%
Displaced Nuclei	0.28%	0	0.4%	0	0.1%	0.5%

We must now digress to histochemical studies which had been going on in our laboratory during the histology work, which offer a separate approach to the color receptor mosaic. After assaying several histochemical techniques, we settled upon the nitroblue tetrazolium-diformazan (NBT-DF) reaction. ENOCH (11) had shown in rat rods and with preliminary experiments on monkeys that the conversion of NBT to the insoluble blue precipitate DF was proportional to quantum capture in the receptors. We (12, 13) reported a series of studies on goldfish retinas by this technique which showed that the morphologically distinct classes of cones shown in Fig. 9 consistently contained one class of photopigment. The response spectra of Fig. 9 were obtained by exposing five different retinas to 30,000 photons/sec/μm^2 of narrow-band 400 nm, 470 nm, 520 nm, 650 nm and 750 nm lights. The procedure was to remove the eye in the dark under infrared illumination, hemisect it, dissect out the retina, remove it to a nylon support continuously perfused with oxygenated teleost ringers receptor side up, expose it to the spectral light for 5 minutes and remove it to an incubation bath of NBT-hydrochloride-succinate in teleost ringers buffered to pH 7.4. After 5 minutes of NBT incubation, the retina was removed to a bath of 10 percent formalin in buffered ringers and after 10 minutes was whole mounted receptor side up on a glass slide and sealed with a cover glass. When the reaction was complete, the relative amount of diformazan in the different morphological classes of cones was assayed with a microspectrophotometer that measured the density at 570 nm - the peak absorption of diformazan. The distributions of absorbances for the five wavelengths shown in Fig. 9 are compared with the absorption spectrum of each class

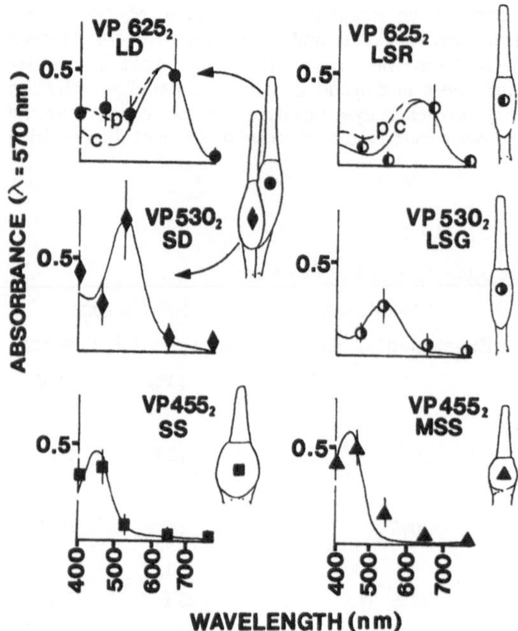

Fig. 9 Response spectra of different morphological classes of goldfish cones obtained by measuring the relative absorption at the peak of diformazan absorption, 570 nm, of the ellipsoids of a number of each class of cones following exposures to 400, 470, 520, 650 and 750 nm lights of 30,000 photons/sec/μm^2. The solid curves - the absorption spectra of goldfish pigments determined by single cone microspectrophotometry by HAROSI and MACNICHOL (14), (from MARC and SPERLING (13)).

of photopigment, adjusted by a constant, as reported by HAROSI and MACNICHOL (14). It is clear that the short single cones contain the blue-sensitive pigment VP 455_2, long single cones contain either the green-sensitive VP 530_2 or the red-sensitive VP 625_2. The double cones consistently show the red-sensitive pigment in the longer member and the green-sensitive one in the shorter member. The ratio of red to green long single cones is 2:1. As additional controls, repeated experiments with white light showed about equal staining in all cones, and retinas kept in the dark showed very little staining. Also, the goldfish retina revealed a very regular repeat pattern of the classes of cones in the mosaic.

Because the cones of the goldfish are morphologically distinct, the goldfish retina provided ideal experimental controls for developing the NBT-Diformazan procedure for cones and establishing that DF density is proportional to quantum catch in cones. We proceeded then to our main objective, to study the primate cone mosaic. At the time, we were unable to obtain rhesus monkeys, so we used baboons (15). Preliminary studies established new exposure times, dissection procedures, and incubation times, but the main features of the procedure in the baboon experiments were very much as described for the goldfish. Unfortunately, however, the baboon retina is much more fragile than the goldfish retina and it was impossible to remove whole retinas without tears in and often complete loss of some regions. The fovea was particularly difficult to remove. Despite these difficulties, we completed successful experiments with blue light exposure on three foveas, and with blue and red lights at 5°, 10°, 15°, 20° and 40° in the periphery. Figure 10 is a photograph of a whole retinal mount of a baboon after blue light exposure and

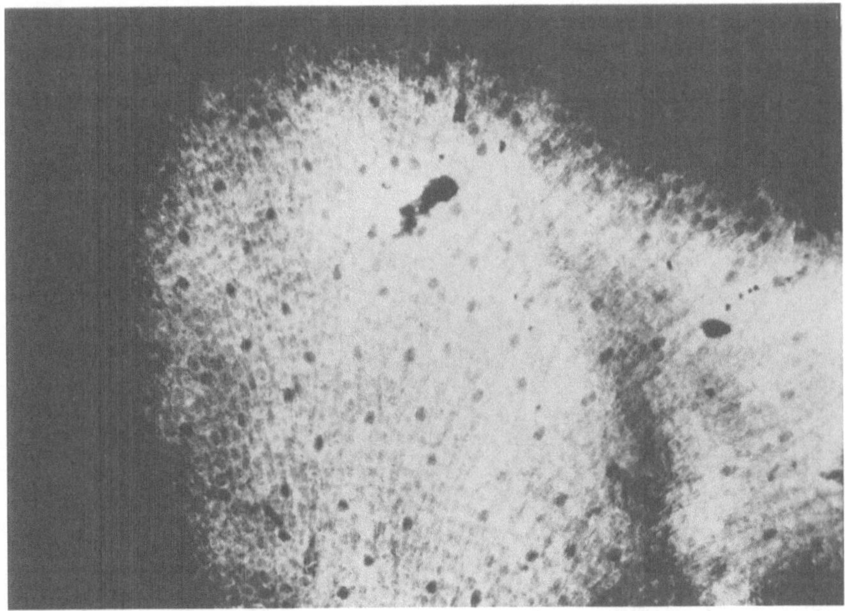

Fig. 10 Photomicrograph (400X) of a whole mount of an adult baboon retina focused for the ellipsoids of the cones in a region 2 mm from the central fovea following exposure to 440 nm blue light of approximately 30,000 photons/sec/ μm^2 and incubation in NBT (from Ref. (15)).

incubation. The dark stained photoreceptors are regularly spaced through this region of the near periphery. At the retinal regions sampled, the blue-sensitive cones consistently appeared regularly spaced. In the foveola, it appeared that they had a rectangular array. The red-sensitive cones appeared more randomly distributed. We were able to reject spatial randomness in the case of blue cones in the periphery of the retina by a statistical, X^2 test. Figure 11 is a summary of the primate evidence to date and a prediction of the distribution of red-, green-, and blue-sensitive cones based on the samplings described above. In the foveola, 2 to 3% of the cones in a regular, widely spaced rectangular array, are blue-sensitive. The frequency of blue-sensitive cones rises to a peak in the outer fovea at about 1.5^o, and then falls in proportion to the red- and green-sensitive receptors which have their highest density in the central fovea. It appears that the most frequent cones are the green-sensitive, about 55% of the total cones. The red-sensitive cones have a frequency of approximately 33% and in the parafoveal to peripheral regions the blue-sensitive cones are approximately 12% of the total cones. Further experiments are underway on baboons and on rhesus monkeys to more completely test this model.

While working on the NBT experiments, we continued exposing rhesus monkeys to intense blue lights. A "blue-blinded" animal was sacrificed at thirty days postexposure and, as shown in Fig. 12, consistently showed a pattern of degenerating

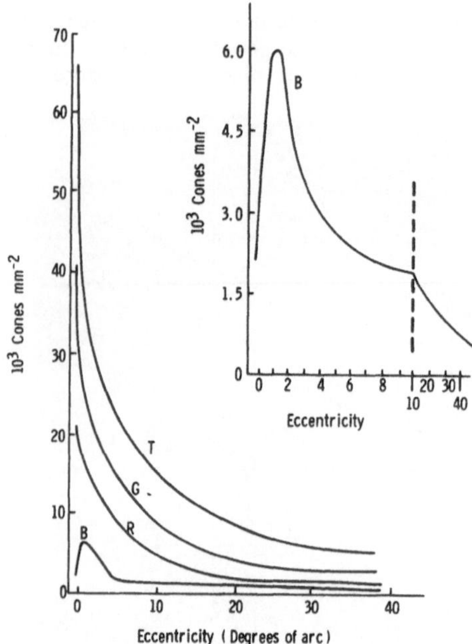

Fig. 11 Estimated density distributions for red (R), green (G), and blue (B) cones, constructed by partitioning the total cone density function for temporal retina (T) according to the relative proportions of red, green, and blue cones in the various retinal areas described. We assumed that the transition between regions containing differing proportions of a given cone type, as determined by discrete measurements, was a smooth, continuous function and that red cones represented 33% of the cones at all loci. (Inset) Expanded blue (B) cone density function. Dashed line indicates change of scale on the abscissa.

cones fairly regularly spaced among normal looking cones. Serial vertical and tangential sections were made from different parts of this retina, and the proportion of damaged cones as a function of position on the retina was estimated. Figure 13 shows the mean (and standard error of the mean) of counts obtained from sections taken from 0.5° to 6.0° along the horizontal meridian in the nasal field of the right eye. The means from the vertical sections are shown as open circles, $S.E._M$ as bars. The open triangles are from counts obtained from tangential sections. It is clear that the distribution of damaged cones peaks a short distance from the center of the fovea in the 1.0° to 1.5° region and then falls gradually towards the periphery, appearing to asymptote at 8 to 10% of the cones at 5° to 6°. In Fig. 14, the distribution obtained from the vertical sections is compared with the percentage of cones that reacted from 0° through 6° with blue light in the NBT-Diformazan experiments. Both studies show a lower frequency of affected cones towards the center of the fovea, a peak near 1.0° to 1.5° and a gradual decline of responding cones towards the periphery. The percentages at the peak are both in the neighborhood of 16% of the cones. The light-damaged cones appear to fall off somewhat more rapidly towards the periphery than the proportion of NBT stained cones. Taken together, these results support the conclusion that the blue-blinding effect with repeated intense blue light exposure is the result of damage to the blue-sensitive cones resulting from differential absorption of energy by the "blue cone" photopigment. The light damage results, in turn, add additional support to the hypothesized distribution of blue cones based on the NBT

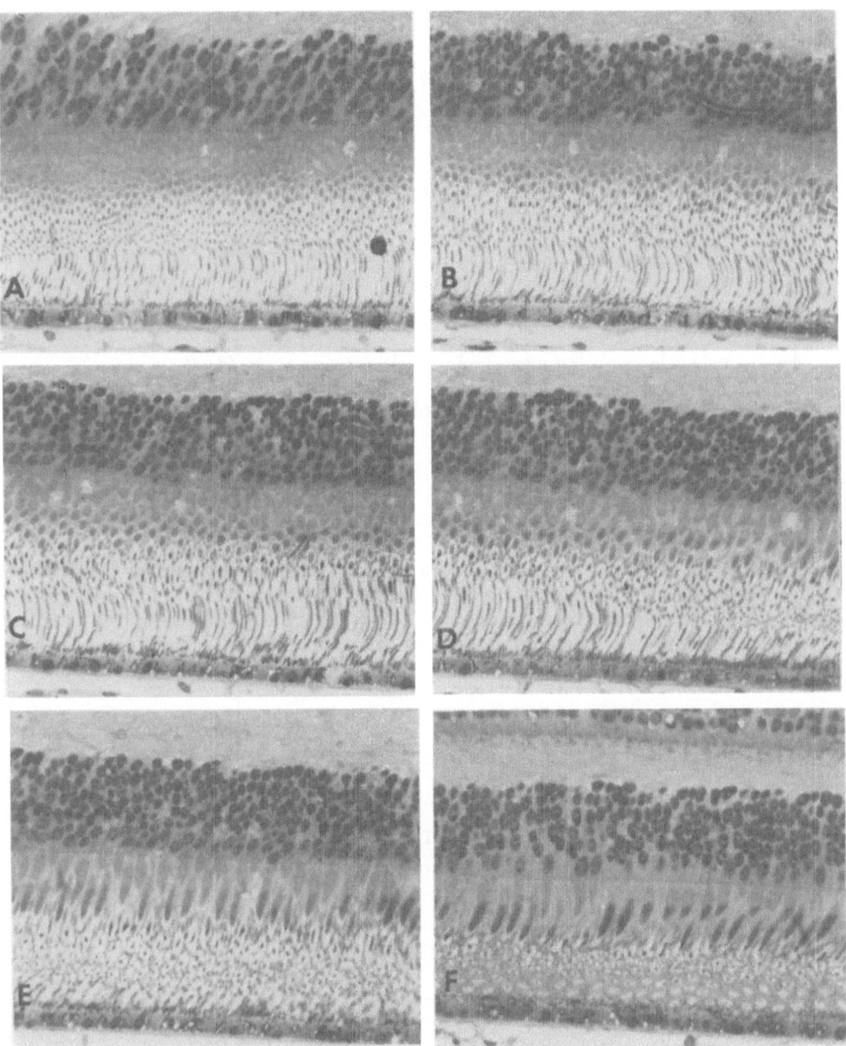

Fig. 12 Photomicrographs of blue-blind monkey retina sectioned 30 days post-exposure 1 µm thick sections at: A - .25°, B - 65°, C - 1.25°, D - 3.0°, E - 5.0° and F - 7.0° in the nasal field of the right eye. Only gross damage seen as gaps in the mosaic were counted in the data of the following figures.

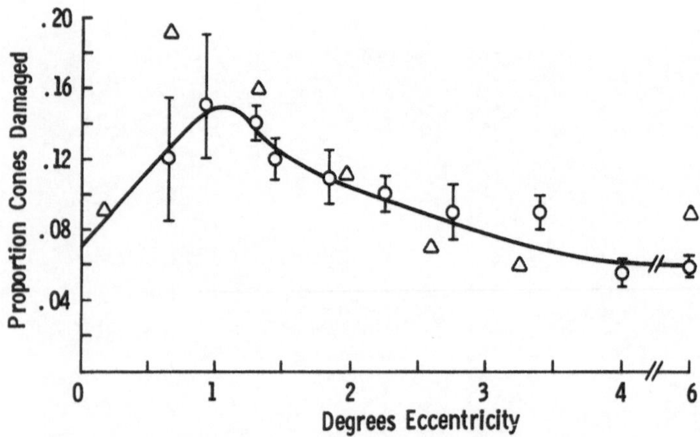

Fig. 13 Proportion of grossly damaged to normal cones in sections of blue-blinded retina at 30 days post-exposure, as a function of distance from the central fovea in degrees. Open circles are mean proportions with the S.E.M. shown as vertical bars from vertical sections. Open triangles are from tangential sections on the same retina. The curve was fitted by eye.

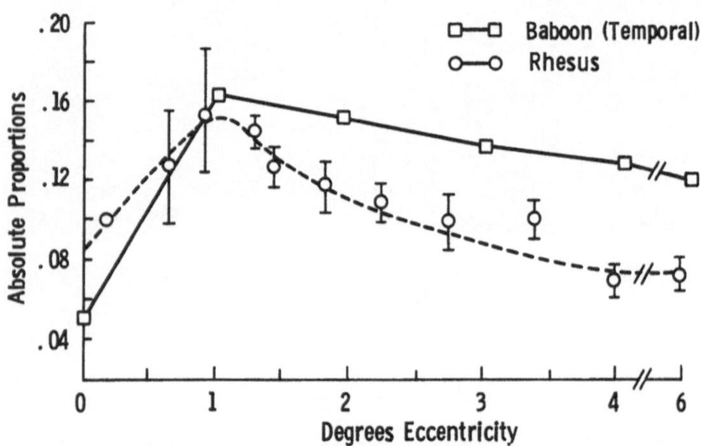

Fig. 14 Comparison of proportion of damaged cones from vertical sections in blue-blinded eye at 30 days post-exposure with proportion of cones stained by the NBT-DF reaction to blue light (see discussion of Fig. 9 above).

studies in primates.

In summary, we have induced irreversible tritanopia in infra-human primates with repeated, brief exposure to blue lights much less intense than lights which produce thermal lesions. We have induced deuteranopia which recovers in 18 to 40 days. We show pathology which clearly points to damage at the receptor level, and we have established a correlation in the case of the "blue-blindness" between the degenerated cones and cones which respond differentially to blue light by conversion of nitroblue tetrazolium to diformazan.

We conclude that the mechanism of the color blinding involves differential absorption of light in the cone photopigments. Since rod vision was found normal in a "blue-blind" eye, we conclude that rods do not serve as blue receptors. The distribution of red- and green-as well as blue-sensitive receptors in the retinal mosaic of primates has been estimated and eventually will be revealed by the NBT-DF reaction.

Acknowledgement

The author is grateful to the following persons for their various contributions to this effort: Norman A. Sidley, Ronald S. Harwerth, Robert E. Marc, Clement Johnson, Donald Prather, David Garrett and M. L. J. Crawford. This research was supported by National Science Foundation Grant NSF BNS 76-07854 and National Institutes of Health Grants EY-00381, EY-01256 and EY-02388.

References

1. M. L. J. Crawford, Vision Res. 16, 117 (1976).
2. N. A. Sidley and H. G. Sperling, J. Opt. Soc. Amer. 57, 816 (1967).
3. W. B. Marks, W. H. Dobelle and E. F. MacNichol, Jr., Science 143, 1181 (1964).
4. H. G. Sperling and R. S. Harwerth, Science 172, 180 (1971).
5. R. S. Harwerth and H. G. Sperling, Science 174, 520 (1971).
6. R. S. Harwerth and H. G. Sperling, Vision Res. 13, 1193 (1975).
7. M. L. J. Crawford, J. Exp. Anal. Behavior 25, 113 (1976).
8. G. Wyszecki and W. S. Stiles, In Color Science, Concepts and Methods, Quantitative Data and Formulas (John Wiley & Sons, Inc., N.Y., 1967).
9. E. N. Willmer, Docum. Ophthal. 3, 194 (1949).
10. G. Wald, Proc. Nat'l. Acad. Sci. U.S. 55, 1347 (1966).
11. J. M. Enoch, Invest. Ophthal. 2, 16 (1963).
12. R. Marc and H. G. Sperling, Science 191, 487 (1976).
13. R. E. Marc and H. G. Sperling, Vision Res. 16, 1211 (1976).
14. F. I. Harosi and E. F. MacNichol, Jr., J. Gen. Physiol. 63, 279 (1974).
15. R. Marc and H. G. Sperling, Science 196, 454 (1977).

Discriminations That Depend Upon Blue Cones

Robert M. Boynton
Department of Psychology
University of California, San Diego
La Jolla, California 92093

Introduction

This report is concerned with discriminations between two parts of a split field for special conditions where the perceived difference depends solely upon the differential stimulation of blue (B) cones[1]. There are three interrelated ways to regard this problem. The first makes use of the concept of chromaticity, as represented for example by the CIE chart of Fig. 1. A second perspective utilizes the hypothesized pathways of an opponent-color model. The third viewpoint pays special attention to the spectral sensitivities of human cones.

Choice of Chromaticities to be Tested

From a chromaticity standpoint, some decision must first be made regarding where, in the domain of all possible chromaticities, discriminations will be tested. Suppose that we arbitrarily pick the point "a" in Fig. 1. Imagine a stimulus having that chromaticity on the left side of a bipartite field; on the right we provide a variable stimulus, one which can be manipulated along a tritanopic confusion line that passes through "a". Provided that luminance is held constant, the level of red (R) and green (G) cone stimulation will be fixed along such a line; otherwise a tritanope would see a difference between the fields as chromaticity varies. On the assumption that B cones do not contribute to luminance, any lack of identity in the appearance of two fields whose chromaticities lie on a tritanopic confusion line must depend solely upon a

[1]These are the shortwave-sensitive cones of the eye whose peak sensitivity is at about 440 nm (See Fig. 4), which also shows the sensitivities of the long- and middlewave-sensitive cones (R and G) according to WALRAVEN (12).

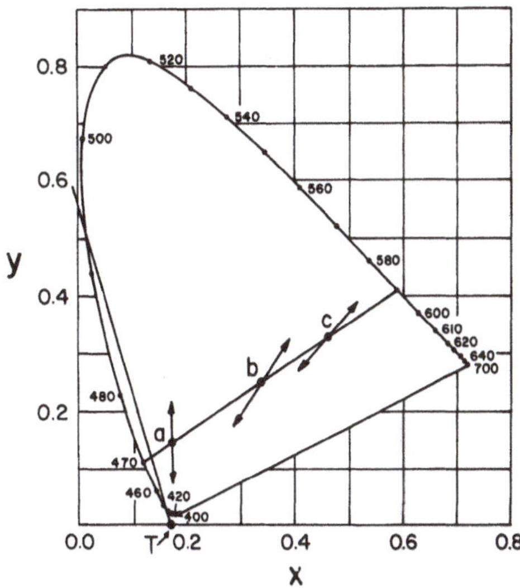

Fig. 1 When plotted on a CIE chromaticity diagram, shown here, discriminations that are mediated solely by B cones are represented along lines that pass through the tritanopic co-punctal point, T. Three examples indicating the directions of such discriminations are shown for stimuli a, b, and c, all of which,at the physical match point, fall along a line connecting spectral stimuli of 470 nm and 590 nm.

chromatic difference that is introduced by B cones.

Opponent-Color Model

The opponent-color model that I wish to use is shown in Fig. 2. Many people have proposed models more or less like this (1, 2, 3, 4). All such models agree that a

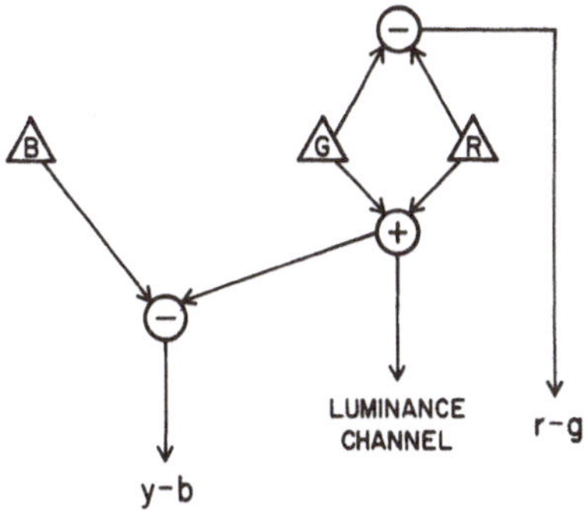

Fig. 2 This simple opponent-color model is used to interpret the experiments reported here. R and G cones (triangles) feed additively into luminance channels. Their signals are also differenced and fed into the r-g opponent-color channels. B cones feed only into the y-b opponent-color channels; these are opposed by signals equivalent to those received by the luminance channels.

difference signal, taken between the outputs of the R and G cones, is transmitted through red minus green (r-g) opponent pathways. A second point of common agreement is that the B cones alone provide the shortwave input to the yellow minus blue (y-b) opponent pathways. A third point of agreement is that the outputs from the R and G cones provide the major input to the luminance channels (sometimes called the achromatic pathways).

Agreement ends about here. Some theorists want to include a minor contribution to luminance from the B cones. Since I assume none, the model that I have in mind allows the B cones to influence perception only through the y-b opponent system. A more contentious point concerns the longwave inputs to the y-b opponent system. Does it come exclusively from the R cones (5), or possibly from the G cones? Or, if it comes from both (1), what are the proportions? I will provide preliminary evidence to show that it comes from both, probably in the same proportions that the R and G cones contribute to luminance. This is what is implied in Fig. 2 by the arrow drawn from the "plus box" in the luminance channel to the "minus" one in the y-b opponent channel.

B Cone Discrimination in the Chromaticity Domain

As one moves in chromaticity space from the shortwave spectral locus toward the longwave one on the other side, the tritanopic confusion lines along which discriminations can be tested slope progressively more, as shown by the lines drawn through points "a", "b", and "c" in Fig. 1. There is nothing fundamentally important about the angles that these lines make with one another, because transformations are possible which would cause all possible tritanopic confusion lines to become parallel. RODIECK, in his textbook (6) has come close to creating such a plot: Fig. 3 shows the well known and justly famous ellipses of MACADAM (7) plotted according to

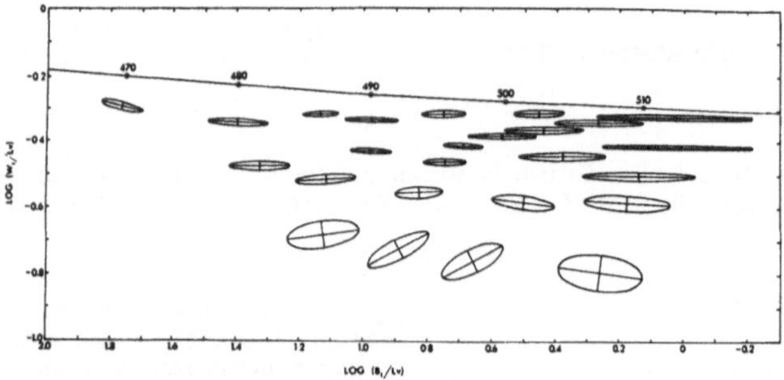

Fig. 3 MacAdam's (1942) ellipses, as re-plotted by Rodieck (6, p. 736) in a special way that confines B-cone variations to directions parallel to the x-axis.

RODIECK'S transformation. This is a very unusual representation where the ordinate represents the fraction of G-cone activation at constant luminance (the remainder being that of R-cones) on a logarithmic scale. Pure G is at the top, and 90% R is at the bottom. The abscissa shows the logarithm of the amount of B-cone activation, also at constant luminance, decreasing rightward.

As you go from left to right on RODIECK'S diagram, the major axes of the skinny ellipses become progressively longer. The figure thereby shows that the relative amount of change of B cone stimulation required for a threshold discrimination increases as the amount of such stimulation decreases. In other words,

Weber's law fails for B cones at low levels of stimulation, much as it does for luminance discrimination (8, p. 264ff) and for STILES' pi-mechanisms (9, p. 578).

Consider now what happens as the chromaticity at which B-cone discrimination is tested shifts away from the shortwave spectral locus. Since "a", "b", and "c" happen to fall along a straight line connecting 470 and 590 nm in Fig. 1, all three chromaticities could be produced by a mixture of these two spectral lights. As one moves from "a" to "b" to "c", the 590 nm component must be increased and the 470 nm component decreased in order to keep luminance constant. In the limit, when the longwave spectral locus is finally reached at 590 nm, there is no longer any B cone stimulation at all. Therefore, one of the factors in B cone discrimination at constant luminance must be the inevitable reduction in B cone activity caused by keeping luminance constant as the threshold of discrimination is tested in various regions of chromaticity space. This is one of the factors that I am studying in my experiments, and it is the one that RODIECK has analyzed using MACADAM'S discrimination data.

Varying Chromaticity (and Luminance) by Adding Longwave Light

It is not, however, a God-given requirement that luminance must be kept constant in such an experiment.[2] One can also shift chromaticity along exactly the same line by adding 590 nm light without altering the intensity of the 470 nm component. Obviously the luminance of such a field will increase as the 590 nm light is added, and the limit of pure 590 nm can never quite be reached. Nevertheless, it should be possible to add enough light at 590 nm to come inside the first discriminable step from 590 nm. In other words, we can anticipate that, if we add enough longwave light, the original discrimination that depends only upon B cones should be adversely affected and, in the limit, totally impoverished. This is the second part of my experiment, and we shall see that the expectation is fulfilled. The question then is why?

How Does Longwave Light Affect B Cone Discriminations?

The hypothesis that the discrimination is affected directly because of 590 nm light being absorbed by B cones can be dismissed, because a large body of evidence suggests that the B cones do not absorb any appreciable amount of such light.[3] A second hypothesis is that the heavy absorption of yellow light by the longwave-sensitive R and G cones unbalances the y-b opponent system in the y direction. To visualize this, return to Fig. 2 and suppose that the longwave input to the y-b channel is very strong because of a large signal from the "plus box" of the luminance channel which is delivered to the "minus box" of the y-b channel. Perhaps this intense signal partially saturates the y-b system, driving its response into a less sensitive portion of its nonlinear input-output curve (10). If so, the use of a variety of longwave lights, none of which is appreciably absorbed by B cones, could reveal the action spectrum to

[2]The decision to keep luminance constant in a chromatic discrimination experiment is not an arbitrary one; if luminance is allowed to vary, the discrimination will no longer be a test of purely chromatic discrimination.

[3]Figure 4 of this paper shows B-cone sensitivity (relative to that of R and G cones), about 2 log units higher than should be if relative probabilities of photon absorption are to be correctly displayed. One basis for this assertion is that spectral sensitivity curves based on absolute threshold show a shortwave "hump" which is very modest compared to what the curves of Fig. 4 would imply. By imagining the B curve of Fig. 4 displaced downward by 2 log units, a proper conception of the negligible sensitivity of B cones to longwave lights can be appreciated.

the longwave input of the desensitized y-b channel. My test with three wavelengths yields results in accord with the model of Fig. 2.

Preliminary Nature of This Research

Before turning to experimental conditions, procedures and results, I wish to emphasize that these are very new experiments. I am reporting here on the very first results, based on only one subject (myself), of what will be a long research program.[4] As you have heard, chromatic discrimination is a very complicated subject. My intention is to look at it one part at a time, hoping eventually to be able to understand the causes of individual differences in chromatic discrimination that are well known to exist, but which are hard to understand from the existing literature.

Choice of Stimulus Conditions: Importance of Cone Sensitivity Curves

I mentioned earlier that there are three ways to view the problem of B-cone discrimination. The chromaticity diagram has afforded one perspective, and an opponent-color model provided another. In order to understand the choice of stimulus conditions for my experiment, we must view the problem from the standpoint of cone sensitivities. The search for the action spectra of human cones has been a long one. To know exactly what they are is especially important for understanding the problems of chromatic discrimination. The reason for this is not hard to understand. As spectral distribution is changed at constant luminance, two things happen simultaneously at the cone level. Although the total stimulation of R+G is held constant (since this is what it means to work at constant luminance), the proportions vary. Usually the activation of B cones also changes. A threshold of wavelength discrimination therefore depends, in the first instance, on a shifting amount of R (or G) cone activity combined with a concurrent change in B-cone output. But this is only the first stage of the process, because the brain probably does not see cone outputs *per se* (11). What the oponent color stage sees, however, can be no more or less than the shifting outputs of the R, G, and B cones.

Choice of Stimulus Conditions: Use of Monochromatic Lights

There exist two sets of R, G, and B functions which especially appeal to me. These are based upon a variety of kinds of evidence that I cannot review here. One of these, first published by VOS and WALRAVEN (12) in 1971, was slightly modified by WALRAVEN (13) in 1974. The other set has been provided by SMITH and POKORNY (14). From the standpoint of this paper, the differences between the two sets are minor. Because WALRAVEN published his curves in tabular form, I have arbitrarily decided to use his functions here; they are shown in Fig. 4. Ideally, the heights of these three curves would reflect the relative probabilities that a photon of the wavelength indicated would be absorbed by each type of cone, but these probabilities are not known. I have positioned the R and G curves so that they cross at 570 nm, which is the least saturated region of the spectrum and very probably close to the balance point for the r-g opponent channels. The B curve is drawn so that its ordinate value at 500 nm is equal to the sum of the R + G curves; this is approximately the wavelength at which the y-b opponent channels are balanced. In other words, the three functions have been positioned so as to describe their relative inputs to the

[4] As of October, 1977, experiments like those reported here had been completed on five observers. The additional data support the conclusions reached here. The newer data show curves, related to those of Fig. 6 and 7 of this paper, which are more nearly linear; the difference is probably related to a change in the method of testing discrimination.

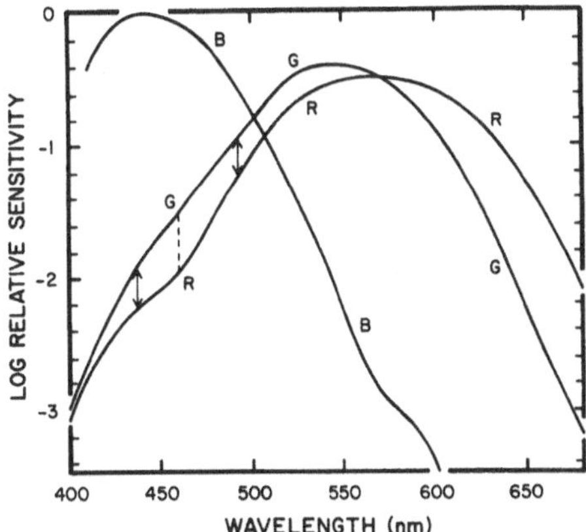

Fig. 4 Walraven's (12) R, G, and B functions which purport to represent the action spectra of the three kinds of cones of the human eye. These have been adjusted vertically without change of shape so that (a) the R and G functions cross at570 nm and (b) R+G = B (in linear sensitivity units) at 500 nm.

opponent channels of the model of Fig. 2. In all probability, however, the B curve is at least 2 log units too high to represent photon absorption probabilities correctly. (This fact implies a large amplification of the B-cone signal.)

In order to test B-cone discrimination alone, one can use two lights that produce equivalent effects, at the same luminance, on both R and G cones. If this condition were met, then the R and G cones would not see an exchange between the two lights. Because any two stimuli that fall along a tritanopic confusion line meet these criteria, there are many such pairs of lights available. Since two such lights stimulate R and G cones each in exactly the same way, the tritanope (who lacks B cones) has no basis for perceiving them as different colors.

Double Silent Substitution

It is advantageous to test discrimination with monochromatic lights because, unlike nonspectral stimuli, these are qualitatively unchanged by pre-retinal absorption. Their use also makes analysis easier. In Fig. 4 it will be noted that the separation between the G and R functions, in the short wavelengths, is maximal at about 465 nm (where the dotted line is drawn). Whether wavelength is lengthened or shortened, the decrease in G to R ratio is monotonic. This means that for each wavelength shorter than 465 nm, there is one (and only one) longer wavelength for which the separation between the R and G curves is the same. For my experiment, I have chosen 439 and 492 nm.[5] The curve separations for these wavelengths are indicated by the arrows in Fig. 4.

Although these two wavelengths are equivalent in their effects upon R and G cones at constant luminance, they are dramatically different for B cones. It can be seen from Fig. 4 that 439 nm is near the peak of the B-cone sensitivity function, whereas sensitivity at 492 nm is about 0.6 log unit less (a factor of four). Moreover, in order to keep luminance constant, about 1.0 log unit (ten times) more of the

[5]These are the same two wavelengths which were used, at the suggestion of Dr. D. I. A. MACLEOD, in our previous research on the "gap effect" (15).

439 nm component is needed, compared to 492 nm, as can be judged from the relative heights of the two arrows in Fig. 4. By exchanging the two components at constant luminance, the stimulation of B cones can therefore be varied by a factor of about 40 without altering what either the R or G cones can see. This is a "double silent substitution" technique.

Viewed in terms of the chromaticity diagram, these same concepts imply that there are lines which can be drawn from the tritanopic co-punctal point that intersect the spectral locus twice in the shorter wavelengths. The line that gives rise to the pair that has been chosen here is shown on Fig. 1.

Conditions for Experiments 1 and 2

The stimulus conditions are schematized in Fig. 5. A five-channel Maxwellian view system is used. The test field consists of reference and variable components. The fixed component is on the left; it consists of an additive mixture of the 492 and 439 nm lights in the luminance ratio of 4:1. The variable component, seen at the

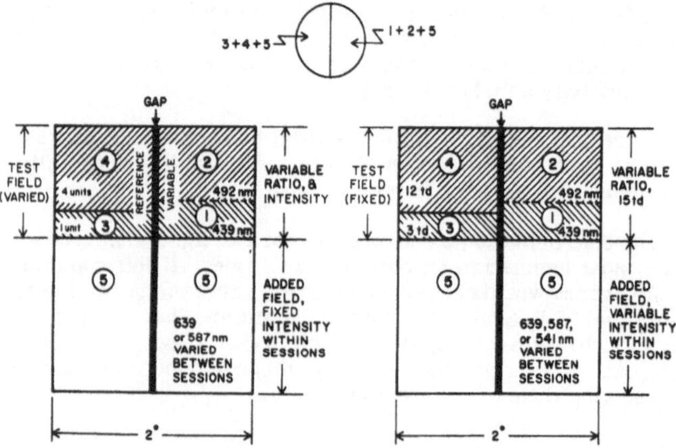

Fig. 5 Top: schematic of the field as viewed by the observer. Channels 1 and 3 of the optical system supply 439 nm light to the variable right-hand field and the fixed left-hand field, respectively. Channels 2 and 4 supply 492 nm light analogously. Channel 5 supplies longwave light to both fields. Bottom: A luminance "cross section" of the fields for Experiment 1 (left) and Experiment 2 (right). Encircled numbers indicate which channel of the optical system is used to supply the indicated component.

right, consists of the same two components, always presented at the same total luminance as the fixed component, but with their ratio continuously variable by the subject. It is therefore possible to make a physical match between the reference and variable components of the test field. The reference and variable components of the test field are deliberately separated by a 6' dividing line, or gap, which is needed in order for B-cone discriminations to be optimal (15).

An added field which is superposed upon the test field consists of two parts, also separated by a 6-min gap and presented in register with the test field. For Experiment 1, the conditions are shown at the left in Fig. 5. The purpose of the experiment is to measure a threshold vs. intensity (tvi) function for B cones alone, based upon unrestricted viewing of a side-by-side field configuration. The test field was superposed upon a weak (c.25 td) added field either 587 or 639 nm. Its illuminance was varied from a maximum of 50 td to a minimum of 0.4 td in steps of 0.3 log unit. In Experiment 2, the test field was fixed at the values shown at the right in Fig. 5; in this case, the added field was varied from 3.2 to 3200 td in 0.6 log unit steps. The purpose of this experiment was to determine how these added fields affect B-cone discriminations even though they are not appreciably absorbed by B cones.

In both experiments, discrimination was specified by the variance of many repeated attempts, under each condition of the experiment, to cause a match between the variable and reference components of the test field. This procedure is identical to the one used by MACADAM in the generation of his original set of 25 ellipses.

Results of Experiment 1

The results of the first experiment are shown in Fig. 6. The data points refer to the axis at the left, where the standard deviation of the attempted matches is shown.[6]

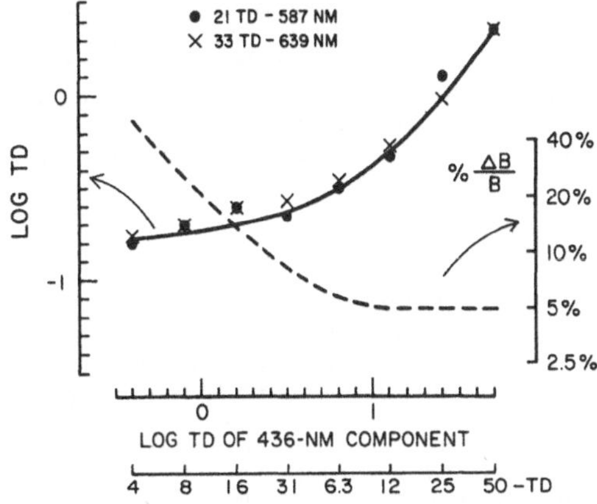

Fig. 6 Results of Experiment 1, where the illuminance of the test field (x-axis) was varied. Left ordinate shows the logarithm of the standard deviation of repeated attempts at matching, expressed in trolands of the 439 nm component of the variable half of the test field. Data points, obtained with added fields as shown, refer to the left ordinate. The results are also shown as B-cone contrast thresholds by the dotted curve, which should be referred to the ordinate at the right.

The abscissa shows the illuminance of the test field at the match point, on a log scale. The result is a rather traditional looking tvi curve. There seems to be no consistent difference between the results for the 587 and 639 nm added fields. The dotted curve in Fig. 6 represents a transformation of the solid curve, expressed in terms of a contrast threshold for B cones (the value of the upper curve divided by its abscissa value). The contrast thresholds for the high-intensity test fields are about

[6]Because almost all of the variation in B-cone stimulation is due to changes in the 439 nm component of the variable field, results are expressed in troland values of that component alone.

5%; this is in good agreement with a similar estimate derived by RODIECK from MACADAM'S data.

At the lowest intensity tested, 0.4 td, the contrast threshold is very high (about 40%). Under this condition, the added field has become very bright compared to the test-field; the latter served merely to slightly alter the basic redness or yellowness of the added field. Much to my surprise I found that, when I occluded the added field under this condition (which I could do without leaving the biting board), the test field was invisible. This was not an instance of transient tritanopia, because the effect was not a temporary one. In other words, a blue light otherwise too weak to be seen has the capacity to alter the appearance of a red or yellow field.

Afficionados of tvi curves will quickly recognize the solid curve of Fig. 6 as beng a respectable-looking specimen, even though it has been determined with the use of side-by-side procedures, and with steady fields. (The usual procedure, of course, is to adapt the eye for a long time with a large conditioning field, and then determine an increment threshold with the use of a small, brief flash.) The most important thing about this result is the demonstration that contrast discrimination which is mediated by B cones (the dotted curve in Fig. 6) grows progressively worse as the luminance of the test field is reduced. Therefore one of the factors that causes B-cone discrimination to become poorer, as one moves progressively closer in chromaticity space to the longwave spectral locus, is the reduction in B cone stimulation that must result if the overall luminance of the field is held constant. As noted earlier, RODIECK came to this same conclusion; in fact he derived a tvi curve from the MACADAM data of Fig. 2. My points fit along a smooth curve better than his, which is perhaps to be expected since my experiment, unlike MACADAM'S, was designed as a direct measure of discriminations along a tritan confusion axis.

Results of Experiment 2

The purpose of this experiment was to determine the effect of added fields *per se*. I chose a test field luminance of 15 td, of which 3 td was supplied by the 439 nm component. This is a rather dim light; it yields a contrast threshold (see Fig. 6) of about 10%. Fig. 7 shows the B-cone contrast threshold, measured as before, plotted this time as a function of the intensity of the added field. The graph indicates that the effect of adding longwave light is definite but minor until the amount of light added is about 30 times or so greater than that of the test intensity (the arrow shows 15 td the luminance of both spectral components of the test field). Above this level the discrimination becomes progressively worse at a faster rate. The graph shows that, over the substantial range tested, the contrast-threshold of the B cones is driven upward by a factor of 10 or more by adding the longwave lights.

Note that there are symbols of three kinds on the graph. Each represents a different wavelength of the added field; their illuminances are all scaled in trolands. When scaled this way, all points seem to fit along the same curve. Had they been scaled instead with units based on the sensitivities of only R or G cones, such fits would be poorer. If this result holds up with further testing,[7] it provides support for the hypothesis that the longwave input to the y-b opponent pathway, as shown in Fig. 2, is in fact the same as the luminance signal derived from the summed activity of R and G cones.

[7] The newer data mentioned in Footnote 4 have been tested using objective curve-fitting procedures. It is not possible to reject the hypothesis that the relative effects of added fields of various long wavelengths are correctly predicted when their intensities are scaled in troland units.

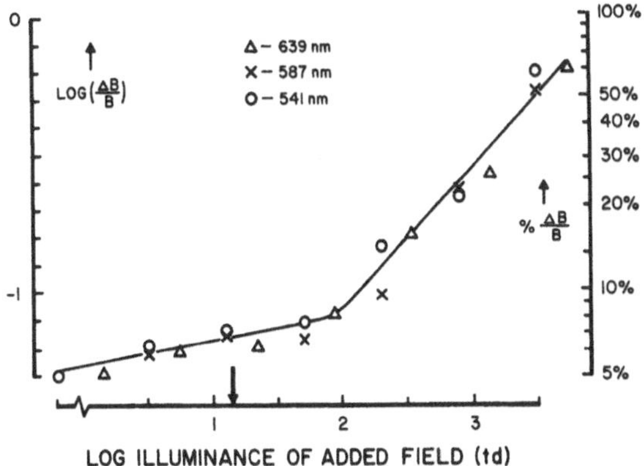

Fig. 7 Results of Experiment 2 where the illuminance of added fields was varied. Left ordinate shows the logarithm of the B-cone contrast threshold for a fixed illuminance of the test field; the scale at the right shows percentage equivalents. Data points are for three wavelengths of added fields, whose illuminances are scaled in trolands.

Summary and Conclusions

The results of the two experiments taken together indicate that the major factor reducing the effectiveness of B cones, as chromaticity is shifted at constant luminance from the blue corner toward the spectral locus of the chromaticity diagram, is the reduced stimulation of the B cones. In addition, however, added longwave light has a deleterious effect upon such discriminations even when the level of B-cone stimulation is held constant. For the particular conditions used in this experiment, the reduction of B-cone stimulation is a much more important factor than is the addition of longwave light. The effect of longwave light *per se* is most likely mediated via the y-b opponent channels. It appears probable that the longwave input to these channels takes the form of luminance signals derived from the summed outputs of R and G cones.

Acknowledgement

This research was supported by Grant EY 01541-03 from the National Eye Institute.

References

1. J. J. Vos and P. L. Walraven, Vision Res. 12, 1327 (1972).
2. S. L. Guth, J. J. Donley and R. T. Marrocco, Vision Res. 9, 537 (1969).
3. R. L. DeValois and K. K. DeValois, in Handbook of Perception (Vol. 5), ed. by E. C. Carterette and M. P. Friedman, (Academic Press, New York, 1975).
4. C. I. Ingling, Vision Res. in press.
5. I. Abramov, J. Opt. Soc. Am. 58, 574 (1968).
6. R. W. Rodieck, in The Vertebrate Retina, (W. H. Freeman, San Francisco, 1973).
7. D. L. MacAdam, J. Opt. Soc. Amer. 32, 247 (1942).
8. Y. LeGrand, in Light, Colour and Vision, Translated by R. W. G. Hunt, J. W. T. Walsh and F. R. W. Hunt, (Chapmand and Hall, London, 1968).
9. G. Wyszecki and W. S. Stiles, in Color Science, (Wiley, New York, 1967).
10. J. Larimer, D. H. Krantz and C. M. Cicerone, Vision Res. 15, 723 (1975).
11. D. H. Kelly and D. van Norren, J. Opt. Soc. Amer. 67, 108 (1977).
12. J. J. Vos and P. L. Walraven, Vision Res. 11, 799 (1971).

13. P. L. Walraven, Vision Res. 14, 1339 (1974).
14. V. C. Smith and J. Pokorny, Vision Res. 15, 161 (1975).
15. R. M. Boynton, M. M. Hayhoe and D. I. A. MacLeod, Optica Acta 24, 159 (1977).

Opponent Chromatic Mechanisms Predict Hue Naming

B. R. Wooten and John S. Werner
Walter S. Hunter Laboratory of Psychology
Brown University

I. Introduction

It now seems certain, as suggested by THOMAS YOUNG (1) some 175 years ago, that human color vision is mediated by three types of receptors, maximally sensitive at different regions of the visible spectrum, but broadly overlapping in responsivity. Although the underlying photopigments have not been isolated, data from microspectrophotometry (2, 3, 4), reflection densitometry (5, 6) and psychophysics (7, 8, 9, 10), converge on estimating the peak sensitivites at about 435, 530, and 562 nanometers (nm). These data also suggest that the shapes of the spectral absorption curves are similar when plotted on a frequency scale. The existence of three receptors of this kind is, of course, consistent with the trivariance of color mixture. The relation between perceived hue and receptor activity, however, appears to be less simple.

As suggested by HERING (11), in the latter part of the 19th century, the perceptual organization of hue seems to demand an opponent schema for the underlying neural processes. His general argument will be familiar to most, but it is worth briefly repeating. HERING noted that yellow and blue are antagonistic in the sense that they cannot exist at the same time and on the same retinal locus, i.e., he could not conceive of a yellowish-blue. When yellow and blue are added together, an achromatic sensation results. He argued along identical lines for the opponency of red and green. HERING further proposed that these simple perceptual relations are isomorphic with neural processes, i.e., that they reflect a basic antagonistic neural organization. Contrary to what is sometimes written, he was quite specific in ascribing this opponency to a neural level, rather than to the receptors. In a distinguished series of papers in the 1950s, HURVICH and JAMESON (12, 13, 14, 15) quantified HERING's model and related it to various color phenomena such as saturation, wavelength discrimination, hue contrast, the Bezold-Brücke hue shift, color naming, and color deficiency. Direct physiological evidence of color opponency was first provided by SVAETICHIN and MACNICHOL (16) working with fish retinas. Somewhat later, DE VALOIS (17) and WIESEL and HUBEL (18) found chromatically opponent cells in the lateral geniculate nucleus of the rhesus monkey, an animal that has color vision closely resembling man's. More recently, GOURAS (19) has shown a similar organization for retinal ganglion cells. Although cortical cells are just beginning to be systematically explored with respect to color (20, 21), the initial evidence also indicates an antagonistic color organization.

In summary, substantial psychophysical and physiological evidence supports the essence of HERING's opponent model of color vision. Yet, at the same time, it is certainly true that many details of the theory remain to be worked out. Perhaps the most fundamental issue, certainly the most challenging, is the question of how activity in the brain gives rise to hue percepts. This sort of question, which is just a specific example of the age-old mind-body problem, seems as elusive as ever. There are, however, smaller questions that are both interesting and currently solvable. One example is whether the opponent channels directly determine hue percepts as reflected by hue naming. Another is how the three receptor types relate to the opponent channels. In this paper, we attempt to shed some light on both of these questions.

II. Opponent Channels and Hue Naming

In HERING's original formulation of the opponent model, he proposed that the yellow-blue (y-b) and red-green (r-g) channels determine hue in a simple, direct manner. At long waves, for example, we experience red tinged with a little yellow because the r-g system is highly sensitive in the red direction, while y-b is slightly sensitive in the yellow direction. Perfect, or pure, yellow occurs at that point in the spectrum where r-g is in equilibrium, while y-b is highly responsive in the yellow direction. To take one more example, a light that looks equally blue and green is found where the y-b and r-g processes are equally activated in the blue and green directions, respectively.

HURVICH and JAMESON (12) quantified HERING's formulation by explicit algebraic expressions that relate perceived hue to opponent channel activation. Their model is as follows:

$$H_{(r,g)\lambda} = \frac{(|r\text{-}g|)_\lambda}{(|r\text{-}g| + |y\text{-}b|)_\lambda} \qquad (1a)$$

$$H_{(y,b)\lambda} = \frac{(|y\text{-}b|)_\lambda}{(|r\text{-}g| + |y\text{-}b|)_\lambda} \qquad (1b)$$

$H_{(r,g)\lambda}$ represents the perceived hue, either red or green, resulting from a given narrow-band light. Similarly, $H_{(y,b)\lambda}$ refers to the perceived yellowness or blueness associated with a spectral light. The terms $|r\text{-}g|_\lambda$ and $|y\text{-}b|_\lambda$ correspond to the absolute amount of opponent channel activity generated by a specific colored light. Thus, the hue of a light is determined by the ratio of the activity in each channel to the total amount of chromatic activity. The percentage of blue, for example, associated with a particular wavelength, can be calculated by taking the ratio of blue activation at that wavelength to the sum of all chromatic activity at that wavelength. In this model, the sum of $H_{(r,g)\lambda}$ and $H_{(y,b)\lambda}$ is always equal to 1.0. Furthermore, since each opponent channel can be excited in only one direction at a time, for a given wavelength, two of the terms in the denominator and one of the terms in the numerator will always be zero.

We set out to directly test HURVICH and JAMESON's expression by determining opponent channel activity and perceived hue in the same observers.

A. Methods and Procedure

Opponent channel activity was ascertained according to the method used by JAMESON and HURVICH (14). Their procedure capitalized on the antagonistic mode of response of the chromatic channels. They assumed that the strength of an opponent channel is reflected by the relative radiance of an added light that brings that channel to equilibrium. For example, the degree of red activity in the r-g channel is proportional to the amount of green that must be added so that the mixture appears neither red nor green. Similarly, the strength of yellow in a yellowish light is assessed by determining the amount of blue light that must be added to just cancel yellowness. The degree of green and blue activation is determined by the amounts of red and yellow, respectively, required to cancel the opponent hue. The blue, green, and yellow cancelling lights were chosen to correspond to each observer's unique (pure) hues. Since unique red is extra-spectral it was not convenient to use it as a cancelling light. Thus, JAMESON and HURVICH used a predominantly red long-wave light for green cancellation. As they point out, it is convenient, but not necessary, to use unique hues as the cancelling stimuli.

In our experiment we presented spectral lights from 400 to 700 nm in 10 nm steps at a luminance level of 2.5 log Trolands. The stimuli were 1 sec in duration with 10 sec between flashes. Following Jameson and Hurvich's procedure, we used each observer's unique (pure) blue, green, and yellow as the cancelling stimuli. The cancelling red was 670 nm. The observers were instructed to turn a neutral wedge in the path of the cancelling stimulus, increasing or decreasing the energy, until they just cancelled out the opponent hue. When cancelling red with green, for example, they were told that the final setting should appear neither reddish nor greenish. The subjects were allowed as many stimulus adjustments as they desired.

A multi-channel Maxwellian-view optical system was used. In addition, an achromatizing lens was used to correct for the chromatic aberration of the eye. The neutral wedges and filters were calibrated at each wavelength. Three color-normal females served as observers.

B. Results and Discussion

In the opponent cancellation experiment, the monochromatic stimuli were all presented, as mentioned above, at the same luminance level of 2.5 log Trolands. Thus, the radiance of the cancelling lights reflect the strength of the opponent channels for stimuli of equal luminance. Expressed in this form, the results are adequate for computing predicted hue according to Eqs. 1a and 1b. It is less convenient for relating the opponent channels to the photopigments, as we do below. Therefore, we followed JAMESON and HURVICH's (14) procedure of adjusting the cancellation values for each wavelength by multiplying them by the C.I.E. photopic luminosity coefficients. With this conversion, the data may be presented for what is effectively an equal-energy spectrum. Note that this procedure does not change the predicted hue from Eqs. 1a and 1b since the luminosity coefficients factor out for any given wavelength. Following JAMESON and HURVICH, we refer to the resulting curves as "chromatic valence functions".

The log chromatic valence functions for our three observers are presented in Figs. 1, 2, and 3. The curves drawn through the data points were fit by eye. The relative heights of the r, g, b, and y curves are not arbitrary. To determine, for example, the height of the blue valence curve, relative to the yellow curve, a method of constant stimuli was used. Unique yellow at 0.25 log Trolands was superimposed on unique blue at 2.25 log Trolands. The luminance of yellow was progressively increased as the luminance of blue was decreased, in 0.10 log unit steps, keeping the total retinal illuminance at a constant 2.5 log Trolands. Six 1-sec trials were presented at each luminance level in a random sequency. The energy of the two optical channels was determined for the point at which the observer's sensitivity to blueness and yellowness was equated. Appropriate adjustments to the height of either the blue or yellow curve were then made. The same procedures were applied to

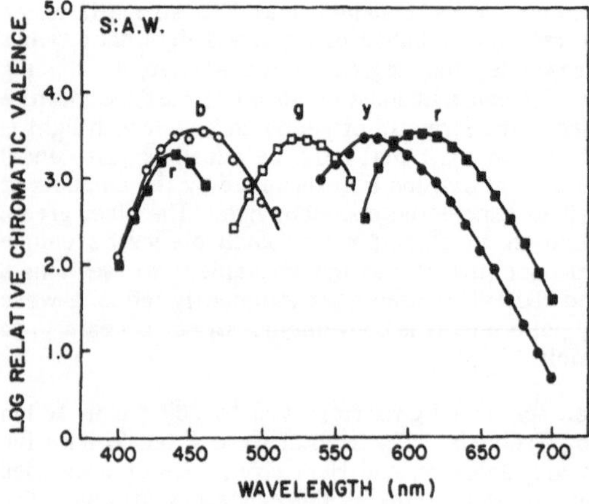

Fig. 1 Log relative chromatic valence of the red-green (squares) and yellow-blue (circles) opponent mechanisms of observer A.W.; plotted for an equal-energy spectrum.

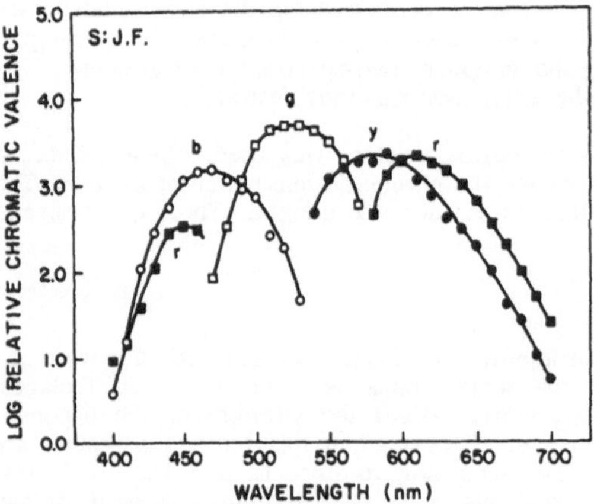

Fig. 2 Log relative chromatic valence of the red-green (squares) and yellow-blue (circles) opponent mechanisms of observer J.F.; plotted for an equal-energy spectrum.

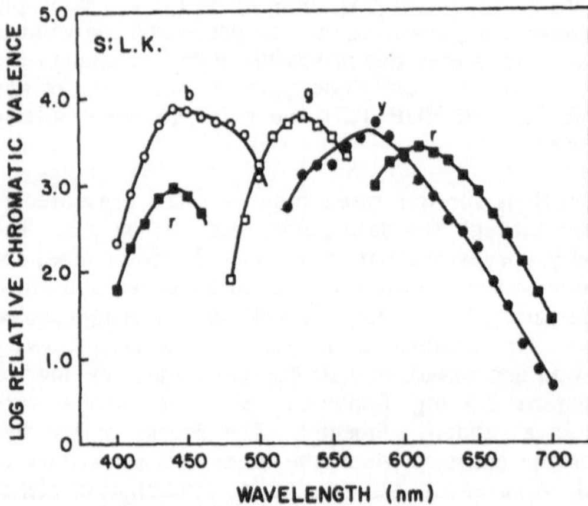

Fig. 3 Log relative chromatic valence of the red-green (squares) and yellow-blue (circles) opponent mechanisms of observer L.K.; plotted for an equal-energy spectrum.

determine the height of the green chromatic valence function relative to the red functions.

The height of the r-g channel relative to the y-b channel was determined post-hoc. In previous studies the r-g channel was pinned to the y-b channel at the point at which red and yellow were described in equal proportions (14, 22). The choice of this locking point wavelength is arbitrary. Therefore, we used all the hue naming data to lock the two channels on the basis of all spectral points. A computer program, beginning with the r-g channel arbitrarily placed below all y-b points, shifted the r-g function relative to the y-b function in 0.05 log unit steps, over a 10 log unit range. At each step it computed hue coefficients and transformed them to percentages in accordance with Eq. 1. The point at which the squared deviations between the hue naming and hue coefficients was minimized was used as the locking point for the two chromatic valence channels.

Individual differences in the log chromatic valence functions are apparent. Most salient is the unusually large green valence of J.F. and the sharply peaked yellow lobe for L.K. This is even more exaggerated on an arithmetic scale, as is shown for the three observers in Fig. 4. There is excellent agreement among observers in the wavelength of the peak sensitivity for the individual lobes of the r-g channel, with the average for the short-wave red curve at 443 nm, green at 525 nm, and the long-wave red lobe at 610 nm. In contrast, the peak sensitivity for the blue and yellow curves is quite variable. For A.W., J.F., and L.K. the blue valence peak is at 460, 470, and 440 nm, respectively; the yellow lobe peaks are at 575, 590, and 580 nm, respectively. The inter-observer agreement in peak chromatic valence may also be paralleled by the agreement among observers in the shapes of the functions. That is, the individual lobes of the r-g channel are fairly comparable among observers, whereas there seems to be more variation in the shapes of the blue and yellow curves, especially that of the yellow lobe.

Our r-g functions are highly similar to those measured for two normal observers by JAMESON and HURVICH (14) and the two normal observers of ROMESKIE (22). The mean λ_{max} for JAMESON and HURVICH, ROMESKIE, and the present study are as follows: short-wave red, 440, 435, and 443 nm; green, 520, 530, and 525 nm; and, long-wave red, 620, 615, and 610 nm. This agreement between studies is not seen for the y-b opponent channel. We found the mean peak in blue sensitivity to be at 457 nm whereas the other studies found it to be at 435 and 440 nm. There are even greater differences in the yellow peak. The mean for HURVICH and JAMESON was 545 nm, ROMESKIE was 555 nm, and in the present study it was 582 nm. It is important to note that this was also the most variable mechanism within each study. For example, the two observers of HURVICH and JAMESON differed by 30 nm. While it is difficult to make quantitative between-study comparisons in the shapes of the individual chromatic valence curves, our informal comparisons indicated that the yellow curves are the only ones for which there is considerable variation.

The fact that there are some individual differences in opponent chromatic response provides a better test of the model which relates opponent mechanisms to hue naming. We ask whether individual differences in the opponent mechanisms are related to individual differences in hue naming.

The square symbols in Figs. 5, 6, and 7 show the hue naming data for each observer. A smooth line has been drawn through the points by eye. The terms red, green, blue, and yellow were sufficient to describe the visible spectrum. Consistent with the HERING model, these observers almost never used the combination of blue and yellow, or the combination of red and green. The unique points of the observers are indicated by arrows in the hue naming figures, in spite of their being determined

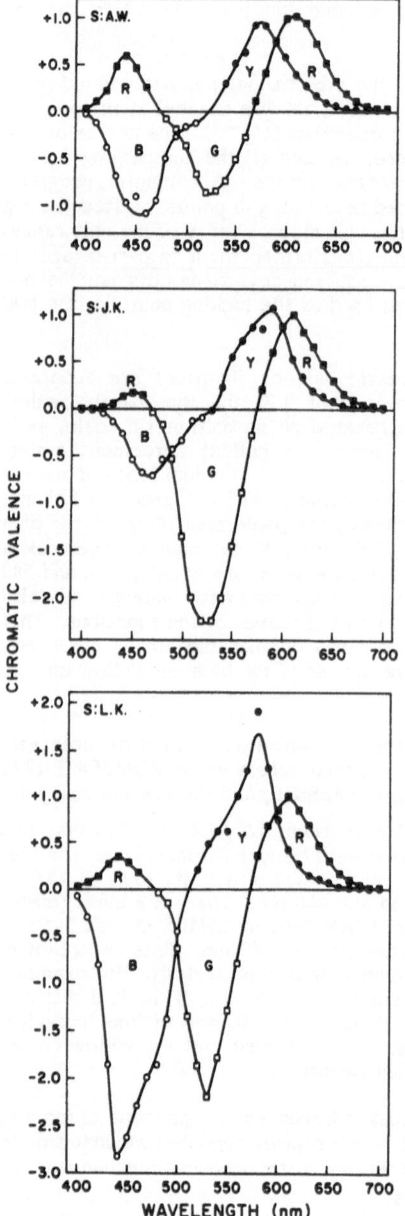

Fig. 4 Chromatic valence functions (arithmetic scale) for each observer as a function of an equal-energy spectrum. Symbols as for Figures 1-3.

by a different method, because within the HERING model they correspond to a hue that is perceived as pure (one opponent channel is at equilibrium).

Values from the log chromatic valence functions were converted to hue percentages by Eqs. 1a and 1b. These predicted points, shown by filled circles, are

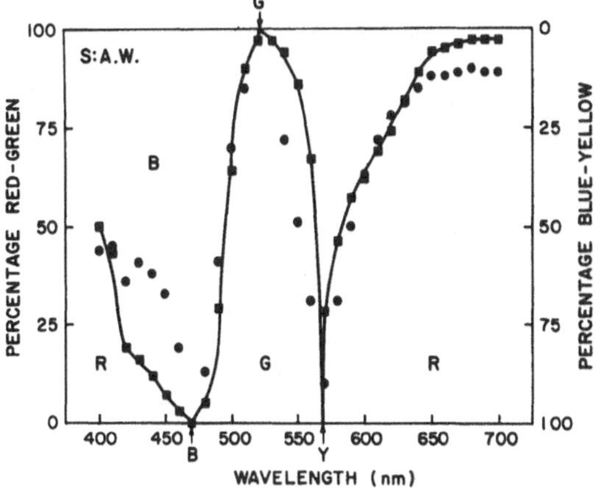

Fig. 5 Hue naming data (squares) and predicted hue naming data (circles) for observer A.W. Red-green is plotted from 0 to 100 percent according to the left vertical axis, while yellow-blue is plotted from 100 to 0 percent according to the right vertical axis.

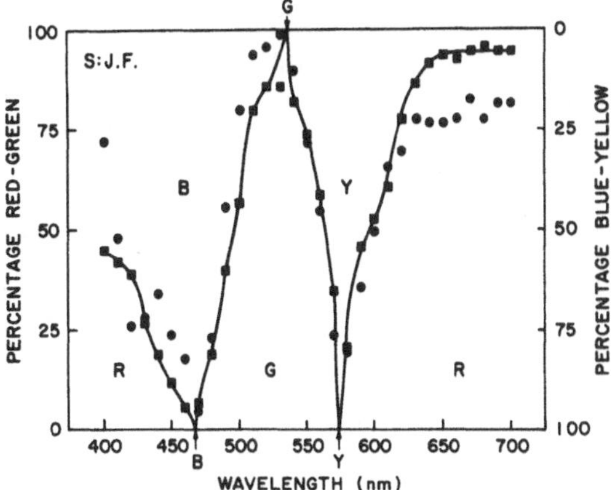

Fig. 6 Hue naming data (squares) and predicted hue naming data (circles) for observer J.F. Red-green is plotted from 0 to 100 percent according to the left vertical axis, while yellow-blue is plotted from 100 to 0 percent according to the right vertical axis.

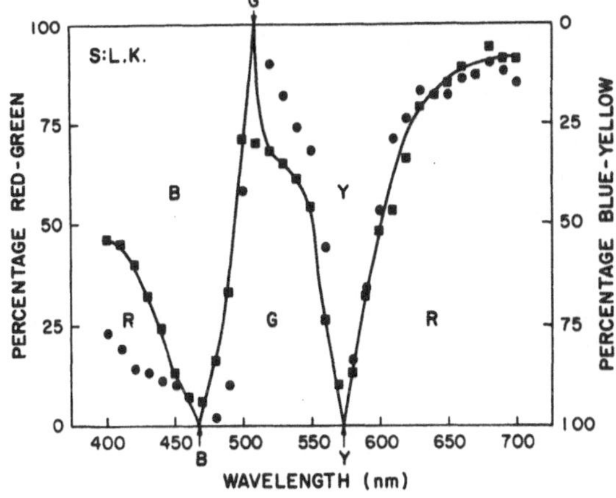

Fig. 7 Hue naming data (squares) and predicted hue naming data (circles) for observer L.K. Red-green is plotted from 0 to 100 percent according to the left vertical axis, while yellow-blue is plotted from 100 to 0 percent according to the right vertical axis.

plotted with the obtained hue naming data. It should be noted that these predictions are not theoretical, but were obtained from opponent cancellation. In general, the predicted hue naming percentages are in good agreement with obtained points. The mean absolute difference between predicted and obtained points was 11.2%. It should be noted that the opponent ratio predictions were obtained for a log scale, but plotted on arithmetic coordinates. Small errors in the log valence functions would be expected to show up as large errors in arithmetic units. In spite of this, when the predicted points were correlated with the obtained points, the result was significant in every case (Pearson product moment correlation, $p < .001$). The proportion of the variance (r^2) in hue naming accounted for by the variance in the opponent mechanisms was about 80% in every case. This only indicates that the two variables are related. To determine whether the relation is 1-to-1, we calculated the slope of the linear equation for obtained hue naming percentages plotted as a function of percentages derived from the opponent ratios. A 1-to-1 relation would yield a slope of +1.00. For observers A.W., J.F., and L.K. the slopes were +0.70, +0.82, and +1.09.

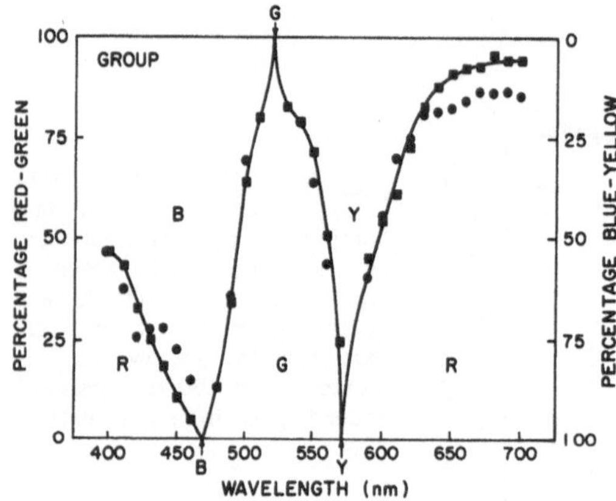

Fig. 8 Average hue naming data (squares) and average predicted hue naming data (circles) for the three observers. Red-green is plotted from 0 to 100 percent according to the left vertical axis, while yellow-blue is plotted from 100 to 0 percent according to the right vertical axis.

The group average of predicted and obtained hue naming percentages is plotted in Fig. 8. Three hue naming points and one cancellation point were omitted because the minor hue component was not the same for all observers. The mean absolute difference between the predicted and obtained points was 5.73%. 96% of the group average variance in hue naming is attributable to the opponent ratios. The only systematic departure between predicted and obtained points is at long wavelengths, where the average opponent response would predict about 9% more yellow than was obtained. We noted over the course of the hue naming practice sessions, prior to data collection, that the reported amount of yellowness progressively increased. Perhaps if more sessions were allowed this trend would have continued and the obtained points would be in closer agreement with the opponent ratios at long wavelengths.

Before JAMESON and HURVICH measured the opponent mechanisms, JUDD (23) proposed theoretical opponent response curves based upon a linear transformation of the C.I.E. tristimulus values (where $(r\text{-}g)_\lambda = 1.0\bar{x}_\lambda - 1.0\bar{y}_\lambda$, and $(y\text{-}b)_\lambda = 0.4\bar{y}_\lambda - 0.4\bar{z}_\lambda$). We used JUDD's formulae to generate theoretical opponent response functions, and then subsequently applied Eq. 1 to obtain a theoretical hue naming function for the standard observer. The mean absolute difference between our hue naming data and the theoretical C.I.E. observer was only 11.1%. The mean difference between the C.I.E. theoretical predictions and the mean predictions derived from the empirical opponent ratios was 7.3%. Thus, our data are consistent with JUDD's transformation for the C.I.E. standard observer. Overall, this suggests that hue naming for the average observer is linearly related to color matching, which, in turn, is linearly related to quantal absorption by the photopigments.

We conclude that to a good first approximation, at this luminance level, opponent chromatic cancellation predicts hue naming. We believe that the code for hue naming is provided by the opponent mechanisms which we measured psychophysically and which others have identified in the receptive fields of single cells at several levels in the primate visual system.

III. Opponent Channels and Photopigments

HERING regarded the opponent chromatic channels as neural processes. Since little was known at that time about the photopigments of color vision, he did not speculate about how these neural systems might relate to the visual receptors. Even now, it must be admitted, we do not know the precise absorption spectra of the human cone photopigments. But,as discussed in the introduction, we do know their approximate shape and spectral position. Thus, it seems worthwhile to ask how our opponent valence functions relate to the probable cone photopigments.

Before considering how quantal absorption in the cones might relate to the hue channels, it is necessary to decide on a specific set of photopigments. Our reading of the psychophysical and photochemical literature leads us to assume values of 435, 530, and 562 nm for the peak absorption locus of the short-, middle-, and long-wave photopigments, respectively. These specific loci can, of course, be debated. It may even be true (24, 25, 26) that there are some individual differences between subjects. As a first approximation, however, we feel that these maxima are adequate for our purpose. For the complete absorption spectra of the photopigments we have assumed the extinction spectrum of iodopsin (27) since it is the only cone photopigment that has been carefully measured by extraction procedures. We have also assumed that DARTNALL's (28) rule can be applied, i.e., that the three pigments have approximately the same shape on a frequency scale. SMITH and POKORNY (9, 10) have shown that these assumptions are good, particularly for the middle- and long-wave pigments. They argue that the human short-wave photopigment is somewhat better approximated by a rhodopsin template, but for our data it makes no detectable difference.

At this point it is, perhaps, worthwhile to comment on the nomenclature used for the three photopigments or receptors of human color vision. Some authors, in the tradition of HELMHOLTZ, refer to them as blue, green, and red pigments or receptors. We feel that this is a mistake, since such terminology inevitably implies HELMHOLTZ's original meaning, i.e., that activation of the red receptor signals redness, and similarly for the green and blue receptors. Yet, it now seems certain that an opponent transformation occurs between the receptors and the perceptual levels that complicates the relation between receptor activity and perceived hue. Some authors will agree with this criticism, but retain HELMHOLTZ's terminology in referring to the receptors. However, by their convention the three receptor types should be called violet, yellow-green, and yellow to correspond to the wavelength peaks of 435, 530,and 562 nm. Even these hue terms ignore the complication of the Bezold-Brücke hue shift. It can be argued that the whole issue is a red-HERING as long as everyone knows what is meant by the various terms. The problem is that confusion inevitably results when hue terms are used to refer to both the receptor and perceptual levels. For these reasons, we have followed JAMESON and HURVICH's (29) suggestion of using the color-neutral terms, α , β, and γ for the short-, middle-, and long-wave photopigments or receptors, respectively.

At the level that we are addressing the question, the basic issue is what mathematical expression adequately relates the estimated relative absorption spectra of the three cone photopigments to the psychophysically determined chromatic valence functions. JAMESON and HURVICH (29) have proposed the simplest possible expression: that the chromatic valence functions are a linear combination of the three cone absorption spectra. Their model is stated as follows:

$$(r\text{-}g)_\lambda = k_{4\alpha\lambda} - k_{5\beta\lambda} + k_{6\gamma\lambda} \quad (2a)$$

$$(y\text{-}b)_\lambda = k_{1\alpha\lambda} + k_{2\beta\lambda} + k_{3\gamma\lambda} \quad (2b)$$

The negative weighting of β in Eq. 2a, and of α in Eq. 2b, is simply a matter of convention, corresponding to plotting y-b with y as positive and b as negative, and the r-g channel with r as positive and g as negative. To evaluate this model with respect to our psychophysically determined opponent valence functions we corrected the absorption spectra of α , β , and λ with the average macular pigment and lens absorption data as given by WYSZECKI and STILES (30). A computer program was used to find the values of the coefficients in Eq. 2 which minimized the sum of the squared deviations between each observer's opponent chromatic response functions and the combination of the three corrected iodopsin nomograms. The best fitting linear combination of the pigments for the r-g opponent response function of the three observers is shown as a smooth line in the left panel of Fig. 9. The data points show chromatic valence on an arithmetic scale, normalized relative to the peak of the long-wave red curve set equal to one. Inspection of these data indicates that each observer's r-g opponent response function is fit fairly well by the linear model. The simplicity of this model and the quality of the fits lead us to tentatively accept Eq. 2a as a reasonable estimate of the transformation of the photopigment input to the r-g channel. This corroborates ROMESKIE's (22) finding with two normal observers whose data were also well fit by Eq. 2a and an iodopsin nomogram.

The best fitting linear combination of the three photopigments to the y-b channel is shown for each observer in the right panel of Fig. 9. It can be seen that the fit is poor. For A. W. the residuals of the linear fit to the r-g channel were about 7% of the residuals for the y-b channel, and for L. K. the r-g residuals were about 11% of the y-b. For J. F. the residuals were approximately equal for Eqs. 2a and 2b; however, the figures show that the problems in fit for the y-b channel were more systematic. Observer J. F. may be somewhat unusual. We note the consistency with which her unique green, and peaks of the yellow and blue valence curves are shifted toward longer wavelengths. This raised the question of whether J. F. is tritanomalous. Additional tests indicated that she was not. She scored normally on the Farnsworth-Munsell 100-hue test and she appeared normal in her metameric matches using the colorimetric equation suggested by HURVICH (31) for detection of tritanomaly (517 nm + 470 nm ≡ 490 nm). We also used a modification of STILES' (32) increment threshold procedure to measure the test sensitivity of J. F.'s π_1. It appeared normal in shape and in the wavelength locus of peak sensitivity. Thus, we have the dilemma of having no independent justification for shifting J. F.'s α photopigment, yet both chromatic valence curves seem to require a shift toward long wavelengths for the optimal fit. Our decision to not make further adjustments to this photopigment, of course, affects all models equally.

For all subjects, the main source of error with the linear model appears to be in the yellow lobe. LARIMER, KRANTZ, and CICERONE (33) have previously reported that they were not able to fit the y-b opponent mechanisms with a linear combination of the VOS-WALRAVEN primaries. Since these primaries are fairly similar to an iodopsin nomogram it seemed reasonable to try to fit our y-b functions with the model that they developed, Eq. 3:

$$(y\text{-}b)_\lambda = -k_{1\alpha\lambda} + k_{2\beta\lambda} \pm k_3 \left|\gamma_\lambda - \beta_\lambda\right|^n \quad (3)$$

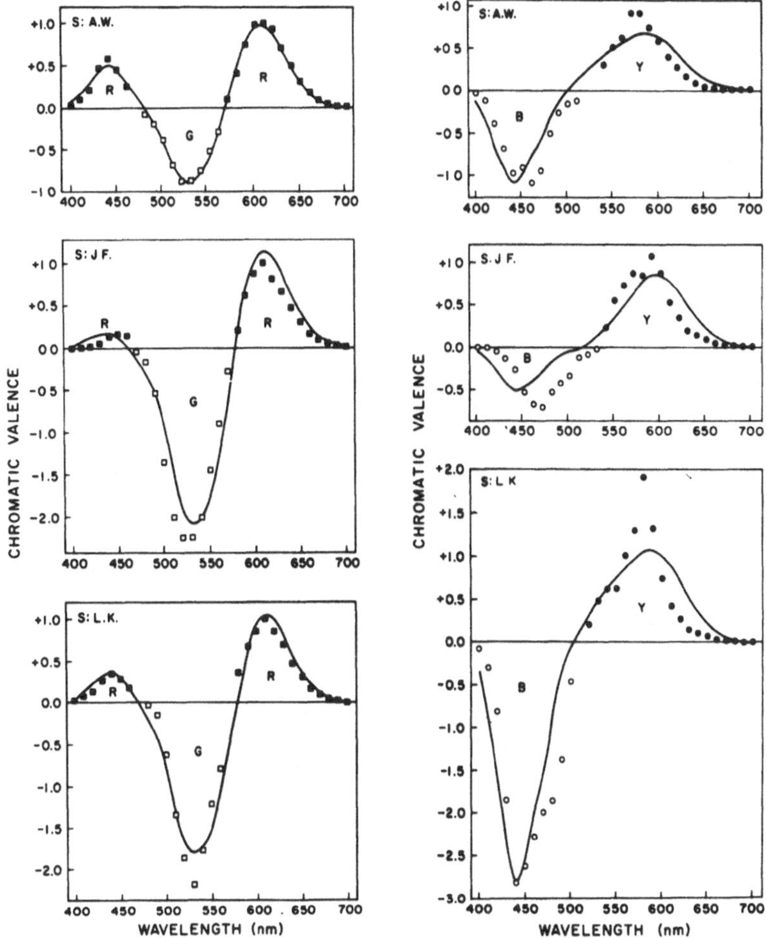

Fig. 9 Red-green (left panels) and yellow-blue (right panels) opponent response plotted for an equal-energy spectrum. The smooth lines represent the best fitting linear (equation 2) combination of the photopigments.

In this model, a compressive transformation is applied after taking the difference between β and γ inputs. This quantity is multiplied by a coefficient which is positive when $\gamma - \beta > 0$, and negative when $\gamma - \beta < 0$. The quantity is then summed with the difference between α and β (each weighted by a constant).

When LARIMER, et al., applied this model to their data the values of n ranged from 0.4 to 0.7 (mean = 0.58). When this model was applied to the present data we obtained comparable values of n; from 0.2 to 0.5 (mean = 0.41). This reduced the average residual relative to the fit of Eq. 2b by about 16%. Unfortunately, the fit with this model was still unacceptable because of a systematic departure between the model's prediction and the obtained data of the yellow lobe. The empirical function was more peaked than the model would predict.

As an alternative we tried Eq. 4:

$$(y\text{-}b)_\lambda \;=\; -k_1\,\alpha\lambda + \left| k_2\beta_\lambda + k_3\gamma_\lambda \right|^n \qquad (4)$$

This model is identical to the linear model, Eq. 2b, except that the sum of β and γ is raised to a power. The main difference between Eqs. 3 and 4 is that the latter model provides for an independent weighting of the β and γ photopigments.

For each observer, the fit with Eq. 4 was markedly better than with Eqs. 2b and 3. Compared to the linear fits, this nonlinear model decreased the residuals for A. W., J. F., and L. K. by 39%, 24%, and 30%, respectively. We feel that the theoretical photopigment fits provided by this model are a reasonable first approximation to the y-b opponent mechanisms. The surprising aspect of these fits is that the exponent required for the optimal fit was large (mean = 3.29). We offer no speculation about what this might imply neurophysiologically. At this point there are no physiological data to either reject or support our model. However, from a curve-fitting point of view, Eq. 4 does fairly well.

We also examined some y-b cancellation functions from previous studies, using the linear and nonlinear models, and iodopsin nomograms that were corrected for average absorption by the preretinal optic media. The extent of the nonlinearity in these y-b data can be summarized in terms of the magnitude of the exponent required for the optimal fit. For example, observer J. of JAMESON and HURVICH (14) required an exponent of 1.28; observers M. O. and D. Y. of ROMESKIE (22) were fit with exponents of 1.06 and 1.65, respectively. Theoretical photopigment fits for these observers' y-b data with the linear model indicated that the nonlinear model provided only a slight improvement. However, a general model accounting for all the y-b data currently available, must include a nonlinear interaction between the β and γ input. Therefore, we conclude that the photopigment input to the y-b opponent channel involves a nonlinear transformation. We further suggest that Eq. 4 describes this tranformation to a good first approximation.

References

1. T. Young, Phil. Trans. Roy. Soc. Lond. 92, 20 (1802).
2. P. K. Brown and G. Wald, Nature 200, 37 (1963).
3. P. K. Brown and G. Wald, Science 144, 45 (1964).
4. W. B. Marks, W. H. Dobelle and E. F. MacNichol, Science 143, 1181 (1964).
5. W. A. H. Rushton, J. Physiol. (Lond.) 170, 10P (1964).
6. H. D. Baker and W. A. H. Rushton, J. Physiol. (Lond.) 176, 56 (1965).
7. G. Wald, Science 145, 1007 (1964).
8. J. J. Vos and P. L. Walraven, Vision Res. 11, 799 (1971).
9. V. C. Smith and J. Pokorny, Vision Res. 12, 2059 (1972).
10. V. C. Smith and J. Pokorny, Vision Res. 15, 161 (1975).
11. E. Hering, Outlines of a Theory of the Light Sense, 1878, trans. by L. M. Hurvich and D. Jameson (Harvard University Press, Cambridge, Mass., 1964).
12. L. M. Hurvich and D. Jameson, J. Opt. Soc. Am. 45, 602 (1955).
13. L. M. Hurvich and D. Jameson, J. Opt. Soc. Am. 46, 416 (1956).
14. D. Jameson and L. M. Hurvich, J. Opt. Soc. Am. 45, 546 (1955).
15. D. Jameson and L. M. Hurvich, J. Opt. Soc. Am. 46, 405 (1956).
16. G. Svaetichin and E. F. MacNichol, Ann. N. Y. Acad. Sci. 74, 385 (1958).
17. R. L. DeValois, Cold Spring Harb. Symp. Quant. Biol. 30, 567 (1965).
18. T. N. Wiesel and D. H. Hubel, J. Neurophysiol. 29, 1115 (1966).
19. P. Gouras, J. Physiol. (Lond.) 199, 533 (1968).
20. P. Gouras, J. Physiol. (Lond.) 238, 583 (1974).
21. J. T. Yates, Vision Res. 14, 163 (1974).
22. M. I. Romeskie, Ph.D. Thesis, Brown University, (1976).

23. D. B. Judd, Handbook of Experimental Psychology, edited by S. S. Stevens (John Wiley and Sons, New York, 1951), Chap. 22.
24. J. K. Bowmaker, E. R. Loew and P. A. Liebman, Vision Res. 15, 997 (1975).
25. M. Alpern, Am. J. Opt. and Physiol. Opt. 53, 340 (1976).
26. M. Alpern and E. N. Pugh, J. Physiol. (Lond.) 266, 613 (1977).
27. G. Wald, P. K. Brown and P. H. Smith, J. Gen. Physiol. 38, 623 (1955).
28. H. J. A. Dartnall, Br. Med. Bull. 9, 24 (1953).
29. D. Jameson and L. M. Hurvich, J. Opt. Soc. Am. 58, 429 (1968).
30. G. Wyszecki and W. S. Stiles, Color Science (Wiley, New York, 1967).
31. L. M. Hurvich, Handbook of Sensory Physiology, Vol. VII/4, edited by D. Jameson and L. M. Hurvich (Springer-Verlag, Berlin, 1972), Chap. 23.
32. W. S. Stiles, Union Internat. de Physique pure et appliquee 1, 65 (1953).
33. J. Larimer, D. H. Krantz and C. M. Cicerone, Vision Res. 15, 723 (1975).

Opponent-Colors Theory and Color Blindness

Martha Romeskie
Department of Ophthalmology
New York Medical College

Introduction

Color vision defects have been studied extensively by visual scientists because they have been considered to be "test cases" for theories of normal color vision. Color defects are assumed to represent modifications of the normal color vision system, rather than totally different systems; therefore, a theory proposed for normal color vision must be able to account for each form of defect by maintaining the same general scheme and merely changing one or more parameter values.

The opponent-process theory that was schematized by HERING and developed quantitatively by HURVICH and JAMESON (1, 2, 3, 4) appears to account successfully for the major phenomena of normal color vision (see the article by WOOTEN, this volume). This paper will describe the opponent-process model for anomalous trichromacy and dichromacy, two major forms of hereditary defect, and some recent data that are consistent with the opponent-process model will be presented.

A central feature of the opponent-colors model is the postulate of two stages of color processing: (1) the cone photopigments, whose absorption maxima occur at 430-450 nm (α), 525-540 nm (β), and 550-570 nm (γ) in normal color vision; and (2) the neural, opponent-response level, which for the normal observer is assumed to comprise the achromatic, or black-white channel, and two chromatic channels, blue-yellow and red-green. Each of these opponent systems is assumed to receive inputs from all three cone types in normal color vision (4, 5). HURVICH and JAMESON (6, 7, 8) argue that it is necessary to postulate alterations at both the photopigment and the neural levels in order to provide a full account of defective color vision.

Anomalous trichromacy is defined on the basis of color matching: persons having this defect require the same number of primaries for color matching as normal observers do, but they use them in different proportions for any given match. In the Rayleigh match, for example, the observer adjusts the proportion of the "red" primary (about 670 nm) and the "green" primary (about 546 nm) to produce a perfect match to the "yellow" standard (about 589 nm). Persons with normal color vision use similar R/G ratios to match the yellow field. The use of significantly more red than normal in this match identifies protanomaly; a higher than normal proportion of green in the match defines deuteranomaly.

The fact that protanomalous and deuteranomalous persons make matches different from those of normals implies that at least one cone photopigment differs from the normal set in each anomalous class. The prevailing view is that protanomalous observers possess the normal α and β pigments and have an abnormal γ pigment whose peak absorption occurs between normal β and γ. Deuteranomalous observers are thought to have the normal α and γ pigments, plus an abnormal β pigment whose absorption maximum occurs between normal β and γ.

The shift of a single pigment in each class of anomaly is commonly considered to be the sole difference between normal and anomalous trichromatic color vision. What HURVICH and JAMESON have pointed out is that a photopigment shift is not sufficient to account for a second characteristic of anomalous trichromacy that is often overlooked: the very wide range of ability to discriminate differences in hue and saturation, which varies from essentially normal to nearly dichromatic in both protanomaly and deuteranomaly. This variation is evident not only in measured wavelength and colorimetric purity discrimination thresholds (9, 10, 11, 12) but also in Rayleigh match data on the range of R/G ratios that an observer accepts as matches to the yellow standard (13, 14). For most normals the Rayleigh match acceptance range is about the same size, and it is very small. But for protanomalous and deuteranomalous observers, that range varies from normal to virtually the entire distance between pure red and pure green.

The critical feature of anomalous color vision is the lack of perfect correlation between the degree of discriminative loss and color match deviance in both classes of anomaly. If one tries to account for anomalous trichromacy solely at the photopigment level, by postulating a single pigment shift, it would be necessary to propose a series of pigment shifts of different magnitudes in order to predict the observed gradation in color weakness. These shifts, in turn, would produce different degrees of color match deviance that necessarily would correlate perfectly with the degree of discriminative loss; that relation is not observed.

In order to account for this relative independence of color match deviance and color weakness, JAMESON and HURVICH (15) have proposed that two mechanisms operate in protanomaly and deuteranomaly: (1) shifts of one or more photopigments in each anomalous class, to explain the color match deviance; and (2) graded reductions of strength of the red-green chromatic system, to explain the discriminative losses. To understand why this second postulate is required, it is necessary to explain in some detail one aspect of the opponent-colors model: i.e., the way in which the model handles color discrimination.

In order to be able to discriminate between two lights that are equated in brightness, there must be a criterion difference in hue or saturation, or both. Since in the opponent colors model, hue and saturation are determined by the relative activities of the chromatic systems, and for saturation by the achromatic system as well, whether or not the two stimuli can be discriminated will be determined by the magnitude of the difference in opponent system outputs for those stimuli. Thus, reduced discrimination implies the reduction in strength of one or more of the opponent systems.

HURVICH and JAMESON have proposed that the reduction occurs in the red-green system for both protanomaly and deuteranomaly. This particular choice was guided at least in part by what is known about these observers' hue perceptions. Descriptions of the spectrum provided by unilateral anomalous observers, who have normal color vision in one eye and anomalous color vision in the other eye and thus can compare the colors seen with the two eyes, suggest that anomalous trichromats see less red and green than normals do (16).

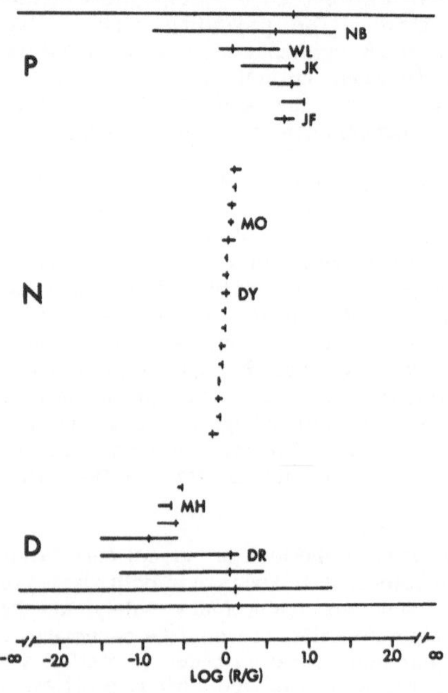

Fig. 1 Rayleigh match data for 16 color-normal, 7 protan and 8 deutan observers. Each line represents the data of one observer: the horizontal line shows the range of accepted matches; the vertical mark indicates the median of 3 free matches. The zero point on the abscissa corresponds to the mean color-normal midpoint.

The experiment described below was designed to test HURVICH and JAMESON's prediction of degrees of red-green system weakness in anomalous trichromacy (for a fuller account, see 17, 18). Chromatic valence functions for the red-green and blue-yellow opponent systems were measured psychophysically for protanomalous and deuteranomalous observers having varying degrees of discriminative loss. The observers for this study were chosen mainly on the basis of Rayleigh match midpoint and range scores, shown in Fig. 1. The match midpoint is defined as the R/G ratio that bisects the accepted range. The large variation in acceptance range in both protanomaly and deuteranomaly contrasts with the characteristically small variation in both midpoint and range for the group of color-normal observers.

The Rayleigh match data of the four protanomalous and two deuteranomalous observers for the chromatic valence function experiment are identified by the observers' initials in Fig. 1. Two color-normal observers (MO and DY) were run both for comparison with the anomalous trichromats and also to validate the current apparatus and procedure by replicating the 1955 data of JAMESON and HURVICH (3).

With the exception of DY, who was a practiced psychophysical observer, none of these observers had previous experience in vision experiments.

The chromatic valence functions were measured with a hue cancellation technique similar to that used by JAMESON and HURVICH (3). This method makes use of the ability of the hue pairs red and green, and blue and yellow, to cancel each other. It rests on the assumption that the response of the chromatic system at a given wavelength is directly proportional to the energy of the light needed to cancel the hue coded by that system. For example, the red function is determined by measuring the relative radiance of a green light needed to cancel perceived redness at each red-appearing wavelength. The higher the radiance of the green light needed for cancellation, the greater the red chromatic response. The cancellation stimuli used in this experiment were each observer's unique blue, green, and yellow wavelengths, which were determined in a separate experiment, and 670 nm for the red stimulus.

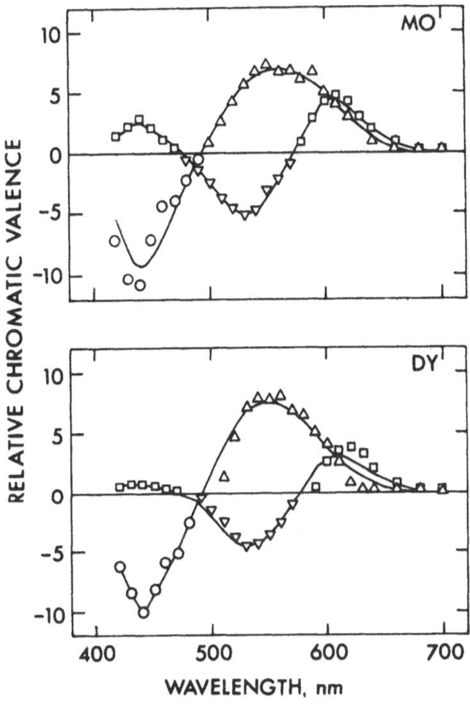

Fig. 2 Relative chromatic valence functions for color-normal observers MO and DY. Each symbol (O : blue; Δ : yellow; □ : red; ∇ : green) represents the mean of 4 measurements. The solid curves are described in the text.

The data for the two color-normal observers are shown in Fig. 2. Each curve represents the energy of the cancellation stimulus required for cancellation of the opponent hue, and converted to an equal-energy spectrum. For all observers, the peak of the blue function has been assigned an arbitrary value of -10. Since the absolute height of this system probably does vary from one person to the next, one cannot compare the absolute strength of any given system across subjects. But since the ratios of the relative heights of the four functions would not be affected by changes in the height of the blue peak, one can compare different subjects on that basis.

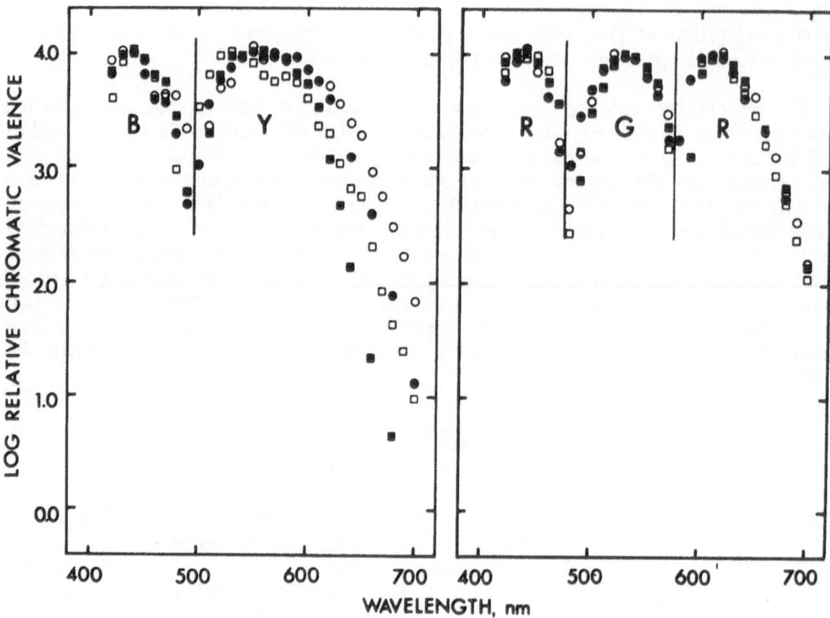

Fig. 3 Comparison of log relative chromatic valence functions for color-normal observers MO (●) and DY (■) with the data reported by Jameson and Hurvich (11) for color-normal observers J (○) and H (□). The maxima of the four chromatic functions for all observers have been arbitrarily assigned the same ordinate value. Data points for MO and DY represent means of four measurements; for J and H, means of twenty measurements.

For the two observers with normal color vision, the spectral shapes of both opponent systems and their relative heights are quite similar overall. The exceptions are the short-wave red branch, which is considerably lower for DY, and the shape of the yellow curve, which for DY has a steeper long-wave slope. These data are compared with the earlier JAMESON and HURVICH (3) data in Fig. 3. Here the data are plotted on a logarithmic ordinate in order to facilitate comparison of the shapes of the functions. The agreement among the data of the four subjects is quite good, which implies that slight procedural differences between the two studies did not have any noticeable effect on the shapes of these curves. The only real variation that is evident here is in the steepness of the long-wave slope of the yellow function.

Figure 4 shows the data for three protanomalous observers arranged, from top to bottom, in order of increasing Rayleigh match range. There is a progressive decrease in the height of the red-green function relative to the blue-yellow function, proceeding from JF, the observer with the small Rayleigh match range, to NB, the observer with the largest Rayleigh match range. There is also a progressive decrease

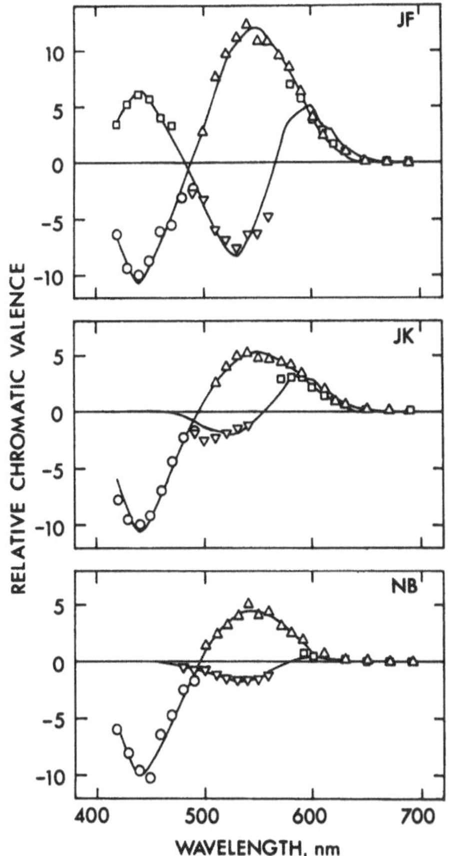

Fig. 4 Relative chromatic valence functions for protanomalous observers JF, JK, and NB. Same as Fig. 2.

in the height of the yellow branch relative to the blue.

The deuteranomalous data are shown in Fig. 5. Here, too, the same trends are evident: comparing the data for the two observers, the red-green system is markedly reduced in strength for DR, and the yellow branch is considerably lower relative to blue.

To summarize these data, the major finding is that the relative strengths of the chromatic opponent systems differ among normal and anomalous observers in a pattern that is correlated with their performance on the Rayleigh match. Increasing acceptance range on the Rayleigh match is associated both with decreasing red-green system strength, which HURVICH and JAMESON had predicted, and also with decreasing strength of the yellow response relative to the blue response.

By assuming a specific relation between the chromatic valence functions and the photopigment absorption spectra, one can generate theoretical curves for the opponent system data, using different sets of photopigments. Then, by examining the goodness of fit obtained with each set, one can make inferences regarding the number

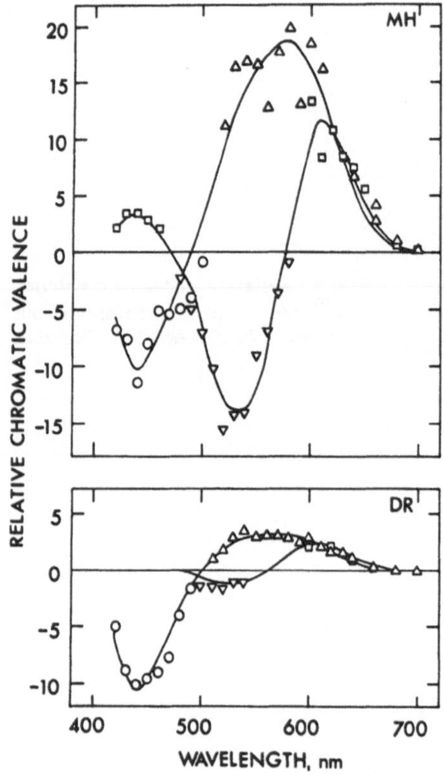

Fig. 5 Relative chromatic valence functions for deuteranomalous observers MH and DR. Same as Fig. 2.

of pigments that differ from normal and the direction of the shifts for each anomalous observer. A simple linear relation between the photopigments and the chromatic opponent systems was assumed for both the blue-yellow and red-green systems, and the method of least-squares deviation was used to generate the weighting constants for the photopigment inputs that produced the best fitting function. The normal pigments were assumed to peak at 435, 530, and 562 nm (quantal absorption at the retina) and to have the shape of iodopsin (19).

The fits obtained to the data of the two normal observers using these three pigments are shown in Fig. 2 by the solid lines. The linear relation produced good fits for both chromatic systems. For the anomalous observers, the normal pigments, and numerous other sets that involved varying amounts of shift of the β and γ pigments were tried. For all three protanomalous observers, the best fit (see Fig. 4) was obtained using the normal α and β pigments, and a γ pigment shifted from 562 to 535 nm. For the deuteranomalous data, a similar-magnitude shift of β toward γ produced a poor fit. The best fits, which are shown in Fig. 5, were obtained using normal α and γ but with β shifted to 535 nm. Thus, with the assumption of a simple linear relation between the photopigments and the opponent-response channels, it appears that these chromatic system data may be consistent with the notion of a single anomalous pigment for both protanomaly and deuteranomaly.

However, it is important to emphasize that this inference is based on testing a limited number of pigment combinations and using data from a rather small group of observers. These data are probably not sufficiently stable to permit a precise identification of pigment peaks; but these results do demonstrate the general type of analysis that probably is appropriate for anomalous trichromats.

Dichromacy

The defining characteristic of dichromatic vision is the ability to match any spectral distribution using only two primaries. Protanopes and deuteranopes accept matches made by color-normal observers but reject each others' matches. To account for the matching data, virtually all color theories have postulated the absence of one of the three normal cone photopigments in each form of dichromacy: loss of the pigment in deuteranopia, and loss of the pigment in protanopia. The two remaining pigments in each class are assumed to be identical to those of normal observers. This is the model of dichromatic vision with which most people are familiar.

The opponent-colors model of dichromacy considers not only the matching functions but also the altered color perceptions of dichromats. Studies of unilateral protanopes and deuteranopes (20, 21, 22) indicate that persons in both classes see only the hues yellow and blue; they do not perceive red and green. In addition, a narrow region of the spectrum that corresponds approximately to the color-normal unique green locus appears achromatic to both types of dichromat; it is termed the neutral point. To explain these features of protanopia and deuteranopia within the framework of the opponent-colors model, it is again necessary to postulate changes at two levels of the visual system: (1) absence of a photopigment; and (2) absence of the red-green chromatic opponent system (1).

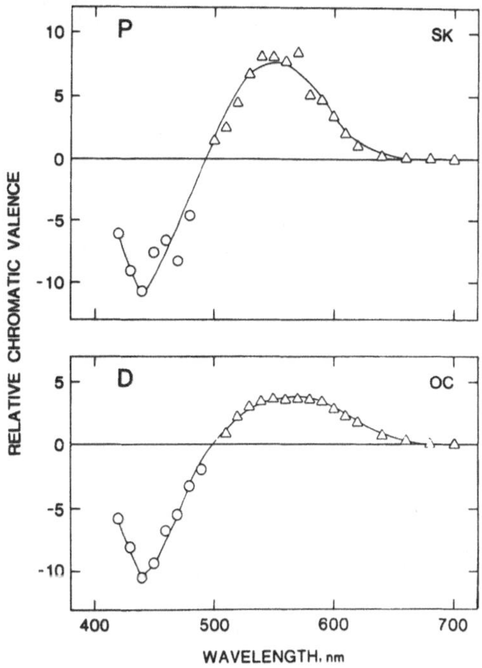

Fig. 6 Relative chromatic valence functions for dichromatic observers SK and OC. Same as Fig. 2.

Figure 6 shows chromatic valence functions determined for two dichromats: SK, a protanope, above, and OC, a deuteranope, below. The cancellation procedure used to obtain these data was similar to that used with the normal and anomalous trichromats but with one important difference: only two cancellation stimuli were required for each dichromat rather than four. As long as the two cancellation stimuli were chosen from either side of the neutral point (470 nm and 600 nm were used for both observers) the mixture always could be adjusted to appear white. Hence only one opponent-response function was obtained for each of these observers (see (23) for details). The important feature of these data is that in both cases, the measured function resembles the blue-yellow function of normal observers. The functions for the protanope and the deuteranope are similar in shape on the short-wave side of the crossover point, which is close to the measured neutral point of each observer. On the long-wave side, the function for the protanope has a considerably steeper long-wave slope.

Photopigment fits were calculated for these dichromatic data, again assuming a simple linear relation between the photopigments and the opponent-response mechanism, and using the method of least-squares deviation. But for these observers, only two pigment inputs were used. A number of different pigment combinations were tried, and the best fits that were obtained are shown in Fig. 6 by the solid lines. The fit to the protanope's data was obtained with α = 435 nm, β = 540 nm; α = 435 nm, γ = 557 nm was used for the deuteranope. These β and γ pigments are close to, but not identical with, the normal β and γ pigments.

In conclusion, these data help to demonstrate the success of the opponent-colors model in accounting for the major features of defective color vision. The model postulates changes both at the photopigment stage and at the neural, opponent-process level of the visual system. For dichromacy, the model predicts absence of one cone photopigment and one of the chromatic channels; for anomalous trichromacy it proposes a shift of one or more photopigments and varying degrees of weakness of one of the chromatic channels.

References

1. L. M. Hurvich and D. Jameson, J. Opt. Soc. Amer. 45, 602 (1955).
2. L. M. Hurvich and D. Jameson, Psych. Rev. 64, 384 (1957).
3. D. Jameson and L. M. Hurvich, J. Opt. Soc. Amer. 45, 546 (1955).
4. D. Jameson and L. M. Hurvich, J. Opt. Soc. Amer. 48, 428 (1968).
5. D. Jameson, In Jameson, D. and Hurvich, L. M. (Eds.) Handbook of Sensory Physiology, Vol. VII/4, (Berlin, Springer-Verlag, 1972).
6. L. M. Hurvich, In Jameson, D. and Hurvich, L. M. (Eds.) Handbook of Sensory Physiology, Vol. VII/4, (Berlin, Springer-Verlag, 1972).
7. L. M. Hurvich, In Color Vision, (Washington, National Academy of Sciences, 1973).
8. L. M. Hurvich and D. Jameson, Mod. Probl. Ophthal. 13, 200 (1974).
9. A. Chapanis, J. Exp. Psychol. 34, 24 (1944).
10. E. Engelking, Klin. Mbl. Augenheilk. 77, 61 (1926).
11. W. M. McKeon and W. D. Wright, Proc. Phys. Soc. 52, 464 (1940).
12. J. H. Nelson, Proc. Phys. Soc. 50, 661 (1938).
13. R. Lakowski, In Color Vision, (Washington, National Academy of Sciences, 1973).
14. M. P. Willis and D. Farnsworth, Med. Res. Lab. Rep. No. 190, 1 (1952).
15. D. Jameson and L. M. Hurvich, J. Opt. Soc. Amer. 46, 1075 (1956).
16. J. L. Kries, Z. Sinnesphysiol. 50, 137 (1919).
17. M. Romeskie, Ph.D. thesis, Brown University, (Ann Arbor, University Microfilms, 1976).

18. M. Romeskie, Vis. Res., in press (1978).
19. G. Wald, P. K. Brown and P. H. Smith, J. Gen. Physiol. 38, 623 (1955).
20. A. L. Hippel, Albrecht L. Graefes Arch. Ophthal. 26, 176 (1880).
21. A. L. Hippel, Albrecht L. Graefes Arch. Ophthal. 27, 47 (1881).
22. L. L. Sloan and L. Wollach, J. Opt. Soc. Amer. 38, 502 (1948).
23. M. Romeskie and D. Yager, Mod. Probl. Ophthal. 19, 44 (1978).

Optical Society Uniform Color Scales

David L. MacAdam
Institute of Optics
University of Rochester

Greetings from the Optical Society of America and congratulations to the University of Houston and to the faculty and students of the School of Optometry on the dedication of this splendid new building.

I am going to talk today about a project that was started 30 years ago. In 1947 a committee of the Optical Society of America was appointed to study what was said by Dr. JUDD, the first chairman, to be the major unsolved problem of colorimetry (1). That was the evaluation of color differences between materials that differ in both luminous reflectance and chromaticity. Data were available for chromaticity differences, alone, for equiluminous reflecting materials. But essentially nothing had been done for the combination of luminous differences with chromaticity differences. That was the problem the committee undertook. Dr. JUDD died in 1972, with the job essentially completed. Before he died, Dr. JUDD asked me to see that color cards designed to illustrate our results were made generally available. They are now available, from the Optical Society. I want to tell you about them.

If we had attempted to make samples of all of the colors that can be distinguished, we would have had to make millions. We had to decide upon a limited number of samples, each of which represents about the same number of distinguishable colors as does any other sample. We finally made 424 colors, each of which closely represents approximately 8,000 distinguishable colors.

To obtain information to guide our choice of colors, the committee conducted a lot of pair-comparison judgements, with 76 color-normal observers. The data were analyzed by statisticians at the U.S. National Bureau of Standards. They reported numbers that indicated the perceptual difference between each of the adjacent pairs in an array of 128 colors that differed in luminous reflectance as well as chromaticity. We also had those colors measured on a spectrophotometer and represented the data on the CIE diagram according to standard colorimetric procedure. We then examined the distances between the points on that diagram that represented those colors and compared the ratios of the distances with the numbers obtained from the judgements of the observers.

For years, we puzzled over the data. We arrived at some disturbing conclusions, as a result of which the committee ultimately had to compromise its principles. In terms of yesterday's discussion I must admit that we plumped for the side of beauty compared to truth. We wanted to complete our job; so in a sense beauty was teaming up with practicality. We wanted to complete the set of colors we had set out 25 years ago to produce. In order to make a set of colors each of which represents the same number of all the millions of colors, say 8,000 each, it is necessary to represent colors in a euclidean space in which equal distances represent perceptually equal color differences. But the committee found that the perceived differences of colors cannot be represented by distance in any euclidean space, at least not for our observers. Our data require a noneuclidean color space. On the other hand, a set of colors of which all nearest neighbors are perceptually equally different cannot be designed if color space is noneuclidean. So we had to make the closest fit we could to our experimental results in terms of euclidean color space (2).

What we wanted to do was select colors such that we could put them together in hexagonal arrays, and pile those hexagonal arrays on top of each other so that around every color there are 12 equally near neighbors, the colors being such that there are 12 equally noticeably different colors around each central color. We wanted to fill euclidean color space with 400 to 500 equally noticeably different colors.

We may describe the arrangement in another way: At the corners of squares in each plane of constant lightness, we have points that represent our colors. In the next lower constant-lightness plane, colors are also arranged in squares, but their corners are exactly under the centers of the squares in the plane next above. In other words, the points in neighboring planes of higher and lower lightness form the apices of regular octahedra (4-sided pyramids base to base) of which the edges and corners touch, but not the equilateral-triangle faces. The feature of this arrangement is that in the next lower (and higher) lightness plane, we do not have the same chromaticities as we have in the central plane, but we have chromaticities that are over and under the centers of the squares in the central plane.

We made the color cards with stable pigments, actually acrylic automobile finishes. Their lightnesses are labelled L, from -7, which is the lowest lightness, up to +5, which is the lightest. In the even-number lightness levels we have even numbers of yellowness, j, including minus values, and even numbers of greenness, g. In the odd-number lightness levels, we have only odd-number yellowness and greenness. The rule that in each lightness plane the notation numbers are all either even or all odd guarantees the arrangement I explained in terms of the regular octahedra that touched each other only along their edges and vertices.

In addition to the 424 full-step colors, the Optical Society will supply 134 extra colors, which split the agreed-upon color differences. Some of our artist and designer friends decided that the color differences we decided on were too big. They said that, especially in the pastel region, the medium-gray region, we need closer colors. Having split the chromaticity levels, the committee also split the lightness levels, in the range from L = -2 to L = +2. So we have plus ½ unit of lightness and a minus ½ unit of lightness and also ±1½ units of lightness. To split all of the color differences in the system would have required a total of about 3300 color cards, which would not be feasible, economically.

A color scale is simply a linear sequence of color samples whose visual qualities change regularly along the sequence. Uniform means that the perceived difference between one color and the adjacent color is the same as between any other pair of adjacent colors in that scale. When we assemble the colors that correspond to any one of the constant lightness planes, we produce a color chart that consists of from three to eight parallel color scales aligned in each of two perpendicular directions (3). In each chart for an even value of L there is one horizontal scale that includes a gray, and one vertical scale that includes a gray. Such scales can be found in other color systems, such as the Munsell System, in which they are called constant-hue scales or series. But none of the other ten to twenty scales on the color chart include a gray. Such scales cannot be found in any other color systems. In the charts for odd values of L, none of the color scales includes a gray. Therefore, none of the scales in those charts can be found in other color systems. In other words, the OSA system is not pinned to the gray scale, to the gray axis. To get uniform color scales of all possible kinds, the committee abandoned the polar coordinate system. It adopted the rectangular coordinate system and revealed all of these new color scales.

There are lots of color scales other than constant-lightness scales. To visualize those, while the paint was still wet, before it was coated onto cardboard for distribution, CARL FOSS dipped a one-inch-diameter lucite ball into each of the paints. I mounted those balls on stainless-steel rods at proper locations and produced a space model in which all kinds of color scales can be seen. In it, the constant-lightness scales are in horizontal planes.

In addition to those planes, there are also vertical planes of color scales. One set of vertical planes consists of scales in which $j + g$ is constant. Different values of the constant correspond to separate, parallel planes. Another set of vertical planes, in which the scales have $j - g$ = constant, are perpendicular to the first set. In each of the vertical planes, two sets of color scales run at 45° to the horizontal, and therefore intersect each other at 90°. All of the color scales in the vertical planes illustrate combined differences of lightness and chromaticness, which was the outstanding problem that the committee, under the leadership of Dr. JUDD, undertook to solve. The job is not yet done. The Optical Society has made the colors available so that all interested persons may assemble the color scales and judge for themselves the validity of the committee's solution. Reports of the results of careful studies of the scales will be welcomed by the Society, and by members of the committee. Anyone who undertakes quantitative analysis of any extensive observational data is advised to remember the compromise that the committee felt impelled to make: i.e., to ignore the almost certain noneuclidean character of real color-perception space.

Finally, there are four different sets of slant planes that consist of color scales. In one of those sets of planes, the scales have $L + j$ = constant. Again, different values of the constant correspond to separate, parallel planes. In each of the slant planes three sets of parallel color scales intersect at 60°. In another set of slant planes, $L - j$ = constant. In the third kind, $L + g$ = constant. In the fourth kind, $L - g$ = constant.

Altogether, there are 420 uniform color scales in the system. They can all be seen in the model, and also in charts in which color cards are arranged in the same way as the balls in the various planes found in the model. In the constant-lightness charts, and in charts corresponding to the vertical planes in the model, square pieces of the color cards are used. In charts that correspond to slant planes, hexagonal pieces of the color cards are best used.

(The lecture was profusely illustrated with color slides of the color charts and the model.)

References

1. D. Nickerson, Opt. News 3, 8 (1977).
2. D. L. MacAdam, J. Opt. Soc. Am. 64, 1691 (1974).
3. M. E. Warga, J. Opt. Soc. Am. 67, 996 (1977).
4. C. L. Sanders and G. Wyszecki, J. Opt. Soc. Am. 48, 389 (1958).

Discussion

Q. Dr. Snodderly: Could you tell us a little bit about how you actually go about choosing pigments and the pigment mixture to get the results you want? From the point of view of natural objects one would like to be able to relate this elaborate set of scales to the kinds of things that might have affected the evolution of color vision. There, one has plant pigments to be concerned with. What do you do if you want to make a red scale? Can you take a single red pigment and mix it through with varying proportions of white - or what?

A. In order to avoid disturbing inconsistencies when these charts are viewed in different kinds of light such as the fluorescent light in this room, each of the 558 colors was made by mixture of not more than four pigments, chosen from a total of about a dozen. The colors were made for viewing in daylight, but I do not think you see any irregularities in the scales. If we had made them in the fashion you suggest, by picking from the hundreds of available pigments the one nearest to the color we wanted, we would have seen very anomalous patches on the charts. They would have looked terrible. The charts would look good only in daylight. In order to avoid that, we used the minimum number of pigments necessary to make the whole set of colors. Another reason for using the minimum number of pigments was that we wanted the most stable pigments, pigments that would not change over the years, pigments that are available commercially so we know their characteristics. Of course we used a white. We used a big-production orange pigment and a big-production red, which were not pigments of maximum available saturations. Likewise, for green, blue, purple and pink. We made as many colors as we could with those few big-production, reliable, stable pigments. Only to make the most saturated colors did we use the most highly saturated, but stable, available pigments. Then we omitted any white but used the nearest moderate-saturation pigment to tone down the extreme-saturation pigment to the color we wanted. By use of four pigments in a mixture, all colors in a tetrahedron in color space can be made. The tetrahedron probably has curved surfaces, depending on the spectral characteristics of the pigments. Within that tetrahedron, we can make all the colors we want by mixing these four pigments together in various proportions.

One member of our committee who makes a business of doing this for pigment and paint manufacturers used a computer to help him decide what percentage of each of the four pigments to use to get each of the colors that I specified in terms of the committee's results. With some set of four pigments he made perhaps ten of the colors we wanted. When he got to one of the four limiting surfaces of the pigment tetrahedron he omitted the pigment that formed the opposite corner and added some highly saturated pigment in the direction he needed to go. Such pigments are very expensive, and perhaps not quite so stable as the others. By use of that new pigment another four-sided space is made available in which all of the colors that I specified in that region could be made. Then he made all of the colors that I specified that he can get with those four pigments. Whenever he approached a surface of the tetrahedron in color space he picked another pigment beyond it, the maximum saturation he could get, and so on until he made all of the colors that I specified.

If you want to select one of the color cards that visually matches any flower or other natural object, the direct thing to do is to do the job visually. Or, you could take the flower to a spectrophotometer, have its spectral reflectance curve measured, then do the CIE calculations and find the CIE specifications. In January of 1978, I will publish the specifications for all of these 558 colors, in the Journal of the Optical Society. In that table, look up the color card that has color specifications nearest to the values of your flower. The color of your flower will undoubtedly be one of the 8,000 near the color of some card. It is not likely to be any one of the 558 colors in the published table. But you can find in this set of color cards a color that is as close to your flower as is possible with any set of 558 color chips. If somebody makes a 1,000 color-chip system, they will charge at least twice as much for it, and maybe you can find in his system a sample that is closer to your sample. But, otherwise, the chances are that with an equal number of color chips you can find in the OSA system a color that is visually as close to any sample you wish.

As for visual effects of color, color vision depends only on the three parameters that are expressed by the CIE system. So if by chance you find a card that visually matches your flower then that card can be used for all visual studies and will be exactly equivalent to your flower. I am not sure this is true for people with anomalous color vision. Presumably, from what we heard yesterday, these color cards will give the same results as any other matching samples will in a test for dichromats.

Q. Dr. Gouras: What are the maximum number of steps in our universe of colors?

A. That number has to be determined by calculation from color-difference experiments. The result is that as many as ten million different colors can be visually distinguished from each other. This means, colors produceable by all means, by fluorescence and by glowing gases, and so on. But with paints made with stable pigments, only about 3 million different colors can be made.

Q. Dr. Gouras: How does this vary from individual to individual?

A. We do not know much about that. We know there was a considerable variation in the judgement made by our 76 normal observers. I do not think the number will differ by a factor of 10, among normal observers.

Q. Dr. Gouras: How does it vary with field size?

A. At least 40 to one. I have done experiments with fields as small as 3 minutes diameter. The maximum differential sensitivity occurs at about 4° visual subtense. In fields larger than about 4° visual subtense, differential sensitivity levels off. The ratio of the numbers of distinguishable colors in fields subtending 4° and 3 min is about 40 to one.

Q. Dr. Castano: I am not sure how you obtained your scales, how you obtained your subjective judgement of the scales of brightness and saturation. I understand you took 76 subjects to judge the closeness of two pairs of colors. Now how about the brightness, you mentioned that.

A. All of our samples were measured in the CIE system. The only observational results that we had, on which the study was based, were the reported ratios of the perceived differences between pairs of colors. In addition to the 43 equal-lightness samples, we had a number of pairs of colors that differed from each other in lightness as well as in chromaticity. Then we attempted to fit all of the ratios in terms of geometrical distances in a geometrical model. We forced those distances to fit into a three-dimensional euclidean space.

Q. Dr. Castano: So you used subjects only to split up the differences for the steps?

A. Not even all of that job. We used a computer to find a highly nonlinear conversion between the CIE specifications, which we got from spectrophotometry and calculation, to a space in which distances were as closely as possible correlated with the ratios of color differences as reported by the observers. No names were used in this color judgement, just which is the biggest difference between two pairs of colors. Having committed himself as to which was the biggest, each observer was asked how big it was compared to the smaller. That had to be a number greater than one. One decimal place was sufficient. The data scattered widely. To fit the data compatibly with their self-consistency, the model had to be such that it implies noneuclidean color space. However, to finish our job, we had to have euclidean space, so we fitted a euclidean space as well as we could to our data, which meant we were plumping on the side of beauty rather than strict truth.

Q. Dr. Miller: Did you ever derive one of the constant-luminance planes in the noneuclidean space that came out of the comparison? Did it resemble the Easter-bonnet-shaped surface that you obtained from NUTTING'S color-discrimination data?

A. No, SANDERS and WYSZECKI (4) did a separate experiment, which I did not take time to describe today. They showed a number of test tiles that were supposed to be equally light to several normal observers and asked each observer to select for each colored tile an equally light gray tile from a very finely stepped gray series of tiles. They published those data in the Journal of the Optical Society. They reported that the lightness-equivalent luminous reflectance was a quadratic function of the x and y coordinates. The committee used that to define the equiluminous surface. We simply ground that formula into our model. If we plot the luminous reflectances vertically, the surface curves downwards from its maximum around gray. It drops quite low for strong reds and blues, which need only about half as much luminous reflectance as a gray in order to appear equally light.

Q. Dr. Wooten: How do these results relate to the CIE uniform chromaticity diagram?

A. They do not relate very closely. They have in common one feature with one of the uniform chromaticity diagrams recommended by the CIE, in that we used a cube-root formula in terms of an opponent color theory. However, unlike the CIE, we transform the CIE tristimulus data to another set of primaries and subtract their cube roots. The chromaticity diagram implied by the committee's formula does not resemble very closely any of the CIE chromaticity diagrams. However, all the committee's differences are big -- they are at least 20 times a just-noticeable difference -- whereas the chromaticity diagrams recommended by the CIE and used by industry are all for one or only a very few JND's, because they want threshold data. Most of the previous work was done with threshold data. We did an entirely different experiment for a different purpose. So the committee emphasized that their data and their formula are not recommended for threshold work, or for color-tolerance work. The formula was devised only in order to make the samples I showed you. We are not pushing our formula for color-difference determinations. I think it has some potentialities, but that is another story.

A Photon Counting Microspectrophotometer for the Study of Single Vertebrate Photoreceptor Cells

Edward F. MacNichol, Jr.
Marine Biological Laboratories
Woods Hole, Massachusetts

Introduction

In order to have a satisfactory description of how the visual system works, particularly a description on which one can base adequate quantitative as well as qualitative theories of visual function, including both normal and abnormal color perception, it's obviously necessary to understand how photoreceptor cells transduce light signals. This paper concerns two techniques that may be useful in improving our knowledge in this area. The first is an improved microspectrophotometer for studying the pigments in single receptors. The second is a microchemical technique for the measurement of light induced reactions in single receptor outer segments. To begin with I would like to go briefly into some of the history of single receptor microspectrophotometry. Until reasonably accurate measurements on single cones in the goldfish were reported by MARKS and myself (1), MARKS (2, 3), and confirmed by LIEBMAN and ENTINE (4), and in human and rhesus monkey receptors by MARKS, DOBELLE and MACNICHOL, (5), and by BROWN and WALD (6), there was no direct evidence that there are three kinds of cones each having a different pigment in organisms known to have trichromatic vision, although much indirect evidence has accumulated over the course of many years from psychophysical and behavioral measurements, the study of defective color perception, and more recently, retinal densitometry. However, for a variety of reasons, the promising work on primate cones of the early 1960's was soon dropped in the two laboratories in which it was done. Although LIEBMAN has reported confirmatory work, neither he nor anyone else to my knowledge has published any more data on single primate cones since then. Thus, such data as are available are based on the measurement of only something like about twenty receptors, and there is disagreement between the work done in the two laboratories both in regard to the wavelength of maximal absorption of each cone type, and the shape of the absorption curves. The number of receptors is so small that we have no good estimate of the proportion of receptors of each type. We also do not know their arrangement in the retinal mosaic. It is possible that the nitroblue tetrazolum (NBT) staining technique developed by JAY ENOCH (7) and applied so elegantly in goldfish and primate retinas by MARC and SPERLING (8, 9) will answer the second two questions, but the answers to the first I think can only be obtained with a microspectrophotometer. The first group of figures summarizes the data that were available at the end of the brief period during which single primate receptors were studied spectrophotometrically. Fig. 1 is a schematic of an instrument typical of the kinds that were used, a double-beam ratio-recording spectrophotometer. It

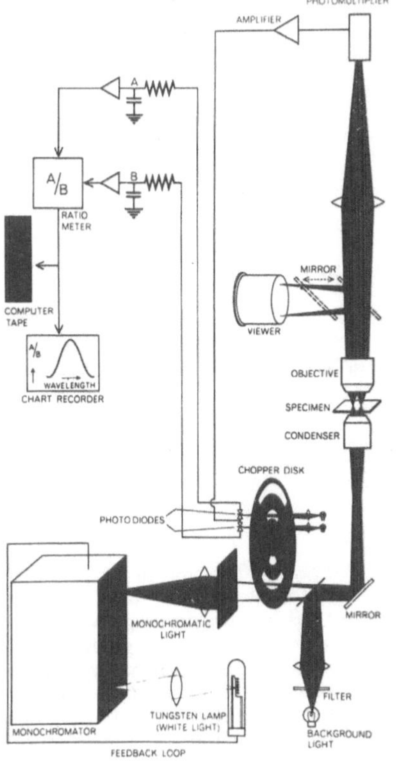

Fig. 1 Simplified schematic diagram of the dual beam microspectrophotometer as used by Marks et al. (From MacNichol, Scientific American, 1964, 211, 48-56). Two feedback loops, one controlling the photomultiplier high voltage to keep the output of the reference channel constant, and the other which controls the lamp current for constant quantum flux throughout the spectrum, have been ommitted.

consists of a broadband light source, a grating monochromator, some kind of a chopper arrangement for alternately switching two beams, one of which passes through the sample, and the other a reference beam which passes through a clear area, or in some instruments through an entirely separate light path. Both beams fall upon a highly sensitive photomultiplier tube, giving pulses of electric charge proportional to the number of photons in the sample and reference beams. An electronic switch, in this case light-activated, sorts out the sample and reference currents and their ratio is recorded while the monochromator scans slowly through the spectrum. Fig. 2 shows how the specimen looked on the stage of the instrument used by MARKS and myself. It illustrates how practically all microspectrophotometers are used. A measuring spot of light passed through a cone outer segment (goldfish) and a reference spot passed through clear area in the retina. Fig. 3 shows raw data from a red absorbing cone obtained directly from the ratio

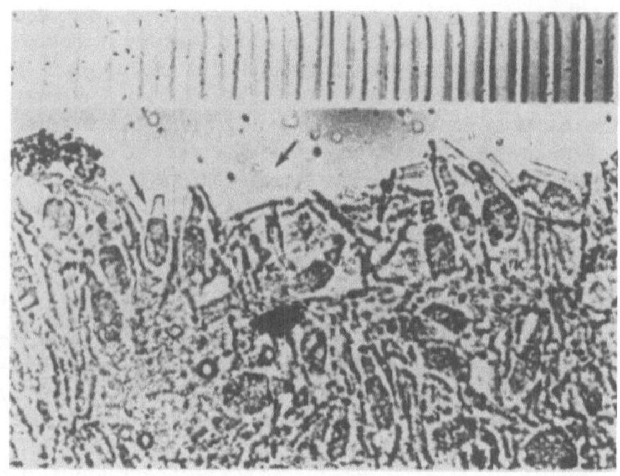

Fig. 2 View of the edge of a goldfish retina in the microspectrophotometer showing the measuring beam passing through a cone outer segment and the reference beam through a clear area (arrows). (From Marks, 1965).

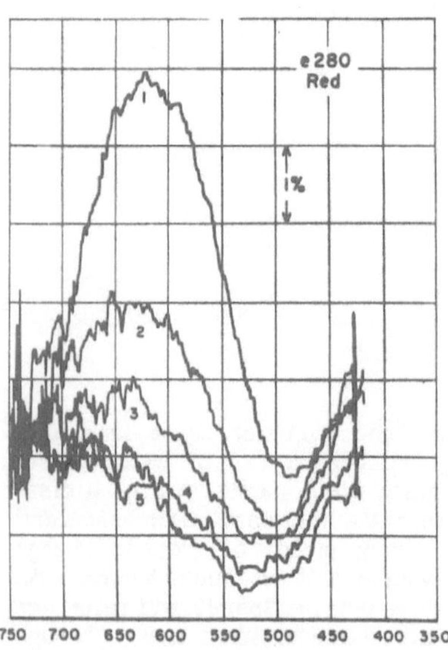

Fig. 3 Tracings from the chart recorder of the MSP used by Marks et al. showing five successive scans of a goldfish red-sensitive cone with a measuring beam strong enough to bleach a large fraction of the pigment during each scan. (From Marks, 1965).

recorder. These curves are successive runs through the spectrum showing that a considerable amount of pigment was bleached during each measurement. MARKS' measurements on a large number of receptors indicated that there are three types with maxima at 455, 530, and 625 nm. In the case of the primates, the small diameter of the cone outer segments makes measurements much more difficult. The work done by BROWN and by MARKS was in the parafoveal region of the retina where the ellipsoids of the cones are large and keep the rod outer segments separated sufficiently to allow a beam passing through the retina perpendicular to its surface to traverse axially only a single cone outer segment. In the fovea the receptors are so closely packed that this kind of measurement is not feasible. The outer segments, although over 40μm long, are only about 0.8μm in diameter. Fig. 4 summarizes essentially all the data on primate cones that MARKS, DOBELLE, and I published back in 1964 showing three groups of receptors having maxima somewhere around 445, 545, and 575 nm. DOBELLE, MARKS and MACNICHOL (10) were also able to

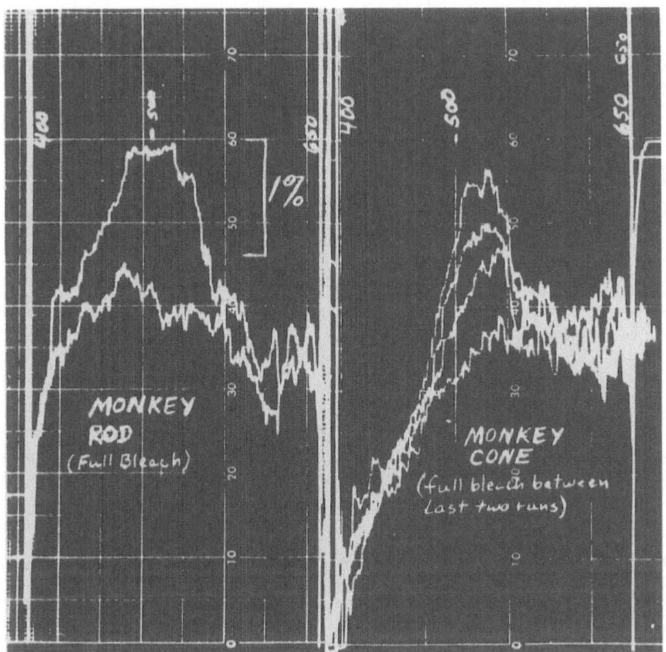

Fig. 4 Computer plots of absorption difference spectra corrected for bleaching of ten primate cones. The points plotted as open parentheses are from human retinas and those represented by numbers are from macaques. (From Marks, 1965.)

record transversely from a few primate rods and foveal cones by teasing out the outer segments and focusing the image of a narrow slit upon them. Because the diffraction fringes are rather broad and light is scattered around receptors of this small size these measurements were too noisy to get an accurate estimate of the wavelength of maximum absorption (λ max.), but they did show that it is possible to make such measurements. Fig. 5 shows the raw records we obtained on a monkey rod and cone. Recently DARTNALL (11) has used a similar technique to measure a large number of primate rods and parafoveal cones, and now has a good statistical sample of both red and green types, though he has found very few blue receptors. Adequate measurements on foveal cones and on single cones of color defective individuals have yet to be made.

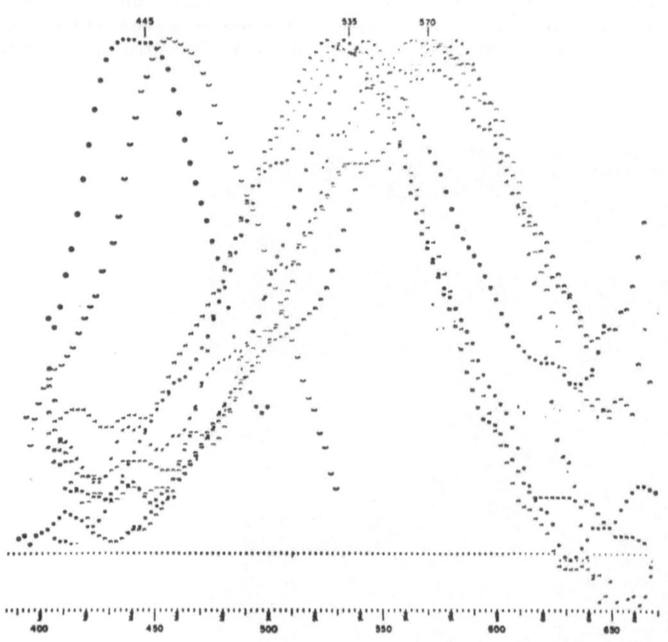

Fig. 5 Recorder Tracings from a single macaque rod and foveal cone measured transversely. (From Dobelle et al. 1969.)

To get better measurements on very small receptors HAROSI and MACNICHOL designed a dichroic spectrophotometer (DMSP) (HAROSI (12); HAROSI and MACNICHOL (13)). Because of the mechanical and optical complexity of dual beam instruments, and the difficulty of balancing the outputs of the sample and reference channels at all wavelengths it was decided to try to design a highly stable single-beam instrument. The initial plan was to use the natural dichroism of the receptor pigment when measured from the side in conjunction with a rapidly rotating polarizing prism to modulate the signal and thereby eliminate the need for a reference beam. The photomultiplier anode current consists of a steady dc component representing the average light intensity upon which is superimposed a sinusoidal ac signal having twice the frequency of the rotating polarizer. The signal is maximal when the electric vector of the polarized light is parallel to the long axis of the receptor and minimal when the electric vector is transverse, when the photopigment absorbs maximally. The ac and dc signals are separated, filtered digitally in a minicomputer which calculates the dichroic ratio. Since the receptors have dichroic ratios of between 3 and 5, the dichroic specturm is quite close to the true absorption spectrum of the pigments. It therefore could be used in place of the absorption spectrum to characterize small receptors that are very difficult to measure in conventional instruments. Furthermore, the instrument scans rapidly and repeatedly through the spectrum summing up the photocurrents, at corresponding wavelength regions (bins).

In effect, it records all wavelengths at nearly the same time so that slow instrumental drifts, while they affect the total size of the measured spectral curve, do not affect the relative measurement of absorption at different wavelengths. Measuring all the wavelengths in rapid succession prevents distortion of the spectral absorption curves due to the progressive bleaching of the pigment during a scan. This is a serious limitation when conventional instruments are used to measure very small receptors. To get a reasonable signal to noise ratio a high photon flux density must be used, which will bleach a considerable fraction of the pigment, which means that the measured absorption will become less and less as a scan progresses. When this happens in the DMSP there is no distortion other than that caused by light-absorbing products of bleaching.

The dichroic spectrophotometer has been a great success but not for its originally intended purpose. HAROSI soon found that it was ideal for measuring changes in the configuration of the rhodopsin molecule subsequent to the absorption of light. He has also used it to study the difference in properties between rod and cone pigments, and between those containing retinal and dehydroretinal. He also made measurements of the deviations from DARTNALL'S relationship. He is currently attempting to explain these deviations in terms of photochemical theory. Thus, the DMSP continues to be a valuable tool in the hands of a gifted investigator. I am delighted that both HAROSI and his instrument are again in my laboratory. However, the episode made me realize that if I wanted to get better primate data, I would have to design, build and run a machine of my own, over which I had complete control. The machine is now running and we have obtained some preliminary results. (MACHNICHOL (14); WILLIAMS, MACNICHOL and JOHNSON (15)). Like HAROSI, I now find that there are so many interesting things to do with it, which are easier than working with very small diameter foveal cones of primates that I too have been distracted at least for the moment. A number of other investigators have also suggested some intriguing problems which are currently being studied.

The Photon Counting Microspectrophotometer

The new instrument for want of a better name is known as the Photon Counting Microspectrophotometer (PMSP). Like the DMSP it is a single beam instrument which scans rapidly and repeatedly through the spectrum, summing the results at corresponding wavelengths in a computer memory. Unfortunately, one soon reaches the point of diminishing returns in improving any instrument. The limit is set (regardless of how the measurement is made) by the number of molecules of photopigments available, and by the fact that every time a photon interacts with a molecule of photopigment it has an approximately 50-50 chance of being bleached. Each receptor contains a limited number of pigment molecules (10^7 to 10^9) to start with, and for every two photons involved in making the measurement one pigment molecule (approximately) is lost. Thus, the efficient use of the photons going through receptor outer segment sets the limit to the signal to noise ratio. To optimise this 1) the optical system must transfer as many as possible of the photons passing through an outer segment to the detector, 2) the detector must have a high quantum efficiency throughout the spectrum, and a dark-current small compared to the photocurrent. The machine has been designed to make it as simple as possible mechanically, optically and electronically, and to get the best possible signal to noise ratio and freedom from drift. Some human engineering has been attempted to make it as easy and fast to operate as possible, to do rapid calculation, store data, and to present the optical density with a minimum amount of man-machine interaction. This allows the accumulation of a great deal of data in a short time, improving statistics of sampling, and minimizing operator fatique. The major departure from previous equipment is that instead of filtering the output of the photomultiplier tube, which is usually used as a detector, then sampling it periodically and feeding the sample to the computer, we amplify the photocurrent in a wide band amplifier which produces an electrical pulse from each photoelectron, pass this through an amplitude discriminator which makes pulses of a fixed size, and count them in a 16 bit counter which is read periodically into our computer. There is of course nothing new about photon counting. In fact, we use a commercially available amplifier and discriminator (Ortec). However, in microspectrophotometry photon counting has some important advantages. In the first place it is an integrating method. The number of counts is equivalent to the integral of the number of photoelectrons liberated during the sampling period. In conventional instruments, in which the photo-current is fed to a computer, it is sampled in a fraction of a microsecond by a "sample and hold" circuit. With a noisy signal like the output of a photomultiplier a sample could be in a noise peak or valley and not be at all representative of the average value during the sample interval. So a filter is usually used. If this filter has a long enough time-constant to do an effective job of smoothing the data it causes a time delay so that the signal being sampled contains some data from previous bins and not all the data from the current bin. Of course it is possible to use an analog integrator which is reset at the beginning of each bin and sampled at the end, but photon counting has two further advantages. The gain of a photomultiplier tube is a power function of the applied voltage (about 0.7 times the number of dynodes) making the photo-current change as about the 7th power of the voltage for a typical P.M. This makes a very highly regulated high voltage supply mandatory. However, when photon counting is used about 90% of the single photoelectron pulses are above discriminator threshold and are counted. Doubling the gain results in only a 5% increase in counts while halving it results in ohly a 10% decrease. In comparison, if photocurrent is used as a measure of photon flux it will be doubled or halved under the same conditions. A further advantage of photon counting is that it is insensitive to the very large random spontaneous pulses occuring in most photomultiplier tubes. These are probably due to natural radioactivity of the materials from which the tubes are made. They cause saturation of the amplifying equipment so that computer programs must be written to recognize and reject them. When photon counting is used they only add a single extra count which is trivial in a usual sample of about 10,000 counts. In addition, when photon counting is done no A/D converter is required: A fast 16 bit counter is easy to build, and makes full use of the precision of the computer. In our computer program only two instructions are required for the counter. The first stops it; the second reads, resets it and starts it.

Description of the PMSP

Figure 6 is a schematic diagram of the PMSP. Light from a tungsten-halogen lamp passes through the adjustable entrance slit of a modified Czerny-Turner monochromator and a manually rotatable Glan-Thompson polarizing prism. It is collimated by a parabolic mirror and falls on a plane reflecting diffraction grating. Light from the grating is collected by another parabolic mirror of longer focal length which focuses the spectrum upon a microscope eyepiece (Ultrafluar projective) acting as the exit slit. This lens in conjunction with an apochromatic objective (Ultrafluar) acting as a condenser focusses an image of a pair of crossed adjustable slits upon the specimen which is held between two cover slips on a Zeiss gliding-rotating stage. The crossed slits forming the field diaphragm are movable by micrometer screws to permit fine adjustment of the region of the receptor being illuminated without touching the stage. A partially reflecting mirror permits the illumination of the

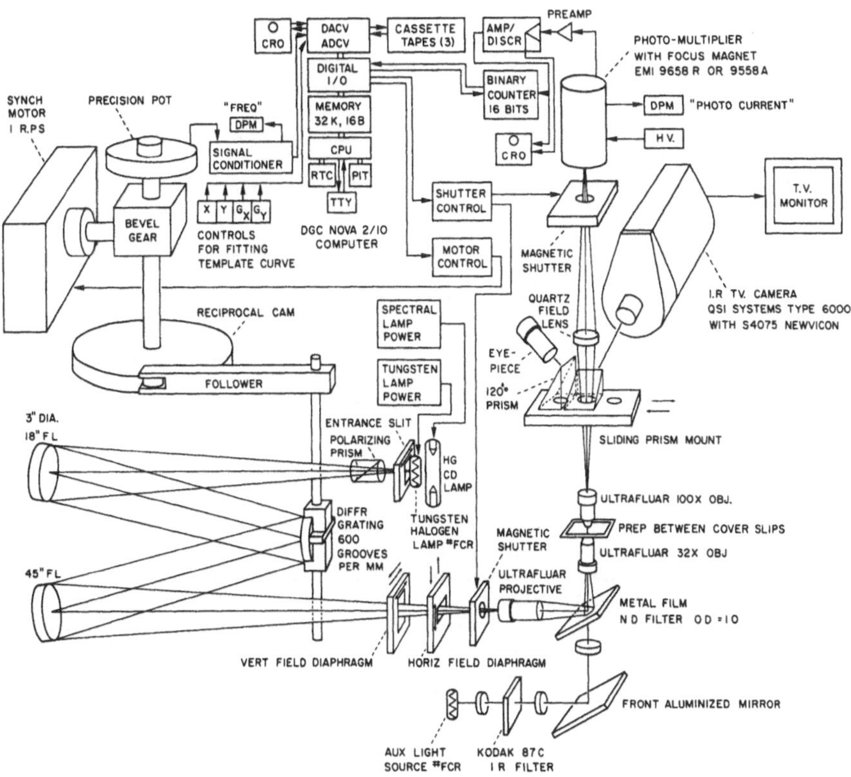

Fig. 6 Schematic diagram of photon counting microspectrophotometer (PMSP).

entire microscopic field by means of an auxilary light source. A large-aperture apochromatic objective focuses the specimen upon an eyepiece, an infared-sensitive vidicon, or upon a quartz field lens which focuses the back focal plane of the objective upon a photomultiplier tube having extended red sensitivity (S20 surface). A permanent focusing magnet reduces the effective diameter of the photocathode to 1 cm which reduces dark counts by a factor of 25. An infared filter is used for T.V. viewing to prevent bleaching of the photopigment. The photoelectron pulses from the photomultiplier tube are amplified, and those larger than a predetermined amplitude converted to pulses of a standard shape by a discriminator. Both the amplified photon-pulses and the discriminator pulses are monitored by a wide-band two channel oscilloscope (Tektronix 465) and counted by a 16 bit binary counter. The 16 bits are read in parallel into the digital input interface of a minicomputer (Data General Nova 2/10) at 9.3 ms intervals under program control. The digital output interface is used to control a pair of magnetic shutters, one of which protects the photomultiplier, and the other of which shuts off the measuring beam between scans. In addition the digital output starts, stops and reverses the synchronous scanning motor. Scanning is done by means of a reciprocal cam and follower which rocks the diffraction grating back and forth so that the scan is linear in frequency rather than wavelength from 400 to 800 fresnels (750 to 300 nm). A frequency scale (1 fresnel = 10^{12} Hz, and 1 nm = 300,000 /fr) is used to make comparison of pigments easier because of DARTNALL'S approximate relation. Also HAROSI (16) has shown that pigment data are nicely fitted by the sum of 3 Gaussian functions when plotted on a frequency scale. The scan occupies 3/4 of the one second rotation period of the cam. Counts are accumulated in 81 five-fresnel bins in the computer memory with two additional bins for dark counts when the shutters are closed. A precision potentiometer mounted on the cam shaft synchronizes the computer with the scan through one channel of an 8 channel A/D converter, and provides a frequency readout via a digital panel meter. A mercury-cadmium lamp is used for calibration.

A small storage oscilloscope (Tektronix 603) is used for plotting data. It is driven by a pair of 12 bit D/A converters and oscilloscope control in the computer. Hard-copy is obtained by photographing the oscilliscope on Polaroid film and printout on the teletype console. Three cassette-tape drives are used to store programs and data. The teletype and cassette tape unit will shortly be replaced by a 60 character/sec keyboard printer and dual diskette drive. These should at least double the rate at which spectra can be measured and stored thus making best use of scarce material such as human retinas.

Programs

The computer has a memory of 32,000 sixteen-bit words. This permits storage of the Extended BASIC programming language operating under Data General's Stand Alone Operating System (SOS), about 50 assembly language subroutines called from BASIC, and still leave adequate room for one of a number of long control programs written in BASIC, and a considerable amount of data. The use of efficient assembly-language subroutines for operating the microspectrophotometer, and for performing rapid calculations, permits these functions to be carried out rapidly, whereas the use of BASIC makes operator-machine interaction very easy, and allows for program modification even in the course of an experiment simply by retyping one or more lines of program.

Three principal programs are used: Data Gathering which contains 257 BASIC statements, Data Analysis (529), and Fast Data Gathering (250).

In normal Data Gathering the operator has the option of collecting new data with the spectrophotometer, storing or recalling data on cassette tape, plotting optical density as a function of frequency on the oscilloscope, or printing out as a table: frequency, corresponding wavelength, average photon counts, and optical density (absorbance) at each frequency. To record: the operator specifies on the console the number of "blank" scans to be made (usually 50) and places the measuring spot on a clear area of the slide. The computer automatically sets the frequency to 400 fr. and opens the beam shutter. When the field has been selected on the T.V. screen pressing the carriage return on the console starts the data collection. When the specified number of scans are completed the computer again sets the frequency to 400 fr. and opens the beam shutter so that the operator can position the measuring spot in a receptor outer segment. The operator then specifies the number of "sample" scans and restarts the scan by pressing the carriage return. After the scans are complete the computer again sets the frequency to 400 and opens the beam shutter. It then calculates the optical density at each frequency bin. O.D.=Log_{10} (Average blank counts-average dark counts) / (average sample counts-average dark counts). It then smoothes the data if desired by averaging each three adjacent points, scales the data to fit the CRT (4000 units full scale), and plots it. It also plots horizontal and vertical tic marks, a numeric identifying code, and the upper and lower limits of the data. When the operator terminates the display the important parameters are automatically typed out. He then has the choice of storing the data, printing it out, or gathering new data. The same blank may be retained in memory and used with successive scans of a number of receptors, which considerably speeds up the data gathering process. A blank and two successive scans through the same receptor can also be made, permitting both an absorbance and a bleaching difference spectrum to be calculated. This is particularly useful in determining the amount of bleaching that has been caused by a measurement in order to calculate the true density the receptor has before the start of the measurement. Fig. 7 is a record obtained from a rod of the Toad, Bufo Marinus. Since these rods are very large a measuring beam of large area was used and the signal to noise ratio is high.

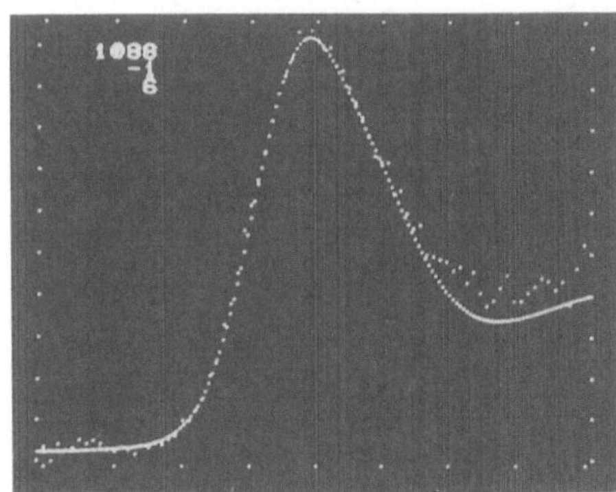

Fig. 7 Spectral absorbance of a single rod of the toad, Bufo marinus plotted with a template curve generated by fitting the sum of three Gaussian components to the average of this and nine other single rod spectra. The computer printout indicated a peak density (Dmax) of .0912 at a frequency (Fmax) of 591.28 fr. (507.4 nm), and a half width (W) of 127.6 fr. Subsequent calibration of the spectrophotometer indicated that Fmax should be corrected to 595.1 (504.1 nm). The corrected average represented by the template curve is 503.6 nm.

The Data Analysis program works entirely with data that have been previously obtained and stored on cassette tape. It permits averaging data obtained from a number of receptors, either scaled or unscaled, plotting data with standard deviation of each point, and stretched both vertically and horizontally by adjusting the zero set controls of 4 of the unused A/D input channels. The template can be made from averaged data, or generated in the computer by calculating 3 Gaussian curves and summing them. The Gaussian parameters are adjusted manually using the A/D zero set controls. The use of Gaussian template permits the accurate determination of the maximum density (Dmax) and its frequency (Fmax). Since the equation of the template is in the computer its maximum value and the frequency at which it occurs is automatically calculated and printed out along with the parameters that specify it. While the method is probably not as accurate for determining the Gaussian parameters as the analytical program (M-lab) used by HAROSI (16) which requires a large computer (PDP-10) it is sufficiently good to obtain Dmax to nearly .001 and Fmax to better than 1 fr. Fig. 8 shows a template curve which has been constructed to fit average data from a number of small cones of the same type. Because the measuring beam had to be very small the scatter in the data points of the individual measurements was large thus making it difficult to determine Dmax by inspection. However, by fitting the template obtained from the average curve to the individual data by adjusting Dmax, Dmin, Fmax and the half width (W) of the template using the zero set controls of four of the A/D inputs the program allows a best fit to be made by eye as shown in Fig. 9. The values of Dmax, Dmin, Fmax and W are typed out as soon as the display is terminated. We believe this procedure is at least as accurate as that used by BOWMAKER, et al. (17) and is much faster and more convenient.

The Fast Data Gathering Program stores individual scans in memory and is being used in collaboration with DR. T. P. WILLIAMS (WILLIAMS, MACNICHOL and

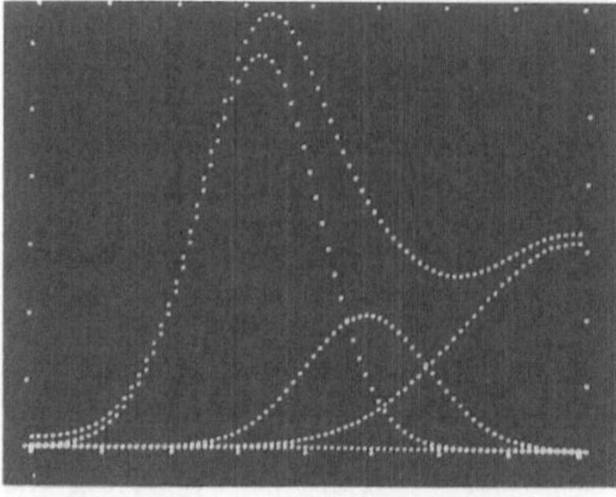

Fig. 8 Three Gaussian components and their sum which best fits the spectral absorbance of the average of eight green-absorbing cones of the tropical fish Cichlasoma citrinellum. Dmax=.043 at 573.6 fr. (523.0 nm) and W=128.5 fr. (Record and data of figures 8 and 9 courtesy of Mr. Joseph Levine).

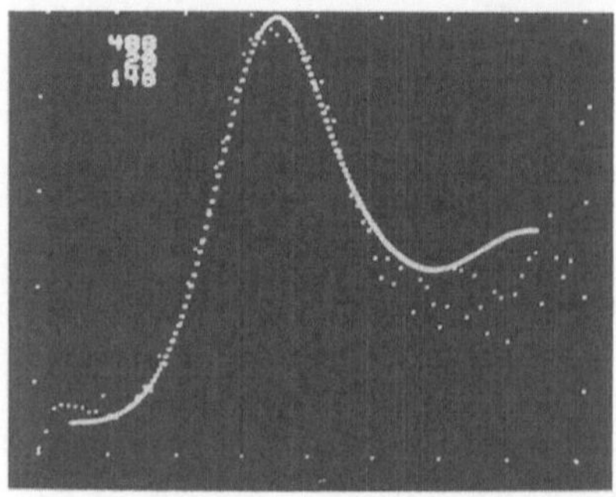

Fig. 9 Template of figure 8 fitted to the spectral absorbance of a single cone of Cichlasoma.

JOHNSON (15)) to study the mobility of rhodopsin molecules in rod disc membranes in a manner similar to that described by POO and CONE (18) and LIEBMAN and ENTINE (19). The advantage of our instrument is that an entire spectrum is recorded each second instead of recording at a single wavelength. This makes the data easier to interpret and allows simultaneous measurement of unbleached pigment and metarhodopsins II and III.

In the current configuration a number of blank scans are made and averaged. Then five scans through a receptor are made and averaged, a bleaching flash given, followed by 25 scans. These are then saved on cassette tape and additional groups of five scans are made at 30 sec intervals and saved on tape for as long as desired. When the data are recalled from tape the five scan groups are averaged and the optical densities displayed. The O.D.s. of 25 scan group can be displayed individually or averaged in groups. Thus, the sequence of events following a bleaching flash can be followed in detail either as changes in absorbance or as bleaching difference spectra. A typical record is shown in Fig. 10.

The Isolated Outer Segment Preparation

In order to study changes taking place in rod outer segments subsequent to exposure to light it is often desirable to isolate single outer segments completely from their surroundings so that they form closed systems. An example of such a procedure is when the dye Arsenazo III is used to measure changes in free calcium and magnesium

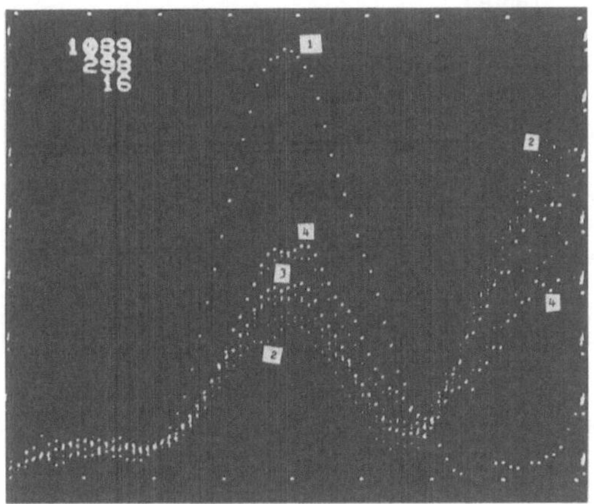

Fig. 10 Absorbance changes of a single Bufo rod using a measuring beam 2 x 30 μ m in cross section after a white light bleach coincident with the measuring beam. A 50 scan measurement (blank) through a clear area was followed by scans at one second intervals through a receptor. After the first five scans a 4.5 sec bleach was given followed by 25 scans at one second intervals. Then two additional sets of five scans were made at 30 second intervals. The scans were averaged in groups of five. The large curve labeled (1) is the average OD of the five pre-bleach scans. (2) shows the first five post bleach scans. Note that the main peak is decreased to about ½, and a large absorbance has appeared in the ultraviolet. (3) Is the average of the last five scans in the 25 scan group, and (4) the final five scans one minute later. Note the steady increase in the main peak and decrease in the UV peak. These changes were due mainly to diffusion of metarhodopsin II out of the measuring beam and unbleached rhodopsin into it and not to the formation of meta III from meta II. This was shown by performing the identical experiment with receptors incubated with glutaraldehyde, then both the main and UV peaks only decreased. (From Williams et al, 1977).

ions. Since both these ions and the dye can diffuse between the receptors and a surrounding aqueous medium the method is not very sensitive. However, if the O.S. are suspended in a strongly hydrophobic liquid each receptor acts as its own microcuvette. Many years ago M. H. KOPAC (20) used silicone and fluorocarbon oils in which to place isolated nuclei and nucleoli during transplantation between unicellular organisms. He reported that these organelles remained viable for long periods in these oils. KUFFLER and YOSHIKAMI (21) have more recently used them to enclose microdrops of fluid containing ACH for determining the sensitivity of motor endplates to known concentrations of this transmitter.

It appeared that it might be possible to incubate receptors with indicator dyes and then suspend them in a heavy oil gradient. Table I shows some of the oils we have tried. All the oils are completely misicible with one another but not with aqueous solutions. For example, after repeated stirring and shaking, and standing for some months with an aqueous solution of 3mM Arsenazo III no color could be observed in

Table I

SPECIFIC GRAVITIES OF SOME SILICONE AND FLUOROCARBINE OILS

Manufacturer	Material	Type	Viscosity	Specific Gravity
Dow-Corning	Methy/Silicone	DC 200	20 cstk.	0.9
Dow-Corning	PhenyMethyl/ Silicone	DC 510	50	0.98
Dow-Corning	Diffusion Pump Fluid	DC 702	medium	1.09
Dow-Corning	Pheny/Methy/ Silicone	DC 710	very high	1.11
Hooker Chemical	FluoroCarbon	FS 5	very low	1.90
Hooker Chemical	FluoroCarbon	MO 10	very low	1.92

the oil phase. Because of the wide range of specific gravities from 0.95 to nearly 2.0 it is possible to produce either continuous or discontinuous gradients by suitable mixing techniques and then to sediment pure fractions of outer segments (spG=1.08) by ultracentrifugation. Our preliminary experiments with discontinuous gradients have given variable results, but we have succeeded in loading some Bufo rod outer segments with Arsenazo III and suspending them in oil. One such preparation mounted on a microscope slide showed no visible deterioration in a Nomarski interference contrast microscope over a period of one week. We have built an apparatus for making continuous linear gradients. It consists of two reservoirs, one cylindrical and the other wedge shaped filled with the heavy and light components respectively. Mercury from a third reservoir is allowed to enter slowly beneath the two oils, maintaining nearly the same level in each because of its much greater specific gravity. The two oils flow from the top of the reservoir into a mixing chamber containing a magnetic stirring bar and then pass through flexible tubing into the centrifuge tube. As the level rises in the wedge shaped reservoir the rate of flow decreases as the cross section decreases, while the rate of flow from the cylindrical tube remains unchanged. By changing the shape of the reservoir, gradients of arbitrary form can be produced. The main problem associated with getting outer-segments into oil has been in getting them to pass through the surface tension to the aqueous to oil interface. Most of them form large clumps including a variable amount of water, and are not easy to separate. However, the technique shows considerable promise and may be applicable to other cells and organelles and to a variety of chemical measurements. Because the oils are excellent electrical insulators it may be possible to study the early receptor potential (ERP) of single rods and cones using external microelectordes, to determine its orientation more accurately and thus to obtain insight into the mechanism of its generation.

Summary and Conclusion

We believe we have developed a microspectrophotometer which has a performance as high as the state of the art will allow, and which is also extremely flexible and easy to use. It also contains a computer which has considerable power, permitting rapid averaging and analysis of data. The system is now being applied to a number of physiological and biochemical problems involving spectrophotometry of receptor outer segments.

The characterization of cone pigments of many species can be accomplished in a relative short time by instruments similar to our PMSP. This should give information on the relationships between visual capacity of an organism to its behavior and habitat selection. It should also yield information in regard to evolution and the development of closely related species. A thorough study of the cones of normal and color defective humans should also be possible.

While most instruments currently being used for single receptors (including ours) approach the theoretical limits of sensitivity set by quantum statistics it is possible that such techniques as fourier transform spectroscopy will have advantages currently unforseen, and that advances in electronic image intensification and processing will permit obtaining spectral absorbance on an entire mosaic of retinal cells simultaneously.

It also appears possible (HAROSI, personal communication) to modify instruments such as the DMSP and PMSP to measure fairly rapid changes in circular dichroism spectra in single receptors in order to obtain new information upon the state of the pigments and the membrane in which they are imbedded, as a result of light absorption.

However, it appears unlikely that there will be any radical improvements in conventional microspectrometry in the forseeable future. On the other hand, two promising and radically different techniques have emerged in the past few years:

Picosecond spectrophotometry, which has already measured the transition time between unbleached pigment and the pre-lumi form (BUSCH, et al. (22)); and

Resonance raman spectroscopy (LEWIS, et al. (23); OSEROFF and CALLENDER (24)). These and greatly improved x-ray diffraction techniques (GRUNER (25)) appear to be the areas in which major advances in instrumentation are likely to result in a greatly increased knowledge of the dynamics of the visual pigments and the mechanism of visual excitation.

References

1. W. B. Marks and E. F. MacNichol, Jr., Biophys. Soc. Abstr. TE2 (1962).
2. W. B. Marks, Biophys. Soc. Abstr. TE2 (1963).
3. W. B. Marks, J. Physiol. 178, 14 (1965).
4. P. A. Liebman and G. Entine, J. Opt. Soc. Amer. 54, 1451 (1964).
5. W. B. Marks, W. H. Dobelle and E. F. MacNichol, Jr., Science 143, 1181 (1964).
6. P. K. Brown, and G. Wald, Science 144, 45 (1964).
7. J. M. Enoch, Invest. Ophthalmol. 2, 16 (1963).
8. R. E. Marc and H. G. Sperling, Vision Res. 16, 1211 (1976).
9. R. E. Marc and H. G. Sperling, Science 196, 454 (1977).
10. W. H. Dobelle, W. B. Marks and E. F. MacNichol, Jr., Science 166, 1508 (1969).
11. Dartnall. Personal communication by J. Lythgoe.
12. F. I. Harosi
13. F. I. Harosi and E. F. MacNichol, Jr., J. Opt. Soc. Am. 64, 903 (1974).
14. E. F. MacNichol, Jr., Invest. Ophthalmol. (Suppl. April 1977) 118 (abstr.)
15. T. P. Williams, E. F. MacNichol, Jr., and H. E. Johnson, Biol. Bull. 153, (1977).
16. F. I. Harosi, J. Gen. Physiol. 68, 65 (1976).
17. J. K. Bowmaker, E. R. Loew and P. A. Liebman, Vision Res. 15, 997 (1975).
18. M. Poo and R. Cone, Nature, Lond. 247, 438 (1978).
19. P. A. Liebman and G. Entine, Science 185, 457 (1978).

20. M. J. Kopac, Ann. New York Acad. Sci. 68, 380 (1957).
21. S. W. Kuffler and D. Yoshikami, J. Physiol. 251, 465 (1975).
22. G. E. Busch, M. L. Applebury, A. A. Lamola and P. M. Rentzepis, Proc. Natn. Acad. Sci. 69, 2802 (1972).
23. A. Lewis, R. Fager and E. Abrahamson, J. Raman Spect. 1, 465 (1973).
24. A. Oseroff and R. Callender, Biochemistry 13, 4243 (1974).

V. Spatial Vision and Form Vision

Spatial Vision in the Cat

Randolph Blake
Northwestern University
Evanston, Illinois 60201

Introduction

During the last several decades the cat has been one of visual science's most generous benefactors, in terms of contributing to our understanding of the neural mechanisms involved in vision. As a result of this generosity, we know a great deal about the receptive field properties of neurons at different stages of the cat's visual nervous system, and we are beginning to see how these properties may be correlated with structural features such as morphology and sites of projection. These anatomical and physiological data, while certainly important in their own right, can take on added significance if their relationship to visual perception can be established. For this reason, as we continue to learn more about the visual system of the cat, it becomes increasingly important to know in some detail the visual capacities of the cat.

For several years now my colleagues and I have been studying spatial vision in the cat using behavioral techniques; our aim has been to examine the extent to which the cat's ability to resolve spatial detail can be related to receptive field properties of neurons within the animal's visual system. This paper provides a progress report on some of our findings. After briefly describing the behavioral approach used in our work, I shall present some psychophysical results which demonstrate the manner in which cat spatial vision depends on stimulus variables such as temporal modulation, background light level and contour orientation. Where possible, an attempt is made to relate the behavioral results to possible underlying neural mechanisms. In some instances I shall be comparing the spatial resolving capacities of the cat with those of human observers. Similarities between cat and human spatial vision would greatly strengthen the argument that the neurophysiological findings in the cat are relevant for understanding the neural bases of human spatial vision.

Spatial Contrast Sensitivity

Traditionally, spatial vision has been expressed in terms of the ability to resolve fine linear detail, such as the minimum separation between high contrast light and dark bars. More recently, though, another approach to this problem of spatial resolution has emerged, an approach which involves measuring the minimum contrast necessary to detect sinusoidal grating patterns of different spatial frequencies. The resulting curve, the so-called contrast sensitivity function, has the advantage of providing a much more complete picture of the spatial resolving capacities of an organism, compared to conventional measures of visual acuity.

To measure contrast thresholds in the cat, we use a behavioral technique known as conditioned suppression (1). With this technique the animal is trained to indicate the contrast value at which it is just able to discriminate a grating pattern from an uncontoured display of the same average luminance; some of the details of the procedure are illustrated in Fig. 1. The grating patterns used in this work are generated electronically on a cathode-ray tube which subtends a 10 deg circular area when viewed from 57 cm, the distance from the cat's eyes to the display. The gratings consist of sinusoidal modulations across space, and the modulation depth (i. e., contrast) of the pattern can be varied in 0.05 log-unit steps. Except where noted the sinusoidal bars are vertically oriented and are viewed at a low photopic light level.

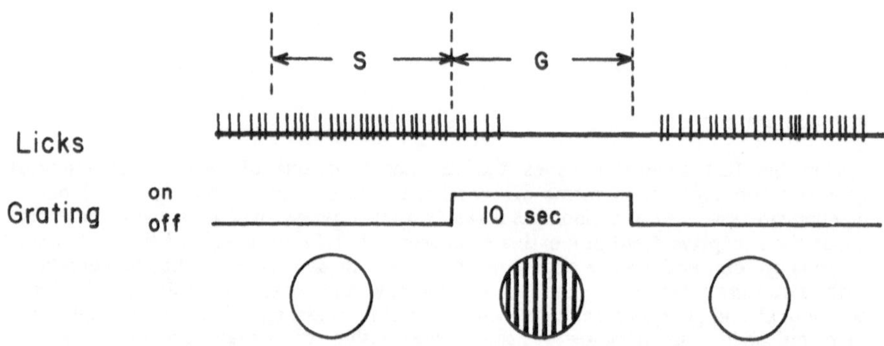

Fig. 1 Contrast thresholds in the cat are measured using a conditioned suppression technique. While housed in a restraining box the cat licks a small tube in order to obtain an occasional food reward. While licking, the animal faces a CRT display which normally appears uncontoured. Test trials involve the replacement of the uncontoured display by a sinusoidal grating pattern; this pattern remains present for 10 seconds, and at the end of this period the grating disappears and the cat receives a brief, mild shock to its paws, through the grid floor of the restraining box. After just a few pairings of the grating and shock, the animal comes to associate the two and typically stops licking whenever the grating appears, in anticipation of the shock. We compare the cat's lick rate during the test period (G) to the rate during a comparable safe period (S) immediately preceding grating presentation. This comparison is illustrated in this Figure. We systematically vary the grating contrast to find the value which produces an arbitrary reduction in lick rate, and define that value as the contrast threshold. This procedure is repeated over a range of spatial frequencies, and the results are plotted in the form of a contrast sensitivity function. Repeated threshold determinations give results within $\pm$ .05 log-units.

Figure 2 shows the typical contrast sensitivity function for a normal adult cat (filled symbols). In our laboratory a total of six ordinary cats have been studied in this fashion, and the resulting curves show good agreement in terms of shape and position along the spatial frequency axis. In general, the visual system of the cat behaves as a band-pass, not a low-pass, filter with peak sensitivity in the vicinity of 0.5 cycles/deg. From the curve in Fig. 2, it appears the passband spans at least a five octave range of spatial frequencies. It is worth noting that this sensitivity profile may be characteristic of vertebrate spatial vision in general, for curves of similar shape have been described for other orders ranging from primates (2,3) to falcons (4).

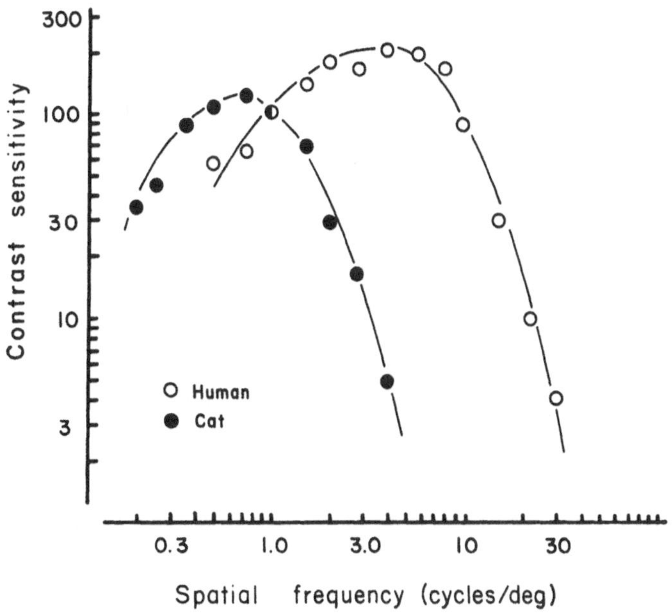

Fig. 2 Contrast sensitivity (reciprocal of threshold contrast) as the function of spatial frequency. These values were measured using a vertical sinusoidal grating which was stationary (i.e., no temporal modulation) throughout the test trial. The values for the human (the author) were taken using a method of adjustment. The small differences in absolute sensitivity (i.e., position of the curve relative to the ordinate) between cat and human may simply reflect differences in threshold criterion. Differences of this magnitude are not unusual between observers of the same species.

For purposes of comparison Fig. 2 also shows the contrast sensitivity curve for a human observer (open symbols) who was tested using the same equipment. Notice that the human and cat curves conform to the same general shape, differing mainly in position along the abscissa. For cats the highest resolvable spatial frequency falls

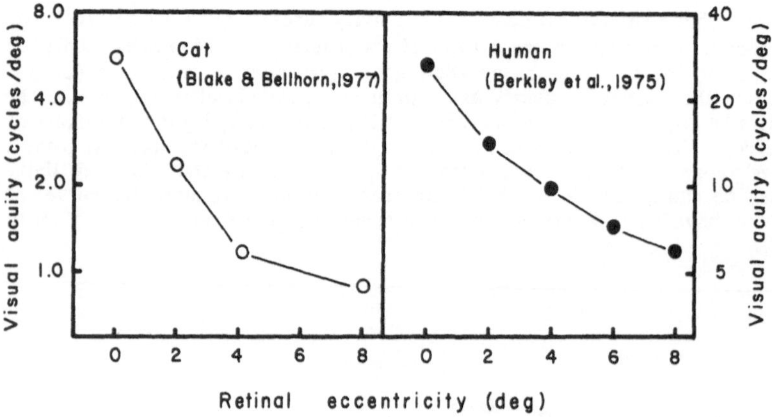

Fig. 3 A comparison of the relationship between acuity (maximum resolvable spatial frequency) and retinal eccentricity for cat and man.

in the neighborhood of 6-8 cycles/deg, while for the human the cut-off frequency comes closer to 40 cycles/deg. These differences in spatial frequency acuity seem reasonable when we compare the retinal networks of the cat and man: in the human fovea the density of cones and ganglion cells is at least an order of magnitude greater than in the area centralis of the cat. Still, despite their differences in absolute acuity, the human and cat show good agreement in terms of the relationship between acuity and retinal eccentricity, as can be seen in Fig. 3. The values for the cat were obtained from animals with bilateral retinal lesions centered on the area centralis (5), and each point represents a different cat; the human results were collected from a normal eye by varying the point of fixation relative to the test pattern (6). Both sets of measurements were performed at comparable light adaptation levels. Now, it is usually assumed that the steady loss in spatial acuity with peripheral viewing reflects a systematic decrease in the receptor-ganglion cell convergence ratio, which results in increasingly larger summation areas. Based on the similarity of the curves in Fig. 3, it would appear that these retinal inhomogeneities are comparably scaled in cat and man.

Flickering Gratings

The threshold curves in Fig. 2 were measured using grating patterns which were stationary, such that the modulation depth remained constant over time (disregarding the initial onset of the grating which occurred gradually over a 1-second period). A rather different picture emerges, though, when an element of temporal modulation is introduced. BLAKE and CAMISA (7) have measured contrast thresholds for the cat using grating patterns which flickered on-and-off sinusoidally at either 1.5 hz or 10 hz, or which appeared continuously (0 hz) throughout the presentation period. At each flicker rate we measured thresholds over a four octave range of spatial frequencies, and the results are given in Fig. 4. There are several features of the curves which are notable:

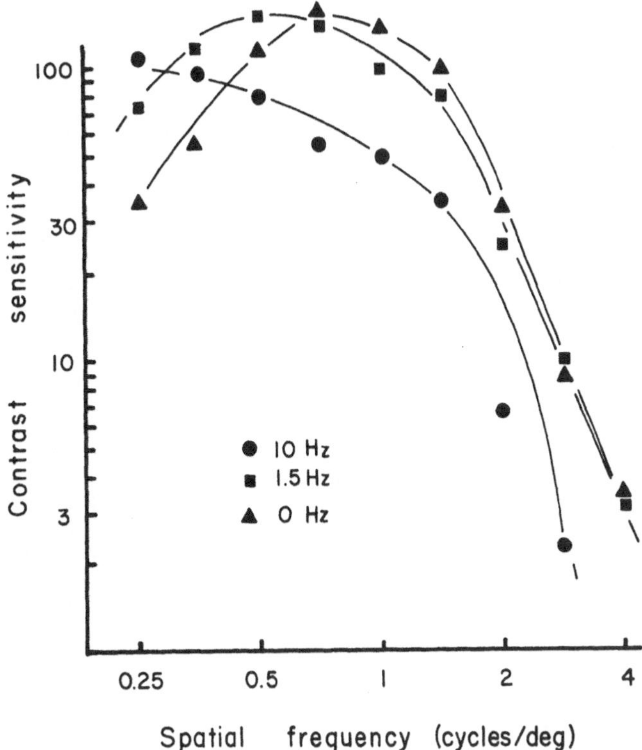

Fig. 4 These contrast sensitivity curves, redrawn from BLAKE and CAMISA (7), show the effects of temporal modulation on the cat's ability to detect gratings. The abscissa has been expanded relative to the ordinate for ease of visualization. The rates of temporal modulation, in this case 'on-off' flicker, are shown in the legend.

1. At higher spatial frequencies, the 10 hz modulation rate produces a significant decrement in contrast sensitivity, shifting the high frequency cut-off point down the spatial frequency axis by about 3/4 octave.

2. The point of maximum sensitivity peaks at progressively lower spatial frequencies as temporal frequency increases.

3. In the low spatial frequency region, increasing flicker rate serves to enhance contrast sensitivity, to the point where at 10 hz there is no evidence for a fall-off in sensitivity with coarse gratings.

We have also performed these experiments using counterphase flicker (wherein the adjacent bars continuously exchange positions) as well as drifting gratings. In all cases, the cat's ability to detect gratings of low spatial frequency is improved when temporal modulation is present, whereas at high spatial frequencies temporal modulation serves to impair grating visibility.

These results demonstrate that grating detection in the cat is mediated by mechanisms whose spatial bandpass characteristics depend upon temporal modulation. With respect to the details of the neural machinery from which these detecting mechanisms are constructed, several alternatives can be imagined, such as a reduction in the influence of the antagonistic surrounds of receptive fields (8). However, we currently favor the hypothesis that the variations in spatial bandwidth reflect the selective involvement of several classes of neurons possessing different spatial tuning and temporal response properties. It is known that the visual system of the cat is composed of different classes of cells which are distinguishable in terms of their receptive field properties, conduction velocities and projection sites. Of these classes, the X- and Y-cells have been studied most extensively (9,10,11) and they appear to constitute a majority of the geniculo-striate cell population. Several differences between X-cells and Y-cells suggest that they are providing the mechanisms for the detection of gratings by the cat. For one thing, the two cell types differ in their selectivity for spatial frequency: X-cells respond to higher spatial frequencies than do Y-cells, and the X- cells show a much more pronounced loss in responsiveness at low spatial frequencies, compared to Y-cells. For another, these two classes of cells differ in their temporal resolution: X-cells respond vigorously to stationary patterns but not to moderate rates of temporal modulation, whereas Y-cells are quite responsive at high rates of flicker or drift. Thus, in view of these receptive field properties, it seems reasonable to propose that the effects of temporal modulation on the cat's contrast sensitivity reflect a shift in the relative contributions of X- cells and Y-cells in grating detection by the cat. Moreover, this tentative conclusion may apply to human vision, too, for temporal modulation produces comparable changes in the form of the human contrast sensitivity function (8).

Does the Cat Show an Oblique Effect?

In addition to their selectivity for spatial frequency and for temporal modulation, neurons in the visual cortex of the cat are noted for their orientation selectivity. The distribution of preferred orientations covers the entire range around the clock (12), but there is some evidence indicating that the horizontal and vertical orientation may be particularly salient. PETTIGREW, NIKARA and BISHOP (13) found among simple cells with receptive fields near the area centralis a slight preponderance of neurons tuned to either vertical or horizontal, compared to oblique orientations. Also, ROSE and BLAKEMORE (14) reported that simple cells preferring vertical or horizontal were more narrowly tuned than neurons selective for obliques; complex cells, on the other hand, showed no such anisotropies in orientation tuning. These physiological findings suggest the possibility that the cat might exhibit meridional differences in visual sensitivity, analogous to those found in humans (15). But in order to observe such a so-called oblique effect in the cat, it might be necessary to employ stimulus conditions which favor simple cells, since the orientation anisotropies found physiologically (13,14) are peculiar to just this case of cortical cells.

Now, it is known that simple cells prefer slower rates of temporal modulation (e.g., movement (16)) compared to complex cells, and it also appears that simple cells respond to higher spatial frequencies than do complex cells (9). With these facts in mind, I set out to determine the contrast sensitivity of the cat for horizontal, vertical and several oblique orientations, testing both at low and high rates of temporal modulation and at several spatial frequencies. The results, which are plotted in Fig. 5, again show that higher rates of flicker enhance sensitivity at low spatial frequencies (in this case, 0.3 cycles/deg) and depress sensitivity at high

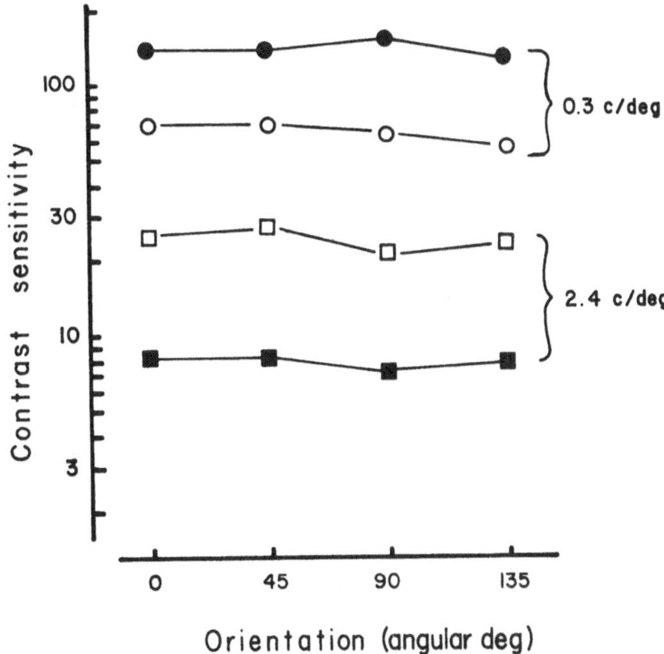

Fig. 5 Contrast sensitivity of the cat for gratings of different orientations: horizontal (0°), oblique right (45°), vertical (90°) and oblique left (135°). Gratings were flickered 'on-off' at either 1.5 hz (open symbols: ○ , □) or 10 hz (filled symbols: ● , ■). Grating spatial frequency is given in the right-hand portion of the Figure.

spatial frequencies (2.4 cycles/deg). But note that the cat's sensitivity for the oblique orientations 45° and 135° is equivalent to that at horizontal (0°) and vertical (90°). Thus, we are unable to measure an oblique effect in the cat, a result which confirms and extends earlier behavioral findings (17).

Of course, to the extent that meridional variations in visual sensitivity stem from selective early visual experience, it is quite possible that the few cats tested so far enjoyed their kittenhood in relatively unbiased, non-carpentered environments. There is, however, some reason to believe that endogenous maturation may play a role in the genesis of the oblique effect in humans (18). In this event, we would be forced to consider the possibility of innate differences in the orientational selective mechanism in cat and man (19).

Adaptation Level

So far I have presented behavioral data on cat spatial vision which were collected at a fixed, photopic level of adaptation. The cat, of course, is an arhythmic animal which often prefers to carry on its business during the evening and at night. Are there reasons to expect the animal's spatial resolving power to change with decreasing light levels? We know that the eye of the cat incorporates several optical (e.g., broad light-gathering cornea paired with enormous pupillary range) and structural (e.g., a rod-rich retina) features which operate to enhance light sensitivity in dim light conditions. Moreover, at low levels of illumination changes occur in the spatial weighting function in the center/surround organization of cat visual neurons; by enlarging the functional size of receptive fields these changes serve to increase spatial and temporal summation (20,21,22), thereby enhancing increment sensitivity at low adaptation levels. At the same time, however, we would expect these structural and neural factors to affect adversely the cat's spatial resolving powers at mesopic and scotopic levels of adaptation.

To test this possibility JOHN CAMISA and I recently measured contrast thresholds in the cat at different luminance levels, where the visual display provided the only source of light. Before each daily test session the cat spent at least 15 minutes adapting to the prevailing light level. Fig. 6 shows the effect upon the cat's contrast sensitivity of reducing mean luminance. Each reduction in background results in an overall depression in sensitivity; at the highest spatial frequency tested, 4 cycles/deg, it proved impossible to measure a threshold except at the two highest luminance levels. We also found evidence for a fall-off in sensitivity at low spatial frequencies at all luminance levels, including the lowest tested. Assuming this portion of the curve reflects the inhibitory contribution of the surround portion of concentric receptive fields (9), these behavioral data indicate that center/surround antagonism operates even at scotopic light levels.

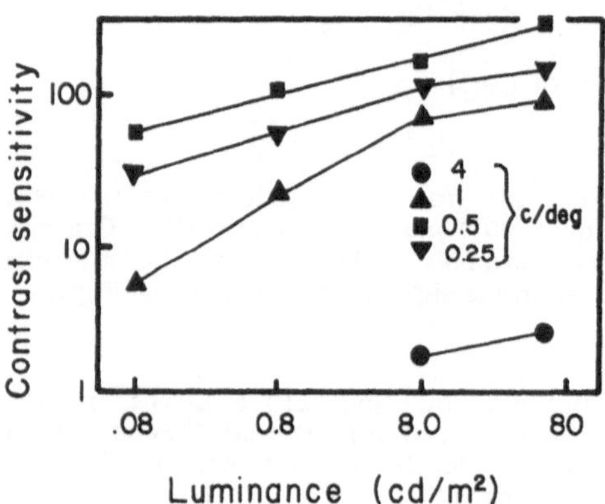

Fig. 6 Cat contrast sensitivity as the function of the average luminance at several spatial frequencies, as shown in the legend. The gratings were flickered in counter-phase at 1.5 hz.

It should be noted that in these experiments the cat viewed the display through its natural pupils, which means we cannot parcel out the relative effects of neural vs. optical factors. Of course, this does not present a problem in the low spatial frequency region, where sensitivity losses must be due to neural factors only. In general, these behavioral data confirm that any enhancement in the cat's sensitivity to light at low luminance levels is purchased at the expense of spatial resolving power. Comparable measurements for humans indicate that a similar trade-off between spatial resolution and increment sensitivity occurs in man (23).

Effects of Visual Deprivation on Cat Spatial Vision

Although most of our work has concentrated on normal adult cats, quite recently we have had the opportunity to study spatial contrast vision in two young cats whose early visual experience was unusual. Both of these animals spent the first six months of life in complete darkness, a form of deprivation which produces abnormal receptive field properties in visual cortical cells (24,25). Following their emergence into a normally illuminated environment, both cats displayed deficits in eye alignment and visuo-motor coordination, conditions which have been described previously (26,27). While there has been some improvement in most aspects of visuo-motor behavior in our cats, one of the two continues to evidence difficulty in tracking moving objects, and this animal remains reluctant to jump from heights which are readily negotiable by normally-reared cats.

Formal behavioral testing of contrast sensitivity in these cats is still in progress, but several interesting results already have emerged. First, these dark-reared cats exhibit an overall depression in spatial resolution, such that they fail to detect low contrast gratings regardless of spatial frequency. As a result the high frequency cut-off point falls at least an octave lower than in normal animals. Of particular interest is the effect of temporal modulation of low spatial frequency gratings. Recall from Fig. 4 that in normal cats increasing temporal frequency produces an enhancement in sensitivity for coarse gratings. For the dark-reared cats, however, we are finding no evidence for such an effect; indeed, for one cat flickering a grating of low spatial frequency actually produces a further loss in contrast sensitivity, on the order of 0.2 log-units. The failure of temporal modulation to enhance sensitivity at low spatial frequencies could reflect selective neural deficits in these dark-reared cats. We are tempted to draw analogies to the loss of Y-cells in binocularly deprived cats (28), but this conclusion must await more comprehensive behavioral testing.

Conclusion

These psychophysical experiments indicate that the cat's ability to detect grating patterns depends upon many of the same stimulus parameters which influence the response properties of neurons within the animal's visual nervous system. This correspondence between psychophysics and physiology serves to reinforce the doctrine that spatial vision is related to the activity of cells in the visual pathways, although we certainly recognize that the details of this relationship remain obscure. Moreover, it is encouraging to see a degree of similarity between human and cat spatial vision, for so many of our current conceptions about the working of the human visual system have been borrowed from cat neurophysiology.

Acknowledgements

Some of the experiments described in this chapter were performed in collaboration with JOHN CAMISA. The project was supported by research grants from NSF (BNS75-17073) and NIH (EY01596).

References

1. R. Blake, S. J. Cool and M. L. J. Crawford, Vision Res. 14, 1211 (1974).
2. R. DeValois, H. Morgan and D. Snodderly, Vision Res. 14 (1974).
3. W. H. Merigan, Vision Res. 16, 375 (1976).
4. R. Fox, S. Lehmkuhle and R. Bush, In preparation.
5. R. Blake and R. Bellhorn, Vision Res. 18, 15(1978).
6. M. A. Berkley, F. Kitterle and D. W. Watkins, Vision Res. 15, 239 (1975).
7. R. Blake and J. Camisa, Exp. Brain Res. 28, 325 (1977).
8. J. G. Robson, J. Opt. Soc. Amer. 56, 1141 (1966).
9. C. Enroth-Cugell and J. G. Robson, J. Physiol. 187, 517 (1966).
10. B. G. Cleland, M. W. Dubin and W. R. Levick, J. Physiol. 217, 473 (1971).
11. L. Maffei and A. Fiorentini, Vision Res. 13, 1255 (1973).
12. D. H. Hubel and T. N. Wiesel, J. Physiol. 160, 106 (1962).
13. J. D. Pettigrew, T. Nikara and P. O. Bishop, Exp. Brain Res. 6, 373 (1968).

14. D. Rose and C. Blakemore, Exp. Brain Res. 20, 1 (1974).
15. S. Appelle, Psychol. Bull. 78, 266 (1972).
16. J. A. Movshon, J. Physiol. 249, 445 (1975).
17. S. Bisti and L. Maffei, J. Physiol. 241, 201 (1974).
18. S. C. Leehey, A. Moskowitz-Cook, S. Brill and R. Held, Science 190, 900 (1975).
19. F. W. Campbell, B. G. Cleland, G. F. Cooper and C. Enroth-Cugell, J. Physiol. 198, 237 (1968).
20. H. B. Barlow, R. Fitzhugh and S. W. Kuffler, J. Physiol. 137, 338 (1957).
21. C. Enroth-Cugell and P. Lennie, J. Physiol. 247, 551 (1975).
22. V. Virsu, B. B. Lee and O. D. Creutzfeldt, Exp. Brain Res. 27, 35 (1977).
23. J. J. Kulikowski, Vision Res. 11, 83 (1971).
24. P. Buisseret and M. Imbert, J. Physiol. 246, 98 (1975).
25. W. Singer and F. Tretter, Exp. Brain Res. 26, 171 (1976).
26. S. M. Sherman, Brain Res. 37, 187 (1972).
27. J. Van-Hof Van-Duin, Brain Res. 104, 233 (1976).
28. S. M. Sherman, K. P. Hoffman and J. Stone, J. Neurophysiol. 35, 532 (1972).
29. R. Blake and D. N. Antoinette, Science 194, 109 (1977).
30. Y. M. Chino, M. S. Shansky and D. I. Hamasaki, Science 197, 173(1977).

Discussion

Q. Dr. Campbell: Human observers report the appearance of two distinct thresholds when confronted with flickering gratings, one the contrast necessary to perceive the spatial structure of the pattern and the other the contrast at which flicker is visible.

A. Yes, that is so. Of course with the cat all we are measuring is the minimum contrast necessary to detect any departure from an uncontoured, nonflickering display. We cannot distinguish whether the cat sees flicker or pattern. I should add, though, that in humans the magnitude of the difference between these two distinct contrast thresholds depends upon whether one employs on-off or counterphase flicker. And we find comparable differences in contrast thresholds for these two forms of flicker in the cat. It will take, however, a more refined discrimination paradigm to answer directly the question of dual thresholds in the cat.

Q. Dr. Kaas: Did you mention Siamese cats?

A. No, but I shall be happy to say a word about their visual performance on this task. Diane Antoinetti and I have measured contrast thresholds in Siamese cats and found some notable departures from the normal curves. Overall sensitivity is depressed, the high frequency cut-off is lower by almost an octave and there is little fall-off in sensitivity at low spatial frequencies. In some respects, the curve for the Siamese cat, using a 1.5 hz flicker rate, resembles the function for the normal cat at 10 hz. When we published our findings (29) we speculated that this breed of cat might have an abnormal proportion of X-cells and Y-cells. This prediction was subsequently confirmed neurophysioloigcally by Chino, Shansky and Hamasaki (30), although they found the major deficits to be in the Y-pathways. This was somewhat surprising in view of the psychophysics, which would have led me to bet on the X-cells.

Q. Dr. Brown: I was trying to think of the behavioral significance of the enhanced sensitivity to low spatial frequencies with high rates of temporal modulation. What are your ideas?

A. I suspect that this reflects the cat's keen sensitivity for movement, which is really a form of temporal modulation. Certainly those of us who have enjoyed cats as pets can attest to their marvelous abilities to detect and capture moving prey, such as a fly. In the laboratory we have measured the cat's sensitivity using moving gratings and the resulting contrast thresholds are on the order of 0.3 log-units lower than with flicker.

Behavioral Analysis of the Role of Geniculocortical System in Form Vision

M. A. Berkley
Department of Psychology
Florida State University
Tallahassee, Florida 32306

J. M. Sprague
Department of Anatomy
University of Pennsylvania
Philadelphia, Pennsylvania 19174

In looking at the title of this session, I am not certain as to how the work I am going to describe fits in. We have been studying the visual system of cats, and, thus it is not clear whether the experiments I will describe are relevant to higher functions of lower animals, or lower functions of higher animals. Basically, the question that we have been asking is a very simple one, and is derived from the recent advances in the anatomy and physiology of the mammalian visual system: which neural structures are important, or perhaps critical for the seeing of shapes. The first experiment I will describe has been done so many times it is difficult to recount all of the people who have performed it, but perhaps it is best to refer to LASHLEY's original demonstrations. In a series of studies, LASHLEY (1, 2) studied the role of visual cortex which he believed to be important in vision. The simple paradigm he used was to estimate its role by testing visual behavior before and after removing visual cortex. As you are all no doubt aware, the animals that he tested were marvelously resistant to revealing deficits after rather extensive lesions. These findings forced LASHLEY to put forward a rather weak hypothesis to account for his results: namely, that vision is mediated by structures that are widely scattered through the brain, perhaps through all of it. While there is nothing basically wrong with this idea, it is not a very satisfying conclusion. Somehow, it seems more reasonable to assume that specific functions are mediated by unique brain parts, and recent discoveries in the physiology and anatomy of the brain tend to support such a view.

To re-evaluate this idea, e.g., specificity of function, we essentially redid the studies performed by LASHLEY using the cat, (instead of the rat) an animal for whom there is a great deal of anatomical and physiological information. The basic experimental paradigm and apparatus is shown in Fig. 1. The cat sits in a small chamber, thrusts its head through a hole into a plexiglas chamber through which it can view various pairs of visual targets that will be used. The outer end of the head chamber has two keys which can be depressed by touching with the nose. The cat's

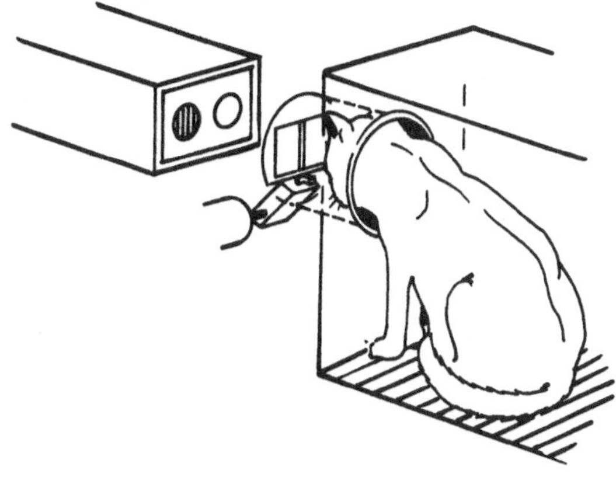

Fig. 1 Schematic representation of two choice simultaneous discrimination testing apparatus employed in testing the visual capacities of cats.

task is simply to depress one or another of the two panels with its nose, indicating its choice (left side or the right side) of one of the stimuli of the stimulus pair. By differentially rewarding the animal for selecting the panel in front of, in this case, the grating, we can communicate with it and instruct it in what it is we want it to tell us about what it sees. To reward the cat, we deliver a little beef baby food for a correct response, while a time out is given for an incorrect response.

I shall return to some of the details of these training procedures later, but for the moment, let me just comment about the general experimental paradigm. Cats are trained on a variety of visual tasks, a portion of their neocortex removed and the animals retested. Specific deficits and residual functions are then noted and correlated with the area and extent of the cortical ablation.

To remind you of what the cat brain looks like and the locus of various cortical areas, I have prepared a summary figure using current anatomical data. Figure 2 shows a pictorial view of the cat brain on which have been superimposed the locations of various cortical areas. At the top is a medial view of the left cerebral hemisphere of the cat; at the bottom, a dorsal view. The figure is a composite of a number of anatomists estimates of the locus of these various areas, and the details of these labels need not concern us. In the present set of experiments, we have confined our experiments primarily to area 17, but in some cases, portions of area 18, e.g., these areas were removed after training cats in a number of visual tasks.

A typical experiment is shown in the next figure (Fig. 3) which shows a reconstruction of a lesion that was placed in one animal (CF10). Comparing the lesion reconstruction with the previous figure reveals that the lesion extends beyond the borders of area 17 and includes most of area 18, and some portions of area 19. Figure 4 shows the pre- and post-operative behavior of cat CF10 on a variety of visual discriminations. Each of the vertical bars on the figure represents a transition from one visual task to another. At the top of each of these zones are indicated the stimuli used in the discrimination task, e.g., a light-dark discrimination; an inverted versus an upright triangle, etc. Prior to surgery, but after acquisition training was completed, the cat was given a two-day test review of all the problems that had been

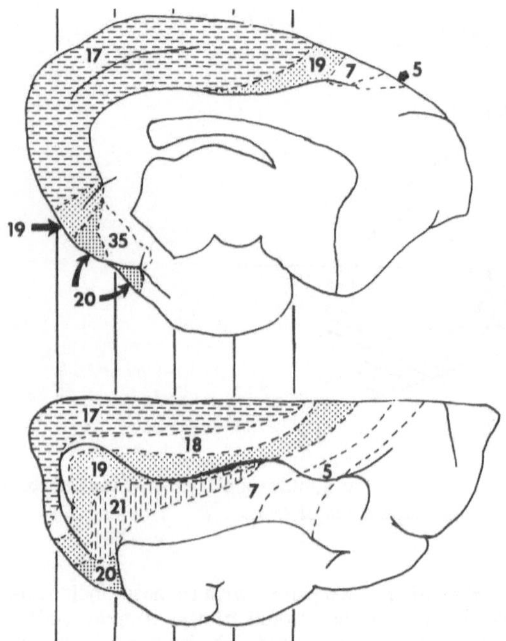

Fig. 2 Diagramatic views of the cat's brain showing the locus of various cortical regions as defined from cyto- and myelo-architectural considerations. Based on ref. (3-7). Top: medial view; lower: dorsal view.

learned to determine whether they were all remembered, etc. Note that in these 2-day tests there is some small decline when compared to terminal acquisition performance levels, but that performance is well above chance. At the end of the review period, the lesion which I showed you in the previous figure was made. After a recovery period, we again used the two-day tests on each preoperatively learned problem. Note that now the cat is performing essentially at chance, except for the light-dark problem which is perfectly retained after the ablation. Assuming that area 17 and 18 are necessary for form vision, these results are exactly what would be expected. However, if training continues, starting with the first preoperative problem, you see that the animal gradually recovers his ability to make contour discriminations. Without a portion of his visual system, it does take quite some time, about as long as it took him to originally acquire the discrimination, but having relearned the initial contour discrimination, he shows transfer of the shapes when brightness cues are changed indicating that shape of the stimuli is the cue. Also, it is not simply a case of recovered memory for these problems, because when we test him on a different problem learned preoperatively but not retrained postoperatively, horizontal vs. vertical stripes, he has to relearn the problem as if it had never been seen before. The conclusion that one is inevitably lead to and which was presented by LASHLEY, is that area 17 is not necessary to mediate many (all?) form discriminations. Given the recent electrophysiological studies showing the strong contour selectivity of neurons in these areas, it is not a very satisfactory conclusion.

The experiment described above has been performed many times. The argument

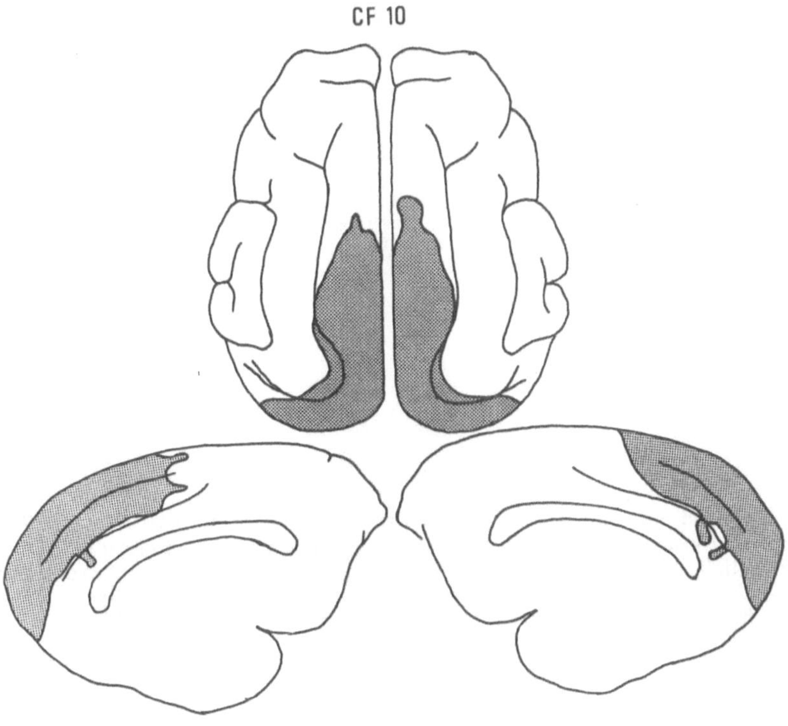

Fig. 3 Reconstruction of cortical lesion placed in cat CF10 showing extent of cortical ablation.

that one can make after examining this kind of data is that there is nothing in it that indicates how the animal is solving these problems. It is quite possible, indeed, it is highly likely, that the way it is solved postoperatively is not the way it was solved preoperatively. Yet the end result, e.g., ability to discriminate the figures, of course, is the same. Several behavioral studies of the cat have suggested that cats may use a variety of cues other than overall shape in form discriminations (Winans (8)).

Let us consider a naive view of the visual system in which it is a simple serial processor of some kind, taking elements, features, whatever you want to call them, and stringing them together somehow to form a visual precept. A diagram of such a scheme is shown in Fig. 5. Clearly any intervention anywhere in such a system would produce total and permanent deficits. It is easy to see, however, that even simple variations open up many possibilities for interpreting our lesion studies. Several variations are depicted in Fig. 6. As can be easily appreciated, interfering with one of several parallel systems still leaves the others intact. If the visual system were organized along these lines, it would not be surprising that one can recover a function

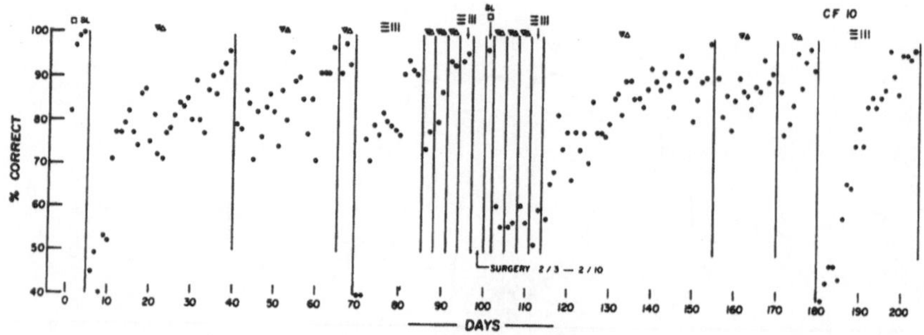

Fig. 4 Pre- and post-operative acquisition and retention performance for cat CF10 on a variety of visual discriminations. Testing days shown on the abscissa; percentage of correct choices shown on ordinate. Vertical lines indicate change in discrimination task. Small figures along the top shows the stimulus pair used for that testing segment. Each dot represents one test day and 200 trials.

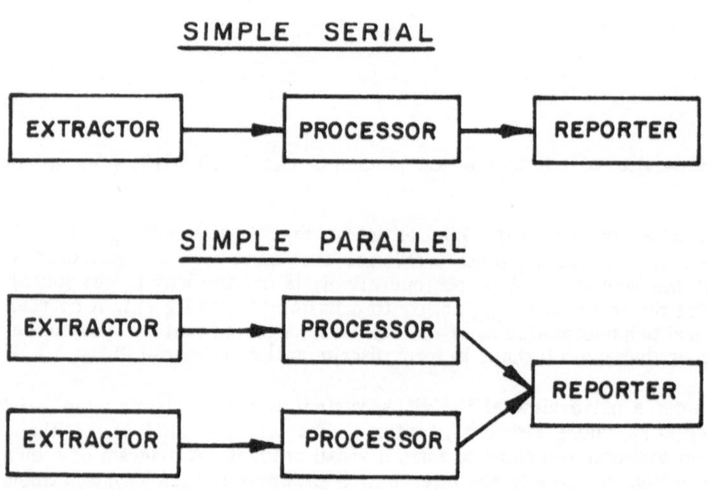

Fig. 5 Block diagram of a simple serial (top) and simple parallel processing model systems.

after an ablation of one portion of the system. By permitting some sharing of features as shown in Fig. 6 you can make these schemes as complicated as you like. The results of the ablation-behavior studies in the cat suggest the operation of such mixed systems. Is there evidence that the cat visual system is organized in such a

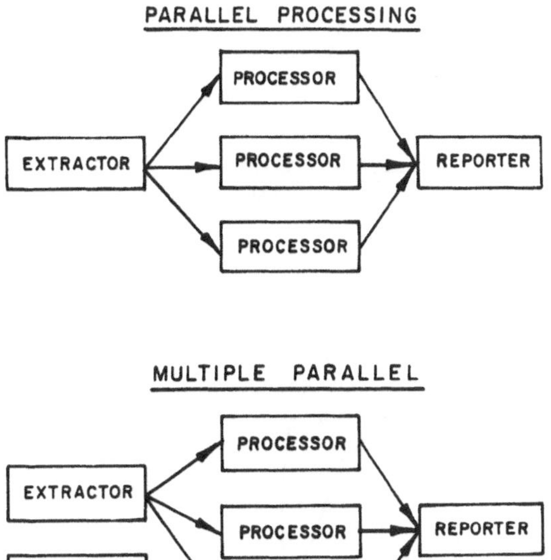

Fig. 6 Block diagram of model processing systems have a combination of serial and parallel pathways.

complex fashion? The next figure (Fig. 7) is adopted from SPRAGUE, et. al. (7) and is a summary of the current knowledge about the anatomical pathways in the cat from retina to cerebral cortex. This figure shows there are many parallel pathways to cortex. Among the many cortical recipient areas, we have been concerned with the one receiving input from lateral geniculate nucleus (LGN) called A17 which receives not only the classic input from the LGN, but from other thalamic regions as well. As can be seen in Fig. 7, the area 17 projection from the LGN actually is only a small part of the visual thalamic projection to cortex. In this context then, it is not too surprising that removal of one small region does not produce very drastic deficits in vision.

How can one deal with a system like this, one that has multiple pathways which share peripheral information? One must begin with the assumption that while each pathway has the common feature of having a neocortical termination zone, they are not equivalent, they cannot be interchanged: that is, they do not perform the same function. How can we determine what each of these pathways and cortical areas are doing? One could guess at various functions and devise tests to evaluate these functions. We chose, however, to proceed in a somewhat different way. We decided

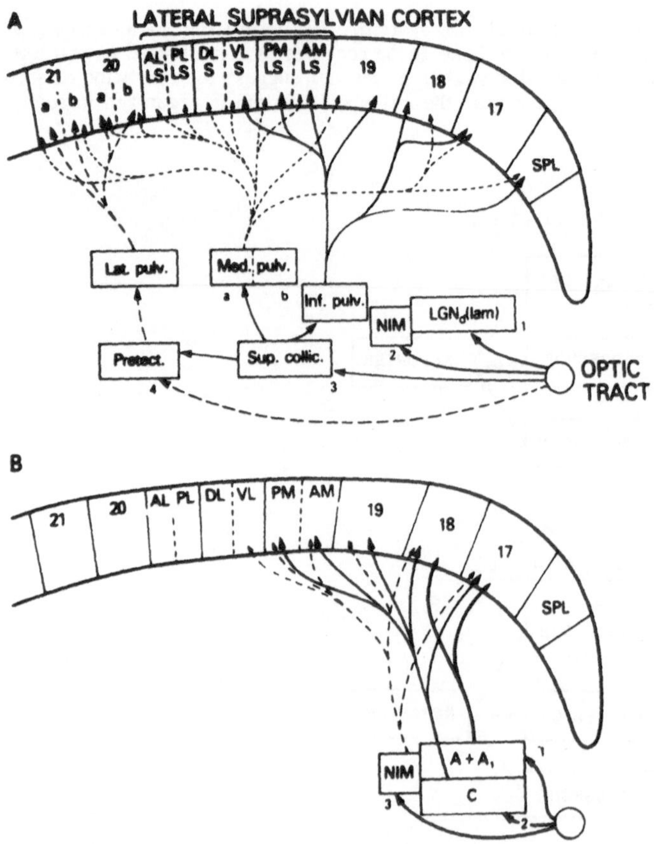

Fig. 7 Summary diagram of anatomical pathways in the cat visual system. A. whole visual system; B. pathways from laminar portion of lateral genciulate nucleus to cortex. Adapted from SPRAGUE, et. al. (7) (with permission).

to limit our studies to areas 17 and 18 and selected the dimensions of vision to test on the basis of the neurophysiological investigations of these areas. Thus, such dimensions as movement, size, orientation and something we have called topographic alignment, all shown to be important aspects of neuron function in areas 17 and 18, were selected for study. Having selected the structure and dimensions, the next requirement was to be certain that the animals were actually attending to the particular dimensions when they were presented. The only way that this can be achieved is to have the animal report whether the dimension presented is the one to which he was attending. This was accomplished by varying the dimension and noting if these changes controlled the animal's behavior. If it does, then we could say with a moderate amount of certainty that that was the dimension to which he was attending.

I would like to stop here for a moment and deal with a number of "housekeeping" difficulties; housekeeping in the sense that there are a number of problems in training cats that interfere with the performance of a simple ablation-behavior experiment. Since we cannot verbally communicate with animal models and tell them which stimulus dimension we want them to ignore, we have to develop a very crude language with which to communicate with these animals. The language consists of essentially two words, Yes and No. To get him to use the language, we make him hungry and pay him for making correct responses with food or to "tell" the animals that what it did was correct. With such a crude language there is a considerable area for misunderstanding between you and the animal, and the animal often misinterprets what it is you want him to tell you. He adopts stratagems that are maladaptive. They often have nothing to do with your tests, and interfere with your testing so that you have to develop counterstrategies to direct the animal, in a simple way, to the task that you want him to perform. This requires careful attention to the animal's behavior and adjustments in training procedures. For example, one of the first things that we noticed in training cats on visual discriminations was that when we first started teaching the animal a problem (of course, we used "easy" problems to get him going) relatively little effort and attention is required on the part of the animal. A quick look at the stimuli, a little jab at the key and a reward is delivered. As the stimuli become more complex, more difficult for him to discriminate, such a stratagem is maladaptive. What is required, as the stimuli become more difficult to discriminate, is that the animal look at them for longer periods of time, to examine them more carefully. How do you teach a cat to do that? First, we wanted to determine what viewing strategy was actually adopted with "easy" and "difficult" problems. For example, we hoped that as the discrimination problems became more difficult, the cats would spend more time looking at the targets. The results of measurements of viewing time during "easy" and "difficult" tasks are shown in Fig. 8. The discrimination task used to make these measurements required that the cats discriminate a moving target from a stationary one. That is, two stimuli are presented, one of which is stationary and one of which is moving. The animal's task is simply to select the moving one. The cat can look at the target as long as it would like, up to two minutes. Figure 8 shows a response latency histogram for 200 responses. The latencies were determined by measuring the amount of time which elapsed between when the time the stimulus was presented and the responses. They are plotted as a histogram with each bin being 200 ms. The latencies observed for two cats on two movement detection problems are shown. The filled columns show latencies for the difficult task (slow movement) while the open columns show the latencies observed while working on an easy problem (fast movement). The mean response latency for one animal is about 800 ms while for the other it is about 600 ms. Human observers tend to look longer at slowly moving targets than they do at faster moving targets. Notice, however, that the animals have not significantly changed their viewing strategies when the stimulus movement was slowed. In one animal, we changed the contingencies so that a mild electric shock was delivered to its feet when an error was made, in addition to delivering a food reward for making a correct response. As you can see from the bottom of Fig. 8, there is a dramatic change in the response latency histogram suggesting a change in observing behavior. The animal is looking at the target longer to make sure he has chosen the right one. These measurements demonstrate that not only must the relevant cue by specified for the animal but a strategy for viewing the targets must be specified. To counter the short viewing time response pattern, we built into our training program a required viewing time so that regardless of the type of target, the animal has to view the target for a minimum time. This procedure has improved the performance of all the animals we have tried it on and is now a standard part of our testing program.

Another problem that came up concerned the position of the stimuli relative to the animals. For example, how far away should the target be placed? This led to another series of experiments to estimate the near point of accommodation of the cat. The results of one of those experiments is shown in Fig. 9. In this case, we made visual acuity measurements as a function of viewing distance. From these data near points of accommodation were derived as shown in the figure. This information allowed us to place the visual targets used in the tests to be subsequently described at

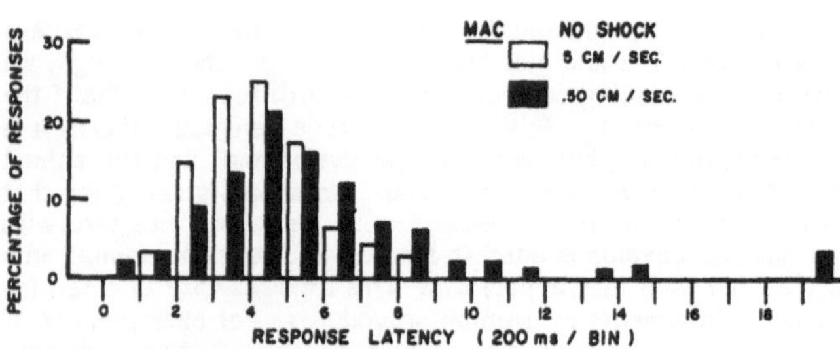

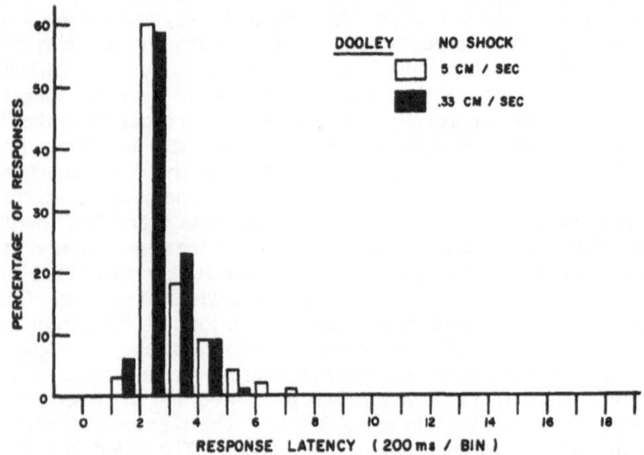

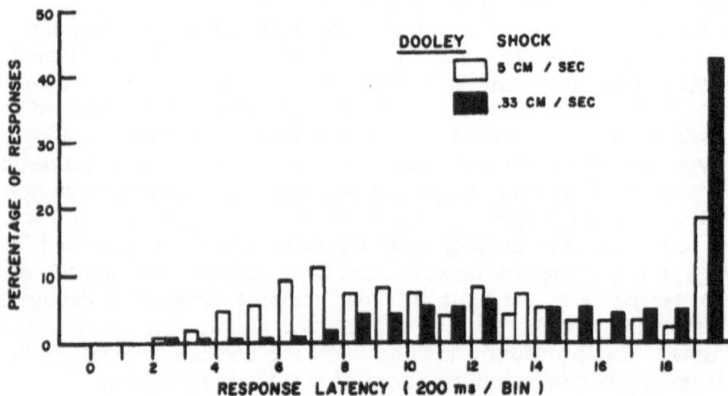

Fig. 8 Response latency histograms for 2 cats (Mac and Dooley) showing the distribution of response latencies, measured from stimulus outset to response, in the simultaneous two-choice paradigm. Discrimination task consisted of a moving and stationary stimulus. Open bars: fast movement; filled bars: slow movement. Lower histogram shows response latency histogram when mild electric shock was delivered to the feet of cat (Dooley) for making errors.

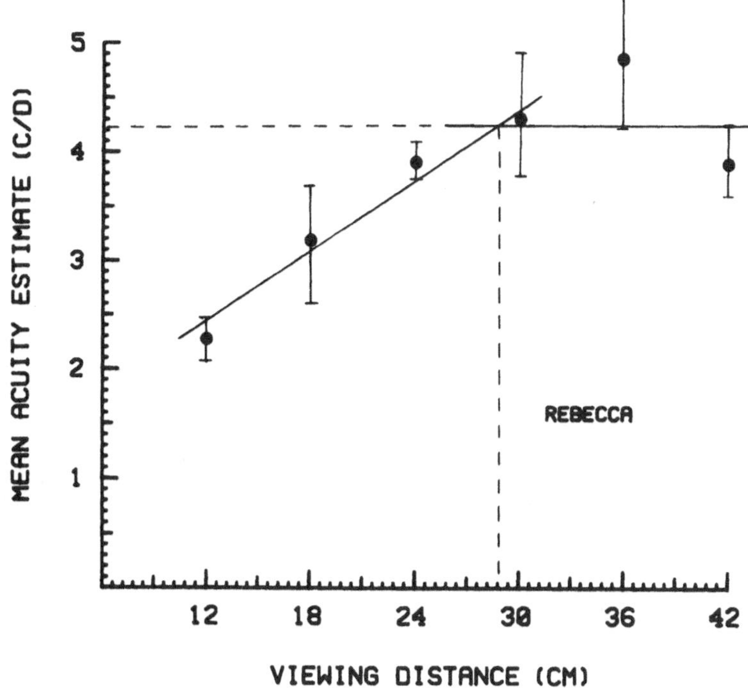

Fig. 9 Plot of grating acuity as a function of viewing distance for one cat. Solid lines are least square fitted functions for a two-component function. The horizontal dotted line intercept indicates maximum grating acuity. The near point of accommodation was taken as the intercept of the two solid line functions (shown by vertical dotted line). From BLOOM and BERKLEY (9) (with permission).

a distance that is appropriate for the cat.

Having established a testing procedure and stimulus presentation arrangement, we proceeded to study the role of area 17-18 in mediating a variety of visual capacities. The first dimension that we studied was movement. Figure 10 shows a reconstruction of a lesion that we made in this particular animal (Agnes). If you recall the location of area 17 and 18 from Fig. 2, you see that the lesion is extensive but incomplete, e.g., portions of 17 and 18 remain. Figure 11 shows the effects of the lesion on movement discrimination. This animal was trained preoperatively on a movement discrimination task (open circles) and you can see that at fast rates of movement, performance is very good, falling off gradually as the movement rate declines. Postoperatively, (filled circles) there is some mild performance deficit, but the threshold value achieved is approximately the same as that achieved preoperatively, it is important to note that one advantage of a psychophysical

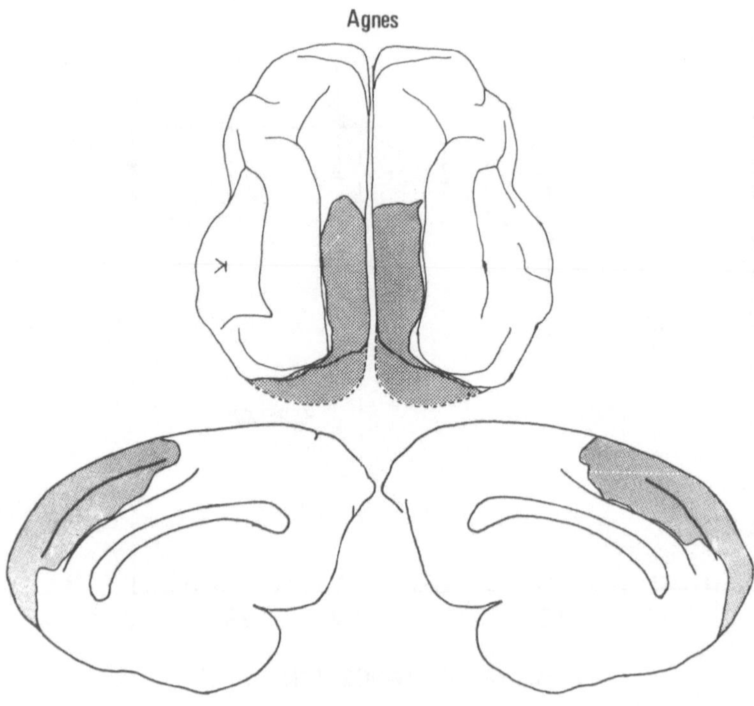

Fig. 10 Reconstruction of cortical lesion placed in cat Agnes showing extent of ablation.

paradigm is that it has built into it a performance evaluation. In other words, if at suprathreshold stimulus values (easily discriminable values of the stimulus) the animal does not show any decrement in performance, then we know that there is no motor impairment, that he can function in the test apparatus, that the rules for working in the apparatus are remembered. Thus, it has in it a self-checking procedure. A second important feature of this type of test paradigm is the demonstration that changing stimulus values changes the animal's behavior, thus demonstrating that the dimension was the one the animal was attending to.

The next dimension of spatial vision that we examined was grating acuity. Figure 12 depicts the extent of the lesion which is somewhat smaller in this animal, but is confined almost exclusively to area 17, with only some minor infringement on area 18. The consequences of the lesion on the ability to discriminate gratings are shown in Fig. 13. In this case, the filled circles are the preoperative performance and the open circles postoperative. Notice that at large grating sizes, the animal is essentially unimpaired and he exhibits only a mild deficit in his grating acuity. A somewhat more dramatic example is seen in another animal. Figure 14 shows the locus and extent of the ablation while Fig. 15 shows pre- and postoperative behavior

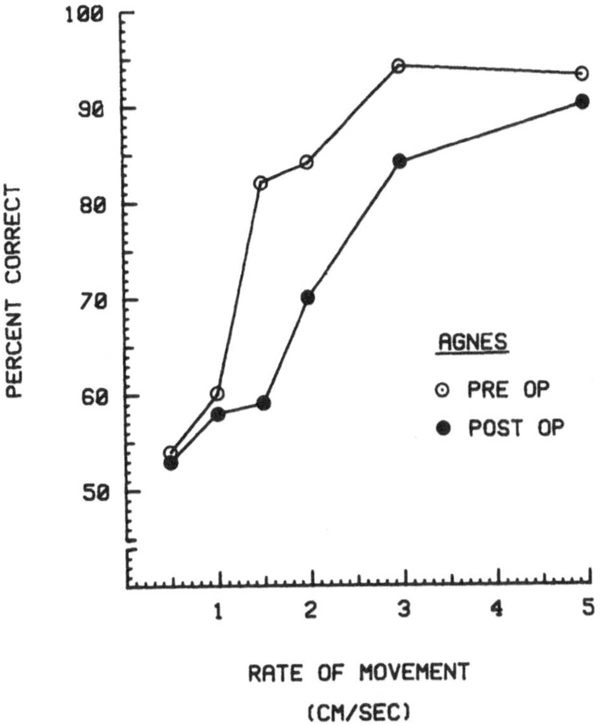

Fig. 11 Performance of two-choice movement discrimination task pre- and post-operatively by cat Agnes. Percentage correct choices shown on ordinate; rate of stimulus movement shown on abscissa.

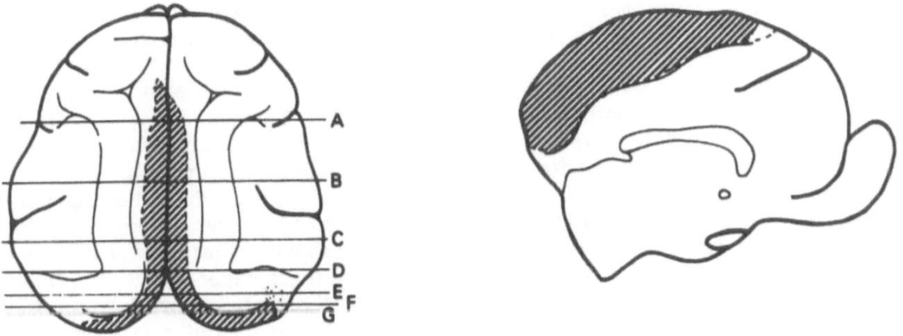

Fig. 12 Reconstruction of lesion placed in cat Zelda showing extent of ablation.

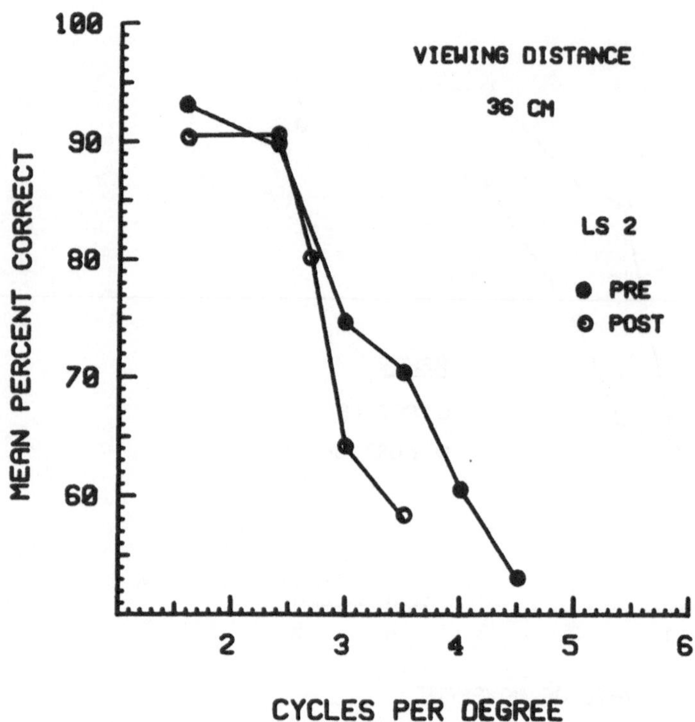

Fig. 13 Pre- and post-operative performance on a two-choice grating acuity task by cat Zelda (LS2). Stimulus pair consisted of a grating and patch of uniform light equal in luminance to the space-average luminance of the grating. Percentage of correct choices shown on ordinate; spatial frequency of grating stimulus shown on abscissa.

on the acuity task. Again, the filled circles represent preoperative behavior, the open circles postoperative. A somewhat larger deficit is observed in this animal. However, the animal is still able to discriminate gratings of about four cycles per degree postoperatively. Not bad for a cat. These findings are surprising because it has been assumed that area 17 is critical for detail vision, and while there is some significant impairment, the animal clearly can see relatively fine gratings. By the way, the animals never recover their preoperative thresholds, regardless of how long you retrain, so the loss is not a temporary one. In doing these experiments, of course, one has to assure oneself that the lesion is specific to the particular task being used and not simply producing a general deficit. Figure 16 shows a similar experiment in which the lesion was not in area 17 and area 18, but lateral to those regions and, as you can see, postoperative performance does not show any impairment.

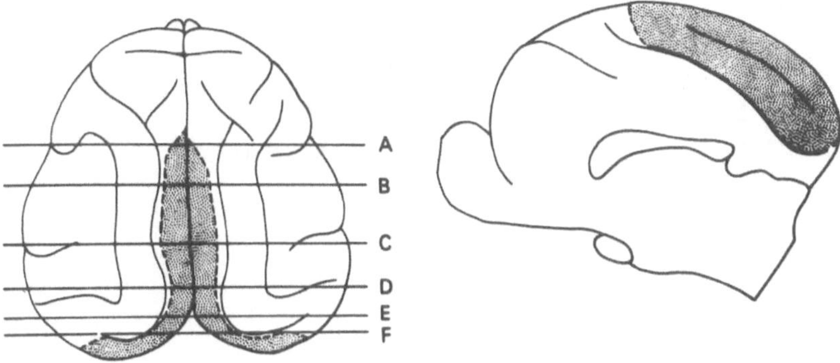

Fig. 14 Reconstruction of cortical lesion placed in cat Scarlett showing extent of ablation.

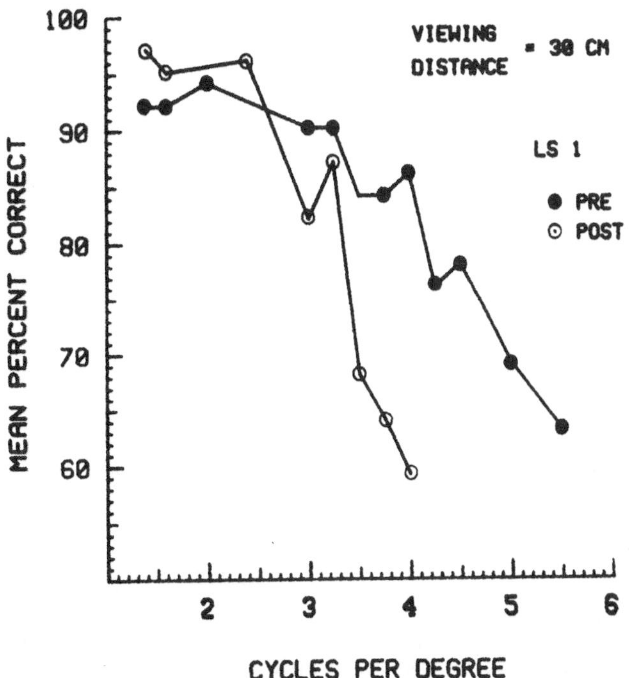

Fig. 15 Pre- and post-operative performance of cat Scarlett (LS1) on a grating acuity task. Other details the same as in Fig. 13.

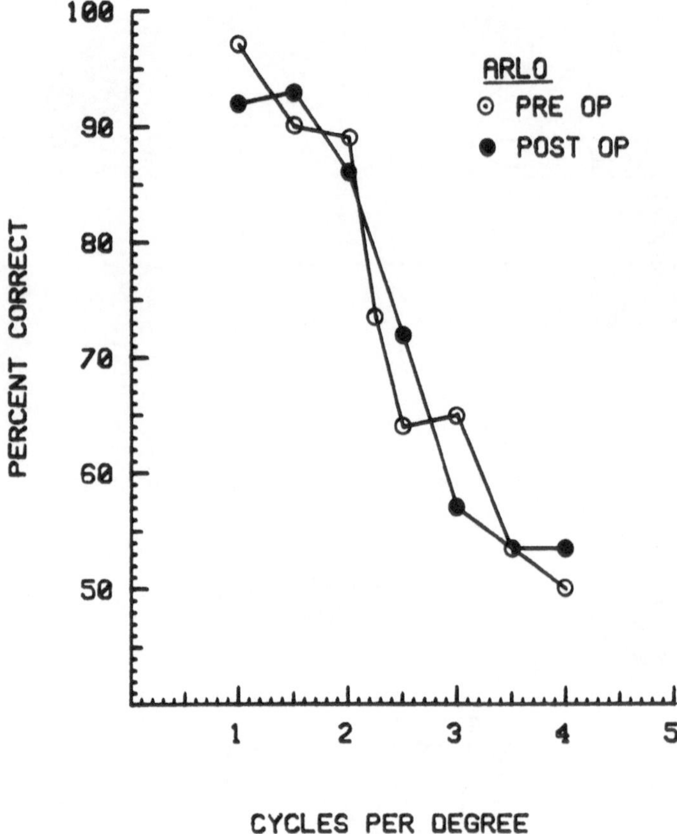

Fig. 16 Pre- and post-operative performance of cat Arlo on grating acuity task. Ablation in this cat was made in cortical region lateral to areas ablated in cats Zelda and Scarlett. All other details the same as in previous figure.

The next dimension that we examined was orientation acuity. The stimuli we used in this task are shown at the bottom of the next figure (Fig. 17). One stimulus was a pair of parallel lines and the other was a pair of non-parallel lines. Their position could be interchanged from trial to trial, as before, etc. This stimulus configuration has a particular advantage in measuring orientation acuity because you can teach the animal to choose the parallel line pair (the concept of parallelness) and then present these stimuli in any orientation. Thus, orientation acuity for different orientations of the stimulus can be measured. The upper portion of Fig. 17 shows another method we have used to teach the animals a stratagem which is adaptive for difficult problems. In this case, the initial discrimination problem is a very large difference in the orientation of the two non-parallel lines. Each time the cat makes a correct response, the angular difference is reduced while each time an incorrect

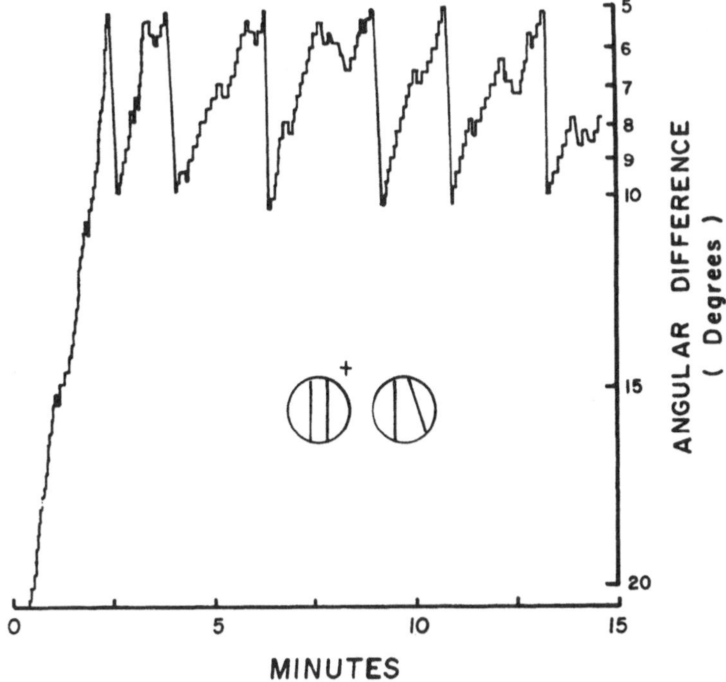

Fig. 17 Recording of discrimination performance for stimulus pair shown at bottom of figure. Pen moved upward with each correct choice and downward with each incorrect choice. Difference in orientation between the non-parallel line pair was reduced after correct responses and increased after incorrect responses. Stimulus was reset to 10° difference when 5° was reached. Orientation difference between the non-parallel line pair shown on the ordinate; time is represented along the abscissa.

response is made, the difference is increased. The upper portion of Fig. 17 records this process; an upward deflection indicating a decrease in the orientation difference while downward deflections indicate an increase. As can be seen, the cat rapidly runs the non-parallel line pair to about a 5° difference without any great difficulty, at which point we reset it and let him do it again. This tracking and resetting procedure is a means of rapidly giving the animal experience with problems that it can easily solve as well as with more difficult ones. In the next figure (Fig. 18) the effects of an area 17-18 ablation on orientation acuity can be seen. As can be seen animals have a threshold of about five degrees, varying somewhat between animals, with an occasional cat with a threshold as low as two or three degrees. Postoperatively, the animal which has the area 17 lesion, shows a fairly substantial deficit in its ability to discriminate orientation. Nevertheless, it does reasonably well. However, if this animal were tested in a classical form discrimination, for example, upright versus inverted triangle or circles versus plus, it would not show impairment in performance.

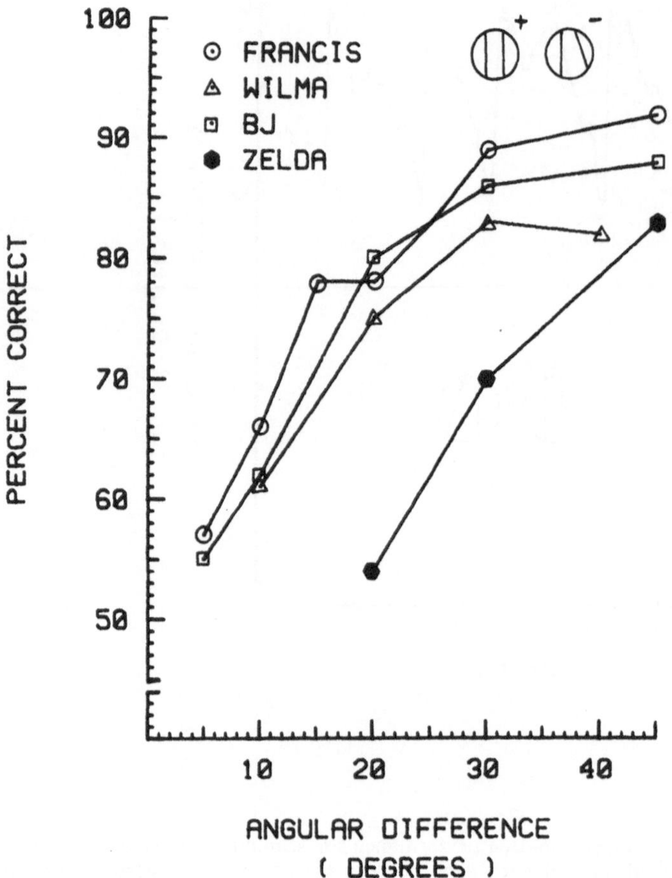

Fig. 18 Performance on line orientation discrimination task for 3 normal cats (open symbols) and one brain-damaged cat (filled symbols). Percentage correct choices shown on ordinate; ordination differences between the non-parallel line pair shown on abscissa. One value of the stimuli to be discriminated is shown at top right of figure.

Such an experiment was shown in Figs. 3 and 4.

The last task which we selected for evaluating the role of area 17 in vision is a task you will recognize as a vernier offset acuity problem and is shown in the next figure (Fig. 19). The targets were generated on an oscilloscope so that the lines were luminous, not an ideal target for this kind of discrimination, but one easy to control. As you may recall, I mentioned earlier that it is often difficult to determine which cue the animal is attending to, and this was particularly true with this problem. When we first tried to teach cats this discrimination, we discovered that it was extremely difficult to teach them to respond to the line offset. We managed to get one cat that

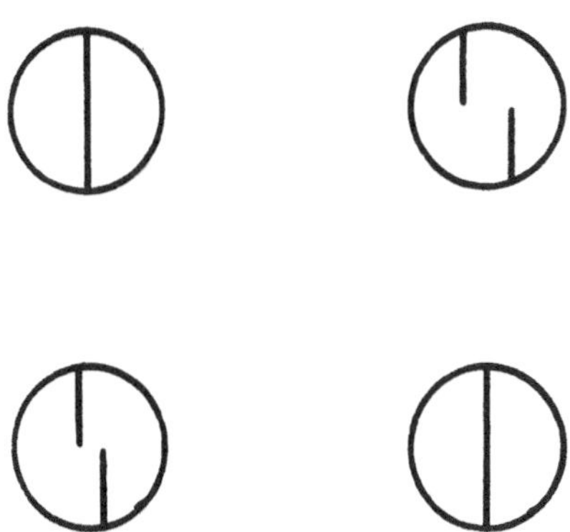

Fig. 19 Two examples of vernier offset stimuli used in vernier acuity task. Magnitude of line offset varied from day to day while left-right position of offset varied from trial to trial within days.

was just showing us a very weak discrimination, being able to discriminate only between the largest offsets. We had an idea that maybe it was not attending to the offset but only to the position of the lines in the viewing aperture. We spent a considerable amount of time trying to devise ways to get him to look at the offset. Finally, we found a simple trick that worked. We placed a small annulus around the offset. The animal then quickly acquired the discrimination at which time the larger annulus was reintroduced.

Figure 20 shows the lesion placed in animal Streak. It has some features worth noting: 1) it is very small, being the smallest of the lesions thus far described; 2) there is a very small portion of area 17 representing the far periphery left. The next figure (Fig. 21) shows the performance of two cats; pre- and postoperatively on the vernier offset task. The animal whose brain is shown in the previous figure (Streak) is represented by open circles. This animal does very well preoperatively having a threshold of near four minutes of arc. Another animal, Earl, is also shown. Postoperatively we could not find any offset large enough that Streak could discriminate. The largest offset that we could produce was about 7 degrees, and yet he could not perform on his task. The other animal, Earl, had a lateral lesion. Postoperatively his behavior is essentially normal. By the way, Streak never recovered after quite some period of time and many attempts at trying to have him relearn his problem were unsuccessful.

What can one conclude from these experiments? Well, there must be an easier way to make a living. But one conclusion seems unescapable and that is that different parts of the visual system, in fact, do do different things. These different functions can be revealed by discovering the proper questions to ask. We have seen how most contour tasks survive A17 ablation. That being the case, what does the

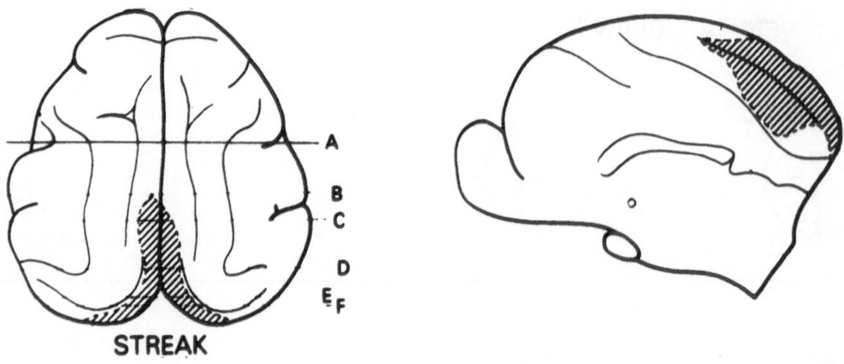

Fig. 20 Reconstruction of cortical lesion placed in cat Streak showing extent of ablation.

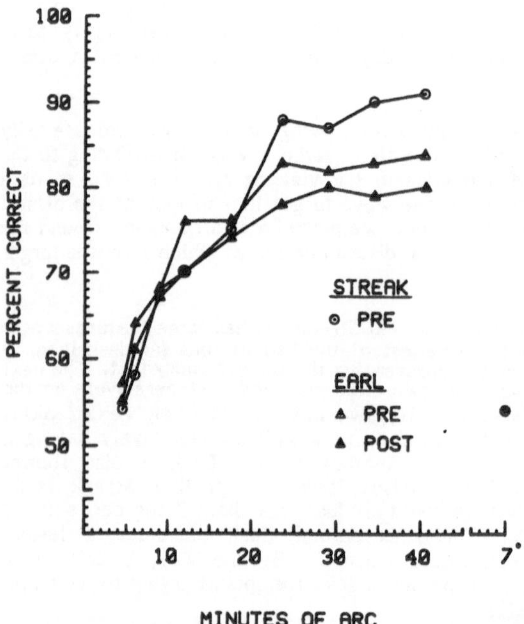

Fig. 21 Pre- and post-operative performance on vernier acuity task for 2 cats, Streak and Earl. Open symbols-preoperative performance level; filled symbols-postoperative performance level. Lesion placed in cat Earl was lateral to area ablated in cat Streak. Percentage of correct choices shown on ordinate; magnitude of line offset in minutes or degrees of visual angle shown on abscissa.

dramatic loss of vernier offset acuity mean? Since there is some evidence that stereoacuity and vernier offset acuity are related, it may not be unreasonable to suggest that the striate cortex may be intimately involved in functions unique to binocular vision, e.g., stereopsis, accurate eye alignment, etc. This is not a very startling conclusion but the present data imply that striate cortex may have relatively little to do with the discrimination of the shapes of things.

References

1. K. Lashley, J. comp. Neurol. 53, 419 (1931).
2. K. Lashley and M. Frank, Comp. Psychol. 17, 355 (1934).
3. E. Jones, The Neurosciences: Third Study Program, ed. by F. Schmitt and F. Worden (MIT Press, Cambridge, 1974).
4. C. Heath and E. Jones, Ergebu. Ant. Entwicklungsgesch 45, 4 (1971).
5. R. Otsuka and R. Hassler, Arch. Psychiat. Z. ges. Neurol. 203, 212 (1961).
6. F. Sanides and J. Hoffman, J. fur Hirnforschung 11, 79 (1969).
7. J. Sprague, J. Levy, A. DiBernadino and G. Berlucchi, J. Comp. Neurol. 172, 441 (1977).
8. S. S. Winans, J. Comp. Phyisol. Psychol. 74, 167 (1971).
9. E. Bloom and M. Berkley, Vision Res. 17, 723 (1977).

The Absolute Efficiency of Human Pattern Detection

H. B. Barlow
The Physiological Laboratory, University of Cambridge
Cambridge CB2 3EG, England

I have a rather complicated story to tell, and, what is worse, it contains statistical arguments that tend to make some people bored and others argumentative. So what I shall do is first describe an experiment, then the result, and then tell you why I did it.

Measuring the Efficiency of Perception

Figure 1 shows examples of targets that a human subject looked at and had to make a decision about. The subject was usually the author or one of his colleagues in this work, AART VAN MEETEREN or BARNEY REEVES. The targets were generated by computer, and the mean and standard deviations of the populations from which these samples were drawn are given below. For a given run of 100 trials the targets were selected at random from two populations, for instance one with, on average, no additional dots on the left side, or one with, on average, 10 additional dots. The subject's task was to decide which population it had been drawn from. The subject's ability to discriminate was expressed as d' in the usual way.

Figure 2 shows the result. As ΔN, the number of extra dots, is increased, the subject's d' rises, in this case rather linearly. Now d' can be regarded as an estimate of the signal/noise ratio of the signals, whatever they are, that enable the subject to discriminate the number of dots. It is an internal signal/noise ratio. But in this situation there is another signal/noise ratio we can talk about, namely the signal/noise ratio of the target. If we had an ideal discriminator and asked it to make the same decision about the targets that we ask the subject to make, the d' of its results would equal the signal/noise ratio of the target. We may call this d'_I to differentiate it from d'_E, the values of d' turned in experimentally by the subject.

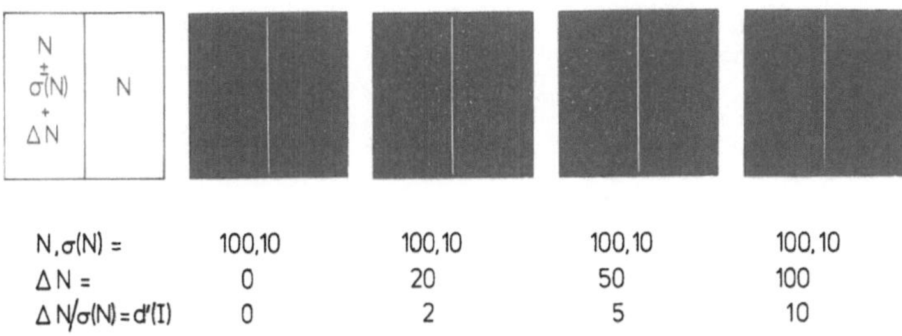

Fig. 1 The four black squares are examples of the computer-generated random dot patterns presented to the subject, and the left hand square shows the plan on which they were constructed. The right half of each square always has exactly N (=100 in this case) randomly positioned dots. The left half has a number of dots randomly selected from a population with mean N and standard deviation σ (N), to which is added exactly ΔN. Below each example the parameters N,σ (N) and ΔN are given for the population from which it was drawn. The subject was presented examples from two populations and his task was to decide from which population it was drawn. Usually one population had $\Delta N = 0$, and the value of d'_I for an ideal discriminator making the discrimination between this population and another is given on the third line below each target.

Now d'_I rises linearly with Δ N in Fig. 2, just as does d'_E, but it rises rather more steeply. Obviously the human subject cannot do better than the ideal discriminator, but it seems he always does worse, and by a roughly constant factor at all ΔN values. This degradation of performance is conveniently expressed as a loss of efficiency: the subject is not using all the information provided to him, and it is "as if" he was only paying attention to a certain fraction F of the dots. Now this fraction F is simply the square of the ratio of d'_E to d'_I:

$$F = (d'_E/d'_I)^2 \tag{1}$$

A more complete account of the theory is given elsewhere (1). The only reasons for using the square are, first that it makes F as here defined exactly equivalent to quantum efficiency (2, 3), and second that it then has the very simple physical interpretation of "the fraction of dots utilized". In Fig. 2 the result is that F = 0.5.

Separating Target from Backgrounds

Now let me explain why I did the experiment. We compare the human performance with ideal performance, and for this task and a wide variety of other similar tasks we know precisely what is necessary to achieve ideal performance. If you are to detect whether there is an increase of dot density in a certain region you must be able to delimit that region precisely and accurately from other regions which contain dots, but which are not part of the target. If you do not have a template to enable you to separate target and background you will be inefficient. Template is perhaps too restrictive a word; you need weighting functions that enable you to apply appropriate weights to dots at various positions, and you need to position these weighting functions appropriately in the visual field.

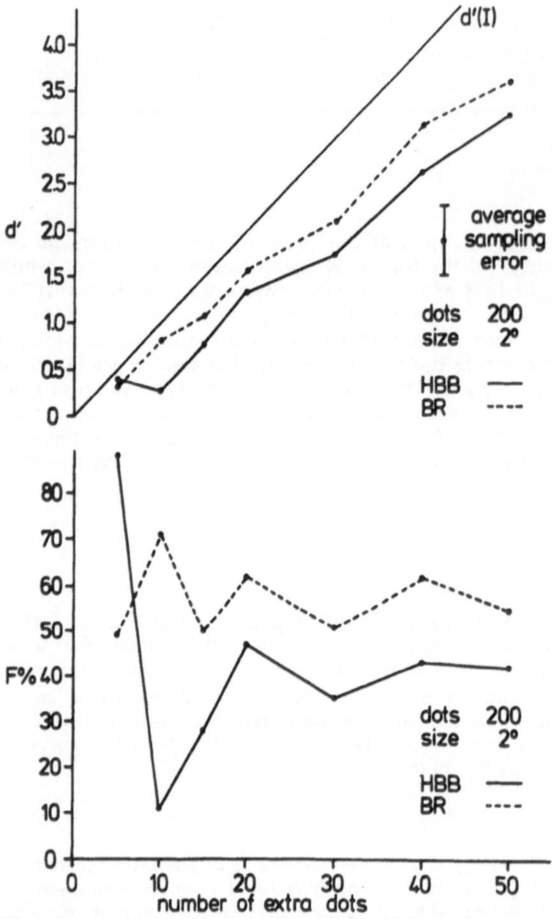

Fig. 2 The top half shows the values of d'_E achieved by two subjects discriminating between populations with $\Delta N = 0$ and ΔN as shown on abscissa (see Fig. 1 for meaning of ΔN). The continuous line shows the value of d'_I for an ideal discriminator. In the lower half the efficiency F is plotted for each point, where $F = (d'_E/d'_I)^2$. Subject BR looked at the targets as long as he wished. For HBB, exposure duration was 200 msec. For both subjects, values of F are unreliable when the number of extra dots is small, but they steady down to a value of 40 to 50% when it is large.

One view of the action of receptive fields holds that they are like weighting functions, and the response of a nerve cell is dependent on the weighted sum of the contributions from the different parts of the receptive field. So a receptive field would be an ideal detector of a perturbation of dot density following a pattern matching the distribution of sensitivity within its receptive field.

Now if we have a technique for measuring efficiency, we should be able to apply this argument in reverse. If we find that human subjects are efficient in absolute terms at detecting certain targets, then that implies that they can delimit or separate out those targets, and they must have weighting functions matching the targets to be detected in the tasks they are good at. So, maybe, we have a superb tool for doing some non-invasive human cerebral neurophysiology.

I must tell you that the first results of applying this tool were very disappointing. I don't regard 50% efficiency, the highest we obtained, as disappointing; I think it is rather good, for if a snap judgement is half as good as careful counting and calculating, it means that it is as good to repeat the observation just once as it is to call in a statistician. Many people would eagerly throw away their statistics book if they only had to repeat the experiment once to get results as trustworthy as those obtained by consulting it! But the trouble from our point of view is that the tasks at which efficiency is high are trivially simple. Also efficiency does not vary enough, and not in the ways that we think it should from the neurophysiology we know.

I shall skip over the effects of varying stimulus duration and target size, for this showed nothing unexpected and simply confirmed the figure of 50% efficiency for a wide range of favourable conditions. We did find higher figures when the average number of dots was reduced, but only when it dropped to the ½ dozen or so that could be counted in a brief exposure. Counting is more accurate than assessing number, as we knew already.

To find out about templates of course the interesting condition to vary is the shape of the target, because our cortical templates are long thin rectangles. In Fig. 3 you see some targets ranging from a square to a long thin rectangle, and Fig. 4 shows the result - a genuinely disappointing one, for squares are actually detected more efficiently than rectangles. CHAMBERS and COURTNEY-PRATT (4) found this in a similar experiment some years ago.

Sinusoidal Templates

Now when these experiments had been done we felt certain we knew what was wrong: we were thinking in the spatial domain instead of the Fourier domain. So, with AART VAN MEETEREN, I started looking at the ability to detect sinusoidal spatial modulations of dot density. The argument is I think sound: if the visual system is performing 2D Fourier transforms, that is the same as saying its templates, or spatial weighting functions, are sets of sinusoidal gratings, and if we match the stimulus to them we should achieve high efficiencies. In particular, if the templates extend over many cycles, then one would only get high efficiencies if one's targets also repeated over many cycles.

Figure 5 shows the result of determining efficiency as a function of spatial frequency. For technical reasons we had to vary the angular subtense of the screen on which the gratings appeared, and we used three window sizes. As you see, efficiency declined with frequency. However it appears to be the number of cycles that is the determining factor, not the spatial frequency. Figure 6 shows the same data replotted to show this. Here efficiency is plotted as a function of number of

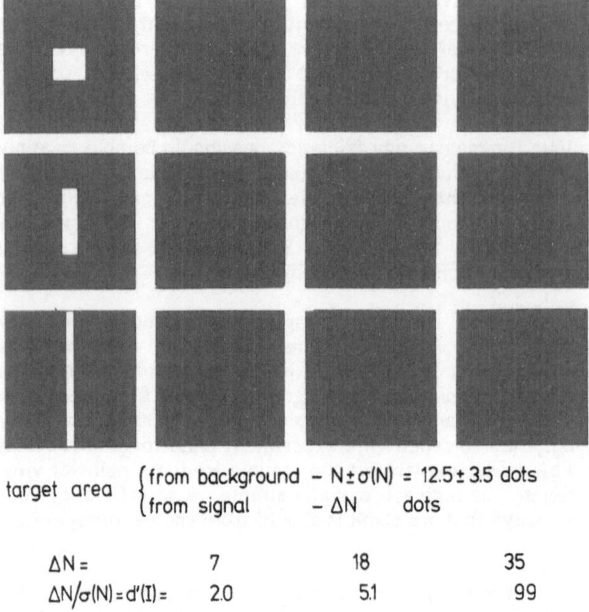

Fig. 3 The effect of shape on efficiency of detecting increments of average dot density. The targets, all of equal area, are shown at left. There were on average 12.5 dots from the background in these areas, with S.D. $\pm$ 3.5. The examples at right show 7, 18, and 35 extra dots in the target areas, giving d'_I values of 2, 5.1, and 9.9. The full height and width of each square is 1°. Experimental values of d'_E are shown in Fig. 4.

cycles. The data from the large (4°) and medium (0.8^o) window coincide. For the smallest window (0.28^o) there is a marked deviation for the highest number of cycles, but 10 cycles in this window corresponds to 35 cycles/deg, and we were almost certainly running into the resolution limit.

There is certainly no improvement with increasing number of cycles, and this was a second major disappointment, for it knocked the Fourier explanation of our disappointment in the first series on the head. Is our superb tool for human cerebral neurophysiology not so sharp after all, or were we using it on the wrong type of task?

Texture, Symmetry and Oriented Pairs

Now our expectations so far have rested quite heavily on the notion that the function of a neurone with a particular receptive field can be satisfactorily represented by supposing that the receptive field is like a spatial weighting function and the cell sums the contributions linearly. The notion is shared by the Fourier view and the view of a receptive field as a weighting function, but perhaps it is not right. I now want to describe some experiments that use the same tool - efficiency measurements, but these aim to explore a different aspect of what the visual system does with the image it receives. I am sure we again have an over-simple viewpoint, but the results

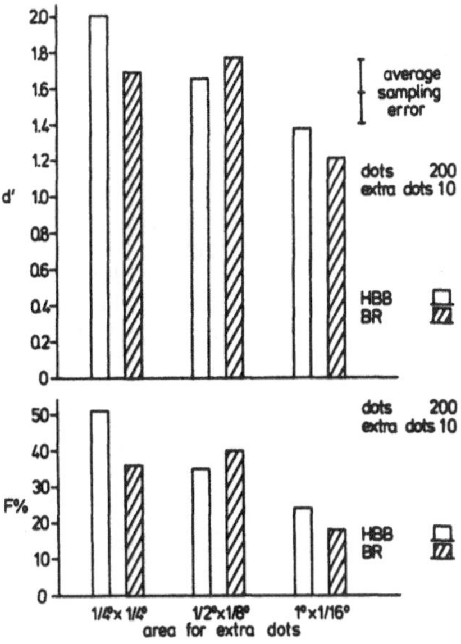

Fig. 4 Values of d'_E obtained experimentally by two subjects for the targets shown in Fig. 3. Δ N values were 0 and 10 for the two populations to be discriminated, giving d'_I = 2.83. Long thin rectangles are not detected more efficiently than squares of equal area.

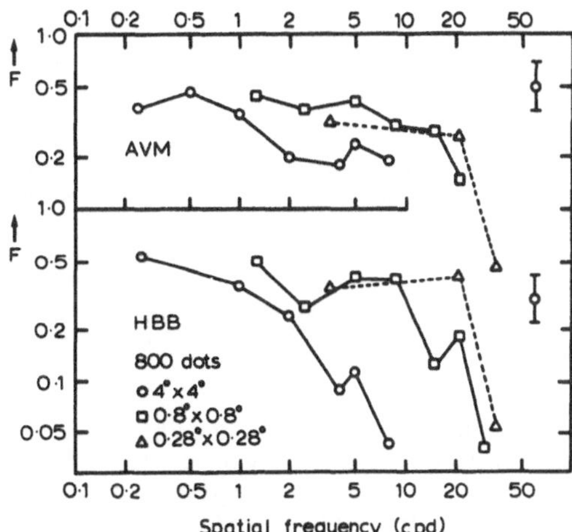

Fig. 5 Efficiency of detecting sinusoidal modulation of the average density of randomly placed dots, plotted as a function of spatial frequency. The experiments used two subjects and three field sizes; the method of determining F was not the same as that illustrated in Fig. 2, and will be described elsewhere (VAN MEETEREN and BARLOW, in preparation). Note that efficiency tends to be highest for the lowest frequency, for each field size.

this time are proving interesting and suggestive.

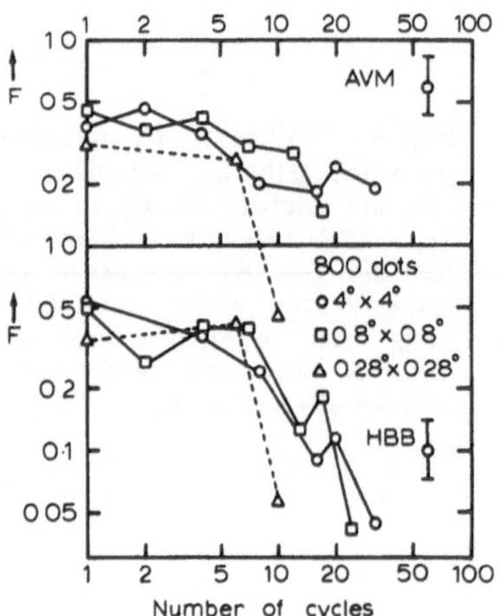

Fig. 6 The same results as Fig. 5 replotted as a function of number of cycles in the target field. The curves superimpose, except for the highest frequency in the smallest target field, where the resolution limit is approached. Note that efficiency is quite well sustained up to about 5 cycles, but is certainly no higher for strongly periodic patterns than for a single cycle.

I shall start by describing the phenomenon we set out to investigate. If you look at the reflection of your own eye in a dusty mirror, what you see is an array of lines radiating from your pupil like rays of light. It looks rather like Fig. 7, top right. Though the lines are formed by the flecks of dust, they continue to be aligned on the pupil even when the mirror is tilted sideways. The flecks of dust appear to reform in lines whenever you do this.

What is happening is that each speck has its own relfection close to it, and this lies on a line between the speck and the line normal to the mirror passing through your pupil, which is the line defining the position of the pupil's own image. So the radiating lines are not lines at all, but pairs of dots. It is hard to believe the radiating lines seen in a mirror are generated by pairs of dots, and when I first saw the effect I suspected multiple internal reflections might contribute to it. But these are very weak and Fig. 7 was generated by a computer unlikely to suffer this particular defect of operation.

This and other related phenomena were described by GLASS and PEREZ (5), and there has been a recent interesting contribution about them by GLASS and SWITKES (6). They are examples of the eye's ability to detect symmetry. One half of the dots has been generated from the other half by a simple operation, in this case radial expansion; though one does not realize this intellectually, what one "sees" is really an idealization of the operation, and the pattern that is operated on is scarcely seen at all. In Fig. 7 it takes close inspection to decide whether the top left pattern is or is not the pattern operated on to form the top right figure. We chose to study in more detail another form of symmetry, namely reflection in a line, or simple mirror symmetry.

Detecting symmetry of any kind is a much more complicated task than detecting changes of dot density. We did not realize at first how much more complicated, because we supposed that it might be a very special operation

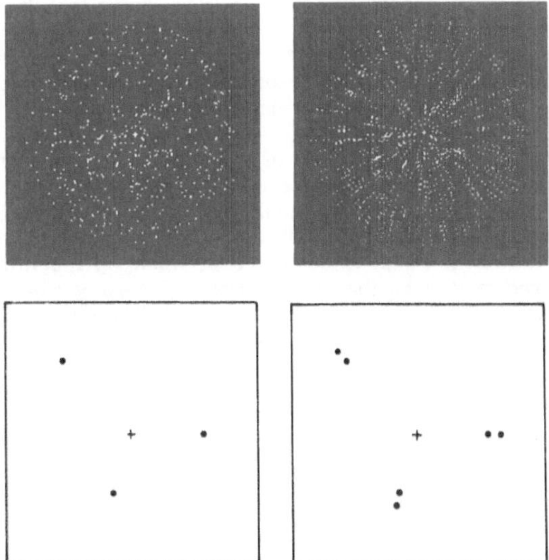

Fig. 7 Each dot in a random array (left) has a satellite dot displaced from it a certain distance radially from the center (+), as shown below. What the eye sees depends on the operation (radial expansion) performed rather than what is operated on. Is the pattern to the left actually the one used to generate the right hand one, or not?

performed in special circumstances. We thought one might only detect perfect symmetry, or only detect it about a vertical axis, and only about the midline of the visual field. None of these restrictions proved to hold. You can detect symmetrical pairs of dots when they are diluted with random, unpaired, dots. You can, for instance, discriminate, with a d' of about 1, between displays containing 100 totally randomly placed dots, and ones containing 60 totally random with 40 in pairs about the vertical midline. You can detect pairs diluted by random dots when the axis of symmetry is horizontal, and even when it is oblique, though not so well in this case. Furthermore you can detect symmetry when the axis is displaced to right or left of the fixation position.

The symmetry detecting mechanism is clearly rather versatile, and it must again be emphasized how much more complex a task it is than detecting changes of dot density. Ideal performance at the latter can be achieved by accurate knowledge of a single quantity - the number of dots in the target area and time. To achieve ideal performance at the symmetry task requires very much more, and this must now be discussed.

Mirror symmetry could be detected by auto-convolution, or a reflection-correlation operation, in which luminances of the scene at corresponding positions on either side of the line of symmetry were multiplied and integrated. It could also be performed on the Fourier transform of the dot-image. But however it is done we shall at some point need to know how accurately a point must be mirrored in order to qualify as a pair. If the mirror points are placed with the full accuracy of our system (1024 x 1024 positions), and our usual number of dots (800) are employed, there will be only very rare mirror pairs other than those deliberately placed, and the ideal system would only very rarely fail to discriminate. But that would be giving the ideal discriminator an unfair advantage over the human, who does not have the exact coordinates of each dot available and cannot be expected to determine them with such accuracy. Before we can measure a valid efficiency for symmetry detection we must know what the positional accuracy of the mirror-pair detecting mechanism is.

To determine this we have done the following type of experiment. Instead of placing the pair to a dot in the exact mirror position we place it at a randomly chosen position within a certain range of the exact mirror position. By varying this range, and observing at what value symmetry detection is impaired, we can determine the accuracy of the system. Figure 8 shows the result of such an experiment. All points were paired, or no points were paired, and when the perturbation was zero or small, there was no difficulty in distinguishing paired from unpaired. However performance deteriorates seriously when the perturbation range is increased to about $\pm$ 0.2^{o} vertically and horizontally.

In Fig. 8 we also show how a quantity, closely related to the auto-convolution or reflection-correlation function, behaves when the tolerance range is increased. The quantity plotted, D, is the expected number of pairs within the tolerance range, a pair being defined as any two dots where one is within the tolerance range of the exact image position of the other. Three dots can thus form two pairs, four dots, four pairs, and N_L, N_R dots on left and right can form up to $N_L \times N_R$ pairs. The lower dotted curve plots D for the zero pair condition, the upper dotted curve is this value increased by the 50 deliberately placed pairs.

Now in deciding whether symmetry is present one cannot do better than count the pairs that qualify under the tolerance rules imposed. The counts obtained will vary from trial to trial: because the occurence of one pair is, to a first approximation, independent of the occurence of other pairs, the values of D will be Poisson distributed with standard deviation approximately $\sqrt{D}$. Ideal detection therefore has S/N ratio $\Delta D/\sqrt{D}$, where ΔD is the number of deliberately placed pairs. In our programmes D is calculated from the tolerance parameters and other values, but there are tricky assumptions behind the calculation and troublesome edge-effects, so we also run Monte Carlo trials to ensure that D and σD are correct.

In Fig. 8 it is clear that D rises as the proportion correct drops, which encourages us to believe that the detection of symmetry is limited by the occurence of spurious pairs within the tolerance range of the mirror-pair detecting mechanism. The lower half of Fig. 8 shows the result of this experiment plotted out in a different form. We have seen how an ideal S/N ratio can be calculated from the values of ΔD and $\sigma(D)$ as tolerance is changed. In the lower half the actual d'_E values are plotted as ordinate against ideal $d'_I(\Delta D/\sigma_D)$ as abscissa. The 100% and 20% efficiency lines are marked and it will be seen that, at this rather complex task, an efficiency close to 20% is reached. I may add that we have obtained figures for efficiency up to about 35% for a different type of symmetry, namely detecting the striated textures formed by linear displacements of paired dots. We think the lower figure for mirror symmetry may result from the fact that the appropriate tolerance range increases when the paired dots are further away from the axis of symmetry; thus the uniform tolerance range we employed was not well matched to the tolerances of the symmetry detecting mechanism.

Possible Neural Mechanisms

Now I said earlier that 50% was quite good for a task which seems ridiculously simple by comparison with the detection of symmetry. How would an ideal device detect symmetry, and how might the nervous system approximate the necessary operations?

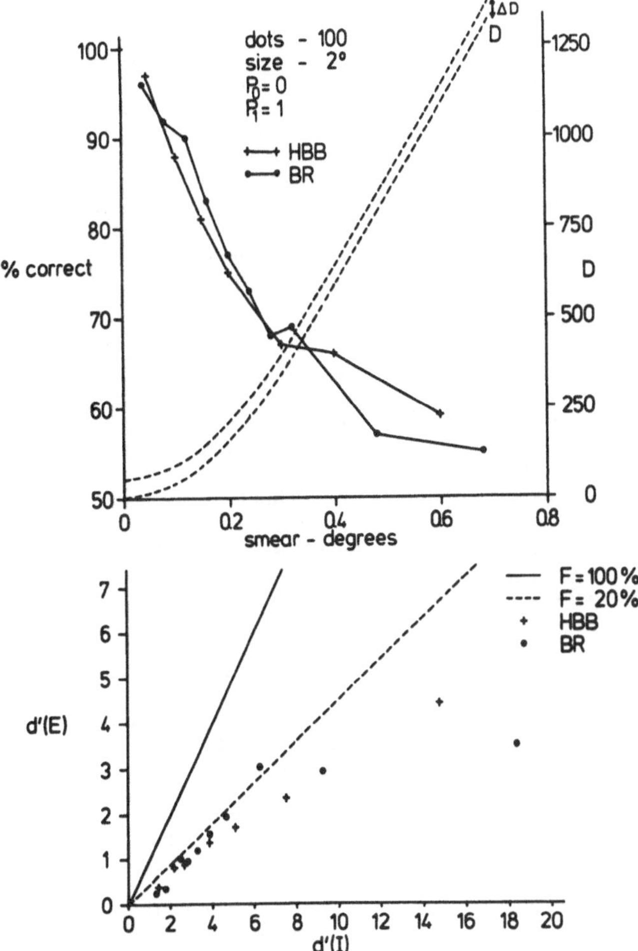

Fig. 8 The effect of tolerance range or "smear" on the percent correct in discriminating between populations of 100 randomly placed dots and 50 paired dots. The dotted curves show the number (right ordinate) of "pairs" (see text) in the random, unpaired, population (lower curve), and paired population (upper curve). The presence of pairs in the random population determines the value of d'_I, and in the lower figure the experimental values of d'_E are plotted as ordinate against these ideal d'_I values. The continuous line shows the performance of an ideal discriminator, the dotted line one operating at 20% efficiency. The subjects approach the latter figure.

One would first think in terms of an auto-convolution or reflexion-correlation operation in which luminances in a visual image at mirror symmetric points are multiplied and integrated. This would have to be done for all possible lines of symmetry. Such an operation may seem utterly different from anything we know that goes on in visual cortex, but one set of experimental facts, together with a reinterpretation of an established model, suggest that it may not be so remote. The set of facts is that many visual properties we habitually assign to lines are in fact possessed by pairs of points. For instance a bend in a line can be detected when its centre is displaced by only a few seconds of arc, but this high positional accuracy is also found when the two ends of the line and its centre are replaced by dots. Similarly many illusions work well when lines are replaced by dots at their junctions and terminations. Are we perhaps wrong, then, in thinking of Hubel and Wiesel's orientationally-selective units as line detectors?

The reinterpretation is to suggest they are "oriented-pair" detectors. The difference between the two views is illustrated diagrammatically in Fig. 9. Formalisations of how nerve cells work are always a bit unsatisfactory, and probably no one is completely happy with the left half as a model of the action of simple cells in the visual cortex. Nevertheless I think it does represent how most of us have thought of these neurones. It says that they sum excitatory influences over a ridge shaped receptive field emerging in a wide saucer of inhibition, with perhaps other lesser parallel ridges on either side. The output is a simple monotonic function of this sum. That is not quite what is needed to detect an "oriented pair"; a small modification, shown to the right, is required. The receptive field here is composed of two parts each of which summate as before, but the output depends on the product, or better the logical conjunction, of these two parts.

I do not think the neurophysiology of cortical neurons is known well enough to exclude this model for their basic receptive fields, but the rather small change makes a big difference to the way we think about visual cortex. The left hand model is a linear filter, while the right hand would perform a logical operation - it would detect the conjunction of excitation in two neighbouring regions. Thinking in these terms one looks at the other types of selectivity found in visual cortex, namely selectivity for direction and velocity of movement, and selectivity for disparity. Movement selectivity, if it is done like rabbit retina (7), depends on detecting pairwise excitation in one region at one moment, and a neighbouring region at another moment; the mechanism probably involves vetoing the unwanted responses by inhibition, but logically it looks like the detection of paired conjunctions. Similarly disparity selectivity involves detecting the paired associations of excitations in the images from the two eyes, or possibly (8) the association of "oriented pairs" in the two images. Again, it would be a logical operation performed on just two inputs. Then again one's mind turns to Julesz' (9) speculation that it is only the second order statistics that are utilized in detecting texture differences (see also (10)).

Finally let me point out that cerebral cortex is an astonishingly uniform structure, and it is unreasonable to suppose that other parts perform operations that differ very radically from those performed in area 17, though of course other parts have totally different inputs. Can we really be content with the idea that cortex does nothing more than linear filtering operations on its inputs, whatever they may be, as the left half of Fig. 9 suggests? The detection of paired associations is a much more attractive proposition as the general function of cortical neurones, and is the proposition we are lead to consider by the remarkably high efficiencies achieved in detecting patterns such as that shown in Fig. 7.

TWO MODELS OF CORTICAL RECEPTIVE FIELDS

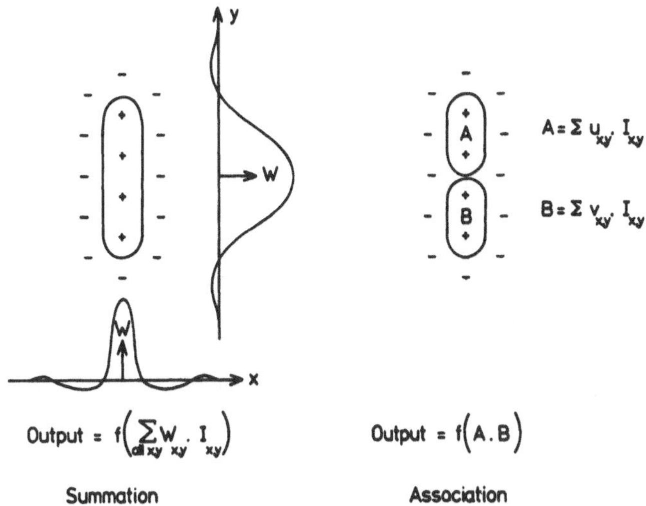

Fig. 9 The left half shows the classical idealization of an orientationally selective simple cell. It consists of an excitatory ridge surrounded by inhibitory areas. The figure below shows the weighting function along a horizontal line through its center, and similarly the curve at right for a vertical section along the ridge. The cell's output is a monotonic function of the weighted excitatory and inhibitory effects of light. As far as its spatial form discrimination is concerned, it is a linear filter. The right half shows the proposed "oriented pair" detector. There are two regions, each of which summates linearly as in the classical model, but the output depends on the product, or better the logical conjunction, of the weighted sums formed by each element of the pair detector. Such a cell can be regarded as detecting a simple example of paired association.

Summary

1) A technique is described for measuring the efficiency of perceptual discriminations on an absolute scale.

2) Many patterns such as squares or spatial sinusoids can be detected with efficiencies up to about 50%; half the statistical information is utilized, half is wasted, as if human subjects utilized about half of the full sample that is available.

3) High efficiencies (20% to 35%) can also be achieved at tasks which would require the detection and counting of pairs of dots for fully efficient performance.

4) It is suggested that selectivity for orientation, for direction and velocity of movement, and for disparity, are all achieved by detecting the conjunction of just two inputs, and that the capacity to detect such paired associations is the basic operation of cerebral cortex.

Aknowledgement

I am much indebted to Aart van Meeteren for planning, performing, and discussing many of the experiments reported here, and to Barney Reeves for relentlessly accumulating data that begin to test some of these speculations.

References

1. H. B. Barlow, Vision Res. (in the press) (1978).
2. A. Rose, J. opt. Soc. Am. 38, 196 (1948).
3. A. Rose, Vision: Human and Electronic, (Plenum Press, New York, 1973).
4. R. P. Chambers & J. S. Courtney-Pratt, Photar. Sci. Engng. 13, 286 (1969).
5. L. Glass & R. Perez, Nature 246, 360 (1973).
6. L. Glass & E. Switkes, Perception 5, 67 (1976).
7. H. B. Barlow & W. R. Levick, J. Physiol. 178, 477 (1965).
8. H. B. Barlow, C. Blakemore & J. D. Pettigrew, J. Physiol. 193, 327 (1967).
9. B. Julesz, Proc. Inst. Radio Engrs., PGIT, IT-8, 84 (1962).
10. B. Julesz, E. N. Gilbert, L. A. Shepp & H. L. Frisch, Perception 2, 391 (1973).

Sources Suitable for Use in Illuminating Visual Acuity Charts

Glenn A. Fry
College of Optometry
The Ohio State University

Charts for measuring visual acuity are generally created by placing black characters on a white background. I am concerned in this paper with the wavelength composition of the light directed toward the eye from the white background of such a chart. The white background may be a diffusely reflecting surface and for our purpose it may be assumed to be nonselective and hence, we can concentrate on the wavelength composition of the source used to illuminate the chart.

I have taken it upon myself to study the role played by the wavelength composition of the source. I am doing this for an NRC Working Group which is attempting to set standards for the measurement of visual acuity. The Working Group wants to designate the CIE source C as the standard source but also wants to specify that certain fluorescent sources are acceptable substitutes.

To be specific, I want to assess the effect of substituting anyone of seven widely used white fluorescent lamps identified as fluorescent sources 1 through 7.

(1) Standard Warm White
(2) White
(3) Standard Cool White
(4) Daylight
(5) Warm White Deluxe
(6) Soft White
(7) Cool White Deluxe

The relative wavelength compositions of these sources have been presented in Table 1.12 by WYSZECKI and STILES (1). Each of these sources has a continuous spectrum and a series of lines. The continuous spectrum is divided into 10 nm bands and the relative amount of flux for each band is specified. The output for each of the lines is specified in terms of even distribution over a band 10 nm wide centered at the wavelength that characterizes the line. The spectral bands at wavelengths below 400 nm and above 730 nm have been ignored.

The relative energy distributions for the seven sources are shown in Figs. 1 and 2.

Because of the axial chromatic aberration of the eye, light of the various wavelengths in a heterochromatic point source will not focus at the same point. Let us assume that the eyes are corrected for astigmatism as is the case when the eyes are being checked for acuity with the best correction. Some of the wavelengths will focus in front of the retina, some behind, and the light of one of the wavelengths will be focused at the retina. The eye is said to be accommodated for this wavelength.

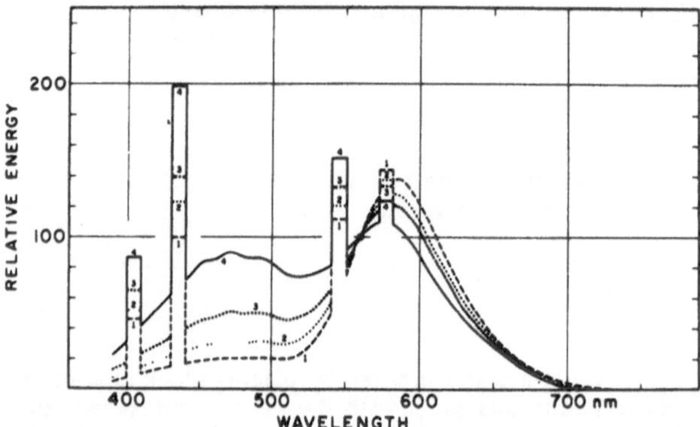

Fig. 1 Relative spectral energy distributions of representative white fluorescent lamps. See Table I to identify the lamps. Reprinted with permission from Ref. (1).

In order to achieve the sharpest image for a given wavelength composition, the eye must be focused for some particular wavelength.

I have tried to determine for each of the sources the wavelength required for best focus and the consequence of not being focused for that particular wavelength.

Although it had nothing to do with the measurement of acuity, I have faced a similar problem in finding out how the use of low pressure and high pressure sodium lamps affects the performance of a task (2). I wanted to know for what wavelength the eye must be focused to achieve the sharpest image. I shall use the same approach for fluorescent sources which I have used for sodium sources.

For each source I have started with the data provided by WYSZECKI and STILES expressed in relative energy per 10 mm band of the spectrum. For my purpose the output of the sources can be described in relative energy and no assumption needs to be made about the absolute values.

The values for relative energy for a given source tabulated by WYSZECKI and STILES have been multiplied by the ordinates $(\bar{y})$ of the CIE relative luminous efficiency. Each element of area of the background of a visual acuity chart may be regarded as a point source. A heterochromatic point source produces a set of images

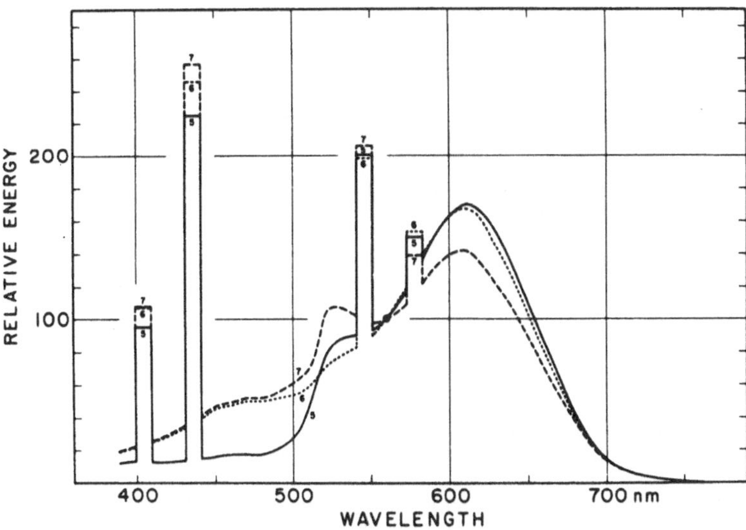

Fig. 2 Relative spectral energy distributions of representative white fluorescent lamps. See Table I to identify the lamps. Reprinted with permission from Ref. (1).

on the retina that correspond to the different bands of the spectrum. Each of these images may be assumed to be radially symmetrical and the total flux (F) in each of these images is proportional to the relative luminous energy given off by the fluorescent lamp for that part of the spectrum.

It has also been assumed that these images are centered at the same point on the retina and that this point falls near the center of the fovea.

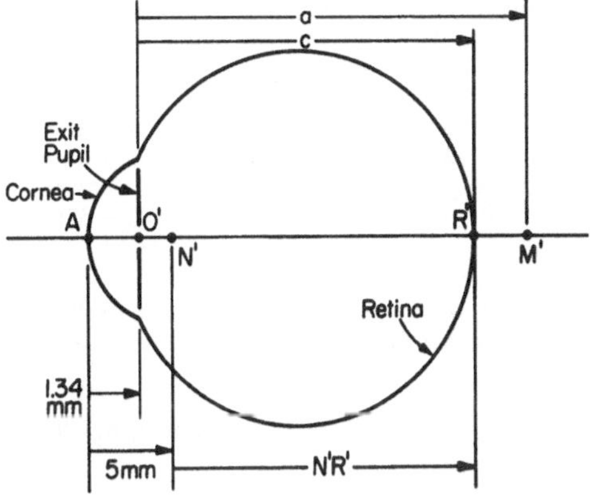

Fig. 3 The Helmholtz schematic eye.

For each point source, the composite image for the entire spectrum is an overlapping of the images formed by the different parts. This distribution is affected by the size of the pupil and the wavelength for which the eye is accommodated.

The HELMHOLTZ schematic eye has been used to assess the amount of blur of the composite image formed on the retina. This eye is illustrated in Fig. 3. The cornea has a radius of five millimeters and the aperture stop, which is also the exit pupil, lies 1.34 millimeters from the cornea. It has been assumed that the point source "M" lies at an infinite distance.

For some particular wavelength the image of the point source will be formed at "R'" at a distance "c" from the exit-pupil. This is the wavelength for which the eye is in focus.

To bring the eye into focus for different wavelengths, I have simply changed the distance from the exit-pupil to the retina.

For any other wavelength, the image falls at "M'" which may lie in front or behind the retina. The symbol "a" represents the distance from "O'" to "M'".

Since the point source "M" lies at an infinite distance from "A", it follows from the conjugate foci formula for a single refracting surface:

that $$\frac{n'}{\overline{AM'}} = \frac{n}{\overline{AM}} + \frac{n' - n}{r} \qquad (1)$$

and $$\overline{AM'} = \left(\frac{n'}{n' - n}\right) r \qquad (2)$$

and, hence $$a = \left(\frac{n'}{n' - n}\right) r - \overline{AO'} \qquad (3)$$

where "r", "$\overline{AO'}$", and "a" are expressed in millimeters. The term "r" represents the radius of curvature, "n" represents the index of the air in front of the eye, and it is assumed that n = 1.

When the eye is in focus for sodium light (589 nm), the value of "c" is 18.669. When the retina falls at some other distance from the exit-pupil, the eye is out of focus for sodium light. The expression (7.142 - 100 n'/c) represents the number of diopters that the eye is out of focus for sodium light. Although one can specify the state of accommodation by giving the distance of "R'" from the exit-pupil, I have chosen to do this by specifying the number of diopters that the eye is out of focus for sodium light. Plus values indicate that the eye is over accommodated.

It is necessary to have a set of values for the index of refraction of the eye-medium for various wavelengths. These have been deduced from the chromatic aberration data of WALD and GRIFFIN and are presented in the form of a table in Ref. 3. There are two minor errors in this table. The value of "n" for "λ" = 578 nm should be 1.3335, and the value for "λ" = 750 nm should be 1.3294.

BEDFORD and WYSZECKI (4) have repeated the measurements made by WALD and GRIFFIN with a large number of subjects and obtained results that are in good agreement.

The data for the relation between "n" and "λ" can be fitted with CAUCHY's equation:

$$n = 1.32546 + 0.002154\,(\frac{10^{-6}}{\lambda 2}) + 0.000176\,(\frac{10^{-12}}{\lambda 4}) \; . \qquad (4)$$

The curve in Fig. 4 represents this equation fitted to the data of WALD and GRIFFIN. The constants are based on the data for three wavelengths, 400 nm, 550 nm and 700 nm, taken from a smoothed curve through the original data.

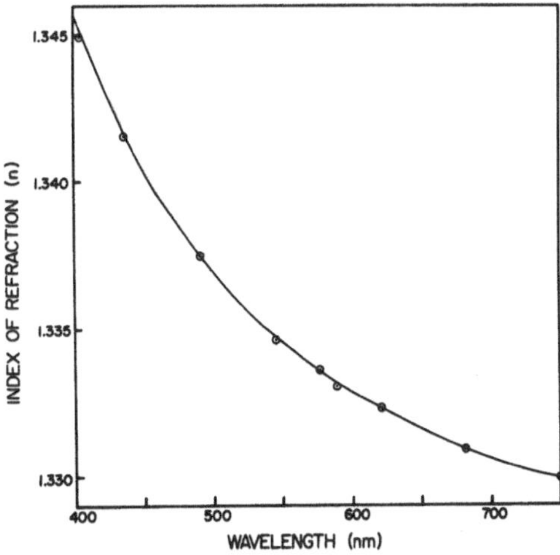

Fig. 4 Dispension curve for the ocular medium.

To assess the amount of blur for the two point sources for various pupil sizes and states of accommodation, we used the Fry-Cobb index of blur. The meaning of this index of blur can be most easily understood in terms of the gradient of illuminance across the image of a border separating a bright from a dark area. If we start with the image of a point source, we must first determine the gradient across a border by convolving the border with the point spread function. The retinal illuminance gradient across the image of a border is illustrated in Fig. 5. The Fry-Cobb index of blur is also illustrated. In a sense, it represents the width of the blur

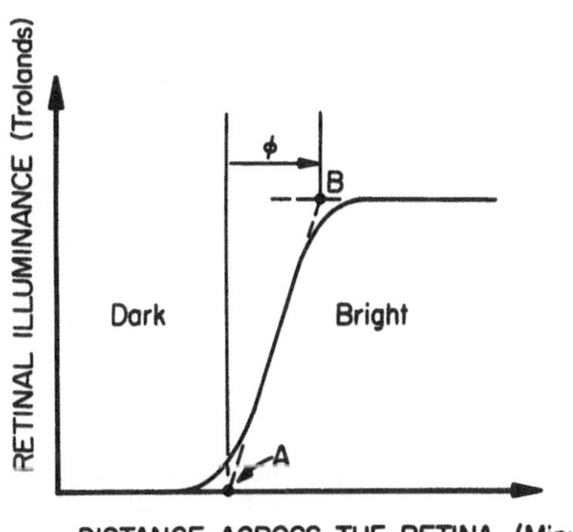

Fig. 5 Fry-Cobb index of blur (ϕ) as related to the gradient of retinal illuminance across of the retinal image of a border between a bright and a dark area.

zone. The points A and B lie on a straight line that coincides with the gradient at the middle of the blur zone. These points lie at levels that represent the illuminance on the two sides of the blur zone. The lateral displacement of B from A measured in minutes of arc at the second nodal point of the eye is the index of blur.

With a monochromatic point source, one can compute the index of blur from Eq. (5) (2, 5).

$$\phi = 3438\ \lambda c/(n'\bar{g}'V\cdot\overline{N'R'}) . \qquad (5)$$

Eq. (5) is based upon the assumption that the eye is diffraction limited, that is, it is free from spherical aberration and astigmatism. The dimensions "c", "$\overline{N'R'}$", and "$\bar{g}'$" are shown in Fig. 1.

c = distance from the exit-pupil to the retina
$\overline{N'R'}$ = distance from the second nodal point to the retina
$\bar{g}'$ = radius of the exit-pupil
λ = wavelength of the light focused at M'
n' = index of the medium

In Eq. (4), "V" is a function of "$\bar{\omega}$ "and is defined as:

$$V = 2\left[\frac{1}{0.5!\ 1.5!} - \left(\frac{\omega}{2}\right)^2 \frac{2!}{2.5!\ 3.5!} + \left(\frac{\bar{\omega}}{2}\right)^4 \frac{4!}{4.5!\ 5.5!} - \ldots\right] \qquad (6)$$

In turn, "$\bar{\omega}$" is a function of the extent to which the eye is out of focus as shown in Eq. (7):

$$\omega = 2\pi\left(\frac{n'}{\lambda}\right)(\bar{g}')^2 \left|\frac{a-c}{ac}\right| \qquad (7)$$

where "a" represents the distance from the exit pupil to the point "M'" at which light of wavelength " λ"is focused (see Fig. 1).

The index of blur for a heterochromatic image is based upon the indices of blur for the separate monochromatic components as shown by:

$$\frac{1}{\phi} = \frac{1}{F}\int F_\lambda\,(1/\phi)_\lambda\,d\lambda \qquad (8)$$

where:

$$F = \int F_\lambda\,d\lambda \qquad (9)$$

Since the values of "F_λ" computed from the energy distributions data represent 10-nanometer wide spectral bands, the integrals in Eqs. (4) and (5) can be evaluated by numerical integration:

$$\int F_\lambda\,d\lambda = \Sigma F_\lambda \qquad (10)$$

$$\int F_\lambda\,(1/\phi)_\lambda\,d\lambda = \Sigma F_\lambda\,(1/\phi)_\lambda \qquad (11)$$

The lines are treated in the same way as 10-nanometer bands.

As explained in a previous paper (2), Eq. (6) cannot be used to compute "V" because the numbers involved have more digits than the computers available can handle, and hence the problem was solved by developing approximate equations for

"V" which can be solved for any value of "ω".

For values of "$\bar{\omega}$" smaller than eight:

$$\frac{1}{V} = .000497\,\bar{\omega}^{-2.5} + 0.58903 \qquad (12)$$

and for values of "$\bar{\omega}$" larger than eight:

$$\frac{1}{V} = C - 0.425 + 0.425 \cos\left[360^{\circ}\,(\bar{\omega}/4\pi)\right] \qquad (13)$$

where

$$C = \frac{1}{4}\left[\sqrt{(\bar{\omega} + 11.87)^2 + 32.7096} - 11.87\right] \qquad (14)$$

The derivation of these equations is explained in Ref. (1).

Figure 6 shows for an exit pupil 4 mm in diameter, the effect on blur produced by throwing the eye out of focus various amounts. The different curves represent the seven different fluorescent sources.

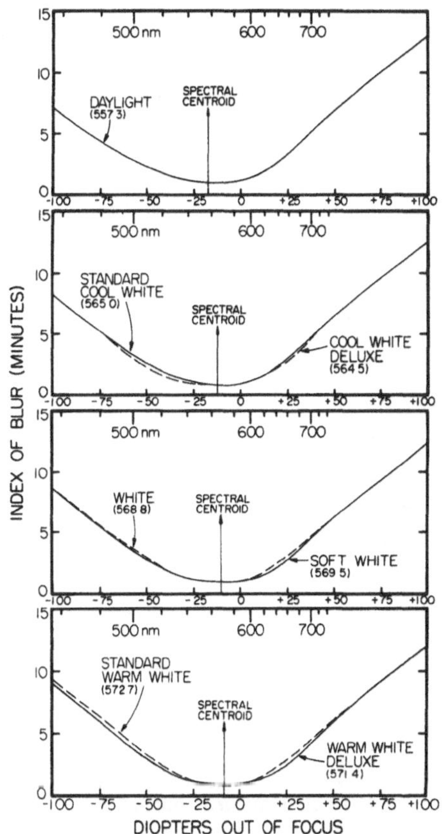

Fig. 6 The effect of throwing the eye out of focus for sodium light (589 nm) on the index of blur. Curves for the seven sources identified in Table I and Figs. 1 and 2. In the lower three graphs the spectral centroids for the two sources practically coincide and are represented by a single line. The upper scale shows the wavelength for which the eye is in focus when the eye is out of focus for sodium light by a given amount.

The curves for Standard Cool White and Cool White Deluxe are nearly the same and are presented in the same graph. The curves for White and Soft White are also nearly alike and presented in a single graph. Finally, the curves for Standard Warm White and Warm White Deluxe are also nearly the same and are represented in a single graph.

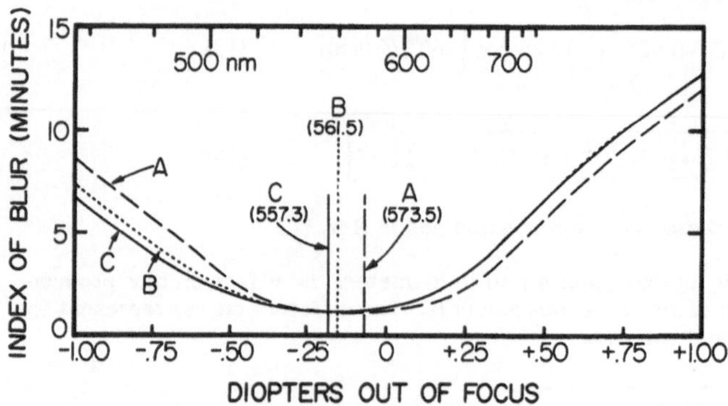

Fig. 7 The effect of throwing the eye out of focus for sodium light (589 nm) on the index of blur. Curves for the CIE standard sources A, B, and C.

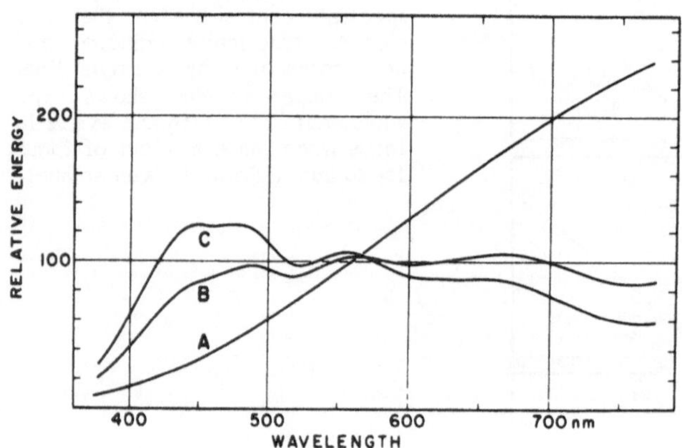

Fig. 8 Relative spectral energy distributions of the CIE standard sources A, B, and C. Reprinted with permission from Ref. (1).

For comparison, a curve for the CIE Standard source C was computed and is shown in Fig. 7. Shown in the same graph are curves for CIE Standard Sources A and B. Source A approximates an incandescent source operated at a color temperature of 2854° K. The correlated color temperature for Source B is 4870° K and that for Source C is 6740° K.

The spectral distributions of relative energy are given by WYSZECKI and STILES in Table 1.9 and the data are presented in graphical form in Fig. 8.

A casual inspection of Fig. 6 or Fig. 7 will reveal how meaningless it is to say that focusing on one specific wavelength will produce the sharpest focus. In the region of best focus there is a considerable range through which the focus can change without producing a noticeable amount of change in blur. If these were analytical expressions for the curves, a minimum could be computed. If the curves were symmetrical, one could establish the axis of symmetry.

The Optical Society of America (6) recommends the spectral centroid of a source as an index of the wavelength to focus for to achieve sharpest vision. The spectral centroid " λ " is a weighted average of the wavelengths in the spectral distribution:

$$\lambda = \int_0^\infty \lambda F d\lambda / \int_0^\infty F d\lambda \qquad (15)$$

The spectral centroids have been computed for the various curves and are shown in the graphs. They represent a useable index and appear to be appropriate.

The spectral centroids for all of the fluorescent sources fall in the small range between the centroids for A and C (see Fig. 7).

The centroids for incandescent sources would also be expected to fall within this range.

The range between A and C represents less than an 1/8 diopter on the dioptric scale.

It is the usual practice for the refractionist to use the same source for measuring the refractive error as for measuring acuity with the best correction and hence the measurement of acuity should not be affected.

Since the average refractionist tends to bias the correction in the direction of leaving the patient ever so slightly myopic, the measured acuity might be improved by switching from Source C to Source A and might be made slightly worse by switching from Source A to Source C. The graphs, however, indicate that the change would be negligible.

If the measurement were made without correction and if the patient were, for example, myopic to the extent of one diopter, there would be a measurable effect in switching from Source A to Source C or vice versa. The amount of change would be equivalent to the amount produced by changing the sphere in the prescription 0.12 D. It would also be equivalent to moving the chart from 4 meters to 8 meters or from 4 meters to 2.67 meters.

Unless the patient were a presbyope the patient would probably accommodate to overcome one diopter of hypermetropia and hence changing the amount of hypermetropia by switching from Source C to Source A or vise versa would have little affect on acuity.

It is recommended therefore that fluorescent and incandescent sources which have spectral centroids that fall between those of the A and C Standard Sources be considered acceptable for the measurement of visual acuity. Since Source C represents daylight, this range includes practically all of the sources (incandescent and fluorescent) which are apt to be used.

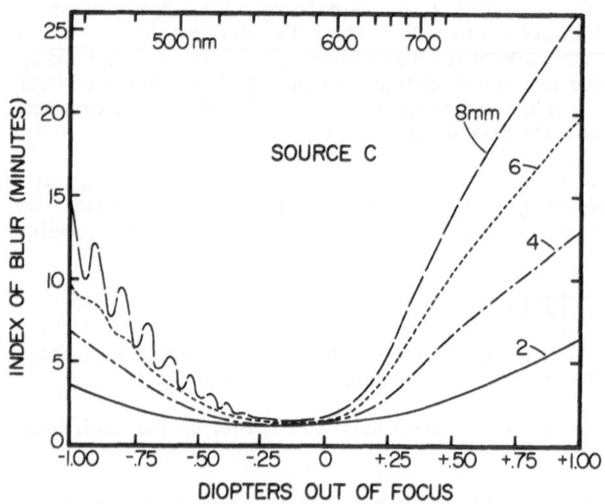

Fig. 9 The effect of varying pupil size on the index of blur curves for the CIE standard source C.

The maximum acuity for a 4 mm pupil is about the same for all of the sources. There are computable differences in the depth of focus but these are negligible in comparison with the effect of changing the size of the pupil. Fig. 9 shows the effect of changing the size of the pupil from 2 to 8 mm in diameter using in each case the CIE Standard Source C.

Addendum

At the meeting Dr. DAVID MACADAM suggested that Source D 6500 which represents one of the phases of daylight should be used instead of Source C as the standard source for the measurement and specification of visual acuity. He indicated that performance with D 6500 would be very similar to that of Source C.

The relative energies for D 6500 have been published by WYSZECKI and STILES (1). Figure 10 which compares the curve for D 6500 with the curve for Source C in Fig. 7 shows that the two sources are nearly identical. The substitution of D 6500 for Source C can be recommended.

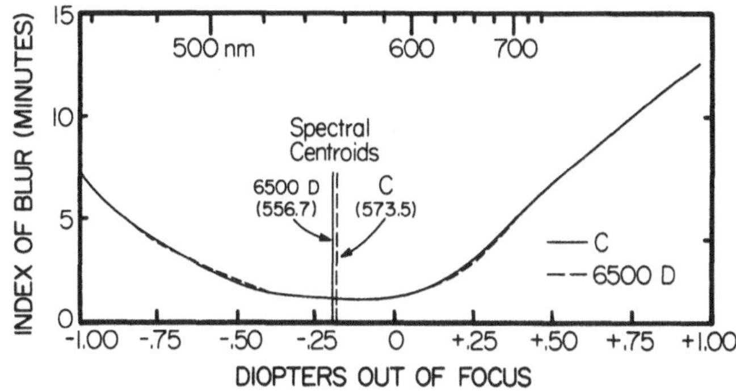

Fig. 10 The effect of throwing an eye out of focus for sodium light (589 nm) on the index of blur. The two curves are for D 6500 and Source C. The data apply to a diffraction limited eye with a 4 mm pupil.

References

1. G. Wyszecki and W. S. Stiles, Color Science, (John Wiley and Sons, New York, 1967).
2. G. A. Fry, J. of IES, 158 (1975).
3. G. A. Fry, Blur of the Retinal Image, (Columbus, Ohio, The Ohio State University Press, 1955).
4. R. E. Bedford and G. W. Wyszecki, J. Opt. Soc. Am. 47, 564 (1957).
5. G. A. Fry, Progress in Optics, VIII, edited by Wolf, E., (Amsterdam, No. Holland, 1970), Chapter II.
6. Colorimetry committee of the Optical Society of America, Science of Color, (New York, Cromwell, 297, 1953).

Current Status of Research on the Spatial Organization of the Human Visual System at Detection Threshold

H. Richard Blackwell, Wunchung Chiou and Bryan R. Blackwell
The Ohio State University

This is the third in a series of reviews of the status of research on visual detection. The first was presented by BLACKWELL (1) in 1963; the second was presented by BLACKWELL (2) in 1972. These earlier reviews included different areas of research on human visual detection. The present review is restricted to aspects of visual detection which reveal the spatial organization of the human visual system. The basic experiment concerns determining the effect of the spatial size and configuration of a stimulus object or target upon the threshold contrast or luminance increment required for bare detection of its presence. The properties of spatial organization of the human visual system are deduced from the results of such experiments.

Early work in this field was summarized by KINCAID, BLACKWELL, and KRISTOFFERSON (3). These authors described what might well be called a first-order theory of the spatial organization of the visual system at detection threshold, and pointed out that the very early studies of FRY and COBB (4) and of GRAHAM, BROWN and MOTE (5) shared with this first-order theory the following implicit assumptions: (a) that there exists a topological mapping of detection targets at some site in the visual neural system which depends upon convolution of an all-positive point spread function (PSF), with the target luminance profile; and (b) that detection occurs whenever a criterion amplitude of neural signal is received at a single site in the visual neural system. Ref. (3) contains a quantitative description of a method for deducing the PSF from detection thresholds of circular disc targets of varying diameters. The detection threshold is characteristically determined under conditions in which a target is presented for a very brief exposure as a luminance increment to a large steadily-exposed field of uniform luminance. The psychophysical data presented in Ref. (3) to illustrate the operation of the method were obtained with the temporal forced-choice psychophysical method first described by BLACKWELL (6). This method encourages the test observer to detect the presence of a target with any possible sensory cue, and results in thresholds of unprecedented smallness.

Ref. (1) contains, among other items, a discussion of this first-order theory in contrast to other theories of the spatial organization of the human visual system. Data are presented which reveal that all-positive PSF's are obtained under a great variety of experimental conditions. However, it is shown that the quantitative character of the PSF depends upon the luminance to which the human visual system is adapted, and the exposure duration and wavelength composition of the target used to measure the detection threshold.

More recently, quantitative evidence has been presented from a multitude of electrophysiological and psychophysical studies suggesting: (a) that the human visual system is organized in terms of center-surround systems or channels which lead to PSF's with positive values in the center and negative values in annular regions of the surround; and (b) that the human visual system consists of multiple functional channels which presumably imply the existence of multiple functional sites in the human visual neural system at which detection could occur. Thus, the two main assumptions of the early theories of visual detection appear untenable today. Let us consider the implications of these new findings for visual detection theory in some detail.

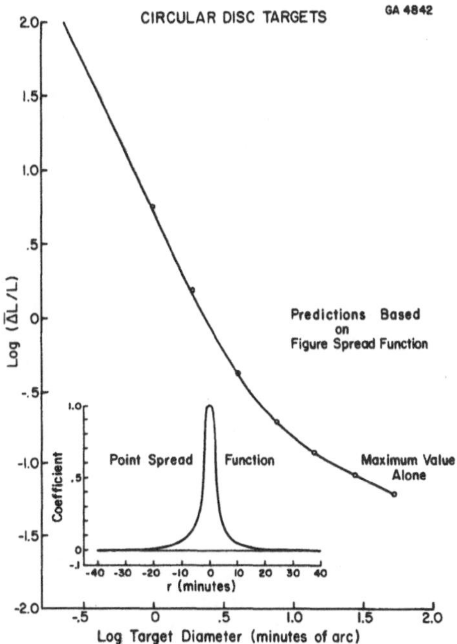

Fig. 1 Data of KRISTOFFERSON (7) revealing the relationship between log contrast threshold and log disc diameter in the solid curve. The point spread function used to predict the curve is shown.

Consideration of the form of the most current theories of visual detection may well begin with a consideration of Fig. 1 (reproduced from Ref. (2)). Here we have experimental data presented first by KRISTOFFERSON (7). Shown are values of the detection threshold contrast, $\Delta L/L$, as a function of the diameter of a circular disc. The PSF shown in the figure was derived from the raw data by the method described in Ref. (3), and then used to predict the solid curve fitted through the data points in the figure. The detection criterion was restricted to a single site. Note that the all positive PSF predicts a monotonic curve, and that the data confirm that there is no doubling back of the curve as would be required if the PSF had the positive and negative values associated with the usual center-surround system.

To be more precise, a non-monotonic curve relating detection threshold contrast to disc diameter would be expected if there were universally a single class of center-surround systems of fixed dimension, or what is usually called a single size-tuned channel, throughout the human visual system. A more complex assumption, and one more in keeping with current thinking, would involve the assumptions first made by THOMAS (8) that the continuous curve in Fig. 1 actually represents the envelope of the size responsivities of a number of different spatially organized mechanisms, each being specialized or tuned to a given range of disc sizes. Each mechanism would presumably involve a positive center, negative surround with dimensions different from each other mechanism. Each of these so-called spatially tuned channels would govern detection over a range of disc diameters to which it gave optimum output, and the unitary seeming monotonic detection curve would in fact represent the operation of a "channel-hopping" process with a number of inconspicuous discontinuities as disc size is varied, produced by the process of hopping from one channel to the next. So far as the data shown in Fig. 1 are concerned, this theoretical model would be difficult to disprove. However, the envelope would of necessity define the relative sensitivity of each of the different sized-tuned channels.

These considerations imply that tests of the relative plausibility of the single site (single channel) theory and the channel-hopping theory must be made in terms of targets other than the circular discs used to derive the PSF in the first-order theory or used to derive the relative sensitivity of the different channels in the channel-hopping theory. This was precisely the notion described in Ref. (3): that the first-order theory should involve deduction of the quantitative form of the PSF from circular disc targets of varying size, then testing of the theory by comparing obtained thresholds for various non-circular targets with thresholds predicted on the basis of the PSF derived from the circular disc data.

Tests of the first-order theory by this basic method have been reported by KRISTOFFERSON (7), by CLEAVER (9), and by CHIOU (10). The data originally presented in Ref. (7) and (9) were summarized in Ref. (2). It was concluded that, whereas in general the obtained thresholds for non-circular targets agreed well with predictions, there were what seemed to be systematic discrepancies in three cases. The data of Ref. (7) suggested that long, thin rectangular targets had somewhat lower threshold contrasts than predicted. Subsequently, analyses reported in Ref. (10) demonstrated that this apparent discrepancy was a mathematical artifact due to truncation errors introduced by the computaton methods used to predict threshold contrasts for the rectangles. The analyses of Ref. (9) suggested that both simple annular targets and targets constructed from patterns of bright bars separated by equal blank spaces (all-positive square wave gratings) were considerably more visible than predicted. Ref. (10) confirmed that this result was not due to mathematical artifacts and argued that it was evidence of the general operation of a second-order mechanism first described in Ref. (1). This mechanism consists of what is called multiple site probability summation (MSPS). The basic idea is that, at least when the forced choice psychophysical method is used, detection will occur whenever a criterion amplitude of output is reached at any of the many potential sites in the human visual neural system. The output of any one site is presumed to vary probabilistically among successive presentations of a given target contrast due to system variability near threshold. Thus, there will be occasions in which a secondary site reaches criterion output when the primary site fails to do so, and detection probability will include all instances in which one or the other reaches criterion so long as this does not occur at the two sites simultaneously. The concept postulates that detection occurs whenever the criterion amplitude of output occurs at one or another site, and that having criterion amplitude at two sites simultaneously is equivalent to having criterion amplitude at only one site. The problem is how to describe what might be called the spatial correlation of temporal variability of different sites in the visual neural system. The description in Ref. (1) did not handle this problem quantitatively. However, the discussion in Ref. (10) did provide the beginnings of a quantitative treatment of the problem, which may be summarized as follows:

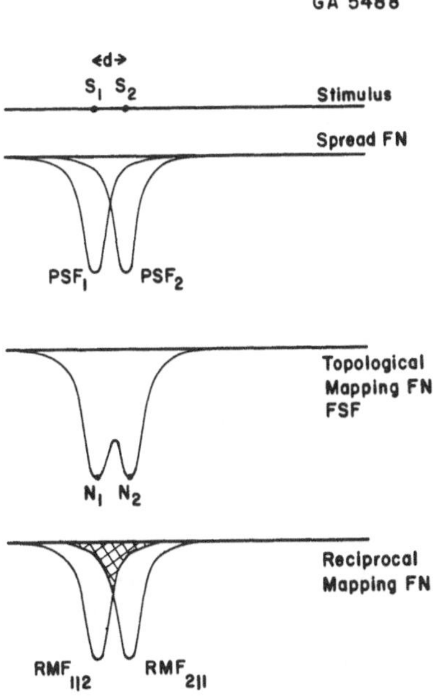

Fig. 2 Construction from CHIOU (10), showing the relations between point spread function (PSF), figure spread functions (FSF), and reciprocal mapping functions (RMF).

Consider first the average amplitude of output at various points in the human visual neural system resulting from a single target consisting of two small bright points of light separated by a rather larger blank space, as shown in Fig. 2. The uppermost layer in the figure represents the optical light stimulus presented to the visual system. The joint action of the image-forming and retinal neural mechanisms may be described in terms of point spread functions PSF_1 and PSF_2 for the two spots as shown in the second layer of the figure. There is a first-order linear process which leads to the topological mapping function designated by Ref. (9) as the figure spread function, FSF. Production of the FSF from the light stimuli can be considered to occur by convolution of the PSF in the usual manner. This is shown in the third layer of the figure.

Examples of FSF's obtained by this process for two annular targets are shown in Fig. 3 and 4 as three-dimensional contourgraphs. Note that in Fig. 3 the FSF corresponding to a small annular target without light at its center, convolved with an all positive PSF such as those shown schematically in Fig. 1 and Fig. 2, has its maximum average amplitude at the center. When the annular target has a larger area at the center without light, the FSF as shown in Fig. 4 has its maximum average amplitude in a circular ridge around the center. However, the average amplitude of the FSF at the center is far from zero. Consider that the neural processing corresponding to the all positive PSF's represents a severe "blurring" of the light stimulus

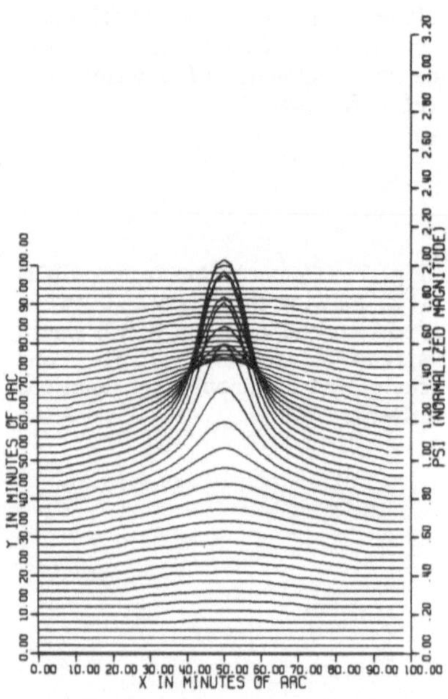

Fig. 3 Contourographic representation of the figure spread function of a simple annulus of width 1 minute and center-to-center diameter of 8 minutes.

pattern. It can be shown that, as unfortunate as this blurring is for good spatial resolution, the broad summative processes it reflects are most effective in improving simple detection of target stimuli.

Returning now to Fig. 2, note that FSF in the third layer is of variable average amplitude across the neural layer at which the mapping exists. Designate two sites, N_1 and N_2 which receive an equal and maximal average output. Reverse the mapping argument and define the reciprocal mapping functions, RMF 1/2 and RMF 2/1 as described in Ref. (10). These functions, shown in the fourth layer in the figure, define the proportions of the total outputs received at the two sites N_1 and N_2 from various spatial points in the neural network, described in terms of corresponding points in the stimulus domain occupied by the target. The cross-hatched area represents the degree to which the two sites N_1 and N_2 share common neural network elements in receiving output. The degree to which the RMF's for the two sites overlap determines the extent to which there is "correlation" between their temporal variability of instantaneous output. As we will see, the concept of "volume overlap" between two neural sites provides the basis for a quantitative theory of multiple site probability summation.

The complete mathematical statement of multiple site probability summation (MSPS) begins with the so-called frequency of seeing curve shown in Fig. 5, obtained by successive presentation of various values of target contrast in the temporal forced-choice paradigm. The ordinate represents the probability of occurrence of detection, after allowance for chance successes made possible by the forced choice method, as described in Ref. (1). The ratio of the two parameters, sigma and mean of the normal

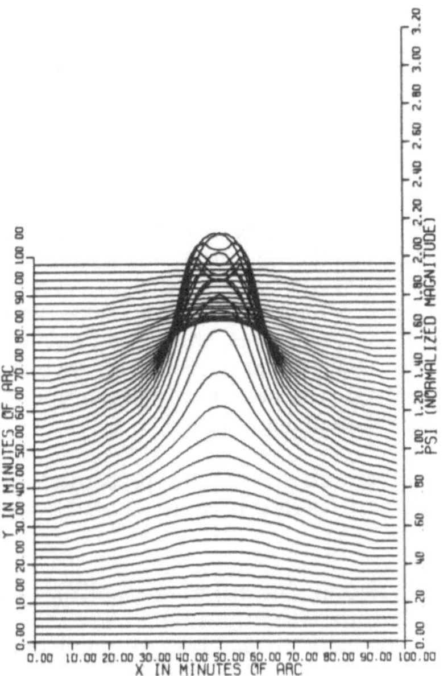

Fig. 4 Contourographic representation of the figure spread function of a simple annulus of width 1 minute and center-to-center diameter of 16 minutes.

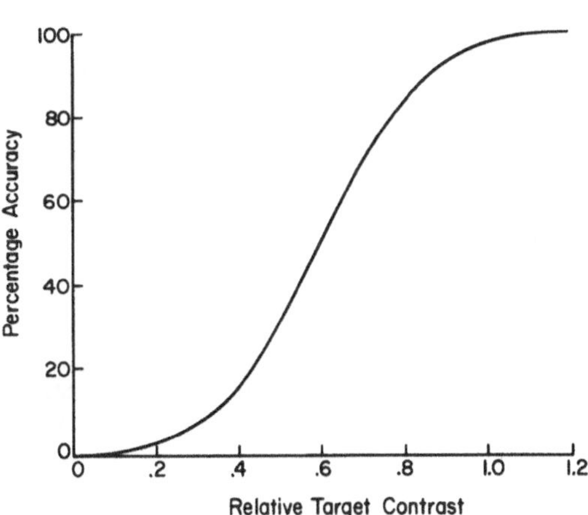

Fig. 5 Frequency of seeing curve obtained in the forced-choice paradigm. Percentage accuracy is computed from the probability of correct forced choice after allowance for successes due to chance.

frequency function depicted in the figure measures the numerical extent of variability in output at each point in the FSF from one moment to the next. Thus, each value of average output shown in Figs. 2-4 must be considered subject to the proportional variability of the human visual system revealed by the frequency of seeing curve. Now assume that a given absolute amplitude of output is required for detection, irrespective of the location at which it occurs. If all neural sites in FSF's such as those shown in Fig. 3 and 4 totally shared the same neural network, amplitude variability from moment to moment would consist of totally correlated waxing and waning of the amplitude of the entire FSF. However, if FSF neural sites such as those shown in Fig. 2 each partially shared neural networks with each other, amplitude variability would be enormously complicated in space through time. Perhaps an analogy will help to clarify the point. Suppose that each neural site receiving variable amplitude of output was used to excite a neon glow lamp, each lamp being set to trigger at an identical value of output. Suppose that detection occurred whenever one or more neon glow lamps were excited. With complex patterns of partially correlated neural sites, the spatial pattern of excited glow lamps at any moment would depend upon the precise target pattern, the PSF and the amount of overall temporal variability. Mathematical description of the probablistic aspects of these patterns presents a challenging problem.

The detailed method used to describe this situation will be presented by BLACKWELL, CHIOU, and BLACKWELL (12). It will suffice here to describe the general features of the method as follows:

Begin by defining the probabilities P_i that each neural site will have amplitude of output equal to or greater than a criterion value. Consider first the simple case shown in Fig. 2 in which there are but two sites and these have equal average amplitude. Then $p_1 = p_2$. The total detection probability, p, is given by the following:

$$p = p_1 + q_1 p_2 \qquad (1)$$

The value of p_1 includes all instances $p_1p_2 + p_1q_2$.

$$\text{Thus, } p = p_1p_2 + p_1q_2 + q_1p_2 \qquad (1a)$$

Assume first that the two sites share no common network components, and hence, that their temporal variability is uncorrelated. Eq. (1) is used directly. For example, if $p_1 = .5$, $p = .75$. Assume next that the two sites have some volume overlap as shown in the fourth layer in Fig. 2. Determing that the proportional volume of the overlap is .2 times the volume of one or the other RMF, then, the four probablistic terms would be written as follows:

$$p_1p_2 = w_{1\bar{2}}p_{1\bar{2}}p_{2\bar{1}} + w_{12}p_{12} \qquad p_1q_2 = w_{1\bar{2}}p_{1\bar{2}}q_{2\bar{1}} \qquad (2)$$

$$q_1q_2 = w_{1\bar{2}}q_{1\bar{2}}q_{2\bar{1}} + w_{12}q_{12} \qquad q_1p_2 = w_{1\bar{2}}q_{1\bar{2}}p_{2\bar{1}}$$

where the bars over 1 or 2 signify "not 1" and "not 2" respectively; and the coefficients, 2, represent proportional volumes relative to the volume of the RMS's. In our example, $w_{1\bar{2}} = w_{2\bar{1}} = .8$ and $w_{12} = .2$. In our simple case of the two point stimuli, $p_{1\bar{2}} = p_{2\bar{1}} = p_1 = p_2$.

In this simple case, eq. (1a) becomes

$$p = p_1 + w_{1\bar{2}}q_1p_2 \text{ (1b).}$$

Then, $p = .70$ when $p_1 = .5$. The 20% volume overlap obviously reduced p from .75 to .70 when $p_1 = .5$.

The method suggested by the above may be used to derive full frequency of seeing curves which include the effect of MSPS. Fig. 6 represents the simple case of uncorrelated sites described by Eq. (1), with the ogive parameters $\sigma/M = .310$. The effect of MSPS in this case is to reduce the threshold contrast by the factor .830. In the case in which the two sites shared .2 of the RMF volume, it may be shown that the effect of MSPS is to reduce the contrast required for detection by the factor .859.

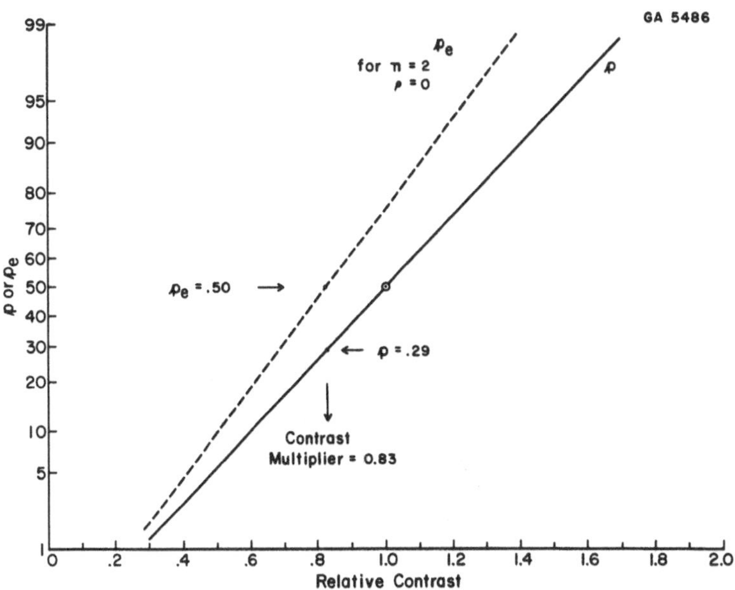

Fig. 6 Probit representation of the frequency of seeing curve for a single site (solid curve) and two possible sites (dashed curve), after allowance for multiple site probability summation.

When a FSF has extended area such as those illustrated in Figs. 3 and 4, the number of sites is potentially infinite. However, it may be shown that little error will be made by stopping the probabilistic analysis at seven sites. Of course, the locations of the seven selected sites must be chosen to obtain the maximum possible value of p for a given vlaue of p_1. Fortunately, sites which optimize p for one value of p_1, optimize values of p for all values of p_1. Site selection proceeds by educated trial and error procedures.

The full MSPS analysis to be described in Ref. (12) involves solution of probabilistic equations analogous to Eq. (1b) expressed in terms of seven sites. There are in all 128 values of p_i to be computed for each value of p_1 of interest, and a total of 128 values of w_i to be computed for each target of interest. The values of p_i each require calcuation of the average amplitude at each site and evaluation of this amplitude in terms of the criterion amplitude implied by each selected value of p_1. The calculations of w_1 each involve computing the volume contained in each overlap category. There will be overlap between each pair of sites, between each triad of sites, etc. In each case, the overlap volume must be computed twice, once with a target assumed to have infinitely large size and then again with an actual target. the method of volume calculation involves in principle cutting slices through each overlap volume at a succession of values of the ordinate y. In each case, x bounds of the volume are given and each of these is used to calculate a slice of volume. The volume overlap consists of the total of these volume slices. The computer program is indeed cumbersome and expensive, but it is capable of computing predicted contrast thresholds for any target of uniform luminance, using a selected PSF. Of course, the above remarks imply that circular disc targets also show multiple site probability summation. Hence, the PSF cannot be derived directly from the threshold contrasts obtained with the disc targets of varying diameter as proposed in Ref. (3) in accordance with the first-order theory.

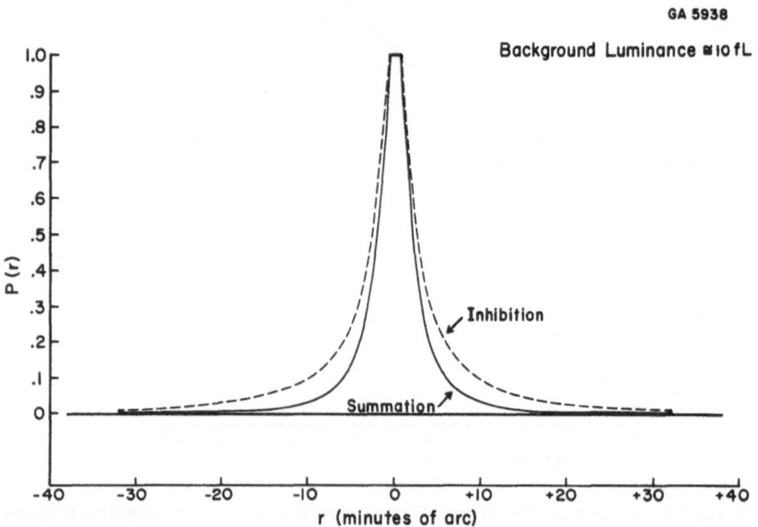

Fig. 7 Hypothetical point spread functions obtained from physiological mechanisms representing spatial summation and spatial inhibition respectively. (see text for explanation).

The full computer program implied by the above descriptions has been used to analyze the threshold contrast data obtained at a background luminance of about 10 foot Lamberts reported in Ref. (7), (9), and (10). Various PSF"s have been used. The PSF providing best fits to the obtained thresholds was constructed in the following manner. It was assumed that the all-positive PSF's which data such as those presented in Fig. 1 seem to demand are the resultant of two underlying physiological processes, one involving lateral summation, the other lateral inhibition. Figs. 7 and 8 demonstrate how an all positive PSF can be obtained from such processes. Fig. 7 depicts hypothetical PSF"s for the two processes, each plotted as positive for direct comparison. Values of r represent distances from the ocular fixational center, in minutes of arc. The PSF's are truncated at ± 32 minutes since targets with larger dimensions than this have deliberately not been studied due to the known non-uniformity of the receptor populations involved beyond these limits. It is assumed that each process has radial symmetry, and that the effects of the two processes are linearly additive. The mathematical functions shown in Fig. 7 were constructed from logarithmic normal frequency distributions, that is, normal frequency distributions expressed in terms of log r, but are plotted here in terms of a linear scale of r. The logarithmic values of σ are .5 for the summative and .6 for the inhibitive processes respectively. Each function was computed by defining x = log 2r and then computing the relative ordinate height of each function from the value of x/σ at each value of r, this quantity being measured in minutes of arc. This definition has the effect of producing a flat top on each PSF extending $\pm$ .5 minutes from r = 0. It is considered that these flat tops can reasonably correspond to the fact that human observers receive light stimuli modified by their dioptric apparatus, which produces image blurring often described approximately by an equivalent blur circle. The flat tops represent the estimated blur circles for the pupil sizes implied by the background luminance used. Since the image blurring effects both physiological processes, the two functions have been given the same flat tops.

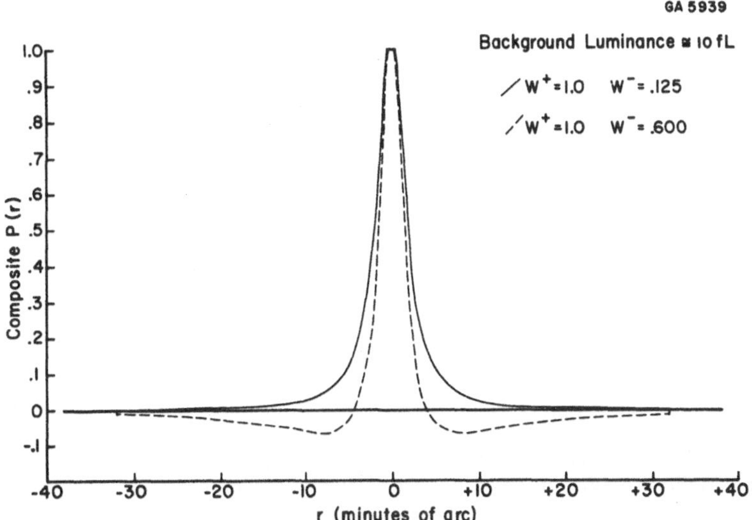

Fig. 8 Hypothetical point spread functions derived from combining the curves from Fig. 7 with the coefficients w^+ and w^- as noted. (see text for explanation).

Figure 8 depicts composite PSF's obtained by adding plus values of P(r) from the summative process and minus values of P(r) from the inhibitive process, with different weights given to the inhibitive process relative to the summative process. Note that when w = .125 and w^+ = 1.0, the composite PSF has all positive values, at least for values of r 32 minutes. This all positive PSF can be said to include hidden or silent inhibition. When the value of w is increased to .600 compared to a value of w^+ = 1.0, the composite PSF has the center positive-surround negative characteristic obtained in electrophysiological measurements and implied by various psychophysical measurements as well. It has been proposed in Ref. (2) that the conditions of the forced-choice detection experiment de-emphasize the role of the inhibitive process in comparison with the conditions under which the psychophysical and most electrophysiological measurements are made. Fig. 8 demonstrates support for this proposal at least to the extent of showing its quantitative plausibility.

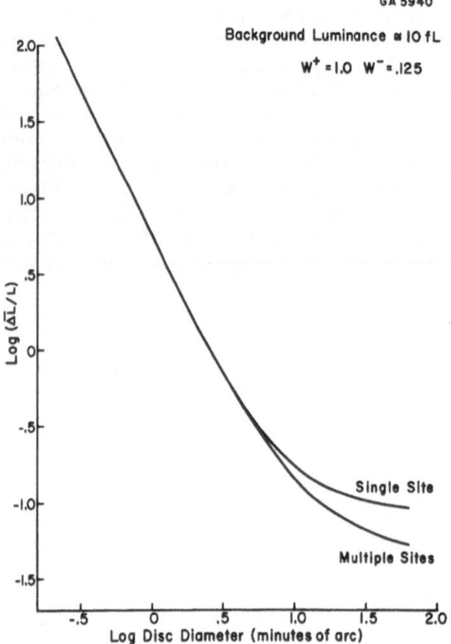

Fig. 9 The relationship between log contrast threshold and log disc diameter derived from the solid curve of Fig. 8 by means of two assumptions: (a) the single-site detection theory; and (b) the multiple site detection theory involving multiple site probability summation.

Figure 9 shows the relation between threshold contrast and disc diameter predicted from alternatively the first-order theory and the MSPS theory, using the solid curve in Fig. 8 as the PSF. It will be shown in Ref. (12) that the curve predicted on the basis of MSPS provides an excellent fit to the combined data obtained in the experiments reported in Ref. (7) and (9) for disc targets. Predicted threshold contrasts for annular and all-positive square wave gratings agree with obtained data within acceptable limits, but only when the MSPS theory is used.

It should be pointed out that the computation methods described here obviously use reciprocal mapping functions (RMF) rather than point spread functions (PSF). The usual assumption is made that RMF and PSF are identical in relative units. The single-site theory which involves only the linear addition of effects due to different elements in each target can be used in either the forward-going (PSF) or backward-going (RMF) directions. However, the non-linear operations involved in MSPS can only be described in terms of the backward-going direction. Thus, we have been constrained to look at the spatial transfer processes as the physiologist does when he measures the receptive field of a neural cell rather than as the physicist does when he convolves a target luminance profile with a PSF. It should be obvious that our RMF is formally identical to the physiologist's receptive field. We designate it the reciprocal mapping function (RMF) to emphasize that we are using a mathematical not a physiological entity and that it is identical in relative terms to the PSF.

Although not explicitly stated, the methods described to this point assume that each neural site has the same RMF. Fig. 10 is duplicated here from Ref. (1) to illustrate that this cannot be the case. Data in the figure were obtained by dividing

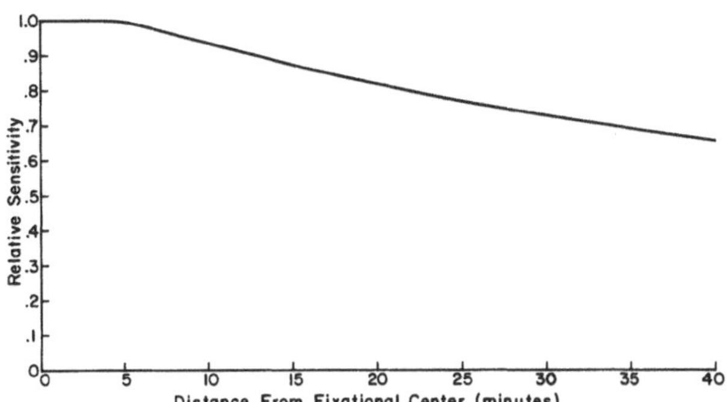

Fig. 10 Data from KRISTOFFERSON and DEMBER (13) showing the reduction in sensitivity as a function of distance from the fixational center, using the forced-choice paradigm with point source targets.

values of the threshold contrast for single point source targets presented at different distances from the fixational center into the threshold contrast for the fixational center. The graph shows that there is a loss in sensitivity when the point source is presented more than about 5 min. from the fixation center, which increases in amount with the extent of separation. These data were obtained with the forced-choice paradigm, with a background luminance of about 10 foot-Lamberts. They seem to indicate clearly that we should decrease the relative ordinate heights of our RMS's whenever a site is assumed to be located the equivalence of more than 5 min. from the fixational center. The computer program allows this change. However, when allowance is made in this way, predicted thresholds no longer agree as well as before with obtained values. Rather, it is necessary for the breadth of the RMF for the eccentric foveal locations to be broadened, while the height is reduced. We are currently in the process of developing the optimum patterns from the RMF broadening to provide the best fit to the obtained thresholds. When included, we may consider that the final detection theory consists of: (a) a linear additive stage; (b) a non-linear stage to take account of MSPS; and (c) a non-linear stage to take account of differences in both point sensitivity and RMF as a function of the spatial location of the neural site or sites involved in the detection of a given target.

Acknowledgements

We are deeply indebted to our colleague Dr. Brian H. P. Tsou for his assistance with modifications of the computer program. We are also most grateful to the Instruction and Research Computer Center and to the College of Biological Sciences, both of the Ohio State University, for their assistance in providing computer time for this study.

References

1. H. R. Blackwell, J. Opt. Soc. Amer. 53, 129 (1963).

2. H. R. Blackwell, Chapter 4 in D. Jameson and L. M. Hurvich (Eds.), Handbook of Sensory Physiology: Visual Psychophysics, Vol VII/4., (Springer-Verlag, Berlin, 1972).
3. W. M. Kincaid, H. R. Blackwell and A. B. Kristofferson, J. Opt. Soc. Amer. 50, 143 (1960).
4. G. A. Fry and P. W. Cobb, Trans. Amer. Acad. Ophthal. Oto-Laryng 423 (1935).
5. C. H. Graham, R. H. Brown and F. A. Mote, J. Exp. Psychol. 24, 555 (1939).
6. H. R. Blackwell, J. Opt. Soc. Amer. 42, 606 (1952).
7. A. B. Kristofferson, Ph.D. Dissertation, University of Michigan, Ann Arbor, (1954).
8. J. P. Thomas, Proc. 76th Annual Convention Amer. Psychol. Assoc. 3, 107 (1968).
9. T. G. Cleaver, Ph.D. Dissertation, Ohio State University, Columbus, (1969).
10. W. C. Chiou, Ph.D. Dissertation, Ohio State University, Columbus, (1973).
11. H. R. Blackwell and W. C. Chiou, J. Opt. Soc. Amer. 64, 1400 (1974).
12. H. R. Blackwell, W. C. Chiou and B. R. Blackwell, (in preparation).

Interactions Among Spatial Frequency Channels In The Human Visual System

Karen Kennedy De Valois
Primate Vision Laboratory
Psychology Department
University of California

Visual patterns consist of variations in two-dimensional space of the intensity and/or the hue of light. Historically, visual scientists have treated pattern perception and color vision as separate topics, and have generally studied the former with stimuli which vary only in luminance, not in color. Although there are practical advantages of using only luminance-varying stimuli, which are easier to produce, to study pattern vision, this has had the unfortunate effect of leading the unthinking to believe that only luminance differences are involved in form perception. This error has doubtless been reinforced by an emphasis on contours in form perceptions, since it is clear that color differences contribute little to sharp contours (BOYNTON AND KAISER (1)). It needs to be emphasized that color variations in a scene also contribute to pattern perception. They may, in fact, be more important than luminance variations in some situations when both are present. We are more sensitive in the low-spatial-frequency range to color than to luminance variations (VAN DER HORST and BOUMAN (2)). In the contribution of low spatial frequencies to pattern vision--a topic ignored in the classic literature--color may well play a dominant role. It would thus be of considerable interest to see to what extent recent theories of pattern perception based on luminance-varying stimuli also apply to patterns defined by color variation.

One of the most useful and provocative recent theories of pattern perception suggests that the visual system contains a number of spatial frequency-specific channels which act as bandpass filters (CAMPBELL and ROBSON (3)). They are presumed to respond to the spatial frequency content of a pattern rather than to more "naturalistic" features such as edges and bars. With the limitations of finite bandwidth, such a system could be described as a crude, patch-wise, Fourier-analyzing mechanism (ROBSON (4)).

A great deal of evidence has accumulated, both psychophysical and physiological, which tends to support this spectral-analyzer model of pattern perception. Among the most convincing is the demonstration that prolonged viewing of a high-contrast, luminance-varying, sinusoidal grating of some frequency, f, is followed by a temporary, orientation-specific, band-limited loss in contrast sensitivity centered at the frequency and orientation of the adaptation grating (GILINSKY (5); PANTLE and SEKULER (6); BLAKEMORE and CAMPBELL (7)). The loss in sensitivity falls to zero by about f $\pm$ 1 octave and has a bandwidth at half-amplitude slightly greater than 1 octave (BLAKEMORE and CAMPBELL (7)). This finding provides strong evidence for the existence of multiple, spatial frequency-tuned channels within the visual system.

It would be of great interest to know whether there exist similar frequency-tuned channels which respond to patterns defined not by luminance, but rather by variations in hue. It is possible that the same channels subserve spatial frequency analysis, regardless of the stimulus dimension on which the pattern is defined. It is also possible that there are multiple distinct sets of channels, some of which respond to luminance and others to color variations. If so, such channels might be completely independent or they might interact in some way. The experiments reported here were initiated in an attempt to answer the first of these questions--namely, are there spatial frequency-specific channels which respond to patterns defined solely by variations in hue. Some unexpected features of the results led to a re-examination of the original BLAKEMORE and CAMPBELL (7) experiment, using luminance-varying patterns; these experiments led to a discovery of important new characteristics of the phenomenon.

Two Tektronix 602 oscilloscopes with a white, P4 phosphor were used to produce hue-varying stimuli. Sinusoidal gratings of identical frequency were produced on the two, but with the pattern phase-shifted 180° on one with respect to the other. The two were then optically combined with a pellicle beam splitter. If the gratings on the two oscilloscopes were equal in contrast and brightness and no filters were used, the result would then be a uniform field. In order to produce a sinusoidal color variation, Wratten color filters were placed between each scope and the beam splitter. When the brightness and contrasts of the two scopes were properly adjusted for the particular color combination, a pattern was produced which varied sinusoidally in hue from, say, red through red-green mixtures to green, without luminance variation. A 2 mm artificial pupil was used to reduce chromatic aberration.

Any of a number of color combinations could be used, but our measurements were limited primarily to red-green patterns. For experiments using luminance-varying black-white gratings only one oscilloscope was used.

Stimuli were generated and presented, and the data were recorded and analyzed by a Nova 1220 computer. The observer was given control of a multi-turn logarithmic potentiometer which the computer read and used to adjust the contrast of the pattern (for the color-varying gratings, both oscilloscopes were, of course, changed in contrast simultaneously). Before-adaptation threshold settings were made for 22 spatial frequencies, covering the frequency range from 0.59 to 22.63 cycles/degree in 1/4 octave steps. Subjects then adapted to a high-contrast grating, either luminance- or hue-varying, for five minutes, then began making the post-adaptation settings at the same frequencies in the same random order as before adaptation. If the observer had not completed his setting in 5 sec, the adaptation grating was interposed for another 10 sec and he was given another chance, etc. When the setting was completed at one frequency, the adaptation grating was presented for 20 sec, then the next test frequency appeared.

Following adaptation to a high-contrast, hue-varying grating, subjects showed a temporary loss in sensitivity to other hue-varying gratings (of the same color combination) at and about the adaptation frequency. The loss in sensitivity fell to zero at about f ± 1 octave, and the shape of the loss describing the adaptation function appeared to be quite similar to that reported by BLAKEMORE and CAMPBELL (7) for luminance-varying gratings. Thus, one can conclude that there exist frequency-specific channels which subserve detection of patterns which vary spatially only in hue. Although the spatial frequency range at which pattern perception operates with color-varying stimuli is shifted to lower frequencies with respect to that using luminance variations, the operation of the color and luminance systems in pattern perception appear otherwise to be very similar.

Although the loss in sensitivity seen with these color gratings at or near the adaptation frequency was quite similar to that BLAKEMORE and CAMPBELL found with luminance-varying gratings, when frequencies far removed from the adaptation frequency were tested an unexpected phenomenon appeared. Frequencies which were removed from the adaptation frequency, f, by more than about 2 octaves showed a significant increase in contrast sensitivity following adaptation! The observer after adaptation was more sensitive in this spatial frequency range than he had been in the normal, unadapted state. Fig. 1 shows an example of the change in contrast sensitivity to red-green gratings of varying frequencies following adaptation to a red-green grating of 1.19 cycles/degree. As with luminance-varying gratings, there is a band-limited loss in sensitivity centered around the adaptation frequency, f, which falls to zero by about f ± 1 octave. (For this adaptation frequency, the low-frequency end of the function did not fall to zero. This was not the usual case, however.) For frequencies which were greater than f by 1-2 octaves, the change in contrast sensitivity was small, but for frequencies removed by more than 2 1/2 octaves there was a significant increase in contrast sensitivity, reaching a peak at 9.51 cycles/degree, 3 octaves above f. By a frequency of 19.03 cycles/degree, 4 octaves above the adaptation frequency, the change in contrast sensitivity had again fallen to zero.

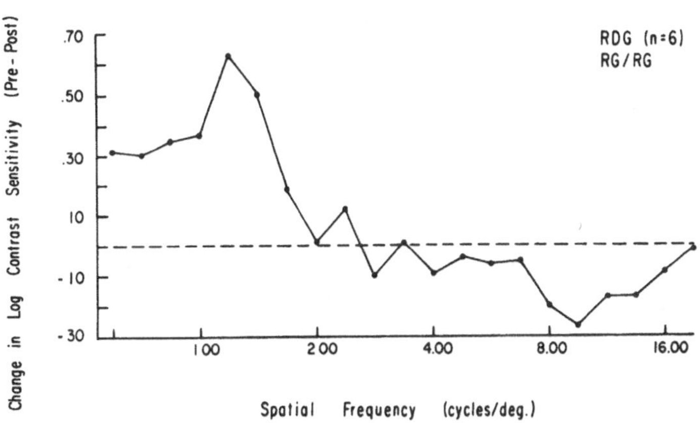

Fig. 1. Change in log contrast sensitivity for red-green gratings of various frequencies following adaptation to a high-contrast red-green grating of 1.19 cycles/degree. Note the significant increase in contrast sensitivity for test frequencies of 8-16 cycles/degree.

The simplest explanation, and the one I would like to propose, for the enhancement effect at frequencies far removed from the adaptation frequency is that there is mutual tonic inhibition among channels (cortical cells) tuned to different frequencies. The reduction in sensitivity due to adaptation of the channel tuned to the adaptation frequency would therefore reduce that channel's inhibition of channels tuned to other frequencies. These other channels, unaffected directly by the adaptation because of their fairly narrow bandwidths, would be relased from their tonic inhibition and thus show an actual increase in sensitivity.

The lack of independence among channels tuned to widely differing frequencies found in this color experiment clearly called for a careful re-examination of the adaptation effect with luminance-varying gratings. The original studies (PANTLE and SECULER (6); BLAKEMORE and CAMPBELL (7)) of frequency-specific adaptation did not discover enhancement to frequencies far removed from the adaptation frequency. In fact, they drew the conclusion that the frequency-specific channels tuned to differing frequencies are independent (BLAKEMORE and CAMPBELL (7)). However, these earlier studies examined only frequencies within about 1 3/4 octaves of the adaptation frequency, so would not have been expected to see the enhancement. Thus it was possible that the enhancement-after-adaptation phenomenon was not just a peculiarity of color gratings. We therefore repeated this experiment using black-white, luminance-varying gratings.

Figure 2 shows an example of the contrast sensitivity function for luminance-varying gratings before and after adaptation to a high-contrast grating with a frequency of 1.19 cycles/degree. As in the earlier color experiment, a large loss in contrast sensitivity is seen for frequencies at and near the adaptation frequency. The loss falls to zero at about f $\pm$ 1 octave, and there is little change in contrast sensitivity for frequencies within the range of about f $\pm$ 1-2 octaves. For frequencies further removed, however, there is a substantial <u>increase</u> in the subject's contrast sensitivity following adaptation.

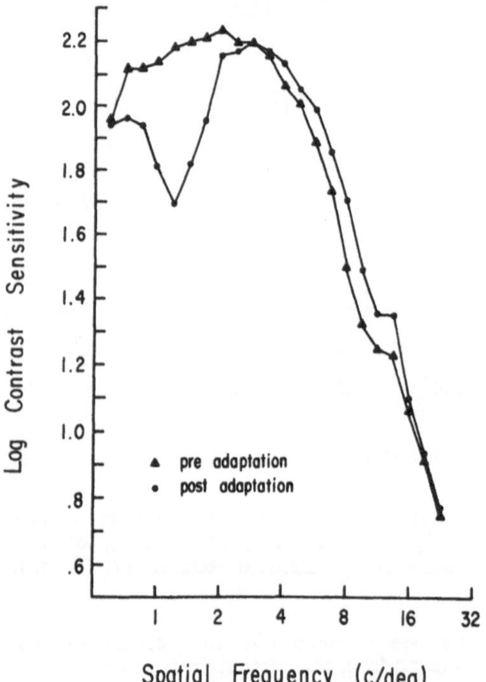

Fig. 2. Contrast sensitivity function for luminance-varying sinusoidal gratings of various frequencies following adaptation to a high-contrast luminance-varying sinusoidal grating at 1.19 cycles/degree. Note the increase in contrast sensitivity after adaptation for frequencies between 4 and 16 cycles/degree.

This adaptation procedure was repeated using three subjects and a variety of adaptation frequencies, ranging from .84 to 13.45 cycles/deg. Enhanced contrast sensitivity for distant spatial frequencies is found after adaptation to either low or high spatial frequencies. The peak of the enhancement generally occurs 2 1/2 -2 3/4 octaves away from the adaptation frequency. Fig. 3 shows how the contrast sensitivity change after adaptation varies with the difference between adaptation and test frequencies, the data from different adaptation frequencies being combined. As can be seen, the adaptational loss is greatest at the adaptation frequency and falls to zero at aobut f $\pm$ 1 octave, as BLAKEMORE and CAMPBELL (7) initially reported. In the region of f $\pm$ 1-2 octaves, there is little change, but beyond two octaves contrast sensitivity is enhanced, reaching a maximum at about f $\pm$ 2 1/2 -2 3/4 octaves.

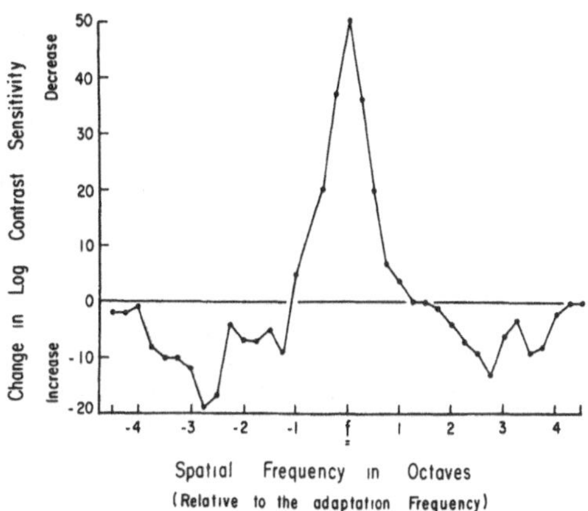

Fig. 3. Change in log contrast sensitivity as a function of the difference (in octaves) between test and adaptation frequencies. Data from 3 Ss and 59 sessions are normalized on the abscissa. Adaptation frequencies ranged from .84 to 13.45 cycles/degree.

The amplitude of the adaptational enhancement is between 1/3 and 1/2 of the amplitude of the adaptational loss; its bandwidth is perhaps slightly narrower. Enhancement can be seen quite easily with either a high or a low adaptation frequency; it is present but more difficult to find when the adaptation frequency is in the middle of the spatial frequency range--say, 3 to 5 cycles/degree. This is presumably because the maximum enhancement would be expected to occur at the extremes of the frequency spectrum where the visual system is least sensitive and measurements least reliable.

The presence of a frequency-specific enhancement following adaptation indicates that the channels detecting these gratings are not independent, but rather tonically inhibit each other. Consequently, in the unadapted state, no channel would show its maximum sensitivity. Only when those distant channels which inhibit it are desensitized by adaptation can a given channel's sensitivity rise to its maximum. This notion is quite contrary to our usual belief that the unadapted system is in the state of maximum sensitivity.

It is at first glance surprising that the maximum enhancement does not occur near the crossover from adaptational loss to enhancement, but rathersome distance away. Many systems which show lateral inhibition have maximum inhibitory interactions between elements which are contiguous or at least very close on some particular scale. (See RATLIFF (8)). In the case of spatial frequency-specific channels, however, it appears that the maximum inhibition--and thus enhancement after adaptation--is between channels tuned to frequencies which are separated by at least 2 1/2 octaves, despite the approximately one-octave estimation of channel bandwidth. However, the characteristics of the cortical cells which are probably involved provide a possible explanation.

Since the frequency-specific adaptational effect is also orientation-specific (GILINSKY (5)), one may safely conclude that it is cortical in origin. R. DE VALOIS, ALBRECHT and THORELL (9) have examined the frequency-specific responses of cells in area 17 of macaque monkeys and found that, while the median bandwidth of all cells studied is about 1.2 octaves, there are great individual differences among cells. Some show spatial frequency bandwidths as narrow as .6 octaves; others have a bandwidth as broad as 2 1/2 octaves.

There thus appears to be a variety of units of many different bandwidths tuned to each small frequency region; if each of these is mutually inhibitory with other units tuned to neighboring spatial frequencies, then the displaced region of maximum enhancement becomes more understandable. Narrowly-tuned units would show an adaptational loss only right around the adaptation frequency, and an adaptational enhancement only a short distance away; broadly-tuned units would show adaptational losses over a wide region about the adaptational frequency and enhancement only a long distance away. It might be expected, then, that in intermediate frequency regions the <u>enhancement</u> produced by the narrowly-tuned units would be cancelled by the <u>loss</u> produced by the broadly tuned ones. As a result, there would be little net change in either direction. Only at distant frequencies, then, would the pure enhancement effect be seen.

The explanation of the enhancement effect on the basis of inhibitory interactions among spatial frequency channels would be more convincing if cortical cells could actually be shown to demonstrate such inhibition. It should be reasonably simple to demonstrate inhibition, if it exists, in cells which show a high maintained discharge level. But since many cortical cells show very little spontaneous activity, it is sometimes necessary to use less direct methods. We have examined inhibitory interactions among cortical cells by two methods. The first involves finding cells which have some maintained discharge and seeing whether they fire to gratings of some spatial frequencies and are inhibited by other frequencies. Such is indeed found to be the case, as shown in Fig. 4. As can be seen, one cell fires to low spatial frequencies but is clearly inhibited by high spatial frequencies. The other cell responds best to high frequencies, and shows maximum inhibition to low frequencies. Both of these cells, thus, show frequency-specific inhibitory responses.

Another example of such inhibition in cells which have measurable spontaneous activity is shown in Fig. 5, in which is plotted the responses to gratings of various frequencies at several contrast levels by a complex cell. For gratings of 3.2 or 4.4 cycles/deg, the response increases as the contrast of the grating increases (up to a saturation level). But to either higher or lower spatial frequencies, 14 or .45 cycles/deg, increasing stimulus contrasts produced a progressive reduction of the cell's response below the maintained level.

The less direct method of demonstrating spatial frequency-specific inhibition in cortical cells involves observing a cell's response to a normally excitatory stimulus in

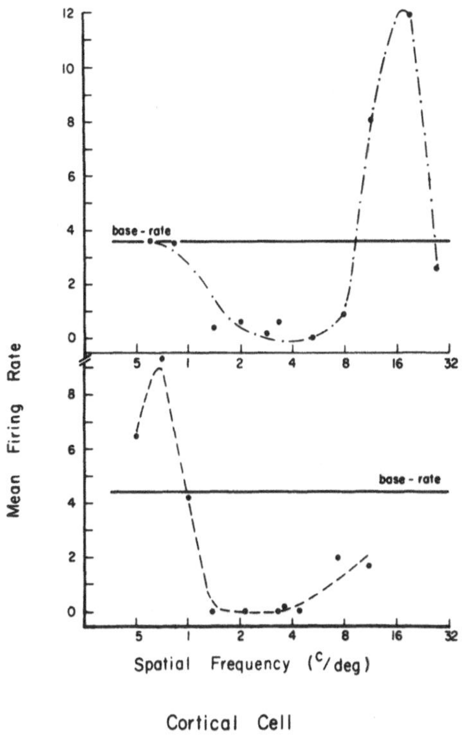

Fig. 4. Responses of two Y cells from area 17 of macaque monkey to sinusoidal gratings of different frequencies. Mean firing rate is plotted as a function of the frequency of the test grating. One cell shows its maximum excitatory response to a 16 cycle/degree grating and inhibition to gratings of 1.3-8 cycles/degree. The other is maximally excited by a frequency of .7 cycles/degree and inhibits to frequencies of 1.3-10 cycles/degree.

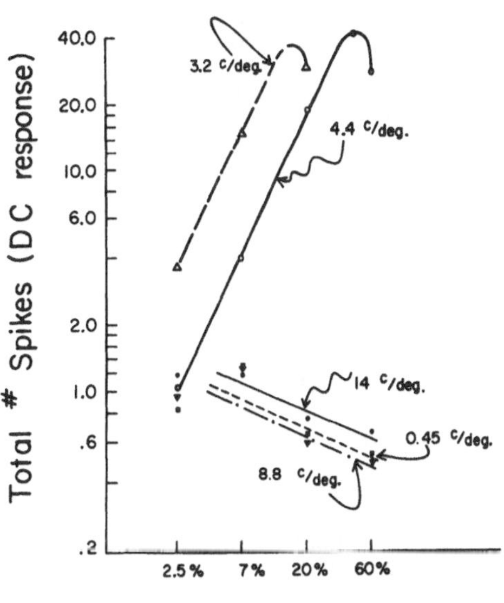

Fig.5. DC response of one cortical Y cell as a function of stimulus contrast for luminance-varying sinusoidal gratings of 5 frequencies. To two frequencies, 3.2 and 4.4 cycles/degree, the cell gives an increasing response to increasing contrasts. The other three test frequencies, 14, 8.8, and 0.45 cycles/degree, produce an inhibition (i.e., decrease in response rate) which is greater in magnitude for higher stimulus contrasts.

the presence and absence of another stimulus which may be inhibitory. We have done this by finding the frequency, f, to which a cell gives its maximum excitatory response. When a grating of four times that frequency, 4f, is presented, a narrowly-tuned cell with no maintained discharge will show no response. But when the two gratings are presented simultaneously the response elicited is significantly less than the response to the grating at frequency f alone. Thus, the presence of the higher frequency grating inhibits the response of the cell to the lower frequency grating. Similar results are often obtained when the second frequency is 1/4 as great as the cell's "best" frequency.

We thus have both psychophysical and physiological evidence for inhibitory interactions among units in the visual system tuned to different spatial frequency ranges. There are several important implications of this finding. The first concerns models which treat the visual system as a crude Fourier analyzer. An ideal Fourier analyzing device would have a very large number of channels of infinitely small bandwidth, each of which was totally independent. As was noted earlier, although estimates of channel bandwidths vary greatly, none approaches infinite narrowness, either in spatial frequency or in orientation. To this limitation, one must now add the restriction that the spatial channels are not independent.

Another implication, and one which is perhaps not immediately obvious, is that increasing the amount of information present in a visual scene may inhibit rather than aid in its recognition. A typical visual scene contains power at a wide variety of spatial frequencies. Generally speaking, the gross characteristics of objects are defined by low spatial frequencies, and the sharp edges and details are defined by high spatial frequencies. Since low and high spatial frequencies appear to interact in an inhibitory manner, it may well be that the presence of both in a scene reduces our ability to recognize either. This might correspond to the old saying that "one can not see the forest for the trees".

The "goodness" of vision has classically been measured by acuity tests. Acuity tests determine one's ability to see contours which are defined by very high spatial frequencies--e.g., the minimum visible separation or displacement of two lines or spots. Indeed, most of the research on spatial vision and theories about the visual system have for years been preoccupied with what we can now see as just the high frequency portion of the contrast sensitivity function. GINSBURG (10), however, has given convincing demonstrations of the overriding importance of low spatial frequencies in object recognition. These low spatial frequencies might thus play a very critical role in our ordinary visual tasks--perhaps more so (except in reading) than high spatial frequencies.

Since we are more sensitive to color than to luminance variations at low spatial frequencies, color variations might well be at least as important for form vision as luminance variations. The neglect of color in studies of form vision has coincided with the ignorance of, or at least the inattention to, the crucial visual role of low spatial frequencies. Given these considerations and the inhibitory interactions between low and high spatial frequencies, perhaps it would be profitable to consider other measures of visual ability which take into account the importance of low spatial frequency stimuli and the inhibitory effects of high frequencies.

Finally, it has been shown that there are some basic similarities between the manner in which the visual system processes information based on luminance variations and information based on hue variations. Both are important in pattern vision, and very likely equally important. We have not learned anything about the nature of the interactions, if any, between them. Any truly comprehensive theory of pattern perception must explore and include these relationships.

References

1. R. M. Boynton and P. K. Kaiser, Science 161, 366 (1968).
2. G. J. C. van der Horst and M. A. Bouman, Journal of the Optical Society of America 59, 1482 (1969).
3. F. W. Campbell and J. Robson, Journal of Physiology 197, 551 (1968).
4. J. G. Robson, In E. C. Carterette and M. P. Friedman (Eds.), Handbook of - Perception, V, Seeing, (Academic Press, New York, 1975).
5. A. Gilinsky, Journal of the Optical Society of America 58 (1968).
6. A. Pantle and R. Sekuler, Science 162, 1146 (1968).
7. C. Blakemore and F. W. Campbell, Journal of Physiology 203, 237 (1969).
8. F. Ratliff, Mach Bands: Quantitative Studies on Neural Networks in the Retina, (Holden-Day, Inc., San Francisco, 1965).
9. R. L. De Valois, D. G. Albrecht and L. G. Thorell, (in preparation).
10. A. Ginsberg, M. A. thesis, School of Engineering, Air Force Institute of Technology, Air University (1971).

Orientation Discrimination[1]

John Lott Brown and Iris M. Kortela
Center for Visual Science
University of Rochester

Introduction

The most carefully considered plans for a series of experiments are rarely followed. Early results often alter the investigator's perception of the problem and require revisions of plans. Our investigations of orientation discrimination are no exception.

It is appropriate to comment at the outset on the stimulus we have opted to use, two small spots of light. While this stimulus is geometrically the simplest we could use for orientation discrimination, its Fourier components are complex. We have chosen two spots for the ease with which we can characterize specific regions of the retina stimulated and their spatial separation at the expense of complexity of the Fourier components.

We arrived at two spots of light as the stimulus for an investigation of orientation discrimination by a diverse path. Several years ago, we were seeking a stimulus that would permit comparison of the results of neurophysiological studies in the cat and psychophysical studies in human subjects. Some attempts have been made to use two small spots of light to drive the excitatory center and the inhibitory surround of receptive fields of the cat, but they have not been very successful. A small spot of light in the surround has not provided sufficient stimulus, particularly in the dark-adapted retina. Neither have two spots of light proven useful for demonstrating inhibitory effects in human vision. SMITH and RICHARDS (1) did devise a method for studying what they believed to be retinal interaction in human subjects using line element stimuli. An interaction effect was observed when the stimulus was a reduction in luminance. The lateral conduction time of an inhibitory process which may have been retinal was calculated from the results. We hoped to achieve a similar result with a simpler stimulus. A requirement that orientation be discriminated was included to permit assessment of the correctness of the response

[1]This is a report of research supported in part by Contract N000 14-67-A-0389 between the University of Rochester and the Office of Naval Research and in part by a National Eye Institute grant, EY 000680, J. L. Brown, Principal Investigator.

and thus to provide some assurance that the stimulus was in fact discriminated. Our two points determine a line, and we can require the subject to tell us whether the line is horizontal, vertical, or at any other angle.

In another experiment, we had investigated critical duration with an acuity target (2). The purpose of that experiment was to determine whether in the completely dark adapted eye, the critical duration would differ significantly if the visual function depended on cones rather than rods. This required a visual task that would depend on the cone process no matter what the state of adaptation of the eye. We could insure this by requiring the subject to report the orientation of a fine grating. A grating target for the study of critical duration has given rise to complicated results, however. When critical duration is measured for visual tasks that involve form discrimination, it may become substantially longer. Critical durations as long as 500 mshave been reported under conditions of photopic vision as contrasted with the value of 10 ms frequently associated with photopic vision and the value of 100 mswhich is typical for simple light detection under scotopic conditions. One of the problems with an acuity grating may be its spatial redundancy. At threshold in a psychophysical experiment, the subject sees fragments of a grating. There is a suggestion of a line or lines in different locations from one determination to the next. The redundancy of the stimulus may be somehow related to extended summation times.

It is clear that to discriminate form or orientation, at least two spatially independent discriminations must be made at the same time or within the same limited time interval. Two light spots which determine a line with no redundancy may therefore afford the fundamental experimental stimulus for the measurement of orientation discrimination.

Experimental Conditions

In the specific conditions that we employed, the nearer of two spots was always located on a 45° meridian, up and to the left, 2° from the point of fixation. The spots were 5 min in diameter. The center distance of the two spots was varied from 10 to 40 min of arc. For lesser separations it becomes difficult to distinguish two 5 min spots. OGLE (3) reported summation of two spots for separations of 3 or 4 min in the central fovea. Summation occurs for larger spot separations with increased distance from the fovea.

The two spots in our experiment were positioned either in a horizontal or a vertical relation. This was achieved by rotating the more distant spot about the stationary spot as a center of rotation. Carefully positioned limit switches established the appropriate locations for the vertical and horizontal relations.

Spots were presented in flashes of 2 ms duration. We investigated various temporal relations between them; one might precede the other by various intervals of up to nearly 1/2 s, or they might appear simultaneously.

The apparatus is shown in Fig. 1: Channels A, B and C of a 4-beam optical system are illuminated by an incandescent lamp L_1. Channel A provides a small red fixation spot for the right eye. An identical spot is provided by Channel C for the left eye. This spot is congruent with the spot in Channel A seen by the right eye. The more distant, moveable test spot is presented via Channel D. A special holder for the field stop which defines this spot permits rotation of the spot about an axis which corresponds to the optical position of the stationary spot provided by Channel D. Channel D is illuminated by a ribbon filament lamp L_2, through a double grating monochromator. The monochromator was set at 575 nm and an interference filter

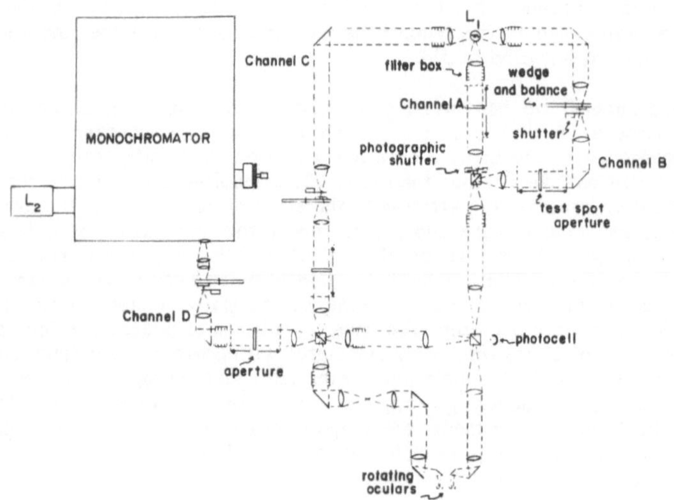

Fig. 1 Schematic illustration of optical system. Three beams, channels A, B and C, were illuminated by a tungsten filament lamp L_1. Channels A and C provided fixation points for the right and left eyes respectively. Stationary stimulus spot provided by channel D, illuminated by ribbon filament lamp, L_2. Moveable test spot provided by channel B.

was located in Channel B to provide spectrally matching illumination. The spot in Channel D could be presented either to the right eye or to the left eye by the appropriate arrangement of baffles. Luminance is controlled by motor driven wedges and by neutral density filters. A motor driven filter changer is located in Channel B. In combination with the wedge in that channel, this permits the automatic adjustment of luminance over a continuous range of 5 log units. Temporal control of stimulation is achieved by Uniblitz shutters(Vincent Associates) in Channels B and D. All experimental procedures, including adjustment of luminance values, timing of stimulus presentations, warning signals to subjects, and recording of threshold values are controlled by a PDP 8 computer. The system is described in more detail elsewhere (2).

Experiments were primarily monocular, with stimulation presented to the right eye. Some haploscopic thresholds were measured, however, with one stimulus spot presented to the right eye and the other to the left. For all experiments, the fixation spot was presented to both eyes against a completely dark background. Subjects found it more comfortable to fixate binocularly and we wished to maintain similar viewing conditions for both monocular and haploscopic threshold determinations.

Procedures

Light detection thresholds were measured for each of the two test spots individually. Independent measurements were made for each of the locations in which the spot illuminated by Channel B was to appear. There was no significant variation in threshold for this spot as a function of location. Threshold was the same for the spots illuminated by Channels B and D.

In a preliminary experiment, luminance thresholds for orientation discrimination were measured starting with both of the test spots at a luminance below threshold and increasing luminance on successive presentations in an ascending method of limits. Threshold was defined as that luminance at which the subject was first able to indicate correctly the orientation of the spots. Stimulus onset asynchronies (SOA's) of 0, 80, 160, and 240 ms were employed, with the spot in Channel B always preceding that in Channel D. Any possible interaction between temporal order and spatial location was ignored. Spacings used were 10, 15, 20 and 25 min of arc, center-to-center distance. Thresholds were measured in two subjects. There was no systematic relation between threshold and spot separation or stimulus onset asynchrony. No evidence of interaction between the two stimulus spots was found for any of the conditions studied. The threshold for discrimination of orientation of the two light spots was a little bit higher on the average than the light detection threshold. Presumably this reflects the fact that it was necessary to detect the two spots independently in order to know their orientation. The luminance necessary to reach threshold probability of detecting both of two spots must be higher than the luminance necessary to reach threshold probability of detecting only one.

In the experiment with which we are primarily concerned, a different procedure was employed. In order to enhance possible interaction effects, the luminance of the stationary spot, illuminated by Channel D, was maintained at a level 0.6 log unit above the light detection threshold. At that level, although it appears quite dim, the spot is seen each time it is presented. The threshold luminance of the other spot for discrimination of the orientation of the two was then measured, as influenced by stimulus onset asynchrony and spatial separation. The spatial separations used were 10, 20, 30 and 40 min, center-to-center. The stimulus onset asynchronies used covered a range of from -420 ms (the suprathreshold spot preceding the spot of adjustable luminance) to +490 ms at intervals of 70 ms. There were thus 14 different stimulus onset asynchronies.

Results

The results of this experiment are presented in Fig. 2 for each of two subjects, JLB and IMK. Data are plotted in terms of luminance in log microlamberts as a function of SOA for each of the stimulus spot separations for which results were available. Each of the points represents an average of the median values for two sessions for both of the stimulus orientations after the application of an adjustment for day to day variability.

The horizontal dash-dot line across each subject's half of the figure represents the light detection threshold. The luminances represented in Fig. 2 are the threshold luminances of the test spot in Channel B for correct identification of the orientation of the two spots as either vertical or horizontal. The most striking aspect of the results is the fact that luminance threshold for orientation of discrimination appears to be systematically lower over much of the range of SOA's than is the light detection threshold. The lowest threshold values appear to occur in the vicinity of SOA's of +70 to +140 ms. There is a suggestion that at extreme SOA values, thresholds may be higher than the light detection threshold, but the differences are small and inconsistent for the two subjects.

The results for the two subjects differ in several ways. Thresholds for IMK reach somewhat lower levels than those for JLB over the range of SOA's from -140 to +210 ms. On the other hand, thresholds for IMK are consistently higher for SOA's from +350 to +490 ms. The data for JLB show two minimum threshold values for the functions presented, one at an SOA of 140 ms and one at an SOA of 420 ms. The maximum value between these two minima is remarkably consistent for all three spatial separations represented. In the data for JLB the minimum threshold appears

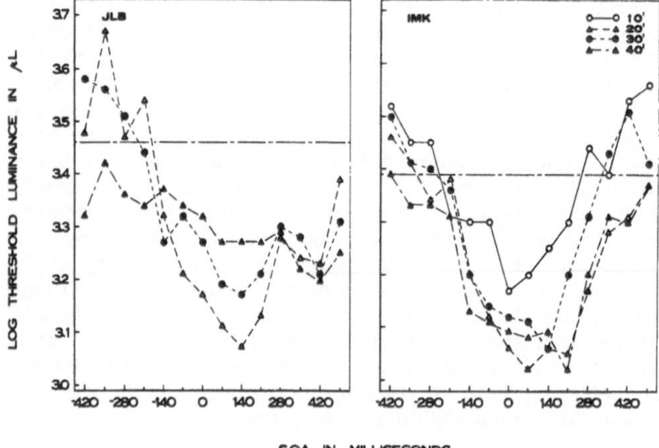

Fig. 2 Test spot threshold luminance for orientation discrimination as a function of stimulus onset asynchrony (SOA). Negative values represent delay of the test spot relative to the fixed luminance spot. Four separations of the stimulus spots are identified by symbols. Subjects JLB and IMK.

to be related to the spatial separation of the stimulus spots; the closer the spots, the lower is the threshold. However, results for JLB with the 10 min separation, although highly variable, suggest that threshold for this separation is somewhat higher than threshold obtained for the 40 min separation. In the data of IMK, the functions for separations of 20 min and 40 min are quite similar. Data for the 30 min separation rise more quickly with increasing SOA values beyond 140 msecs. The data for the 10 min separation represent the highest threshold values of all.

Statistical Analysis

The data were subjected to several statistical analyses. An analysis of variance was performed independently on the data for each of the two subjects. For the purpose of these analyses, data for adjacent pairs of SOA values were pooled. This procedure eliminated any confounding of day to day shifts in overall threshold level with SOA. The classifications employed were SOA blocks, spatial separation and stimulus orientation. The influence of SOA on threshold is highly significant for both subjects. Spatial separation of the test spots is highly significant for IMK, but not significant for JLB. Although the functions for different separations are dissimilar for JLB, the average threshold values over SOA are quite similar for the three separations investigated. In the case of IMK, the functions for three of the separations are similar in form, but differences in the overall threshold values for the different separations are apparently sufficient to render the spatial separation of test spots a highly significant classification. On the other hand, the different form of the functions for different separations is associated with a significant interaction between SOA and separation for JLB. This interaction is not significant for IMK. The orientation of the stimulus spots proved to be significant for both observers, at a slightly lesser level for JLB than for IMK. There was a substantially greater number of errors for horizontal positioning of the stimulus spot for IMK and a slightly greater number of errors for the vertical position for JLB. The interaction between stimulus orientation and SOA was not significant for either observer. The interaction between orientation and separation was significant for IMK, reflecting a substantially greater difference in threshold for horizontal and vertical orientations for the two wider separations than for the 10 and 20 min separations of the test spots.

The apparent deviation from an SOA of zero of the minimum threshold in the data of Fig. 2 indicates that the test spot for which luminance threshold was being determined, the dimmer of the two spots, must be presented ahead of the fixed luminance spot for minimum threshold. This finding, if correct, is of importance in the interpretation of these results. In order to test the statistical significance of this deviation from zero SOA, a procedure described by WILLIAMS (4) was employed. A polynomial equation was fitted to the function in Fig. 2 for each of the subjects. The differential of the fitting equation is determined, within which a minimum or maximum value may be expected at various chance levels. On the basis of this analysis, the deviations from zero of the minima for JLB were significant at the 1% level of confidence for both the 20 min and 30 min separations. The 40 min function for JLB in Fig. 2 does not provide sufficient definition of the minimum value for the technique to be applied. The best reasonable fit of this function is a straight line of negative slope. The function is thus in accord with the conclusion that threshold values are lower when the dimmer stimulus spot precedes that of fixed luminance.

In the data of IMK, displacement of the minimum threshold toward a positive SOA value was statistically significant at the 1% level for separations of 20 and 40 min and at the 5% level at a separation of 30 min.

Haploscopic Study

The results of threshold determinations with the variable luminance spot presented to the right eye and the fixed luminance spot presented to the left eye are shown in Fig. 3. Only the 30 min separation was employed. Both subjects showed a decrease in

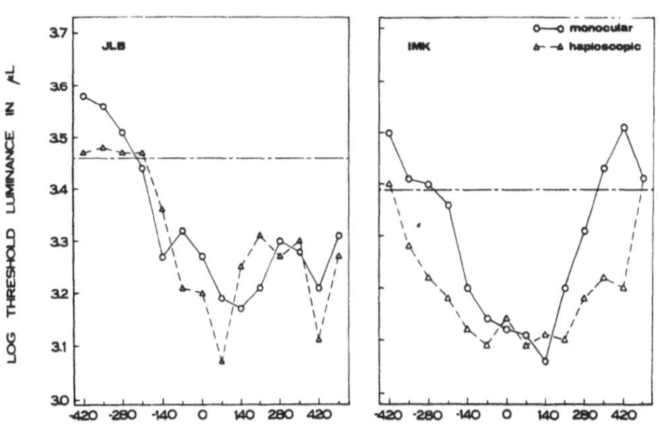

Fig. 3 Same as Fig. 2, but for 30-minute stimulus separation only with test spot presented to right eye, fixed-luminance spot to left eye. Monocular data for 30-minute separation from Fig. 2 are included for comparison.

threshold similar to that shown for the monocular data and in the same SOA range. The two minima which are shown in the monocular data for JLB are also found with this procedure, along with the substantially higher thresholds for negative SOA than for positive SOA values. The results for IMK are more symmetrical with respect to SOA than the results for JLB as was the case with her monocular data. The haploscopic results differ from the monocular in showing a broader region of SOA values in which threshold is substantially lower than the light detection values, but the minimum threshold values are remarkably similar for the haploscopic and monocular conditions. The qualitative results obtained monocularly are replicated haploscopically, both with respect to the general finding of lowest threshold in the middle region of SOA values and with respect to those characteristics of the results unique to each of the individual subjects.

Method of Constant Stimuli

An additional experiment was carried out to determine whether the result was dependent on the psychophysical method employed and to examine the slope of the frequency of seeing function for orientation discrimination thresholds. Light detection thresholds were redetermined for JLB and two additional subjects with the method of constant stimuli. Two spots separated by 30 min were then presented with one suprathreshold by 0.6 log unit as before and the other at each of 3 luminances ranging from the light detection threshold to a luminance 0.4 log unit lower. Only two SOA's were employed, -350 and +140 ms. These were selected as representative of the maximum difference in threshold found in the original experiment (Fig. 2). A forced-choice procedure was used; subjects were required to choose either "horizontal" or "vertical" after each stimulus presentation.

Results were the same for all three subjects. The light detection threshold based on 50% affirmative response in the constant stimulus procedure was approximately equal to the orientation discrimination threshold for an SOA of -350 ms. The orientation discrimination threshold for an SOA of +140 ms was significantly lower. The results for JLB are illustrated in Fig. 4. The original results are thus confirmed with a different procedure and with two additional subjects. It is of interest to note that the slope of the function for the lower threshold, +140 ms SOA, is less steep than that for the higher threshold, -350 ms SOA. This was true for all three subjects.

Discussion

The results of this experiment suggest that information concerning the brief illumination of a small region of the retina may be utilized more effectively when that illumination occurs in conjunction with stimulation of an adjacent region. The spatial orientation of two test spots in the visual field can be discriminated reliably when one of the spots is illuminated at a level below that required for its detection in isolation. This effect is optimum when a low luminance test spot for which a threshold value is being determined precedes by from 70 to 200 ms a stimulus spot of fixed luminance at approximately four times its light detection threshold level. One possible implication of this result is that the interpretation of the visual stimulus depends upon a cortical process which utilizes information from multiple retinal stimulation most effectively when resulting cortical events occur simultaneously. Thus, lower luminance stimulation, which is associated with a longer delay between retinal excitation and the subsequent resulting cortical activity, must precede higher luminance stimulation by a compensating amount of time under the circumstances. A stronger argument for dependence of the effect on a cortical process is provided by the replication of results with haploscopic stimulation. The lowered threshold can not be attributed to retinal interaction effects under these circumstances.

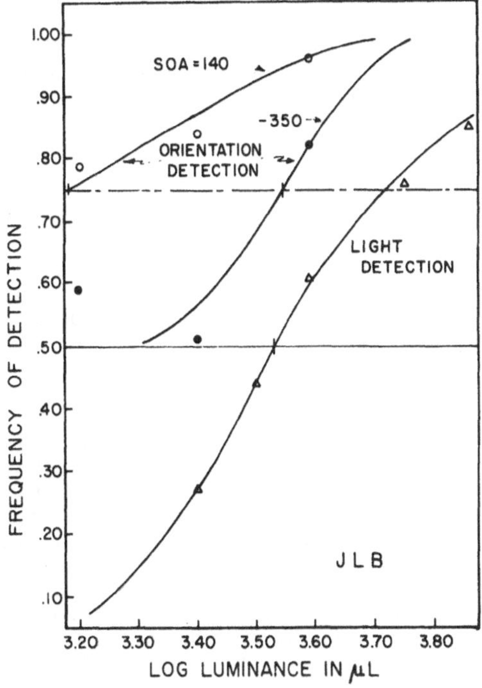

Fig. 4 Frequency of light detection or of correct detection of orientation as a function of the logarithm of test spot luminance. Stimulus onset asynchronies of -350 and +140 ms. Subject JLB.

Quantum Thresholds

There is a continuing debate as to the minimum number of quanta which must be absorbed by retinal receptors for a stimulus to be consciously perceived. It has been widely accepted for some time (5) that between 5 and 7 quanta are sufficient to produce sensation, although others have argued that as few as 2 may be sufficient (6). An argument favoring the larger number of quanta has been linked to the advantage this would afford in eliminating confusion from spontaneous retinal activity. Breakdown of individual rhodopsin molecules in the retina occurs spontaneously, but these events do not seem to result in spurious visual sensations.

An important element in this debate is the nature and stability of the criterion of threshold which a subject employs for detection of a visual stimulus. In complete darkness with no physical stimulation of the retina, one becomes aware of faint visual sensations which have no correlation with light stimulation. Such spontaneous effects and those resulting from stimulation can not be discriminated with confidence at low levels of physical stimulation. With forced choice procedures, however, subjects are able to detect very low levels of physical stimulation at significantly better than chance levels.

SAKITT (7, 8) has argued for the adequacy of a single quantum for stimulation of the visual system. She has demonstrated that under circumstances where subjects are very uncertain as to the presentation of a test flash, their ratings of the probability of presentation in each of several possible intervals are significantly better than chance (8). There is thus evidence that information is transmitted from retina to cortex under circumstances in which subjects are unwilling to accept the resulting effect as positive with respect to some threshold criterion in an ascending method of limits experiment.

In our experiment, when those stimulus spots which were subthreshold with the criteria of threshold we adopted for light detection were presented in conjunction with another suprathreshold stimulus spot in the right temporal relation, they were very often readily discriminable. The criteria of threshold were obviously different for light detection and orientation discrimination, but neither procedure involved forced choice.

In the later experiment when a forced choice technique was employed in orientation discrimination, the same facilitative effect was clearly shown for the +140 ms SOA. We conclude that the effect is not in some way an artifact of our psychophysical method. The less-steep slope of the frequency of seeing function for the lower threshold, +140 ms SOA condition, as compared to the -350 ms condition is in accord with the idea that fewer quanta are required at threshold (5).

Mechanism of the Effect

It is of interest to speculate what the mechanism for this effect may be. Both the nature of the SOA under optimum conditions and the replication of the effect haploscopically argue for a cortical locus of the mechanism. It is possible that cortical cells which respond much more vigorously to line stimuli than to point stimuli provide a basis for summation of any stimulus elements presented on the retina as long as these elements lie along the appropriate line. The visual acuity for detection of a dark line against a light background is remarkably high. A line subtending a visual angle of less than 1 s of arc can be discriminated if it is sufficiently long. A short line segment or a black spot with a diameter equivalent to the line width could never be detected. The explanation of this type of visual acuity must rest with some sort of summation process occurring in the cortex where a cell or cells receive converging signals from two or more retinal ganglion cells, the receptive fields of which are arrayed along the line of retinal stimulation.

Another possible mechanismwhich occurred to us was one involving cortical motion detectors. When two light spots at an appropriate separation are presented at the proper time interval, there may be the perception of motion. If specialized cortical cells are aroused by such stimulation, their activity might enhance the detection of the orientation of two spots. The subject might not see two spots clearly but would be aware of motion in a given direction and the direction of motion would define the orientation of these spots. The conditions under which minimum thresholds were found in the present experiment did not produce any significant apparent motion. Apparent motion was reported for larger SOA's, however, at near threshold levels. Therefore, movement discrimination mechanisms do not seem to provide a likely explanation of the results. We conclude that the facilitation effect may indeed depend upon some mechanism involving cells that respond to specific stimulus orientations.

Electrophysiological Studies

In an attempt to test some of our speculations as to the mechanism of the effect we have reported,we have undertaken electrophysiological studies of single cells in the cortex of the cat. This work is continuing and we will not report it in detail here. Some of the results are of interest, however, and some of the advantages of the procedure are worthy of mention.

We have found single cells in striate cortex that will respond to stimulus lines in a specific orientation. We have stimulated simple cells with small, stationary test spots presented individually at various positions along a line corresponding to the optimum position of the line stimulus used for original definition of the receptive field. Pairs of stimulus spots have then been employed in simultaneous stimulation. For several of the cells we have examined, two spots elicited a response similar in strength (number of spikes in a selected time interval) to the response to a line of light. The response to two spots was larger than the sum of the responses to the spots presented individually. More often, however, the response to two spots presented simultaneously is less than the sum of the individual responses.

The use of two spots with the more responsive cells permits an examination of the relation between optimum temporal and luminance conditions at the retina. Where two spots elicit a vigorous response, a reduction in luminance of one of the spots may require that it precede the higher luminance spot for optimal response. The relation between luminance and relative conduction time from retina to cortex for specific cell types can thus be assessed from investigations of this kind.

For complex cells, we have employed moving spots, swept across the receptive field in various positions. Here again, two spots may produce a vigorous response.

Additional Psychophysical Studies

We are presently conducting studies to determine whether the oblique effect may be manifested in our orientation discrimination paradigm. One might suppose that the oblique effect could lead to a reduction in the facilitation effect we have reported for two spots oriented obliquely relative to each other. This would be expected if the facilitation effect actually involves a number of line element detection cells and there are fewer such cells available for oblique as contrasted with horizontal and vertical orientations. Preliminary results for our investigations of horizontal and vertical orientations of the spots as compared with oblique orientations show no difference in the amount of the facilitation effect as a function of orientation.

References

1. R. A. Smith and W. Richards, J. Opt. Soc. Amer. 59, 1469 (1969).
2. J. L. Brown and J. E. Black, Vision Res. 16, 309 (1976).
3. K. N. Ogle, J. Opt. Soc. Amer. 52, 1035 (1962).
4. E. J. Williams, Regression Analysis (Wiley, New York, 1959).
5. S. Hecht, S. Schlaer, M. H. Pirenne, J. Gen. Physiol. 25, 819 (1942).
6. M. A. Bouman, in Sensory Communication, edited by W. Rosenblith (Wiley, New York, 1959).
7. B. Sakitt, J. Physiol. Lond. 223, 131 (1972).
8. B. Sakitt, Vision Res. 16, 782 (1976).

Discussion

Q. Dr. Blackwell: I have a suggestion. I would predict that you might discover that if instead of having the first spot present at .6 log units above its threshold, you were to put a ring around the place where your spots are presented, the threshold for both spots would be reduced. You had a visible spot in the same location on each presentation as part of the stimulus for orientation discrimination. The fact that it was involved in the orientation discrimination task may have had nothing to do with the facilitation effect you have shown. STAN SMITH showed some time ago that if you identify a locus of an off axis point, it increases sensitivity for that point. A kind of sensitizer, or attention mechanism undoubtedly is present. Something visible in the critical part of the retina tells the subject where to direct attention in order to detect the stimulus and facilitates a lowering of threshold.

A. It is an interesting idea.

Q. Blackwell: Incidentally, we have found something absolutely astonishing about off-axis threshold. If we had not been anxious to determine the upper asymptote very carefully, that is, get to 100%, we never would have found this. You can not just get a few data and assume you are going to get an ogive that goes to 100%. If you look in the center, and take a single Landolt ring, one degree off to the side, so it is a classical off-axis experiment, you find that it does not go to 100% when the contrast is high. For the two degrees, well, there is a couple of percent difference if you look carefully for it. It goes 98, not 100. Go to three degrees, go to four degrees, go to five degrees, and lo and behold, you can increase contrast to a maximum and you do not get 100% detection probability.

A. Is this not just further evidence that peripheral vision is imperfect?

Q. Blackwell: No, I do not think that is what it is at all. Because if you take the psychophysical curve you get, which is asymptoting at say 90%, and simply divide each probability by .9, that is, express it as a ratio if its own maximum in probability terms, you get the kind of ogive you are used to. No that is consistent with the idea you have a mechanism like blinking, only it is suppression, where the system shuts off and it shuts off a certain fraction of the trials regardless of the contrast.

It does not have to be peripheral vision if you have saccades, you get the same kind of thing more centrally than that, closer into the line of sight. I do not think it is peripheral at all; I think it is actually a general principle in the visual system. So you might want to look at the upper asymptote of your curves. You might find you are not getting 100%.

A. We have not done so as you saw in Fig. 4.

Follow-up Experiment

Dr. H. R. Blackwell suggested that the suprathreshold spot may serve to focus attention on the region of the visual field involved, or to enhance for a brief interval the sensitivity of cortical mechanisms which serve that region. The facilitation effect then would not depend upon having the suprathreshold spot and the facilitated spot both involved as elements in the discrimination of orientation. The suprathreshold spot would serve merely as a cueing signal.

As a test of this possibility, the stimulus pattern was changed. The suprathreshold spot was presented in the same location and at the same luminance. Two additional five-min spots were presented. The first of these was on the same 45° meridian as the suprathreshold spot, but 30 min further away from the fovea. The second was 30 min away from the first, either horizontally to the left, or directly above. These two spots were always presented simultaneously and at the same luminance, either 140 ms before or 350 ms after the suprathreshold spot. It was assumed that if the effect we have found can be explained as a kind of attentional effect or cueing effect of the suprathreshold spot, not specifically dependent upon orientation discrimination, then the threshold luminance for discrimination of the orientation of the two more distant spots might be influenced by their temporal relation to the suprathreshold spot; the facilitation effect should be found with a -140 ms SOA. This proved not to be the case. Threshold luminance for discrimination of the orientation of the two spots was approximately equal to their individual light detection thresholds at both SOA's. Our failure to get the effect under these circumstances may have resulted from the fact that the more distant spot was relatively far from the cueing spot. It must be sufficiently far so that our facilitation effect does not occur, however, or the orientation of the two low luminance spots would be revealed in terms of the orientation of each of them relative to the cueing spot. We used a spot instead of an annulus because we wished to maintain a stimulus condition as like that of the initial experiment as possible. It remains a possibility that the effect may be found with a cueing annulus.

The requirement for orientation discrimination could be eliminated and a single test spot could be located in the center of a cueing annulus. Temporal relations between the cueing annulus and the test spot like those in our experiment could be investigated. If a facilitation effect were found, it could not be attributed to orientation discrimination, but it might result from the sensitization by the annulus of a number of line element detectors with receptive fields passing through the location stimulated by the test spot.

A New Approach to Perceptual Grouping

Barbara Gillam
State University of New York, State College of Optometry
100 East 24 Street
New York, New York 10010

I will be talking about a very old problem in visual perception - which goes under various names such as grouping, coupling, organization and unit formation. Theories of such effects have tapped multiple levels of visual processing. My talk will add to what is known about the factors contributing to these effects.

It is generally assumed that underlying all perception there must be some initial process which imposes a structure on which later processes such as recognition and depth perception, certainly monocular depth perception, depend. For example, in order to recognize, one has to define the object of recognition. A depth cue must have a range of operation. (Not all information in the field has depth information and those parts of the field that have depth information must extend it to parts that do not.) This structuring of the input, which must underlie later processes, has been called "grouping" by the Gestalt psychologists and others. You have all seen pictures in elementary textbooks illustrating the Gestalt grouping principles and yet, in fact, very little is known about them beyond these demonstrations in line drawings. Very little quantitative work has been done. The reason for this is partly that it is very difficult to specify a criterion or even an adequate definition for grouping. It's very difficult, therefore, to find a measure of it. Detection and discrimination tasks require too little of the subject and recognition tasks require too much to really get at what we experience as grouping. So, we're left with a rather vague feeling that there's something to investigate there and we don't quite know what it is.

I want to describe a new approach to this subject involving a very precise definition of grouping - a definition which does yield quantitative measures which are sensitive to parametric manipulations. I have to digress a little to explain what this method is. Just consider an oblique line rotating around a central vertical axis (Fig. 1a). Consider further that one is not observing the line itself, but a parallel projection of the line. In a parallel projection, when a line rotates clockwise around a vertical axis, its edge moves sinusoidally in a horizontal path across the screen. However, because the light is parallel, if the same line were to rotate counterclockwise at the same speed, its edge would describe an identical sinusoidal motion across the screen. In a parallel projection of a line rotating into depth, the

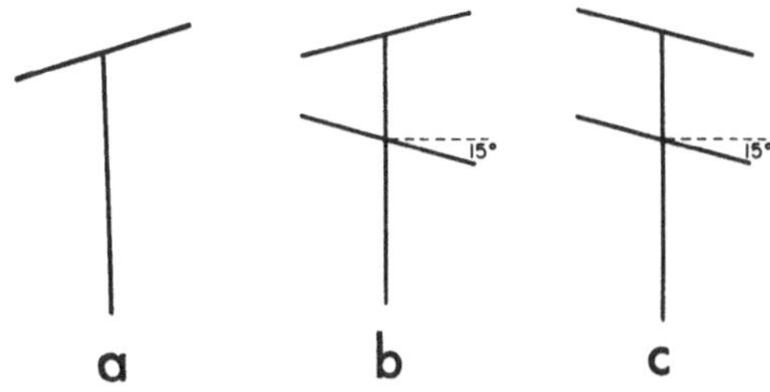

Fig. 1a.)
1b.) Some figures described in text.
1c.)

change in angle (see Fig. 2) creates a kinetic depth effect, that is, a powerful impression of rotation, but the stimulus is completely ambiguous with respect to the direction of rotation and so there are frequent reversals of apparent direction of rotary motion.

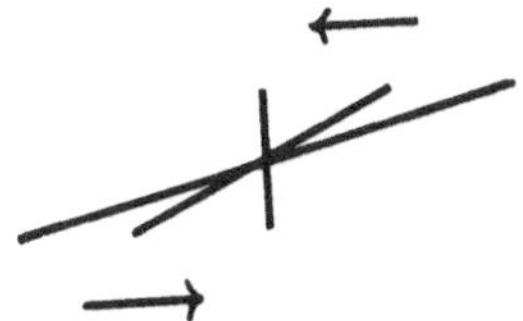

Fig. 2 Parallel projection of an oblique line rotating around a central vertical axis.

Now for my definition of perceptual grouping; consider more than one line rotating in a parallel projection. Let us assume that because of the ambiguity, the probability for each line of clockwise motion is 0.5 and of counterclockwise motion is also 0.5. Then the a priori probability of both lines appearing to rotate in the same direction at any time is 0.5 assuming that they are processed completely independently by the visual system, with respect to their motion in depth. In other words, if they're independent, they'll appear to be going in the same direction about half the time. Motion in the same direction more than half the time would signify a non-independence in the resolutions of the motion directions of the 2 lines. It is this lack of independence which I equate with grouping. Grouping can be defined as a coincidence in the resolutions of ambiguous state for two elements, more frequently than would be predicted from adding the resolutions obtained for them independently. In fact, as you shall see, in most of the experiments to be described we used an empirical rather than a theoretical baseline against which to evaluate the effect on grouping of various parameters.

Grouping in this sense of a common response to elements in ambiguous states is not uniquely confined to dynamic arrays such as I have used. Figure 3 shows the ATTNEAVE TRIANGLES (1). These triangles all appear to point in a particular direction. If you keep looking at them, they all suddenly appear to point in a different direction. They can point toward any one of three apices. From my point

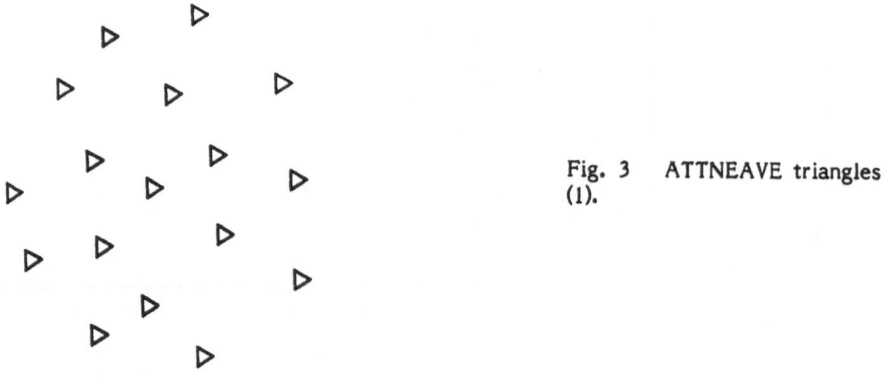

Fig. 3 ATTNEAVE triangles (1).

of view, the interesting thing about these triangles is that they are not responded to independently. When one reverses apparent direction of pointing, so do all the others.

In our experiments we had subjects view two or more lines on the same vertical plane rotating in depth around a vertical axis for ten revolutions and viewed in parallel projection. Subjects had to press a switch whenever the lines appeared to be going in opposite directions from one another. This time, accumulated over the ten revolutions, was called the "fragmentation time." We assumed it to be a negative function of grouping strength, i. e. the tendency for unitary processing of the lines. All the figures in our experiments subtended a visual angle of less than two degrees so that they were all clearly visible within one single glance.

With an operational definition of grouping and a quantitative measurement, we were ready to investigate its nature. In a sense the most interesting question about grouping is the extend to which it is entirely a high level "cognitive-type" process. In recent years, since the mechanisms proposed by Gestalt theory have been discredited, theories of grouping have tended to be couched in terms of the imposition of rules about line relationships and intersections (artificial intelligence approach) or the imposition of a minimum information principle to find the most redundant solution to any input (2, 3). On the other hand, physiological research raises the possibility that there are aspects of cortical neural coding which might influence grouping; for example, the nature of the receptive fields and the organization of line detectors in areas 17 and 18. Initially, we tried to investigate variables which might reveal more low-level topological influences.

The first major finding was that the grouping of two lines is strongly influenced by their angular similarity. Figures 1b and 1c show two of the basic figures for this experiment. The upper line was varied in 5 steps between its position in 1a and 1b. Fragmentation time as a function of this manipulation is shown in Fig. 4. So, whatever kind of aggregation process or dependency there is in the processing of these pairs of lines, it does decrease monotonically with a difference in their angle. I do not want to go into detail about this effect which I have dealt with elsewhere (4). Today I shall concentrate on the two extreme cases shown in Figs. 1b and 1c and ask what stimulus manipulations can influence the independence of the component lines of these figures. First of all, changing the orientation of either figure by 90° results in a very marked reduction in fragmentation time. That is, the figures behave much more like a unit when converging north or south (symmetric about a vertical axis) and rotating on a horizontal axis than when converging east and west (symmetric about a horizontal axis) and rotating on a vertical axis. In the case of Fig. 1b this axis effect

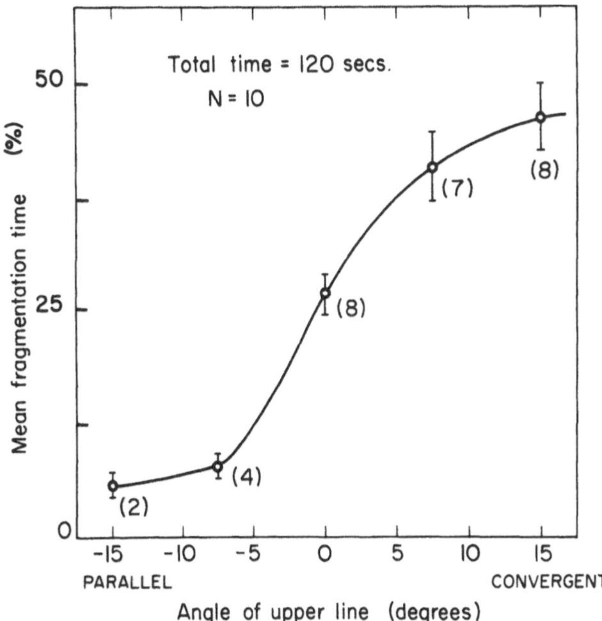

Fig. 4 Mean percent fragmentation time as a function of the angle of the upper line (lower line fixed at -15 deg).

may be owing to the greater saliency of symmetry around a vertical axis, as shown by BARLOW, JULESZ, ROCK and others. However, this would not account for the fact that we obtained the same effect of 90° rotation with figures that are not symmetric, such as Fig. 1b. We do not at present know the reasons for the axis effect. However, we do know something about it. It is determined entirely by figure axis relative to the retina, not relative to the environment or to gravity. In other words, it shows no orientation constancy. Table I shows the results of an experiment in which there was a complete reversal of the means for the two environmental axes when the subject was placed in a horizontal instead of a vertical viewing position. This lack of orientation constancy is interesting because it clearly differentiates nonconfigurational grouping in our sense from recognition processes with simple figures which do show orientation constancy (5). So here we have a determinant of grouping, which cannot be accounted for by cognitive-type processes.

Another candidate for a possibly early topological determinant of grouping would seem to be proximity, which in our figures is defined as line separation. We took Fig. 1b (converging lines) and Fig. 1c (parallel lines) and measured the fragmentation time as a function of line separation along the vertical axis for each figure. The results are shown in Fig. 5. You can see that for Fig. 1b fragmentation time is a monotonic function of line separation. The parallel lines (Fig. 1c) also showed an effect of separation. In fact, in later experiments the effect of separation on the parallel lines tended to be somewhat greater than is shown in this early experiment. So, here we have another effect which is difficult to account for by processes of a cognitive nature, like a minimum information principle or any kind of process involving schemata or meaning. It may be possible to relate this result very directly to neural coding in the visual areas. To see if this is feasible, it was

Table 1

1 2 3 4 5 6

	Vert. Axis Targets		Horiz. Axis Targets	
S Vertical	#1:	23.45	#4:	5.39
	#2:	19.02	#5:	6.01
	#3:	31.45	#6:	7.27
	X =	24.64	X =	6.22
S Horizontal	#1:	2.89	#4:	25.50
	#2:	2.33	#5:	12.36
	#3:	8.47	#6:	19.85
	X =	4.56	X =	19.23

important to determine whether or not the separation effect showed size constancy. If the distal (real) separation, rather than the retinal separation, were critical, then one could assume that the separation effect operates at a post-constancy level and presumably beyond area 18.

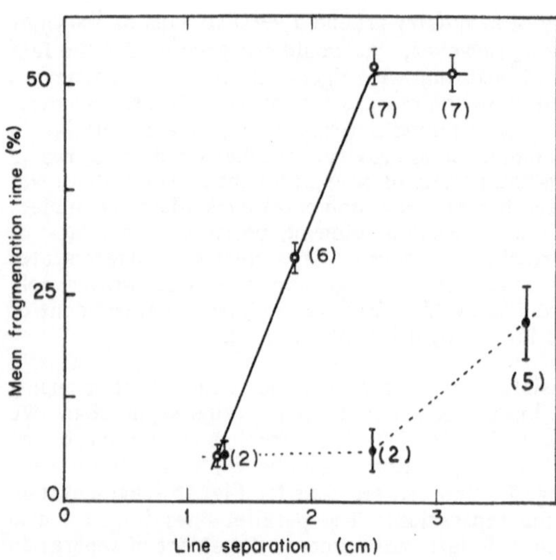

Fig. 5 Mean percent fragmentation time as a function of line separation for converging lines (open circles) and parallel lines (filled circles). N = 16, N = 14, respectively.

We presented pairs of parallel lines at three different separations with a ratio 1:2:3 at three different distances from the subject also in the ratio of 1:2:3. The results are shown in Fig. 6. Whereas separation along the axis had a large and

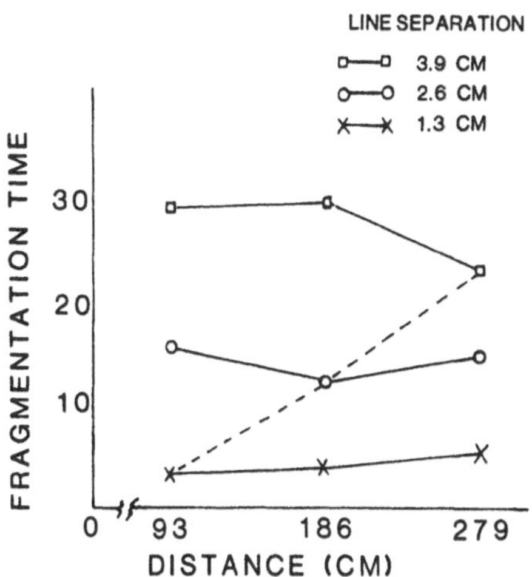

Fig. 6 Fragmentation time in secs for parallel lines as a function of viewing distance with line separation as a parameter. The dotted line joins line pairs with equal visual angle separations.

significant effect, distance had no significant effect although it produced equivalent retinal separations to the separation variable (see Fig. 6). So, retinal separation is clearly not what is important here. Either the distance is taken into account in evaluating separation for grouping, or alternatively the critical factor is the ratio of the separation of the lines to their overall magnitude (their length). To separate these two possibilities, we varied the overall retinal size of the stimuli (a) by changing distance and (b) by changing stimulus size directly and equivalently on the CRT screen. Again we found that there was a large and significant effect of changing line separation, but there was no effect of changing the overall size, either directly or by varying stimulus distance. The results of this experiment are shown in Table 2. It now seemed clear that grouping depends on separation relative to length. This was confirmed in a further experiment in which we varied the line length for a constant separation. Indeed, fragmentation time increased by a factor of 3 as length decreased by a factor of 3.

Clearly, we have an effect here which is not a constancy effect in the sense that distance cues are evaluated and distance taken into account, but neither is the critical separation retinal. The proximity effect depends on size relations within the configuration and has the effect of maintaining perfect size constancy. It would be odd if the grouping of contour elements into units did vary with their distance. If so, grouping would not reflect the actual dependencies of lines in the environment which, of course, do not depend upon distance. Although it is surprising that distance is not "taken into account", dependence on separation relative to the dimensions of the

Table 2 Fragmentation times for 3 pattern shapes viewed under different conditions.

		PATTERN				
		A	B	C		Means
	140cm (x=25cm)	32.76	17.68	26.71		24.72
VIEWING DISTANCE	140cm (x=12.5cm)	28.54	20.68	25.48	(half-size)	24.9
	280cm (x=25cm)	34.78	15.77	22.69		24.41
	Means	32.00	18.04	24.96		

Note: A, B, and C refer to the shape of the pattern as labeled in Fig. 2. "x" refers to the mean vertical distance between the oblique lines of the pattern.

separated elements would give a roughly veridical result in the grouping of elements into objects and surfaces and probably would result in a faster process than would an analysis of depth cues. ROCK & EBENHOLTZ (6) and others have considered the possibility that the relation between size of object and size of surround is the basic mechanism for size constancy. Our data are in line with that idea.

Most of what I have said so far about the nature of the separation variable as it affects grouping is based on data obtained using parallel lines. The situation is not so clear for converging lines - that is, lines of different orientations. In this case, any processes that organize those lines into a unit do seem to depend on distance to some extent. This applies to Fig. 1b and also to figures like Fig. 7 which have collinear

Fig. 7 A figure which exhibits closure (using the present grouping criterion) as a function of gap.

attachments which "group" (see ref. 7). It is possible that grouping of these lines represents an entirely different process from the grouping of parallel lines. These and other differences in the behavior of converging lines and parallel lines have led me to postulate two grouping processes. The grouping of parallel lines can be thought of as a process of aggregation similar to what you saw happen with the ATTNEAVE TRIANGLES. In the case of aggregation, a resolution that is applied to one element is automatically duplicated for other similar elements. In the case of the parallel lines, aggregation would refer to the situation in which one line mimics the other at any moment but what both lines do is not different in nature from what either line alone would do. However, when the converging lines are organized as a unit, we seem to have a somewhat different process going on. Here we have something which is much more like what the Gestalt psychologists meant by organization, that is, that the whole is more than the sum of the parts in the sense that what the pair of lines together do when grouped (usually oscillate) is different from what either line would do alone. The response emerges from the unified figure; it is not just a duplication of the responses to the individual elements. This latter process of organization seems to depend on distance in a way that aggregation does not. This distinction needs a good deal more development in experimental terms. We are currently looking at variables which differentially affect grouping in the sense of aggregation and grouping in the sense of organization.

So far, the determinants of grouping that I have mentioned appear to be operating at a relatively early stage in perceptual processing. They have been mostly either topological or, at the most, relational effects. They have all been obtained using only two lines. However, like many perceptual effects grouping is influenced by processes at different levels of abstraction in the visual process. The set of factors I shall next describe appear to arise from a higher level in the perceptual process where such factors as the minimum information principle could apply. This work is also of interest because more than 2 coplanar lines were used, which allowed us to investigate grouping factors which apply to multiple lines. This can be conceived of as research on the determinants of perception of a surface. The goal of these experiments was to take a readily fragmenting pair of lines (Fig. 1b), and to investigate the effects on the grouping of such lines of other lines placed between them (as in Fig. 8). The outer lines were constant and the inner lines varied in

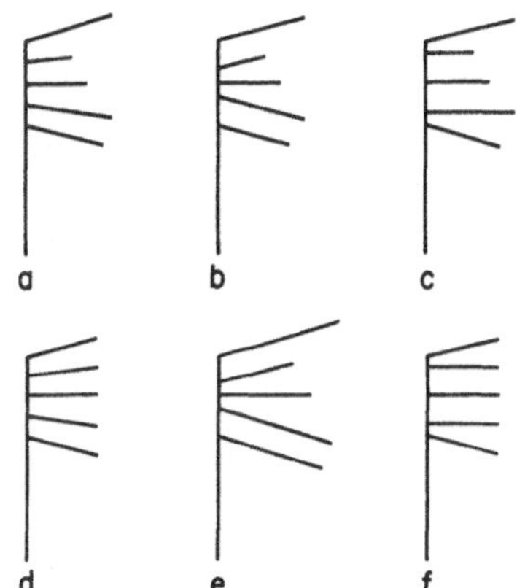

Fig. 8 Some examples of figures used to investigate surface perception.

number, angle, separation and length in various combinations in different experiments.

Another way of putting the question is to ask whether a dynamic resolution can be transmitted from one line to another across intervening lines and, if so, what properties must the intervening lines have to optimize this effect. In these experiments the subject was asked to press the button if any one of the lines appeared to rotate in a different direction from the rest. Figure 9 shows the figures used in the first experiment with their fragmentation times. Notice that the fragmentation time was drastically reduced as the intervening lines were placed between the outer lines. That is, there was a very powerful grouping elicited by placing the intervening

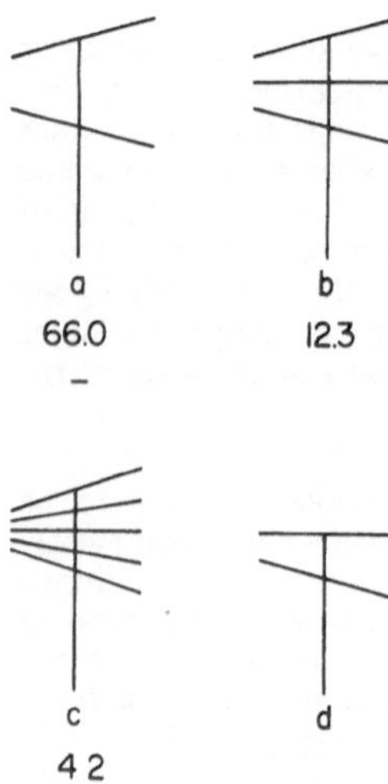

Fig. 9 Figures with varying number of "inlines" showing fragmentation times.

lines there.

The first question that arises is whether this is just simply owing to the greater ease of transmission of a depth resolution across neighboring elements when the interline separation is smaller. We showed that this is not a sufficient explanation by comparing fragmentation time for the two lines of Fig. 9d with fragmentation time for all three lines (Fig. 9b). Despite the greater opportunity for fragmentation in the three-line stimulus (Fig. 9b) fragmentation time was much greater for the two line case. It can be concluded that the effect of adding intermediate lines is not simply a matter of reducing the interline separation of successive lines. A second possible explanation of the results shown in Fig. 9 is that grouping is related to the redundancy with which an implicit vanishing point is defined. The greater the redundancy, the greater the grouping. We indeed found that intervening lines that formed an implicit vanishing point with the outer lines showed a greater degree of grouping than intervening lines of any other orientations. Another strong effect was an approximately two-fold decrease in fragmentation time when a set of lines formed an implicit colinear edge (Figs. 8d and f). However, even for the lines which did not form an implicit collinear edge, an implicit vanishing point facilitated grouping. Furthermore, the addition of intervening lines to the outer lines of Fig. 1b increase grouping even when the lines did not define a vanishing point or a collinear edge. This suggests that the reduction in interline separation per se contributed to the effect.

The vanishing point factor and the collinear edge factor (which are new grouping principles) can very well be explained by the minimum information principle, or by schemata theories. However, it is possible that a more elementary principle could be invoked, perhaps involving the Fourier components of the various patterns of lines. I have not looked into this possibility. Other effects in multiple patterns such as the grouping produced in a pair of lines by decreasing the separation of successive lines placed between the pair may depend on earlier processing stages.

Let me summarize what I have said today. We have devised a technique which I think allows us to measure unambiguously and with a good degree of reliability and sensitivity to parametric manipulation, a lack of independence in the depth processing of lines. Both apparently structural, topological aspects of the relationship between the lines as well as some apparently high level, more abstract, properties of the relationship seem to determine the grouping effect. Such grouping, particularly the kind we have called organization, would provide a basis for recognition. Grouping also provides for considerable economy in depth processing for it provides a basis for redundancy. Resolutions for one element could be applied automatically to other elements which would lead to a good deal of economy in the system. This applies to both the organization and the aggregation varieties of grouping.

To what extent the grouping process which we've discovered is general, I'm not sure. It seems likely that it is. However, in any case, dynamic three-dimensional displays have more ecological validity than static two-dimensional displays, so the effects are of great interest in themselves.

Note:

N. B. A film of many of these effects was shown at the talk and is available from the author.

References

1. F. Attneave, Am. J. Psychol. 81, 447 (1968).
2. F. Attneave, Psychol. Rev. 61, 183 (1954).
3. J. Hochberg and E. McAlister, J. Exp. Psychol. 46, 361 (1953).
4. B. Gillam, Per. Psy. 11, 99 (1972).
5. I. Rock, Orientation and Form, (Academic Press, N. Y., 1973).
6. I. Rock & S. Ebenholtz, Psychol. Rev. 66, 387 (1959).
7. B. Gillam, Per. Psy. 17, 521 (1975).

Optical Illusions and Visual Functions

R. A. Weale
Institute of Ophthalmology
London

The optical illusions I should like to discuss are non-trivial in the sense that they may be explicable in physiological terms. In other words, they may have significant objective elements.

The first illusion on my list is known as Mach bands. These can be observed at a luminance step between two uniform fields, when the edge of the darker of two fields appears even darker, and vice versa. The bands disappear when the boundary between the fields is occluded. MACH (1), their discoverer, attributed them to interaction between the stimulated retinal areas. In point of fact the bands were shown to exist by the Venetian painter MANTEGNA approximately in 1480. This is important because it establishes once and for all that visual science began before the discovery of America. The painting of the bands appears on one of MANTEGNA'S Triumph of Caesar cartoons at Hampton court near London: the artist painted not the stimulus but the perception, as good a documentary piece of evidence as one is likely to get for a scientific subtlety datable to the fifteenth century.

Elsewhere in this volume GLENN FRY showed a Fourier representation of a target, and it contained an illustration of what I wish to examine in connection with Mach bands, namely GIBB"S phenomenon (2). In this connection we ought to remind ourselves that the eye is a low pass filter (3). If it were an ideal one, it would be possible to characterize it with a single cut-off frequency $_c$: it would pass spatial frequencies without attentuation below V_c (= $_c$/ 2), and completely absorb all those higher than this. In fact, the eye is diffraction limited. If we assume that, to some degree of approximation, the theory of the step function (2) can be applied to two adjacent photometric fields, then it is easy enough to calculate the unit response given by an ideal low-pass filter (Fig. 1). Note that the angular gap R between the under-and overshoots is inversely proportional to the cut-off frequency: in fact, it is easily shown to be

$$R = 1/V_c$$

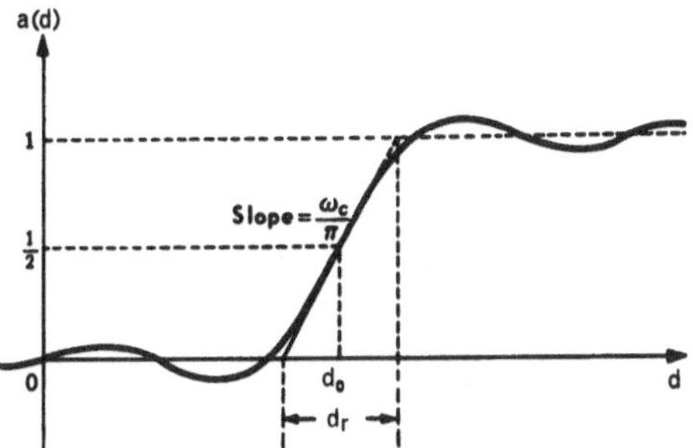

Fig. 1 The unit step response of an ideal low-pass filter.

In order to test the notion whether Mach bands may not perhaps reflect the overand under-shooting of the response, one can use results (4) for Mach bands spacings as a function of luminance on the one hand, and contrast sensitivity data (5) to provide cut-off frequencies on the other. These are defined as the frequency of semi-maximal response (6). Fig. 2 illustrates the comparison between observed and predicted values, allowance being made for the fact that the Dutch workers used different pupil sizes for the small and large fields respectively. No shifting of data or curve-fitting was needed to obtain the measure of agreement shown. Admittedly, it depends in part on how we define V_c; but other tests of the above relation can be based on physiological experiments in which retinal interaction is abolished, e.g., in Lumulus preparations.

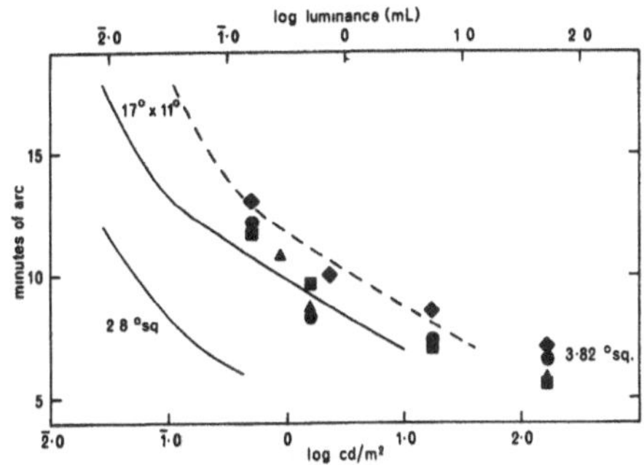

Fig. 2 Ordinate: angular distance between the principal Mach-bands; abscissa; luminance. The continuous curves are calculated from equation (1) and the data from (5). The points are taken from (4). The dashed curve corrects the large-field function for equal retinal illumination.

The other non-trivial illusion I wish to consider is of the geometric type (Fig. 3); ORBISON'S, LUCKIESH'S and ZÖLLNER'S phenomena are versions (7) probably of a single phenomenon (8). The basic effect was first described by MONTAIGNE (9), the famous French essayist, as long ago as in the sixteenth century. In stressing how unreliable our sense organs are, he says that some effect or other reminds him of the ring made of feathers held together with two hoops, which appear to be converging as you turn them, yet in fact stay parallel. Because MONTAIGNE discovered this, the illusion (Fig. 3c) is named after ZÖLLNER, a matter for little surprise in the history of visual science. The magnitude of the apparent convergence of the parallel lines can be measured (Fig. 4) by comparison with pairs of lines converging at known angles. It can be shown that slight blurring of the background, e. g., with a

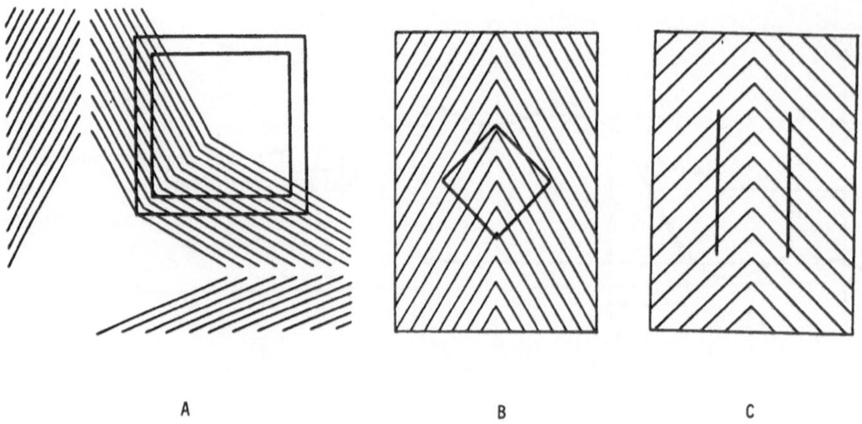

Fig. 3 The Luckiesh, Orbison, and Zöllner patterns.

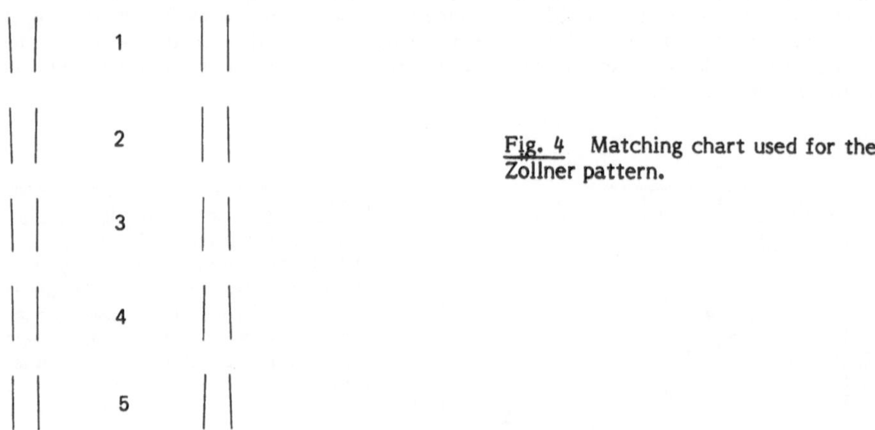

Fig. 4 Matching chart used for the Zöllner pattern.

cylindrical lens (8) with its axis perpendicular to the direction of the parallel, or test pair, affects the illusion. This is shown in Fig. 5. The magnitude of the illusion is determined as a function of the cylinder through which the test is projected, three angles B between the background (the perturbing lines) and the test lines serving as parameters. The lower part of Fig. 5 shows that the illusion was effectively abolished with lens powers much too low to affect the visibility of the perturbing lines to a significant extent. Following a suggestion of HORACE BARLOW'S, we repeated the experiment with spherical instead of cylindrical lenses, and found that the illusion was abolished with predictably weaker powers. It is well to remind guests of this College of Optometry that optometrists can do untold harm by providing us with acute (high spatial frequency) vision: they stop us from seeing the world as it really is.

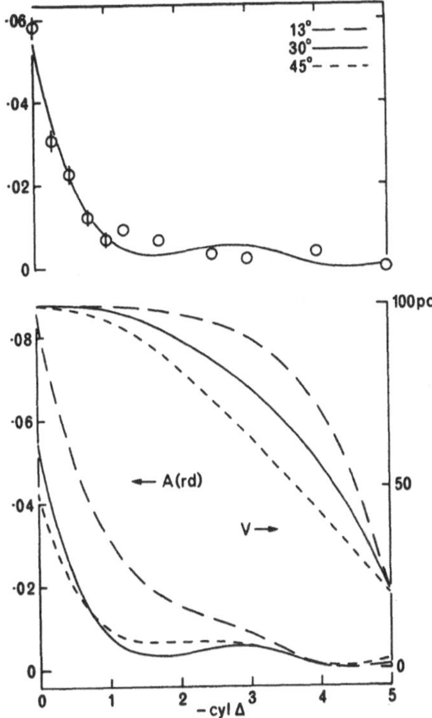

Fig. 5 Top: magnitude of ZÖLLNER illusion is radians (ordinate) as a function of the cylinder added to the projecting lens with the cyl axis perpendicular to the pairs of parallel lines. The curve is the best-fitting quartic curve. The first five points are shown with their standard errors to indicate goodness of fit.

Bottom: similar data were obtained for the three targets with distorting angles of 45°, 30°, and 13° respectively. Only the quartic functions are shown. The V-curves represent visibility scores of the herring-bone pattern. Note that the magnitude of the illusion response is low even when V is still relatively high.

Another way of destroying this illusion is by reducing its contrast (8). If this is changed from 70-80% to below 10% (Fig. 6) then ZÖLLNER'S and related illusions are abolished: illuminating engineers and printers clearly sin in the same sense as do optometrists.

Figure 7 shows that contiguity between perturbing and test lines is important. If a break between them does not exceed 6' of visual arc, the magnitude is unimpaired (Fig. 8), but it drops for larger values (10,11). This critical angle is significant as will appear in a moment.

It is hardly surprising that the magnitude of the ZÖLLNER illusion varies with over-all angular subtense. We found in some preliminary tests that both this and other illusions increase when the angular size of the perturbing grating is reduced (at constant physical contrast) in the range between threshold resolution frequency and 1/3 - 1/4 this value. However, WALLACE and CRAMPIN (10) observed smaller illusion magnitudes with larger spacings, i.e., lower frequencies. Accordingly, we extended our range, and found that the magnitude decreases again for large spacings (Fig. 9). The maximum of the free-hand curve drawn through the points --- each of which represents the mean for 4 - 6 observations per different observer (n_{Obs} = 6) --- peaks at 6' - 7' of arc.

Fig. 6 Reducing the contrast (of Fig. 3c) reduced the magnitude of the illusion.

A very simple semi-empirical theory (8) suggests that the angle B between the perturbing background lines and the test-lines affects the magnitude of the illusion, if we assume that it is subject to the inhibitory effects which BLAKEMORE, et al. hypothesized as acting between cortical orientation detectors. The lower part of Fig. 9 takes this into account; the experimental points have been scaled down by the factor of $1 + \cos^2 B$ and this reduces their scatter.

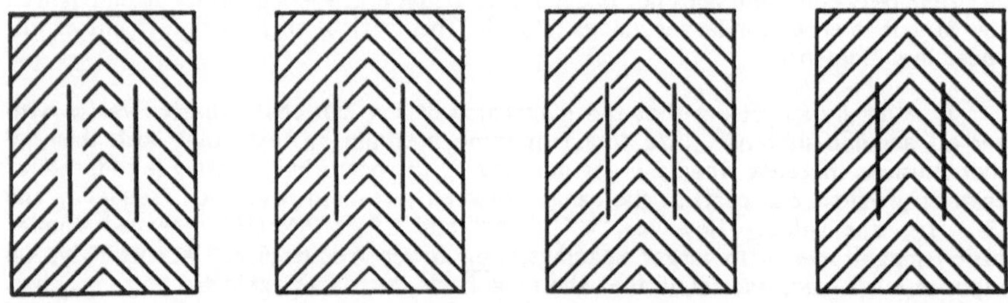

Fig. 7 For details see text.

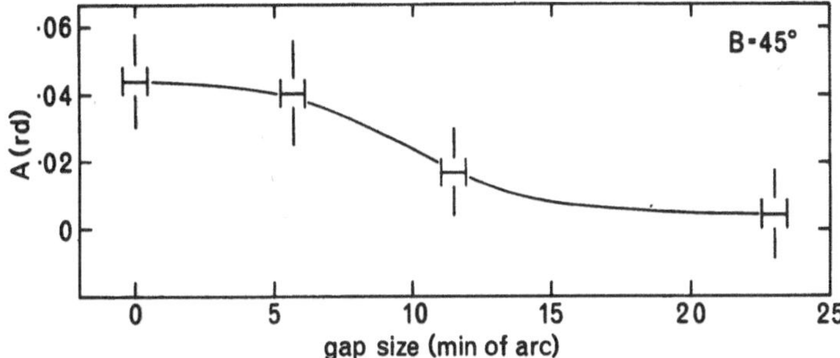

Fig. 8 Magnitude of the Zöllner illusion in radians (of Fig. 7) as a function of the angular gap size between the parallel lines and the background pattern. The vertical lines indicate the standard deviation (one test for each of six observers).

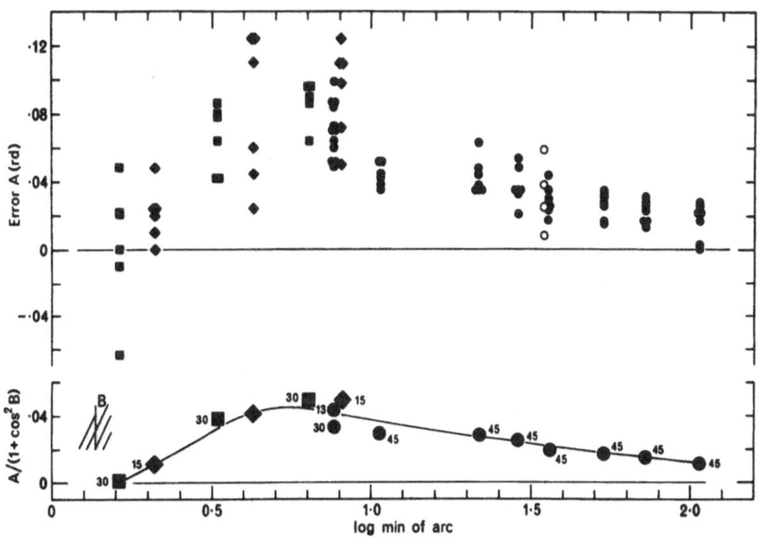

Fig. 9 Ordinate: apparent vergence of the parallel test lines (Fig. 3c) in radians. Abscissa: angular spacing of background lines (minutes of arc in the visual field). Every point in the upper part represents the mean of 4-6 observations for any one observer. The empty circles are taken from (10,11). The numbers against each point in the lower part represent B in degrees, and were obtained by the transformation shown along the lower ordinate. The curve is free-hand.

I may mention parenthetically that, in the second series of tests, one of the observers assured us that she could see the test-figure equally well at all test-distances (the variation in which controlled that of the angular subtense), and declines to wear her glasses. On seeing her results we persuaded her to submit to a repeat session wearing her correction. The two sets of results are shown in Fig. 10: the difference between them is statistically significant (p = 0.001).

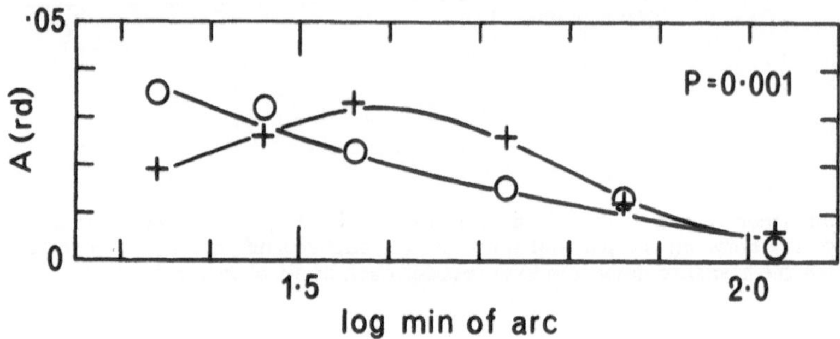

Fig. 10 As for the upper part of Fig. 9: data obtained for one observer with (0 0 0) and without (+ + +) glasses.

In conclusion, I should like to speculate briefly on a possible physiological link between these and other data. The receptive field size of rhesus orientational units projected to cortex from the fovea for both simple and complex cells is less than 15' (13). DOW (14) narrowed this estimate down to 6', a value which is consistent with the estimates in Figs. 8 and 9 above. this comparison lends new support to the idea that there is a physiological basis for the type of distortion shown in Fig. 3.

Achkowledgements

I thank Miss G. M. Villermet for technical assistance, numerous anonymous observers for their goodwill, and the Binstead Fund for financial assistance.

References

1. E. Mach, Sitzber. d. Wien. Akad. Wiss. 52, 303 (1865).
2. H. P. Hsu, Fourier Analysis. (Simon and Schuster, New York, 1970).
3. F. W. Campbell and D. G. Green, J. Physiol. 181, 576 (1965).
4. Celeste McCollough, J. Exp. Psychol. 49, 141 (1955).
5. A. van Meetersen and J. J. Vos, Vision Res. 12, 825 (1972).
6. F. F. Kuo, Network Analysis and Synthesis, Second edition, (John Wiley and Sons Inc., Londong, 1966).
7. J. O. Robinson, The Psychology of Visual Illusion, (Hutchinson, London, 1972).
8. R. A. Weale, Vision Res. (In press).
9. M. de Montaigne, Essays, Apology for Raymond Sebond,Chap. XII, ed. W. Hazlitt, (John Templemen, London, 1842).

10. G. K. Wallace and D. J. Crampin, Vision Res. 9, 167 (1969).
11. G. K. Wallace, Percept. Psychophys. 5, 261 (1969).
12. C. Blakemore, R. H. S. Carpenter and M. A. Georgeson. Nature 228, 37 (1970).
13. D. H. Hubel and T. N. Wiesel, J. Physiol. 195, 215 (1968).
14. B. Dow, This Volume.

VI. Binocular Vision and Stereopsis

Binocularity and Stereopsis in the Evolution of Vertebrate Vision

Robert Fox
Department of Psychology
Vanderbilt University
Nashville, Tennessee

The investigation of the visual system forms a major part of the effort to compose a more comprehensive picture of the evolution of brain and behavior in vertebrates. The major thrust of this chapter is to consider the emergence of binocular vision and stereopsis as it is now revealed by quite recent anatomical, behavioral, and electrophysiological evidence.

Binocular vision and stereopsis are frequently discussed together because it is generally acknowledged that binocular vision, which permits a common segment of visual space to be viewed by the two eyes simultaneously, provides the necessary condition for stereopsis. But it cannot be assumed that all animals with binocular vision are automatically endowed with the capability for stereoscopic depth perception. Over the years, two major hypotheses have been advanced to account for the development of binocular vision during the course of vertebrate evolution. The first hypothesis, which might be termed the special or elitist hypothesis, maintains that the mechanisms for binocular vision and stereopsis have been closely associated with the emergence of mammals and, in particular, have been highly elaborated and refined during the evolution of primates. The second hypothesis, which might be termed the general or proletarian hypothesis, maintains that stereopsis is present in all animals with binocular vision.

Let us briefly consider the several lines of evidence in support of the elitist hypothesis, which until recently had gone virtually unchallenged.

First, it has been known since 1838 that, in humans, binocular vision provides stereoscopic depth information--information that is not available to either eye alone and that would be of great adaptive significance in such activities as predation and locomotion in environments where other depth cues are absent. Second, there are prominent features in the visual systems of man and monkey that seem to have evolved precisely for the promotion of binocular vision. These include frontal placement of the eyes as well as semidecussation of the optic tract, which permits wholesale interaction of inputs from corresponding parts of each retina. There is also an elaborate system of eye-movement control, which acts to ensure that objects stimulate corresponding retinal points.

A third class of evidence, more recently developed, comes from single-unit recording, which reveals cells optimally tuned for retinal disparity produced by discrete contours (1-4). And, finally, there is behavioral evidence indicating that animals with this kind of neural machinery possess stereopsis. Evidence for stereopsis in the monkey has been reported by BOUGH (6), COWEY, PARKINSON, and WARNICK (6), and SARMIENTO (7). For stereopsis in the cat, see FOX and BLAKE (8), PACKWOOD and GORDON (9), BLAKE and HIRSCH (10), and LEHMKUHLE, FOX, and BUSH (11).

In addition to the positive evidence supporting the elitist hypothesis, there is also evidence suggesting that nonmammalians do not have the requisite machinery. Their eyes are more laterally placed, providing them with less binocular overlap. Their eyes can move independently, rather than being yoked. But, most important, there appeared to be no site for extensive interaction between the eyes. The principal target of each eye is the contralateral optic tectum, and classic anatomical methods could reveal no pathways by which the two eyes could communicate.

The evidence for the elitist hypothesis has been compelling, and it is not surprising that it has enjoyed widespread acceptance (e.g., 12, 13).

What evidence, then, has there been for the proletarian hypothesis? First, from naturalistic observation it appears that at least some nonmammalians have excellent depth perception. Second, anatomical specializations have evolved to provide binocular overlap. The most conspicuous of these is the temporal fovea, seen in predatory birds, fish, and lizards. But the capacity for binocularity does not mean that stereopsis is present, and one reasonable alternative explanation for binocular depth perception in nonmammalians has been described as stereopsis by triangulation (14). This refers to an absolute or egocentric depth localization, whereby an object at a fixed distance from an animal would simultaneously stimulate both eyes and thus trigger a reflexive response. This is not comparable to the precise discrimination of relative depth that is the essential feature of stereopsis as we know it.

What has been said so far describes the situation as it was about fifteen years ago, but since that time more incisive anatomical methods have yielded evidence favorable to the proletarian hypothesis. The major discovery has been that in many classes of vertebrates there are two parallel pathways that transmit information from retina to the telencephalon. One pathway involves a direct connection between retina and thalamus, and the second a direct connection between retina and tectum (15). Many investigators are engaged in working out in detail the connections that form the parallel pathways in a broad spectrum of animals, including turtle (16), frog (17), and shark (18). But for now I would like to emphasize the investigations of the parallel pathways in the bird visual system--investigations that to a large extent have been carried out by HARVEY KARTEN and his colleagues (19). Figure 1 shows the main structures comprising the two pathways in the bird system, specifically in pigeon and in owl. The pathways are distinct and widely separated. The tectofugal pathway proceeds from retina to tectum to the rotundus nucleus of the diencephalon and upward to the telencephalic structure known as the ectostriatum. The thalamofugal pathway proceeds from retina to a group of thalamic nuclei known collectively as OPT and then upward to layers in the hyperstriatum, the region known generically as the visual Wulst. It has been suggested that the Wulst is a structure analogous in function to primary visual cortex in mammals, while the ectostriatum serves as an analog of extrastriate visual cortex (20). Of special interest for the question of binocularity and stereopsis is evidence that there are ipsilateral and

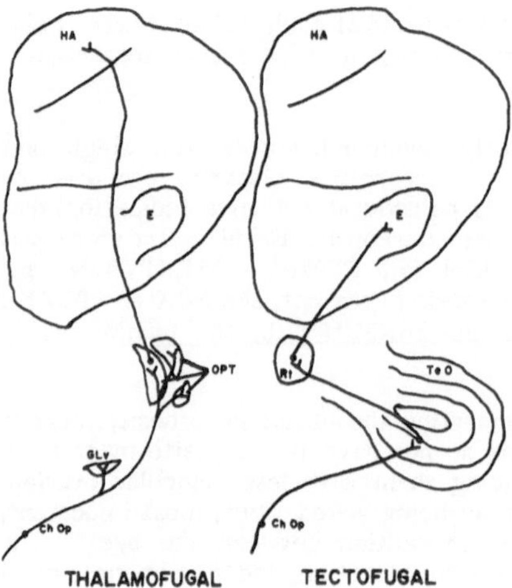

Fig. 1 See text for explanation.

contralateral projections to the Wulst. This is illustrated schematically in Fig. 2, where crossing over to the ipsilateral eye is accomplished by the DSO pathway or superoptic decussation.

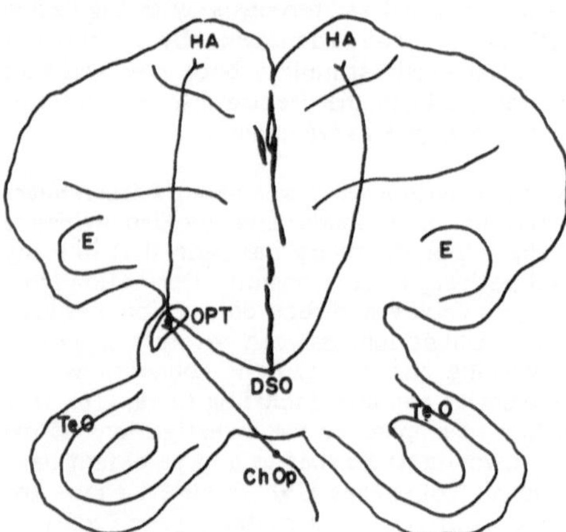

Fig. 2 See text for explanation.

Motivated by these data, we undertook a behavioral investigation of stereopsis in the falcon. I will describe the results of that investigation, but let me reverse the chronology by first describing anatomical investigations of the falcon visual system that we have recently begun. This work has been done in collaboration with my colleagues Dr. VIVIEN CASAGRANDE and Dr. STEVE LEHMKUHLE. In brief, using autoradiographic methods we have found that the falcon visual system is composed of the same structures that are found in pigeon and owl. In particular, we have found evidence for both contralateral and ipsilateral projections in the visual Wulst. This is shown schematically in Fig. 3. The right eye of the falcon was injected with a 3000-microcurie dose of proline three times (separated by 24-hour periods), and then the animal was allowed to survive for ten days. The shading in the schematic indicates very heavy labeling in the contralateral tectum and transynaptic transport of label in both contralateral and ipsilateral Wulst. In both projections the band appears continuous.

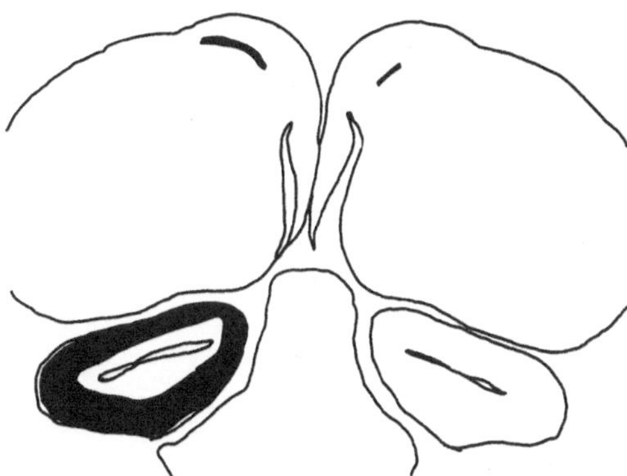

Fig. 3 See text for explanation.

Turning now to our behavioral investigation of stereopsis, our subject was an American kestrel, which weighs about 110 grams and is slightly smaller than a dove. Yet despite its small size it is a true falcon with all the attributes of its larger colleagues.

The method of testing we used is the classic two-choice discrimination task. Fig. 4 shows a schematic of the appartus. The basic idea is quite simple. The bird sits on a perch and views two visual displays, one with the correct stimulus and one without. It is trained to fly to the correct stimulus in order to obtain a food reward.

The key element to testing stereopsis in animals is a stereopsis display that contains no monocular cues, even though an animal might try various strategies, such as closing one eye or making rapid head movements to obtain parallax information. A display devoid of these cues is provided by random-element stereograms, in which the disparity information is camouflaged by a random matrix of thousands of minute dots (21). We used random-element stereograms consisting of large matrices of red and green dots generated on modified color television receivers. When the displays are viewed through appropriate red and green filters the red and green matrices stimulate separate eyes and the conditions for stereoscopic viewing are met. This is, of course, the well-known anaglyph method of dichoptic or stereoscopic stimulation. The elements of the combined matrix appear to be in motion because all the dots are randomly replaced every 16 ms. This apparent motion does not alter the characteristics of the stereoscopic form, but it does contribute to the elimination of any potential monocular cues. In our experiments, the stereoscopic form consisted of a vertical rectangle that could be varied in depth as well as in orientation.

To provide for separate eye stimulation, the bird was trained to wear a goggle

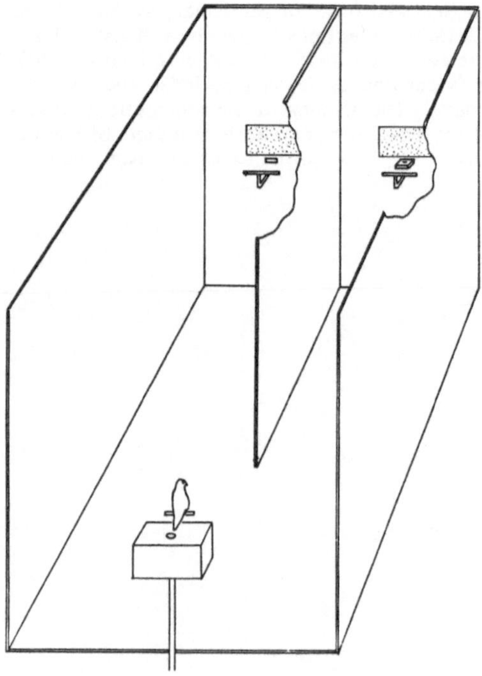

Fig. 4 See text for explanation.

device containing red and green filters (see Fig. 5). Now, imagine the bird wearing the goggles and viewing two displays, one with a stereoscopic form and one without a form. Through a gradual series of steps the bird was trained to fly to the stereoscopic form. We found that the bird could make this discrimination on the order of 80% correct under a variety of conditions. This itself is very strong evidence for stereopsis. However, we carried out additional experiments to test more stringently for stereopsis and to explore some of its characteristics.

One standard test of stereopsis is to measure performance under monocular viewing conditions, since stereopsis requires the simultaneous interaction of both eyes. Since the bird would not fly with one eye occluded, we carried out the equivalent of monocular occlusion by placing filters of the same color before both eyes. This procedure permits only one dot matrix to stimulate the eyes. These 'monocular' control sessions were introduced several times during testing, and typical results are shown in Fig. 6. It can be seen that discrimination falls to chance when the eye filters are of the same color. The solid black dots indicate performance during stereoscopic stimulation. The split dots indicate performance when both eyes were covered by either red or green filters. Note that in these cases performance falls to the 50% chance level. The sessions shown are consecutive ones. In general, a session consisted of 20 to 25 trials, with two sessions each day, one in the morning and one in the afternoon. The bird's entire daily food ration was earned in these sessions.

Fig. 5 See text for explanation.

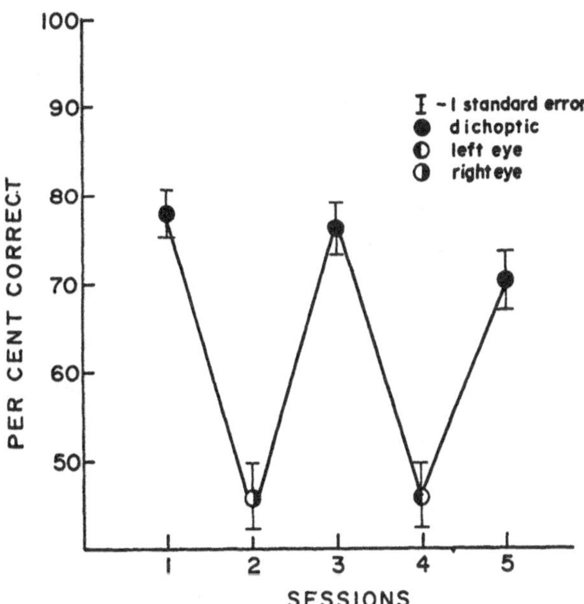

Fig. 6 See text for explanation.

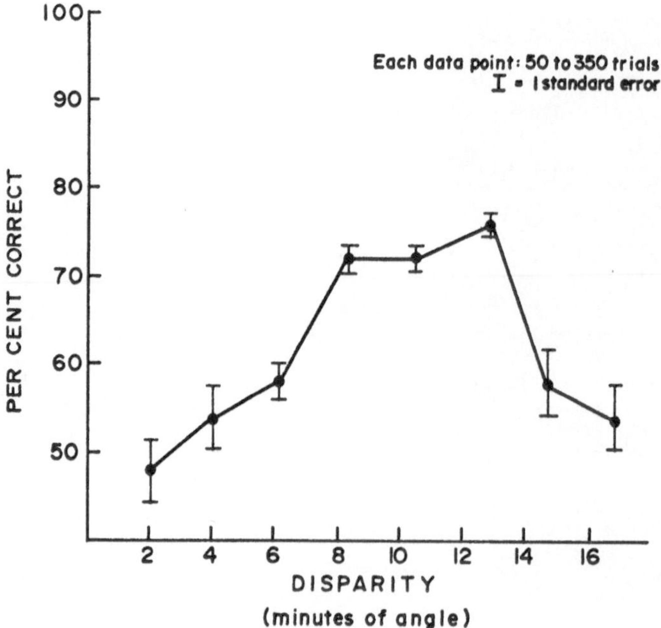

Fig. 7 See text for explanation.

Figure 7 shows how discrimination varies as a function of disparity. As disparity increases and the stereoscopic form stands further out in depth, performance increases. But performance declines at the largest disparities (see Fig. 7). Human observers also report the loss of stereopsis at these large disparities, presumably because some fusional limit has been exceeded. Although performance as a function of disparity cannot be directly compared with humans, the data do suggest that the falcon is sensitive to the same magnitude of disparity as humans.

Figure 8 shows the effect of changing the orientation of the stereoscopic form. Perhaps the bird is responding to 'something' in depth rather than to the configuration. If so, then a change in configuration would not alter performance; but if the configuration or shape was used to make the correct discrimination, a change in configuration would initially impair discrimination because of negative transfer. To test the sensitivity to configuration, we rotated the stereoscopic form by 90 degrees. The result of this change, as shown in Fig. 8, is that performance falls abruptly upon introduction of the horizontal rectangle and then gradually improves with repeated exposure. This disruption in performance strongly suggests that the bird was responding not only to the depth of the form but also to its configuration.

Together, the results lead us to conclude that the falcon possesses stereopsis, a conclusion that complements nicely the physiological evidence for binocular interaction within the bird visual system. Note also that the results match well with the evidence for disparity-tuned neurons in the visual Wulst of the owl, reported by PETTIGREW and KONISHI (22) and discussed elsewhere in this volume by Dr. PETTIGREW.

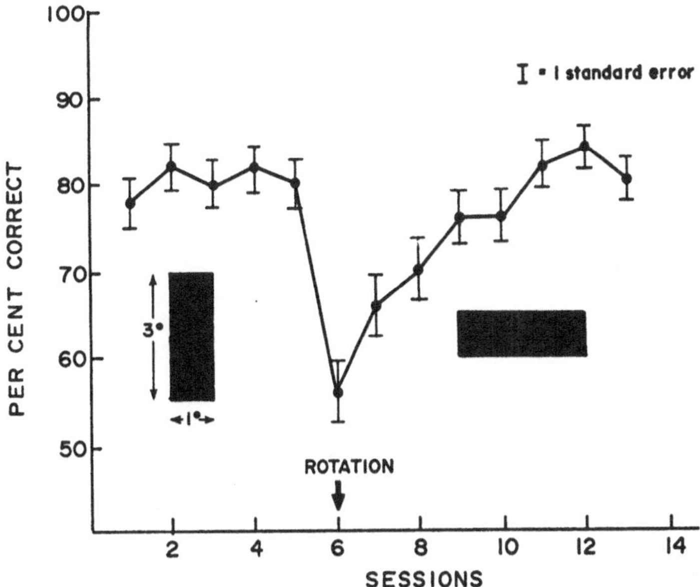

Fig. 8 See text for explanation.

The existence of stereopsis in the falcon is, of course, congruent with the proletarian hypothesis. A comprehensive evaluation of that hypothesis, however, requires that behavioral tests of stereopsis be performed on binocular animals occupying a representative spectrum of evolutionary positions. But the training of animals for behavioral testing is quite time-consuming and difficult, especially in the case of exotic species. For these reasons, we have been looking for objective tests of stereopsis that would minimize the need for exensive training. This has led us to investigate the induction of optokinetic nystagmus by stereoscopic or cyclopean moving contours formed from dynamic random-element stereograms (23). We used the same kind of color television generator described above with the generation program modified to produce an array of vertical contours standing out in depth and appearing to move continuously in a horizontal direction. The control condition consisted of contours produced on the television screen by the suppression of segments of the dot pattern, resulting in an array of nonstereoscopic contours, moving in the same direction and at the same speed, defined solely by differences in luminance. Fig. 9 shows a typical EOG record from a human observer.

The results of a factorial experiment with human observers in which contour velocity and disparity were varied are shown in Fig. 10. Note that nystagmus frequency for the luminance contours is consistently higher than that for cyclopean contours. Analysis of the EOG records reveals that this difference is due to the slower pursuit movements associated with cyclopean contours. Presumably this difference is related to the fact that cyclopean contours are not formed until cortical levels, and, in a sense, skip more peripheral neural stages. The increase in duration might reflect the time required to form contours and send signals to the eye-movement control centers.

While these data have implications for models of eye-movement control, the main point is that cyclopean contours can elicit robust nystagmus. Since nystagmus

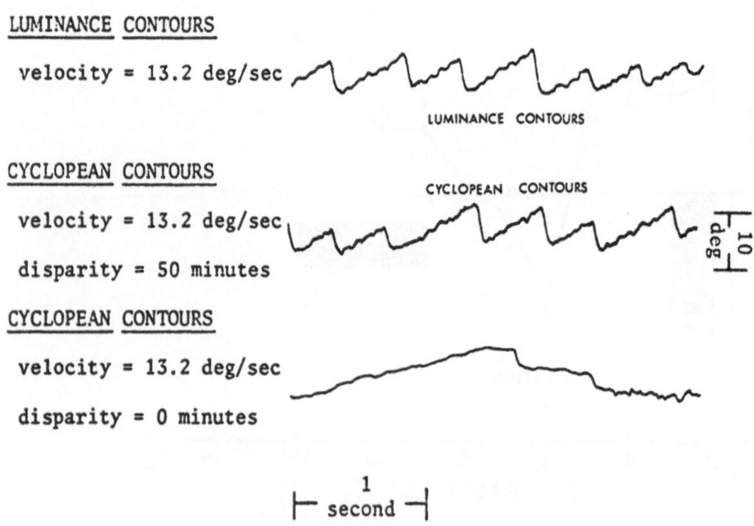

Fig. 9 See text for explanation.

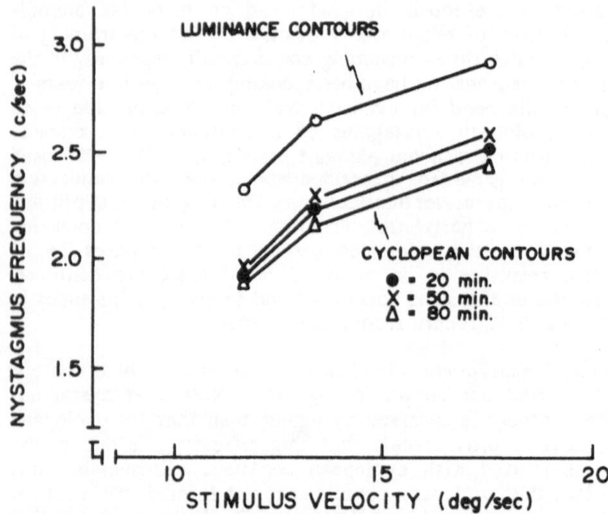

Fig. 10 See text for explanation.

reactions are reflexive and can be obtained in a wide variety of animals, it is possible that the induction of cyclopean nystagmus could be used as a relatively rapid technique for testing stereopsis. Although such a test would be indirect, the results would be quite convincing--the induction of nystagmus by cyclopean contours formed from random-element stereograms requires perception of the contours, and these can be seen only if the observer has stereopsis.

About 130 years have intervened between the demonstration of stereopsis in humans and its investigation in animals. The delay is not due to an absence of interest. Rather, only recently have techniques emerged that make testing possible. These include the extension of operant conditioning principles to psychophysical inquiry and the introduction of random-element stereograms that are inherently free of irrelevant cues. These methods, perhaps with the aid of derivative ones such as cyclopean nystagmus, may make it possible to determine the prevalence of stereopsis. And this will permit us to decide whether stereopsis is a fundamental attribute of vertebrate vision or a unique capability bestowed upon a small number of elite animals.

References

1. H. B. Barlow, C. Blakemore and J. D. Pettigrew, J. Physiol. 193, 327 (1967).
2. P. O. Bishop, The Neurosciences: Second Study Program, ed. by F. O. Schmitt, (Rockefeller University Press, New York, 1970).
3. J. D. Pettigrew, Nature, New Biology 241, 123 (1973).
4. J. D. Pettigrew, T. Nikara and P. O. Bishop, Exp. Brain Res. 6, 391 (1968).
5. E. W. Bough, Nature 225, 42 (1970).
6. A. Cowey, A. M. Parkinson and L. Warnick, Quar. J. Exp. Psy. 27, 93 (1975).
7. R. F. Sarmiento, Vis. Res. 15, 493 (1975).
8. R. Fox and R. R. Blake, Nature New Biology 233 (5314), 55 (1971).
9. J. Packwood and B. Gordon, J. Neurophy. 38, 1485 (1975).
10. R. Blake and H. V. B. Hirsch, Science 190, 1114 (1975).
11. S. Lehmkuhle, R. Fox and R. C. Bush, Paper presented to ARVO (1977).
12. W. E. LeGros Clark, Antecedents of Man (Edinburgh University Press, Edinburgh, 1959).
13. S. Ramon-y-Cajal, Die Struktur des Chiasma opticum nebst einer allgemeinen Theorie der Kreuzung der Nervenbahren (J. A. Barth, Leipzig, 1899).
14. A. Linksz, Physiology of the Eye, II. Vision (Grune-Stratton, New York, 1952).
15. D. Ingle and G. E. Schneider (Eds.), Brain Behav. 3 (no. 1-4), 1 (1970).
16. A. M. Granda and W. N. Hayes (Eds.), Brain Behav. Evol. 5 (no. 2-3), 89 (1972).
17. D. Ingle, Science 181, 1053 (1973).
18. T. A. Duff and S. O. E. Ebbesson, Science 182, 492 (1973).
19. H. J. Karten, W. Hodos, W. J. H. Nauta and A. M. Revzin, J. Comp. Neurol. 150, 253 (1973).
20. R. Fox, S. W. Lehmkuhle and R. C. Bush, Science 197, 79 (1977).
21. B. Julesz, Foundations of Cyclopean Perception (University of Chicago Press, Chicago, 1971).
22. J. D. Pettigrew and M. Konishi, Science 193, 675 (1976).
23. R. Fox, S. Lehmkuhle and L. E. Leguire, Vision Research (in press).
24. R. Blake and R. Fox, Per. Psy. 14, 161 (1973).

Questions

Q. If you do not get nystagmus it does not necessarily follow that the animal does not have stereopsis, because the animal would have to make the proper vergence movements for the disparity to be detected.

A. True, a negative result would not be informative, only a positive result. And there are good reasons why we may not get nystagmus. The animal must have binocular overlap and some capability for convergence. But the opportunities for convergence, I suspect, are widespread among animals.

Q. That is not quite my point. You must get the animal to use the convergence it has.

A. In the case of our human observers we simply asked them to attend to the display and, being compliant humans, they attended and made the appropriate convergence movements. For animals, it would be necessary to train them to look at the display--to capture their attention, so to speak. But it is easier and much less time-consuming to train animals to attend to a display than it would be to teach them a complete set of discrimination responses, as we had to do for the falcon.

Q. The only clear gain from binocular vision is stereopsis. I have never heard of other benefits of two frontal eyes, and a lot is lost, such as panoramic vision. Therefore, if an animal has frontal vision it probably has stereopsis.

A. That expectation would seem to be consistent with the proletarian hypothesis, although the situation might be more complicated for animals that have both panoramic and frontal vision by virtue of a double fovea.

Q. Is it possible that an animal could gain information from two frontal eyes without having stereopsis?

A. It has been proposed that binocular nonmammalians possess stereopsis by triangulation, which refers to an absolute or egocentric depth localization. Imagine two fixed eyes that intersect at a point in space. Whenever a prey animal crosses the intersecting lines of sight some response would be triggered--the tongue leaps out, or a tentacle uncoils, and the prey is snatched. This would not be stereopsis as we know it. Stereopsis is relative depth localization with respect to two points, and there is a continuous range of depths. One point lies on the horopter and the second lies off the horopter, and we see one point relative to the other because we are comparing the depth positions of the two points. But with egocentric depth localization, the animal--say, the frog--simply is stuck with an absolute 2 inches or whatever it might be. In this way it would be possible for an animal to have binocular vision and to acquire visual information, but it would not have stereopsis.

There are, of course, some advantages of binocular vision relative to monocular. For humans, on several tasks, binocular performance is superior to monocular by an amount greater than that anticipated by probability summation (for a recent review of the psychophysical literature, see ref. (24). For instance, Dr. Campbell has demonstrated that contrast sensitivity at threshold for grating patterns is lower by a factor of 1.4 for binocular viewing relative to monocular. But the superiority of binocular viewing does not seem sufficiently great to justify, by itself, the evolutionary pressure for the maintenance and elaboration of binocularity.

Q. It is possible with Julesz-type random-element patterns to detect the area of decorrelation--where the stereoscopic form is located--by superimposing the dot matrix seen by one eye upon the matrix seen by the other eye. Is it conceivable that the falcon was doing something like that?

A. The dichoptic conditions of stimulation ruled out the operation of any kind of physical superposition, but of course it is conceivable that some superposition process occurred at the neural level. However, the fact that the dots in the matrix were being replaced at random every 16 msec to produce a dynamic random-element stereogram would make it more difficult for any kind of superposition process to be effective.

Q. What about vertical disparity? Have you run any controls using vertical disparity?

A. Our control conditions, taken together, are somewhat more powerful than introducing vertical disparity. As I understand it, the vertical control condition would involve rotating the stereogram 90 degrees. The assumption underlying that test seems to be that an observer might be basing a response on the detection of a vertical line produced by horizontal disparity, and if so this strategy would be revealed by rotation--or that perhaps through some process such as superposition the stereoscopic form could be seen even though it does not appear to lie in depth. Presumably, the rotation to vertical disparity would produce chance performance if the discrimination were based on stereopsis and successful performance if it were based on a nonstereopsis cue. But the results might be misleading, at least in the case of discrimination by animals. Recall that when we rotated the vertical stereoscopic form by 90 degrees performance initially fell to chance and then recovered. The initial deficit in performance can most likely be attributed to the fact that an animal's performance tends to be disrupted whenever a new element in the stimulus configuration is introduced. So, if complete reliance were placed on the vertical disparity control alone, it would be possible to conclude that discrimination was based on stereopsis when indeed it was not. A more convincing test of the hypothesis that the stereoscopic form was seen in depth would be to vary the depth position by systematic variation of disparity. Performance should be impaired at near-threshold disparities, rise to a maximum at intermediate disparities, and then fall again at disparities that exceed some fusional limit. This is what we did, and that is the result we obtained. In terms of the extensive literature on animal discrimination, the tactic can be thought of as testing all values of the stimulus dimension presumed to control performance. If performance covaries with the manipulation, and all other stimulus variables are held constant, it is difficult to avoid the conclusion that the animal is responsive to the relevant stimulus dimension.

Comparison of the Retinotopic Organization of the Visual Wulst in Nocturnal and Diurnal Raptors, with a Note on the Evolution of Frontal Vision

John D. Pettigrew
Beckman Laboratories of Behavioral Biology, Division of Biology 216-76,
California Institute of Technology
Pasadena, California 91125

Introduction

Anatomical studies by KARTEN and his coworkers (1), more recent physiological recording experiments (PETTIGREW and KONISHI (2,3)), and behavioral studies (e.g., see FOX, this volume), all indicate that the visual Wulst of birds is analogous in its visual processing capabilities to the striate cortex of mammals. Similarities in function between the two structures include the following: (1) Both have laminar organization with monocular information from both eyes arriving in the granular layers (cf. KARTEN, et al., (1), and PETTIGREW and KONISHI (2,3) with HUBEL and WIESEL (4)); (2) Binocular convergence of excitatory information in both systems occurs here first (PETTIGREW and KONISHI (2, 3, 5), versus HUBEL and WIESEL (6, 7, 8)); (3) In both structures, binocular excitatory convergence appears to occur at the same level of processing as the elaboration of orientation selective neurons, since the monocular, thalamic afferents have concentrically-organized receptive fields which contrast with the orientation-selective receptive fields of the binocular cortical neurons (cf. PETTIGREW and KONISHI (2,3), with HUBEL and WIESEL (7, 8)); (4) Both structures are primarily concerned with binocular visual processing and neurons are found in each structure which appear to be selectively tuned to stereoscopic depth cues (e.g., cf. PETTIGREW and KONISHI (2,3) with PETTIGREW, NIKARA and BISHOP (9)). This last point is controversial with respect to the monkey because of the technical difficulties involved in the control of the relevant stimuli which, based on psychophysical considerations, may have the order of seconds of arc in primates. Note however that there are now positive reports of disparity-selective neurons in monkey striate cortex (FISCHER and POGGIO (10)) to balance the earlier negative one (HUBEL and WIESEL (11)); (5) The functioning of both is sensitive to visual experience in the neonatal period (PETTIGREW and KONISHI (3); HUBEL and WIESEL (11)).

All of these points of similarity underline the potential significance of the study of the avian Wulst for an understanding of our own visual system, in addition to the intrinisic value of such study. For these reasons I have begun a comparison of the Wulst and its pathway in a variety of bird species with different visual adaptations. Of particular interest are the diurnal raptors which have two foveal specializations in each eye, a temporal one for binocular vision and a central one directed more peripherally (POLYAK (12); WOOD (13)). The availability of such species for study is strictly limited and I report here preliminary observations on the visual Wulst of a bifoveate, diurnal raptor, the kestrel, Falco sparverius. These observations suggest that the visual Wulst of the kestrel has a representation of only the temporal fovea and that the main projection of the central fovea is therefore into another visual pathway. The same conclusion appears to apply to the vulture, another diurnal raptor with a central fovea in addition to the temporal retinal specialization.

Some questions about the evolution of frontal eyes in birds were raised by this comparison of diurnal and nocturnal raptors, and speculative answers to these questions are included in the Discussion.

Methods

The work presented here is based upon studies of 9 barn owls, Tyto alba, 2 kestrels, Falco sparverius, and one black vulture, Coragyps atratus. An interesting comparison is thereby provided since the eye of the barn owl has a single temporal area centralis, with no other area of retinal specialization, the eye of the kestrel has both a temporal and a central fovea and the eye of the vulture has a single central fovea and a temporal area centralis (see below).

All birds were anesthetized with Ketamine (6-12 mg/kg) and additional thiopentone (2-5 mg/kg) was given to the vulture. Paralysis was not necessary in the owl whose strictly limited eye movements could be monitored as previously described (PETTIGREW and KONISHI (2)). Eye movements were marked in both kestrel and vulture and even allowed some observations to be made upon the patterns they exhibited. Therefore, for single unit recording the kestrel and vulture were paralyzed with d-tubocurarine (0.2-0.5 mg/kg) and artificially respired by continuous passage of warm, moist air (5-20 cm H_2O pressure) through a cannula inserted into the posterior air sac.

Retinal landmarks were projected with an ophthalmoscope to a tangent screen 57 cm in front of the bird. In the case of the barn owl the only readily visible landmark is the pecten and so the projection of the center of the area centralis was obtained after the experiment by examination of a stained retinal whole mount (see below). In the kestrel both the temporal and central foveae are visible as small pits, darker than the brightly-colored, refractile surrounding retina and best located by the bright blue, concentric light reflexes which encircle them. In the vulture there was a readily visible central fovea pit with surrounding blue reflex. The vulture's temporal area centralis was not distinguishable with the ophthalmoscope, so its position in relation to both pecten and central fovea was determined after the experiment from retinal whole mounts.

The corneas were protected with plastic contact lenses (radii of curavature 6-7 mm). The refractive state was determined by both retinoscopy and by subjective ophthalmoscopy. No corrective lenses were necessary to bring the temporal retina in focus on the tangent screen at 57 cm but the nasal retina was consistently (in all birds studied) 1-3 diopters more hypermetropic than the temporal retina. This difference in refractive state was also observed after paralysis of the accommodative mechanism with curare, so it appears to reflect the structural asymmetry of the avian eye, to which attention has already been drawn in the case of the pigeon (BLOUGH (14).

Single unit recording was carried out from the visual Wulst with tungsten-in-glass microelectrodes in a closed chamber. Techniques of visual stimulation have already been described (PETTIGREW (15); BLASDEL, MITCHELL, MUIR and PETTIGREW (16); and PETTIGREW and KONISHI (2)). Retinal whole mounts were prepared as described by others for the cat (e.g., HUGHES (17)) with the added precaution that the birds were dark adapted for an hour or two before perfusion. This step,plus a soak of the posterior globe in water overnight before the retina is dissected off, help overcome the problem of adherence of the pigment layer to the photoreceptor layer, a problem which makes the dissection much more difficult for diurnal bird eyes than for the cat.

Maps of the density of retinal ganglion cells were obtained by direct counting under the microscope at regular intervals. These maps (PETTIGREW and WATHEY, in preparation) show a vertically elongated region of increased ganglion cell density in the temporal retina of the barn owl. In the kestrel both foveae were readily visible on the retinal whole mount, as they had been in vivo. The vulture's central fovea could be directly visualized and a temporal area centralis of increased ganglion cell density defined from a density map.

Results

Barn Owl. As already described (PETTIGREW and KONISHI (2, 3, 5)) the representation of the barn owl's visual field within the Wulst was like that of the cat's visual field within Area 17. The vertical meridian was mapped upon the lateral margin of the Wulst and receptive fields recorded from more medial locations moved systematically into the periphery of the contralateral hemifield. Recording sites located more anteriorly yielded receptive fields which moved into the inferior visual field and posterior penetrations yielded superior field positions. Also in close similarity to the map of visual space in the cat Area 17, the map in visual Wulst had a marked over-representation of both the area centralis region and the inferior visual field. More than half the total area of the Wulst was concerned with the visual field within 20° of the vertical meridian. In consequence only a very tiny strip along the medial boundary of the Wulst was concerned with the monocular crescent at the periphery of the owl's visual field. In line with the binocular properties characteristic of most of the neurons found there, the visual Wulst thus appears to be concerned primarily with binocular visual processing.

In contrast to the organization of cat visual cortex, the visual Wulst does not have a mirror image representation of the visual field abutted along its lateral border in the same way that Area 18 abuts Area 17 in the cat (e.g., HUBEL and WIESEL (18); TUSA, PALMER, and ROSENQUIST (19)). Neurons recorded just off the lateral margin of the Wulst were not visually responsive. The properties of some neurons recorded within the Wulst, such as an absolute requirement for binocular stimulation and a preference for dark, terminated-bar targets, suggest that some layers of the Wulst carry out processing which is similar to that taking place in cat Area 18. If this is the case it may be possible to think of multiple visual field representations stacked, in register, within the Wulst, whose 5-6 mm thickness may thus be equivalent to more than one visual area in the 2 mm thick cat cortex.

Kestrel. (a) Eye movements: Before describing the results of recording from the Wulst of the kestrel I will briefly describe some qualitative observations made upon the kestrel's pattern of eye movements. These observations were made before paralysis in a bird lightly anesthetized with ketamine in which there was clear visual following of targets. Since the head was fixed an ophthalmoscope could be used to estimate, on occasions, the direction of gaze based upon the projection of either the temporal fovea or the nasal fovea. In these circumstances it was possible to observe two quite distinct patterns of eye movements. The more common pattern consisted of independent movements of each eye. For example the right eye was observed to track the movement of a person in the laboratory some 50°-60° to the right of the bird's vertical merdian while the left eye was motionless. Similarly, if a finger was passed in front of the bird from the far left peripheral field to the far right, the left eye would first track the target alone, then there was a brief period where both eyes tracked the target, followed by the left eye's ceasing to track while the right eye followed the target around into the right peripheral field. Observations with the ophthalmoscope indicated that the bird was using the region of the central fovea, and not the temporal fovea, for these independent tracking movements.

The second eye-movement pattern was more difficult to elicit and consisted of coordinated movements of both eyes. This pattern was like the familiar mammalian one of conjugate gaze and stood in marked contrast to the bizarre, chameleon-like pattern described above. The conjugate pattern was usually seen if there was no movement in the peripheral visual field of either eye (dark "blinkers" fitted to the kestrels seemed to increase the chances of observing this pattern) and if there was a small target (1-2°) moving around in the binocular visual field directly in front of the bird. Ophthalmoscopic observations carried out in this phase showed that the region of the temporal fovea of each eye was directed at the target. This alignment of the axes of the temporal foveae stands both in contrast to the varying degrees of divergence of these axes which are seen during the phase of independent eye movements and in contrast to the wide divergence of these axes in paralysis (see Fig. 1).

These observations suggest that there are at least two separate eye movement mechanisms in the kestrel. One mechanism permits independent motion of either eye and appears to be mediated for each eye separately by the region of the central fovea. The other mechanism appears to involve the coordinated use of both temporal foveae to examine binocular visual space.

b) Visual Wulst: Despite the fact that the kestrel has two foveae in each eye and a retinal ganglion cell density higher than that found in the owl (unpublished observations), the kestrel's visual Wulst is proportionally smaller than that found in the owl. The difference in size is particularly noticeable in the medio-lateral dimension which maps the horizontal meridian of visual space.

Physiologically, the properties of the limited sample of units so far obtained from kestrel visual Wulst are like those recorded from the owl. Most neurons are orientation-selective, except for some in the granular layers receiving thalamic input which have concentrically-organized receptive fields. Most neurons were binocularly-activated and in the superficial laminae there were disparity-selective neurons with an absolute requirement for binocular stimulation like those found in the superficial laminae of the owl Wulst (PETTIGREW and KONISHI (2,3)).

The visual field representation within the kestrel's Wulst was also like that of the owl. The anterior, posterior, medial and lateral margins of the kestrel visual Wulst represented respectively the inferior, superior, far peripheral and midline visual fields.

There was also a marked over-representation of the binocular visual field with the kestrel's Wulst. This is shown in the experiment of Fig. 1 where it can be seen that almost the whole of the lateral-to-medial extent of the Wulst is traversed before receptive fields could be found which were not in the region of binocular overlap. The central visual field around the central fovea's projection and the peripheral visual field are barely represented at all within the Wulst, which therefore appears to receive projections from the temporal foveae but very little from the central fovea.

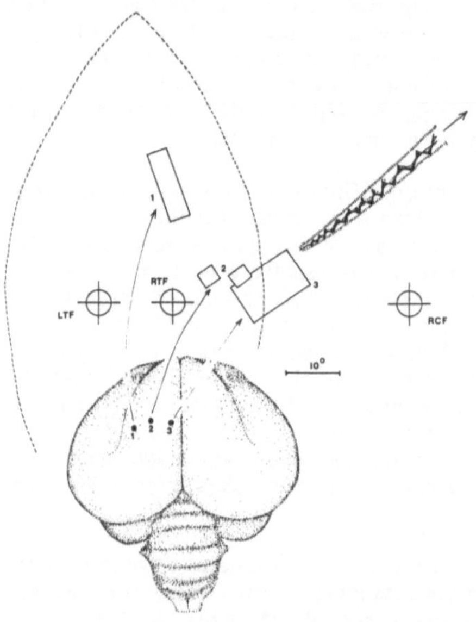

Fig. 1 Single unit recording from three points along the medio-lateral dimension of the dorsal surface of the left visual Wulst in the kestrel. Dotted line indicates the region of binocular overlap on the tangent screen at 57 cm. LTF - projection of temporal fovea of the left eye: RTF - projection of temporal fovea of right eye: RCF - projection of central fovea of right eye. The outline above RCF and to the right of the receptive fields is the projection of the base of the pecten of the right eye. Only the lower half of this projection is shown since the upper half of the projection extends overhead and cannot be shown on the tangent screen. Note that, as in the visual Wulst of the owl, there is a representation of the contralateral hemifield of visual space; receptive fields recorded from neurons in the left Wulst are located to the right of the vertical meridian passing through the right temporal fovea (RTF). Also like the arrangement in owl visual Wulst is the representation of the vertical meridian at the lateral margin of the Wulst and the representation of more peripheral parts of the visual field as penetrations are made more medially. There is also a markedly increased representation of the region of visual space close to the vertical meridian like that found in the owl. This is to be expected in view of the high density of retinal ganglion cells in the temporal fovea. Unexpected is the apparent lack of an adequate representation of the central fovea within the Wulst, despite an even higher ganglion cell density in this part of the retina. Even the most medial penetrations made in the Wulst of the kestrel yield receptive fields which are close to the region of binocular overlap.

Discussion

The most surprising finding in this comparison of owl and kestrel is the absence in the kestrel Wulst of a stong representation of the central fovea. This finding has been confirmed in another diurnal raptor, the black vulture, Coragyps atratus, whose visual Wulst also lacks a representation of the central fovea and is concerned largely with the temporal area centralis (unpublished observations). In the case of the vulture the lack of a strong representation of the central fovea is all the more striking in view of the very much larger ganglion cell density found in this region of the retina in comparison to the temporal retina, which lacks a fovea and yet commands the greater proportion of the area of Wulst.

An obvious question raised by these findings concerns the destination of most of the axons from the ganglion cells in the central fovea, since their poor representation in the visual Wulst suggests that few of them can project from the eye to this forebrain structure. A fairly obvious answer, but one which has yet to be tested because of the scarcity of bifoveate experimental material, is suggested by the formulation of KARTEN and his colleagues (1). They described two large visual pathways in birds. The first corresponds to the pathway through the bird's "LGN" to the visual Wulst described here. This KARTEN, et al. called "thalamo-fugal," but it is perhaps better called "geniculo-striate" because of its now well-established functional relationship to the pathway of the same name in mammals and because the second pathway also passes through a thalamic nucleus (n. rotundus). The second pathway, "tecto-fugal," passes from eye to optic tectum to nucleus rotundus (roughly equivalent to mammalian pulvinar) to ectostriatum (probably equivalent to the "third tier" of extrastriate visual cortical areas in mammals (ALLMAN (20)). This latter tectofugal path is known to be of the greatest significance for a pigeon's visual ability (HODOS and KARTEN (21)), and I now suggest that it utilizes information largely derived from the central fovea. Notice that this conclusion is derived by exclusion, since, on the basis of the present findings, the central fovea does not appear to be playing a major role in the geniculo-striate pathway. This does not mean that I think the optic tectum will prove to have a representation of the central fovea to the exclusion of the temporal fovea. On the contrary, since there is a strong projection from the visual Wulst to tectum (KARTEN, et al. (1)), I would expect to find in the tectum a strong representation of the binocular visual field (temporal fovea) in addition to the representation of the central fovea, like that already described in the optic tecta of some teleosts (SCHWASSMANN (22)). Initial anatomical work on the pathways concerned is at least consistent with the formulation presented, since the retinotectal pathway is relatively much larger in the kestrel than it is in the owl (unpublished observations).

The division of function I am proposing in the kestrel between the geniculo-striate pathway and the tecto-fugal pathway suggests an intriguing relationship to the two patterns of eye movements shown. Since the optic tectum of all vertebrates, with the exception of primates, has a representation of the complete, contralateral retina (cf. ALLMAN (20)) it is possible to envisage each tectum in independent control of the opposite eye, as suggested by the first pattern of movements described above. On the other hand, visual processing in the geniculo-striate pathway seems primarily to be binocular and would be appropriate for the control of the fine conjugate movements of the second pattern, which one imagines is brought into action only when the bird is close enough to the prey to be able to estimate its distance stereoscopically.

A Note on Frontal Vision

An important evolutionary question is raised by the comparison of the kestrel's visual system with that of the owl. Baldly stated the question is as follows: Why, in the process of evolutionary adaptation to the nocturnal niche, has the owl not retained the capability of sophisticated laterally-directed vision mediated by the central fovea, when both binocular vision and lateral vision are carried out in the diurnal kestrel? The question can be put another way, which makes it more pertinent to man's own visual system: Why have frontally-placed eyes when both lateral vision and binocular vision can be highly developed with the eyes in a more laterally-placed position, such as that found in the kestrel?

The answer to these questions appears to be linked to the special demands made upon the visual system by the nocturnal niche and may therefore be quite relevant to the understanding of man's "frontal eye syndrome," since it now appears to be generally agreed that the primate evolutionary line began with a nocturnal form (ALLMAN (20)). I believe that the answer is concerned with the optical properties of a nocturnally-adapted eye. Such an eye is a "fast" lens system, with a small 'f' ratio (owl eyes have 'f' ratios around 1 (MARTIN (23) and PETTIGREW and WATHEY, in preparation) which have limitations such as very critical focussing (small depth of field) and a high degree of optical aberration for off-axis light paths (SOULE (24)). The latter property of a nocturnally-adapted optical system may mean that it is not possible to achieve good image quality with those rays entering very far from the optical axis. In contrast the "slower" optical system of the kestrel, which is optimal for use in higher levels of illumination, may have much better image quality away from the optical axis than the owl eye and may help account for the fact that, in the kestrel's eye, the temporal visual axis (through the temporal fovea) can be situated 45° away form the central visual axis (through the central fovea) and the optical axis (which lies close to the central visual axis) (see Fig. 1). In other words, the nocturnal animal may have to settle for one visual axis close to the optical axis. One might expect the one chosen to be the temporal visual axis, since this is the part of the system participating in binocular vision, where the tiny retinal image disparities to be detected will demand the best image quality available.

These considerations lead to a testable prediction, viz., when it is possible to make a direct comparison between two closely-related species, one of which is diurnal and the other of which is nocturnal, then the latter will have more frontally-directed eyes.

This "natural experiment" can be performed in two cases of which I am aware and in both cases the results conform to the prediction. The first concerns the nocturnal, swallow-tailed gull of Galapagos (_Larus_ _(Creagrus)_ _furcatus_) which does indeed have eyes which are both "faster" (in the sense of a lower 'f' ratio) and more frontally-directed than those of diurnal, closely-related gulls (PETTIGREW, in preparation). The second case concerns the kakapo, or owl parrot, _Strigops habroptilus_ of New Zealand. This species has close psittacine relatives, like the kea _Nestor mirabilis_, which are diurnal and would afford a close comparison. Although the kakapo is an endangered species and the opportunity for study of its visual system may not arise again, the available information suggests that it may also conform to the prediction. The kakapo does appear to have more frontally-placed eyes than other parrots, according to G. R. WILLIAMS of the New Zealand Wildlife Service, who is one of the very few people to have first-hand knowledge of the bird (personal communication). In addition, the fundus of the kakapo is so un-parrot-like and so like an owl's fundus (even to the extent of having a well-developed temporal fovea) that WOOD (13) was led to declare, on ophthalmoscopic grounds, _Strigops_ to be a member of the owl family rather than a member of the parrots!

Further work along these lines should lead to valuable information about the origins of the "frontal eye syndrome" which characterizes our own visual system. Of particular interest will be further tests of the suggestion that frontalization of the eyes is associated with changes in central visual organization, such as increasing specialization of the temporal foveal region, and its geniculo-striate pathway.

Acknowledgements

This work was supported by The Spencer Foundation, and by grants MH 25852 and EY 01909 from the U.S. Public Health Service.

References

1. H. J. Karten, W. Hodos, W.J.H. Nauta and A. M. Revzin, J. Comp. Neurol. _150_, 253 (1973).

2. J. D. Pettigrew and M. Konishi, Science 193, 675 (1976).
3. J. D. Pettigrew and M. Konishi, Nature (Lond.) 264, 753 (1976).
4. D. H. Hubel and T. N. Wiesel, Nature (Lond.) 225, 41 (1970).
5. J. D. Pettigrew, Proc. Roy. Soc. B, In press (1978).
6. D. H. Hubel and T. N. Wiesel, J. Physiol. (Lond.) 148, 574 (1959).
7. D. H. Hubel and T. N. Wiesel, J. Physiol.(Lond.) 160, 106 (1962).
8. D. H. Hubel and T. N. Wiesel, J. Physiol. (Lond.) 195, 215 (1968).
9. J. D. Pettigrew, T. Nikara and P. O. Bishop, Exp. Brain Res. 6, 391 (1968).
10. B. Fischer and G. Poggio, J. Neurophysiol. 40, 1392 (1977).
11. D. H. Hubel and T. N. Wiesel, Nature (Lond.) 225, 41 (1970).
12. S. Polyak, Vertebrate Visual System, 2d. impression, (Chicago University Press, Chicago, Illinois, 1968).
13. C. A. Wood, The Fundus Oculi of Birds Especially as Viewed by the Ophthalmoscope, (The Lakeside Press, Chicago, Illinois, 1917).
14. P. Blough, Vision Res. 12, 978 (1972).
15. J. D. Pettigrew, J. Physiol. (Lond.) 237, 49 (1974).
16. G. G. Blasdel, D. E. Mitchell, D. W. Muir and J. D. Pettigrew, J. Physiol. (Lond.) 265, 615 (1977).
17. A. Hughes, J. Comp. Neurol. 163, 107 (1975).
18. D. H. Hubel and T. N. Wiesel, J. Neurophysiol. 28, 229 (1965).
19. R. J. Tusa, L. A. Palmer and A. C. Rosenquist, J. Comp. Neurol, in press (1978).
20. J. A. Allman, in Progress in Physiological Psychology, Vol. 7. James Sprague and Alan Epstein (Eds.), (Academic Press, New York, London, 1977).
21. W. Hodos and H. J. Karten, Brain, Behav. Evol. 9, 165 (1974).
22. H. O. Schwassmann, Vision Res. 8, 1337 (1968).
23. G. R. Martin, Nature (Lond.) 268, 636 (1977).
24. H. V. Soule, in Electro-optical Photography at Low Illumination Levels, (Wiley Inter-Sciences, New York, 1967).

Orientation and Position Disparities in Stereopsis

P. O. Bishop
Department of Physiology
John Curtin School of Medical Research
Australian National University

Binocular depth perception or stereopsis is possible because of the brain's ability to discriminate differences in the images in the two eyes. Although it is generally agreed that horizontal position disparities are the main cue for binocular depth discrimination it is possible that orientation disparities may also provide important cues. The term orientation disparity refers to the angular difference between line images in the two eyes. These angular differences or orientation disparities are a very common occurrence in everyday life and they assume particular importance when we look downwards as in reading or when we view the ground ahead of us while walking.

Consider a subject seated at a table (Fig. 3) viewing a straight line XY (Fig. 1) on the table top. By looking down from above on the subject's eyes and table we get a plan view as in the top part of Fig. 1. I will assume that the line on the table lies in the subject's median plane and that he is fixating the center point of the line. You can see that the two retinal images are at an angle to one another. They are convergent downwards because the extremity X forms its image lower and more towards the median plane than the extremity Y.

In Fig. 2, I have made the two retinae exactly congruent so that the two foveolae and other geometrically corresponding retinal points are accurately superimposed. When this is so, the two retinal images of the line will intersect at the point conjugate with the fixation point which is, in this case, the midpoint of the line on the table. The angle between the two images is the orientation disparity of the line. An understanding of the position disparity of an end of the line is rather more complex. The position disparity of the end Y would normally be expressed with respect to the fixation point and would be given in terms of the difference between the two angles that the line joining the nodal points of the two eyes subtends respectively at the fixation point and at the end Y. It can be shown, however, that the difference between these two angles is proportional to the length of the line

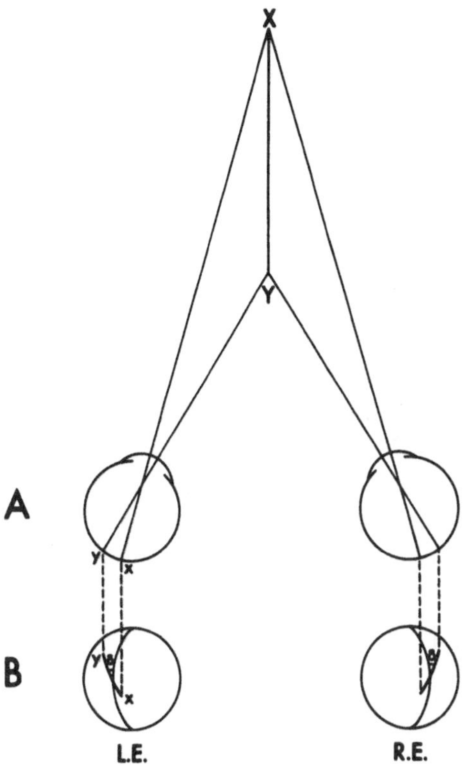

Fig. 1 Orientation disparity of retinal images of a line XY on a table top as seen by a seated human subject showing plan (A) and elevation (B) views of the subject's eyes. The orientation disparity is the sum of the angles $\delta_L + \delta_R$ (from ref. 1).

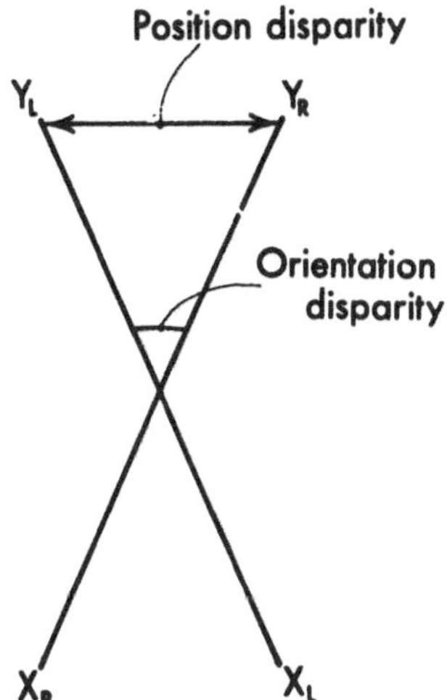

Fig. 2 Superimposed retinal images Y_LX_L and Y_RX_R of the line XY in Fig. 1 obtained by making the two eyes exactly congruent. The position disparity of the ends Y_L and Y_R is the angle that the line Y_LY_R subtends at the nodal point of the imaginary "superimposed" eye.

$Y_L Y_R$ that joins the ends of the two superimposed retinal images in Fig. 2. This angular difference is, in fact, equal to the angle that the line $Y_L Y_R$ subtends at the nodal point of the imaginary "superimposed" eye. Hence the position disparity of the end Y is given by the length of the line $Y_L Y_R$ expressed in terms of a visual angle.

In the literature orientation disparities are usually referred to as torsional disparities or cyclodisparities. I do not like either of these terms because they seem to imply that orientation disparities are associated with the torsional or cyclorotary movements of the eyes round the anteroposterior axis. Orientation disparities certainly occur without any torsional rotations of the eyes.

It is also important to appreciate that orientation disparities always occur in association with position disparities but the reverse is not necessarily the case. Position disparities commonly occur without any associated orientation disparities. Any part of the line XY in Fig. 1 has a position disparity but the most important position disparities are always the end points. In normal viewing, of course, the two disparate retinal images of the line are fused into a single line and this single line is seen as tilted in depth with the end X further away than the end Y.

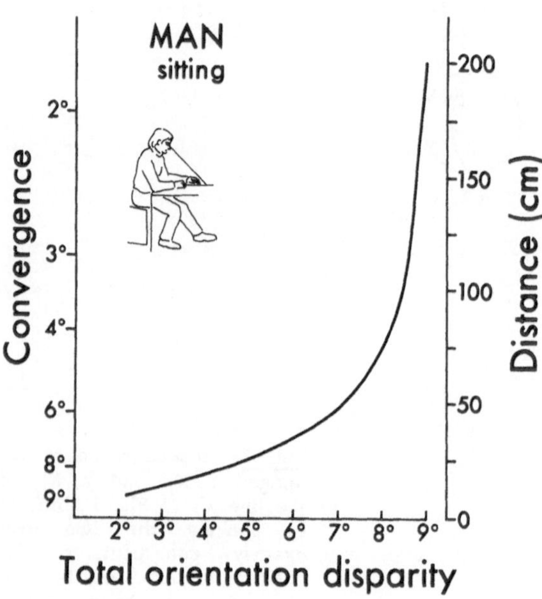

Fig. 3 Orientation disparity as a function of convergence and viewing distance for a seated human subject when, with the nodal points of his eyes 40 cm above the table top, he views a line in his median plane on the table top. The ordinate scale gives horizontal distances from the nodal points measured in the plane of the table top.

The main question I want to consider here is the role of orientation disparities in binocular vision and particularly the extent to which they contribute to binocular depth perception.

For the graph in Fig. 3, I have calculated the orientation disparities a subject would experience if, when seated at a desk with the nodal points of his eyes 40 cm from the desk top, he views a line in his median plane. At a typical reading distance of 32 cm (i.e., horizontal distance of 25 cm) the orientation disparity is approximately 5°. The distances given by the right hand ordinate scale in Fig. 3 are horizontal distances from the nodal points measured in the plane of the desk top. So all the time you are reading under these conditions you are continually fusing orientation disparities of the order of 5°. Elevating the eyes to view the line at a greater distance increases the orientation disparity of the line. There is thus a continuum of orientation disparities which vary systematically with viewing distance. The curve in Fig. 3 asymptotes to about 9° and, since we do not see double at viewing distances beyond 1 to 2 m, the brain is apparently still able to fuse orientation disparities of that order.

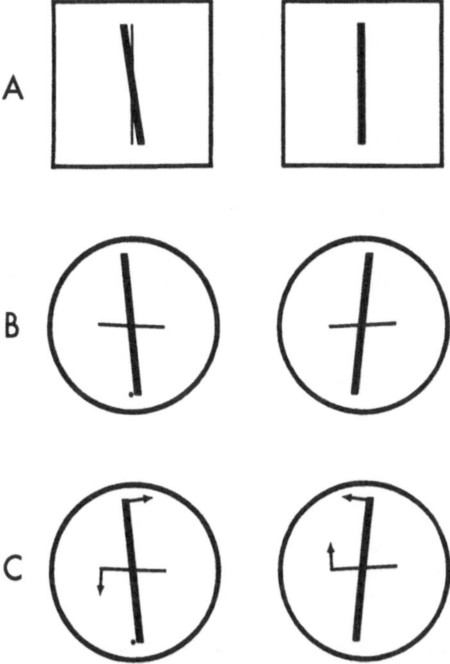

Fig. 4 Stereopair (B) modified from WHEATSTONE (2) (A) to demonstrate that fusion by cyclovergence would require impossible eye movements (C).

As long ago as 1838 WHEATSTONE (2) demonstrated that lines whose images do not fall on corresponding points of the two retinae may still appear single. In Fig. 4A I have redrawn one of WHEATSTONE'S (2) original stereopairs. The thin line in the left hand target and the thick line in the right hand one are both vertical. When viewed in a stereoscope these two dissimilar lines will presumably fall along corresponding points in the two retinae whereas the two thick lines, which are at an angle to one another, will give rise to an orientation disparity. Nevertheless, when viewed binocularly, the two thick lines, each seen by a different eye, fuse into a single line tilted out of the plane of the square frame while the thin line retains its original vertical position and appearance in the plane of the frame. WHEATSTONE (2) noted that there had been a fusional displacement of the thick vertical line to combine with the other thick line which is at an angle to the vertical. Besides demonstrating the fusion of orientationally disparate retinal images WHEATSTONE (2) concluded that "this experiment affords another proof that there is no necessary physiological connection between the corresponding points of the two retinae, - a doctrine which has been maintained by so many authors". And maintained, I might add, by many authors even to this day.

WHEATSTONE'S (2) paper, published in the Philosophical Transactions of the Royal Society, is a superb paper and must be one of the most original that has ever been written. It is without any scientific antecedents and is still very well worth reading because most of the elements of the psychophysics of binocular vision are already present in it.

Figure 4B is a simple but instructive adaptation of WHEATSTONE'S (2) original stereopair. I have omitted the thin vertical line in the left hand target and altered the two thick lines so that they are both now inclined to the vertical and at an angle of 11° to one another, convergent downwards. I have also added two short approximately horizontal lines across the centers of the two thick lines. These two short lines are symmetrically placed about the horizontal and at an angle of 4° to one another. Viewed stereoscopically I have no difficulty in fusing the stereopair so that it appears as a single figure with the cross member still located in the plane of the circular frame but now accurately horizontal. The thick line is now vertical but tilted about its center point out of the plane of the frame. The dot in the left hand target appears to the left of the fused line, despite the fact that, in the other eye, its image falls to the right side of the line (3).

If cyclorotation of the eyes is to be responsible for the fusion of the disparate retinal images in Fig. 4C, then fusion of the approximately vertical lines would require an incyclovergence, as shown by the arrows in fig. 4C, whereas an excyclovergence would be required for fusion of the horizontal lines. Since both these vergences cannot occur simultaneously, cyclovergence cannot be responsible for the fusion. There are, of course, many examples in the literature where fusion would require impossible eye movements of this kind. That rotation of the eyes was not the basis of the fusion of orientation disparities was doubtless already known to WHEATSTONE (2) since the stereopair in Fig. 4C is essentially the same as that in Fig. 4A. If incyclovergence had been the basis of the fusion of the thick lines in Fig. 4A it would necessarily have led to a doubling of the square frame.

There has recently been controversy concerning the role of cyclorotary eye movements in the fusion of orientation disparities and, indeed, controversy also whether fusion occurs at all (4). However from the work of KERTESZ and his colleagues (5, 6) we now have, I believe, quite convincing evidence that fusion of orientation disparities can occur and that it operates by a central neural mechanism without any significant movements of the eyes.

So far I have described orientation disparities as experienced by the human subject. Since I want to go on to describe the neural mechanisms that my colleagues, DR. NELSON and DR. KATO, and I have been studying in the striate cortex of the cat (1) I thought I should give you some idea of the orientation disparities the cat is likely to experience under natural conditions.

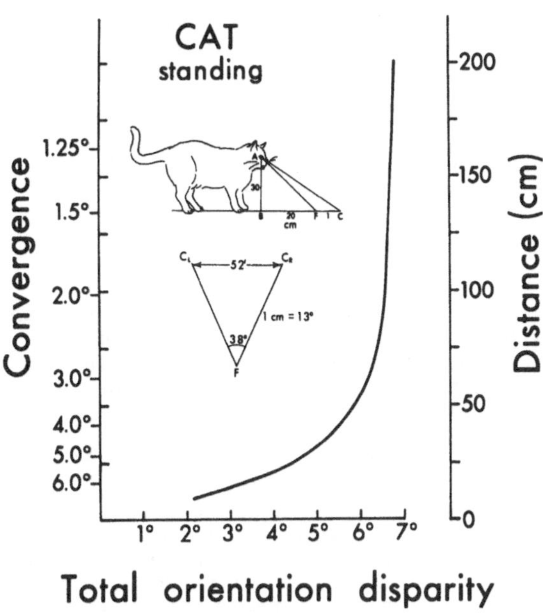

Fig. 5 Orientation disparity as a function of convergence and viewing distance for standing cat when, with the nodal points of its eyes 30 cm from the ground, it views a line on the ground in its median plane. Inset: Orientation (3.8°) and position (5.2°) disparities for a line FC = 1 cm located 20 cm (BF) horizontally from the nodal points.

Consider a cat standing with the nodal points of its eyes 30 cm from the ground and looking down on the ground at a line in its median plane. In Fig. 5 I have calculated the orientation disparities the cat would experience at different viewing distances. As in Fig. 3, the values given by the right hand ordinate scale are horizontal distances. You can see that the orientation disparities for the standing cat are much the same as those for a man sitting at a desk. The disparities in Fig. 5 asymptote to about 7° for viewing distances beyond 1-2 meters. When the cat assumes a semi-prone position of sphinx-like recumbancy, rather larger orientation disparities will be experienced and the curve now asymptotes to about 20° for the longer viewing distances. The closer the eyes are to the ground and the longer the viewing distance the greater the orientation disparity. Thus, depending upon the posture it adopts, a cat will be required to fuse orientation disparities ranging up to about 20°.

In the literature until relatively recently the aspect of retinal image orientation disparities that has dominated consideration has been the problem of fusion in the interests of single vision. Recently however BLAKEMORE, FIORENTINI and MAFFEI (7) carried out experiments that led them to propose that the detection of orientation disparities by striate neurons might operate as a second neural mechanism for depth discrimination. This is an appealing idea because it would provide a very economical mechanism for the perception of the tilt of a line. It would be economical because it would require the discrimination of only a single orientation difference rather than a series of position disparities.

If binocular cells in the visual cortex are to serve as orientation disparity detectors in a binocular depth discrimination mechanism then they should display two essential properties. First, there should be a population of cells that show a sufficiently wide range of different orientation disparity preferences as to cover the range of disparities likely to be experienced in normal life. And second, any given cell in this population should be sharply tuned to detect its particular disparity preference.

In order to investigate the first of these requirements we set out to prepare quantitative orientation tuning curves for each eye from as many binocular cells in the striate cortex as we could in the one animal. The 14 pairs of orientation tuning curves in Fig. 6 were recorded from 13 simple and one hypercomplex I cell in the one cat. In the same animal, but not illustrated here, we also recorded tuning curves from 6 complex and one hypercomplex II cell, making a total of 21 cells in the one animal. The data points in Fig. 6 were recorded in orientation steps of 3° but, in order to avoid confusion, the points are shown for only one curve in each pair. All the tuning curves for the right eye have been arbitrarily brought into line and the whole of the orientation disparity in each case has been transferred to the left eye curve. If each cell had the same optimal orientation for the two eyes (i.e., zero orientation disparity) then all the left eye curves should also line up. But this is clearly not the case. First of all the left eye curves are bodily shifted to the right. Without exception all the curves fall to the right of the 0° line. Then, in addition, there is a scatter in the peaks of the curves to either side of the mean orientation, the mean being indicated by the two vertical arrows.

On the assumption that striate cells have a zero mean orientation disparity, we believe that the net rightward shift of the left eye curves is a consequence of the incyclorotation of the eyes that results from the anaesthesia and the paralysis of the extraocular muscles. Both of these procedures are, of course, an essential part of the experimental method. Over a series of 9 cats we have found that this incyclorotation averages about 10°, so that presumably for each eye there is a medial rotation of the upper end of the slit pupil by about 5°. Many years ago, using the same general experimental procedures but a quite different method for measuring eye rotation, we also observed an incyclorotation of the two eyes that totalled 10°. However, at the time, we felt that we could not rely on the accuracy of our method (8). In the same year HUBEL and WIESEL (9) also concluded from their observations on the anaesthetized and paralyzed cat that "the combined inward rotation of the two eyes seldom exceeded 10°".

Of more immediate interest is the scatter in the optimal orientation disparities. For the 21 cells that we recorded from the cat used for Fig. 6 there was a scatter in the optimal disparities of about 10° to either side of the mean or zero disparity. When we pooled the data from 74 cells recorded in 9 cats we found a scatter that totalled nearly 20° to either side of zero disparity. So there is a striking parallel

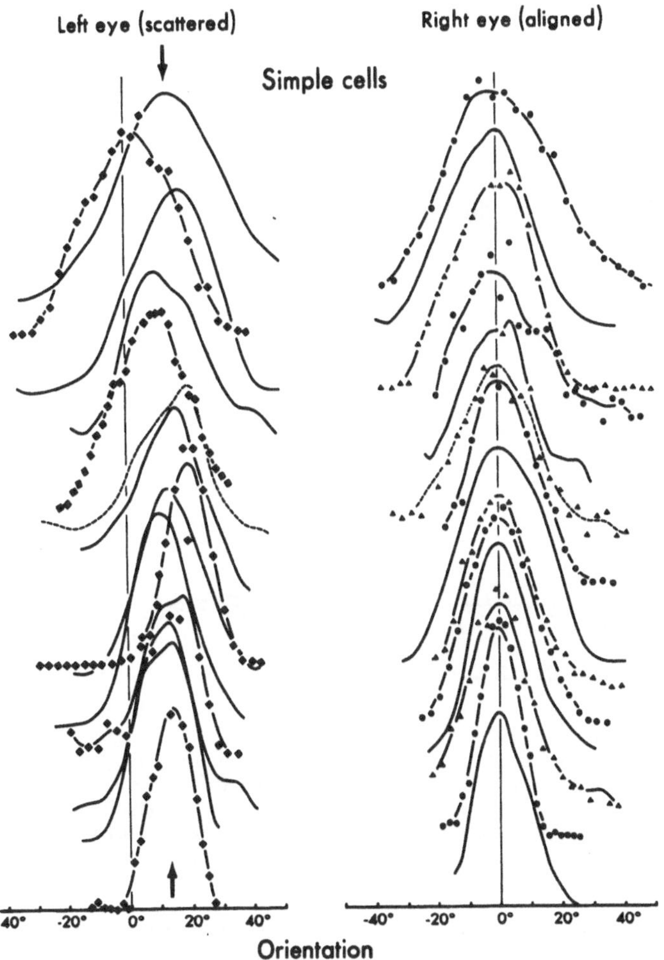

Fig. 6 Scatter in monocular stimulus orientation tuning curves for 14 binocular striate cells (13 simple, continuous lines; 1 hypercomplex I, dashed lines) in one cat. All the tuning curves have been normalized to the same height, ordered according to half-width at half-height, and spaced vertically. To avoid confusion, data points are shown for only one curve of each pair (from ref. 1).

between the range of preferred orientation disparities that we found for cells in the cat striate cortex and the range of retinal image orientation disparities likely to be experienced by the animal under natural conditions. I might add that we very carefully monitored any eye rotations that occurred during the course of an experiment by a technique that was at least an order of magnitude more accurate than the optimal orientation disparities that we sought to measure.

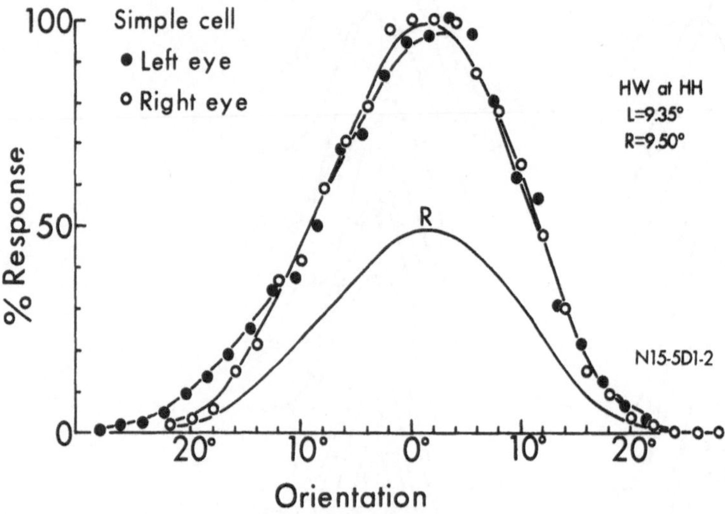

Fig. 7 Monocular stimulus orientation tuning curves for each of the two eyes from a binocular simple cell in the striate cortex with the curve (R) for the right eye normalized to the same height as the curve for the left eye. The abscissa zero is arbitrarily located so that the curves are approximately symmetrical about it. HW at HH: half-width at half-height of the left (L) and right (R) eye tuning curves.

It is a curious fact that, while the optimal orientation disparity varies widely from cell to cell over a range of nearly $\pm 20^{\circ}$ so that the match that any given cell achieves between the preferred stimulus orientation for each of the two eyes tested monocularly is relatively poor, the match the cell achieves between the sharpness of its monocular orientation tuning curves is extremely good. Thus the half-widths at half-height of the two monocular orientation tuning curves for the simple cell in Fig. 7 differ by only 0.15°. In all we prepared 80 monocular orientation tuning curves from 40 cells in the simple family (simple and hypercomplex I) and the maximum difference between the half-widths at half-height for any given pair was 8°, with a mean difference of only 2.5°. I have included Fig. 8 because it so happened that the difference between the half-widths at half-height of the two monocular curves for this cell was exactly the same as the population mean, i.e., 2.5°. By contrast with these observations, the half-widths at half-height of the curves varied widely from cell to cell over a range of about 37° from 7° to 44°. Thus the two curves from any given cell are always closely matched for sharpness whether or not the cell itself is finely tuned for orientation discrimination.

As I have already remarked, the close matching between the two monocular curves also does not extend to their optimal orientations. Thus cells having widely different preferred orientations for the two eyes are just as closely matched for sharpness as those with closely similar preferred orientations. Though closely matched for sharpness, the stimulus orientation disparities for the two cells in Figs. 7 and 8 were 7.8° and 5.5° respectively.

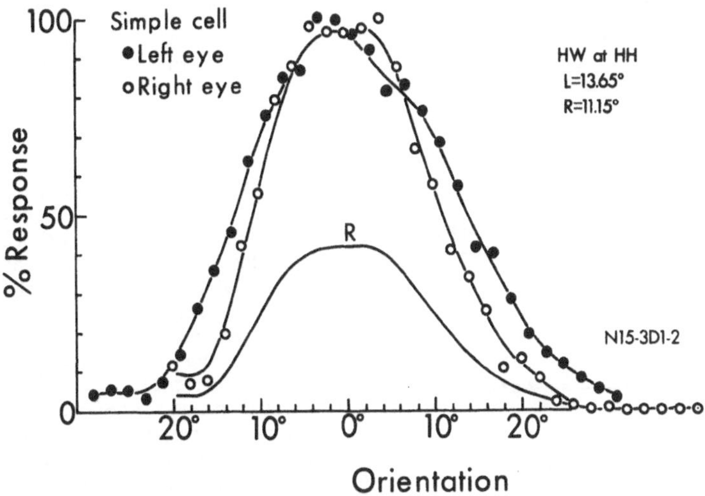

Fig. 8 Monocular stimulus orientation tuning curves as in Fig. 7 but from another simple cell from the striate cortex of the same cat. The difference in the half-width at half-height (HW at HH) of the two curves (2.5°) is the same as the mean difference in the curves from the two eyes for 37 simple and 3 hypercomplex I binocular cells recorded in 9 cats.

Furthermore the close matching between the two monocular curves from a given cell is independent of its ocular dominance preference. Thus the two cells I have used for Figs. 7 and 8 both showed a marked preference for the left (contralateral) eye with a response from the right eye about half that from the left. In both figures the curves for the two eyes have been normalized to the same height but the relative amplitude of the response from the right eye is shown in each case by the continuous curve labelled R.

From these observations we can conclude that, despite Nature's ability to match the sharpness of the two monocular orientation curves of a binocular cell, there has apparently been no great evolutionary pressure to reduce the range either of the sharpness of the tuning curves themselves or of their optimal orientation disparities. Indeed it may even be the case that the scatter in preferred orientation disparities has provided an evolutionary advantage.

The first requirement for a depth discrimination mechanism based on orientation disparities was that there should be a population of cells preferring a range of different orientation disparities. Such a population of striate cells certainly exists.

I now propose to turn to the second requirement for the depth discrimination mechanism, namely the requirement that each of the cells concerned should be sharply tuned to detect its particular preferred disparity. The key question I want to consider is whether the striate neurons in the cat are capable of discriminating the position and orientation disparities that seem from behavioral(10) and other considerations (1) to be required of the animal in normal life. By way of answering these questions we prepared single cell tuning curves for both position and orientation disparities. We did this by diverging the two eyes with prisms so that they could be stimulated at the same time, though each independently of the other, by separate but identical stimuli (1).

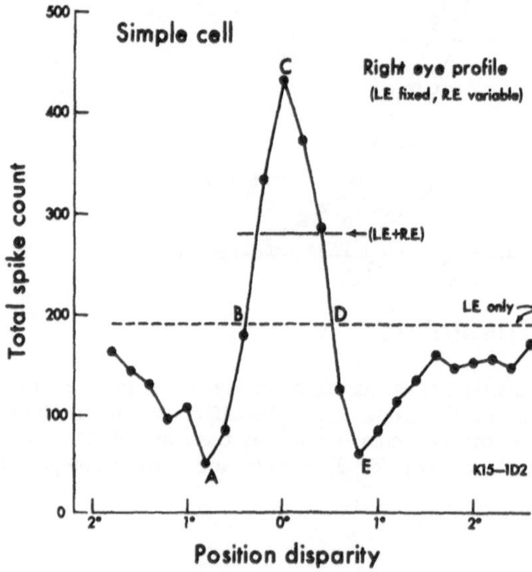

Fig. 9 Position disparity tuning curve from Fig. 10 corresponding to the optimal stimulus orientation (86°) for the variable (right) eye. Dashed line: level of the monocular response from the left eye at the fixed (optimal) stimulus orientation of 93°. short continuous line: summed monocular responses from the two eyes with both stimulus orientations optimal (from ref. 1).

Figure 9 shows a position disparity tuning curve obtained from a simple cell. The maximal binocular response is much greater than the sum of the responses from each eye stimulated on its own. This maximal response occurred when the two eyes were stimulated independently, each with the optimal stimulus, but at exactly the same time. This situation is precisely equivalent to what happens in normal life when a single object feature stimulates the two receptive fields of the cell at that depth in space where the two fields are in accurate spatial register - that is, a 0° of position disparity. Some years ago we found that the optimal binocular response from simple cells averaged 76% greater than the sum of their monocular responses (11). But as soon as a position disparity in either direction was introduced to the cell in Fig. 9 by stimulating the two receptive fields at slightly different times the binocular response fell off very rapidly indeed. The cell in Fig. 9 was, however, not particularly sharply tuned for position disparity. Quite a few of the cells we have examined switch from maximal response to complete inhibition for a position disparity increment of less than 0.5 prism diopters (17 min arc) (11). In general, therefore, simple cells are very sensitive to position disparity.

Now what about their sensitivity to orientation disparity? An orientation disparity tuning curve is produced by recording the binocular response from a cell with the stimulus for one eye fixed at the optimal setting while varying the stimulus orientation for the other eye. You have just seen that simple cells are very sensitive to position disparity and considerable care has to be exercised when recording an orientation disparity tuning curve so as to avoid introducing unintentional position disparities. The presence of an unwanted position disparity can have a powerful distorting effect, making the cell appear to be much more sensitive to orientation disparity than it really is. We avoided position disparity errors by allowing the cell in question to pick its own preferred position disparity at each orientation disparity setting. We did this by preparing a complete position disparity tuning curve, like the one in Fig. 9, at each orientation disparity setting. In other words we presented a range of position disparities at each combination of stimulus orientations. In this way we were able to extract a tuning curve for one kind of disparity when the other kind of disparity was zero (i.e., optimal).

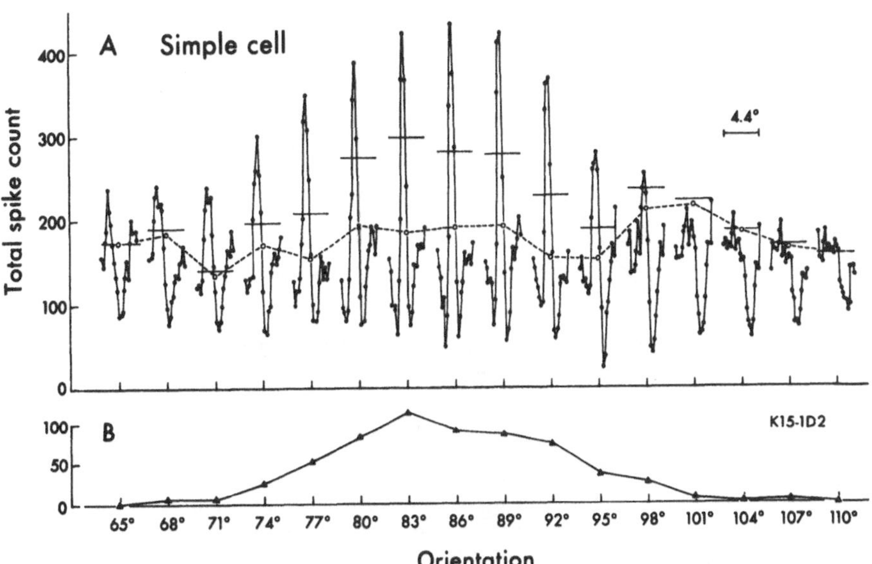

Fig. 10 A. Family of 16 position disparity tuning curves from a simple cell recorded over a 45° range of orientation disparities at increments of 3°. All orientation and position disparity changes occur in the right eye with the left eye's stimulus held constant at 93° throughout. Dashed lines (open circles): level of control monocular responses from the fixed orientation (left) eye. Short continuous lines: level of the summed monocular responses from the two eyes with the stimulus orientation for the variable (right) eye as indicated. B. Monocular orientation tuning curve for the right eye from the same cell (from ref. 1).

All the data used for Fig. 10 were obtained from the same simple cell as for Fig. 9. In Fig. 10 there are 16 position disparity tuning curves recorded over a wide range of orientation disparity settings in orientation increments of 3°. The position disparity tuning curve at orientation 86° is the one used for Fig. 9. The curve given by the peaks of the 16 position disparity curves is the true orientation disparity tuning curve for this simple cell. This is so because the curve drawn through the peaks of the various position disparity tuning curves shows the effect of changing the orientation disparity when the position disparity is maintained at its optimal setting.

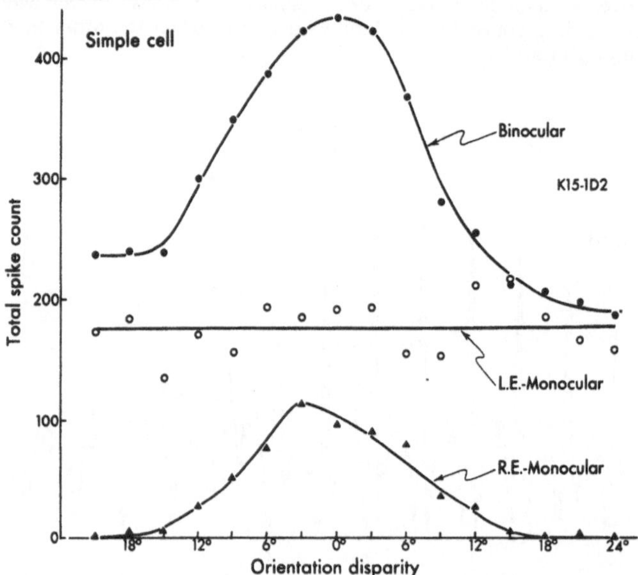

Fig. 11 Simplified version of Fig. 10. Filled circles: binocular orientation disparity tuning curve. The data points correspond to the peaks of the 16 position disparity tuning curves in Fig. 10. Open circles: control monocular responses from the fixed orientation (left) eye. Filled triangles: control monocular orientation tuning curve from the right eye. The three curves share a common abscissa and ordinate but the data for the control monocular responses only correspond to the orientation disparity settings given by abscissa scale.

For our present purposes I have considerably simplified the presentation of the data in Fig. 10. Fig. 11 is the simplified version. I would particularly like you to note that the three curves in Fig. 11 share common abscissa and ordinate. Orientation disparity is, of course, a binocular parameter so that the data for the two control monocular curves only correspond to the orientation disparity settings given by the abscissa scale. The top-most curve is the binocular orientation disparity tuning curve, the data points being the peaks of the 16 position disparity tuning curves in Fig. 10. The control monocular responses given by the two lower curves are, for the most part, very much smaller than the binocular responses. The open circles are the monocular responses from the eye with the fixed stimulus, the straight line drawn through the data points giving the mean level of the response. The bottom-most curve is the monocular orientation tuning curve for the other or variable eye.

This is not the place to discuss the nature of the binocular orientation disparity tuning curve in detail (1) but you can see that it is very broad. In fact, in all the cells that we have examined, it is always broader than the corresponding monocular tuning curve. Even over the very wide range of orientation disparities that we used to test the cell in Fig. 11 the binocular response barely fell to half of what it was at the optimal orientation disparity setting - that is at 0° orientation disparity. There are apparently no novel binocular mechanisms to sharpen up the discrimination of orientation disparities.

Before concluding I would like to consider once again the orientation disparities that the cat would experience under natural viewing conditions. The inset to Fig. 5 shows a standing cat viewing a point F and a line FC 1 cm long on the ground 20 cm ahead of the animal. If we take the distance from eye to ground as 30 cm then the orientation disparity of the two retinal images of the line FC is 3.8° and the position disparity of the end-points C_L and C_R of the line images is 5.2 min arc. A position disparity of 5 min arc is probably very little above threshold for the cat (1, 10). Our conclusion is that striate neurons in the cat would probably be able to detect this position disparity but would be unable to detect directly the orientation disparity associated with it.

Thus orientation disparity detection is unlikely to operate as a second neural mechanism of binocular depth discrimination as proposed by BLAKEMORE *et al.* (7). Nevertheless a striking feature of the range of optimal orientation disparities of striate neurons ($\pm 20^{\circ}$) is that it is approximately the same as the range of retinal image orientation disparities that the cat is likely to experience in normal life. Thus for any tilted line that the cat observes there stands available a receptive field pair whose preferred stimulus orientation disparity matches the orientation disparity of the line concerned. Not only does this ensure maximum efficiency in a tilt-detecting mechanism based on position disparities but it is also doubtless the basis of the cyclofusion needed for binocular single vision. The separate contributions from the two eyes lose their identity when they come together to give rise to a single binocular discharge from the cortical cell. Furthermore, when the stimulus is optimal, that is when it is located at the depth in space preferred by the cortical cell, the binocular discharge is nearly always significantly greater than the sum of the individual responses from the two eyes under the same conditions of stimulation (11). With optimal binocular stimulation, both eyes make an excitatory contribution to the common response from the cortical cell and suppression of one eye by the other does not occur. This combined response now conveys information, not about retinal image disparities, but about the depth in space of a particular single object feature.

Concluding a recent interchange of *Letters to the Editors* in *Vision Research* KAUFMAN and ARDITI (12) questioned "the adaptive purpose, if any, of central cyclofusion". They went on to remark that "we have tried to imagine situations in nature which would require the evolution of a central mechanism which would compensate by as much as 13° for the lack of ability of the eyes to counter-rotate. Such a central mechanism must be complicated. Our own failure to define the biological usefulness of such a mechanism is a major reason for the doubts we have expressed concerning central cyclofusion". Quite apart from the question of binocular single vision, I have argued that central cyclofusion has another important adaptive role, which is to make the position disparity detecting mechanism insensitive to the relatively large orientation disparities that occur under natural conditions, and I have outlined the way this orientation disparity mechanism might operate.

Acknowledgements

I wish to thank my colleagues Dr. J. I. Nelson and Dr. H. Kato for their collaboration in carrying out most of the experimental work. The following assistance is gratefully acknowledged: Ms. E. Elekessy with computer programming, Mr. K. Collins with the illustrations and Miss J. Livingstone with the preparation of the typescript.

References

1. J. I. Nelson, H. Kato and P. O. Bishop, J. Neurophysiol. 40, 260 (1977).
2. C. Wheatstone, Phil. Trans. R. Soc. Lond. 128, 371 (1838).
3. J. I. Nelson, J. Theor. Biol. 49, 1 (1975).
4. L, Kaufman and A. Arditi, Vision Res. 16, 535 (1976).
5. A. E. Kertesz and R. W. Jones, Vision Res. 10, 891 (1970).
6. A. E. Kertesz and M. J. Sullivan, Vision Res. 16, 545 (1976).
7. C. Blakemore, A. Fiorentini and L. A. Maffei, J. Physiol. London 226, 725 (1972).
8. P. O. Bishop, W. Kozak and G. J. Vakkur, J. Physiol. London 163, 466 (1962).
9. D. H. Hubel and T. N. Wiesel, J. Physiol. London 160, 106 (1962).
10. R. Blake and H. V. B. Hirsch, Science 190, 1114 (1975).
11. P. O. Bishop, G. H. Henry and C. J. Smith, J. Physiol. London 216, 39 (1971).
12. L. Kaufman and A. Arditi, Vision Res. 16, 551 (1976).

Stereoscopic Depth Channels for Position and for Motion

D. Regan, K. I. Beverley, and M. Cynader
Department of Psychology, Dalhousie University
Halifax, Canada

Summary

In this chapter we summarize psychophysical and single-neuron evidence that information as to movement in depth and information as to position in depth are processed in different psychophysical channels and possibly by different neural mechanisms. One item of evidence is that inspecting a target oscillating in depth reduces visual sensitivity to depth motion, but only for a restricted range of trajectories close to the adapting trajectory. This suggests that there are several binocularly-driven neural filters, each of which is preferentially sensitive to a different direction of motion in depth. Further evidence is the existence of areas of the visual field blind to motion in depth but where positional (static) depth perception is normal. Supporting evidence obtained by single-neuron recording in cat cortex is the existence of two classes of neuron that emphasize information as to the direction of motion in depth rather independently of position: one class prefers trajectories passing between the eyes while the other class prefers trajectories that miss the head. A third class of neuron is well known. This class is sensitive to position in depth and is best driven by binocularly-viewed sideways motion.

Introduction

When a real object moves in depth its retinal image size changes. Conversely, changing size is a cue to motion in depth.

We propose that information as to changing size is processed in a psychophysical channel separate from both motion and flicker information. Our evidence is that inspecting a changing-size stimulus reduces visual sensitivity sepcifically for changing-size stimuli: movement adaptation has no such effect. This adaptation is indifferent to contrast (i. e. transfers from bright to dark stimuli). This finding is consistent with the existence of neural filters sensitive to increasing size and to decreasing size respectively. We also find that, following inspection of an increasing-size stimulus, a static square appears to move continuously away from the observer, and vice versa. This negative aftereffect supports the notion that the changing-size filters drive the neural mechanism that underlies the perception of motion in depth.

We should emphasize that we have restricted both our physiological and psychophysical experiments to the component of motion in the horizontal plane. The vertical component is also important in everyday life, but cues other than the ones we have considered must be used to judge the direction of motion in the vertical plane.

1. Dynamics of Binocular Depth Perception: Psychophysical Sensitivities to Oscillating Disparity and to Pulsed Disparity

Since 1838 when WHEATSTONE invented the stereoscope and demonstrated to the Royal Society that retinal disparity by itself is sufficient to generate the perception of depth (1, 2), most research into stereopsis has been concerned with positional (static) depth perception and binocular fusion, with disparity as the sole cue to depth.

Most early experimenters paid little attention to the dynamics of depth perception, and still less was the perception of motion in depth treated as a problem distinct from the question of positional depth perception. For most subjects, psychophysical sensitivity to disparity oscillations is greatest at low frequencies falling off rapidly for oscillation frequencies greater than about 1 to 2 Hz and failing entirely at a frequency of 3-5 Hz (3,4): the dynamics are slightly different for crossed and uncrossed disparities (5). Such measurements are easily made at superthreshold levels, for matched depths methods may then be used. At threshold, however, the subject is attempting to detect any change, and monocular cues can be confounded with cyclopean cues: depth oscillations and sideways oscillations may not look different. This threshold problem has been tackled by stimulating the left and right eyes with sinewave movements of slightly different frequencies, thus dissociating the time courses of retinal image movements and disparity changes (6). Again, senstivity was greatest at low frequencies, falling off above 0.6 Hz or so (Fig. 1). No depth oscillations at all could be seen above a frequency of 3 to 5 Hz. As illustrated in Fig. 1, sensitivity to disparity oscillation was restricted to a much lower frequency band than sensitivity to sideways oscillations (i. e., oscillations in the frontoparallel plane).

If the left and right eyes view identical bars that oscillate from side-to-side with the same amplitudes, then if the motions are in phase the binocularly-fused bar appears to oscillate sideways, whereas if the motions are opposed (antiphase) the binocularly-fused bar appears to oscillate in depth.

The question whether sensitivity to sideways oscillation is greater or less than sensitivity to oscillation in depth cannot be answered unequivocally without carefully controlling binocular tracking and binocular convergence (5). The reason for this proviso as shown in Fig. 2, is that the relative sensitivity to depth oscillation (STEREO) and binocularly-viewed sideways oscillation (BINOC) depends on whether the mean stimulus disparity is near-zero ($\pm$ about 7 min arc).

Sensitivity to depth oscillations can be greater than to sideways oscillations in some stimulus situations (mean disparity less than $\pm$ 7 min arc, frequency less than about 0.8 Hz) (5), but in most situations the reverse is true (7). Pulsed presentations

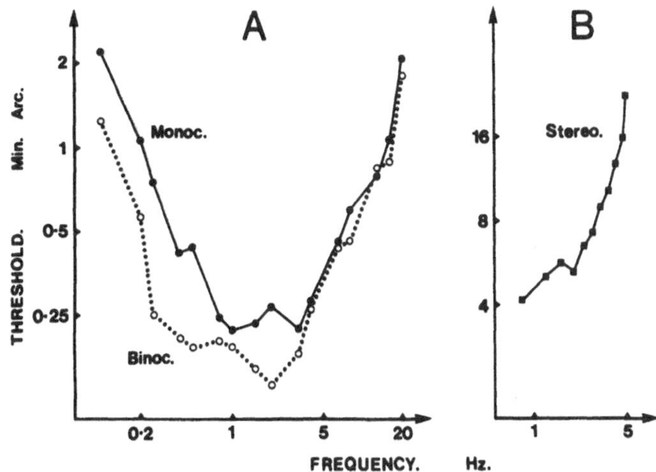

Fig. 1 A. Visual sensitivity for sideways oscillations. Threshold amplitudes of oscillation are plotted versus the frequencies of sine wave displacements of the target. Continuous line - monocular viewing with right eye occluded; chain line binocular viewing. Log-Log axes. B. Visual sensitivity for depth oscillations. The left eye viewed a target that oscillated at (F+ Δ F)Hz where F was a constant small value (0.1 Hz). Depth oscillations waxed and waned at a rate of 0.1 Hz, and the subject set the smallest amplitude of oscillation (ordinates) for which he could just see the 0.1 Hz periodicity at various rates of depth oscillation (F + Δ F)Hz plotted as abscissae. Same subject as in Fig. 1A. Log-Log axes. Modified from (6).

(flashes) of left and right retinal images are ineffective in generating a depth impression when the end of the first presentation occurs more than 100-180 msec before the start of the second presentation (8, 9, 10). Such experiments give a basis for estimating the "memory" of the depth mechanisms. With simultaneous movements of the left and right retinal images, a depth movement grows progressively easier to see as the duration of the disparity pulse is progressively lengthened (11), though Bloch's law is not strictly obeyed (12). Curiously, the eye seems to integrate disparity change over longer durations when information as to the moment of disparity change is cyclopean only: monocular timing information seems to restrict integration (13).

2. Psychophysical Adaptation to Motion in Depth

It is well known that visual sensitivity to static disparity, although very acute, is not adequate to give appreciable discrimination of positional depth at distances of more than 10 meters or so. However, it does not follow that accurate judgements of the direction of motion in depth need necessarily be restricted to short range. For

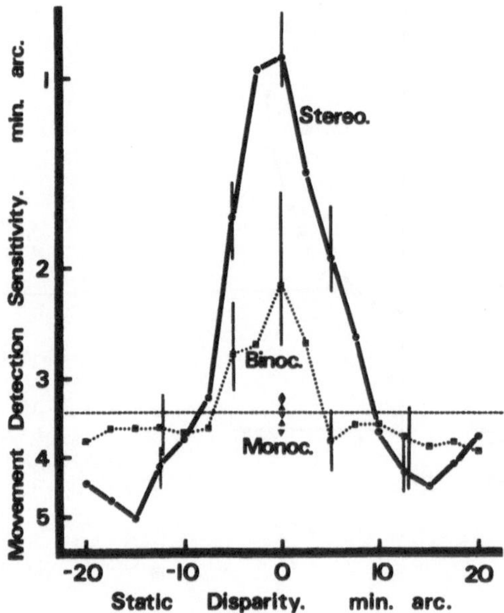

Fig. 2 The effect of position in depth (i. e. mean disparity) on sensitivity to motion. Ordinates are threshold amplitudes of oscillation; amplitudes are mean disparities. Positive (crossed) disparities are nearer than the fixation point, while negative (uncrossed) disparities are farther away. In the stereoscopic case the targets viewed by the left and right eyes oscillated in antiphase, while in the binocular case they oscillated inphase. In the monocular case, either the right (upright triangles) or left (inverted triangles) eye was occluded. The horizontal line is drawn through the mean monocular threshold. Oscillation frequency was 0.1 Hz. The bars show total range of 5 settings. From (5).

example, the visual system might process information as to motion in depth by responding directly to the relative velocities of the retinal images, thus dissociating visual responses to the direction of motion in depth from visual responses to position in depth (see below, Section 5).

We proposed that information as to motion in depth and information as to positional (static) depth are processed in different psychophysical channels in order to explain our finding of directionally-specific adaptation to motion in depth. This adaptation effect can be observed in the following way.

Two identical random-dot targets were viewed, one by the left eye and one by the right eye. The targets were oscillated independently from side to side by sinewaves of the same frequencies. Viewed in binocular fusion the stimulus appeared to be a single target oscillating along a straight line in depth. The direction of this trajectory was determined by the relative velocities of the left and right eyes' targets. Illustrating this point, Fig. 3 shows how the relative velocities of the left and right retinal images provided a sensitive cue to the direction of motion in depth.

For example when the left and right images moved in opposite directions (i.e. in antiphase) then the real object viewed would be moving along a line directed between the eyes. If the left image moved slower than the right, then the trajectory passed closer to the left eye than the right, and vice versa (Fig. 3). Note that in real space

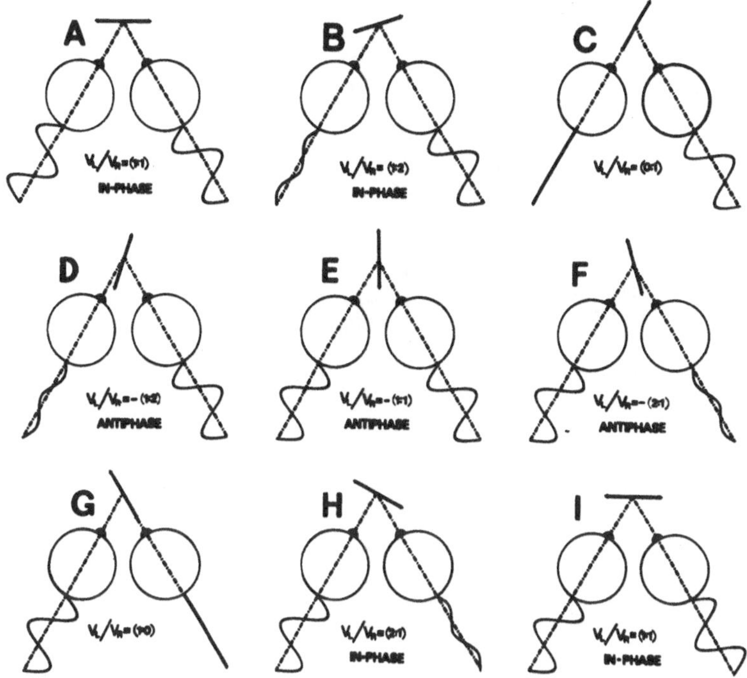

Fig. 3 Schematic diagram illustrating the motion of the images on the left and right retinae of a binocularly fixated stimulus which oscillates sinusoidally in various directions in three-dimensional space. The ratio V_L/V_R is the physiological cue to the direction of motion in depth. From (14).

the range of directions in depth for all possible antiphase motions could be very narrow (e. g. 2.5° for a viewing distance of 57 inches assuming an interpupillary separation of 2.5 inches)[1]. An object moving along a line passing wide of the head caused the left and right retinal images to move in the same direction, and again the relative speeds indicated the exact direction (Fig. 3).

1. Some authors have expressed surprise at the very narrow directional tuning of these binocular motion filters. Therefore we should emphasize that this very narrow selectivity is merely a geometrical consequence of computing the direction of motion in depth directly from relative retinal image velocities. This point is illustrated by the plots of Fig. 4, obtained by calculation rather than by physiological experiment.

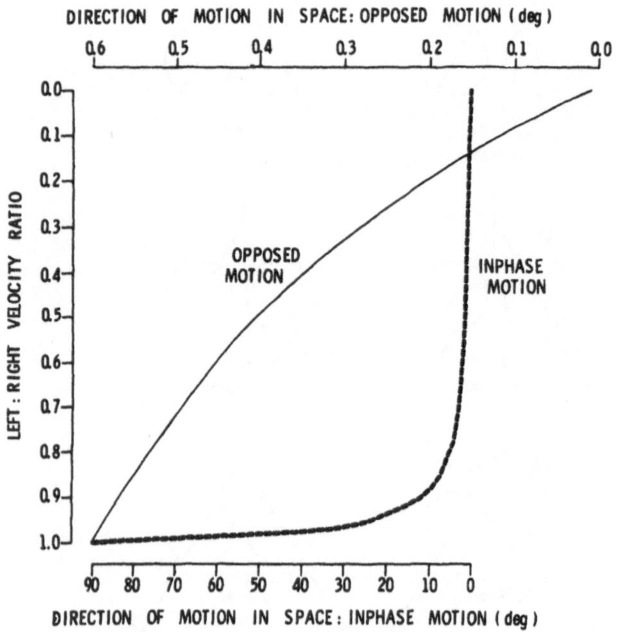

Fig. 4 Ratio of target speeds seen by the left and right eyes (V_L/V_R) (ordinates) versus the direction of motion in real space (abscissae) when the two motions are oppositely directed (i.e. the trajectory passes between the eyes, continuous line), and when the two motions are in the same direction (i.e. the trajectory misses the head, dashed line). An angle of 0° signifies a trajectory that passes through one eye. The data were calculated geometrically (for cat) with an interpupillary separation of 3.0 cm and viewing distance of 145 cm. From (23).

Subjects viewed a binocularly-fused target that appeared to oscillate in depth through 20' arc. We found after an inspection time of about 15 min the initially-large depth oscillations appeared small or even totally absent, although the perceived amplitude of side-to-side oscillations (parallel to the frontoparallel plane) was comparatively unaffected. Figure 5 illustrates our central finding that this sizeable adaptation to movement in depth was direction-specific. In particular, after adapting to motion along a line inclined in depth and passing to the left of the nose, visual sensitivity to movement in depth was reduced for all trajectories passing to the left of the nose, but was unaffected for all trajectories passing to the right of the nose. Conversely, adapting to movement in depth directed to the right of the nose reduced visual sensitivity for all trajectories passing to the right of the nose, but had no effect on sensitivity to trajectories passing to the left (14).

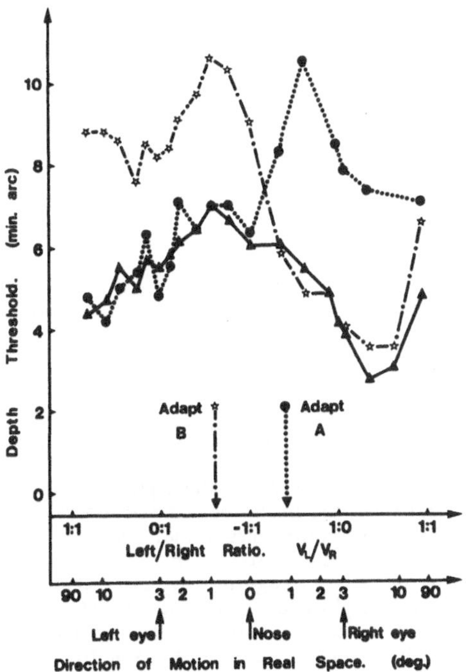

Fig. 5 Evidence for directionally-selective binocular motion filters. Threshold amplitudes of disparity oscillation (ordinates) are plotted versus V_L/V_R where V_L and V_R are the left and right retinal image velocities. Full line, before adaptation at B. Oscillation frequency 0.8 Hz. Negative V_L/V_R means left and right retinal images always moved in opposite directions; positive sign means images always moved in same direction. Directions in real space for the viewing distance of 63 cm are shown in the lower abscissae.

This finding led us to postulate that there are at least two binocular filters in the visual pathways, sensitive to motion along trajectories passing to the left and the right of the nose respectively.

In further experiments we asked whether there were more than two classes of binocular filters. We repeated the measurements of Fig. 5 for an additional range of 11 adapting directions: in all we separately adapted to 13 directions of motion in depth. However, this did not produce 13 different curves of the type shown in Fig. 5. All the data could be described in terms of only 4 underlying sensitivity curves, and these are shown in Fig. 6. Two of the sensitivity curves are concerned with trajectories passing between the eyes, and are so selectively tuned that they indicate no response at all when the direction of motion departs by more than 1.5° from the preferred direction. The "missing the head" curves are much more broadly tuned, and deal with the remaining 177° of visual space.

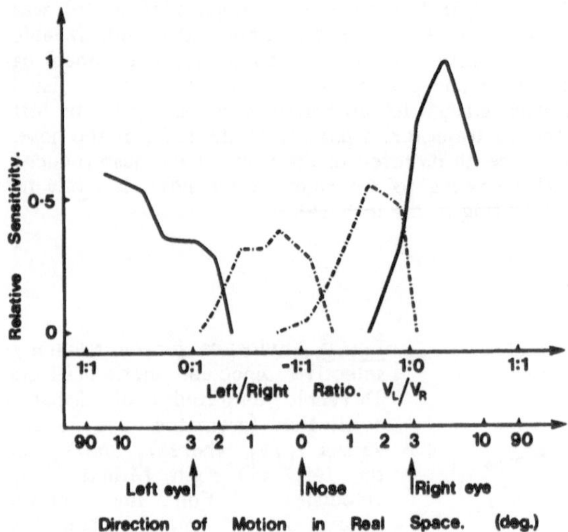

Fig. 6 Sensitivity curves of four hypothetical directionally-tuned binocular motion filters. From (4).

We proposed that the curves of Fig. 6 describe the directional selectivities of 4 classes of binocular motion filters. These filters can be modelled as shown in Fig. 7 (14, 15). The left monocular input separately signals left-to-right retinal image motion and right-to-left retinal image motion. Similarly for the right monocular input. These four signals are segregated and form the basis for a broad directional tuning to four directions in depth (Fig. 7A). Narrower directional selectivity is achieved on the basis of velocity tuning. Each of the four filters of Fig. 7A is divided into two parts, one more sensitive when the left eye's image moves faster than the right eye's and the other more sensitive when the right eye's image moves faster than the left eye's. This gives a total of eight binocular filters, four sensitive to movement towards the head and four sensitive to movement away from the head (Fig. 7B). Of these eight filters four are sensitive to trajectories passing between the eyes (two towards, two away) and four to trajectories missing the head (two towards, two away). This model can explain our findings on direction-selective adaptation (14, 15).

Movement-in-depth filters of the type proposed in Fig. 7 might explain the characteristics of two anomalous subjects described by RICHARDS (16). These subjects were most sensitive to depth oscillations of frequencies near 1.0 Hz with little sensitivity at higher or at low frequencies.

Independent evidence for the existence of neural mechanisms sensitive to unidirectional motion in depth directed towards and away from the head was provided by evoked potential recording (17, 18).

3. The Relation Between Discrimination and Sensitivity in the Perception of

Direction of Motion in Depth.

Figure 6 shows the directional tuning curves (i. e. sensitivity curves) for our proposed binocular motion filters. These filters alone would be sufficient to provide a physiological basis for discriminating between different directions of motion. If we assume that discrimination is entirely mediated by the filters of Figs. 6 and 7, then we can use the following argument to predict the way in which discrimination will depend on the direction of motion in depth.

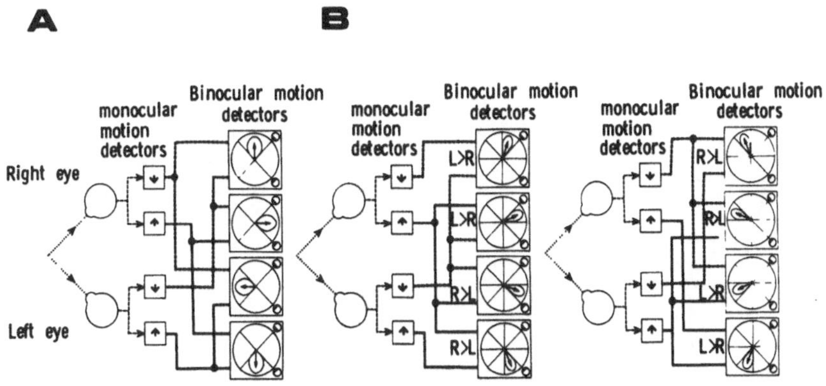

Fig. 7 Monocular responses to motion are supposed to be directionally-specific, either right-to-left or left-to-right. Segregation of these monocular signals creates coarse directional selectivity (Fig. 7A). This coarse selectivity is sharpened by velocity tuning (Fig. 7B). Each binocular motion detector in 7A is composed of two parts. One part is preferentially sensitive when the target's speed seen by the right eye exceeds the speed seen by the left eye (R L), while the other part is more sensitive when the right eye sees a slower speed than the left (R L). Theoretical directional tuning curves for each of the eight predicted binocular motion detectors are graphed in polar coordinates.

Clearly, if only one filter is stimulated, then the direction of motion must be somewhere within the range of directions to which that filter is sensitive. Similarly, if two filters are stimulated, then the direction of motion must lie somewhere within the range of directions to which one or the other filter is sensitive. Therefore, the just discriminable difference would be very large for trajectories wide of the head (even up to roughly 98°). Discrimination would fall to minima of very roughly 1°-2° for trajectories midway between the nose and the left and right eyes, and there would be a submaximum of very roughly 3° for a trajectory cutting near the nose.

In order to test this prediction we measured discrimination in the following way (experimental details and precautions to avoid ocular tracking are described in ref. (15). A stimulus bar was located 1° to the right of the fixation point. The bar appeared to execute one sinusoidal oscillation in depth along a predetermined

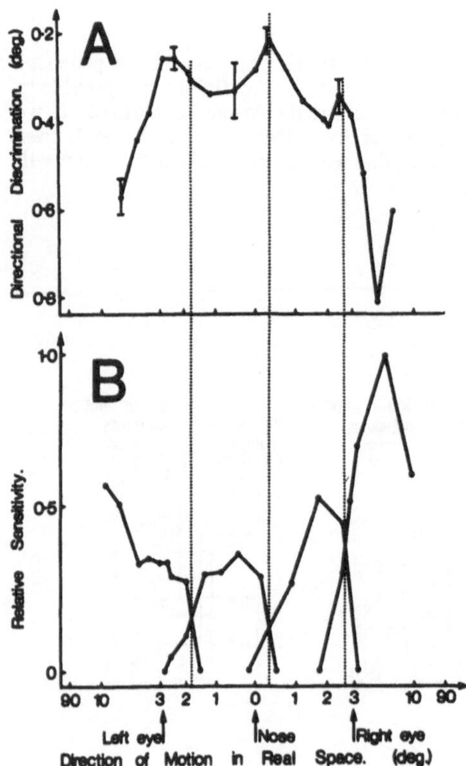

Fig. 8 Discrimination of the direction of movement in depth. Discrimination is plotted as ordinates, the directions of the target's trajectories as as abscissae. Standard deviations are shown for one measurement of 200 judgements; each point is based on between two and six such measurements. B. Sensitivity curves of four hypothetical groups of binocular mechanisms tuned to different directions of movement in depth. From (15).

direction (the comparison direction) and then a second sinusoidal oscillation along one of eight possible test directions. The subject was asked to judge whether the test direction was to the left or to the right of the comparison direction. Data obtained in this way by the method of constant stimuli was subjected to Probit analysis so as to obtain the just discriminable angle between test and comparison directions. The experiment was repeated for fifteen comparison directions.

Figure 8A shows a plot of just discriminable angle versus comparison direction. The shape of this plot is quite different from the prediction above. Therefore the experiment did not support the notion that discrimination is entirely mediated by the directionally-tuned binocular motion filters. Best discrimination did not coincide with the peak sensitivities of the filters (i. e. midway between the nose and the left or right eye). Also, the best discrimination was unexpectedly acute (about 0.2°).

Our proposed explanation for the discrimination data links the shape of the discrimination curve (Fig. 8) with the shape of the filter sensitivity curves (Fig. 6)

(15). Our proposal is as follows. The outputs of adjacent filters are compared (e. g. subtracted) at a more central stage. This would have the effect of enhancing discrimination between different directions of motion in depth so as to explain our finding of acute (0.2^{o}) discrimination. However this comparison process would have, as its "signature", characteristic maxima and minima in the discrimination plot. Suppose that discrimination is determined by the difference between the slopes of adjacent sensitivity curves. Then each time this difference passes through a maximum, discrimination will pass through a maximum. The sensitivity curves of Figs. 6 and 8B pass through three such maxima close to the crossover points. These are marked by dotted vertical lines in Fig. 8. It can be seen that the discrimination curve of Fig. 8A has three maxima. Furthermore, these fall close to the dotted lines.[2]

Thus, the shape of the discrimination curve supports the notion that a comparison stage exists, central to the filter stage.

4. Visual Field Defects for Motion in Depth and for Positional (Static) Depth.

Stereofields were measured in a way analogous to conventional perimetry according to RICHARD'S method. The left eye saw a bar and the right eye saw a similar bar. The two bars oscillated from side to side with opposed motion, so that in binocular fusion a single bar was seen to oscillate in depth. If this stimulus was too far into the periphery of the visual field no depth movement could be seen, but as it was slowly moved towards the center of the field a sensation of depth movement was suddenly elicited and the fused bar appeared to oscillate in depth. RICHARDS and REGAN plotted the region of the visual field over which depth oscillations could be seen (19).

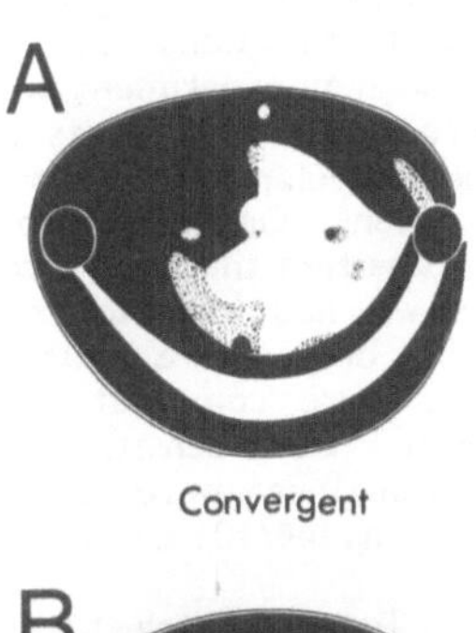

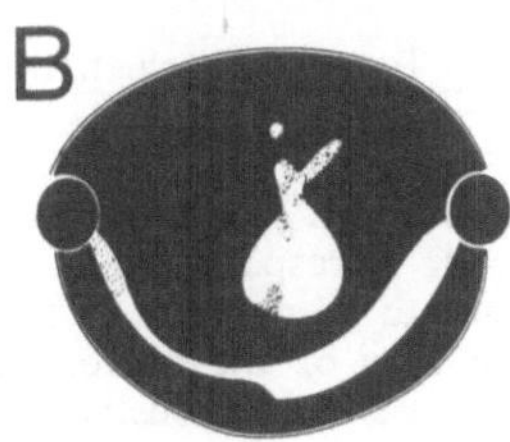

Fig. 9 A. Stereo field plot for a convergent stimulus disparity, obtained using a 1° bar oscillating back and forth from 0 to 0.4^{o} disparity twice per second. The two ovals on the horizontal axis are the blind spots. The dark regions inside the double-line perimeter indicate regions of no depth. (The field was not explored beyond the perimeter shown). The dotted areas are regions of amblyopia, where unstable depth sensations could occasionally be elicited. The cross corresponds to the central fovea. Subject DR. B. Stereo field plot for a divergent stimulus disparity for subject DR. Except for reversing the bars presented to each eye (a sign reversal), the stimulus is identical to that used to obtain the field shown in Fig. 1. From (19).

2. This argument is analogous to the well-known argument that relates the shapes of cone action spectra and the wavelength discrimination curve in colour vision theory.

Figure 9 shows such stereofield plots. White areas indicate where oscillatory motion in depth could be seen, black areas indicate where depth motion could not be seen and dotted areas mark regions of unstable perception. The stimulus was always in front of the frontoparallel plane in the upper plot, and behind the plane in the lower plot. Visual field defects for depth motion are evident in both plots. Since these defects differed for crossed and uncrossed disparities they were due, not to monocular defects, but to impaired processing at cyclopean level 3 (i. e. to defects after the signals from the two eyes had converged). Conventional perimetry showed that there was no loss of visual acuity in the defective regions. Interestingly, these stereo defects were not evident when positional depth perception was tested by flashing the bars.

Thus, there exist regions of the visual field blind to motion in depth but where positional (static) depth can be seen normally. In retrospect this finding can be taken as further support for the notion that information as to motion in depth and as to position in depth are processed by different neural mechanisms.

5. Single-Neuron Evidence that Information as to Motion in Depth and Positional (Static) Depth are Processed by Different Neural Mechanisms.

5.1 Introduction

It has been firmly established that in cat and monkey visual cortex there are single units whose binocular responses depend on the positional (static) disparity between stimuli viewed by the left and right eyes. These are the so-called "binocular depth units" (20, 21, 22). However, until recently there has been little known about binocularly-driven neurons sensitive to motion in depth as distinct from position in depth. In order to reveal binocular interactions that depend on the direction of motion in depth it is necessary to search for them by simultaneously stimulating both eyes with targets that can move in either the same or in opposite directions and can move at different relative speeds (c.f. Fig. 3). The results of such an experiment are summarized in Figs. 10 and 11. CYNADER and REGAN recorded from single units in area 18 of cat visual cortex (23). By comparing monocular and binocular responses we were able to measure the strength of purely binocular interactions. On the basis of the strength and nature of these binocular interactions we classified the units into four types. A substantial proportion were of the type we had expected on psychophysical grounds, being sensitive to motion in depth but only over a limited range of directions in depth (29/101 cells). These cells were comparatively insensitive to position in depth, or indeed to position in space. Cells sensitive to position in depth were fairly common (17/101 cells). The remainder were either monocular (11/101 cells) or showed only weak binocular interactions (44/101 cells).

5.2 Cells that Preferred Opposed Motion (i.e. Trajectories Passing Between the Eyes).

The responses of a representative unit of this type are plotted in Fig. 10. When only the left eye was stimulated this unit behaved like a simple cell. It responded best to a bar oriented near vertical and moving rightward. Responses to stimulation of the right eye were absent or extremely weak. However, when bars of the preferred orientation, direction and velocity were presented to both eyes simultaneously, the unit's responses were markedly depressed (up to tenfold) compared with monocular responses to left-eye stimulation.This depression is represented by the marked trough in Fig. 10A illustrating that responses to binocular stimulation (open circles, dashed line) were much less than the sum of responses to monocular stimulation of the left and right eyes (open squares, fine continuous line). Fig. 10A shows how this binocular depression held over a wide range of stimulus disparities, in other words over a large

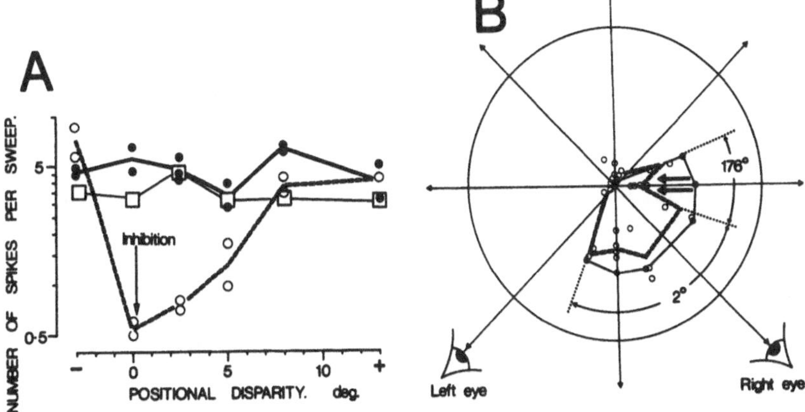

Fig. 10 A. A unit that was preferentially sensitive to motion directed between the eyes. The number of spikes per sweep in 20 stimulus cycles at 37 deg/sec is plotted logarithmically as ordinates versus the positional disparity between the left and right bars (plotted linearly). In this, as in all subsequent Figs., the zero of disparity is arbitrary. The fine continuous line plots the linear sum of response to separate stimulations of the left and the right eyes. The heavy dashed line shows that responses were inhibited tenfold below this linear prediction when the left and right eyes were simultaneously stimulated with identical movements (i.e. binocularly-viewed sideways motion, c.f. Fig. 1). The arrow indicates the depth of this inhibition which clearly extended over a broad range of disparities. The full line shows that no such inhibition occurred when left and right eyes were stimulated with identical speeds in opposite directions (i.e. binocularly-viewed motion directed along a line midway between the eyes, c.f. Fig. 3). Thus, over a large volume of space, an inhibitory mechanism caused this unit to favour motion direction between the eyes over sideways motion. The preferred orientation was 15° clockwise from vertical. B. Binocular selectivity for the direction of motion plotted in polar coordinates looking down onto the left and right eyes. The number of spikes per sweep for 20 stimulus cycles at 37 deg/sec is plotted radially on a linear scale (dashed line, open circles). The fine continuous line (filled circles) shows the linear sum of responses to separate stimulations of the left and right eyes: each point is based on separate empirical measurements. This figure shows how the inhibition for binocularly-viewed sideways motion (arrowed) had the effect that appreciable responses were generated by only a narrow (2°) range of directions of motion for which the target would either hit or narrowly miss the animal's head. The number of spikes per sweep for 20 stimulus cycles is plotted radially on a linear scale. The ratio between the speeds of the left and right retinal image is plotted linearly round the circumference of the circle. For explanatory purposes, the corresponding angular directions in real space are also shown, in this case for a viewing distance of 145 cm and interpupillary separation of 3.0 cm. Note that this angular scale is very nonlinear, and emphasizes the 2.4° range of directions that pass between the eyes at the expense of the remaining 357.6° range of directions that miss the head. From (23).

volume of space. It will be recalled from Fig. 3 that this stimulus condition corresponded to binocularly-viewed sideways motion. Quite different responses were obtained when the bar traversed a path that led it directly towards and away from the nose (filled circles and heavy continuous line in Fig. 10A). Note that the only difference between the binocular stimulus conditions was the phase relation between the left and right eye's stimuli: monocular stimulation was the same in both conditions.

Figure 10A shows that over a large range of disparities (i. e. a large volume of space) motion along a trajectory passing through the nose gave much greater responses than sideways motion. This difference was achieved entirely by binocular inhibition.

Figure 10B plots in more detail this unit's preferential sensitivity to the direction of motion in depth. We obtained these data by setting the binocularly-viewed stimuli at the positional disparity marked as 0° in the abscissa of Fig. 10A where inhibition to sideways movement was maximal. We then systematically varied the relative velocities and directions of stimuli in the two eyes so as to produce the directional tuning curve of Fig. 10B. This figure shows that the unit was sharply tuned for the direction in which the stimulus moved in depth. Responses to all directions of movement from sideways to 88° on either side of it were markedly inhibited. Only over a narrow range of directions in depth (about 2° for the viewing distance used in these experiments) could a strong response be elicited. As can be seen, the strongest responses were obtained for stimuli moving toward the animal along trajectories that would result in a collision with his head.

The unit described above was an example of a cell type that was not uncommon in our sample from the cat parastriate cortex. These units were distinguished by appearing to be solely monocularly driven, or nearly so, when tested with conventional visual stimuli. When binocular stimulus conditions were employed, an inhibition for sideways motion was observed. This binocular inhibition was relaxed for only certain left-right velocity ratios resulting in the tuning for direction-in-depth seen in Fig. 10B.

A striking feature of the opposed-motion-selective units was a clear tendency to appear in aggregates during our perpendicular penetrations. While our findings do not constitute proof of a columnar organization for these units, they are certainly consistent with this notion. It is worth noting that columnar organization for stereoscopic responses is a feature of monkey area 18 (21) and that columns of cells receiving excitation from one ear and inhibition from the other can be found in cat auditory cortex (24). Our findings add to those of earlier workers (25, 26) in providing evidence for a cortical mosaic with groupings of cells according to highly-specific functions.

5.3 Cells that Preferred Trajectories that Missed the Head.

Units which responded optimally to stimuli moving in depth but along a line that missed the head were enountered less frequently than those responding to binocularly-opposed movement (note that our classification required that these cells responded best to motion at an angle to the frontoparallel plane). We found that all such units were characterized by facilitatory interactions in which responses to stimulation of the two eyes together were much greater than the linear sum of responses of the two eyes stimulated alone. Data from a representative cell responding to depth movement missing the head is illustrated in Fig. 11. This unit was binocularly activated (ocular dominance group 3) and responded best to stimuli oriented near 11:00 o'clock moving right and up. Responses from each eye independently were

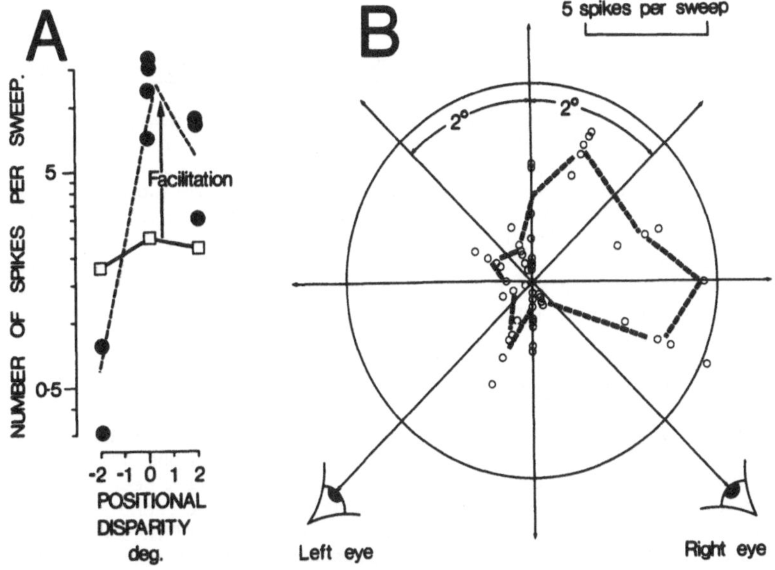

Fig. 11 This unit illustrates the class of neuron that responds selectively to motion along trajectories that pass wide of the head. A. The number of spikes in 20 stimulus cycles plotted logarithmically as ordinates (dashed line, filled circles) versus mean disparity plotted linearly. The continuous line plots the linear sum of responses to separate stimulations of the left and right eyes. Facilitation for binocularly-viewed sideways motion is indicated by the arrow. Unit orientation: 30° anticlockwise from vertical. B. Binocular directional selectivity. Best responses are to directions away from the head along a line wide of the head. From (23).

rather weak, but simultaneous presentation of stimuli having the same orientation and direction to the two eyes resulted in strong facilitation. The firing rate for the binocular response in Fig. 11A was approximately five times that which would be predicted from linear summation between the two eyes. This facilitation was present in this case over a rather restricted range of stimulus disparities (2°-3°). In most cases, units with strong facilitatory interactions showed a more restricted range and sharper tuning for retinal disparity than did units characterized by strong inhibitory interactions.

In order to obtain the polar plot of direction selectivity shown in Fig. 11B, we aligned the binocular stimuli at a position corresponding to 0° on the left-hand side of the figure (the location in depth where facilitation was at a maximum) and then varied the relative speeds and direction of stimuli presented to the two eyes. The results for this unit revealed a broad directional selectivity for movement in depth with responses over a range of more than 90°. Such broad tuning for direction of movement in depth seems characteristic of this type of unit. It is clear that, despite the broad tuning, these units responded differently not only to targets moving toward or away from the animal but also to targets that would have missed or would have hit the head.

5.4 Cells that Preferred Sideways Movement (Parallel to the Frontoparallel Plane)

In most cases, units with strong facilitatory interactions responded best for binocular sideways movement parallel to the frontoparallel plane (i. e. when the two stimulus bars were moving at the same speeds and directions in the two eyes, Fig. 3). When the retinal speeds differed (i. e. motion was at an angle to the frontoparallel plane) responses became weaker and this distinguishes these units from the "missing the head" units described above. The binocular responses could be more than 100 times greater than the sum of monocular responses. In general, these units were more tightly tuned for positional disparity than were the opposed-motion units described earlier and therefore they may emphasize positional disparity more than movement information. We called these "positional disparity" units.

5.5 Comparison with Other Studies

Only when both eyes were stimulated simultaneously with different relative speeds and directions was it possible to observe the binocular interactions we describe. Neurons sensitive to oppositely-directed monocular stimulation have been described in monkey cortex (27). The binocular interactions that we describe cannot, however, be observed with monocular stimulation as used by ZEKI. PETTIGREW (28) has described neurons in cat area 18 which bear some resemblance to our opposed-motion selective units. His units were, however, exceedingly rare (4 out of 200 units studied) and appear to be a subclass of our opposed-motion units which work by facilitation rather than inhibition. We have observed only one unit like those he described during this investigation.

The two classes of neuron tuned to the direction of motion-in-depth that we report above cut across the established categories of cortical neurons (29, 30). To conventional testing many cells seemed unexceptional and of familiar types (simple, complex, etc.).

5.6 Possible Functions of Neurons Tuned to the Direction of Motion in Depth and Neurons Tuned to Static Disparity

The volume of three-dimensional space over which these motion-in-depth neurons retained their directional selectivities was rather large. Thus, they emphasized directional motion selectivity at the expense of signalling information as to position. Now static depth perception is very dependent on ocular convergence. Therefore, rather than the neurons whose activities underlie static depth perception, an answer to the question "will that moving object hit me?" might better be signalled by a neuron sensitive to the direction of motion in depth whose performance was comparatively little affected by the object's position and disparity. These are exactly the characterizing properties of our two classes of neurons tuned to the direction of motion in depth.

If these neurons are, in reality, responding to motion relative to the head, then they might also be involved in visually-guided locomotion. For example, strong activation of opposed-motion neurons would signal that one is moving directly towards a particular object in one's field of view.

In this sense our third class of neuron had complementary properties. These neurons showed strong binocular facilitation for trajectories parallel to the frontoparallel plane. However, facilitation was appreciable over a somewhat narrower range of static disparities than for the other two types of unit. Thus, these neurons fired strongly only when the binocularly-viewed stimulus was located in a rather smaller region of three-dimensional space, and so might signal position in depth. Of course, units with much sharper tuning for positional disparity have previously been described in cortical area 17 (20, 22). One function of such units may be to signal a particular value of depth and their activity is widely supposed to underlie stereoscopic depth perception.

6. Changing Size as a Cue to Motion in Depth

6.1 Neural Filters Sensitive to Changing Size

So far we have discussed only one cue to movement in depth, namely the relative velocities of the left and right retinal images. Of course there are a number of other cues. Among them is changing size, and in this section we describe experiments in which changing size was the only cue to motion in depth.

In these experiments we distinguish between the static size of a known object (e. g. a human figure), a well-recognized cue to position in depth (31) and the dynamically changing size of an object that signals motion in depth, just as in previous experiments we distinguished between static disparity (a cue to position in depth) and the relative velocity of the retinal images (a cue to motion in depth).

Our chief aim was to find whether information as to changing size is processed in a separate psychophysical channel. Our main experiment was to stimulate the eye with a bright square whose opposite edges oscillated to and fro with a sinewave motion (32). In one condition opposite edges moved in the same direction at the same time, so that the square appeared to oscillate as-a-whole along a diagonal without changing size. (Fig. 12A, movement condition). In the other stimulus condition, opposite edges moved in opposite directions at the same time so that the square oscillated in size without changing position (Fig. 12A, oscillatory size condition). In order to avoid any effect due to the eye's tracking the moving square, two squares were used that oscillated towards and away from the fixation spot (Fig. 12A).

The central point of this experiment is that movements of the edges were identical in both oscillatory motion and oscillatory size conditions. Only the relationship between movements of the opposite edges differed in the two stimulus conditions.

Figure 12B shows the results of adapting to oscillating size and Fig. 12C the results of adapting to oscillatory motion.

The continuous line in Fig. 12B shows the percentage elevation of visual threshold to size oscillations caused by adapting to sinusoidally oscillating size. The sensitivity loss was clearly very marked, reaching a threshold elevation of 520% near the adapting frequency of 2 Hz. This was significant at the 0.001 level.

Of course, this finding is not sufficient to support a proposal that there is, in the human visual pathway, an information-processing channel for changing size. Three plausible alternative hypotheses spring to mind. One alternative explanation for our finding that does not invoke the notion of a changing-size channel is that the perception of threshold oscillations of size is mediated entirely by motion detectors. This explanation would have the merit of parsimony, since there is already strong psychophysical evidence for the existence of motion channels in the human visual pathway (33, 34, 35). We checked this explanation by comparing visual sensitivities to oscillatory motion and oscillating size before and after adapting to the oscillating size stimulus. (In fact, the two test stimuli were alternated one after the other

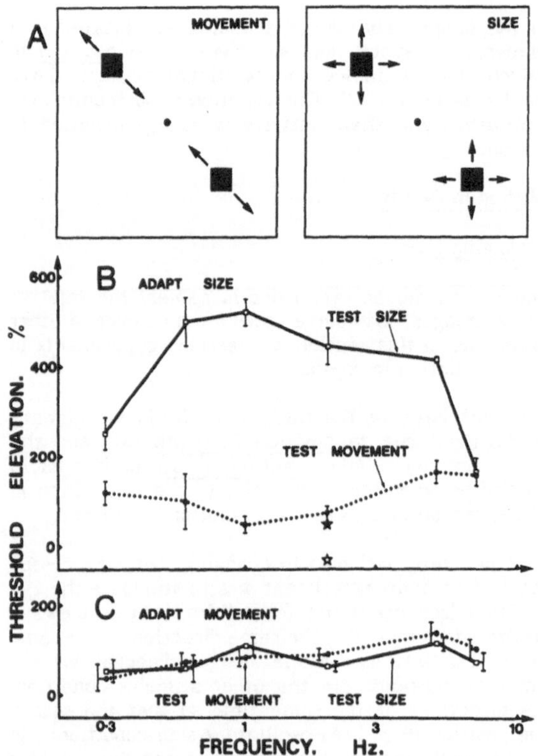

Fig. 12 Adapting to oscillating size preferentially depressed visual sensitivity to oscillating size. The stimulus was two identical squares on either side of a fixation spot (A). Ordinates plot percentage elevations of visual threshold produced by 25 mins adaptation to 2 Hz oscillating size (B) and 2 Hz oscillatory motion (C). Test frequencies are plotted as abscissae. Vertical lines show $\pm$ 1 SE. The large filled star and large open star show threshold elevations produced by adapting to 2 Hz flicker upon sensitivity to oscillating size and oscillatory motion, respectively. From (32).

during every experiment.) The dotted line in Fig. 12B clearly shows that adaptation to oscillating size depressed visual sensitivity to oscillatory motion much less than it depressed visual sensitivity to size oscillations up to 7 Hz. This experiment shows that the depression of visual sensitivity to size oscillations cannot be explained in terms of motion channels.

There is a second and independent argument that adaptation of motion channels cannot explain the depression of visual sensitivity to size oscillations shown in Fig. 12B. This argument is based on the data of Fig. 12C. Here we measured visual sensitivity to size oscillations and to oscillatory motion both before and after 25 mins adaptation to oscillatory motion. Figure 12C shows that visual sensitivity was little depressed for both oscillatory motion and size oscillations, and that the percentage depressions were not greatly different.

A comparison of Figs. 12B and 12C demonstrates our main finding: the effects of adaptation to oscillating size were quite different from the effects of adapting to oscillatory motion.

A rather trivial, explanation of our finding might be couched in terms of the known effect of adaptation to flicker (36, 37). Size oscillations must necessarily be accompanied by luminance flicker (if total light flux is kept constant) or by flicker of the total light flux (if luminance is kept constant) or by some mixture of the two. As a control we carried out an adaptation experiment similar to that of Fig. 12 excepting that we adapted for 25 mins to stationary squares flickering at 2 Hz. Visual sensitivities to size and motion test stimuli were little affected, confirming that the selective depression of sensitivity to oscillatory size shown in Fig. 12 could not be explained in terms of adaptation to flicker.

In separate experiments we asked whether the threshold elevation for changing size (Fig. 12B) was specific to the sign of spatial contrast. In brief, monocular adaptation to a bright square whose size oscillated sinusoidally produced approximately the same threshold elevation in an oscillating-size test square, independently of whether the square was bright or dark (and similarly for a dark adapting square).

Secondly, adapting to a bright square whose size increased with a ramp waveform elevated threshold for both bright and dark test squares whose sizes <u>increased</u> with a ramp waveform: thresholds were not elevated for <u>decreasing</u> size. Similarly, adapting to a dark square whose size increased with a ramping waveform elevated thresholds for both dark and bright test squares whose size increased with a ramping waveform whereas threshold were not elevated for decreasing size. Again, threshold elevations caused by adapting to decreasing size were qualitatively independent of the sign of test or adapting contrast.

On the basis of these experiments we proposed that there are channels within which psychophysical information as to dynamically changing size is preferentially processed, and that these channels are distinct from the known movement channels. Our ramp data support the notion that increasing and decreasing size are processed in different channels.

More tentatively, our evidence is consistent with the existence, within the human visual pathway, of neural mechanisms (for example, classes of single neurons) preferentially sensitive to increasing or to decreasing object size, respectively.[3] That a substantial proportion of these neurons would be binocularly-driven is indicated by our finding that adaptation to changing size shows interocular transfer. This would locate the neurons either in colliculus or at, or more central to, primary visual cortex. On this point, it is interesting to note that ZEKI (27) has reported single neurons sensitive to changing size in monkey visual cortex, although it is not clear how unequivocally his neurons would signal size change.

Several explanations at the single-neuron level are ruled out by taking together our findings that (a) <u>independently</u> of whether contrast is positive or negative, thresholds for ramping size changes are elevated when adapting and test ramps have the same direction while thresholds are depressed when adapting and test ramps have opposite directions, and (b) adaptation to changing size is quite distinct from adaptation to flicker.

---

3. The results of subsequent microelectrode experiments support this suggestion.

The first class of explanation that can be discounted is cast at the level of the lateral geniculate body in terms of concentrically-organized receptive fields. For example, a centre-on neuron would indeed be excited by increasing the size of a bright square falling within its field centre. On the other hand, increasing the size of a dark square would have the opposite effect. But we find that the threshold elevation effect transfers from a ramping increase in the size of a bright square to a ramping increase in the size of a dark square rather than to a ramping decrease in the size of a dark square. Similarly, for a ramping decrease in size, the threshold elevation transfers independently of the sign of contrast. A second argument against this geniculate-level explanation is that adaptation to flicker does not transfer to an oscillating-size test square, and vice versa.

Turning to cortical level, many cells (simple cells) respond better to movement than to flicker (26). Explanations in terms of their centre-surround organization are again ruled out by our findings for positive and negative contrasts.

One mechanism that might correspond to our changing-size filters would be a type of neuron or neural orgnaization that received inputs from two different regions of the visual field. When these two regions were simultaneously stimulated by contrast borders (either light-dark or dark-light) that moved in opposite directions the neuron would respond, either by excitation or inhibition. According to BISHOP, COOMBS and HENRY (38) some 50% of simple cells in cats respond to one direction of motion independently of contrast, and complex cells with this property are also well known. Perhaps such cells might converge onto the changing-size neural mechanism. Stimulation of either region alone or simultaneous stimulation by similarly-directed movements would have comparatively little effect. These two regions that together constituted a hypothetical "receptive field for changing size" might be separated by a distance of up to several degrees, and the neuron's special sensitivity to changing size would be revealed only by simultaneously stimulating the two regions with oppositely-directed motion.

6.2 Motion-In-Depth Aftereffect and Nulling of Aftereffect

Using his right eye a subject gazed steadily for 20 minutes at a dark fixation point in the center of a 1° bright square while the square's side length increased at a fixed rate (e. g. 24 min arc/sec) with a ramp time of 1.0 sec. During the ramp changes of square size, the square appeared to be moving in depth. Subjects noted a compelling negative movement aftereffect: the subsequently-viewed static test square appeared to be moving continuously away from the head along the line of sight. Conversely, after adapting to decreasing size, a static test square appeared to move continuously towards the head. (As with the classical movement aftereffect, the illusory motion was not accompanied by changing location, the necessary accompaniment of real motion).

We attempted to null this illusory motion in depth by causing the test square to continuously change size in the same sense as the adapting stimulus. It did indeed prove possible to null the aftereffect in this way. The rate of change of size required to null the illusory depth motion provided a quantitative measure of the aftereffect (39).

Figure 13 shows the relation between the strength of the adapting stimulus and the size of the aftereffect as measured by nulling. Over a considerable range the relation was approximately linear on log-log axes. Thus, for all but the weakest adapting stimulus the aftereffect obeyed the equation $N_V = kV_A{}^n$ where V_N was the rate of size change to null the aftereffect, V_A was the adapting rate of size change, and k and n were constants (rates of change of side length are in min arc/sec). For subject K.B., constant n was equal to 0.92 and 0.94 for adaptation to increasing and decreasing size respectively, so that approximately $V_N = kV_A$. Constant k was equal to 0.77 for increasing size and 0.56 for decreasing size. For subject D.R. the corresponding values for n were 0.86 and 0.70, and for k were 0.52 and 0.84.

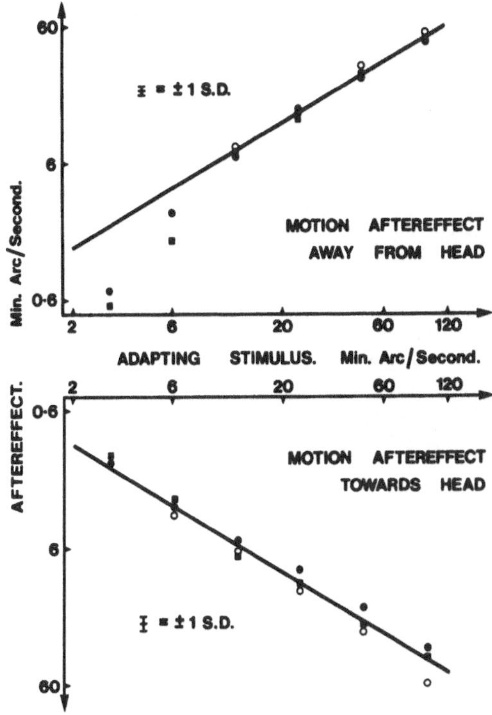

Fig. 13 Rates of size change required to null motion in depth after-effect versus the adapting rates of size change (both axes log min arc sec^{-1}). Upper adaptation to increasing size. Motion aftereffect away from head: Lower adaptation to decreasing size. Motion aftereff 0.25 sec blanking interval between successive ramps. Three test ramp times of 0.25 (open circles), 1.0 (filled squares) and 3.3 sec (filled circles) all with 0.25 sec blanking intervals were used. The mean $\pm$ 1 standard deviation for all points is shown: all SDs were similar on log axes. From (32).

There is a qualitative difference between our aftereffect and the adapting stimulus that generates it. This is an important distinction from previously-known aftereffects such as, for example, the Waterfall effect and the tilt aftereffect since it implies that the motion-in-depth aftereffect cannot be explained along the lines proposed by SUTHERLAND (40) and BARLOW and HILL (41). In order to account for the observation that our aftereffect was of movement in depth rather than changing size we propose that the outputs of the changing-size filters can drive the neural organization that underlies the perception of motion in depth. Evidence against more parsimonious explanations based on the classical motion aftereffect is that our aftereffect builds up and decays over durations completely different from the buildup and decay of the classical aftereffect (39).

Acknowledgements

K.I.B. was supported by a Killam Postdoctoral Fellowship and by the N.R.C. this work was supported by the Canadian N.R.C. (to D. R., A-0323; to M. C., A-9939) and Canadian M. R. C. (to M. C., MA-5201).

References

1. C. Wheatstone, Phil. Trans. R. Soc. 13, 371 (1838).
2. C. Wheatstone, Phil. Trans. R. Soc. 142, 1 (1852).
3. W. J. Richards, J. Exp. Psychol. 42, 376 (1951).
4. W. Richards, J. Opt. Soc. Amer. 62, 907 (1972).
5. D. Regan & K. I. Beverley, Vision Res. 13, 2369 (1973).
6. D. Regan & K. I. Beverley, Vision Res. 13, 2403 (1973).
7. C. W. Tyler, Science 174, 958 (1971).
8. R. Efron, Brit. J. Ophthal. 41, 709 (1957).
9. K. N. Ogle, J. Opt. Soc. Amer. 53, 1296 (1963).
10. C. L. Godek & R. B. Lawson, Psychol. Rev. 23, 243 (1973).
11. K. N. Ogle, Researches in Binocular Vision (Saunders, Philadelphia, 1950).
12. D. Regan & K. I. Beverley, Vision Res. 14, 175 (1974).
13. K. I. Beverley & D. Regan, Exp. Brain Res. 14, 175 (1974).
14. K. I. Beverley & D. Regan, J. Physiol., 235, 17 (1973).
15. K. I. Beverley & D. Regan, J. Physiol. 249, 387 (1975).
16. W. J. Richards, Air Force Report AFDSR-TR-73-0439 (1973).
17. D. Regan & H. Spekreijse, Nature 225, 92 (1970).
18. D. Regan & K. I. Beverley, Nature 246, 504 (1973).
19. W. Richards & D. Regan, Invest. Ophthal. 12, 904 (1973).
20. H. Barlow, C. B. Blakemore & J. D. Pettigrew, J. Physiol. 193, 327 (1967).
21. D. H. Hubel & T. N. Wiesel, Nature 225, 41 (1970).
22. J. D. Pettigrew, T. Nikara & P. O. Bishop, Exp. Brain. Res. 6, 391 (1968).
23. M. Cynader & D. Regan, J. Physiol. 274, 549 (1977).
24. T. S. Imig & J. F. Brugge, Neurosci. Abstr. 2, 19 (1976).
25. J. I. Mountcastle, J. Neurophysiol. 20, 408 (1959).
26. D. H. Hubel & T. N. Wiesel, J. Physiol. 160, 106 (1962).
27. S. Zeki, J. Physiol. 242, 827 (1974).
28. J. D. Pettigrew, Nature 241, 123 (1973).
29. D. H. Hubel & T. N. Wiesel, J. Neurophysiol. 28, 229 (1965).
30. F. Tretter, M. Cynader & W. Singer, J. Neurophysiol. 38, 1099 (1975).
31. C. Graham, Vision and Visual Perception (Wiley, New York, 1965).
32. D. Regan & K. I. Beverley, Vision Res. 18, 415 (1978).
33. A. Wohlgemuth, Brit. J. Psychol., Monogr. Suppl., 1 (1911).
34. A. Pantle & R. Sekuler, Vision Res. 8, 445 (1968).
35. R. Sekuler, A. Pantle & E. Levinson, Handbook of Sensory Physiology, ed. by R. Held, H. Leibowitz and H. I. Teuber, Vol. 8 (Springer Verlag, Berlin,1976).
36. R. A. Smith, Vision Res. 10, 275 (1970).
37. A. Pantle, Vision Res. 13, 943 (1973).
38. P. O. Bishop, J. S. Coombs & G. H. Henry, J. Physiol. 219, 625 (1971).
39. D. Regan & K. I. Beverley, Vision Res. 18, 209 (1978).
40. N. S. Sutherland, Quart. J. Exp. Psycho. 13, 222 (1961).
41. H. Barlow & R. M. Hill, Nature 200, 1345 (1963).

Neural Mechanisms Underlying Stereoscopic Depth Perception in Cat Visual Cortex

Max Cynader, Jill Gardner and Robert Douglas
Department of Psychology
Dalhousie University
Halifax NS
B3H 4J1

Since the basic discoveries of WHEATSTONE (1) in the last century, it has been known that the neural integration of the slightly different images seen by the two eyes is sufficient for the perception of depth. In the intervening period, an impressive body of research has emerged which has explored such topics as the relationship between retinal disparity and perceived depth, the relationship between monocular and binocular recognition of visual stimuli and the relationship between disjunctive eye movements and stereopsis. It is only in the last decade that neurophysiological methods have been applied to problems in this area. The pioneering research of HUBEL and WIESEL (2) showed that single cortical cells could be influenced by visual stimuli presented through either eye. It was soon established (3-6) that single cells in the cat and monkey visual system responded differentially to moving visual stimuli as a function of their retinal disparity.

All of these studies have dealt with the detection of spatial disparity by the visual system. Our purpose in this paper is to propose a different mechanism operating in the visual cortex which may also underlie stereoscopic vision. We will contrast our new formulation with existing models and explore some of its strengths and weaknesses.

There are two well-known aspects of binocular vision that are central to our reasoning. The first is the ocular dominance distribution of the monkey visual cortex shown in Fig. 1 (Redrawn from (7)). The abscissa of this figure is the ocular dominance scale devised by HUBEL and WIESEL (2), in which the numbers from one to seven represent the relative excitory contributions from the contralateral and ipsilateral eyes. Cells in ocular dominance group 1 receive exitatory input exclusively from the contralateral eye while cells in group 7 are driven exclusively

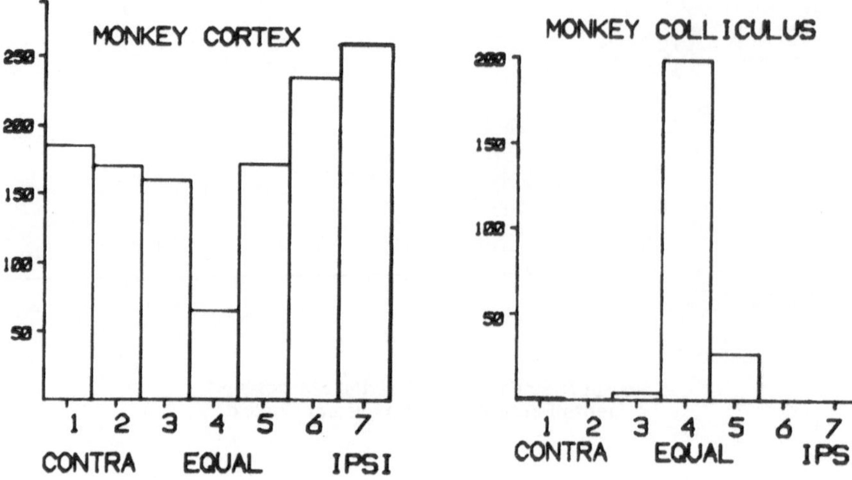

Fig. 1 The distribution of ocular dominance in the visual cortex (L.H.S.) and superior colliculus (R.H.S.) of the rhesus monkey. On the abscissa, the numbers from one to seven represent a trend from exitatory influence solely by the contralateral eye to influence solely by the ipsilateral eye. Units in group 4 receive equal strength inputs from the two eyes. Units in group 2, 3, 5, and 6 are binocularly excited, but one eye's input is clearly stronger. The ordinate represents the number of cells encountered in each ocular dominance category.

through the ipsilateral eye. Units in ocular dominance group 4 are driven equally well via either eye. Cells in ocular dominance categories 2, 3, 5 and 6 receive exitatory input from both eyes, but with unequal strength. One might imagine that units in groups 1 and 7 are important for monocular viewing and cells in group 4 for binocular vision but units in ocular dominance groups 2, 3, 5 and 6 appear to lack a functional role. It is not immediately apparent why these neurons should be encountered with any frequency. One possibility which should be considered is that these cells simply represent biological variance. It maỷ be very difficult to arrange the inputs from the two eyes so that they impinge equally on single cells. The right hand side of Fig. 1 shows, however, that single units in the Superior Colliculus, another visual structure with binocular input, are nearly all driven equally through the two eyes (8). This suggests that activation with unequal strength in the two eyes among cortical cells is not due to biologic variability, but rather that these cells in ocular dominance groups 2, 3, 5 and 6 may have a specific functional role.

A second consideration is the Pulfrich phenomenon (9), which is illustrated in Fig. 2 taken from R. L. GREGORY'S introductory textbook "Eye and Brain" (10). In this situation the subject wears a neutral density filter over the left eye. The neutral density filter causes the signals from that eye to be delayed relative to those of the other eye (11, 12). As the pendulum swings from left to right and back again, it appears to move in an elliptical path, first behind the fixation plane and then in front. It is generally thought, (and GREGORY'S last statement in the figure caption illustrates this point of view) that the brain converts the <u>time difference</u> between the two eyes into a <u>spatial difference</u> which then produces a sensation of depth.

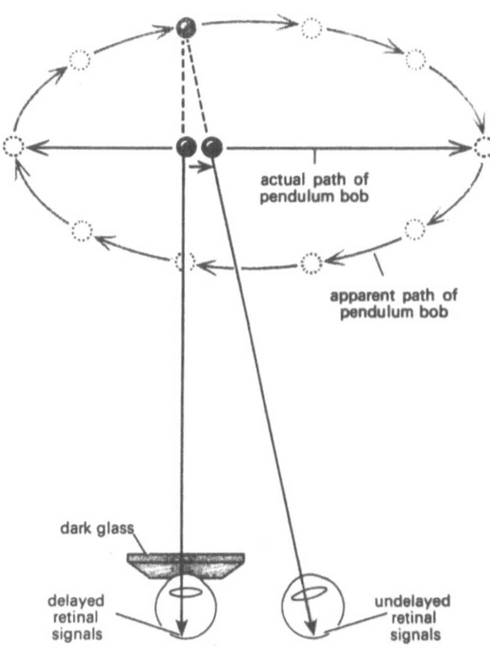

Fig. 2 The usual explanation of the Pulfrich phenomenon, taken from GREGORY (10). One eye views through dark glass and the other eye views normally. The darkened eye responds with longer latency. It is generally thought that the visual system converts the time difference into a space difference to give the sensation of depth. It is clear however that the stimulus conditions are such that information from corresponding points in the two eyes arrives at the visual cortex at different times.

This is, however, an assumption which cannot automatically be accepted. In fact, the Pulfrich phenomenon represents a condition in which information from corresponding points in the two eyes reaches the central nervous system at different times. This may be sufficient in itself for stereoscopic depth perception.

The results presented below attempt to connect the two topics illustrated in Figs. 1 and 2. We have found that cells in ocular dominance groups 2, 3, 5 and 6 which are activated binocularly but with unequal strength are also activated with unequal latency through the two eyes. Neuronal responses to stimuli presented through the stronger eye consistantly occur with shorter latencies. In a not-so-metaphorical sense, cells in ocular dominance groups 2, 3, 5 and 6 are seeing the Pulfrich phenomenon all the time. This observation leads directly to a time-difference-based theory of stereoscopic depth perception.

Our general procedures for preparing cats for neurophysiological experiments and for recording single cell responses were conventional and have been described elsewhere (8, 13). Most of these experiments were performed in cat area 18, but we have studied some units in cat area 17 as well with similar results.

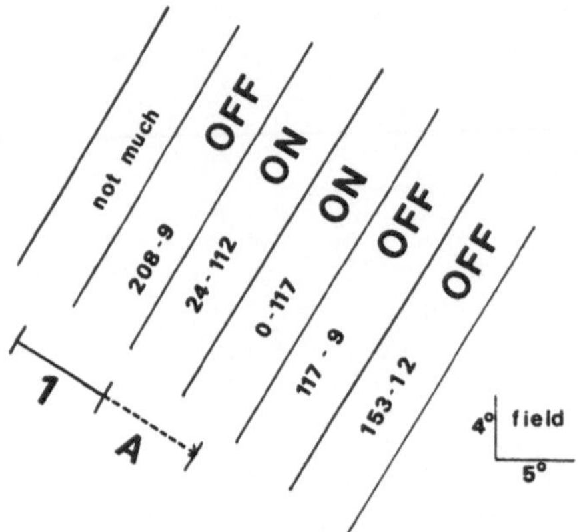

Fig. 3 Receptive field plot of an area 18 simple cell. Numbers to the left and right respectively indicate the number of action potentials elicited by the onset and offset of a light slit. Spikes were tabulated over a 10 second period. Response latency was determined for this unit by jumping the stimulus to position "A" from position "1" in order to summate the ON and OFF responses.

Stimulus delivery and data collection were handled by a PDP 11/10 computer. In characterizing the response of a cell to stimuli as a function of disparity, orientation, contrast, etc., the stimuli were always presented in an interleaved, randomized order and the neural responses collected, analysed, and graphed by the computer. Flashed stimuli were used throughout. Fig. 3 illustrates the number of spikes evoked in a cell by an optimally-oriented slit flashed "on" and "off" at several positions within the receptive field. Since the receptive field of this cell could be divided into separate "on" and "off" regions it was classified as a simple cell (2). To determine the latency for simple cells, our procedure was to first place the slit onto the "off" region of the receptive field and then "jump" from this region to the "on" region of the field. For complex cells, which lack separate "on" and "off" areas, latency was determined by jumping the stimulus between two positions near the center of the receptive field. Neuronal spike responses were collected while this procedure was repeated 64 to 256 times. Post-stimulus time histograms like that of Fig. 4 were compiled. Typically, there was an early large and sharp peak, with little or no indication of any preceding response. The time to this first peak was used as the measure of response latency.

The response latency of single units was determined in this fashion for stimuli presented to each eye separately. In some cases, we presented stimuli at different contrast levels above a constant background (0.86 cd/m^2). The data from one such cell is shown in Fig. 5. This unit was found in Area 17 and was classified as a simple cell, ocular dominance group 2. Tested through the stronger (contralateral) eye, the unit responded with a latency of 45 ms when the stimulus was 1.5 log units above background. As the stimulus was dimmed, response latency increased (as would be expected from earlier classic studies (14, 15)). Tested through the weaker eye, the units responded with a latency of 65 ms to a stimulus 1.5 log unit above background. This 20 ms difference in response latency through the two eyes was maintained over a rather broad range of stimulus contrast.

It can be seen that the latency of response through the weaker eye is the same as that evoked through the stronger eye when the stimulus is dimmed by about .7 log units. Thus it may be useful to imagine this cell as wearing a .7 log unit neutral density filter over the non-dominant eye.

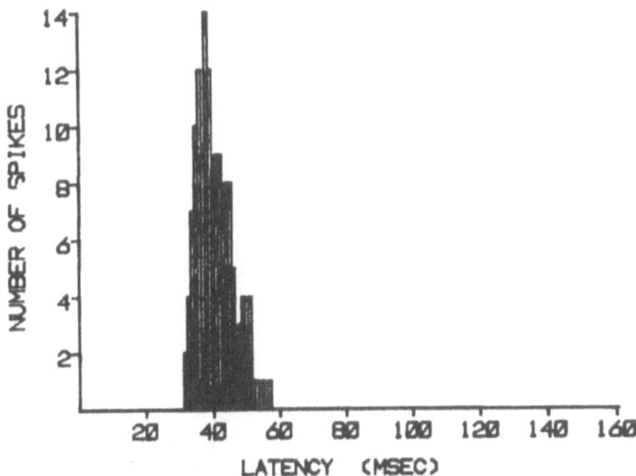

Fig. 4 A histogram illustrating the summed responses of a representative area 18 cell to 64 flashed presentations of an optimally oriented stimulus. Response amplitude is determined by counting the total number of action potentials while response latency is determined by noting the bin in which the largest number of spikes fall.

Figure 6 illustrates the distribution of interocular latency differences for 101 cells in the cat visual cortex as a function of ocular dominance. The hatched rectangles in this distribution represent cells in ocular dominance group 4. These units, driven with equal strength through the two eyes were activated with the same or similar latency through each eye. The mean absolute interocular latency difference for these cells was only 2.2 ms. The unfilled rectangles represent cells in ocular dominance groups 3 and 5. These units could can be driven well through either eye, but one eye was clearly stronger. For cells in ocular dominance group 3, the latency was consistantly shorter through the contralateral (stronger) eye. For cells in ocular dominance group 5, stimulation of the ipsilateral eye resulted in shorter latency activation. The mean interocular latency difference for units in ocular dominance groups 3 and 5 was 7.1 ms. The filled rectangles represent units in ocular dominance groups 2 and 6. In these units, stimulation of the dominant eye evoked much more vigorous responses than those evoked through the other eye. The mean interocular latency difference for these units was 13.1 ms although latency differences of over 20 ms between the two eyes were not uncommon.

The foregoing data demonstrate that single cells in the visual cortex can be activated with different latencies through the two eyes. Fig. 7 shows how these latency differences might serve as a basis for a stereoscopic depth mechanism. The illustration represents a binocularly-driven cell, preferring the left eye, with receptive fields located on corresponding points in the two eyes. We can assign the cell a response latency of 50 ms through the right eye and a 40 ms latency through the left eye. As a stimulus moves leftward along the frontoparallel plane it

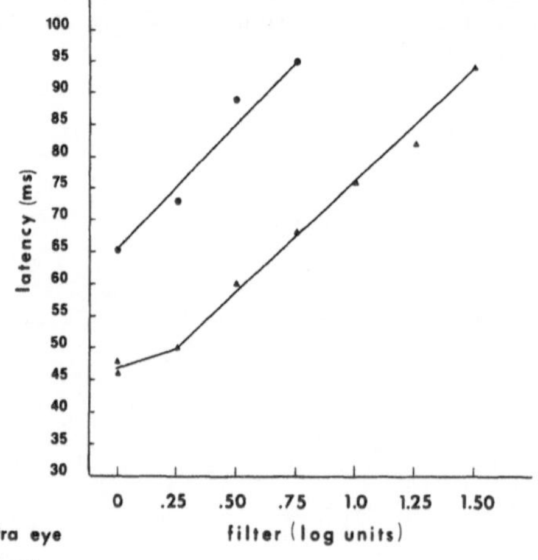

Fig. 5 Contrast response functions taken through both eyes of a simple cell, OD group 2. As indicated by the legend, the response latency of the contralateral eye is represented by the lower graph and the response of the ipsilateral eye is shown by the upper curve. The latency-contrast function is similar in both eyes, with the inter-ocular latency difference remaining relatively constant across the range of stimulus contrasts which were examined. The contralateral response was considerably more sensitive than the ipsilateral response yet the latency at threshold was approximately the same through both eyes. It is worth noting that this cell would appear to be monocularly-driven if tested near the threshold for the contralateral eye.

intersects the lines of sight of the two eyes at the same time. However, as seen from the cortical neuron, the events do not appear simultaneous, since the response from the left eye arrives 10 ms before that of the right eye. We assume that summation of the combined responses from the two eyes will not be optimal. Now, if one considers instead a leftward-moving stimulus with a crossed disparity (i.e., in front of the fixation plane) it can be seen that the stimulus will intersect the line of sight of the right eye first and then only later intersect the line of sight of the left eye. If the stimulus intersects the line of sight of the left eye 10 ms later, the built-in interocular latency difference will be neutralized. This will result in optimal summation of the inputs from the two eyes and hence maximum response from the cortical cell.

Similarly, it can be seen that a leftward-moving stimulus beyond the fixation plane (i.e., with an uncrossed disparity) would intersect the line of sight of the left eye first and only later the line of sight of the right eye. In this situation, responses evoked from the two eyes would be even more asynchronous than the case in which stimuli moved along the fixation plane.

The time-based stereopsis mechanism described above thus provides cells with receptive fields on corresponding points with sensitivity to crossed (or uncrossed) disparities, depending on which eye is dominant. It is evident that the value of the crossed disparity which will result in synchronous activation through the two eyes

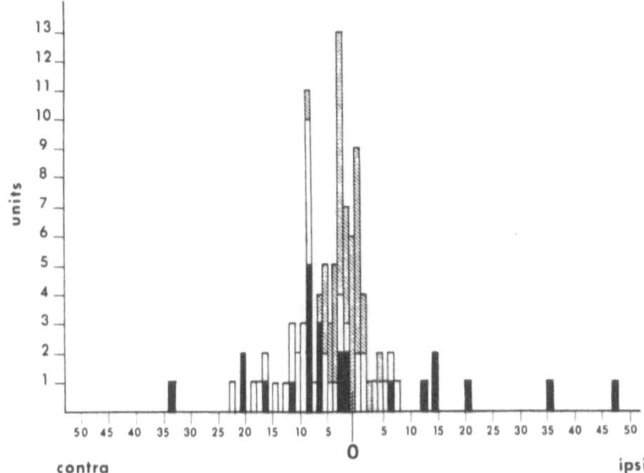

Fig. 6 The distribution of interocular latency differences for 101 units from areas 17, 18 and the 17/18 border. Most of these units were encountered in area 18. Numbers on the abscissa indicate the time difference between the two eyes (msec.) with units that were ipsilateral or contralateral dominant being shown to the right or left of zero respectively. Units classified as group 4 are represented by stippling, units in OD groups 2 and 6 are grey and units in OD groups 3 and 5 are white.

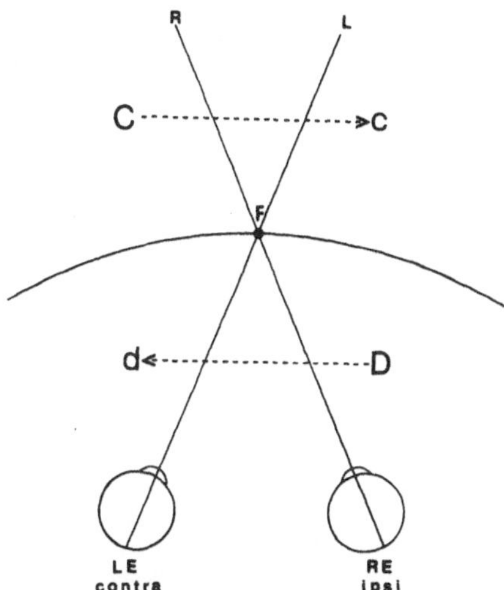

Fig. 7 The proposed mechanism for stereoscopic depth perception based on interocular delay. Explanation in text.

depends on the stimulus velocity. A stimulus velocity of 10°/sec and an interocular latency difference of 10 ms yields a preferred crossed disparity of 6 minutes of arc. Higher velocities result in larger preferred disparities for the same interocular latency difference.

Several problems are posed by this formulation of stereoscopic mechanisms. While it is clear that the cell of Fig. 7 might respond optimally to leftward-moving stimuli with crossed disparities, the same argument predicts that the cell should respond just as well to stimuli with uncrossed disparities moving rightward. This ambiguity in signaling the sign of disparity must be considered as a serious potential weakness for the proposed mechanism. A solution to this problem is provided, however, by the observation that over 80% of the cells encountered in the visual cortex exhibit direction selectivity. Thus the cell in Fig. 7 might respond only to leftward-moving stimuli. This preference for leftward movement would prevent the unit from responding to stimuli with uncrossed disparities. This provides us with a rather different perspective on the functional role of cortical directional selectivity, namely that it renders unambiguous the value of retinal disparity. We have devised a simple rule which summarizes the relationship between a unit's preferred direction, ocular dominance and preferred disparity: If the preferred eye and the preferred direction are the same, the unit should respond optimally to uncrossed disparities. If the preferred eye and preferred direction are opposite, then crossed disparities should provide optimal stimulation.

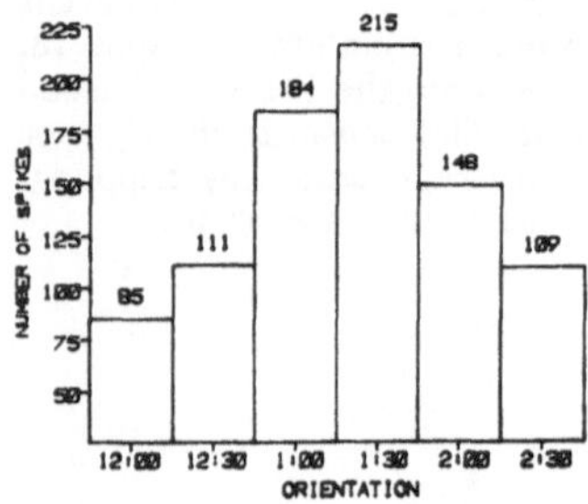

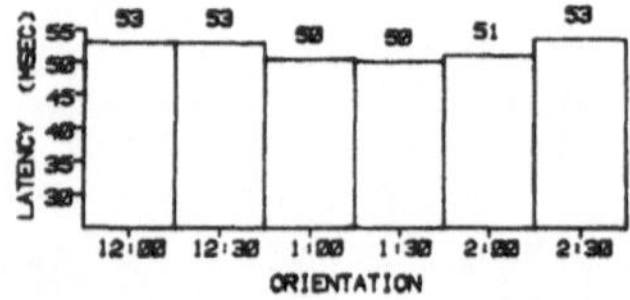

Fig. 8 Response amplitude and response latency in an area 18 complex cell. Six orientations were tested, ranging from 12:00 to 2:30. The preferred orientation of the unit was 1:30 and so this stimulus orientation elicited most responses. The top histogram slows the spike count for each orientation, and the lower histogram shows the response latency. Latency of response was somewhat shorter for the best orientation, but the effect is a small one.

A second problem for a time-based stereoscopic mechanism is a possible contamination of response latency by response strength. It is well known in many sensory systems that decreased strength of response is associated with increased latency of response (14). One obvious example occurs with decreasing stimulus contrast (Fig. 5). The observed interocular latency differences, in which the weaker eye has the longer latency, are also consistent with the general premise that response amplitude and response latency are negatively correlated. If this were always the case, then latency of response in cortical neurons would be heavily dependant on the particular stimulus configuration used. For example, it is well known that cortical cells respond differentially as a function fo stimulus orientation. If latency were to change markedly with orientation, then orientation disparity (16, 17) would inevitably contaminate the temporal disparity mechanism described here. We have found however, that changes in stimulus orientation have very little effect on response latency. Fig. 8 illustrates the response of a complex cell to stimuli of various orientations. The upper bar graph shows that response amplitude increases markedly at the best orientation. Despite the marked change in response amplitude with orientation, latency of response remained nearly constant. Even the weak responses at the worse orientation occurred at nearly the same time as those to the best orientation. This observation, which has been a consistent one in 30 cells studied thus far, indicates that response strength and response latency are <u>not</u> inevitably correlated.

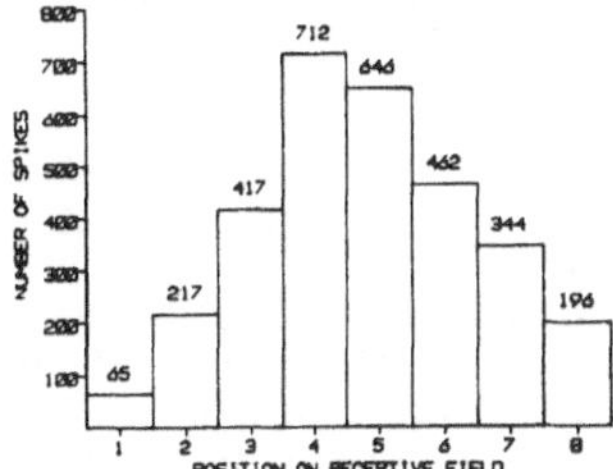

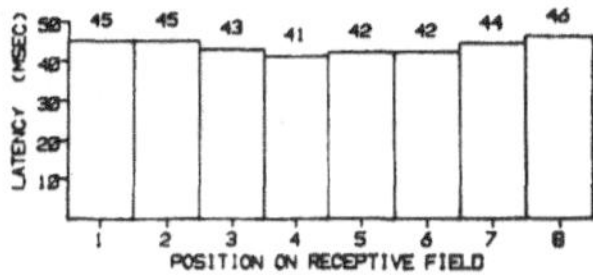

Fig. 9 Number of spikes and latency of response as a function of stimulus location across the receptive field of an area 18 complex cell. The upper histogram shows the total number of spikes elicited from 8 spatial positions across the receptive field and the lower histogram shows the response latency. Each position represents approximately 1° of visual angle. Response strength is much greater near the center of the field, but latency varies by only 5 milliseconds across the receptive field.

A second potential contaminant for a temporal disparity mechanism would be the existence of marked latency changes across the receptive field. If the latency of visual response were to vary markedly across the cell's receptive field, then interocular latency differences might themselves be a function of the spatial disparity of the stimulus. Fig. 9 shows response amplitude and latency of a representative complex cell to stimuli flashed onto different positions of the receptive field. The response amplitude increases near the center of the receptive field and fell off at the margins. The lower bar graph shows that the latency of response varied only slightly with stimulus position on the receptive field. As with orientation, it is possible for the neuron's response to become a good deal weaker without marked changes in the timing of that response. It appears as if cortical mechanisms maintain the latency of response under tight constraints despite wide variations in response strength.

It may be instructive at this point to compare the status of the time-based disparity mechanism which we have proposed with the earlier formulations based on spatial disparity. BARLOW, BLAKEMORE, and PETTIGREW (3) argued that the mechanism underlying stereoscopic depth perception was the scatter in the locations of binocular receptive fields they observed in the visual cortex. Their data suggested that some cortical cells should be best activated by stimuli on non-corresponding points in the two eyes since some receptive fields were so located. They provided evidence for a range of interocular misalignment in the spatial domain. In subsequent experiments, they and others (3, 4, 5) showed that cortical cells responded differentially to moving stimuli as a function of their retinal disparity. These two findings, namely evidence for spatial scatter in the binocular receptive fields and evidence indicating selective preferences for particular spatial disparities in single cells, are the major underpinnings of the currently-accepted spatial disparity mechanism.

The data presented here provide evidence for interocular scatter in the temporal domain (i.e., in the time of arrival of stimuli through the two eyes, rather than the location of the fields). We have shown that these temporal disparities between the two eyes are maintained under a wide variety of stimulus conditions. It is now necessary for us to show that single cells are sensitive to temporal disparities between the two eyes just as has been claimed for spatial disparity.

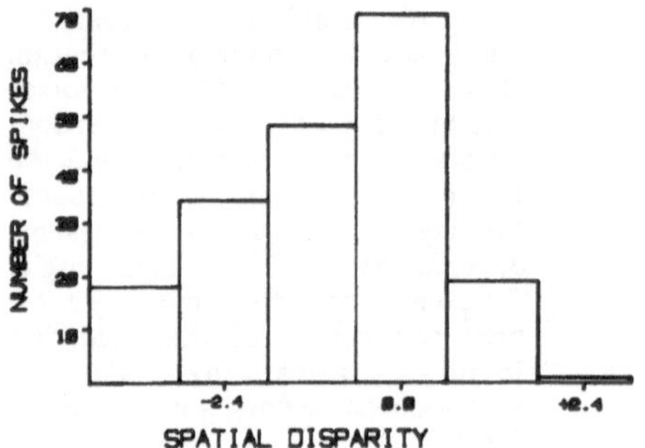

Fig. 10 Responses to flashed stimuli as a function of spatial disparity. Stimuli were presented to each eye simultaneously while the spatial disparity of the stimulus was varied. The ordinate represents the number of spikes evoked by 64 presentations of the stimulus at various spatial disparities. The data indicate marked differential responses as a function of spatial disparity. Abscissa values are in degrees.

Before describing our results on temporal disparity tuning in single cortical cells, it is worth noting that all previous studies which have purported to demonstrate selectivity for spatial disparity among cortical cells have in fact confounded the contribution of spatial and temporal mechanisms to the selectivity of the unit. Since these experiments have all employed moving visual stimuli at one velocity, the roles of temporal and spatial selectivity cannot be distinguished. One way to distinguish the contributions of spatial and temporal selectivity would be to vary the velocity of the stimulus. Another was is to present flashed stimuli to the two eyes at different retinal positions and then to measure the responses separately as a function of spatial or temporal disparity. We have carried out this experiment, and Fig. 10 taken from a representative unit shows that single cortical cells do indeed respond differentially to stimuli as a function of their spatial disparity. To demonstrate the effect of temporal disparity on the responses of this unit we presented stimuli at the position marked "O" in Fig. 10, but varied the time of stimulus presentation to the two eyes. Fig. 11 illustrates the responses of this cell to stimuli with various interocular delays. As can be seen, the unit's response varied markedly with a maximum at the point of simultaneous arrival on input from the two eyes.

In general, the maximal response was obtained with simultaneous input from

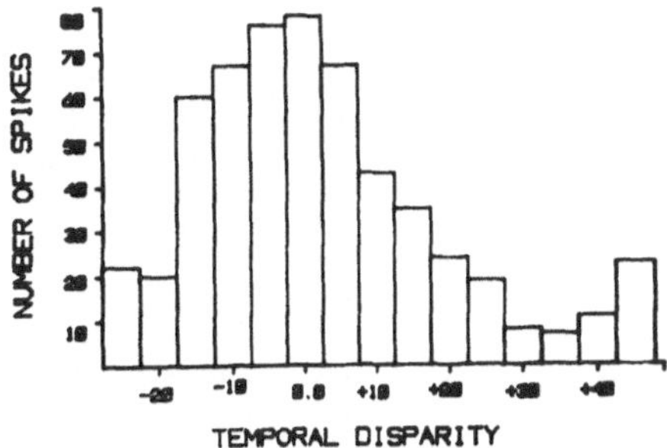

Fig. 11 Responses to temporal disparity. The cell shown in Fig. 10 was tested with stimuli which were presented at the same spatial location throughout, but with varied time relationships between the two eyes. Minus signs on the abscissa indicate prior stimulation through the left eye and plus signs prior stimulation through the right eye.

the two eyes, as predicted by the monocular latencies, rather than with simultaneous stimulus presentation. In many cells, the selectivity for temporal disparity was quite sharply tuned, with stimuli presented only 10 or 15 ms away from optimum interocular synchrony evoking much weaker responses. Fig. 12 illustrates the responses of another cortical unit to stimuli of varied temporal disparities. In this case, we tried to see how small a time difference between the two eyes could be discriminated in the cell's response. As can be seen, this cell's firing rate varied by a factor of 3 for 10 ms of temporal disparity. Each 2 ms change in temporal disparity evoked a discernibly different response from the cortical cell. These data demonstrate that the firing rates of single cortical cells can be greatly affected by temporal disparities which are well within the range of the observed (Fig. 4) latency differences between the two eyes.

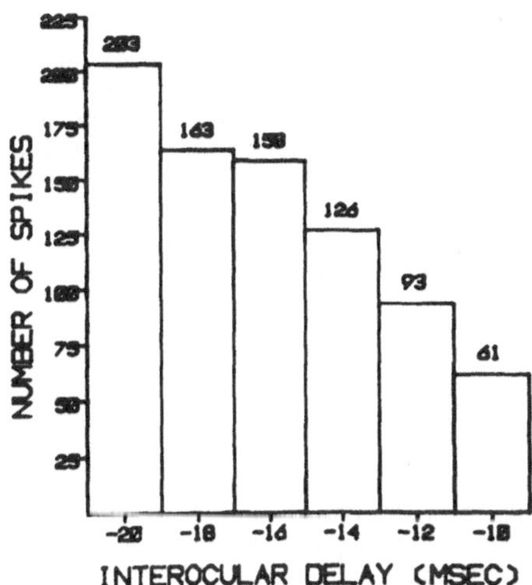

Fig. 12 Histogram of the temporal responses of a simple cell, OD group 1, found on the 17/18 border. Flashed stimuli were presented to both eyes with an interocular delay ranging from 10 to 20 ms. The number of spikes elicited for each temporal delay is indicated above the relevant bin. The data for each position represent 64 stimulus presentations at a frequency of 2 Hz. The monotonic decrease in spike count seen with a decrease in binocular delay shows that this unit was highly sensitive to the timing of input from the two eyes.

The data presented in Figs. 11 and 12 were gathered by choosing one position on the receptive fields in the two eyes and studying the consequence of varied interocular delay. Conversely, the data of Fig. 10 were gathered by picking a particular temporal disparity and varying the spatial disparity of visual stimuli. Neither of these procedures provides us with a complete description of the cortical cell's responses to stimuli with varied spatial and temporal disparity. To achieve a more complete picture of a cell's response to stimuli over a range of both spatial and temporal disparity, we have presented stimuli in an interleaved fashion at several different spatial and temporal disparities. Fig. 13 shows what we call a "space-time" plot of the binocular response of a cortical cell. We presented stimuli at 7 different spatial disparities and 7 different temporal disparities to generate 49 data points. Interpolated vectors were computed. The data of Fig. 13 show that this cell was highly sensitive to both spatial and temporal disparities between the two eyes. Only when both spatial and temporal stimulation was appropriate did the unit respond most vigorously. Inappropriate spatial or temporal disparities evoked far fewer spikes.

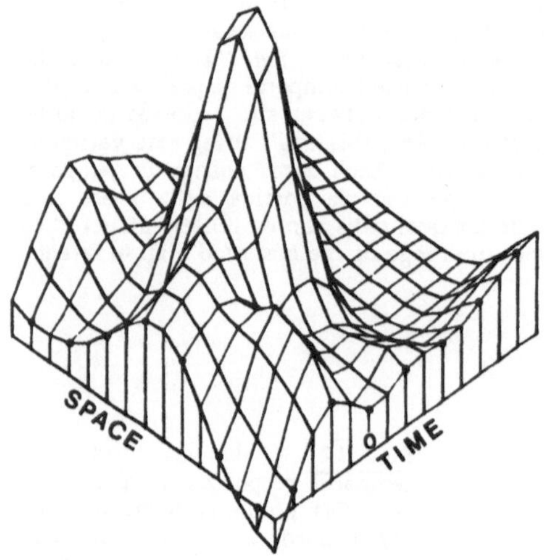

Fig. 13 Space-time plot of an area 18 cell which responded poorly to monocular stimulation but showed strong binocular facilitation. Around the maximum response peak, this facilitation dropped off with ± 15 ms of interocular delay. For this unit, 49 histograms were simultaneously compiled, representing seven spatial locations and seven interocular delays. Interocular delays ranged from -20 ms to +40 ms around synchronous stimulation ("0") and spatial positions covered 4.8° of visual angle. The procedure for binocular stimulation was similar to that previously described. The best response occurred when one eye was stimulated 10 ms ahead of the other eye.

Figure 14 shows the responses of a simple cell to stimuli with various spatial and temporal disparities. One can clearly distinguish the exitatory and inhibitory parts of the spatial tuning curve, but once again these spatial interactions occur only if the temporal disparities are appropriate. The data indicate that both spatial and temporal characteristics of the cell's binocular responses must be considered to derive a complete picture of its behavior.

It should be emphasized that the selectivity of cortical units to stimulus disparity, both spatial and temporal, varies markedly among different cells. We have found in the study of over 100 cells that some units appear rather insensitive to time-of-arrival differences over a range of ± 100 ms. Other cells are clearly more sensitive than those reported in this paper. The detailed nature and location of these highly sensitive units are subjects for further study.

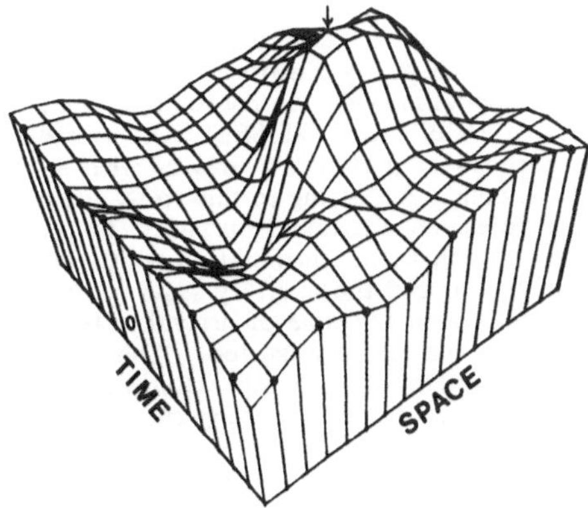

Fig. 14 Space-time plot of an area 18 simple cell. This unit was OD group 2 and showed an interocular latency difference of 10 ms. Stimuli were presented to both eyes simultaneously and 64 histograms representing the space and time variables were compiled. The position of the stimulus in the contralateral eye was held constant as the spatial location of the second stimulus was varied across the receptive field of the ipsilateral eye. The eight positions on the "space" axis represent 8.8° of visual angle. At each of these eight positions, eight interocular delays were tested, ranging from -30 ms to +40 ms around the point of synchronous stimulation ("0"). Binocular interactions were seen with interocular delays of ± 20 ms and showed maximum excitation and inhibition around the presumed point of simultaneous excitation, from the two eyes rather than the point of simultaneous stimulation.

To summarize, the data presented show that single cells in the visual cortex may be activated binocularly, with different latencies in the two eyes. This interocular latency difference appears to remain rather constant with changes in stimulus contrast, orientation, or location. Single cells in the visual cortex appear to be highly sensitive to the timing of input from the two eyes, as well as being influenced by the spatial disparity of stimuli.

The new formulation of stereoscopic depth mechanisms described above has both strengths and weaknesses. A potential difficulty with a time-based stereoscopic mechanism is that the disparity tuning of single cells should depend on the velocity of the stimulus. Higher stimulus velocities should result in both a change in preferred disparity (for cells with interocular latency differences) and also a broadening of the the disparity tuning curve (for all cells). This effect would be minimized in cortical cells with sharp velocity preferences (18, 19), but it may be necessary to have an independent signal concerning the stimulus velocity to correctly interpret the exact disparity value of the stimulus.

There are at least three strengths of this mechanism which may be considered. First, a time-based stereoscopic mechanism provides a raison d'etre for cells in ocular dominance groups 2, 3, 5, 6. These cells are a numberical majority in the monkey striate cortex (7) and they are frequencly encountered in cat striate and parastriate cortex as well (2, 20). It is now well known that cells with similar ocular dominance such as 2, and 3 or 5, and 6 aggregate in parallel slabs as viewed in tangential section. As a direct result, temporal disparity sensitive cells may be laid out in a similar regular manner. Thus our model provides an anatomical basis for stereopsis in the ocular dominance slabs of HUBEL and WIESEL (21). A second virtue of a time-based stereoscopic mechanism is that it provides us with a new meaning for direction selectivity in cortical cells. According to this view, direction selectivity has, in addition to its former roles, an additional purpose: to render unambiguous the sign of temporal disparity. Third, this formulation points out an essential commonality between the visual system and the auditory system. In both systems, inputs from paired end organs are compared and time-of-arrival differences between these end organs are employed for spatial localization. The auditory system used these differences to locate stimuli along the frontoparallel plane while the visual system uses a similar mechanism to locate stimuli along the orthogonal plane.

References

1. C. Wheatstone, Phil. Trans. Roy. Soc. Lond. 11, 371 (1838).
2. D. H. Hubel and T. N. Wiesel, J. Physiol. 160, 106 (1962).
3. H. B. Barlow, C. Blakemore and J. D. Pettigrew, J. Physiol. 193, 327 (1967).
4. T. Nikara, P . O. Bishop and J. D. Pettigrew, Exp. Brain Res. 6, 353 (1968).
5. D. H. Hubel and T. N. Wiesel, Nature, 225, 41 (1970).
6. N. Berman, C. Blakemore and M. Cynader, J. Physiol. 246, 595 (1975).
7. D. H. Hubel and T. N. Wiesel, J. Physiol. 195, 215 (1968).
8. M. Cynader and N. Berman, J. Neurophysiol. 35, 187 (1972).
9. Von C. Pulfrich, Naturewissenshaften, 10, 553 (1922).
10. R. L. Gregory, Eye and Brain (McGraw-Hill, New York, 1966).
11. M. Alpern, Psych. Rev. 75, 260 (1968).
12. B. J. Rodger and S. Anstis, Vis. Res. 12, 909 (1972).
13. M. Cynader, N. Berman and A. Hein, Exp. Brain Res. 25, 139 (1976).
14. E. D. Adrian and R. Matthews, J. Physiol. 63, 378 (1927).
15. H. K.Hartline, J. Cell Comp. Physiol. 5, 229 (1934).
16. C. Blakemore, A. Fiorentini and L. Maffei, J. Physiol. 226, 725 (1972).
17. P. O. Bishop, this volume (1978).
18. J. A. Movshon, J. Physiol. 249, 445 (1975).
19. J. A. Orban and M. Callen, Exp. Brain Res. 30, 107 (1977).
20. D. H. Hubel and T. N. Wiesel, J. Neurophysiol., 28, 229 (1965).
21. D. H. Hubel and T. N. Wiesel, Proc. Roy. Soc. Lond. B. 198, 1 (1977).

Mechanisms for Stereopsis

Whitman Richards
Massachusetts Institute of Technology
Department of Psychology
Cambridge, Massachusetts 02139

Two Visual Systems

In 1966 GERALD SCHNEIDER completed a thesis at M.I.T. proposing that visual function was modular and that there were two visual systems, not one. One system dealt with recognition and was cortical and the other resided in the midbrain and was concerned with localization. The cortical system answered the question "What is it?", whereas the midbrain system asked, "Where is it?". Although SCHNEIDER'S work (1) was based upon experiments with the hamster, many of us then began to examine our own work more closely to determine whether the visual problems we were studying could also be subdivided into two main components designed to answer the "where?" and "what?" questions.

Our first step toward making this distinction for stereopsis was to compare the stereoscopic response to a moving stimulus versus one that merely changed position. Large movements in depth occur during normal locomotion, or when objects independently approach and recede. The analysis of these movements helps to determine where things are in the environment. On the other hand, when objects are foveated and examined in detail, each new fixation causes abrupt positional changes in disparity that provide information useful for a three-dimensional representation of shape and size.

To characterize simply these two types of stereoscopic stimuli, smooth movement versus abrupt change of postion, two vertical bars were made to oscillate either sinusoidally or as a square-wave. When seen through appropriate polarizers that delivered one bar to one eye and the second bar to the other eye, the stimuli appeared to move either smoothly in and out, toward or away from the observer (sinusoidal) or to jump from one position to another either closer or farther from the observer but in the same binocular direction (square-wave). By varying the rates of movements or positional change, the sensitivity profiles for those two quite different stereo scopic stimuli could be measured.

Figure 1 shows the form of these sensitivity functions. The circles represent sinuosidal movement, whereas the squares are the abrupt displacements. The maximum disparity for all points in the figure was the same: 1.0 degree. The ordinate shows how much perceived depth could be elicited for this constant disparity as the modulation rate was varied from 1/8 to 8 c/sec. Note that the positional stereoscopic analysis fails at lower modulation rates than the movement in depth, but at very low rates of oscillation (i.e., slow movement), the movement analysis is impaired. This observation generally holds for other stimulus disparities, both larger and smaller, except when the disparity amplitudes are very small and are below 1/4 degree where both types of modulation yield similar results.

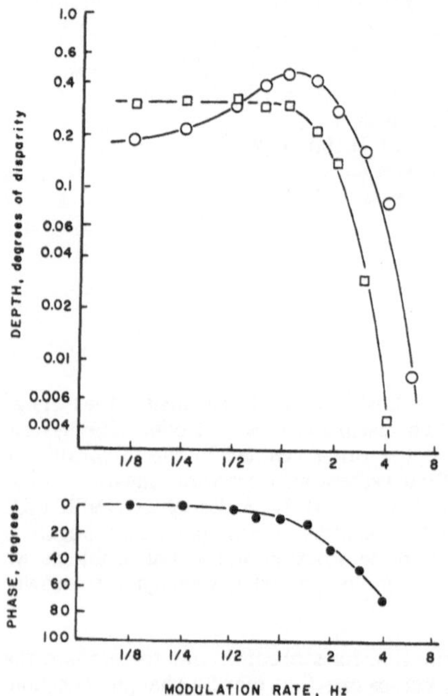

Fig. 1 Apparent depth generated by a bar oscillating back and forth through the fixation plane, with the maximum amplitude held constant at $\pm$ 1.0 deg of disparity. Sine-wave modulation, circles; square-wave modulation, squares. The lower portion of the graph shows the apparent phase angle for sinusoidal modulation. (From RICHARDS (2)).

To further reinforce the distrinction between the positional and movement stereo systems, it is desirable to isolate each separately. One method of isolation is to demonstrate that some observers lack one system but not the other. Figure 2 is an example of a subject whose positional mechanism is greatly attenuated (5-fold for this 1° disparity). Yet his vergence tracking is normal, failing only near 6 Hz.

Subjects can also be found that behave in the complementary manner, where positional changes are processed similar to the "normal" as in Fig. 1, but whose movement analysis is grossly attenuated, especially at low modulation rates (2). Hence we have a double dissociaiotn between two rather different kinds of stereo losses--one concerned with positional changes, the other reflecting sensitivity to large amplitude, slow motion in depth.

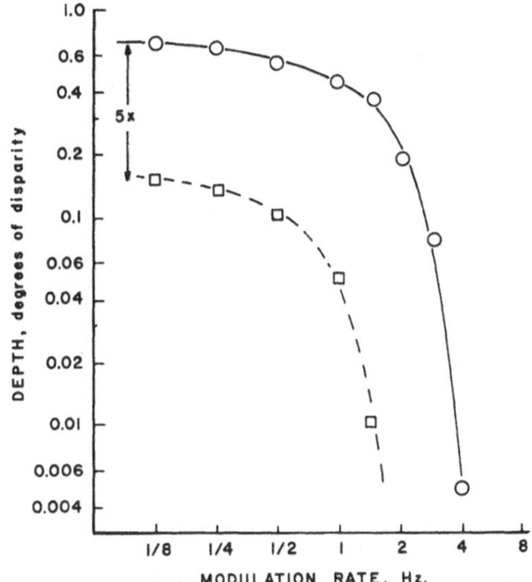

Fig. 2 Depth-response curves for an observer possessing only a divergent disparity mechanism. Modulation amplitude is ± 1.0 deg of disparity. Open circles, sine-wave modulation; squares, square-wave modulation. Compare with Fig. 1. (From RICHARDS (2)).

Further Feature Filters

Positional change (dz) and movement (dz/dt) are not the only cues that stereoscopic systems may use. These two cues only serve to highlight a distinction between two separate modes for using stereoscopic information. Within these two broad categories, additional cues and features are utilized, and the main categories themselves can be further subdivided.

Three "Pools"

The most important subdivision is probably the breakdown of the disparity dimension into three regions: convergent, near-fixation, and divergent. Both psychophysical and neurophysiological results support this subdivision (3, 4, 5). For human stereopsis, the results are quite clear. When a large population of subjects are analyzed in terms of their ability to process convergence or divergent dispairites, a sizeable percentage (around 30%) confuse all divergent (or all convergent) stimuli with monocular stimuli that carry no stereoscopic information.

Figure 3 illustrates the nature of these confusions. In the lower left column, the relation between disparity (abscissa) and perceived depth (ordinate) is characterized for an observer that does not confuse stereoscopic with monocular disparity. This non-monotonic relation between depth and disparity is typical of most observers. When pressed to make forced-choice judgements as to whether stimuli are convergent (crossed) or divergent (uncrossed), a high level of performance is achieved over a wide range of disparities (upper graph, left column). Other observers may be found, however, who fail to discriminate divergent disparities, for example, from monocular control stimuli (middle column, upper graph). In these cases, the relation between perceived depth and disparity is also altered, and the depth seen for disparities not discriminated is roughly the same as that elicited by a monocular stimulus (middle column, lower graph).

Three major classes of residual stereoscopic mechanisms can be shown to underly all cases of stereoblindness to positional changes in disparity. One class processes only divergent disparities, while the second processes only divergent disparities (i.e., the data are mirror-symmetric to those in the middle column of Fig. 3). In both cases, when a loss occurs, it extends over the full range of disparities from as small as 1/16 degree to as large as 4° (6). In addition to these two types of reduced stereopsis, there is also a third reduction where the observer appears to process disparity information in the neighborgood of the horopter, but confuses the sign of the disparity (Fig. 3, right column). Just as the reductions in color blindness were taken as representations of the three modes of normal color vision, we make the same inference about normal stereopsis.

The relation between these three types of disparity "pools" and the two broad categories of stereopsis (i.e., where? and what?) is still not clear. There is a high correlation between fixation disparity errors, however, and the convergent, divergent or "on" stereoscopic losses (7). Thus, one is led to believe that the fine alignment of the eyes for detailed pattern vision may be under a positional control mechanism driven by the interrelated activities of the three stereoscopic "pools" (8). For the initiation of large vergence movements, however, although there is a relation between the losses in positional stereoanalysis, the correlation is not as high (9). Perhaps this is in part due to the possibility that the normal initiation and maintenance of large vergence movements may require velocity or directional information lacking in a disparity pulse (10). Such information would be more appropriate for the "second" visual system, presumed to reside in the midbrain.

The different pattern of losses between the initiation of large vergence movements and phoria, as well as the related losses in movement and positional stereomechanisms, need not imply two entirely independent mechanisms. More likely is an interaction between the two modes of viewing, or even a "collapse" of the "three pool" classes in the case of the large amplitude vergence system addressing itself to the question "where?".

Two sets of data suggest that a collapse of the "three pool" categorization may occur. First, we know from anatomical studies in the Siamese (as well as the ordinary "wild-type" cat who is cross-eyed) that aberrations in the visual pathway occur both in the geniculo-striate and midbrain routes (11, 12). This suggests a common retinal origin for at least part of the disruption in the integration of binocular information, whereas cortical studies reveal additional disruptions (13, 14, 15). Secondly, psychophysical data obtained from patients suffering damage to visual cortex show that the remaining stereoscopic processing collapses to merely a distinction between binocular versus monocular stimuli, typical of one of the three basic stereo-mechanisms (Fig. 3, right), but without complete loss of movement sensitivity (16, 17), or eye movement control (18).

Thus, we conclude that although the cortex may pool disparity information into either convergent, divergent, or "on" (i.e., near the fixation plane), in the midbrain the three pools may appear to collapse into two ("far off" versus "near" the fixation plane), especially if additional motion information is excluded.[1] Both systems could still have a common retinal basis for setting up each mechanism.

Directionality

Although usually associated with motion parallax, directional movement provides a powerful cue for stereoscopic analysis. Its importance has already been emphasized for the second (midbrain) visual system concerned with flow patterns and localization. Yet directionality is also important for the finer analysis of object relations, particularly in the presence of small head movements. Then the relative velocities of different portions of the image plane serve to isolate both the objects themselves as well as their relative distance relations.

Psychophysical techniques can again be used to illustrate that this cue is set up separately from the positional cue. For example, blindness to motion in depth can be observed in regions of the visual field where positional stereoscopic information is intact (19). REGAN (20) discusses further examples of the directional nature of stereoscopic analysis, using selective adaptation procedures. More recently, SMITH (21) has also shown that directional components can be assigend to either convergent or divergent disparity movements simultaneously, suggesting that both cues can combine to constrain the stereoscopic analysis[2]. These data are consistent with the known behavior of simple cortical feature analyzers, which are sensitive to directionality and disparity, as well as to orientation and stimulus width.

Spatial Frequency

Still another feature dimension along which stereoscopic analyzers may be classified is bar width or spatial frequency (Fig. 4). Several results show that small displacements of disparity are processed by the "small-bar detectors", whereas the larger disparity analysis is carried out by "large-bar detectors" (6, 22, 23). This relation between bar size, or spatial frequency and disparity is independent of the pooling along the dispairty dimension, as well as of directionality. Yet to be determined is whether bar height is also still another independent dimension for constraining stereoscopic analysis.

Contrast

Whether the correlated pairs of stereo elements are black or white has an effect on the nature of the depth perception (24, 25). Although complex, the results imply that a change in contrast may change the sign of the depth signal, but perhaps only in cases where the disparity signal may be slightly ambiguous. It has been known for some time that positive and negative contrast is processed by two separate classes

[1] The introduction of directionality should restore the three pool distrinction although in terms of (dz/dt) rather than (dz) alone. In this case, both systems could still have a common retinal basis for setting up each mechanism.

[2] Although not yet demonstrated, directionality may be the critical cue that allows complex stereograms to be built up, perhaps by binding binocular direction vectors to correlations in the two image planes as the eyes scan the three dimensional pattern, with the assigments dependent upon the z direction of the scan.

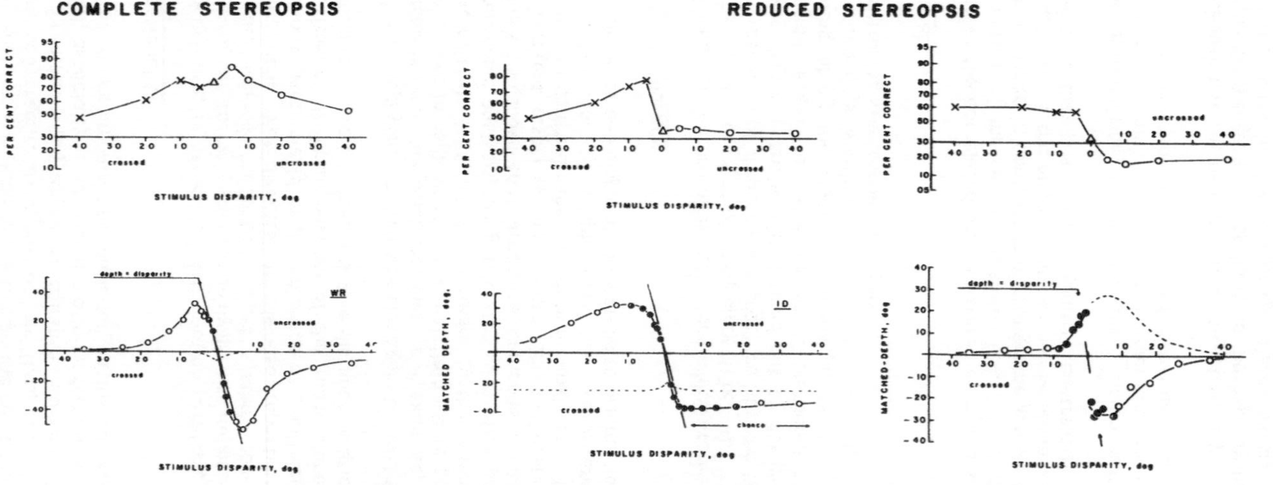

Fig. 3 Upper row of figures show percent of correct responses to stimuli of different disparity. Each graph is based upon about 150 trials. Lower row of figures show the perceived depth elicited from a bar flashed with different disparities relative to the fixation position. A filled circle indicates that the binocular stimulus appeared as single; the open circles show which stimuli appeared double. The dashed lines show the apparent depth of the monocular stimuli. The regions indicated by the arrows show where the anomalous observers respond to chance on the forced-choice test. (From RICHARDS (3)).

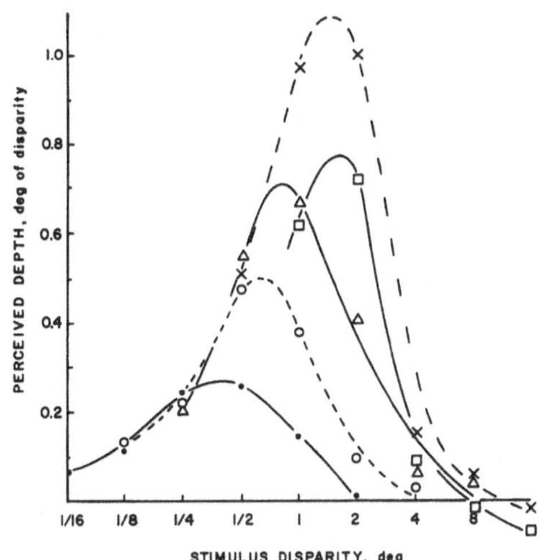

Fig. 4 Relative perceived depth vs stimulus disparity averaged for three observers. Crossed (convergent) disparities only. Each curve represents a separate bar width; with the bars presented as brief flashes. Dots, 0.05°; open circles, 0.1°; triangle, 0.2°; cross, 0.4°; square, 0.8°. Bar height was fixed at 0.5°. (From RICHARDS and KAYE, (6)).

of feature analyzers (21), and more recent psychophysiocal results support this dichotomy in processing for man (27, 28). Thus it is not surprising that our stereo system tries to pair dark bars together and light bars together rather than mixing the two (29). What is surprising is that the sign of the contrast can influence the nature of the stereoscopic signal. Perhaps because "white" objects are generally superimposed on backgrounds, whereas "black" objects are not (i.e., they are often "holes"), the sign of the contrast may be used only as a parity check for the sign of the disparity.

False Targets: A Pseudoproblem?

The property that several cues may be read independently by the same stereoscopic analyzer raises questions about whether "false targets" occur often enough in the real world to be considered a serious threat to image processing. MARR and POGGIO (30) were the first to recognize that the stereoscopic correspondence problem could in principle be solved by using the property that the high spatial frequency content of an image only conveys information about small disparities, whereas lower spatial frequency "masks" (i.e., channels) are required to analyze larger disparities. By trading off resolution with range of disparity, false targets can be eliminated first by analyzing the larger elements of the pattern where no false pairings are made, and then proceeding to a succession of finer image and disparity analyses.

But the spatial frequency content of the pattern is only one of several cues that can help solve the false target problem. Once directionality, orientation, element height, and disparity "pooling" is added, then the constraints upon possible correspondences plummet drastically. Thus, we might well have a system that in fact fails for stimuli having many possible false correspondences but no cues to eliminate them. An approximation to such a stimulus is JULESZ' random dot stereogram (31). When such a stereogram is presented in a flash so that eye movements are eliminated, then stereopsis is indeed markedly impaired for most and may collapse entirely for some individuals (32). The false target problem, therefore, may be solved by the multiple cue assignments that can be made with natural images.

Summary

Stereopsis is a problem that must not be considered in isolation, but rather with respect to the questons it is designed to answer. Two such questions are "where are things in our three dimensional world, and what are these things?". Within these two broad categories, stereopsis is only one of several mechanisms brought into action. All of these mechanisms tend to reinforce or deny the assertions made by one another, and hence interact. Important constraints upon stereopsis are imposed by directionality, spatial frequency, contrast and orientation. Each of the two major stereo systems may place special emphasis on one rather than another of these cues, depending upon the task at hand. Table I is an attempt to highlight some of these distinctions at several conceptual levels.

Table 1

	What?		Where?
Primary Cue	Position (dz)		Motion (dz/dt)
"Opponent" Channels	Front-Back	Binoc.-Monoc.	Approach-Avoidance
Oculo. Cor.	Fixation		Tracking (Pursuit)
Dominant Site	Cortex (LGN)		Midbrain (Pretectum)

References

1. G. E. Schneider, Psychologische Forsch. 31, 52 (1967).
2. W. Richards, J. Opt. Soc. Amer. 62, 907 (1972).
3. W. Richards, J. Opt. Soc. Amer. 61, 410 (1971).
4. P. G. H. Clarke, I. M. L. Donaldson and D. Whitteridge, J. Physiol. 256, 509 (1976).
5. T. Poggio and B. Fischer, J. Neurophysiol. (in press) (1977).
6. W. Richards and M. G. Kaye, Vis. Res. 14, 1345 (1974).
7. W. Richards, Amer. Acad. Opt. Annual Meeting. Columbus, Ohio, Dec. See also Chapt. 10 in Handbook of Perception, Vol. V. ed. by E. C. Carterette and M. P. Friedman (Acad. Press, N. Y. 1975).
8. W. Richards, Neurosciences Research Program Work Session on The Visual Field, Dec. 9-11, 1973 (to appear in the Neurosciences Research Program Bulletin, 1977).
9. R. Jones, J. Physiol. 264, 621 (1977).
10. C. Rashbass and G. Westheimer, J. Physiol. 159, 339 (1961).
11. R. Kalil, S. R. Jhaveri, and W. Richards, Science, 174, 302 (1971).
12. W. Richards and R. Kalil, ARVO spring meeting, Sarasota, Florida, (1976).
13. D. H. Hubel and T. N. Wiesel, J. Phsyiol. 218, 33 (1971).
14. S. J. Cool and M. Crawford, Vis. Res. 12, 1809 (1972).

15. J. Kass and R. Guillery, Brain Res. 59, 61 (1973).
16. W. Richards, Exp. Brain Res. 17, 333 (1973).
17. E. Poppel and W. Richards, Exp. Brain Res. 21, 125 (1974).
18. E. Poppel, R. Held and D. Frost, Nature (Lond.) 243, 295 (1973).
19. W. Richards and D. Regan, Invest. Ophthal. 12, 904 (1973).
20. D. Regan, (1978) (Chapter in this volume).
21. R. A. Smith, Vis. Res. 16, 1507 (1976).
22. T. B. Felton, W. Richards and R. A. Smith, J. Phsyiol. 225, 349 (1972).
23. C. Blakemore and B. Hague, J. Phsyiol 225, 437 (1972).
24. S. M. Anstiss and B. J. Rogers, Vis. Res. 18, 957 (1975).
25. W. Richards, Am. J. Opt. and Arch. Am. Acad. Opt. 50, 853 (1973).
26. S. W. Kuffler, J. Neurophys. 16, 37 (1953).
27. K. DeValois, Vis. Res. 17, 209 (1977).
28. S. Purks and W. Richards, J. Opt. Soc. Amer. 67, 765 (1977).
29. H. L. Helmholtz, Treatise on Physiological Optics. (J.P.C. Southall (Trans.) Dover, N.Y., 1962).
30. D. Marr and T. Poggio, (to be submitted to Proc. Roy. Sco., Lond.)(1977).
31. B. Julesz, Foundations of Cyclopean Perception (Univ. Press: Chicago, 1971).
32. W. Richards, Vis. Res. 17, 967 (1977).

Spatio-Temporal Aspects of Binocular Depth Discrimination

Alfred Lit
Department of Psychology
Southern Illinois University at Carbondale
Carbondale, Illinois 62901

Introduction

When I originally planned my presentation for this Symposium, convened to dedicate the new Optometry Building in Houston, I had intended to discuss some of our experiments on temporal factors in binocular vision. These experiments, many of which are still unpublished, deal with four interrelated classes of binocular functions: (i) the effects of exposure duration on stereoscopic thresholds; (ii) the effects of interocular delay on binocular depth discrimination; (iii) depth-displacement effects obtained in studies on the Pulfrich effect (1) in relation to correlative psychophysical and electrophysiological studies on simple visual reaction time and on visual evoked potentials; and (iv) depth-displacement effects obtained in the so-called Mach-Dvorak phenomenon (2). As I began to prepare and organize the necessary slides for my presentation, it became quite obvious that I could not effectively deal with more than one of these topics in the limited time alloted. I decided, finally, to limit my presentation to a discussion of some of my recent work on the Mach-Dvorak phenomenon, performed with the aid of several of my student assistants[1].

This interesting depth-displacement effect, first reported about a century ago by V. DVORAK (2), who was then a student in Ernst Mach's laboratory in Prague,

[1]The first systematic study of the Mach-Dvorak effect in our laboratory was performed by Joel Brauner in 1973 as a Doctoral Dissertation in the Department of Psychology in which I served as chairperson of the Dissertation Committee. I am most pleased to acknowledge the many contributions made by Arthur Menendez to all phases of our more current experiments on the Mach-Dvorak effect; Dan Libke has been most helpful in the organization, analysis, and presentation of the data; Mary Ann Myskowski has most effectively assisted as experimenter during the data-collection phase.

provides a unique experimental situation for systematically studying some of the basic spatio-temporal factors in binocular depth discrimination. The study of this depth effect will not only serve to advance our basic knowledge of the underlying binocular physiological processes involved, but will also provide the test situation for assessing the integrity of the binocular integrating system (that is, the stereoscopic fusion mechanism) in a wide variety of possible clinical applications.

My presentation this morning aims to elucidate the complex geometric relationships in the Mach-Dvorak phenomenon and to identify the various classes of stimulus variables that can be fruitfully investigated. Special emphasis will be given to the task of relating the theory and data of the Mach-Dvorak effect to those of binocular space perception and of clarifying the simularities and differences underlying both the Mach-Dvorak effect and the Pulfrich stereophenomenon. Finally, I shall discuss some experimental results which we have obtained on the Mach-Dvorak effect. Because of space limitations, I shall give little consideration in this report to an historical account of the previous experimental work done on these two stereoscopic phenomena. I hope this omission will be forgiven by my colleagues; the individual contributions of the many investigators of the Pulfrich and the Mach-Dvorak effects and of related studies will be fully documented in future articles.

The Pulfrich Stereophenomenon

Let me begin by contrasting the Mach-Dvorak effect with the more familiar Pulfrich stereophenomenon.

The Pulfrich stereophenomenon, you will remember, involves a depth effect resulting from a stereoscopic parallax condition which occurs when a target that is oscillating from left to right and from right to left in a frontal plane is viewed with unequal binocular retinal illuminance, produced, for example, by placing a neutral density filter in front of one of the eyes. Under these illumination conditions, the oscillating target will appear to rotate out of its plane of motion toward the observer for one direction of stroke and away from the observer for the return stroke, in either a clockwise or counterclockwise rotation, depending on which eye is wearing the filter.

I have discussed the geometric relations existing for the apparent depth-displacement effects in the Pulfrich phenomenon in previous papers (see, e.g., LIT (3); LIT and HYMAN (4); LIT (5)). These relationships may be summarized with the aid of Fig. 1, which depicts the apparent near and far depth-displacement positions of the target, localized, respectively, at P'_N and P'_F in the observer's median plane when the observations are made with a filter placed in front of the right eye. In this case, an apparent counterclockwise target rotation is produced, as seen from above. The apparent near and far depth displacements, C_N (= OP'_N) and C_F (= OP'_F), can be accounted for on the basis of an hypothesized differential time delay in the signaling of the target locations in the two eyes resulting from the produced difference in binocular retinal illuminance. The delay in the signaling of the target positions is presumed to be greater for the dimmer eye than for the unfiltered one. Accordingly, when the target is moving from left to right with constant linear velocity, V, and is actually at the position P_{LR}, synchronously arriving cortical impulses from the two eyes will have been initiated by stimulation from different target positions along the target path, W_1W_2, located in a frontal plane at a distance, d, from the two eyes. The arriving cortical impulses from the left eye at the moment the target is actually located at P_{LR} are presumed to have been initiated when the target was located at position B, a small distance behind P_{LR}. The exact magnitude of the distance BP_{LR}, for a given target velocity, V, will

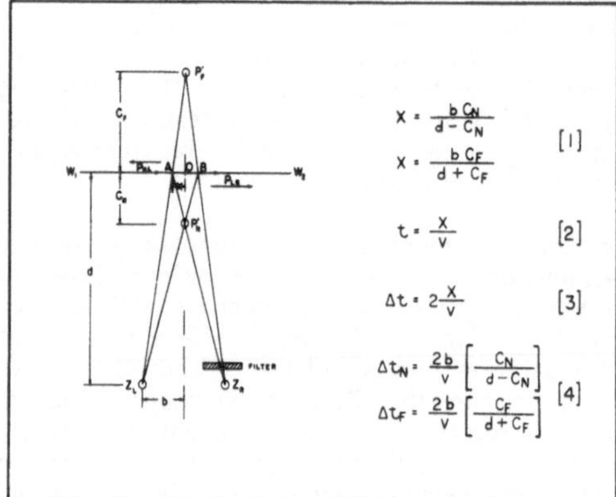

Fig. 1 Geometrical representation of the apparent near (C_N) and apparent far (C_F) depth displacements in the Pulfrich stereophenomenon and their corresponding computed near and far latency differences, Δt_N and Δt_F. For a filter placed in front of the right eye, the direction of target motion, as seen from above, is counterclockwise.

depend upon the magnitude of the absolute latent period of the left eye under the given conditions of illumination. Synchronously arriving impulses to the cortex from the dimmer eye are presumed to have been initiated earlier than those from the left eye (e.g., when the target was at point A, located at a slightly greater distance behind P_{LR}). The exact magnitude of the distance AP_{LR}, for a given target velocity, V, will depend on the magnitude of the (larger) absolute latent period of the right eye with filter. According to the laws of binocular depth perception, the intersection of the lines of sight of the two eyes, Z_L B and Z_R A, will locate the apparent near position of the target at P'_N; the magnitude of the apparent near depth displacement is then specified by the distance C_N (= OP'_N). By similar reasoning, when the target is moving from right to left and is actually located at P_{RL}, the lines of sight through points A and B will now locate the target at the apparent far position, P'_F; the magnitude of the corresponding apparent far depth displacement is specified by the distance C_F(=OP'_F).

It should be noted that the distance AB (=2X) represents the prevailing linear stereoscopic parallax existing between the two unequally illuminated eyes. The time required for the target to move from A to B or from B to A (t = 2X/V) measures the latency difference, Δt_N or Δt_F, existing between the two eyes. Both the spatial and the temporal measures of AB (linear stereoscopic parallax and binocular latency difference) progressively increase, within limits, as the binocular difference in illumination is increased. If the filter over the right eye is now transferred to the left eye, the oscillating target will appear to reverse its rotation to a clockwise motion, similarly as seen from above, but with unchanged amplitude of displacement. From geometric considerations, a given latency difference (Δt_N = Δt_F = K), will result in an apparent far depth displacement, C_F, that is always greater than the corresponding apparent near depth displacement, C_N, particularly for latency differences of large magnitude.

The relationship existing between the depth displacements, C_N and C_F, and the corresponding latency differences, Δt_N and Δt_F, can be established by simple geometry, as shown by the four groups of equations derived in Fig. 1. If the magnitudes of C_N and C_F are determined experimentally by an observer whose interpupillary separation is 2b, the magnitudes of the corresponding latency differences, Δt_N and Δt_F, can be readily computed from "1" (Fig. 1) when the observation distance, d, and the constant linear target velocity, V, are also given. Eq. (1) is taken from "4" in Fig. 1:

$$\Delta t_N = (2b/V)\,(C_N/(d - C_N))$$

and (1)

$$\Delta t_F = (2b/V)\,(C_F/(d + C_F))$$

The remarkable capability of the Pulfrich phenomenon to transform the presumed time differences into readily discriminable binocular depth differences can be demonstrated with the aid of Eq. (1). Thus, for a given unequal binocular illumination that produces an experimentally measured apparent near depth displacement of 1 mm (i.e., C_N = -0.1 cm) and an apparent far depth displacement of 1 mm (i.e., C_F = +0.1 cm), the average computed latency difference, $\overline{\Delta t}$ (= $(\Delta t_F - \Delta t_N)/2$), is only 0.4 msec under conditions where 2b = 6.5 cm, d = 100 cm, and V = 16.5 cm/sec (9.37 deg/sec). That is, for each msec of latency difference, the predicted depth difference between the far and near points of target localization, $P'_F - P'_N$, is about 5 mm under these specified conditions of testing.

We and many other investigators, following Pulfrich's pioneering efforts, have extensively studied (both theoretically and empirically) a wide variety of stimulus variables which influence the magnitude of this fascinating binocular depth effect, a phenomenon that is based on a visual-latency hypothesis which assumes that the magnitude of the absolute visual latent period in each eye is an inverse function of the level of stimulus illumination. We have frequently urged that serious consideration be given to the application of this phenomenon to clinical practice, particularly for the early detection of unilateral pathology in the visual centers or pathways and for assessing the quality of binocular depth discrimination.

The Mach-Dvorak Effect

Geometric Theory

The Mach-Dvorak phenomenon similarly involves a binocular depth-displacement effect for oscillating targets, but the required linear stereoscopic disparity, AB, is produced directly, without filters, by exposing the oscillating target for a short exposure duration (ED) along its path of motion, first to one eye by means of a shutter, and then after a brief interocular delay (ID) only to the other eye for the same short exposure duration by means of a separate shutter. The delayed target presentations to each eye are given in repeated pairs of flashes by means of the two shutters at a given repetition cycle (RC) as the target moves from left to right and from right to left in its frontal plane of motion. If the monocularly delayed, paired targets are presented successively along the path of target motion with an appropriate repetition delay (RD) between successive pairs, the intermittently presented targets are reported to produce nearly as "smooth" a target movement as is obtained in the case of the Pulfrich effect where binocular viewing occurs under continuous presentation of the oscillating target. That is, smooth apparent movement occurs in the Mach-Dvorak situation provided the time interval between successive flashes to the same eye, i.e., the repetition cycle (RC), lies within the range of values required to obtain "good" apparent movement for spatially separated targets that are presented non-simultaneously, as has been demonstrated in the case of the classical "phi phenomenon" by WERTHEIMER (6).

Fig. 2 may be helpful in clarifying the geometric relations existing for the apparent depth displacements obtained in the Mach-Dvorak phenomenon for an observer with interpupillary separation, b. Figure 2 illustrates the situation for three successive pairs of target flashes, A' and B', A and B, and A" and B", as the target moves with constant linear target velocity (TV) from left to right in a frontal plane (W_1W_2) at an observation distance (d). As in the case of nearly all of the experiments we shall report here, the right eye for each paired member is always

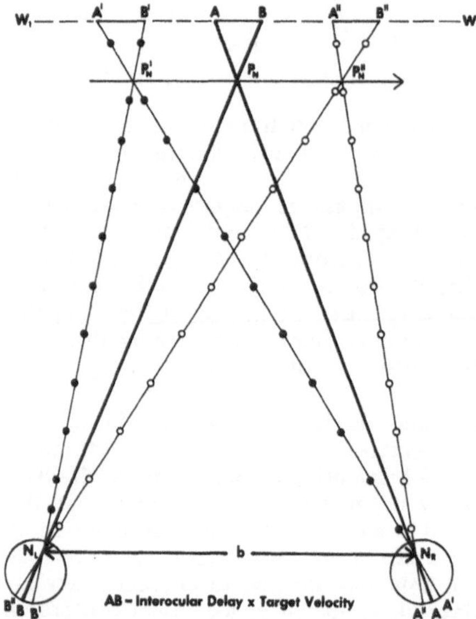

Fig. 2 Geometrical representation of the apparent near depth displacement in the Mach-Dvorak effect for target motion from left to right with the right eye serving as the "lead" eye.

stimulated first. Note in Fig. 2, for target movement from left to right, that point A, the position of target stimulation (and also that of target localization) for the "lead" (right) eye is always depicted to the left of point B, the position of target stimulation (and localization) for the paired "lag" (left) eye. In accordance with the laws of binocular space perception, the amount of linear stereoscopic parallax produced is measured by the magnitude of AB. Also, the intersection point of the appropriate lines of sight in the two eyes drawn through A and B locates, theoretically, for this direction of target movement, the predicted apparent near displacement position, P_N, (and positions P'_N and P''_N), if binocular fusion obtains for each of the paired flashes. The magnitude of AB, as indicated, is given by the product of the interocular delay (ID) and the target velocity (TV); theoretically, the greater the value of ID or TV, the greater the magnitude of the apparent near depth displacement for target flashes of very short duration.

The corresponding situation for target motion from right to left is shown in Fig. 3, where, again, the right (lead) eye is stimulated first at all three successive paired target positions, now designated as B" and A", B and A, and B' and A'. Note in Fig. 3, for target movement from right to left, that the point of target stimulation and localization for the lead (right) eye (now designated as B) is always depicted to the right of the target position of stimulation and localization for the paired lag (left) eye (now designated as A). The appropriate lines of sight through the respective target positions A and B, will now intersect beyond the plane of target motion to locate the apparent far displacement position of the binocularly fused target at P_F (and P'_F and P''_F). As in the previous figure, AB = ID x TV.

Thus, for conditions in which the right eye serves as the lead eye, the resulting direction of target rotation for the total to-and-fro movement will be counterclockwise as shown in Fig. 4, where the depth-displacement effect for target motion in both directions is depicted in the observer's median plane.

If, under otherwise identical conditions of stimulation, the left eye is now made to become the lead eye, then the oscillating target will appear to reverse its direction of rotation to a clockwise motion, but with unchanged amplitude of

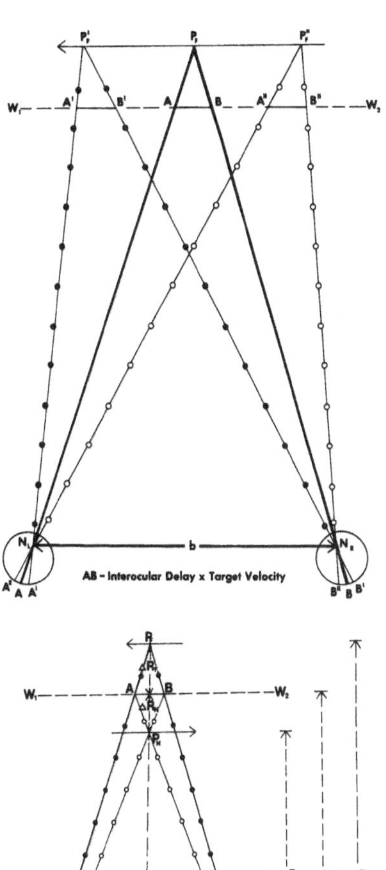

Fig. 3 Geometrical representation of the apparent far depth displacement in the Mach-Dvorak effect for target motion from right to left with the right eye serving as the "lead" eye.

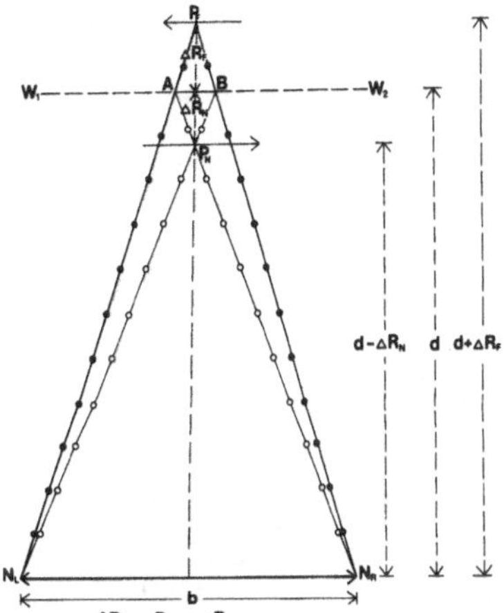

$$\frac{AB}{b} = \frac{\Delta R_N}{d-\Delta R_N} = \frac{\Delta R_F}{d+\Delta R_F} \quad [1]$$

$$AB = ID \times TV \quad [2]$$

$$\Delta R_N = \frac{d(ID)(TV)}{b+(ID)(TV)} \quad [3]$$

$$\Delta R_F = \frac{d(ID)(TV)}{b-(ID)(TV)} \quad [4]$$

Fig. 4 Geometrical representation of the apparent near (ΔR_N) and apparent far (ΔR_F) depth displacements in the Mach-Dvorak effect. For the right eye serving as the "lead" eye, the apparent direction of target motion, as seen from above, is counterclockwise.

displacement. As mentioned earlier, a similar target-reversal effect occurs in the case of a filter transfer from one eye to the other in the Pulfrich stereophenomenon. But the reversal in direction of apparent target rotation in the two phenomena occurs for different reasons. In the Pulfrich stereophenomenon (see Fig. 1), the motion reversal is the consequence of a reversal in the magnitude of the absolute latent period of the two eyes resulting from the filter transfer. As the target moves from left to right, the left eye, (now wearing the filter) has the greater absolute latent period so that its point of target stimulation (A) lags behind the corresponding point of target stimulation (B) for the right eye. This will produce target localization at the more distant position, P'_F, where the new lines of sight, $Z_L A$ and $Z_R B$, will now intersect. Conversely, for target movement from right to left, the filter transfer, i.e., the latency reversal, will produce target localization at the near position, P'_N, where the new lines of sight $Z_R A$ and $Z_L B$ will now intersect. Thus, with the filter placed over the left eye, the apparent target rotation will now be clockwise in direction. On the other hand, the reversal in apparent target rotation in the case of the Mach-Dvorak phenomenon (see Fig. 4) is produced by an exchange of the lines of sight through the points of target stimulation A and B, because of the modification in the lead eye-lag eye spatial relationship. Thus, for target motion from left to right, the line of sight of the left (now the lead) eye passes through point A (i.e., $N_L A$) and the line of sight of the right (now the lag) eye passes through point B (i.e., $N_R B$). The two new lines of sight will now intersect at the far depth-displacement position, P_F. Conversely, for target motion from right to left, the new line of sight of the left (lead) eye, $N_L B$, will intersect that of the right (lag) eye, $N_R A$, at the near depth-displacement position, P_N. Accordingly, the change in the lead eye-lag eye relationship will now produce an apparent clockwise target rotation. It is important to note that the target-rotation reversal in the Mach-Dvorak phenomenon does not involve any considerations dealing with the visual-latency hypothesis as is required in the case of the filter transfer in the Pulfrich effect.

The four equations which accompany Fig. 4 have been derived by simple geometric considerations involving proportional sides of similar triangles. The formulas in "2", taken from "3" and "4" in Fig. 4, specify, respectively, the predicted magnitude of the apparent near and far depth displacements out of the plane of target motion, here designated as ΔR_N and ΔR_F, rather than by the symbols C_N and C_F, used in specifying the magnitude of the depth displacements in the Pulfrich effect. Eq. (2) applies to an observer of interpupillary separation, b, who is exposed to very brief, paired target flashes which are presented with a given interocular delay, ID; also the target is presumed to be oscillating with constant linear velocity, TV, in a frontal plane located at a distance, d, from the eyes.

$$\Delta R_N = (d \times ID \times TV)/(b + (ID \times TV))$$

and (2)

$$\Delta R_F = (d \times ID \times TV)/(b - (ID \times TV))$$

As an illustration of the predicted relative magnitudes of ΔR_N and ΔR_F for different values of AB, consider the conditions of the previously discussed case where b = 6.5 cm, d = 100 cm, and TV = 16.5 cm/sec. With the aid of Eq. (2) it can be shown that for a very small value of AB (say, e.g., for AB = 0.1 cm), the computed theoretical value of the average depth displacement, $\overline{\Delta R}$ (= (ΔR_F - ΔR_N)/2) is very large (15.4 mm). This means that for a linear parallax of only 1 mm, the predicted depth difference between the far and near positions of target localization, $P_F - P_N$ in Fig. 4, is slightly over 3 cm under these specified conditions of testing.

It is generally preferred that the linear magnitude of the computed depth displacements, ΔR_N, ΔR_F, and $\overline{\Delta R}$, be converted into their equivalent angular stereoscopic parallax values, $\eta_{\Delta R_N}$, $\eta_{\Delta R_F}$ and $\eta_{\overline{\Delta R}}$, expressed in sec of arc. This is accomplished by substitution of the appropriate numerical value in each of the terms contained in Eg. (3), as discussed by GRAHAM (7):

$$n_{\Delta R_N} = (206265\,(b)/R_S)\,(\Delta R_N/R_V)$$

and (3)

$$\eta_{\Delta R_F} = (206265\,(b)/R_S)\,(\Delta R_F/R_V)$$

where R_S = d, represents the fixed distance (100 cm) of the oscillating target from the observer's eyes; R_V represents, respectively, the average distance of the apparent near or the apparent far position of the oscillating target, obtained empirically by groups of equidistance settings performed in the observer's median plane; b represents the observer's interpupillary separation; ΔR_N and ΔR_F represent, respectively, the magnitude of the apparent near (-) or the apparent far (+) depth displacement values; and 206265 is the numerical constant that converts $\eta_{\Delta R_N}$ and $\eta_{\Delta R_F}$ from radians into sec of arc when all distances are expressed in the same linear unit.

It can be readily demonstrated with the aid of Eq. (3) that for b = 6.5 cm and d = 100 cm, $\eta_{\Delta R_N}$ will equal -13.4 sec of arc and $\eta_{\Delta R_F}$ will equal +13.4 sec of arc for a depth displacement of $\pm$ 1 mm (i.e., for ΔR_N = -0.1 cm and for ΔR_F = +0.1 cm). The apparent near and far target positions in that case would be localized by the observer at 99.9 cm and 100.1 cm, respectively, as measured from the observer's eyes. In the previous illustration, where the value of AB = 1 mm was shown to give rise, theoretically, to an average linear depth-displacement value of $\overline{\Delta R}$ = 15.4 mm, the average value of the corresponding parallax angle, $\eta_{\overline{\Delta R}}$ (= $(\eta_{\Delta R_F} - \eta_{\Delta R_N})/2$), as computed by use of Eq. (3), is about 206 sec of arc, the same theoretical value as would be obtained by multiplying the average depth displacement $\overline{\Delta R}$ in mm (=15.4) by 13.4 sec of arc.

Because the magnitude of the Mach-Dvorak effect is essentially a stereoscopic and not a latency measure, the experimental values of $\overline{\Delta R}$ will generally be computed in terms of the corresponding angular stereoscopic parallax, $\eta_{\overline{\Delta R}}$ (in sec of arc), rather than in terms of the computed latency difference, Δt (in msec), as in the case of depth-displacement measures, C_N and C_F, in the Pulfrich effect.

Apparatus Implications for Stationary Targets

It should be emphasized that a moving target is not essential for obtaining depth-displacement effects in the Mach-Dvorak phenomenon, unlike the case for depth effects in the Pulfrich stereophenomenon. Binocular depth-displacement effects produced by "apparent movement" of paired stationary targets presented in stereoscopic parallax can be accomplished by spatially arranging the members of a bank of stationary lights to form many closely spaced pairs at fixed intervals along the observer's frontal plane; that is, by using a series of paired lights placed in stationary positions such as at A' and B', A and B, A" and B", etc., as depicted in Figs. 2 and 3. By use of an appropriate timing mechanism, each simultaneously presented A and B pair can be flashed in sequential order at discrete positions, progressing first from left to right and then from right to left at a rate designed to produce optimal apparent movement for each binocularly fused pair. It will, of course, be necessary in this arrangement to have one member of each of the paired lights presented to only one eye and the other member of each pair presented to only the other eye. The magnitude of the apparent near and apparent far depth displacements produced by these sequentially flashing stationary lights will depend, in accordance with the laws of binocular space perception, on the spatial separation, AB, of the paired members (i.e., on the magnitude of the linear stereoscopic parallax). In this situation, it is clear that the same depth-displacement effect will be obtained whether the spatially separated members of each flashed pair, A and B, are presented simultaneously or with slight interocular delay. If the lights at each paired position, A and B, are binocularly fused, then the depth effect predicted by the intersection of the lines of sight to each of the respective A and B positions) will depend upon only the spatial separation in the frontal plane of each flashed AB pair. The magnitude of AB and, consequently, that of the apparent depth effect will remain constant even if the temporal delay existing between one flashed pair and the succeeding ones (i.e., as the repetition cycle, RC) is varied by large amounts within the range which yields good apparent movement. It should also be obvious that the depth-displacement effects in this arrangement of stationary targets should not be affected if the retinal-illuminance levels of the members of the binocularly fused pairs are unequal. Although the binocular latency relations will be altered by unequal retinal illuminance, the conditions of illumination should not alter the effective spatial separation, AB, of the fixed, paired target lights. Accordingly, for increasing unequal binocular illumination, produced by decreasing or increasing the illumination to one eye, the points of intersection of the lines of sight through the respective points A and B will remain unaltered until binocular fusion is destroyed by suppression of vision in the dimmer eye. In that event, the depth-displacement effect will disappear and (monocular) apparent movement in a single frontal plane will be reported (i.e., $\Delta R = 0$).

Apparatus Implications for Moving Targets and Some Additional Theory

The general preference in Mach-Dvorak experiments for using moving targets which are presented to the observer in successive paired flashes by means of a binocular shutter mechanism--one member of each stimulus pair always presented only to the lead eye and the second member always presented only to the lag eye--is probably based on the relatively simple manner in which many of the governing stimulus variables can be systematically controlled. Thus, when an oscillating target whose constant velocity can be varied over a wide range of values is used in conjunction with an electromechanical shutter mechanism that can systematically vary the time delay between target onsets for the right and left eyes in each paired presentation (i.e., that can introduce controlled variations in ID values), then any desired magnitude of linear stereoscopic parallax (AB) can be produced. For more complete stimulus control, the shutter mechanism should also be capable of varying the repetition delay (RD) existing between the end of stimulation of the lag eye of each paired target presentation and the beginning of stimulation of the lead eye in the succeeding paired presentation. This capability would also allow control of the magnitude of the repetition cycle (RC) which specifies the time between the onset of stimulation to each eye in one paired presentation and the corresponding onset to each eye in the succeeding paried presentation. Also required to be specified is the inter-stimulus interval (ISI) value existing in each cycle between target offset for the lead eye and target onset for the lag eye, for the given exposure duration of the lead (right) eye (ED_1) and the lag (left) eye (ED_2). Finally, of possible theoretical importance is the total duration of binocular stimulation for the paired-target presentation, i.e., the time from the onset of stimulation in the lead eye to the offset of stimulation in the lag eye within the same cycle.

A summary of the definitions and specifications of some of the temporal variables mentioned above are presented in Fig. 5. Each specified variable is given in terms of its equivalent spatial relationship, based on an assumed constant target velocity. These variables will generally be analyzed in terms of their spatial rather than their temporal relationships.

STIMULUS SPECIFICATION: DEFINITIONS

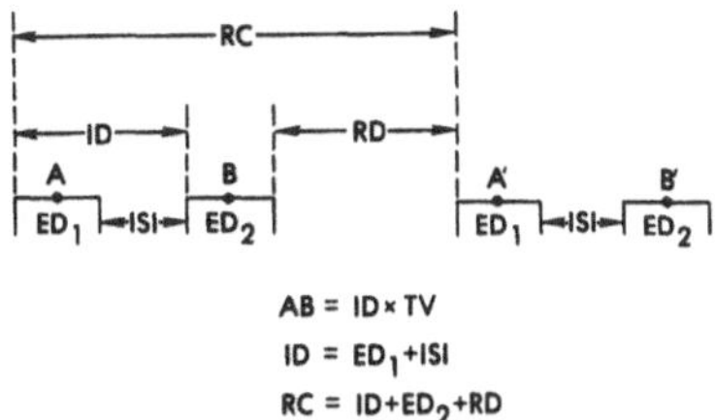

Fig. 5 Specification of the relations among the basic spatio-temporal variables in the Mach-Dvorak effect.

Three additional points should be mentioned regarding the geometrical relations presented in Fig. 5. First, for short target exposure durations of the lead and lag eyes (ED_1 and ED_2), at even the lowest target velocities used, the "points" A or B must each now be represented by a substantial spatial region of stimulus presentation whose magnitude is defined by the product of the temporal value of ED_1 or ED_2 and the target velocity(TV). That is, target positions A and B can no longer be designated as points as in the idealized cases depicted in Figs. 2, 3, and 4. The effective magnitude of the linear parallax, AB, will now depend precisely on where the spatio-temporally integrated target positions A and B are localized by the observer's visual system within these respective broad spatial regions of target exposure. For purely descriptive convenience, target localization points A and B in Fig. 5 have been arbitrarily placed in the center of their respective regions of target exposure rather than at corresponding stimulus-onset positions for the two eyes, although both sets of presumed localization positions give identical linear parallax (AB) measures. It will be fruitful to investigate how closely the magnitude of the average depth effect ($\overline{\Delta R}$ or $\eta_{\overline{\Delta R}}$) as determined empirically for both equal and unequal exposure durations to the two eyes will compare with the corresponding predicted displacement values computed by Eqs. (2) and (3), on the assumption that the effective linear parallax, AB, is measured by the distance from the point of onset of target stimulation for the lead eye to the point of onset of stimulation for the lag eye (i.e., by the distance represented by the product of ID and TV). We will designate any such discrepancy between empirical and predicted values of $\overline{\Delta R}$ or $\eta_{\overline{\Delta R}}$ as a measure of the "exposure-duration" (E-D) effect, an important non-geometric variable in the Mach-Dvorak phenomenon.

The second point to be mentioned regarding the geometrical relations in Fig. 5 is the possible effects on the magnitude of the depth displacements, $\eta_{\overline{\Delta R}}$, of variations in conditions of equal and unequal binocular retinal illuminance. As stated earlier, the level of target illumination should have little effect on $\eta_{\overline{\Delta R}}$ for a slowly moving target having very short monocular exposure durations and no effect on $\eta_{\overline{\Delta R}}$ for a horizontal row of paired stationary point targets, where each pair is presented in stereoscopic parallax as described in the previous section in timed sequence to produce "good" apparent movement. However, for a moving target having long monocular exposure durations, conditions of target or background illumination could systematically influence $\eta_{\overline{\Delta R}}$ by producing a lateral shift in the monocular positions of target localization, A and B, in a directon either toward or away from that of target motion, the exact amount and direction of shift depending, respectively, on the amount of increase or decrease in target or background illumination introduced in each of the two eyes. Theoretically, if the net effect of the monocular or binocular change in illumination is to increase the stereoscopic parallax, AB, then the observer should report seeing an increase in depth displacement; a net decrease in the linear separation between A and B should produce a reduction in perceived depth. It is probably more appropriate to state this relationship in converse form: An increase, decrease, or reversal in perceived depth (i.e., $\eta_{\overline{\Delta R}}$) implies, respectively, an increase, decrease, or reversal in the magnitude of the hypothesized linear parallax, AB. We assume that the time course of such target-localization shifts, confined within the respective monocular regions of target presentation, could be influenced by changes in total binocular-time duration between the onset of neural activity in the lead (right) eye and the offset of neural activity in the lag (left) eye. We also assume that an increase or decrease in the illumination to one eye (i.e., an increase or decrease in the level of monocular neural activity) will, within limits, systematically shift the position of target localization within the region of target presentation for that eye respectively in the direction of target motion or in the opposite direction. The longer the monocular exposure durations, and thus the wider the spatial extent of the monocular regions of target stimulation (i.e., the greater the value of ED x TV in each eye), the greater could be the shift in the positions of A and B with time. Each monocular target-localization position would have a specified time course, functioning under conditions of either overlapping or separated regions of target stimulation, i.e., for $ID < ED_1$ or for $ID \geq ED_1$, respectively, or, equivalently, for $ISI < 0$ or for $ISI > 0$. To establish whether conditions of illumination have a differential effect on the time course of the shift in A and B, for both equal and unequal exposure durations, ED_1 and ED_2, will require an extensive experimental research program. We have initiated a series of such investigations, for both equal and unequal binocular background illuminations, designed to test and quantify these assumed relationships. Some preliminary results of these studies and a more detailed analysis of the effects of illumination will be given in a later section of the report.

It should be emphasized that the latency hypothesis of the Pulfrich phenomenon does not apply when accounting for depth-displacement effects produced by unequal binocular illumination in the Mach-Dvorak phenomenon. An increase in monocular latency (by use of a filter) for, say, the lead (right) eye should not, as such, result in a shift of the region of target stimulation for that eye in a direction opposite to that of target motion and by an amount specified by the product of the magnitude of the change in the absolute visual latent period of the right eye produced by the filter and the prevailing target velocity. Such lateral shifts in target localization are correctly presumed to occur for points, lines, or even wide target surfaces when viewed under conditions of continuous motion as in the Pulfrich phenomenon and as depicted for point A in Fig. 1. In the Pulfrich phenomenon the assumption is made that the lateral "backward" shift in the point of target localization for the right eye with filter (point A in Fig. 1) would be of such magnitude (Δt x TV) as to result in synchronously arriving impulses at the visual cortex from the retinal image of points A and B. However, in the case of the Mach-Dvorak phenomenon where the target is continuously moving only in intermittently presented regions of monocular exposure for each of the two eyes, an increase in time delay produced by the filter placed over the lead (right) eye should, indeed, delay the perception of the fixed onset (or offset) position of the region of target stimulation for that eye, but the delay, as such, should not affect the spatial localization of that onset (or offset) position. Spatial displacements in target localization could occur within the fixed regions of target stimulation as an indirect result of monocular latency changes as, for example, from possible effects of variations in the total duration of binocular neural excitation for each paired target presentation. As will be discussed later in greater detail, such variations could affect the "updating" process of target localization--a process which is presumed to occur throughout the total course of binocular excitation for each paried stimulation. In any case, it does not appear meaningful to assert (see, for example, HARKER and JONES (8)) that a predictable portion of the depth displacements obtained in the Mach-Dvorak phenomenon under conditions of unequal binocular illumination consists of a depth component attributable to the Pulfrich effect. I prefer[2] to analyze the depth effects in the Mach-Dvorak phenomenon which result from variations in conditions of unequal binocular retinal illuminance without alluding to so-called "depth components produced by the Pulfrich effect."

The third point to be mentioned in regard to the geometrical relations presented in Fig. 5 deals with the effects of reducing the magnitude of RD, the repetition delay, a delay which also affects the temporal and corresponding spatial magnitude of the repetition cycle, RC. For large values of RD, the distance between the paired points of target localization, A and B, in one cycle will be smaller than the spatial separation between point B in that cycle and point A' in the succeeding cycle. If RD is progressively reduced, a condition will be reached in which the distance between points A and B will be greater than that between points B and A', i.e., AB > BA'. If the assumption is made that the target members which become paired and binocularly fused are the ones having the smaller spatial separation (pair BA' in this condition of reduced RD value), then the lines of sight of the right and left eyes should now be depicted as passing respectively through A' and B rather than through the previous points A and B. The intersection point of the appropriate lines of sight in the two eyes ($N_R A'$ and $N_L B$) for target motion from left to right should now lie beyond the plane of target motion, thus giving rise to a reversal in the direction of target rotation from a counterclockwise to clockwise motion. That is, the left eye should now become the lead eye. The experimental results confirm these predictions based on the above-mentioned assumption that the stimulus members having the smaller of the two spatial separations become paired for binocular fusion. At the value of RD where AB = BA', the detectable depth effect remains, but the observer has difficulty in stating whether the direction of apparent target motion is clockwise (if $N_R A'$ and $N_L B$ serve as the binocular lines of sight as occurs when AB > BA') or counterclockwise (if $N_R A$ and $N_L B$ are now the lines of sight as occurs when AB< BA'), as can be understood with the aid of Fig. 4. Similar modifications in the magnitude and direction of apparent target rotation can be achieved for a constant RD value by progressively increasing the magnitude of ID. Thus, as ID is increased from zero, the counterclockwise depth effect will progressively increase until ID = RD. For this condition (AB = BA'), the direction of apparent target motion, despite the large depth effect, will be reported to alternate between clockwise and counterclockwise rotation. As ID is further increased (AB> BA'), an initially large and steady clockwise rotation (i.e., a reversal in direction of target rotation) will be reported. The magnitude of the depth effect for this reversed target rotation will start to decrease and will continue to decrease as ID is still further increased until BA' = 0, when the depth effect will disappear because the lead eye and the lag eye are being simultaneously stimulated (i.e., ID = 0).

We have experimentally studied this interesting class of target reversal effects (see, LEE (9) as an example of an earlier systematic study of this situation), but I will not report our findings on this occasion.

Some Experimental Results on the Mach-Dvorak Effect

This final section of the presentation will summarize several preliminary experiments which we have performed to assess the effects of three classes of

[2] My rejection of the notion which asserts that (under varying conditions of unequal binocular illumination), the depth displacements in the Mach-Dvorak phenomenon contain depth components whose magnitudes are predictable from measurements of the Pulfrich stereophenomenon, obtained separately under the same given conditions of binocular illumination, represents a modification in the viewpoint expressed in my Symposium presentation.

stimulus variables previously discussed: 1) the geometrical effects of varying the magnitude of the linear stereoscopic parallax (AB) by using various ID-TV stimulus combinations; 2) the effects of varying both the monocular and the binocular exposure durations, ED_L and ED_R; and 3) the effects of equal and unequal binocular variations in retinal-illuminance levels, E_L and E_R.

Apparatus and Procedure

The basic apparatus[3] we used in obtaining data on the Mach-Dvorak effect is the same (see Fig. 6) as the one used in our experiments on stereoscopic acuity (see, e.g., LIT (10) and (5); LIT, FINN and VICARS (11)) and on the Pulfrich effect (see e.g., LIT and HYMAN (4); LIT (12) and (13)). Three additional components had to be

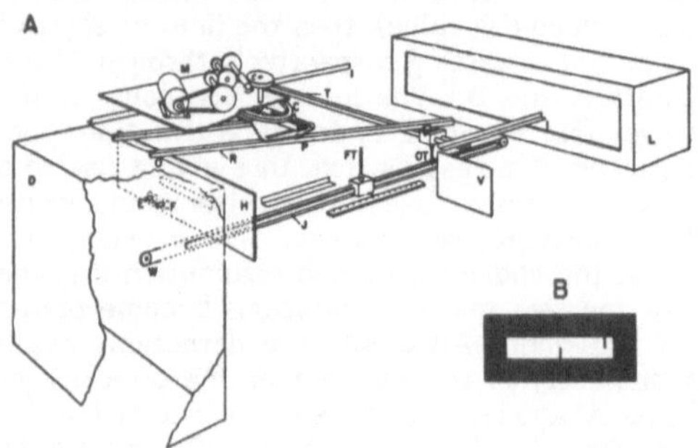

Fig. 6 A. Schematic representation of the basic apparatus used for measuring depth displacements of the oscillating target in the Pulfrich stereophenomenon and in the Mach-Dvorak effect. B. View of the upper oscillating (standard) target and the lower fixation (comparison) target as seen by the observer against the uniform white background produced by the light box.

added to the basic device: 1) a guillotine shutter mechanism in front of each eye to vary stimulus exposure duration, 2) an electronic timing system for independently controlling each of the spatio-temporal variables involved in the Mach-Dvorak effect, and 3) a fixation device for preventing eye movements between stimulus presentations.

The oscillating (standard) target (OT) is a vertical, black metal rod, 5.79 mm (0.33°) in diameter. It is located in the upper half of the observer's visual field at an observation distance (d) of 100 cm. It can be made to oscillate in the frontal plane by means of a motor driven (M), cam-regulated mechanism (C) which produces reciprocating linear velocity over a wide range of values, with the central 90% of stroke at constant speed. Unless otherwise specified, the linear target velocity (TV) is kept constant at 16.5 cm/sec (9.37 deg/sec) at the fixed observation distance (d = 100 cm) in each of the experiments to be reported here. The fixation (comparison) target (FT) is identical with the oscillating target and is located in the lower half of the observer's visual field. It can be moved by the observer along his median plane in a direction either towards or away from his eyes by means of a pulley wheel (W) located in the dark room. The distance of the adjusted comparison target from the observer's eyes can be estimated to within 0.1 mm by the experimenter with the aid of a vernier scale. The bottom end of the upper oscillating target and the upper end of the lower comparison target lie in the observer's horizontal plane of fixation, as shown in (B) of Fig. 6.

[3]The basic apparatus was originally constructed at Pupin Laboratories, Columbia University, partially through funds from a research grant-in-aid provided by the American Academy of Optometry. The three additional components needed for measuring the Mach-Dvorak effect were constructed at Southern Illinois University by Charles Popp, through funds provided by the Graduate School.

The observer is seated in a dark room (D) and binocularly views the targets through a pair of circular artificial pupils (E) that are 2.5 mm in diameter and adjustable for interpupillary separation. A mouthbite containing the observer's dental impression is used together with a head rest to keep the eyes aligned with the artificial pupils. Uniform background illumination is provided by a light box (L). Neutral density filters, placed in the pair of filter boxes (F) located on the outside dark room wall, control the level of background illumination presented separately to each eye. The constant horizontal rectangular field of view (21.6° x 4.2°) is established by means of horizontal (H) and vertical (V) screening units.

In performing depth-displacement (i.e., equidistance) settings in the Mach-Dvorak effect, the observer continuously fixates the upper end of the comparison target and adjusts this target in his median plane first by moving it away from his eyes until it appears to be located directly below the apparent near path of the upper oscillating target and then directly below the apparent far path; the apparent far and near paths are again located by now moving the fixation target towards his eyes. Usually six pairs of equidistance settings are made during an experimental session under each given value of the variable under investigation: six settings to locate the apparent near depth-displacement position and six settings to locate the apparent far depth-displacement position. Typically from three to five replications are held for each experimental session, scheduled in counterbalanced order.

The electronic timing system consists of a train of four, one-shot multivibrators whose periods are continuously variable. Each multivibrator independently controls one of the four temporal variables within a single cycle of target presentation that are involved in the Mach-Dvorak effect: 1) the exposure duration of the lead eye (ED_1); 2) the exposure duration of the lag eye (ED_2); 3) the interocular delay (ID) between the onset of stimulation of the lead eye and that of the lag eye; 4) the repetition delay (RD) between the offset of stimulation of the lag eye in one cycle and the onset of stimulation of the lead eye in the succeeding cycle. Both the repetition cycle, RC (i.e., the time between the onsets of stimulation of the lead eye in successive cycles), and the inter-stimulus interval, ISI (i.e., the time between the offset of stimulation of the lead eye and the onset of stimulation of the lag eye in a given cycle), become fixed in magnitude when the time-period values are preset in the dials which control each of the four multivibrators.

A clearer understanding of the timing sequence may be obtained with the aid of Fig. 7. At the termination of the preset time period of RD from the previous cycle, the onset of both ED_1 and ID occur simultaneously at the start of cycle 1. The durations of ED_1 and ID are determined by the preset values in dials T_1 and D_2, respectively. At the end of the preset ID period, the onset of ED_2 occurs, and ED_2 remains active for the preset period of dial T_2. The offset of ED_2, in turn, serves as the onset signal for RD, for the time period preset in dial D_1. The termination of RD in cycle 1 initiates the timing sequence in cycle 2, a sequence which duplicates that which prevailed in cycle 1. The device can produce specified time periods for each of the four variables ranging from a few milliseconds to several seconds.

The exposure duration for each eye is mechanically accomplished by means of a guillotine shutter which is mounted in the housing of each artificial pupil. A lever system connecting each shutter to a stepping motor is used to drive the shutters. The exposure duration of each eye is thus determined by the action of its one-shot multivibrator. The onset of each multivibrator steps the motor in one direction to open the shutter, and the offset of each multivibrator steps the motor in the reverse direction to close the shutter.

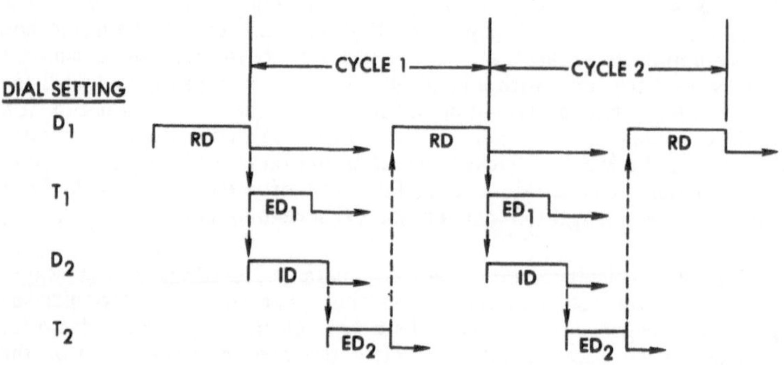

Fig. 7 Schematic representation of the stimulus timing sequence for measuring depth displacements in the Mach-Dvorak effect. The repetitive timing is accomplished by four separate one-shot multivibrators and their associated preset dial-setting timers.

During the time interval between stimulus presentations, the observer is provided with a pair of dim red fixation lights which are physically located behind the head, one for each eye. By means of an inclined side mirror for each eye, combined with a beam splitter in the housing of each artificial pupil, the fixation lights are reflected into the observer's eyes and fused to form a single luminous point which appears to be located in depth at the upper end of the fixation target at a distance of 100 cm. When the shutters are open, the binocularly fused dim fixation points become invisible against the prevailing illuminated background. Thus, binocular fixation is kept nearly constant both during and between stimulus exposures.

Effects of Covarying Interocular Delay and Target Velocity

As mentioned earlier, the magnitude of the linear stereoscopic parallax (AB) can be systematically controlled either by varying ID for a fixed TV value, or by varying TV for a fixed ID value. Specific, constant values of AB can be produced by using a variety of equivalent ID-TV combinations, the number of AB values available depending on the limitations of the apparatus. When the exposure duration of each eye is extremely short, so that the positions of target stimulation, A and B, involve essentially a pair of points rather than a pair of spatial regions, the magnitude of AB (= ID x TV) can be geometrically specified with relatively great precision. Accordingly, the empirically determined depth effect, $\eta_{\overline{\Delta R}}$, for any given AB value, should show very good agreement with the corresponding theoretical $\eta_{\overline{\Delta R}}$ value as computed with the aid of Eqs. (2) and (3).

Experiments of this type have been initiated in our laboratory using a wide range of ID values in combination with both constant and varying TV values. The confirming experimental results which were contained in JOEL BRAUNER'S Dissertation on the effects of varying ID, for a fixed TV value at each of six photopic background illumination levels, have been reported (BRAUNER and LIT (14)) and are being submitted for publication, so the details of that study will not be discussed here. Instead, the results of more recent confirming studies in this laboratory on the effects of varying ID for a fixed TV value (16.502 cm/sec = 9.4° per sec) are shown in Fig. 8. The data were collected at an observation distance, d =

100 cm, for a background illumination of about 1.5 log td. The exposure duration of each eye was kept constant at 10 msec. To help clarify the depicted stimulus configuration in this and all later experiments, the parallel lines drawn within each region of target stimulation are slighlty slanted in the appropriate direction to indicate which eye is being exposed to the given member of the paired target presentations. The two observers (CJ and KB) had the same interpupillary separation, b = 6.15 cm. The ID values used ranged from 5 to 75 msec. However,

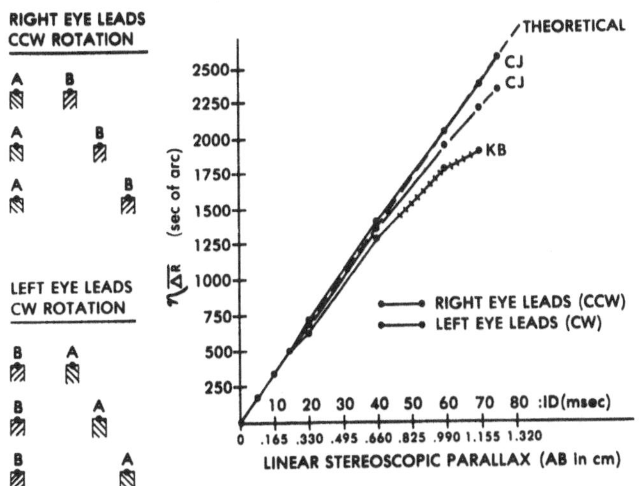

Fig. 8 The average angular magnitude of the depth displacements, $\eta_{\overline{\Delta R}}$, in the Mach-Dvorak effect plotted as a function of the prevailing linear stereoscopic parallax, AB (=ID x TV). Target velocity held constant at 16.502 cm/sec; also ID values are indicated on the abscissa axis. Data representing conditions in which the right eye leads and in which the left eye leads are presented separately. The hatched segment of the curve for observer KB represents ID values under which binocular fusion became impaired.

observer KB reported having problems in maintaining binocular fusion at ID values of 60 to 70 msec; he was not able to make any depth settings at the highest (75 msec) ID value. The portion of his curve in Fig. 8 which presents the $\eta_{\overline{\Delta R}}$ data obtained at the two large ID values under which binocular fusion was disrupted is accordingly depicted by hatchmarks. Data were collected on observer KB under conditions of increasing ID values, but only when the right eye served as the lead eye. Two sets of data were collected for observer CJ: one set in which the right eye served as the lead eye and the other set in which the left eye was the lead eye. As predicted from an analysis of the drawings on the left side of the graph in Fig. 8, in which A lies in the region of target stimulation of the right eye and B lies in that of the left eye (for left-to-right target motion), a counterclockwise (CCW) apparent target rotation occurs when the right eye is the lead eye and a clockwise (CW) apparent rotation occurs when the left eye is the lead eye. The obtained experimental results in Fig. 8 for the computed values of n_{R} are plotted for each observer as a function of AB (= ID x TV) for each of the indicated ID values. The dashed line without data points represents the locus of the predicted $\eta_{\overline{\Delta R}}$ values for all AB conditions used, as computed by Eqs. (2) and (3). The data on CCW target rotation obtained from observer CJ falls almost precisely on the theoretical line. The corresponding data obtained from observer KB deviate only slightly below the theoretical line for small AB values, but the deviations become quite marked when binocular fusion is impaired, as indicated by the large downward deviations for the hatchmarked segment of his curve. The data on CW target rotation for observer CJ do not fall as precisely on the theoretical line as do her data on CCW target rotation. Additional studies are planned to identify specific measures of the quality of binocular vision (e.g., phorias, ductions, eye dominance, etc.) that may correlate with discrepancies between the empirical and the theoretical magnitudes of the depth displacements, $\eta_{\overline{\Delta R}}$. The results of such planned investigation should have important clinical implications for assessing and training binocular depth discrimination.

A second class of experiments which we have recently initiated is designed to test the validity of the hypothesis that linear stereoscopic parallax, AB (=ID x TV), is the underlying basis of the depth-displacement effects in the Mach-Dvorak phenomenon for very short exposure durations. In these experiments, data were collected on the effects of systematic variations in target velocity (with ID serving as parameter) and of systematic variations in ID (with TV serving as parameter). Fig. 9 gives partial results obtained from three additional observers for the indicated ID-TV combinations which produce seven constant AB values ranging from 0.00 to

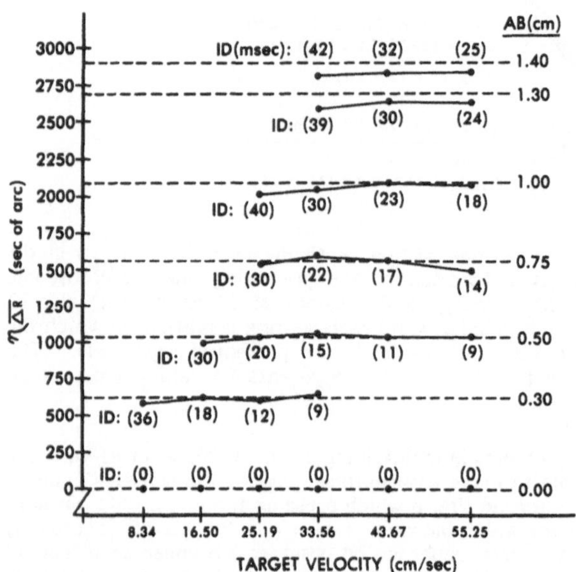

Fig. 9 The average angular magnitude of the depth displacements, $\eta_{\overline{\Delta R}}$, in the Mach-Dvorak effect plotted as a function of target velocity, with linear stereoscopic parallax, AB, serving as parameter. The dashed horizontal lines give the predicted $\eta_{\overline{\Delta R}}$ values for the seven indicated linear parallax conditions. The prevailing value of ID(in msec) is indicated for each data point.

1.40 cm. The exposure duraton of each eye was held constant at 20 msec. The predicted $\eta_{\overline{\Delta R}}$ values are indicated by the dashed horizontal lines. The goodness of fit, except for the two highest AB values, is remarkably satisfactory, considering that the data for each point on a given AB line were obtained on different days. The data for $\eta_{\overline{\Delta R}}$ at the two highest AB values are constant in magnitude as predicted, but the empirical data fall on lines which are each displaced slightly below their respective theoretical lines. Additional experiments are being planned to assess the effects of variations in TV on $\eta_{\overline{AB}}$ under conditions of binocular target exposure duration (ED_L and ED_R) in which the magnitude of the region of target presentation for each eye will be kept constant under each of the varying target velocities for a given AB condition. For the data of Fig. 9 in which ED_L and ED_R were kept constant at 20 msec, the respective regions of target stimulation in each eye increased proportionately with the magnitude of TV. The positions of target localization (A and B), and hence the depth-displacement effects can be systematically influenced by such variations in the size of the regions of target stimulation. This would hold particularly for large monocular regions of target stimulation as are produced either by high TV values or by large ED_L or ED_R values for a fixed TV. The systematic depth-displacement effects obtained by increasing the regions of target stimulation by varying ED_L and/or ED_R for a constant TV, will be demonstrated in the following section which discusses some of our experiments that were specifically initiated to assess the so-called "exposure duration" (E-D) effect.

Effects of Equal and Unequal Binocular Target Exposure Duration

Although target exposure duration is not a geometric variable in the Mach-Dvorak phenomenon, depth-displacement changes produced by variations in exposure duration have been of sufficient magnitude (see, e.g., HARKER (15)) to justify conducting a systematic experimental analysis of this important variable. Fig. 10 presents some preliminary results on observer CJ which we have recently obtained on monocular and binocular E-D effects under conditions of simultaneous binocular onsets (i.e., for ID = O). Target background illumination was kept constant at 1.5 log td. Three separate cases are depicted in Fig. 10: (i) ED_R remains constant at 10 msec, while ED_L is progressively increased from 10 to 75 msec; (ii) ED_L remains constant at 10 msec, while ED_R is progressively increased from 10 to 75 msec; and (iii) ED_L and ED_R are equal and both are progressively increased from 10 to 75 msec.

In the two cases of monocular change, cases (i) and (ii), a small, constant exposure duration (10 msec) was respectively selected for ED_R and ED_L in order to ensure that any obtained E-D effect could be safely ascribed primarily to shifts in target localization which occur in only the opposite eye, the one subjected to the progressively increasing exposure durations. The assumption is made that for the two monocular cases, the respective positions of the points of target localization, A for the lead (right) eye and B for the lag (left) eye, will remain relatively unchanged because of the small constant ED value selected. In case (iii), the exposure durations for both eyes are increased equally to test whether or not the points of target localization for each eye shift equally in the same direction within their respective regions of target stimulation.

If increases in exposure duration were to have no effect on depth displacements, then the results for all three cases depicted in Fig. 10 should be identical: No observer will detect any depth effect, since ID = O. The actual results show that for case (iii), the points of target localization, A and B, for the right and left eyes, respectively, fall on correspondingly displaced positions in the direction of target motion within their regions of target stimulation, including the target-onset positions. Thus, in the stimulus configuration above the graph in Fig. 10, the depicted magnitude of the linear stereoscopic parallax, AB, is zero in case (iii) for all values of exposure duration. For convenience, the midpoints of the regions of target stimulation have been selected to represent the shift in the positions of target localization from the corresponding target-onset positions in the two

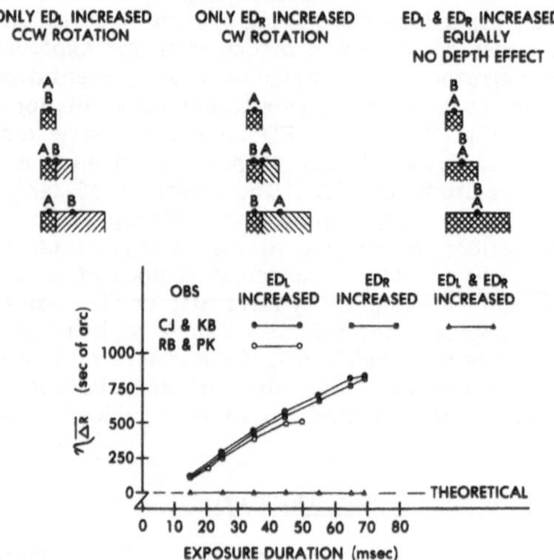

Fig. 10 The exposure-duration (E-D) effect in the Mach-Dvorak phenomenon. The average angular depth displacement, $\eta_{\overline{\Delta R}}$, is plotted as a function of monocular and binocular variations in exposure duration, ED_L and ED_R, for the condition of zero interocular delay. For ID = 0, both the equal and unequal conditions of binocular exposure duration produce displaced monocular target localizations, A and/or B, in the direction of target motion for the eye(s) undergoing the increasing target exposure duration.

completely spatially-overlapping regions. The data in Fig. 10 for the two cases of increasing monocular exposure duration, cases (i) and (ii), show that a systematic (E-D effect is obtained: Increasing the magnitude of ED_L produces an apparent counterclockwise target rotation of progressively increasing amplitude, whereas, increasing ED_R produces an apparent clockwise target rotation, but of similar progressively increasing amplitude. (The combined data for RB and PK were taken from J. BRAUNER'S Dissertation for a background illumination of 0.97 log td.) The reversal in direction of target rotation for the two monocular cases can be understood with the aid of the stimulus configurations depicted above the graph in Fig. 10, for the condition of target motion from left to right.

It should be emphasized that the results for all three cases depicted in Fig. 10 are qualitatively consistent with an assumed spatio-temporal principle of directional localization which is proposed to predict the changes in the monocular localization of a target that is moving in a frontal plane under conditions of monocular or binocular increasing ED values: The monocular point(s) of target localization (either A or B or both A and B) are laterally displaced in the plane of target motion in the direction of target motion. The positions of each of the spatio-temporal integrated points of target localization, A and B, as indicated in the stimulus configurations drawn above the graph in Fig. 10, have been depicted in accordance with the predictions implied by the above-stated directional-localization principle for moving targets. It should also be emphasized that the indicated direction of target localization for the E-D effect in the Mach-Dvorak phenomenon is consistent with the notion that the spatio-temporal, neurally integrated activity in each eye, which serves to specify the hypothesized points of target localization (A and B) during a total cycle of binocular target stimulation, progressively "decays" with time, starting at the onset portion of the respective retinal regions of target stimulation and progressively moving in the direction of target motion toward the offset portions of the respective retinal regions of target stimulation. The resulting effect, in each eye, is then a continually shifting point of target localization, as a function of time, always moving (within limits) in the region of target stimulation in the direction of target motion.

A neural mechanism consistent with the spatio-temporal integration of the type proposed here can account not only for the specified direction of the E-D effect in the Mach-Dvorak phenomenon, but possibly also for the monocular, inhibitory-spatial effect produced in the vicinity of the onset portion of the "growing" retinal image formed by a moving target in the so-called Fröhlich effect. FRÖHLICH (16) observed that when a moving vertical slit of light passes behind a horizontal opening in a screen, it does not seem to come into view at the entrance edge, at which point the eye remains fixated, but rather, is first perceived a short distance within the border of the screen aperture. He attributed this target displacement to the visual latent period of the eye (die Empfindungszeit). It may, however, be more fruitful to analyze the monocular lateral displacements in the Fröhlich effect in terms of the notions proposed above which for this case would hypothesize a very rapid initial decay in neural activity involved in directional localization near the onset position of the region of monocular target stimulation. The task still remains, of course, either to develop a theoretical model which would allow a quantitative assessment to be made of the E-D effect in the Mach-Dvorak and other related spatio-temporal phenomena or to derive an empirical equation based on an extensive experimental program designed to establish the quantitative relationship between the magnitude of the depth displacements ($\eta_{\overline{\Delta R}}$) and the prevailing values of the binocular target exposure durations.

Figure 11 provides some additional exploratory data on the E-D effect obtained on one observer (CJ) for both binocular and monocular variations in exposure duration. The experimental design is essentially similar to that used to obtain the experimental data presented in Fig. 10, except that the ID values systematically varied from 15 to 75 msec in each of the three cases. The depth-displacement measures, $\eta_{\overline{\Delta R}}$, were obtained not only for a repetition delay (RD) of 160 msec, as in the case of the data for Fig. 10, but also for a higher (double) RD value of 320 msec in order to assess the possible complicating depth effects which arise under testing conditions involving large exposure durations in which the spatial separation between the onset of stimulation for the lead eye and the offset of stimulation for the lag eye in a given cycle approaches the value of RD as expressed in spatial terms (i.e., RD x TV). This possible complicating depth effect for large exposure durations is presumed to be analogous to the complicating depth effect previously discussed for short exposure durations under the critical condition in which ID approaches RD. It should be noted with the aid of the stimulus configurations depicted above the graph in Fig. 11 that both for the case of binocular exposure-duration variations and that for the monocular case in which only ED_R for the lead eye was increased, the ISI value was kept constant at 5 msec. The dashed line in Fig. 11 gives the theoretical value of $\eta_{\overline{\Delta R}}$ for each ID value used, based on computations performed with the aid of Eqs. (2) and (3). As previously shown in Fig. 8, all of the control data obtained on observer CJ for $ED_L = ED_R = 10$ msec fell very close to this theoretical line.

Consider first the uppermost pair of curves in Fig. 11 for the monocular case in which the exposure duration for only the lag (left) eye is increased. The $\eta_{\overline{\Delta R}}$ data for both the 320 and 160 RD curves lie above the theoretical line indicating, as

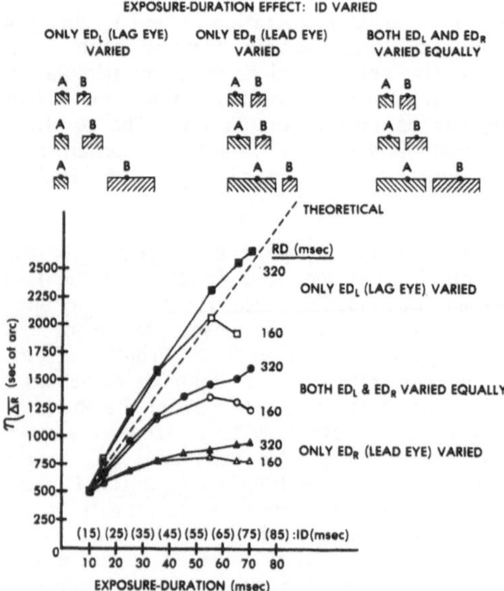

Fig. 11 Schematic representation of the predicted exposure-duration (E-D) effect in the Mach-Dvorak phenomenon for both equal and unequal binocular exposure durations.

expected, the operation of an E-D effect for the left eye, as depicted in the configuration above the graph. Depth settings obtained at RD = 160 msec were reported by CJ as being very difficult to make at the two highest ED_L levels used; only those obtained for ED_L = 75 msec were omitted in Fig. 11 because of their particularly high variability. Also, as expected, an E-D effect occurs in the lowest pair of curves in Fig. 11 for the lead (right) eye in the monocular case where the exposure duration for only that eye (ED_R) is progressively increased. In this monocular case, the E-D effect for the right eye should shift the target localization point A in the direction of target motion (while the corresponding point B for the left eye remains fixed because of the short exposure duration, i.e., small region of target stimulation, for that eye), thus producing a progressive decrease in the linear parallax, AB. The rapid leveling-off of the experimental curves in this monocular case implies that the distance of A from the changing offset position in the progressively increasing regions of target stimulation for the right eye remains constant as ED_R is increased under this testing condition in which ISI is held constant at 5 msec. An expected E-D effect also occurs in the binocular case (the middle pair of curves in Fig. 11) where the exposure durations for both eyes are equally increased, as depicted in the configuration shown above the graph in Fig. 11. The linear parallax, AB, in the binocular case for each ID value is always greater than the corresponding AB value obtained in the comparable monocular case in which the small, constant exposure duration of the lag (left) eye (ED_L = 10 msec) could not allow any appreciable E-D effect to become manifest for that eye. It is important also to account for the fact that, unlike the findings for the binocular case in Fig. 10 for ID = 0, the data for the binocular case in Fig. 11 imply that A and B are not equally displaced in the direction of target motion for ID > 0. The binocular data in Fig. 11 imply that in addition to the equal shift of both A and B in the direction of target motion, an additional progressive shift of A for the lead (right) eye occurs in that same direction to decrease the linear parallax, AB, that is, to increase the discrepancy between the higher theoretical values of $\eta_{\overline{\Delta R}}$ and their corresponding lower empirical values. This implication is in complete qualitative agreement with our earlier proposed notion that an increase in the total time of binocular neural excitation should, within limits, shift the position of target localization for the lead eye in the direction of target motion.

Finally, it is to be observed for all three cases in Fig. 11 that RD has a consistent effect of increasing $\eta_{\overline{\Delta R}}$, particularly at the critical condition where the spatial separation between the onset of stimulation for the lead eye and the offset of stimulation for the lag eye approaches the value of RD, expressed in spatial terms. The increased $\eta_{\overline{\Delta R}}$ values result in extending the linearity of the experimental curves for higher ED values. A systematic investigation of the effects on $\eta_{\overline{\Delta R}}$ of increasing the values of RD will be undertaken to establish the range of stimulus conditions under which RD exerts no influence on the magnitude of the depth displacements. Clearly, such information regarding RD is critical in designing experiments attempting to quantify the E-D effect when other basic variables in the Mach-Dvorak phenomenon are being investigated.

The exploratory data in Fig. 12 were collected on observer CJ to assess E-D effects occurring under conditions of "overlapping" binocular target stimulation. Thus, as depicted in the configuration accompanying the graph in Fig. 12, the exposure duration of the lag (left) eye was kept constant (ED_L = 10 msec) while that

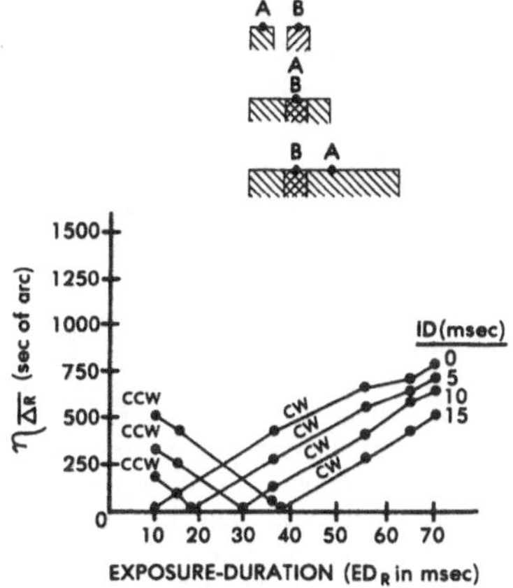

Fig. 12 Measures of the exposure-duration (E-D) effect in the Mach-Dvorak phenomenon. $\eta_{\overline{\Delta R}}$ is plotted as a function of equal and unequal binocular exposure durations for varying magnitudes of ID. The depth-displacement data ($\eta_{\overline{\Delta R}}$) were collected at each of two ID values: 160 msec and 320 msec.

of the lead (right) eye was progressively increased from 10 to 70 msec. Depth-displacement settings for the series of ED_R conditions were obtained under each of four ID values: 0, 5, 10, and 15 msec. The experimental data are plotted to show how $\eta_{\overline{\Delta R}}$ varies as a function of the exposure duration of the lead eye (ED_R), for each of the four indicated ID values serving as parameter. The results for all four ID conditions show that $\eta_{\overline{\Delta R}}$ progressively decreases in linear fashion as ED_R is increased until the counterclockwise apparent target motion disappears (zero depth effect); $\eta_{\overline{\Delta R}}$ then increases linearly to produce clockwise apparent target motion of increasing amplitude as ED_R is further increased. It is noted that the curve segments for the four ID conditions are parallel. The larger the prevailing ID value, the larger the corresponding ED_R value at which the reversal in the direction of apparent target motion first occurs. These results are in good qualitative agreement with predictions based on the presumed functional characteristics of the visual mechanism underlying the E-D effect as described in earlier sections of the paper and with the experimental results on the E-D effect obtained in the other two experiments described in this section of the paper. The experimental results obtained for the condition of target "overlap", as well as the results of the other two experiments on the E-D effect in which $\eta_{\overline{\Delta R}}$ continues to change systematically as the exposure duration of the lag eye is progressively increased, strongly support the conclusion that the process of target localization which serves to specify the integrated directional positions A and B within their respective regions of target stimulation does not terminate until the offset of stimulation, or more properly stated, until the offset of neural activity occurs for the lag eye. Clearly, considerably more experimental data, collected under a wide variety of conditions of monocular and binocular exposure duration and of total time of binocular neural excitation, are required to yield an equation (or possibly a set of equations) that adequately describes the quantitative relationships involved in the E-D effect at very many ID values and conditions of binocular retinal illuminance.

Effects of Equal and Unequal Binocular Retinal Illuminance

Although it can be readily demonstrated that conditions of illumination systematically affect the magnitude of $\eta_{\overline{\Delta R}}$, it is conceptually inappropriate for reasons discussed in an earlier section of the paper, to assert that a component of the depth effect in the Mach-Dvorak phenomenon obtained under conditions of unequal binocular illumination is attributable to latency effects in the Pulfrich phenomenon and that the magnitude of this depth component can be predicted by the geometry of the Pulfrich stereophenomenon as specified by the equations in Fig. 1. Differences in binocular illumination could, indeed, produce shifts in monocular target localization for points A and B within their resepctive regions of target stimulation because of at least two presumably interrelated processes of target-localization shifts: 1) A systematic shift in target localization in a given cycle for the same eye(s) as that whose illumination is being varied, and 2) An additional systematic shift in target localization for only the lead (right) eye as a result of variations in the total time of binocular neural excitation produced by the effect of the change in level of illumination on the monocular latent period for each eye during a single cycle. As previously indicated in the report, the changes in target localization resulting from these two interrelated effects can be specified as follows: 1) Any monocular increase in illumination (i.e., increase in level of neural excitation) in a cycle of target stimulation should result, within limits, in a shift of target localization for that eye in the direction of target motion; a decrease in monocular illumination should result in a shift, within limits, in a direction opposite to that of target motion for that eye. To account for the difference in the direction of target-localization shift between incremental and decremental illumination changes we make the further assumption that the higher the level of neural activity, the more rapidly does the neural "decay" occur across the retinal region of target stimulation in the direction of target-image motion. Hence, the neural contribution of the onset retinal segment of the moving image to the specification of the integrated, localized position of the target should be progressively less effective under conditions of increasing illumination so that the point of target localization, A or B, will be progressively shifted, within limits, in the direction of target motion, thereby reflecting the greater contribution to directional target localization made by the neural activity occurring in the offset region of the monocular retinal stimulation. Contrariwise, a reduction in level of illumination should decrease the rate of the monocular neural decay so as to produce a "backward" shift in target localization in the opposite direction as that of target motion for that eye. Thus, in terms of the effects of monocular illumination changes in the case of non-overlapping binocular regions of target stimulation, an increase in the illumination in the lag (left) eye or decrease in the illumination in the lead (right) eye should result in an increase in the prevailing linear parallax, AB, and hence, an increase in the magnitude of the CCW depth effect as measured by $\eta_{\overline{\Delta R}}$; conversely, a decrease in $\eta_{\overline{\Delta R}}$ should occur if the illumination in the lag (left) eye is decreased or if the illumination in the lead (right) eye is increased. In the case of the overlapping binocular regions of target stimulation, it should be possible to produce a reversal in apparent direction of target motion by means of relatively small changes in monocular illumination. 2) Any increase in illumination (i.e., decrease in visual latent period) in the lag (left) eye, in the case of non-overlapping binocular regions of target stimulation, should reduce the total time of binocular neural excitation; a decrease in illumination in the lag eye should increase the total binocular neural excitation time. Conversely, increasing or decreasing the illumination in the lead (right) eye should respectively increase or decrease the total binocular neural excitation time. As previously stated in the report, the assumption is made that the longer the duration of the total binocular neural excitation period, the later does the "updating" process or stage of target localization occur for the lead eye. Accordingly, in terms of the monocular latency effect, the point of target localization (A) for the lead (right) eye should be progressively shifted, within limits, in the direction of target motion, the greater the increase in the total time of binocular neural excitation, produced either by increasing the illumination in the lead (right) eye or by decreasing the illumination in the lag (left) eye. Thus, in terms of the effects of changes in the total duration of binocular neural excitation, an increase or decrease in depth effect ($\eta_{\overline{\Delta R}}$) should occur depending on whether the change in monocular latency increases or decreases the prevailing linear parallax, AB. In any set of experimental conditions involving unequal binocular changes in illumination, the resulting depth displacement, $\eta_{\overline{\Delta R}}$, should reflect the combined influence on target localization, of both forms of illumination effects, as each is analyzed in terms of its modification of the prevailing linear parallax, AB.

In concluding this section of the report, I would like to present some preliminary results for equal and unequal binocular background illumination on depth settings obtained to assess the E-D effect for monocular variations in exposure duration under simultaneous target onsets, i.e., for ID = 0.

Figure 13 presents some preliminary data obtained on observer CJ under the two conditions of stimulation depicted in the configurations above the graphs. In both cases, the exposure duration for only the left eye was progressively increased (from 10 to 80 msec) while that for the simultaneously presented right eye was kept constant at 10 msec. In one series, as indicated in the left-most section of Fig. 13, a comparison is made between depth settings obtained under equal binocular illumination ($E_L = E_R$ = 1.00 log td) and those obtained when the illumination level of the left eye was increased to 3.00 log td. The right-most section of Fig. 13 deals with the corresponding series which compares the obtained depth settings when the binocular illumination level was equal (but now at a higher level, 3.00 log td) to those when the illumination of the left eye was decreased to 1.00 log td. The configuration above each graph, as usual, represents the condition of target motion from left to right.

EXPOSURE-DURATION EFFECT (ED_L VARIED) WITH EQUAL AND UNEQUAL BINOCULAR VARIATIONS IN ILLUMINATION FOR ID=0

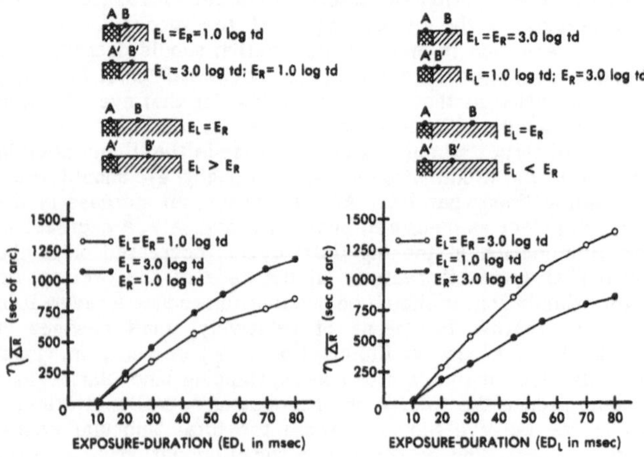

Fig. 13 Effects of equal and unequal binocular background illumination on the magnitude of the E-D effect under conditions of simultaneous target onsets, i.e., for ID = 0. The monocular increase and the binocular decrease in illumination were introduced in the same eye (left) as the one undergoing the progressive increase in exposure duration, ED_L. These conditions of testing, as designed, produce apparent target rotation in a counterclockwise direction.

The results show, as can be seen with the aid of the configurations depicted above each graph, that monocular changes in illumination systematically influence the magnitude of the depth displacements ($\eta_{\overline{\Delta R}}$) obtained under conditions in which ED effects were produced by monocularly increasing the exposure duration for the left eye, resulting in apparent CCW target rotation: Increasing the illumination in the eye whose exposure duration is being increased enhances the magnitude of the depth effect, $\eta_{\overline{\Delta R}}$; decreasing the illumination in that eye reduces the depth effect. That is, an increase in E_L is presumed to shift the point of target localization B for the left eye in the direction of target motion to B', as previously discussed; the shift from A to A' for the right eye is presumed to be minimal because of the small (1.65 mm) region of target stimulation produced by the short flash duration for that eye. The predicted overall effect is thus to increase the linear parallax, AB, at each ED_L value. Conversely, a decrease in E_L presumably shifts B, as previously discussed, in the direction opposite to that of target motion to B', thus resulting in a reduction of the linear parallax, AB, at each ED_L value.

It is to be noted that for the two curves representing equal binocular illumination, the value of $\eta_{\overline{\Delta R}}$ obtained at the higher illumination level for any given ED_L is consistently larger than the $\eta_{\overline{\Delta R}}$ value obtained at the lower illumination level. This illumination effect confirms the expectations previously discussed that binocular level of illumination should systematically influence the magnitude of the depth effects when the constant exposure duration for one eye (ED_R) is presumed to be too short to produce a shift in target localization (A) for that eye despite any variation in illumination level. In that event, an increase in binocular illumination should shift only B in the direction of target motion to produce an increase in linear parallax, AB. Unaccountably, level of equal binocular illumination was not found to have any systematic influence on the E-D effect for the two observers in JOEL

BRAUNER'S Dissertation (BRAUNER and LIT (14)).

The preliminary data presented in Fig. 14 on the same observer (CJ) represent the corresponding results for the condition in which the exposure duration for the right eye was progressively increased from 10 to 80 msec, while that for the

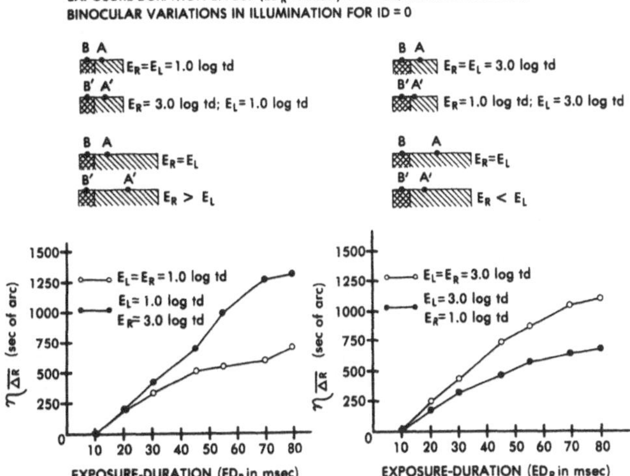

Fig. 14 Effects of equal and unequal binocular background illumination on the magnitude of the E-D effect, as in the case of the data in Fig. 13, except that the increase and decrease in illumination were introduced in the right eye, the one undergoing the increase in exposure duration, ED_R. These conditions of testing, as expected, produce an apparent clockwise target rotation.

simultaneously presented left eye was held constant at 10 msec. The left-most section allows a comparison to be made between depth settings under conditions of equal binocular illumination and those obtained when the illumination level in the right eye was increased. The right-most section allows a similar comparison when the illumination in the right eye was decreased. For both cases in Fig. 14, a clockwise apparent direction of target rotation is produced, as expected.

The results in Fig. 14 on apparent CW target rotation confirm the findings in the previous figure on apparent CCW target rotation. The results in both figures demonstrate the predicted systematic illumination effect on target localization for an eye whose exposure duration is sufficiently long to produce a large region of target stimulation within which a substantial shift in target localization could occur for that eye. Variations in illumination are assumed to have a negligible effect for the eye having the small, constant exposure duration. As in the previous case, the results show that increasing the illumination in both eyes, or only in the one whose exposure duration is being increased, produces an increase in the depth effect; reducing the illumination only in that eye, or in both eyes, decreases the depth effect.

Because of the short exposure duration (10 msec) which always prevailed for one of the eyes in these experiments on illumination effects, any shift in target localization for that eye, resulting from illumination changes that alter the total period of binocular neural excitation, cannot be adequately assessed. That is, shifts in target localization for that eye (from A to A' in Fig. 13 or from B to B' in Fig. 14) must be restricted to positions lying within only very narrow regions of target stimulation. Different experimental designs will be required to separate the magnitude of the direct target-localization shifts produced by variations in monocular or binocular illumination from the magnitude of any additional target-localization shift that occurs in only the lead eye by variations produced in the total period of binocular neural excitation as a result of these monocular or binocular illumination changes. Thus, for example, new experiments are needed using progressively longer exposure durations for the lead eye under various conditions of monocular and binocular variations in illumination, while the offset of stimulation for the lag eye is systematically delayed, using very short exposure durations for that eye in each case and for a wide variety of constant ID values.

Finally, it should be mentioned that a control experiment was performed on observer CJ in which depth settings were made under the same conditions of unequal binocular illumination (E_L = 3.0 log td and E_R = 1.0 log td; E_L = 1.0 log td and E_R = 3.0 log td) as prevailed in the experiments described in Figs. 13 and 14, but for continuous target viewing as in the usual case of Pulfrich-effect measurements. The results showed that the average depth displacement ($\overline{\Delta R}$) in each case was 7.11 cm and 8.04 cm, respectively. The corresponding latency-difference values ($\overline{\Delta t}$) were 26.3 msec and 29.3 msec and the corresponding angular stereoscopic parallax values ($\eta_{\overline{\Delta R}}$), as computed by Eq. (3) from the obtained ΔR_N and ΔR_F measures, were 896 and 1000 sec or arc, respectively, which represent an average value of 948 sec of arc.

If the conceptually erroneous view is accepted that a component of the depth displacements obtained under the conditions of unequal binocular illumination is comprised of a "Pulfrich-latency effect", then the experimental data in Figs. 13 and 14 are in sharp conflict with this (meaningless) prediction. Adding 948 sec of arc to each of the $\eta_{\overline{\Delta R}}$ values obtained at each ED_L and ED_R for the curves representing the condition ED_L = ED_R = 3.00 log td, very poorly predicts the empirically obtained data of the curves representing the condition of unequal binocular illumination. The depth differences between the paired curves in Figs. 13 and 14 are not constant at all values of ED_L and ED_R, as that questionable notion would predict. Also, even the largest depth difference which was obtained for the four paired curves at an exposure duration of 80 msec is only 582 sec of arc; the smallest depth difference at this largest exposure duration for the four paired curves is a low value of 328 sec of arc.

Summary

I would like to conclude this presentation on the theory, methods, and data of the Mach-Dvorak effect by emphasizing the following four main points.

1) The Mach-Dvorak effect is a stereoscopic effect that can be produced by introducing binocular parallax for either stationary or moving targets presented intermittently under sequentially-timed conditions that yield good apparent movement. For an oscillating target presented intermittently with interocular delay in binocularly paired sequences in the form of very brief flashes, the magnitude of the depth displacments ($\eta_{\overline{\Delta R}}$) in angular terms can be accurately predicted on the basis of simple geometric considerations derived from the classical theory of stereoscopic vision.

2) When monocular or binocular target flashes of long duration are used,

particularly in conjunction with a fast moving target which combines to produce a wide region of target stimulation in one or both eyes, then the specification of the prevailing linear stereoscopic parallax (AB) given simply in terms of the distance between the position of target onset (A) for the lead eye and the position of target onset (B) for the lag eye, is no longer adequate in predicting the magnitude of the Mach-Dvorak effect. The empirical results imply that variations in monocular exposure duration produce systematic effects in shifting the hypothesized position of target localization (A or B) for that eye in the plane of target motion, always in the direction of target motion. The depth effects of monocular or binocular variations in target exposure duration are designated as the exposure-duration (E-D) effect. The theoretical and empirical relationship between $\eta_{\overline{\Delta R}}$ and the magnitude of the binocular exposure durations are under current investigation in our laboratory. A general neurophysiological model is presented in the paper to account for the E-D effect.

3) Conditions of binocular illumination also have a systematic influence on the magnitude of the Mach-Dvorak effect. Increasing the illuminaton level in one eye will displace the hypothesized position of target localization (A or B) in that eye in the direction of target motion; decreasing the illumination will displace the position of target localization in the opposite direction to that of target motion.

An additional shift in target localization occurs for only the lead eye when conditions of binocular illumination produce a change in the total period of binocular neural excitation (from the onset of cortical neural activity resulting from stimulation of the lead eye to the offset of cortical neural activity resulting from terminaton of stimulation of the lag eye). Increasing the total duration of binocular neural excitation will additionally shift the position of target localization (A) for only the lead eye in the direction of target motion; decreasing the total binocular excitation time will now additionally shift only A in the direction opposite to that of target motion. The proposed general neurophysiological model can be readily expanded to incorporate the lateral-displacement effects of equal and unequal binocular illumination for the two eyes.

The magnitude of the Mach-Dvorak effect under conditions of unequal binocular illumination does not involve any consideration of a depth-displacement component which can be related to the latency hypothesis underlying the Pulfrich stereophenomenon. That is, the change in linear parallax (AB) resulting from a change in monocular illumination is not given by the product of the change in latent period produced for that eye and the target velocity, as would be predicted by the latency hypothesis of the Pulfrich stereophenomenon. The theoretical and empirical relationship between $\eta_{\overline{\Delta R}}$ and conditions of equal and unequal binocular illumination is also under current study in our laboratory. Experiments are planned to assess the effects of white and colored targets presented under various target-background contrasts on the magnitude of the Mach-Dvorak effect.

4) The proposed basic studies on the Mach-Dvorak effect will help to establish optimal conditions of testing and will provide initial normative data necessary for clinical application in the field of binocular depth discrimination. Quantitative assessments of the correlation between specified performance on the Mach-Dvorak effect under "stressful" stimulus conditions of testing and the results on clinical tests of binocular functioning should prove useful for both diagnostic and vision training purposes. The results of the proposed basic studies on the Mach-Dvorak effect, as in the case of the basic studies on the Pulfrich stereophenomenon, will make a direct contribution to the theory and data of many basic visual functions, such as stereoscopic vision, monocular directional localization, the visual latent period, intensity discrimination, color vision, and visual neurophysiology.

Acknowledgement

Partial support for our studies presented here was provided by a research grant (EY 00383) from the Eye Institute of the U.S. Public Health Service awarded to Professor Alfred Lit.

References

1. C. Pulfrich, Naturwissenschaften 10, 553 (1922).
2. V. Dvorak, Sitzber d. k. Böhm. Gesellsch. d. Wiss, in Prague 65 (1872).
3. A. Lit, Am. J. Psy. 62, 159 (1949).
4. A. Lit and A. Hyman, Am. J. Opt. 28, 564 (1951).
5. A. Lit, Optom. Weekly 59, 42 (1968).
6. M. Wertheimer, Z. Psychologie 61, 161 (1912).
7. C. H. Graham, In Vision and Visual Perception, Edited by C. H. Graham (John Wiley and Sons, New York, 504 1965).
8. G. S. Harker and P. D. Jones, Bulletin of the Psychonomic Society 6, 434 (1975).
9. D. N. Lee, Vision Res. 10, 65 (1970).
10. A. Lit, J. Opt. Soc. Am. 50, 321 (1960).
11. A. Lit, J. P. Finn and W. M. Vicars, Vision Res. 12, 1241 (1972).
12. A. Lit, J. Exp. Psy. 59, 165 (1960).
13. A. Lit, J. Opt. Soc. Am. 50, 970 (1960).
14. J. D. Brauner and A. Lit, Bulletin of the Psychonomic Society 6, 443 (1975).
15. G. S. Harker, Vision Res. 13, 1041 (1973).
16. F. W. Frohlich, Die Empfindungszeit, Ein Beitrag zur Lehre von der Zeit-Raum und Bewegungsempfindung. Jena, 1929.

VII. Neurophysiology of Visual System Function

Discharges of Visual Neurons in Eye Movements

Hiroharu Noda
Brain Research Institute, University of California at Los Angeles
School of Medicine
Los Angeles, California

In animals with sharp vision, only a small portion of the retina is specialized for detailed vision. This portion, the fovea, has a high density of photoreceptors and in our eyes we can count as many as 7000 cones in a small area which covers only two degrees of the visual field. The fovea has outstanding spatial sensitivity. For the rest of the retina, however, the spatial sensitivity is extremely poor. Because of such a difference in capabilities we have to move our eyes in order to bring the image onto the fovea; therefore, the eye movement is especially well developed in animals having a discrete fovea. The function of eye movements is to acquire a visual target and then track it so that the image stays on the fovea. For that purpose the brain has to select a visual target and transform its spatial coordinates into a motor signal necessary for the production of accurate eye movements by the oculo motor system. The information about the location of the target could be derived from either branch of the visual system, namely, the midbrain pathway terminating primarily in the superior colliculus or the geniculo-cortical pathway terminating chiefly in the striate cortex. There is no doubt that the control of eye movements is strongly dependent on the feedback of sensory information from the retina. The primary pathway for visual information is by axons of retinal ganglion cells from the eyeball to the lateral geniculate nucleus and then through the optic radiations to the visual cortex. Some fibers of retinal ganglion cells also travel to the midbrain, terminating chiefly in the superior colliculus and the pretectal region. In this session on "Neuronal Substrates of Eye Movement," I have been asked to discuss the properties of neurons in the afferent pathway of the eye movement control system. In order to study response properties of visual neurons during eye movements, I recorded single unit activity from the optic tract and the lateral geniculate nucleus in the chronic cat. The cat had permanently implanted, nonpolarizing EOG electrodes, with which both horizontal and vertical EOGs were measured. The cat was fully awake and making frequent eye movements. A stainless steel microelectrode was advanced by a mechanical drive until it reached the target structure; thereafter, it was driven by a hydraulic system. Action potentials of retinal ganglion cells were recorded from the optic tract, a few millimeters posterior to the optic chiasm. In the second series of experiments, spikes were recorded from relay cells of the lateral geniculate nucleus.

The first question to be answered was, "Do all ganglion cells of the retina respond to image motion induced by saccadic eye movement?" In most experiments on anesthesized and paralyzed animals, the speed of the moving visual object is relatively slow. We know that the peak velocity of saccadic eye movement in awake cat may exceed 300°/sec. No information was available to answer this question, because this speed is too high for the speed commonly used in acute experiments. The answer is shown in Fig. 1. When saccadic eye movements occured in a lighted environment, the discharge pattern of retinal cells, though considerably different from cell to cell, fell into one of the three classes exemplified in Fig. 1. In each record, the upper trace is the discharges of the ganglion cells, while H and V represent the horizontal and vertical EOGs, respectively. When both EOGs are flat, that means both eyes are stable, while sudden shifts in EOGs mean saccadic eye movements. The first class of cells, as represented by the example shown in the upper record, is characterized by a burst of discharges for each saccadic eye movement. The background activity during fixation is fairly constant. The cells in this class are tentatively called "T-units", because they show transient discharges to saccades. The second class of cells, as represented by the middle unit, did not show transient responses to saccades. Instead, they showed different position-dependent levels of sustained firing during steady fixation and, consequently, are referred to as "S-units". The third class of cells, as represented by the bottom unit, also shows a transient burst to saccades, but, in addition, exhibits a change in the level of tonic activity depending upon the fixation point. Since cells of this class showed mixed responses of both transient and sustained components, they are called "M-units".

Figure 2 shows the time course of the transient responses to a saccade seen in an "on" center T-unit. In the raster, spikes are represented by dots which are aligned to the start of saccades. The superimposed traces at the bottom show horizontal EOGs during 21 consecutive saccades. As you can see, the cell showed a vigorous burst to the image motion of the saccade with the firing within a burst reaching levels more than ten times greater than the background activity. Latencies from the start of the saccade was about 20 to 30 ms and the burst lasted for about 100 ms. A point we should notice here is the fact that the duration of the burst is considerably longer than the duration of image motion during a saccade. The duration of medium size saccades of the cat is about 40 to 50 ms, so that the burst responses are almost twice as long as the stimulus duration.

Before I explain what these transient responses are and how they are produced in certain classes of ganglion cells, I would like to show you the response properties of the class of ganglion cells which do not show transient responses to the image motion induced by saccades. Figure 3 shows examples of ganglion cell activity not charaterized by transient responses to saccades. Since the cell shows sustained changes in activity as the eyes were directed to different points, this is an S-type unit. Now, I have sufficient evidence to conclude that this type of cell is equivalent to the X-cells observed in acute experiments. The conduction velocities for the S-units, estimated from the latencies of antidromically elicited axon spikes by stimulation of axons and terminals within the lateral geniculate nucleus, ranged from 5 to 25 m/sec which corresponded fairly well to the value for X-cells. In order to compare response properties of different classes of cells, a standard visual stimulation was used throughout the experiments. The visual field was comprised of a stationary grating of vertically oriented, 5° wide dark stripes separated by 5° wide light stripes. This grating was projected on the rear of a tangent screen. The same

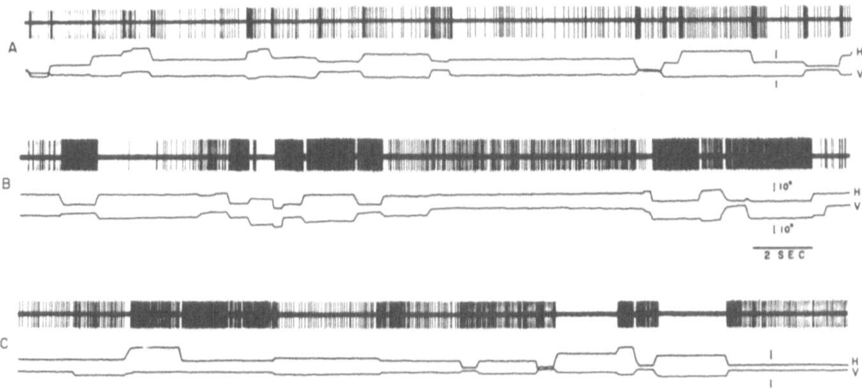

Fig. 1 Discharges of 3 types of retinal ganglion cells during free eye movements scanning the stationary grating pattern. A: firing of a T-unit, showing transient responses to saccade, indicated by a potential shift in EOGs. When the eyes were stationary, indicated by flat EOGs, it showed a constant level of firing. B: firing of an S-unit, showing sustained firing to direction of gaze, reflecting local luminance in the receptive field. C: firing of an M-unit, showing both transient and sustained responses. Horizontal (H) and vertical (V) EOGs are shown. Upward deflection of the H and V tracings represent eye movements to the right and upward, respectively.

grating was moved horizontally by reflecting its image by an optical scanner, when I wanted to stimulate a cell with a moving pattern. Record A shows the discharge of the cell when the cat was looking at the stationary grating. The cells showed sustained firing which was related to the position of the eyes with respect to the light or dark stripe of the grating. The fact that the firing level was dependent on the local luminance is proven by the next experiment shown in B. Here diffuse light covering the tangent screen was switched on and off. It is clear from a comparison of cellular activity with the record from the photocell (S), that the cell increased firing when the light was on. The sustained level of activity continued as long as the luminance level was maintained. Record C was taken 5 minutes after the diffuse light was turned on and the activity in D was recorded 5 minutes after the cat was immersed in complete darkness. In E is shown the response of the same cell as tested by a moving grating. The grating consisted of 20 alternating dark and light stripes, 5° in width covering a total of 100° of visual angle. The signals from the photocell registered the time when a stripe passed over the center of the screen as shown in Fig. 3E, F. When the grating was moved at low speeds, for example, 20° or 30°/sec, the cell showed responses to each cycle of the grating and it continued to respond as the target velocity was increased until the speed was approximately 300° to 400°/sec. This is the maximum speed for this cell, above which the modulated response was replaced by an unmodulated response which persisted only during movement of the grating. As a matter of fact, the unmodulated response continued only for the duration of this stimulus movement, and the firing rate within the response reflected the average luminosity of the grating (2).

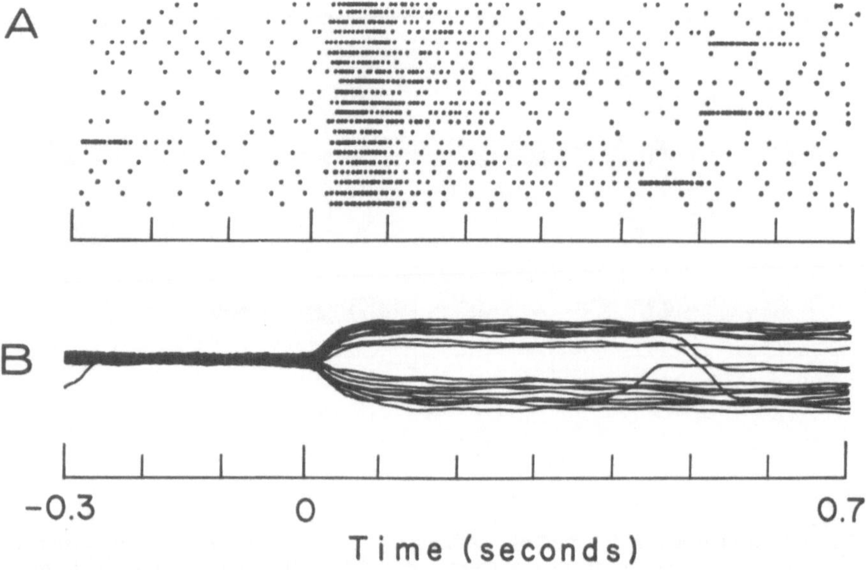

Fig. 2 A: a raster showing the time course of the transient responses of an on-center T-unit to saccadic eye movements across the stationary grating pattern. Unit discharges were represented by dots aligned to the start of saccadic eye movements. B: horizontal EOGs showing 21 consecutive saccades. EOG calibration, 10°.

Now, I would like to return to the ganglion cells showing transient responses following saccades and explain the nature of their burst responses. Figure 4 shows an example of a T-unit which, I believe, corresponds to a Y-cell. The estimated conduction velocity for this class of cells ranged from 20 to 50 m/sec. When saccades occurred in the presence of the stationary grating as in A, the firing of the cell was initially suppressed as indicated by an arrow. Then it exploded in a burst. As seen in G, when three parallel light stripes were moved over the tangent screen, discharge was clearly suppressed three times, indicating that the cell had an off-center. When diffused light was switched on and off on the screen, the cell showed a similar transient response to light, showing a stronger response to the off-set of the light. In the experiment shown in C, vertical stripes were placed on the periphery of the screen, 30° away from the center, the cell showed transient responses, regardless of the fact that eye movements were limited usually to the central 20° of the visual field. For the transient responses, however, visual contrast was essential and when the cat was tested in darkness as in D, the transient response to a saccade was completely eliminated. Another important feature was that the cell showed an almost constant level of activity during fixation. The firing level during eye movement in the light (C) or dark (D) or in the presence of a pattern (A), stayed almost constant. Thus, this cell did not respond to local luminance. When the grating was moved across the tangent screen, two substantially different types of responses, i.e., a primary and a secondary response, were observed. The primary response appeared when the target's motion was sufficiently slow, while the secondary response appeared as a burst when the speed of the target motion was high. Both

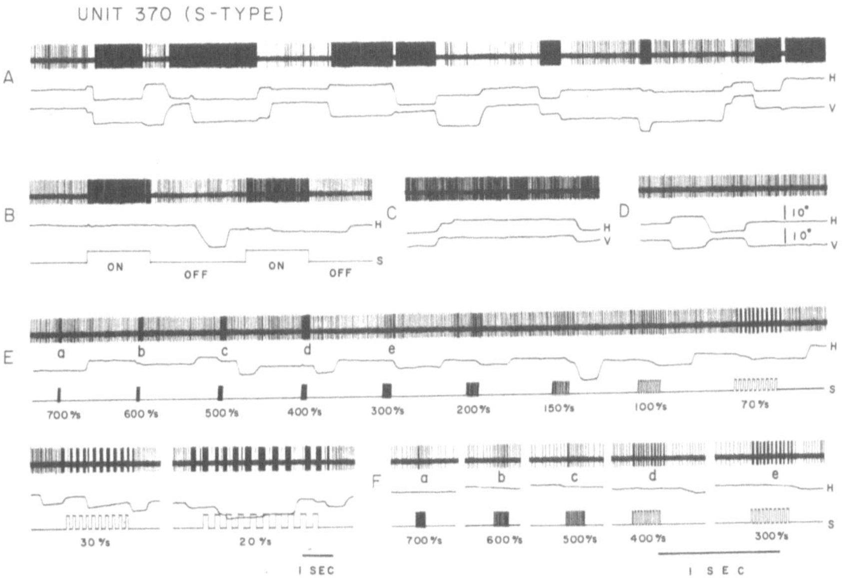

Fig. 3 Responses of an S-unit to saccadic eye movements in the stationary grating field (A) and to the onset and offset of diffuse light (B). Firing pattern after prolonged illumination with diffuse light (C) and in complete darkness (D). Note that firing level was dependent upon the luminance (A, B, C, D). The maximum velocity for the primary response was approximately 400°/sec (d of F), and above that speed the response was saturated into a burst firing (a, b, c of E and F). H: horizontal EOG. V: vertical EOG. S: stimulus marker representing potentials from a phototransistor. The time calibration at the end of E applies also to A, B, C, and D.

responses are seen in records E and F, with the latter being an expanded version of the former and each alphabet corresponding. When the movement of the grating was too fast, as in a-e, the primary response did not follow the stimulus. Only a remnant suppression appeared at the beginning of the passage of the stimulus, followed by a burst. The point that I want to emphasize here is the nature of the transient response to a saccade. When we compare the burst responses during saccades (A) with the secondary responses, typically seen with fast target motion (a-e of E), it becomes clear that the bursts are analogous to the secondary responses and certainly not equivalent to the primary responses. The secondary responses were quite non-specific. Similar responses also appeared when the diffuse light was switched on or off (B) or even in response to flashes of light. Almost any stimulus which causes a sudden change in luminance seemed to be effective in eliciting the secondary responses.

Another important feature of the secondary responses was that they could be

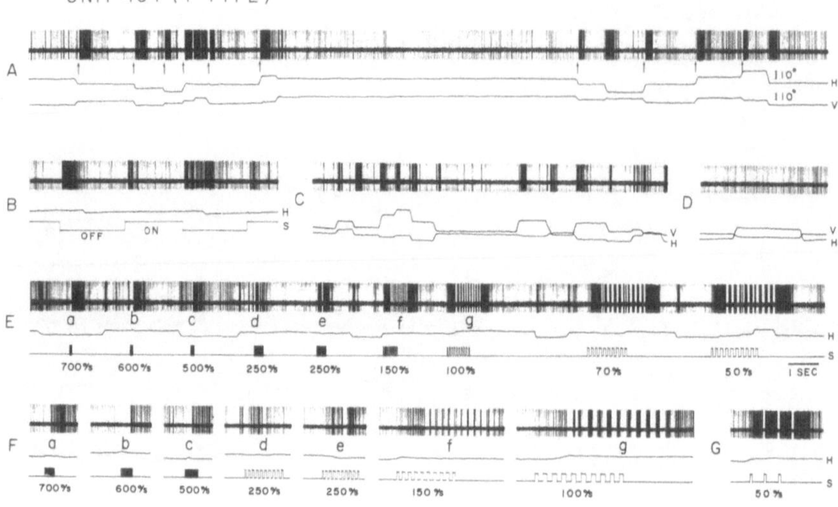

Fig. 4 Responses of a T-unit to saccades in the stationary grating field (A) and to the onset and offset of diffuse light (B). Firing pattern after prolonged illumination with diffuse light (C) and in complete darkness (D). Note that the background firing rate during fixation is almost identical in A, B, C and D, indicating that this unit did not reflect the luminance. E and F: responses to moving grating at different speeds as indicated below the stimulus marker (S). The responses in F are portions of E and the sweeps were expanded in F. This unit has an off-center receptive field and showed evidence of the periphery effect (G). Time calibration at the end of E applies also to A, B, C and D.

elicited by a moving stimulus presented as far out as 30° from the receptive field center of the cell. We can see this in the 50°/sec record (right end of E). The increased activity started almost half a second before the first stripe entered the receptive field and it lasted for a certain period after the last stripe passed over the receptive field. Since the stripes were moved at 50°/sec, the fact that the response started half a second before the entry of the stimulus indicates that the cell started responding even when the stripes were 25° away from its receptive field center. The secondary responses were, therefore, the periphery effect which was first described by MCILWAIN (3). CLELAND, DUBIN and LEVICK (4) found later that it was obvious and strong in Y-cells but weak or absent in X-cells. The secondary response is very sensitive to the attentiveness of the animals. The best response was obtained when the cat was making a slow tracking eye movement, during which I assume that both convergence and accommodation are functioning perfectly. The response became less obvious when the cat became less attentive or drowsy which was indicated by EOGs showing a slow drift and by a marked reduction in the speed of saccades. In this state, although the eyes may still be open, they are not fixating on the tangent screen. The secondary response disappears almost completely with anesthesia, as originally discovered by MCILWAIN (3).

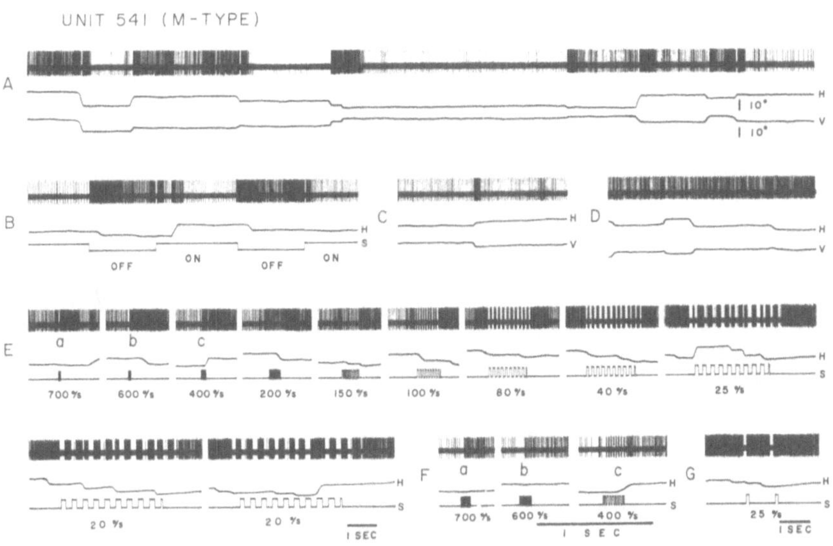

Fig. 5 Responses of an M-unit to saccades in the stationary grating field (A) and to the onset and offset of diffuse light (B). The discharge rate in complete darkness (D) was higher than in diffuse light (C), but showed a burst discharge when eye movement occurred in the light. This cell was an off-center unit (G) and showed a remarkable secondary response (a, b, c and E). There was evidence of periphery effect as can be seen by an increased firing preceding and following the moving grating.

Another class of cells which showed a transient response to saccadic eye movement had receptive field properties of both S-units and T-units. Since they had mixed properties, they are called M-units and are found in about 25% of the ganglion cells. Their conduction velocities fell in between S- and T-units. An example of an M-unit is shown in Fig. 5. When the cat was facing the stationary grating (A), the cell showed a burst discharge in response to saccades. This was characteristic of T-units and was not seen in S-units. However, the level of activity depends on the position of the eyes, showing the characteristic of S-units. When tested with diffuse light switched on and off on the tangent screen (B), the cell showed a transient response to the onset of the light and overall activity during the off period was higher than during the on period. The response to saccades were apparent here also. The record in C was taken 5 minutes after the diffuse light was projected, and that in D was taken 5 minutes after the cat was immersed in complete darkness. You can notice the difference in the tonic level of activity. Note too, that the burst response seen in light was completely wiped out when saccades occurred in complete darkness. When tested with moving gratings we can see response properties of both S- and T-units. In response to slowly moving gratings the cell showed a modulated primary response and in response to fast motion, secondary burst responses are observed.

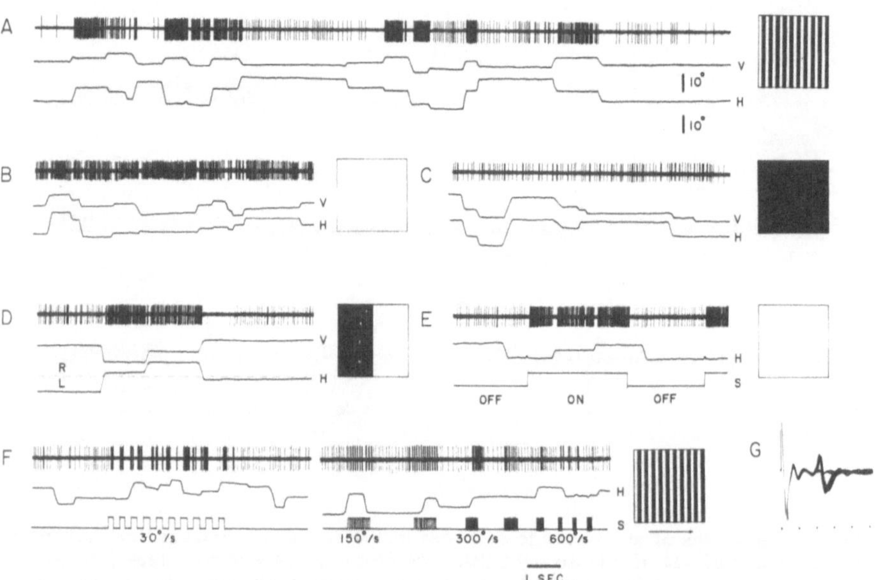

Fig. 6 Responses of an S-cell of the lateral geniculate nucleus. Discharges with eye movements in the stationary grating field (A), in diffuse light (B), in complete darkness (C) and in a visual field where only the right half of the tangent screen was diffusely illuminated (D). The dashed line in horizontal EOG in D represents the level of EOG potential when the eyes were directed to the center of the screen. Responses to the onset and offset of illumination (E) and to moving grating (F). The velocity of the moving grating is indicated below each response. Spike responses to optic chiasm stimulation are shown in (G).

It is generally accepted that in higher mammals, midbrain structures play a fundamental role in visual orientation, while the visual cortex is intimately involved in form discrimination. It is also well known that a part of the retinal output is destined for the superior colliculus. It would be very simple and reasonable to assume that the S-cells are related to the geniculo-striate system and that T-cells are related to the retinotectal system. For the next series of experiments, I asked whether or not geniculate cells could be classified as S-units or T-units in the same way as retinal ganglion cells and whether there is any preference for a certain class of cells projecting to the lateral geniculate nucleus. Keeping these questions in mind, I (5) have recorded soma spikes from relay cells of the lateral geniculate nucleus. Each cell was identified by the latency of spikes in response to optic chiasm stimulation. I employed the same tests as used for retinal ganglion cells. Relay cells also could be classified into S-, T- and M-types, and their response properties are surprisingly similar to those of corresponding classes of retinal ganglion cells. I would like to show you the responses of these cells just to prove how identical they are.

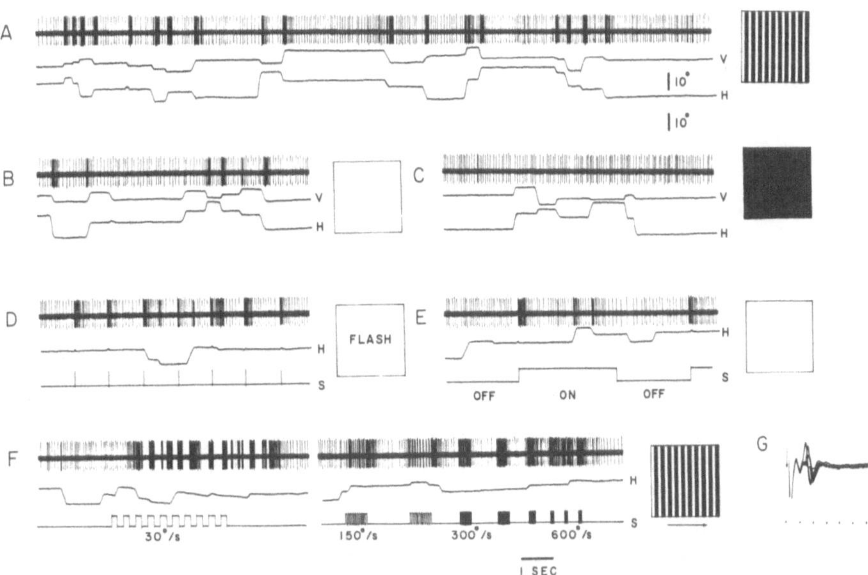

Fig. 7 Responses of a T-cell of the lateral geniculate nucleus. Discharges with eye movements in the stationary grating field (A), in diffuse illumination (B) and in complete darkness (C). Responses to flashes of light (D), to the onset and offset of illumination (E) and to moving grating (F). G: spike responses to optic chiasm stimulation.

Figure 6 shows an example of S-relay cells of the lateral geniculate nucleus. The cell responded to stimulation of the optic chiasm with a latency of 1.0 msec. When tested in the presence of a stationary grating pattern, the cell showed sustained responses sensitive to local differences in luminance. The maintained level of activity in diffuse light (B) was higher than that in complete darkness (C). In a visual field where the right half of the screen was illuminated, the cell increased activity when the eyes were directed to the right, as indicated by the horizontal EOG rising above the level of the dashed line (D). The same was true when diffuse light was switched on and off over the tangent screen (E).

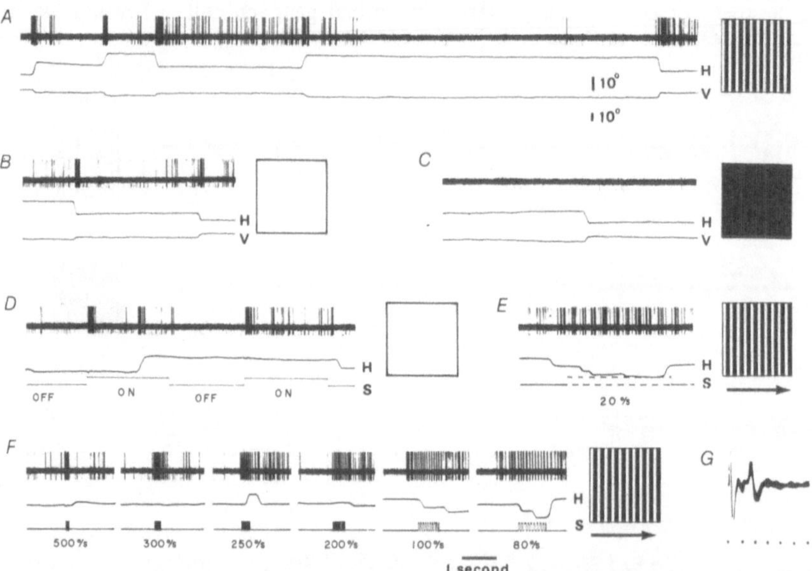

Fig. 8 Responses of an M-cell of the lateral geniculate nucleus. Discharges with eye movements in the stationary grating field (A), in diffuse illumination (B) and in complete darkness (C). Responses to the onset and offset of illumination (D) and to moving grating (E and F). G: spike responses to optic chiasm stimulation.

Figure 7 shows an example of T-relay cells of the lateral geniculate nucleus. The cell responded to stimulation of the optic chiasm with a latency of 1.1 msec. Analogous to its response to saccades in the presence of step changes in luminosity (A), the cell showed responses to the frame of the screen (B) located about 30° from the center; but these burst responses disappeared when tested in complete darkness (C). A similar transient response was elicited by flashes of light (D) or by the onset of diffuse light (E).

Figure 8 shows an example of M-relay cells. Both sustained and transient responses were seen in the presence of the stationary grating pattern (A). The maintained level of activity was higher in the light (B) than in complete darkness (C). Transient and subsequent sustained responses were seen with the onset of diffuse light (D). The periphery effect was also seen in this unit.

In Fig. 9 are presented frequency histograms of spike latencies to optic chiasm

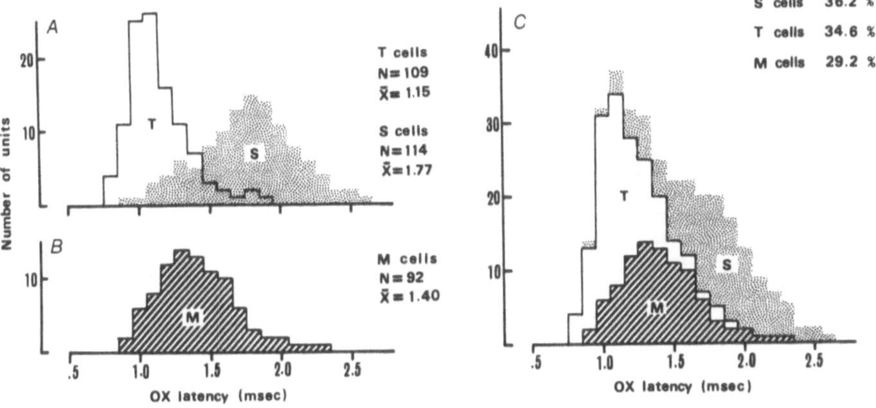

Fig. 9 Frequency histograms of spike latencies for 315 relay cells of the lateral geniculate nucleus to optic chiasm stimulation. A: latency distribution for 109 T-cells (T) and 114 S-cells (S). B: latency distribution for 92 M cells. C: latency distribution for the whole population of 315 relay cells.

stimulation for a total of 315 relay cells. T-cells responded at shorter latencies than S-cells. The average latency for S-cells was 1.77 msec and for T-cells it was 1.15 msec. Latencies of M-cells had a distribution in between those of S- and T-cells. The distribution of latencies for the whole population of lateral geniculate cells was unimodal. The distribution corresponds fairly well to that of HOFFMAN, STONE and SHERMAN (6), except that these authors found a sharp dip at 1.5 msec and I did not.

There is considerable evidence that X- and Y-cells are distributed very differently in the retina. First of all, the majority of cells in the area centralis are X-cells. Compared with Y-cells, X-cells have a smaller receptive field. According to the data of STONE and FUKUDA (7), the average diameter of the receptive fields of X-cells remains less than one degree, even 30° away from the center, while the average diameter of Y-cell receptive fields increases up to 2°. Besides the larger receptive field center of Y-cells, these cells can be activated by moving stimuli applied as far out as 30° from the receptive field. This indicates that the peripheral retina is covered with a huge "antenna" composed of densely overlapping Y-cells receptive fields. As seen in the T-unit of the retinal ganglion cell, the Y-cell is very sensitive to the transient visual phenomena. However, because of its "fuzzy" receptive field organization, Y-cells are not ideal cells which can inform the brain of the details of spatial contour of visual objects. In response to the image motion of a saccade or to a change in luminance or to a quick shift of the visual field or even to a flash of light, T-cells exhibited almost identical burst discharges. The signal was always in the form of a strong burst of impulses. This signal may be useful for the reflex arc of the goal directed saccadic eye movement. For example, the saccade for the foveation mechanism which was proposed by DR. SCHILLER (8).

Acknowledgement

Supported by NIH Grant EY 01051.

References

1. H. Noda, Brain Res. 84, 515 (1975).
2. H. Noda and W. R. Adey, Brain Res. 70, 340 (1974).
3. J. T. McIlwain, J. Neurophysiol. 27, 1154 (1964).
4. B. G. Cleland, M. W. Dubin and W. R. Levick, J. Physiol. (Lond.) 217, 473 (1971).
5. H. Noda, J. Physiol. (Lond.) 250, 579 (1975).
6. K. P. Hoffmann, J. Stone and S. M. Sherman, J. Neurophysiol. 35, 518 (1972).
7. J. Stone and Y. Fukuda, J. Neurophysiol. 37, 722 (1974).
8. P. H. Schiller and F. Körner, J. Neurophysiol. 34, 920 (1971).

Q. Dr. Barlow: I found these results very fascinating because it showed the difference in function between the X- and Y-cells which is far from obvious during the ordinary acute experiment where you can find differences between the X- and Y-systems and have little difficulty in classifying the cells but there is nothing there that seems to make functional sense and here you have shown that during ordinary eye movements there is a radical difference in the way they behave which to me is fascinating. We for a long time thought that the shift effect described by Fisher and Kruger might be caused by stray light and therefore thought that some of the results you have shown here might also be caused by stray light, but you have now proved to our satisfaction and I think even to Robinson's satisfaction that this is not so. And that what you say about the Y-cells being excited from the extreme periphery of the retina is absolutely correct.

A. Thank you for your kind comment, Dr. Barlow. Since the content of my presentation wasn't exactly what Dr. Stark expected, in closing, I would like to comment briefly on a mechanism in the eye which might be related to the psychophysical phenomenon known as saccadic suppression. We have some data that tested the excitability of relay cells to the stimulation of the optic chiasm. During saccadic eye movement, we found that the excitability which is measured by the firing probability of relay cells in the lateral geniculate nucleus to the orthodromic impulses is markedly suppressed for about 100-150 msec. This means that the information transmitted from the eye to the primary visual cortex is inhibited during, and the period immediately following, the saccade. However, when the test was conducted in complete darkness, such a decrease in the excitability was completely eliminated. Furthermore, when the visual field was shifted as a whole in a saccadic fashion, similar depression in the excitability in the geniculate cells was observed in the absence of eye movements. So we concluded that the effect of producing the decreased excitability in the geniculate cells was produced by visual input coming from the retina.

The Primate Superior Colliculus and its Sensory Inputs

Peter H. Schiller
Department of Psychology
Massachusetts Institute of Technology
Cambridge, Massachusetts

Let me begin by showing you a simplified diagram of the rhesus monkey visual system. The retinal fibers, as shown in Fig. 1, project to the superior colliculus and to the lateral geniculate nucleus. The geniculate in turn projects to visual cortex. From layer 5 cortical cells project down to the colliculus. What I want to emphasize is that the superior colliculus receives a dual projection; one directly from the retina and the other indirectly from visual cortex. If you look at this diagram do not take it as a literal proportional relationship of what the brain looks like. The superior colliculus is drawn big here simply because it is my point of focus. In reality, it is the smallest of these structures; in monkey its surface is only about five or six mm in diameter.

My talk divides into two parts; in the first I will describe what we have found when we studied the superior colliculus; in the second part I will describe the inputs to this structure from the retina and from visual cortex.

When we first began this work, our task was to study the characteristics of single cells in the superior colliculus both with respect to vision and to eye movement. Therefore, we had to develop a preparation in which you could do both of these things. This is a problem, because when the eye moves it is very hard to map receptive fields and to establish where they are located on the retina. There are several ways of solving this problem. BOB WURTZ solved it by training monkeys to fixate (1). We solved it by immobilizing one eye and letting the other eye move freely (2).

Eye immobilization is accomplished by transection of the third, fourth and sixth cranial nerves. Around the normal eye electrodes are implanted in the orbital

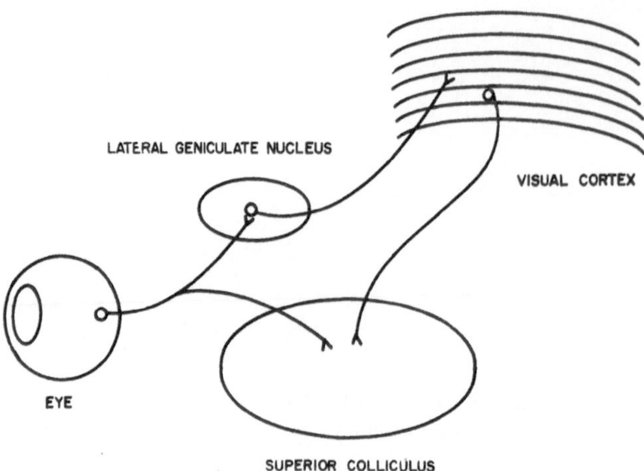

Fig. 1 Basic wiring diagram of the monkey visual system.

bone so that one can record eye movements. During the experiment the head is restrained. By occluding the moving eye, it is possible to map collicular receptive fields through the eye which is immobilized. If one now occludes this eye, it is possible to look at the saccadic or pursuit eye movements of the moving eye.

The work I will describe to you has been done over the past six or seven years at MIT, in collaboration with a number of individuals which include NANCY BERMAN, MAX CYNADER, FRITZ KOERNER, JOE MALPELI AND MIKE STRYKER.

The first point to be made about the superior colliculus is that it is a structure upon which the visual field is laid out in a beautiful, topographic fashion with the left superior colliculus representing the right visual hemifield and the right superior colliculus representing the left hemifield. The anterior part of the structure deals with central vision and the posterior part handles peripheral vision. Furthermore, the superior colliculus is a layered structure; the response characteristics of cells are different in the superficial and the deep layers. When one records in the superficial layers single cells have clearly defined, neat receptive fields. In central vision, they are small, but get larger with increasing retinal eccentricity. The response properties of these cells in the monkey are not particularly exciting. They do not seem to care about the orientation or the shape of the stimulus, or its direction of movement. By contrast, in the cat, you find many directionally selective cells in the superior colliculus. Cells in both of these animals respond both at the onset and the offset of a flashing stimulus. The response is always more vigorous to small stimuli than to large ones. An example of this appears in Fig. 2.

A series of discs of different sizes was presented centered on the receptive field. The stimulus was on for half of the histogram and off for the other half. What you can see is that a 1° stimulus elicited the optimal response. When a large stimulus was used, there was practically no response at all. The other point to be made is that the responses are quite transient; a brief burst was obtained both when

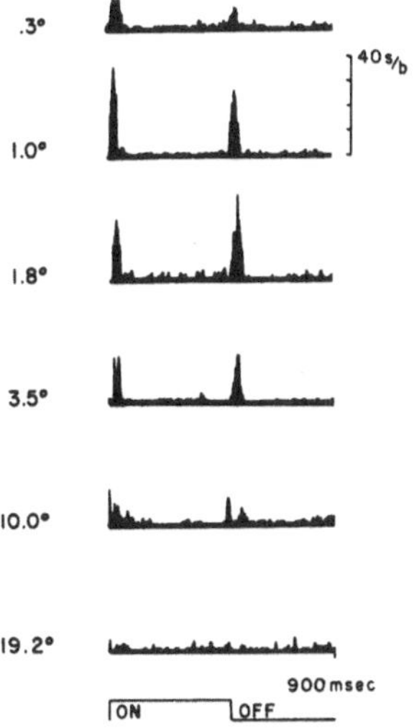

Fig. 2 Stimulus response histograms obtained from a collicular cell to stimuli of varying diameter. Stimulus duration was 500 msec, presented once every 1,300 msec. Response field was 13° located 18° from fovea. s/b: number of spikes per bin.

the stimulus came on and when it went off. It may be stated then, that superficial cells respond consistently, have well defined receptive fields and give both on and off responses. The last thing I should mention (which is not obtained from this preparation but from flaxidilized monkeys) is that almost all of the cells in the binocular portion of the visual field of the superior colliculus can be activated from either eye.

As one advances the microelectrode into the intermediate layers of the structure, a rather dramatic shift occurs in the response properties of single cells. One begins to see more and more cells which respond specifically in relationship to eye movement. An example of such a cell appears in Fig. 3. The first four segments of this figure show a brisk response which begins prior to the small saccade and reaches a crescendo just as saccade appears. Fig. 3C shows that the other saccades, even if they are in the same direction but have a larger amplitude are not preceded by a burst. What we can see in this figure then is that the discharge of the collicular cells in the intermediate layers is rather specific for the size and direction of saccades.

If one takes a long record of this sort and scores each of the saccades in terms of its amplitude and its direction, a motor map can be generated. A map of this sort is shown in Fig. 4. The discs represent those saccades which were preceded by a burst of discharges; saccades during which the unit remained silent are represented as circles. As you can see, this unit fired only when a saccade occurred which moved the eyes leftward a few degrees. Clearly, this particular unit is quite specific, and DAVID SPARKS, (this volume), the next speaker will present more data regarding their exact characteristics.

Some of the eye-movement units have a dual property in that one can also elicit visual responses from them. The record in Fig. 3D was obtained when we occluded the monkey's moving eye so that only his immobilized eye saw the world.

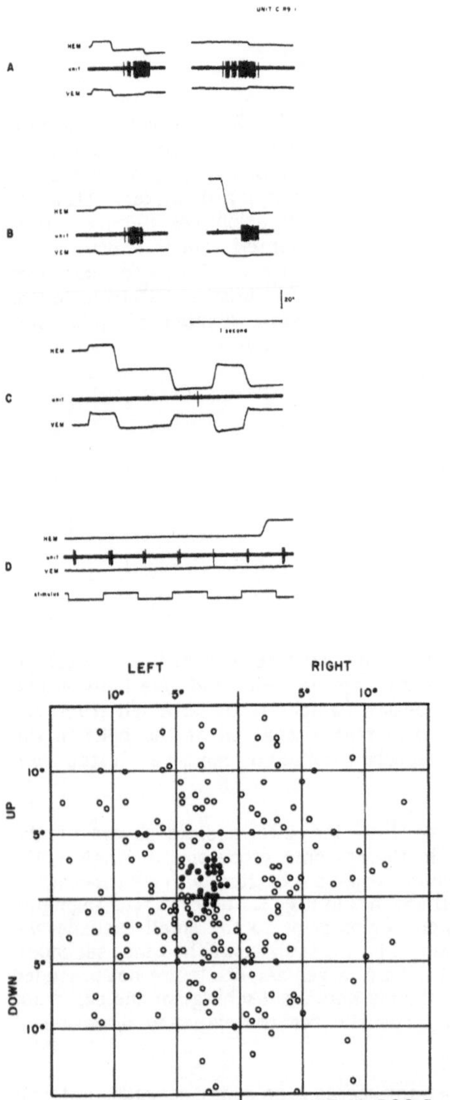

Fig. 3 Discharge characteristics of a unit related to eye movement. A, B, and C: unit discharge and eye movement in the light, moving eye unoccluded. D: response to a 0.25° light spot moved back and forth within the receptive field of the immobilized eye with a square wave; moving eye occluded. Stimulus marker displaced up and down represents onset and offset of light.

Fig. 4 Saccade-associated unit response. Each mark represents the size and direction of a saccade. Circles represent saccades not associated with unit activity. Discs show saccades which were preceded by a burst of spikes. Direction and size of saccade shown by quadrants designated left, right, up and by degrees within these areas.

We looked for the receptive field of this unit and after finding it, we exposed it to a flashing spot. This elicited a weak but consistent response even though there were no discernible eye movements. On the basis of this one can conclude that some of these units have a dual property. On the one hand they fire prior to saccades and on the other they have a visual receptive field. Now the next important thing to determine is the location of the visual receptive field. Figure 4 is a motor map, but one can readily convert it into a sensory map since the coordinate system is the same, with the foveal gaze located in the center of the map. The receptive field of this unit on the converted map falls into the region of the filled circles. This finding suggests that when the unit discharges with a high frequency, a saccadic eye movement occurs which brings the eye into that part of the visual field where the receptive field of that particular unit was located prior to the saccade. On the basis of this one can conclude that in the superior colliculus resides a saccadic guidance mechanism which enables the system to bring the fovea to various parts of the visual field.

To subject this hypothesis to further scrutiny, we thought it desirable to employ another approach. To do this we resorted to electrical stimulation of the superior colliculus through microelectrodes. This method enables one to do two things. On the one hand, one can record and find the location of the receptive fields of the recorded cells. Next, one can electrically stimulate this area (3).

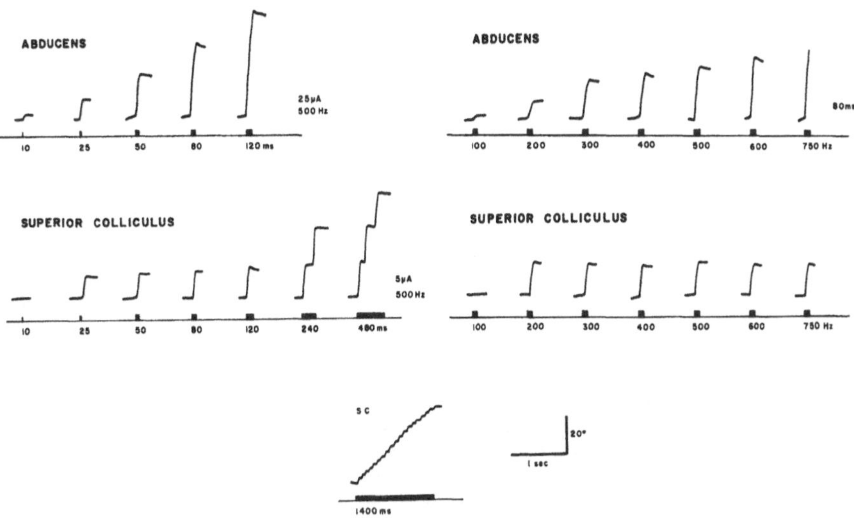

Fig. 5 Effects of electrical stimulation in the abducens nucleus and the superior colliculus as a function of burst duration and frequency. The long staircase of saccades shown at the bottom of the figure was elicited by stimulating within the anterior tip of the superior colliculus. All eye-movement records are horizontal with saccades going to the left.

To appreciate what electrical stimulation does in the superior colliculus, it is instructive to compare two structures: the colliculus, and the abducens nucleus, which as you know, is the structure which contains the motor neurons of the final common path to the lateral rectus muscle. We will only be concerned with the left side of Fig. 5. In the abducens, increasing the duration of the stimulation increased saccade size. This makes good sense, of course, because the longer the burst, the more the muscle will contract. But now if you do the same thing in the superior colliculus, a completely different result is obtained. Once above threshold, no matter how long the stimulus burst, the saccade size remains the same. If a certain duration is exceeded, one gets a staircase of identical saccades. At the bottom of the figure we have a record which was obtained while stimulating the anterior portion of the superior colliculus for 1400 msec. The result is a long staircase of small saccades.

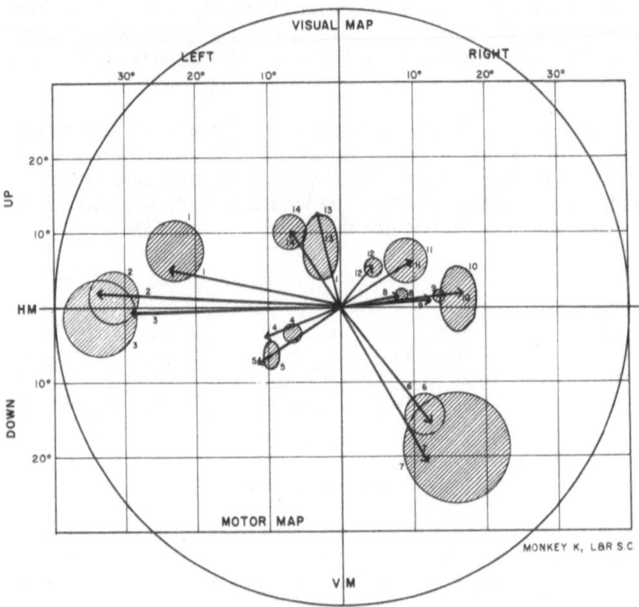

Fig. 6 Effects of recording and stimulation in the superficial layers of the superior colliculus. The visual map with the receptive fields of 14 units is superimposed on the motor map with its arrows representing the electrically elicited saccades at each of the 14 sites. The length of each arrow represents the mean length of 8-14 stimulation-elicited saccades; the direction of each arrow represents the mean direction of saccades. HM=horizontal meridian, VM=vertical meridian.

What about the size and direction of saccades? What determines that? As you have seen in Fig. 5, it is not a function of stimulation parameters. The size of the saccade and its direction is contingent upon where one stimulates in the colliculus. In the anterior part, tiny saccades are obtained. In the posterior one gets large saccades. Knowing this, we can take the next step. We can find out where the receptive fields of cells are located in the colliculus near the electrode tip and we can then stimulate and see what kind of saccade occurs. We have done this for many sites and a sample of the results appears in Fig. 6. Here we have superimposed the motor map and the sensory map; the receptive fields are shown as cross hatched circles. Saccades elicited by stimulation are shown as arrows. For example, at site

No. 1, we found the receptive field of the unit through the immobilized eye; we then occluded this eye and stimulated the colliculus. The size and direction of the ensuing saccade is shown by arrow No. 1. It is generally the case that stimulation brings the fovea into the receptive field of the stimulated cells. This method supports the idea of some sort of guidance system in the colliculus which enables the organism to move the eyes in a reflex fashion to various locations in the visual field.

There are of course many factors which determine whether or not the saccade should occur. This is necessary, for with each glance, many new stimuli impinge on the retina, and one of those must be chosen for the next saccade. This decision process must take time. We know, for example, that it takes about 200 ms to perform a saccadic eye movement in reaction to a light stimulus. Yet when you stimulate electrically in the superior colliculus, saccade latency is only about 20 or 25 ms. The length of time it takes for the visual information to reach the superior colliculus from the retina may take 50 to 100 msec. The second loop around from cortex down takes just a little longer. There remain almost 100 msec unaccounted for, and this probably represents the time required to make a decision about the size and direction of the ensuing saccade. We have here a rather complex series of events. I do not want to imply that the colliculus is a robot, which triggers an eye movement in response to any incoming light stimulus. An eye movement can only occur when there is a lot of activity in a particular region of the deeper layers of the colliculus. In order for this activity to occur there must be controlling or grating inputs from other areas.

The last point I want to make about the colliculus before looking at its inputs is this: if it is a structure which is involved in the initiation of saccadic eye movments, one would expect that its ablation should produce deficits in eye movement activity. It turns out that this is by and large true, but it is much clearer for animals lower on the phylogenetic scale. Thus, for example, in an animal like a tree shrew, the ablation of the superior colliculus has devasting effects; CASAGRANDE has a beautiful film of this. When you take a mealworm on the end of a pair of forceps and move it around rapidly in front of the treeshrew, this animal has beautiful head and eye movements. It follows the stimulus rapidly and when close enough, it will snatch the worm. When the colliculus is removed, there is no orientation and no eye movement; the animal is incapacitated and cannot get moving mealworms; it has to be fed differently. In the cat there is also serious deficit but there is considerable post-operative recovery. In the monkey, the deficits are less pronounced. Colliculectomy produces an increase in the latency of the eye movement initiation, a decrease in accuracy, and a decrease in spontaneous saccades. The effects are transitory, and over a period of many weeks the animal improves.

This then comprises the first part of my presentation. I will now proceed to talk about the nature of the inputs to the superior colliculus from the retina and from the visual cortex. The method we employed to study of the retinal inputs to the superior colliculus appears in Fig. 7 (4). We recorded in the eye from retinal ganglion cells extracellularly by inserting microelectrodes through the eyeball. Then we stimulated electrically the superior colliculus and also the optic chism with single pulses, enabling us to identify those cells which go to the superior colliculus by the antidromic activation. This is a difficult experiment, because one can activate only a very small portion of the cells in the retina from the superior colliculus.

It is important to note at this point that several distinct classes of retinal ganglion cells have been identified in mammals. This was first discovered in the cat by ENROTH-CUGELL and ROBSON (5). They discovered two classes which they called the X- and Y-cells. Subsequently STONE has added a third class of cells

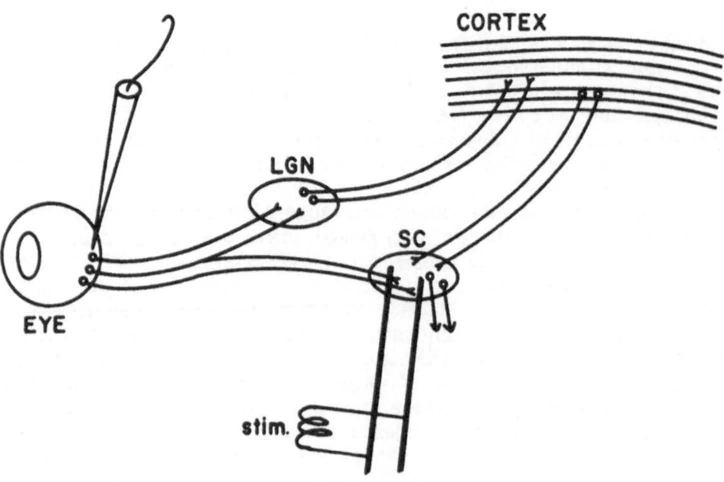

Fig. 7 Method for recording retino-tectal ganglion cells.

which he called W-cells, the very slowly conducting small ganglion cells of the retina (6). Work on the monkey by PETER GOURAS has also disclosed that there are several subclasses of cells (7). When we did our work we confirmed these findings. In our work on the monkey we divided retinal ganglion cells into three classes. Two of these are very distinct in the monkey and the third one is sort of a grab bag. The first two are the color-opponent cells and the broad-band cells as originally named by GOURAS. Color-opponent cells are those which respond to one wavelength in the center and to another wavelength in the surround and do so in an opponent fashion. So, for example, you may have a cell which in the center gives an "on" response to red light and in the surround gives an "off" response to green light. The cells are quite numerous in the retina. The second class of cells are called broad-band because they do not seem to be particularly selective for color. The antagonistic center-surround organization is independent of wavelength. When one looks at the details of the organization of these two classes of cells, they appear to be very similar to the X and Y categorization of the cat. There are several criteria; Fig. 8 shows some of these. On the left-hand side we have cells which are color opponent and on the right we have cells which are broad-band. I will simply tell you what the distinguishing features are in our experience. First of all, we found that color-opponent cells are specific for color while broad-band cells are not. That is the first criterion and really the defining criterion because it is the easiest to use. The second is that color-opponent cells tend to respond in a relatively sustained fashion as you can see on the left hand side of Fig. 8. By contrast, the broad-band units fire an initial, sharp high frequency burst of about 100 msec duration. After this, the response frequency rapidly declines to the baseline level. The third criterion is the conduction velocity. The broad-band units conduct much more rapidly than do the color-opponent cells.

The third population of retinal ganglion cells seem to resemble STONE'S W category. The basic characteristic of these cells is that they are small and conduct very slowly. They tend to respond more sluggishly. The receptive-field characteristics of these cells are not homogeneous. That is why I said that this

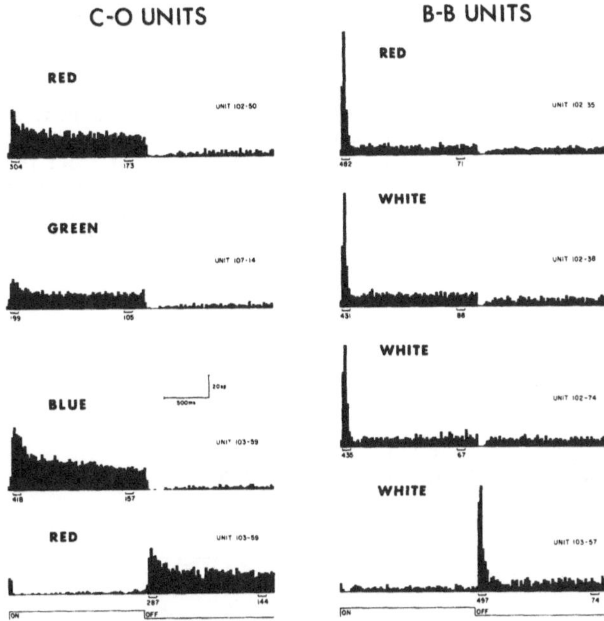

Fig. 8 Stimulus-response histograms for three color-opponent and four broad-band cells. Stimuli were small, stationary discs of the color indicated, centered on the receptive field. The bottom two histograms on the left side were obtained from the same cell and demonstrate the color-opponent response. Twenty repeated measures per histogram.

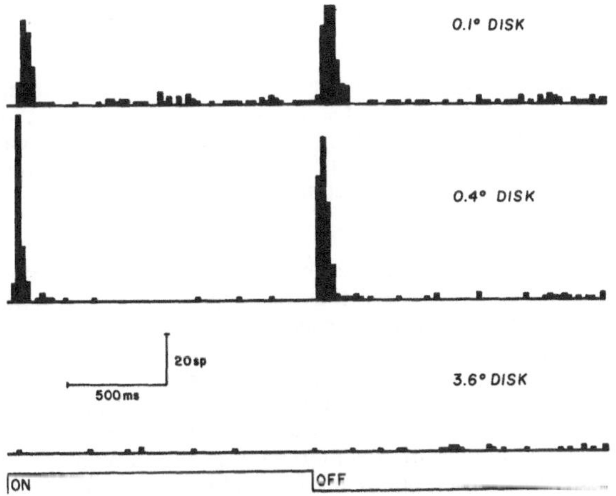

Fig. 9 Response of retino-tectal cell to flashing, stationary stimuli of different sizes. Twenty repeated measures per histogram.

category is sort of a grab bag. An example of a cell falling in this category is shown in Fig. 9. This particular cell gives both an on and off response in the center of the receptive field. When a large stimulus is used there is no response at all. This cell then has very strong center-surround antagonism. It is not color selective. This particular cell was antidromically activated from the superior colliculus. In general it is true that cells falling into the third category project to the colliculus extensively. In addition, the superior colliculus also receives an input from the broad-band cells but apparently in much smaller numbers. The color-opponent cells do not go to the colliculus at all; they go instead to the lateral geniculate nucleus. The broad-band cells actually go both to the colliculus and the lateral geniculate nucleus, and in the latter they appear to terminate mostly, if not exclusively in the magnacellular layers (8, 9).

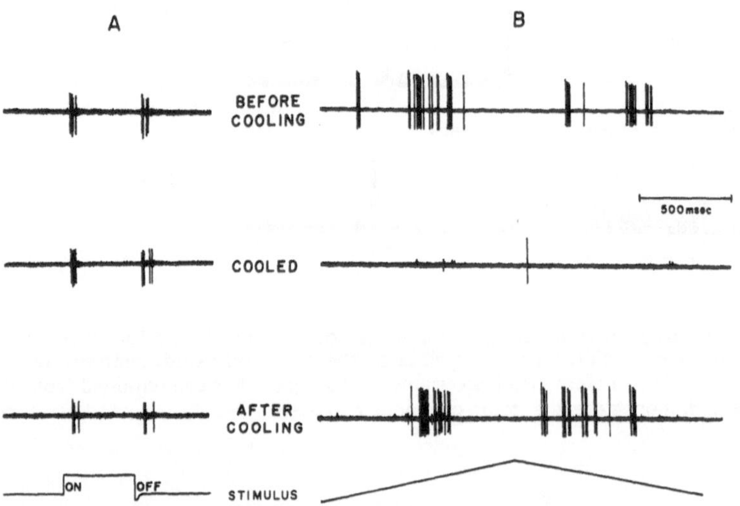

Fig. 10 Response characteristics of two units in the superior colliculus before, during, and after cooling of visual cortex. A: unit 220 below the surface of the colliculus. The lower trace represents shutter voltage. Stimulus: flashing spot centered in receptive field. B: unit 400 below the surface of the colliculus. Lower trace is mirror galvanometer voltage. Stimulus: sweeping spot.

I now come to the final part of my talk: what is the nature of the input to the superior colliculus from visual cortex? To answer this question, several different techniques can be employed. One of these is rather crude and let me tell you about it first. This method involves the disruption of the cortico-tectal pathway while recording in the colliculus. Disruption can be accomplished readily in two ways. One is to ablate visual cortex and see what happens in the colliculus. The other is to place a thermoelectric cooling plate on the cortex to cool it until the cells in the cortex become unresponsive. The advantage of this method is that it is reversible. Collicular cells can be studied before, during and after cortical cooling. When one does this, the effects are most dramatic, but only in the deeper layers. Recording in superficial layers of the colliculus shows practically no effect at all. By contrast, in the deeper layers every cell we have studied stopped responding to visual stimuli when cortex was cooled. This is shown in Fig. 10. It appears then that except for the superficial layers of the colliculus where the retinal input is the heaviest, collicular cells depend extensively on the cortical input. Without the cortico-tectal downflow, visual information does not seem to make it to the deeper layers (10).

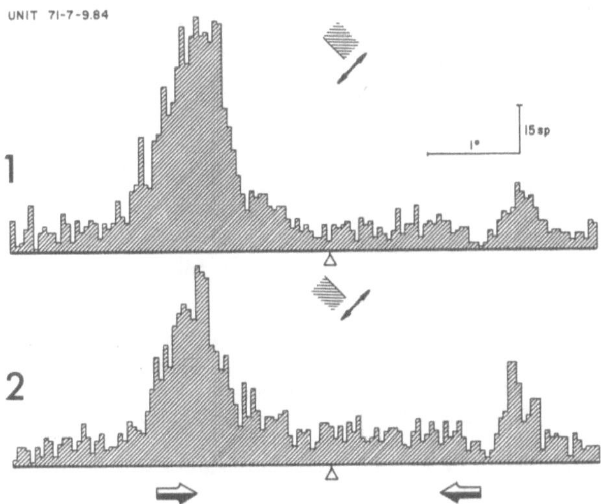

Fig. 11 Stimulus-response histogram of a cortico-tectal cell to moving edges; 30 repeated presentations at 2°/s velocity. 1: light edge in up direction, dark in the other. 2: dark edge up in direction, light in the other.

The second approach we used in studying the cortico-tectal input was to examine the characteristics of those cells in area 17 which project to the colliculus. The method is similar to the one we used in the retina. Instead of sticking the electrode into the eyeball, we place it in cortex and we stimulate the superior colliculus to antidromically activate the cortico-tectal cells. This is a masochistic experiment because we had to examine thousands of cells before collecting 50 which were activated antidromically. These cells are located almost exclusively in layer five of visual cortex. Functionally, they fall into the so called complex category of the HUBEL and WIESEL classification system. An example of such a cell appears in Fig. (11). The cell responds both to light and to dark edges. The size of the receptive field as assessed by how long the cell discharges as a stimulus traverses the field is rather large, at least a degree across. The cell is directional, and the discharge to repeated stimulation is vigorous and consistent. Some cortico-tectal cells are directional some are not; some but not all such cells are also orientation specific. One may ask: since we do have oriented and directional cells which project to the superior colliculus, how come the colliculus itself lacks orientation and direction specificity? There are at least two possibilities. One is that these cells converge extensively upon a collicular cell, thereby losing their specificity. The other possibility is that the cortical input has some sort of modulatory or gating function on the collicular cells rather than providing them with their receptive field properties.

In summary, the superior colliculus is a structure which appears to be involved in saccadic guidance operations. There are cells in the intermediate and deeper layers which are selective to saccadic eye movement, and respond prior to certain saccades. The motor responses are arranged in a topographic fashion and are in a register with the sensory map above it. When the colliculus is ablated, there are deficits in visually guided eye movements. The colliculus receives rather specific input from the retina from broad-band cells and from the third class of sluggish small cells. It does not receive an input from color-opponent cells. From visual cortex there is a very important downflow because when this pathway is disrupted, the deeper collicular cells no longer respond to visual stimuli. From area 17 the projection is from the complex cells located in layer 5.

References

1. M. E. Goldberg and R. H. Wurtz, J. Neurophysiol. 35, 542 (1972).
2. P. H. Schiller and F. Koerner, J. Neurophysiol. 34, 920 (1971).
3. P. H. Schiller and M. Stryker, J. Neurophysiol. 35, 915 (1972).
4. P. H. Schiller and J. G. Malpeli, J. Neurophysiol. 40, 428 (1977).
5. C. Enroth-Cugell and J. G. Robson, J. Physiol. 187, 517 (1966).
6. J. Stone and Y. Fukuda, J. Neurophysiol. 37, 722 (1974).
7. P. Gouras, J. Physiol. 204, 407 (1969).
8. B. Dreher, Y. Fukuda and R. W. Rodieck, J. Physiol. 258, 433 (1976).
9. P. H. Schiller and J.G. Malpeli, The functional specificity of the lateral geniculate nucleus laminae of the rhesus monkey. In press.
10. P. H. Schiller, M. Stryker, M. Cynader and N. Berman, J. Neurophysiol. 37, 181 (1974).
11. B. L. Finlay, P. H. Schiller and S. F. Volman, J. Neurophysiol. 39, 1352 (1976).

Properties of Saccade-Related Unit Activity In the Monkey Superior Colliculus

David L. Sparks, Jay G. Pollack, and Lawrence E. Mays
Department of Psychology and The Neurosciences Program
University of Alabama in Birmingham
Birmingham, Alabama 35294

The foveation hypothesis of SCHILLER and KOERNER (1) maintains that the superior colliculus (SC) is part of a neural system which acquires visual targets for foveal viewing. The SC is thought to be involved in coding the location of an object relative to the fovea and in eliciting saccadic movements which produce foveal acquisition of the object. A major line of evidence offered in support of this hypothesis is the finding that neurons in the intermediate and deeper layers of the SC discharge maximally prior to eye movements with a particular direction and amplitude (1-9). These neurons have a movement field - that is, the neurons discharge prior to a range of movements of similar directions and amplitudes. However, the temporal relationship between the spike discharge and saccade onset has not been examined carefully. Furthermore, it has been suggested that the discharge of SC neurons reflects an attentional mechanism rather than a command to initiate a particular saccade (10,11).

The first experiment to be described was based upon the following reasoning. If the foveation hypothesis is correct and neurons in the SC are involved in initiating saccadic movements, then one should be able to isolate neurons which display the following characteristics:

1) for visually-elicited movements the latency of the neuronal response should be tightly linked to the latency of the saccade; and
2) in a situation in which the occurrence of a visual stimulus sometimes elicits a saccade and sometimes fails to elicit a saccade, some component of the neuronal activity should be related to the probability of saccade occurrence - that is, the probability of the neural response should be highly correlated with the probability of the behavioral response.

We examined these hypotheses in the following manner. Four rhesus monkeys were trained to track a visual target presented on an oscilloscope with a CRT viewing area of 11 by 15 in. The horizontal and vertical positions of the target, a dot subtending a visual angle of less than one-tenth degree were controlled by a small laboratory computer. Horizontal and vertical eye movements were measured using a magnetic field, search coil technique (12).

Using standard chronic microelectrode recording techniques, the movement field of a SC neuron was plotted. Then, four types of trials, illustrated in Fig. 1, were

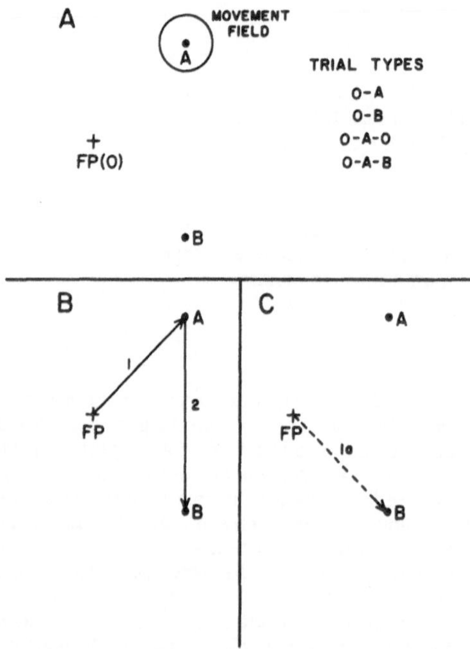

Fig. 1 The eye movement task for experiment 1. See text for details.

presented randomly. During 0-A trials, monkeys were required to fixate a dot at the center of the oscilloscope screen (0) and after a variable interval the target was moved to a position near the center of the movement field of the neuron being studied (A) (See Fig. 1A). If the animal acquired the target within 400 msec and maintained fixation for 2 sec, a liquid reinforcement was presented.

During 0-B trials, after a variable duration of fixation of point 0, the target was moved to point B. A saccade to acquire the target at point B was not preceded by a neural discharge, i.e., saccades to acquire target B were not in the movement field of the neuron being investigated.

0-A-B trials required a fixation of point 0. Then the target was moved to point A and remained in that position for a chosen interval, T_1. At the end of interval T_1, the target was moved to point B and remained at that position until acquired by the subject. At certain durations of T_1, the monkey would sometimes make a saccade to position A and then a second saccade to position B (Fig. 1B). On other trials with the same duration of target A, the saccade to acquire the target at position A did not occur and a single saccade to position B was observed (Fig. 1C).

Similarly, during some 0-A-0 trials, with a given duration of target A, a saccade to acquire A would occur followed by a saccade back to 0. On other trials, with the same duration of target A, a saccade to acquire the target at position A failed to occur and the monkey maintained fixation of point 0.

From recordings taken during 0-A-0 and 0-A-B trials, we were able to determine the degree of relationship between the probability of a neural response and the probability of saccade occurence. The relationship between the latency of the behavioral response and the latency of the neural response was determined from recordings taken during 0-A trials.

We have isolated one type of SC neuron, hereafter referred to as "saccade-related burst neurons", which generates a pulse of spike activity beginning approximately 20 msec before the onset of saccades with particular amplitudes and directions. Figure 2 illustrates typical spike acitvity of two such neurons. Examination of the instantaneous frequency record reveals that for both neurons there is a relatively discrete pulse of spike activity which occurs approximately 20 msec prior to the onset of the saccade.

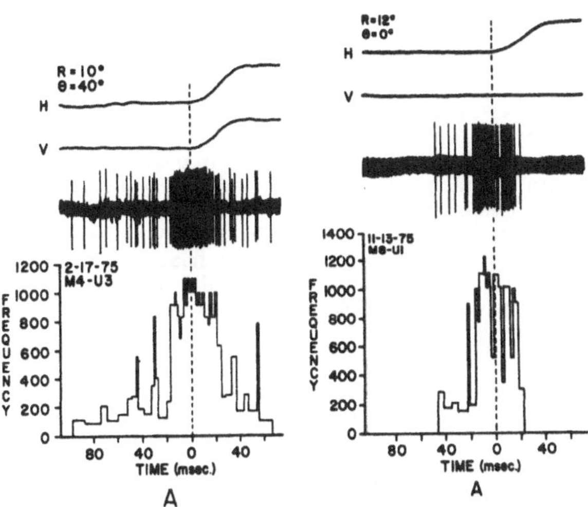

Fig. 2 Saccade-related discharge patterns recorded from two SC neurons. H: horizontal eye position. V: vertical eye position. Middle tracing: spike activity. Bottom graph: instantaneous spike frequency. The dotted line represents the onset of the eye movement.

As Fig. 3 illustrates, for this class of SC neuron, the pulse of spike activity is tightly coupled to saccade onset. Spike burst and saccade latencies were measured for 38 saccades to the same point in the movement field for the neuron illustrated on

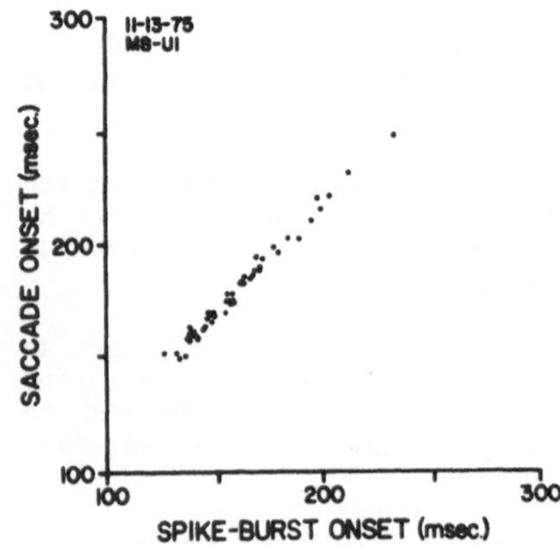

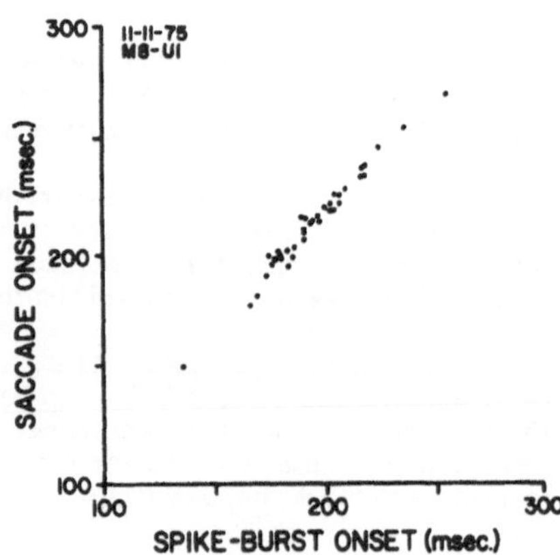

Fig. 3 Relationship between spike burst latency and saccade latency for two neurons. The abscissa represents the interval between target onset and the onset of the spike pulse. The ordinate represents saccade latency.

the left and for 50 saccades for the neuron illustrated on the right. The high degree of association between the pulse of spike activity and saccade onset is apparent. Thus, the functional properties of this type SC neuron meet one criterion for participation in the initiation of visually-elicited saccades - the latency of the neural response is tightly linked to the latency of the saccade.

Also, for saccade-related burst neurons there is a high degree of relationship between the occurrence of the pulse of spike activity and the occurrence of the saccade associated with this activity. Figure 4 illustrates typical results for this type neuron. The top left segment of the figure illustrates a 0-A-B trial in which the monkey made a saccade to acquire the target at position A and a second saccade to acquire the target at position B. The instantaneous frequency record of this trial shows a typical response - some pre-pulse activity followed by a fairly discrete pulse of spike activity. The top right segment of the figure shows a trial with the same duration of target A, but on this trial a saccade to acquire A did not occur and a single saccade to acquire target B was observed. Some build-up of spike activity was observed, but a sharp pulse of activity was not present.

The bottom left segment of Fig. 4 illustrates a 0-A-0 trial in which a saccade to acquire target A and a saccade to reacquire target 0 occurred. The saccade to acquire target A was preceded by a pulse of spike activity. The bottom right segment of the figure illustrates a 0-A-0 trial, with the same duration of target A, in which a saccade to acquire A did not occur and the eye remained fixated at position 0. Once again, some build-up of spike activity was observed, but a discrete pulse of activity was not present.

Results of the first experiment are interpreted as supporting the foveation hypothesis of SC function. A response of one class of SC neurons has been observed which is tightly coupled to the onset of appropriate saccades. Furthermore, the probability of occurrence of this response is highly correlated with the probability of saccade occurrence. It should be noted that these results refer to a homogeneous category of cells - neurons with a discrete burst of activity preceding the onset of an

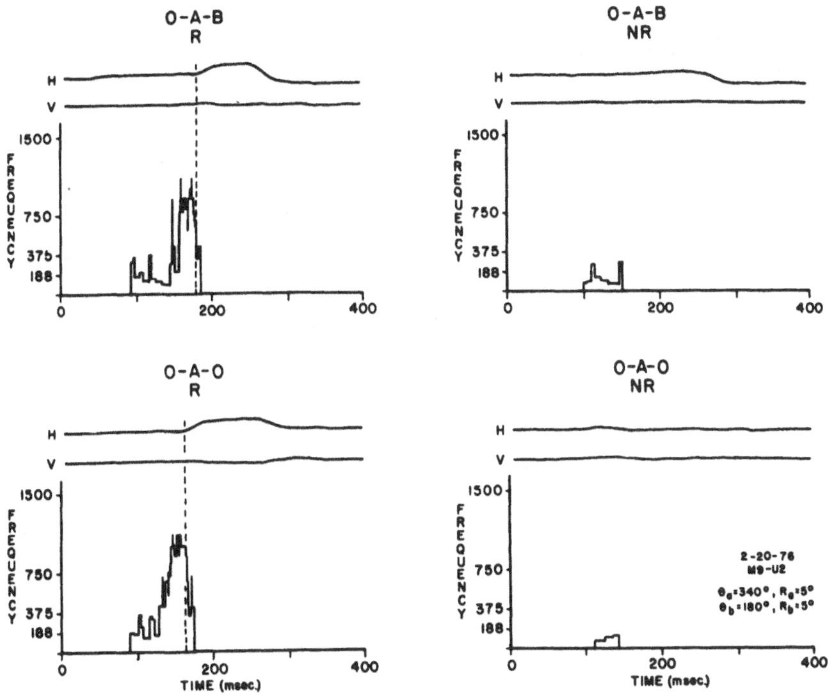

Fig. 4 Top left: 0-A-B trial with a saccade to A and a second saccade to acquire B. Top right: 0-A-B trial with a single saccade to acquire B. Bottom left: 0-A-0 trial with a saccade to 0 and a saccade to reacquire 0. Bottom right: 0-A-0 trial in which the monkey maintained fixation of 0.

appropriate saccade. Other SC neurons display different patterns of saccade-related activity which are not characterized by such tight coupling.

The second experiment to be described is concerned with the linkage between neural activity in the superficial layers of the SC and neural activity in the intermediate and deeper layers. These experiments are still underway and the data reported are preliminary findings.

As a microelectrode is lowered through the SC of an alert monkey, three major types of cells are encountered. As the electrode first enters the SC, neurons with visual properties are encountered. Then, as the electrode reaches a depth of 1.5 - 2.5 mm beneath the surface of the colliculus, neurons with both visual and eye-movement related spike activity are encountered. Finally, as the electrode goes deeper, neurons with spike activity related more exclusively to eye movements are seen.

Although it has been commonly assumed that the progression of cell types from visual to visual-motor to motor represents a translation from the encoding of visual information to the generation of efferent commands, it seems unlikely that this translation occurs completely within the SC. Indeed, very little is known concerning the intrinsic connections of the primate SC or concerning the functional linkage between visually-related activity in the superficial layers of the SC and eye movement-related activity in the deeper layers of the SC.

We have attempted to look at this linkage by training monkeys to make identical saccades in response to a single target in one half of the visual field or to two brief targets in the other hemifield. This paradigm is illustrated in Fig. 5.

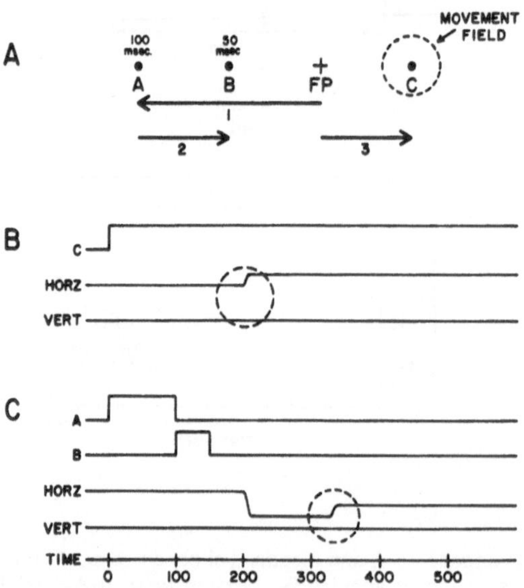

Fig. 5 The task for experiment 2. FP: fixation point. C: target in the receptive or movement field of a SC neuron. A and B: targets out of the receptive or movement field. Horz: horizontal eye position Vert: vertical eye position.

A tone signalled the appearance of the target at the center of the oscilloscope screen (FP). Monkeys were permitted 500 msec to acquire the target and then were required to maintain center fixation for a variable period (1-3 sec). On single saccade trials (Fig. 5B) the offset of the center target was coincident with the onset of target C in the receptive or movement field of the neuron under observation. If target C was acquired within 400 msec and fixated for 2 sec, a water reinforcement was delivered. On double saccade trials (Fig. 5C), the offset of the center fixation target was followed by successive presentations of targets A and B, neither of which were in the neuron's receptive or movement field. The total duration of targets A and B was less than the reaction time of the monkey (i.e., both A and B were off before a saccade occurred). Nevertheless, the computer program made reward contingent upon a saccade to position A followed by a saccade to position B. A critical feature of this paradigm is that a saccade from position A to position B has the same amplitude and angle as a saccade from FP to C. Only the origins of the movements differ.

Figures 6 and 7 present, schematically, predictions of the response of SC neurons during performance of this task. The right visual field is mapped upon the

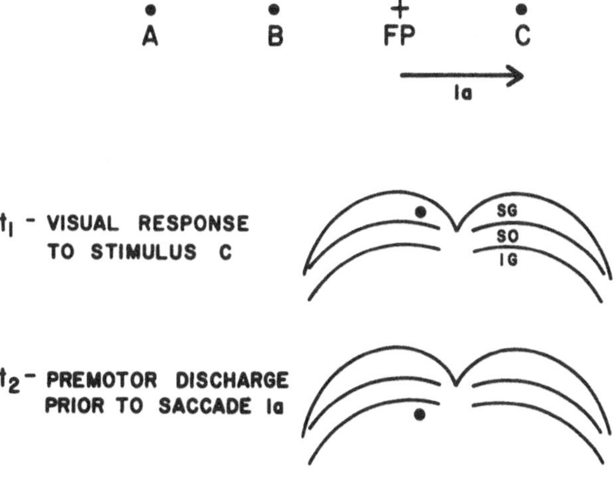

Fig. 6 Hypothetical example of SC neural activity occurring during single saccade trials

superficial layers of the left SC and saccade-related unit activity in the left SC occurs before saccades to the right. On single saccade trials, with center fixation, the onset of target C in the right visual field will elicit a visual response in the superficial layers of the left SC (Fig. 6, t_1). Later, a premotor discharge will be observed in the intermediate layers of the left SC prior to the saccade to acquire target C (Fig. 6, t_2).

On double saccade trials, the onset of target A in the left visual field will elicit a response from neurons in the superficial layers of the right SC (Fig. 7, t_1). The onset of target B, also in the left visual field, will elicit a visual response from a different population of SC neurons in the superficial layers of the right SC (Fig. 7, t_2). Next, a neuronal discharge will occur in the intermediate layers of the right SC before the left saccade to A (Fig. 7, t_3). Finally, a discharge of neurons in the intermediate layers of the left SC will occur before the rightward saccade from A to B (Fig. 7, t_4). The neuronal discharge in the deeper layers of the left SC will occur, presumably, in the absence of visually-related neural activity in the overlying superficial layers of the left SC.

Preliminary results confirm our expectations. SC neurons in the intermediate layers discharge prior to saccades elicited by the two brief visual targets. Figure 8 (top section) shows the saccade which occurred to acquire the target presented in the left visual field. Also shown is the burst of spike activity which preceded the onset of this saccade. The pulse of spike activity preceds the onset of the saccade by approximately 25 msec. Figure 8 (bottom section) also illustrates the results when two brief stimuli are presented. Both are terminated before the eye leaves center fixation. The monkey made a right saccade to the location of target A and then a left saccade to the location of target B. The saccade from A to B is of the same

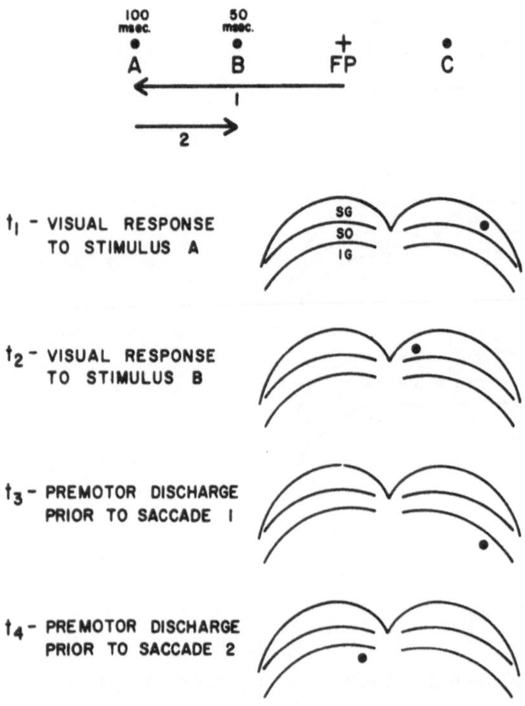

Fig. 7 Hypothetical example of SC neural activity occurring during double saccade trials.

distance and direction as the saccade from the fixation point to C. A vigorous burst of spike activity preceds the saccade from A to B. The onset of pulse or spike activity preceds the onset of the eye movement by approximately 20 msec; a time comparable to the control condition.

The spike burst preceding the saccade from A to B may occur in the absence of visually-elicited neuronal activity in corresponding regions of the superficial layers of the same colliculus. Figure 9 (top) illustrates the transient "on" response and the subsequent sustained discharge evoked by the onset of stimulus C in the receptive field of a SC neuron in the superficial layers. Figure 9 (bottom) illustrates the absence of neural activity during double saccade trials.

We believe these observations have important implications for attempts to describe the role of the SC in the control of eye movements. Current models of the saccadic eye movement system (13,14) assume that the major error signal is retinal error - the distance of a retinal image from the fovea. Neural models usually assign to the SC a major role in the computation of retinal error. There is a topographically coded signal of retinal position in the superficial layers of the SC. Aligned with this visual map in the superficial layers is a motor map in the deeper layers of the SC. It is assumed that activity in the superficial layers becomes translated into activity in corresponding regions of the deeper layers. Thus a saccade with a particular

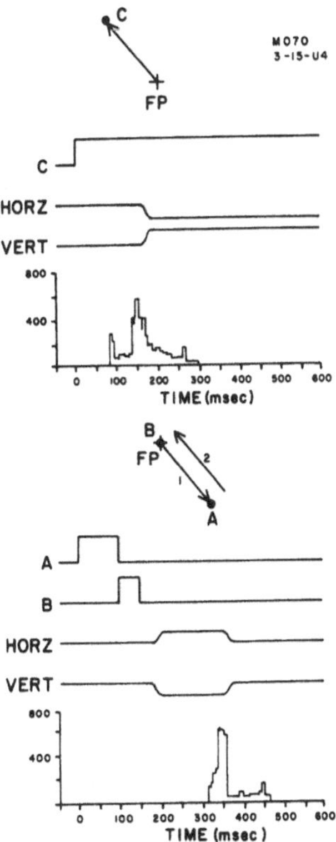

Fig. 8 Discharge pattern recorded from a SC "burst" neuron during single and double saccade trials.

direction and amplitude is produced which will bring the foveal projection onto that part of the visual field occupied by the receptive fields of those neurons which initiated the movement. This would occur regardless of the initial position of the eye in the orbit.

While this scheme may be valid for most saccades, these preliminary results suggest that this is not the only mode of triggering saccade-related activity in the SC. Vigorous activity may occur in the superficial layers which does not become translated into saccade-related discharges in corresponding regions of the deeper layer cells. Furthermore, saccade-related discharges may occur in the deeper neurons without prior activation of superficial layer neurons receiving a direct retinal input. Thus, activity of neurons in the superficial layers receiving direct retinal input is neither necessary nor sufficient to produce discharges of saccade-related neurons in corresponding regions of the deeper layers of the same colliculus.

What are the signals which enable the saccade-related acitvity of the deeper SC neurons in the double saccade task? At this point we do not know. The vectors for the two saccades may be precomputed based solely upon the relative position of the two visual targets and current eye position. Or, the vector for the second movement may be computed based upon the location of the second target with respect to a future position of the eye. Or, the vector of the second movement may be computed

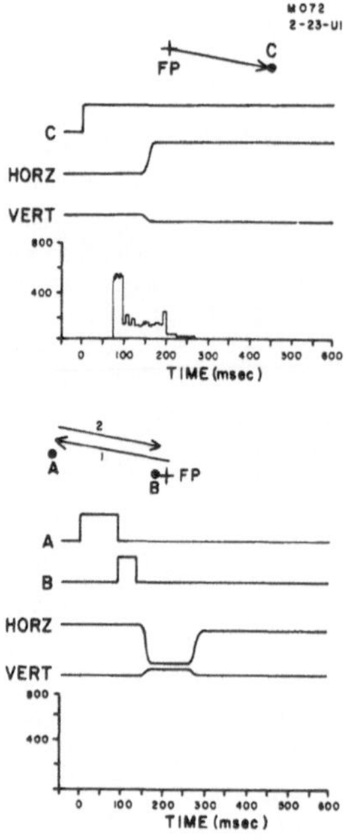

Fig. 9 Discharge pattern recorded from a SC "visual" neuron during single and double saccade trials

after the eye reaches the first target position based upon a memory of the location of the second target. Actually, the animal may use several of these strategies or even other strategies to solve the task. However, our data indicates that the SC participates not only in the elaboration of saccades elicited by a conventional retinal error signal, but also in the saccades elicited by these more complex, higher-order, relative signals. An important implication of these findings is that the brain areas performing these higher-order computations must have topographically arranged inputs to the SC in order to activate particular subsets of the saccade-related collicular units.

Acknowledgements

We thank Dick Holland, Sally Marcus and Richard Sheetz for technical assistance.

This work was supported by NIH Grant EY 01189. Jay G. Pollack was supported by a NEI Postdoctoral Research Fellowship 1 F32EY05112. L. E. Mays was supported by a NIMH Training Grant, 1 T32MH14286-02.

References

1. P. H. Schiller & F. Koerner, J. Neurophysiol. 34, 920 (1971).
2. A. Arduini, R. Corazza & P. Marzolla, Brain Res. 73, 473 (1974).
3. D. L. Robinson & D. C. Jarvis, J. Neurophysiol. 34, 925 (1971).
4. P. H. Schiller, in: Cerebral Control of Eye Movements and Motion Perception, ed. by J. Dichgans, & E. Bizzi (S. Karger, 1972), pp. 122-120.
5. P. H. Schiller & M. Stryker, J. Neurophysiol. 35, 915 (1972).

6. D. L. Sparks, Brain Res. 90, 147 (1975).
7. D. L. Sparks, R. Holland & B. L. Guthrie, Brain Res. 113, 21 (1976).
8. M. Straschill & F. Schick, Exp. Brain Res. 27, 131 (1977).
9, R. H. Wurtz & M. E. Goldberg, J. Neurophysiol. 35, 575 (1972).
10. R. H. Wurtz & M. E. Goldberg, J. Neurophysiol. 35, 587 (1972).
11. C. W. Mohler & R. N. Wurtz, J. Neurophysiol. 39, 722 (1976).
12. A. Fuchs & D. Robinson, J. Appl. Physiol. 21, 1068.
13. D. A. Robinson, Kybernetik 14, 71 (1973).
14. D. A. Robinson, Basic Mechanisms of Ocular Motility and Their Clinical Implications, ed. by G. Lennerstrand and P. Bach-y-Rita (Pergamon, New York, 1975), pp. 337-374.

Potentials Accompany Eye Movement

John C. Armington
Northeastern University

In past years visual theory has attributed several important functions to the small saccadic movements of the eye; but more recently, a succession of experiments has challenged much of this. Stabilized image experiments have shown that no component of eye movement including microsaccades is responsible for the excellent acuity of the human eye (1). When the visual field is stabilized, it fades, but saccades are not essential to prevent this from happening furthermore, saccades may not be needed to restore the position of the eye when it has slipped off the target (2). They contribute to counting behavior only under restricted conditions (3). Thus, it now seems that the principal function of saccades may be merely that of scanning the visual field in the vicinity of fixation, and it has been suggested that microsaccades actually interfere with vision (4). Yet, despite all of these negative indications regarding specific roles for small saccadic eye movements, it remains true that they are accompanied by significant electrical activity at the brain and at the retina. This activity is the subject of the present discussion.

Recording Method

A combination of techniques that are used for eye movement recording, evoked potential recording, and electroretinography is needed to investigate the potentials that are linked to saccadic movement. A sketch of a suitable recording assembly is presented in Fig. 1. The main parts are a stimulator, an eye movement recording system, and an electrophysiological system for recording electroretinograms and cortical potentials. In addition, a tape recorder for storing the recordings and a small on-the-line computer for processing the recordings are included in the system.

Depending upon the objectives of the experiment, the stimulator is capable of presenting the subject with a range of stimuli extending from single fixation points to checkered stimulus fields of high contrast. Luminance, field size, and other characteristics of the stimulus are easily varied. For most experiments, the stimulus is a steady grating that does not move, flash, or alternate. However, a mechanism, that is not shown in this figure, makes it possible to record responses to phase alternation of the pattern (5). Responses produced by this conventional method of stimulation can thus be compared with those produced by saccades.

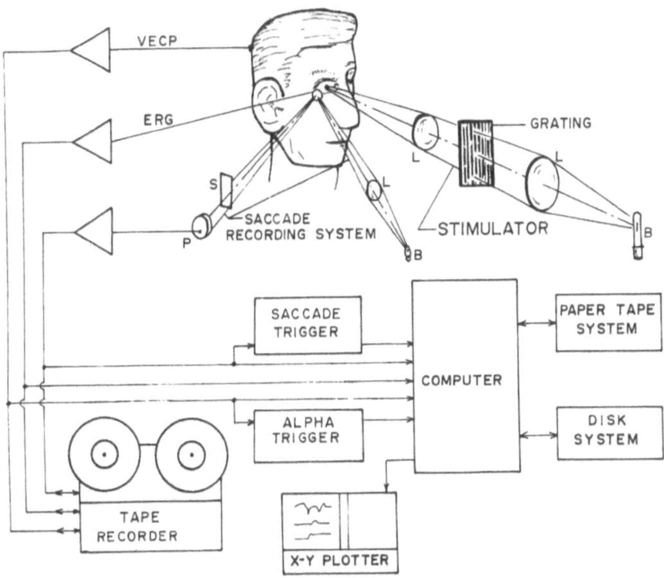

Fig. 1 Sketch of recording system. The following components of the stimulator and the eye movement recording system are identified by letters: B, tungsten lamp; L, lens; S, straight edge stop; P, photocell. The apparatus is explained further in the text.

The eye movement recording system is of the optical lever type employing a mirror on a tight fitting contact lens and a photoelectric cell (6). A beam of light, reflected from the mirror moves back and forth with the eye. It is intercepted by a straight edged stop. The amount of light that passes to the photoelectric cell is, thus, proportional to the position of the eye. The eye position signal is amplified and recorded directly on magnetic tape. Most of the present experiments also require that the eye movement signal be fed to a special analog computer device, the saccade trigger. This device sends synchronization pulses that signal the occurrence of saccades of specified magnitude and direction to the computer.

An electrode also supported by the contact lens, is used for obtaining the electroretinogram. Scalp electrodes pick up the evoked potentials. Amplifiers, suitable for human physiological research, increase the amplitudes of these signals so that they may be sent to the computer, the tape recorder, or both. Usually signals are recorded on tape and played into the computer subsequent to actual recording. In addition to all these devices, the system also includes a second analog device, the alpha trigger. It signals the computer whenever alpha, defined as filtered ten per second electroencephalographic activity, exceeds a preset magnitude (7).

The computer may be programmed to accomplish a variety of functions. Its main activity is that of response averaging. Synchronization pulses delivered from the saccade trigger pace this action. Furthermore, because the computer samples the recording continuously, averaging can be performed prior to saccadic movement of the eye as well as after it. The alpha trigger places an additional contingency upon response averaging; potentials that occur when alpha is present in the EEG can be separated from those that occur when it is not. Finally, the computer is used in another configuration to plot the data and to fit curves to the results (8).

An illustration of averaged waveforms obtained with this system is shown in Fig. 2. An average saccade for a shift in gaze to the right (marked "8" in Fig. 2) is

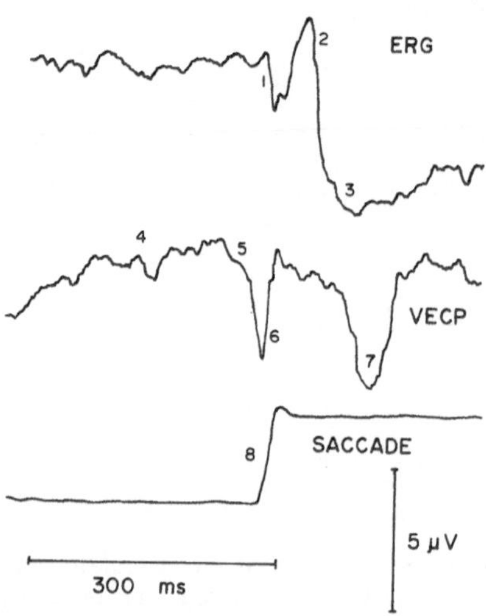

Fig. 2 A typical recording of potentials accompanying microsaccades. The components marked with numbers are explained in the text.

shown at the bottom of this figure; the evoked potential and the electroretinogram that accompanied it are shown above. Certain features of these waves will be identified. The electroretinogram is characterized by a positive B wave "2" and a late negative potential "3". The retinal A wave "1" is confounded with an eye movement artifact. The evoked potential also has a complex waveform. Most conspicuous is a positive peak "7". Small potentials that precede the saccade are evident in the scalp recording. These are an early slow negativity "4", an ensuing positivity "5" and a positive spike "6" that reaches its peak at the onset of the saccade. It must be emphasized that this figure is shown merely to identify the principal features of the recordings. As would be expected, the waveforms, particularly those from the scalp, differ according to electrode placement, stimulus conditions and a variety of other recording conditions. Certain properties of these potentials will now be considered in brief detail. Detailed consideration will not be given to the effects produced by the alpha trigger. It will merely be noted that results obtained with this device in operation indicate that alpha potentials do not contribute to the VECP, but instead can be regarded as an interfering noise (7). The discussion will first consider response potentials and then will turn to the potentials that precede microsaccades.

Response Potentials

One question that might be asked is how well response potentials that follow microsaccades compare with those elicited by conventional methods. The answer is that they seem to be virtually the same. Fig. 3 compares electroretinograms and evoked potentials that accompany saccades with those obtained by the now well-known method of stimulus alternation (5). The waveform is similar at both recording

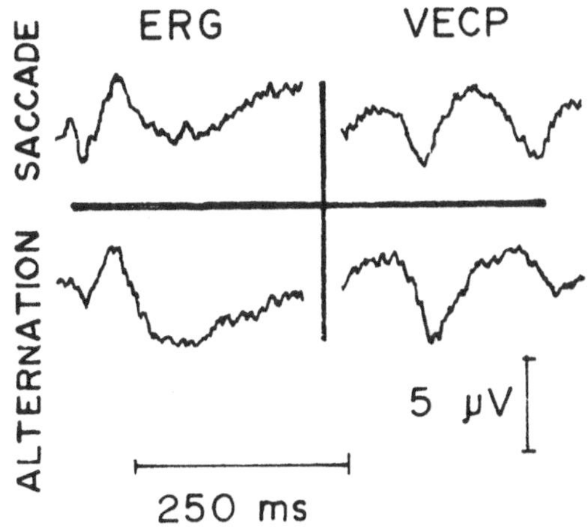

Fig. 3 A comparison of the responses that follow saccadic movement with those produced by stimulus alternation.

stations. The similarity is all the more striking when account is taken of the differences in the rates at which responses were produced for the two conditions. Alternating stimuli produced responses at the regular rate of 4 hertz, but saccades occurred irregularly triggering the computer once every several seconds.

It has also been found that the responses obtained with the two recording methods have similar stimulus properties. For example, in both cases there is a regular growth of response amplitudes with stimulus luminance. More importantly, the response waveforms recorded by both methods show similar additive properties (8).

Since the following paragraphs will review results testing additivity, it will be explained further. Additivity in the present case refers to response waveforms only. Stimulation of any very small area of the retina is deemed to produce "elemental" responses--both at the eye and at the cortex. When larger areas are stimulated, a number of elemental responses are produced simultaneously. These add together to produce the recorded result (9, 10, 11). Several experiments have shown that while additivity does not hold in an exact sense, it often holds within the limits of practical usefulness (8). In one experiment (12), responses were recorded as the subject made spontaneous saccades that swept his eye across gratings of several different spatial frequencies. It was found that the size of the responses depended on the spatial frequency of the stimulus; the largest responses were obtained when the visual angle of the stimulus stripes matched that of the saccades. In other words, the largest response was obtained both at the retina and at the cortex when the maximum number of receptors within the stimulus field was stimulated as a result of the saccadic movement.

Additivity may be tested more directly in another way. Responses are obtained when alternation of the stimulus grating is not complete and different fractions of a stripe width are stimulated. Results may be compared for potentials initiated by saccades with those produced by stimulus alternation. In the case of the eye movement method, the system is set to respond only to saccades of specific preset size. The data are recorded on tape and replayed into the computer with different presettings, each of which corresponds to a specific excursion across the test pattern: the larger the saccade, the greater the proportion of stripe width traversed and the greater the number of receptors stimulated. In the case of stimulus alternation, the grating pattern is displaced different fractions of a stripe width. The apparatus for producing phase alternation is not shown in Fig. 1. The procedure is, however, to displace the stimulus image back and forth with a sharp square wave motion. A small electromagnetically driven mirror is used for this purpose. The mirror may be driven to produce displacements that are preset proportions of a full stripe width.

Amplitude measures for the b wave and for the VECP recorded for saccades of several sizes across a fixed grating are shown in the upper part of Fig. 4. The

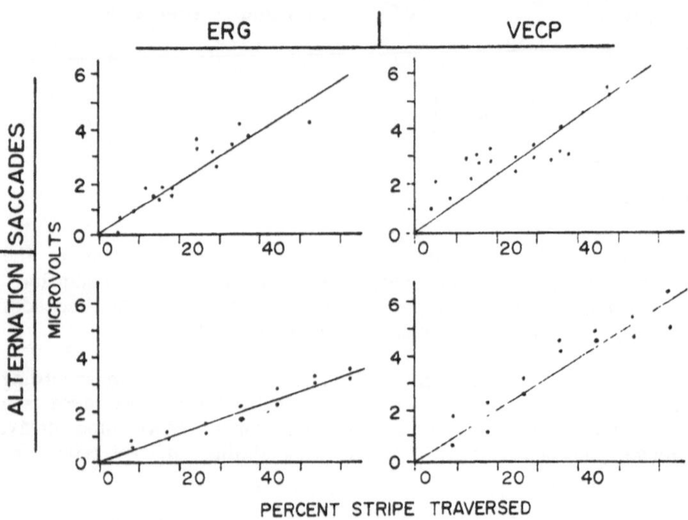

Fig. 4 A comparison of amplitude measures of responses obtained by alternation and by the saccadic eye movement method for different amplitudes of movement (per cent of stripe traversed).

straight line, fitted to the points, shows that the response is proportional to the size of the eye movement and hence, to the proportion of the stripe traversed. Similar measures of responses produced by displacing the stimulus pattern are shown in the bottom of the figure. A close linearity (particularly so for the ERG) between response amplitudes and the proportion of a stripe traversed is again in evidence. As far as additivity is concerned, the important feature in Fig. 4 is the proportionality between stimulus and response. The eye movement and alternation data were recorded on different occasions and with differing stimulus dispositions so that no comparisons of absolute response amplitudes are justified.

The analysis of responses released by microsaccades may be carried a step further by taking the spatial frequency of the stimulus pattern into account. If microsaccades of only a single amplitude are considered and if the spatial frequency of the stimulus pattern is made an experimental variable, an increase of response amplitude with frequency is to be anticipated because more receptors are then stimulated (13). New data showing this effect and collected with alternating, checkered stimulus patterns is presented in Fig. 5. The response of the electroretinogram, perhaps because of optical degradation of the retinal image, falls off as the stimulus becomes finely textured (to the left) but is maintained for

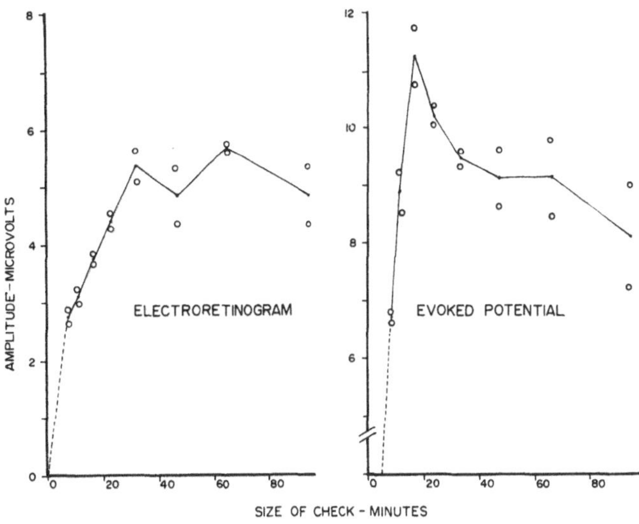

Fig. 5 Dependence of the amplitude of responses on the spatial frequency of the stimulus.

relatively coarse patterns. The visually evoked cortical potential, on the other hand, shows maximum response for stimuli whose elements subtend a little less than 20 minutes of arc. Results that are published in detail elsewhere show that when corrections are made for the effects shown in Fig. 5, both the electroretinogram and the visually evoked cortical potential show an increase in amplitude with spatial frequency that is consistent with the concept of response additivity (14).

YARBUS has speculated that blinking may trigger visual activity (15). An attempt to observe this activity was made by triggering the response averaging equipment with blinks rather than with microsaccadic movements. Blinks were detected in the form of electrical potentials picked up with standard electroencephalogram electrodes attached to the skin close to the eyes and were fed to the saccade trigger which then initiated the averaging process. Sample recordings obtained with this procedure are shown in Fig. 6. An averaged blink potential is shown on the top line. Activity obtained with occipital electrodes is shown below for a "no stimulus" condition and for three test luminances. A prominent potential appears at the same time as the blink is made. It may contain some response activity, but if so, there is a heavy contamination with a blink artifact. Then, after some delay (200 ms) a gradually rounded potential, whose amplitude depends on stimulus luminance, appears. It is a more likely candidate for response activity. This potential appeared only when the subject made very prominent blinks, however. Thus, the data suggest that evoked responses associated with blinks are not as strong as those associated with saccades. Blinking seems less effective than microsaccades in producing vision. Some reservation must be made to this conclusion, however. It is not possible to examine the records during the actual course of the blink. Furthermore, blink potentials do not rise so sharply as do the signals that describe saccades. It is not possible to synchronize on them with the same degree of temporal accuracy, but the regularity of the recordings suggests that this was not responsible for the smallness of the averaged potentials.

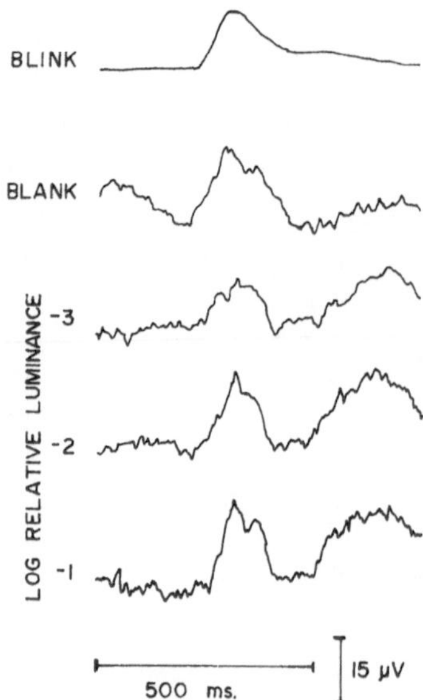

Fig. 6 Potentials accompanying eye blinks.

In addition to emphasizing the usefulness of the concept of addivitity in accounting for mass action waveforms, the present experiments indicate that microsaccades do play a useful role in vision. They produce strong response activity as the eyes scan local areas of the visual field. The experiments do not define any specific role for saccades. It is reasonable to speculate that the visual system will take advantage of any response activity however it may be produced.

Potentials that precede microsaccades

Several waves that precede microsaccades can be seen in the cortical recording shown in Fig. 2. They seem to be the same as the waves that are reported as anticipatory to large saccades. The broad potential (numbered 4) has been identified with contingent negative variation (16). It is recorded easily from some subjects but not others. It appeared in the present case when saccades were paced between the two fixation points by an auditory or tactile signal. It disappeared when the saccades were self-paced. These properties found with small saccades, suggest that the broad negative potential is the same one that has been described for large eye movements. Under some circumstances this potential has been seen as early as several seconds before large saccades (16). No attempt has yet been made to look more than 300 ms in front of small saccades.

The positive complex that immediately precedes the saccades also has been described for large eye movements. It seems to consist of two components. The early gradually descending phase of this complex has been given the descriptive name of pre-motor positivity (16). It gives way to a sharper potential or spike. There is some controversy regarding its function. Thus, it has been called, "the rectus potential", a term which relates it to the action of the extraocular muscles (16); and in another context, it has been suggested to be an electrophysiological correlate of "correlary discharge" (17). It has been given the descriptive name of "positive spike" here.

Recordings comparing potentials obtained with horizontal left and right self-paced movement are shown in Fig. 7. The recordings extend before and after the

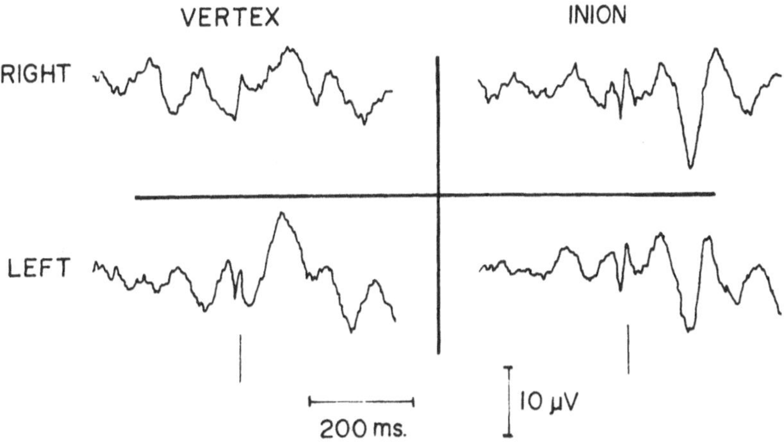

Fig. 7 Potentials accompanying right and left microsaccades.

saccades whose time of occurrence is indicated by thin vertical lines beneath the recordings. The data were obtained simultaneously from two electrode sites. Response potentials that follow the saccade may be seen in all the tracings, but they are clearest from the inion where they take the well-known "W" form (18). The positive spike that occurs almost simultaneously with the saccade is evident in all recordings, but other possible premotor potentials are difficult to discern because of the presence of activity in the alpha frequency range.

It is clear that the recordings are characterized by substantial variability. To meet this problem the technique illustrated in Fig. 8 has been adopted. The upper section of the figure shows averaged recordings from five experimental sessions, all taken from the vertex electrode placement and all produced by 25' saccades to the right. Recordings in the left column show the full epoch that was sampled. The records on the right are expanded sections of those on the left. The times that the saccade trigger "fired" is shown by the vertical line at the end of the time scales. The expanded sweep makes it possible to examine the recordings immediately in front of the saccades for waveform details. Grand averages from all five sessions combined are shown in the bottom row. Premotor positivity can be seen in the expanded grand average because much of the interfering background (chiefly the alpha rhythm of the EEG) has been cancelled out. The data presented below for the spike potential are all based on grand averages.

Spike Potential

If the spike potential is related to a tensing of the rectus muscles, it might be expected that the amplitude of the spike potential would vary in regular fashion with the excursion of the saccade that follows. To test this, an experiment was made to see whether the size of the spike increases with the size of the ensuing saccade. Recording was performed while saccades were made between fixation points that were spaced at intervals ranging up to 77 minutes. Typical recordings for left and right saccades are compared in Fig. 9. The spike potential seen at the right of these

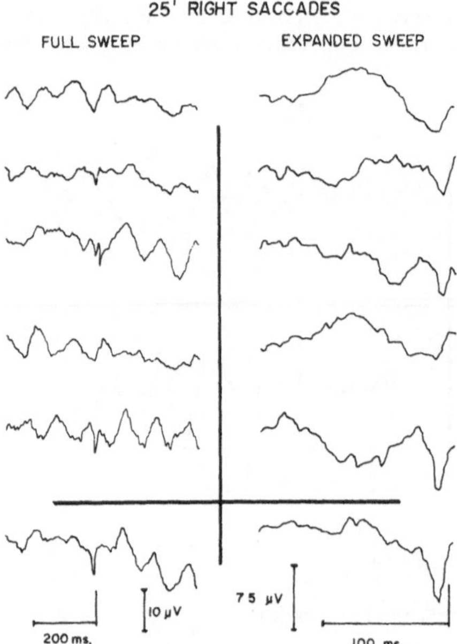

<u>Fig. 8</u> Technique of averaging responses from repeated trials.

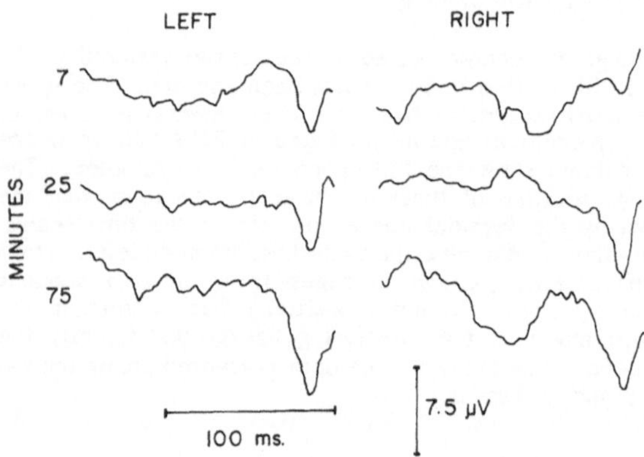

<u>Fig. 9</u> Positive spike potentials accompanying right and left microsaccades of several amplitudes.

tracings (displayed with an expanded sweep) is similar for left and right movements. Clearly, it grows with the size of the saccade. This is seen better, however, in Fig. 10 where the amplitude of the positive spike is plotted against the size of the saccade. There is a consistent increase in the size of the potential with eye

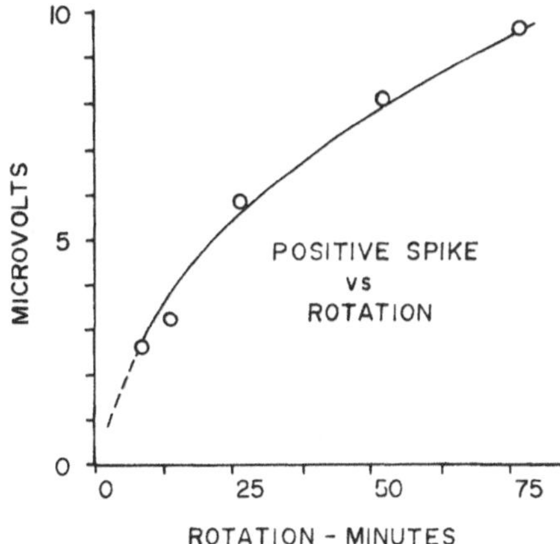

Fig. 10 Dependence of the amplitude of the spike potential on the amount of rotation of the eye.

movement, but it is not a linear one. In fact, the literature suggests that even with very large movements, the potential does not become much greater than seen here (16). Because it does not change in polarity with the direction of movement, this spike cannot be an electro-ocular artifact. Furthermore, because it is not directly proportional to the size of the saccade, it may not be a simple muscle tightening that precedes actual eye movement.

Conclusion

The present recording techniques and the results are at an early stage of development. As is true of all mass action potentials, the physiological bases is not well understood. The strength of the methods is that they permit recording from the human subject with minimal encumbrance. The visually evoked cortical potential has already demonstrated its value for interpreting psychophysical phenomena. The results reviewed here do suggest that saccades do have an important role in vision. Significant activity precedes their occurrence. They change the eyes' line of regard efficiently, and in so doing, they initiate sensory activity. A question for the future is not whether vision is possible without saccades but rather how saccades contribute to vision.

References

1. U. Tulunay-Keesey, In R. A. Monty and J. W. Senders, editors, Eye Movements - and Psychological Processes, (Erlbaum, New Jersey, 1976).
2. J. Nachmias, J. Opt. Soc. Amer. 50, 569 (1960).
3. E. Kowler and R.M. Steinman, Vis. Res. 17, 141 (1977).
4. R. M. Steinman, G. M. Haddad, A. A. Skavenksi and D. Wyman, Science 181, 810 (1973).
5. E. P. Johnson, L. A. Riggs and A. M. L. Schick, In Clinical Electroretinograph, ed. by H. N. Burian and J. H. Jacobson (Pergamon Press, N. Y., 1966) p. 75.
6. J. C. Armington, Behavior Research Methods and Instrumentation 4, 61 (1972).
7. K. G. Jones and J. C. Armington, Vis. Res. 17, 949 (1977).

8. J. C. Armington, In Visual Evoked Potentials in Man: New Developments, ed. by J. E. Desmedt, (Clarendon, Oxford, 1977).
9. E. P. Johnson, AMA Arch. Ophthal. 60, 565 (1958).
10. A. Troelstra and N. M. J. Schweitzer, J. Neurophys. 31, 588 (1968).
11. P. B. C. Fenwick and C. Turner, Electroencephalography and Clinical Neurophysiology 43, 74 (1977).
12. J. C. Armington, K. Gaarder and A. M. L. Schick, J. Opt. Soc. Amer. 57, 1534 (1967).
13. J. C. Armington, T. R. Corwin and R. Marsetta, J. Opt. Soc. Amer. 61, 1514 (1971).
14. J. C. Armington and M. B. Bloom, J. Opt. Soc. Amer. 64, 1263 (1974).
15. D. L. Yarbus, Eye Movements and Vision, (Plenum, New York, 1967).
16. W. Becker, O. Hoehne, K. Iwase and H. H. Kornhuber, Vis. Res. 12, 421 (1972).
17. P. Kurtzberg and H. G. Vaughan, Jr., In Desmedt, J. E., Editor, Visual Evoked Potentials in Man: New Developments, (Clarendon, Oxford, 1977).
18. H. Emrich, Vis. Res. 10, 1155 (1970).

A Model of Function at the Outer Plexiform Layer of the Cyprinid Retina

Leo E. Lipetz
Department of Biophysics
The Ohio State University

Introduction

This talk concerns behavior. The behavior of animals is thought to depend on activity of their nervous systems. Certainly, if the nervous system is altered by damage, the behavior is altered. Nevertheless, we would feel more confident of that dependence of behavior on the nervous system and would feel we had a better understanding of the nervous system if the actions of the nervous system could be described cell by cell and shown to lead to the observed behavior. Many scientists are working toward this end.

I would like to tell you about one attempt to explain the neural basis of a piece of visual behavior. It concerns goldfish and closely related fish, like carp, bream and tench. These are members of a group of fish called "cyprinids". These fish have been found able to distinguish shapes visually. For example, they can be trained to swim to a black circle displayed on a white background, and they can be trained to swim away from a black cross on a white background. They also can distinguish different colors. Thus, they can be trained to swim to a red light in preference to a green light, or vice versa. This behavior depends on the action of light on the eye; if the eyes are covered, this behavior ceases. But, light is light. How is it that one time the light causes one behavior--swimming toward--, and another time, the light causes a reverse behavior--swimming away? Red light elicits the reverse behavior to that elicited by green light. A particular distribution of white and dark, the cross, elicits the reverse behavior to that elicited by a different distribution of the light and dark on the retina, the circle. How does this occur?

In my laboratory I was studying the visual behavior of animals like the frog, the horseshoe crab, and the turtle. But it just so happened that the crucial data to understanding was collected for the cyprinids before it was collected for these animals. I saw the opportunity to use that data along with the other existing data to explain a portion of visual behavior. So, the studies on cyprinids I am going to describe were not done by myself or my collaborators.

The usual attempt to explain away behavior, such as visual discriminations, is that "it is done in the brain". Researchers have examined the signals sent by the eye to the brain and found these to consist of the firing of nerve impulses by the optic nerve fibers that connect the retina to the brain. In 1960, WAGNER, MACNICHOL and WOLBARSHT (1) measured the firing of single optic nerve fibers in the goldfish. What they found was that an optic nerve fiber had either an "on" response (firing during the illumination) or an "off" response (firing when the lights went off). An interesting result when the fiber was tested with monochromatic light was that many fibers had the maximum sensitivity of their off response in the green spectral region. Those same optic fibers had their maximum sensitivity for an on response in the red spectral region. This is clearly a reversal of response as the wavelength was changed from green to red. And this reversal was present in the output of of the retina before it reached the brain.

Another interesting finding was that if they moved a spot of white light across the receptive field of the optic fiber, that in the center of the field the greater sensitivity was for an on response, while in the periphery, the greater sensitivity was for an off response. The type of response was reversed as the position of the light on the retina was changed. Thus, it has been demonstrated that the information needed for the two forms of visual behavior shown by the fish already exists in the output from the retina to the brain. Some mysterious activity of the brain is not required-- the retina has already made the necessary discriminations of the visual inputs. The question then becomes one of how the retina gives different responses depending on the wavelength or the position of the input light. I will try to show the mechanism of this retinal functioning in the form of a model derived from existing data and reasonable assumptions.

The Functional Basis of the Model

The model involves only the first layers of retinal neurons which partake in the response to light, the photoreceptors, the horizontal cells, and the bipolar cells. This model is limited to cone-dominated activity in the retinas of cyprinid fishes. It provides a parsimonious explanation of published findings regarding the electrical responses of cones, horizontal cells, and bipolar cells. The starting functional data are the establishment in 1965 (2) that each cone contains only one light-sensitive pigment, and that in the goldfish there are three such pigments; a blue-sensitive, a green-sensitive and a red-sensitive. The next critical piece of data was the finding by TOMITA and his collaborators (3) that each cone responded to illumination with a negative-going change in its internal electric potential. The action spectrum of this voltage response of the cone was similar to that of one of the light-sensitive pigments found in the cones of such fish. At all wavelengths, each of the cones recorded showed only a negative-going response. That certainly does not provide the basis for the two types of behavioral and retinal responses, so those must involve interactions with other cells of the retina.

Each cone connects to a number of cells at its pedicle, and these include horizontal cells and bipolar cells (Fig. 1). It was found (4) that in a cyprinid there were three types of horizontal cells functionally. In one type, the most common, the cell's internal potential went more negative during illumination of the retina, and it went most negative in response to the red wavelengths. This was called a monophasic response, meaning that it had only one peak of voltage response as a function of wavelength. A second type of horizontal cell had a negative-going peak in the green region of the spectrum and a positive-going peak in the red region. This was called a biphasic response. A third type of horizontal cell had a large negative-going peak in the blue region, a positive-going peak in the green region, and had a small negative-going peak in the red region. This was called the triphasic type of horizontal cell. All of these responses of horizontal cells are, for historical reasons, called S-potentials. The crucial data were obtained by SPEKREIJSE and NORTON in 1970 (5).

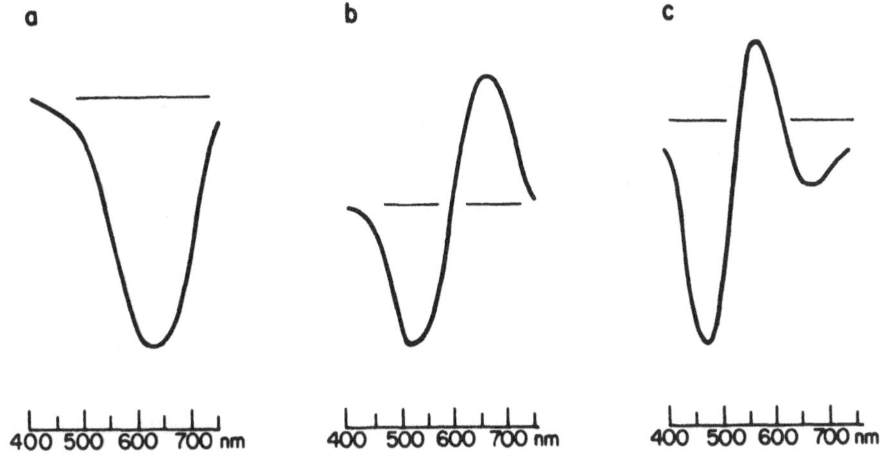

Fig. 1 The common types of S-potentials of carp recorded as a function of wavelength: (a) monophasic, (b) biphasic, (c) triphasic. In all records of this figure a positive potential change (depolarizing) at the intracellular electrode is upward and a negative potential change (hyperpolarizing) is downward. The horizontal lines indicate the potential level in the absence of a light stimulus. All three spectral scans were recorded from short to long wavelengths over a five second period. Adapted from (5).

TABLE I

S-Potentials of Cyprinid Fish

	Component		
S-Potential Type	Red	Green	Blue
Monophasic			
polarity	-		
delay, msec	25		
Biphasic			
polarity	+	-	
delay, msec	50	25	
Triphasic			
polarity	-	+	-
delay, msec	75	50	25

They used transfer function determinations to measure the time delay between a change of light intensity on the retina and the electric voltage change at the horizontal cell. What they did was to stimulate the retina with the wavelength which evoked a peak response in the type of the horizontal cell, from which they were recording. Thus, in a monophasic horizontal cell, they used just one wavelength; in the biphasic cell they used first the wavelength for the green peak and then the wavelength for the red peak; and in the triphasic cell, they used successively the wavelength for the blue peak, the green peak, and the red peak. They modulated the light intensity and measured by transfer function analysis the delay between the moment of maximum illumination and the moment of maximum response.

Their results are summarized in Table 1. What they found was that the largest response for each of the three types of the horizontal cell was always negative-going and always had the same minimum delay. The two positive-going responses found in the biphasic and triphasic horizontal cells had twice that minimum delay. The small negative-going response to red found in the triphasic horizontal cell had three times the minimum delay. The regularity in these delays and polarities of the responses suggests a simple regularity of interconnections of the horizontal cells and the cones. The possible patterns of connection are basically of two types. In one type, there are parallel connections from the different cones to the different horizontal cells (see Fig. 2). The second type involves series connections among these cells (See Fig. 3).

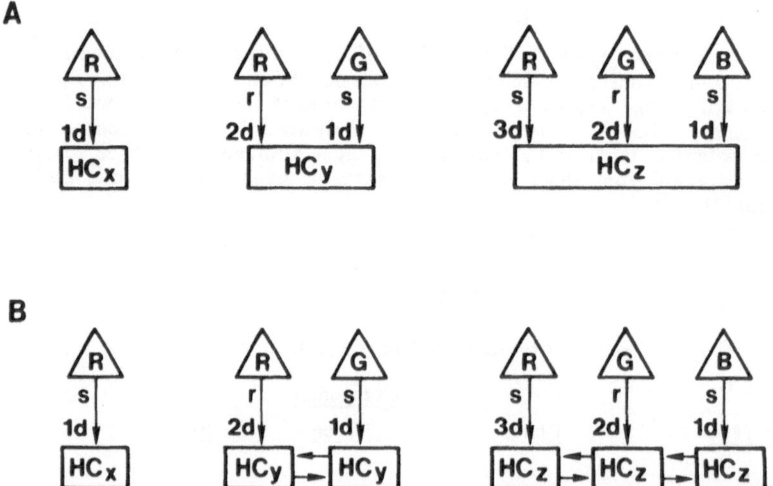

Fig. 2 Two possible parallel arrangements of connections from the cones to the horizontal cells which would give the observed polarities and relative delays of the S-potentials' components in cyprinid fish. R, G, B, red, green, and blue cones, respectively. HC, horizontal cell; the subscripts X, Y or Z indicate that its S-potential is monophasic, biphasic, or triphasic, respectively. Signal flows (of voltage changes) are indicated by arrows. The flows from each cone are labeled s or r to indicate that the voltage change is transmitted with the same or reversed polarity, respectively. The relative transmission delay in units of least delay, d, is also marked adjacent to the arrow for each flow. A, each biphasic and triphasic horizontal cell connects to more than one type of cone. B, each horizontal cell connects to only one type of cone. The biphasic horizontal cells all interconnect with each other, as do all the triphasic horizontal cells, by synapses which have small transmission delay and preserve the polarity of the transmitted voltage change.

A

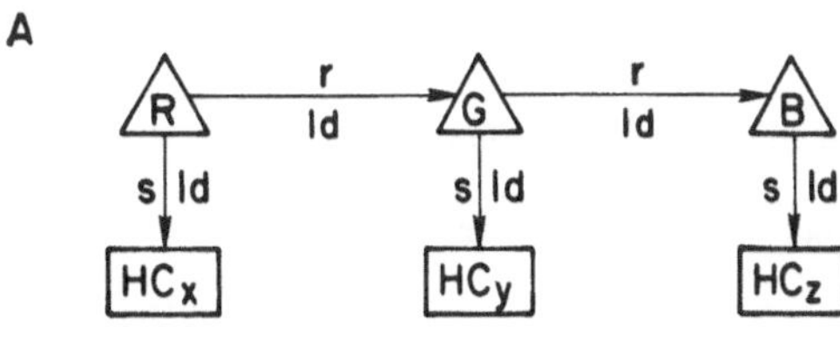

B

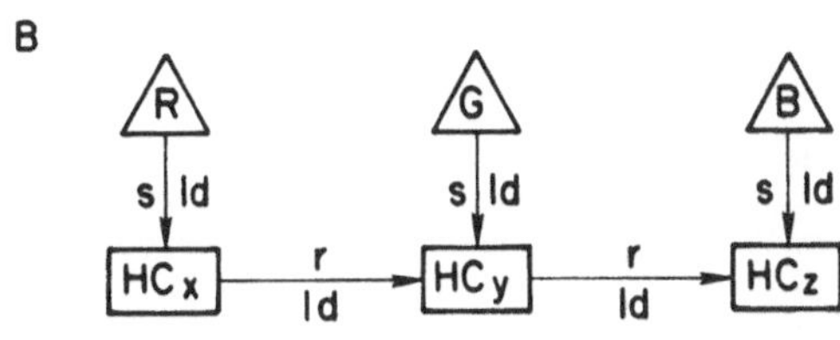

C

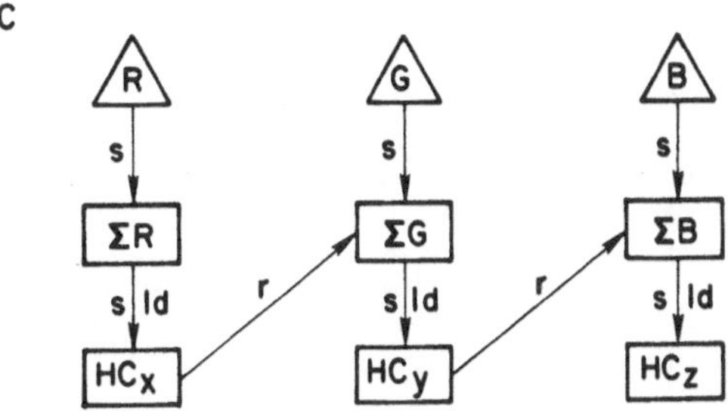

Fig. 3 Possible series arrangements of connections among the cones and horizontal cells of cyprinid fish which would give the observed polarities and delays of the S-potentials' components. Labels as in Fig. 2. Σ, combining site for voltage changes transmitted from a cone pedicle and from a horizontal cell. The transmission delay from cone pedicle to combining site and from horizontal cell to combining site is taken as negligible. The cone's ribbon synapse is probably the combining site.

Let us start with the parallel connections (Fig. 2). One way to produce the monophasic response would be to have the red cone, (i.e., red-sensitive cone) transmit directly to the monophasic horizontal cell via a link having a single unit of delay and giving the same polarity of response in the horizontal cell as in the cone. Thus, the response is negative in both the cone and the horizontal cell. For the biphasic response, the horizontal cell must receive from the green cone via a polarity-preserving link having a single unit of delay, and receive from the red cone via a polarity-reversing link having two units of delay. For the triphasic response, the horizontal cell must receive from the blue cone via a polarity-preserving synapse of one unit delay, from the green cone via a polarity-reversing synapse of two units delay, and from the red cone via a polarity-preserving synapse of three units delay. This set of parallel connections is unsatisfactory in that it requires synapses having two and three units of delay, and such delays are not ordinarily observed. It is also unsatisfactory because it would require three different kinds of synapses (s/1d, r/2d, and s/3d).

A more parsimonious arrangement is the set of series connections (Fig. 3). To produce the the monophasic response, the red cone sends to the monophasic horizontal cell via a polarity-preserving synapse having a single unit of delay. The monophasic horizontal cell's response is entirely negative-going. The monophasic horizontal cell is connected to the biphasic horizontal cell in such a way that the monophasic cell's response shows up in the biphasic cell as a positive-going voltage change. The green cone, also, sends to the biphasic horizontal cell via a single unit delay, polarity-preserving synapse. The biphasic horizontal cell, in turn, sends to the triphasic horizontal cell via a polarity-reversing synapse of a single unit delay. That would cause the green cone's signal to be reversed in polarity as it goes from the biphasic horizontal cell to the triphasic horizontal cell and to show up at the triphasic horizontal cell as a positive-going voltage change in the green region. The red cone's signal would be reversed twice in getting to the triphasic horizontal cell, so it would show up as a negative-going response at the triphasic horizontal cell. Thus, the set of series connections takes care of all the delays and all the polarities observed in the S-potentials. It is more reasonable than the parallel connections because it does this more parsimoniously than the parallel connections because it requires only two types of synapses; one is polarity-preserving and the other is polarity-reversing, and so I much prefer the series connections.

To decide among the possible series connections (A, B and C of Fig. 3) some anatomical data can be applied. The connections of Fig. 3A require unidirectional polarity-reversing synapses from red cones to green cones and from green cones to red cones. While functional connections have been found between cones in fish (7), these connections have been observed to be polarity-preserving and cannot provide the circuit of Fig. 3A.

The connections of Fig. 3B require unidirectional, polarity-reversing synapses from monophasic to biphasic horizontal cells, and from biphasic to triphasic horizontal cells.

It is known that in the cyprinids, the connections from horizontal cell to horizontal cell are predominantly gap junction electric synapses. (A few chemical synapses are seen.) Electrical synapses can be either polarity-preserving or polarity-reversing, but the structural arrangements of the synapses between horizontal cells in cyprinids would be expected to be only polarity-preserving. Furthermore, the gap junctions, with but two known exceptions (6) are bidirectionally symmetric. Thus, in two ways, they cannot provide the connections required for Fig. 3B.

The Model

What is the anatomical pathway of the series connections?

The horizontal cells all have processes extending into the cone pedicles and different horizontal cells can be seen connecting into the same cone pedicle. Therefore, this seems a good candidate for their point of series interconnections, and is modeled in Fig. 3C. I propose to locate the combining site of the cone and horizontal cells' signals at the ribbon synapse of the cone pedicle. The red cone, via its synapse sends a signal to the monophasic horizontal cell. The monophasic horizontal cell has an input connection to the ribbon synapse of the green cone, and it is this connection which is polarity reversing. The output from the green cone's ribbon synapse is to the biphasic horizontal cell. The biphasic horizontal cell has a polarity-reversing synapse into the ribbon synapse of the blue cone. The output from the blue cone's ribbon synapse is to the triphasic horizontal cell. Thus, the series connections of Fig. 3C explain the major features of the spectra, the polarities, and the response delays of the monophasic, biphasic, and triphasic horizontal cells, and do so by synapses consistent with the known anatomy.

There is one other piece of data that I have to include in any model, and this is the observation that a change of voltage can be evoked in any of these horizontal cells by illumination of any region of the retina. Naka and Rushton (8) found that the voltage change decreased as the light spot was moved further from the horizontal cell being recorded from, and it also decreased as the retinal area illuminated was made smaller. They showed that this lateral transmission could best be explained as the spread of electric current from the illuminated region to the horizontal cell being measured. In my model, I will postulate that each horizontal cell has low resistance electric synapses to neighboring horizontal cells of the same type. This means, monophasic horizontal cell to monophasic, biphasic to biphasic, and triphasic to triphasic. The synapses have to be restricted in that way because otherwise the responses of the horizontal cell types would get mixed up and there would no longer be three separate types of responses. Electric synapses have been observed anatomically in the cyprinid fishes and they have been demonstrated functionally in the shark by Kaneko (9).

There is a sixth group of data to be explained and these are the findings of Kaneko and his collaborators on bipolar cells in these fish (10, 11, 12, 13). Some bipolar cells were found to give the same polarity of voltage change in response to illumination of any part of the receptive field. But there were other bipolar cells which gave opposite polarity changes of voltage to illumination of the receptive field center as against illumination of the periphery. They found still other bipolar cells which gave changes of voltage of opposite polarity depending on the wavelength. Thus, for one type, red illumination tended to give, in this case, a positive-going change, and short wavelength illumination tended to give a negative-going response. The sensitivity to the long wavelengths was greatest near the center of the receptive field and the sensitivity to the short wavelengths was greatest in the periphery. Thus, you can already see at this bipolar cell the reversing of response with wavelength and the reversing of response with illumination position on the retina that was also observed at the optic nerve fibers.

In cyprinids bipolar cells are known to each contact the ribbon synapse of several cones. As part of my model, I am going to postulate that each particular type of bipolar cell connects only to the ribbon synapses of a particular type of cone. Thus, for the green cone's ribbon synapse, there is a type of bipolar cell that connects only to it. For the blue cone's ribbon synapse, there is a type of bipolar cell that connects only to it. And similarly for the red cone's ribbon synapse. Then, these bipolar cells should show different properties in their responses. And, in fact, would give the kinds of response properties found by Kaneko.

There is one more piece of information that should be included. Kaneko and others (10, 11, 12, 13, 14) found that some bipolar cells had positive-responding receptive field centers but negative-responding peripheries, while other bipolar cells had negative-responding centers and positive-responding peripheries. Therefore, I postulate that there are two sets of bipolar cells. One set has polarity-preserving inputs, the other set has polarity-reversing inputs. There is functional evidence for these two types of connections to bipolar cells from studies of amphibian retinas (15, 16). That postulate completes the model.

Tests of the Model

This model was developed in 1971 and was circulated by private paper to leading visual researchers around the world.[1] WILLIAM STELL and his group in Los Angeles

1.An updated version of this paper, giving in detail the data and the reasoning on which the model is based, is available for handling costs from The Institute for Research In Vision, The Ohio State University, Columbus, OH 43214, USA. Ask for Publication No. 16, "Information Processing in the Outer Plexiform Layer of the Retinas of Cyprinid Fish". The preparation of the original paper was partially supported by grant EY467 from The National Eye Institute and grant GN534 from The National Science Foundation.

decided to test the anatomical connections predicted by this model; for this I am most grateful. Their results were published in a series of papers starting in 1975. These results, as well as work done on the cyprinid bipolar cells by SCHOLES, provide a test of the model. All these studies start with the finding that there are six anatomically distinct types of cones in the cyprinid retina. There is a double cone with one long and one short member, and two long single cones with a little difference in squatness between the two, and two short cones, the short single and the miniature single. By a variety of techniques, various experimenters have shown that a particular visual pigment is associated with each anatomical form of cone (2, 17, 18, 19, 20, 21, 22). They showed that the principal cone (the long double) contains the red-sensitive pigment, the accessory cone (the short double) contains the green-sensitive pigment, most of the long single cones contain a red-sensitive pigment and there are a small percentage of them (possibly the squatter ones) that contain a green-sensitive pigment. The short cones (the miniature and the short single) both contain a blue-sensitive pigment. By tracing the anatomical connections of these different cone types to the different horizontal cells, it is possible to predict the spectral inputs to those different horizontal cells.

Stell and Lightfoot (23) found three anatomical types of horizontal cells in cyprinids. They traced the connections to cone types by light microscopy of Golgi-stained horizontal cells followed by electron microscopy to determine in which cones each terminated. The largest of the horizontal cells, H3, contacted only the blue-sensitive cones. The H2 horizontal cell contacted both the blue-sensitive and the green-sensitive types of cones, and the H1 horizontal cell contacted red-sensitive, green-sensitive and some blue-sensitive cones.

They still needed to know which of these connections were inputs and which were outputs. That was established in a further study (24). It was already known that at the ribbon synapse of the cone there are lateral elements which come from horizontal cells and there are central elements. Sometimes these central elements come from bipolar cells, but what Stell and his collaborators found was that many times they come from horizontal cells. Again using Golgi staining and light and electron microscopy they traced the terminations of each horizontal cell type. They found that the H3 horizontal cell has only one type of connection, a central element connection to the ribbon synapses of the blue cones. The H3 cell has to have an input because it has a response, so they concluded that the central element is the input element to the horizontal cell. Now, if the same holds true for all other central elements, and if all the lateral elements are outputs from the horizontal cell to the cone, then their findings are that the H2 type of horizontal cell is receiving its input from the green cone and sending its output to the blue cone. Similarly, the H1 horizontal cell is receiving its main input from the red cone and lesser input from the green cone and has its main output to the green cone, with some output to the blue cone. Fig. 4 summarizes these connections as found anatomically. It can be seen that they agree quite well with the connections predicted in my model from strictly functional data.

Next, to extend these anatomical studies to the bipolar cells, Scholes in 1975 (19) stained bipolar cells and traced their connections to the various types of cones. For example, he showed that wide-field rod bipolar cells connected to rods, but they also connected to cones, and the only cones they connected to were the red-sensitive ones. He found at least ten different anatomical types of bipolar cells, of which only six were rather common. Of these six; two connected to rods and to the red cones, one connected to the green cones only, one connected to the blue cones only, one connected to red and green cones plus some rods, and one connected to blue and green cones plus some rods. I had predicted in the model that there would be a different bipolar cell type connecting to the ribbon synapse of each cone type--three

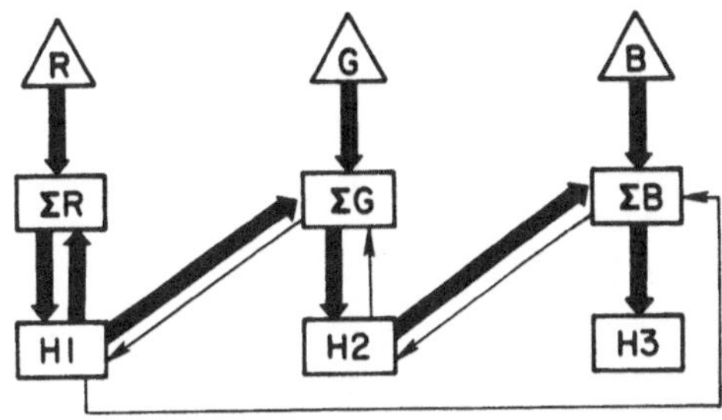

Fig. 4 Diagramatic representation of the synaptic connections in the outer plexiform layer of goldfish as determined by anatomic techniques. From data of (23, 24), R, G, B, red, green, and blue cones, respectively. H1, H2, H3, cone horizontal cell types. ΣR, ΣG, ΣB, ribbon synapses of the red, green and blue cones, respectively. The less frequently found synaptic paths are indicated by the thinner arrows.

bipolar cell types. It turns out that there were more than that. There were also two bipolar cell types that each connected to the ribbon synapses of two different types of cones. This and the rest of the anatomical data are summarized in Fig. 5.

There is a feature of the model I have not mentioned--the organization of the receptive fields of the bipolar cells. Each bipolar cell receives its input via its dendritic connections to cone ribbon synapses. The dendrites of each bipolar cell spread over a certain region of retina and connect to a number of cones in that region. Illumination of that retinal region will stimulate the cones (say, the green cones) that directly feed to a particular bipolar cell (say, the B3 of Fig. 5) and we can call that the central field of that bipolar cell's receptive field. Since each of the cones connects with the bipolar cell via, let's say, polarity-preserving synapses, then the negative-going response of the cones will result in a negative-going response of the bipolar cell. Next, consider the effect of illumination at a distance from this central region. It will stimulate other receptors (say, red cones) at a distance that do not connect directly to the (B3) bipolar cell. These cones, in turn, will transmit through their ribbon synapses to monophasic horizontal cells. These connect to neighboring monophasic horizontal cells all the way across the retina. Some of them are going to transmit into the ribbon synapses of the green cones that transmit directly to the bipolar cell. But each feeds into the ribbon synapse with a polarity-reversing connection, so the response at this (B3) bipolar cell then will be positive-going when the illumination is out on the periphery of its receptive field. This explains how the central versus peripheral responses of a bipolar cell can be of reverse polarity. We have already explained how the series connections of the model cause the difference in spectral response of the cones to produce different polarity responses of the horizontal cell for different wavelengths. In the same way, it produces different polarity bipolar cell responses for different wavelengths. Thus, the model has accounted for the two features of response reversal that we set out to explain.

There are certain discrepancies in the model. From the known spreads of the bipolar cell dendrites you can predict the size of the cell's central receptive field and compare it with what is found functionally. When you do that, you find that there is a discrepancy. The central receptive field response occurs over an area that is two times the diameter of the dendritic spread, and I simply do not know the explanation. So that is one discrepancy with this model. From Schole's data as to the types of cones contacted by each of the different bipolar cells, one can predict the spectral responses of each cell's central and peripheral receptive fields. Since there are six

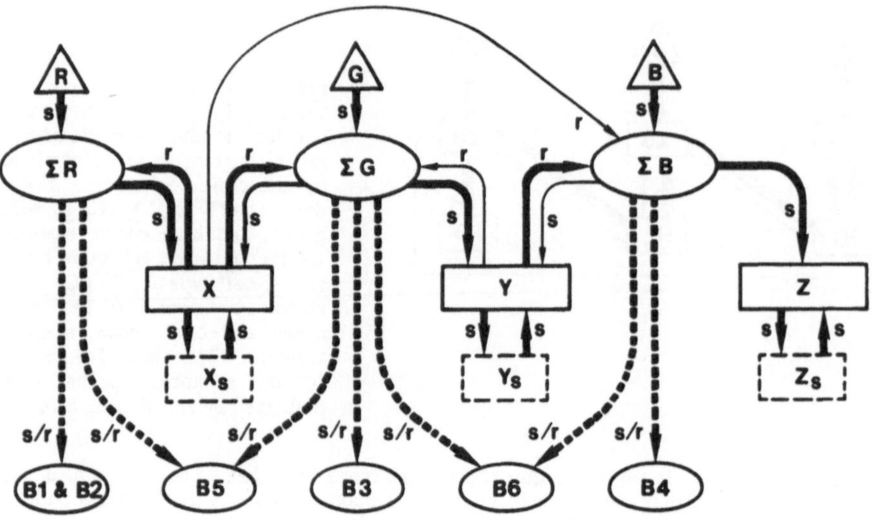

Fig. 5 Diagram of synaptic connections in the outer plexiform layer of the retina of cyprinid fish. The solid lines indicate connections determined anatomically in goldfish (23, 24). The dashed lines indicate connections determined anatomically in rudd (19). The heavier lines indicate connections considered significant in the model developed in this paper from functional data. R, G, B, red, green, and blue cones, respectively. ΣR, ΣG, ΣB, ribbon synapses of red, green, and blue cones, respectively. X, Y, Z, horizontal cells defined functionally in this paper's model. They correspond to the H1, H2, and H3 cone horizontal cells defined anatomically by STELL and LIGHTFOOT. Note that each type of horizontal cell synapses with neighboring cells of its type, across the entire retina. B1 through B6, bipolar cells as defined anatomically by SCHOLES. B1, narrow field rod bipolar; B2, wide field rod bipolar; B3, accessory cone bipolar; B4, single cone bipolar; B5, RG mixed cone bipolar; B6, BG mixed cone bipolar. s indicates that the signal is transmitted with the same polarity, r with its polarity reversed. s/r indicates that the synaptic path may be of s type for some cells and of r type for other cells. There is a relative delay of one unit each time the signal passes through a ribbon synapse.

types of bipolar cells, there should be six sets of spectral responses. I have compared the observed sets with those predicted by the model. For three of the bipolar cell types, the two types of rod bipolar cells, and the (B5) mixed red-green bipolar cell, the predictions match the spectral sensitivities of the observed electrophysiological responses. There should be three other sets of bipoplar cell responses, but they have never been observed. Here is the other discrepancy. It may be that such responses were not observed simply because they are harder to record. The cells that have the largest bodies are the two types of rod bipolar cells. The bipolar cells that are most common aside from those two are the red-green mixed bipolar cell type. Therefore, it is at least reasonable that these three types of bipolar cells have been recorded from simply because they are the largest or the most numerous, and the other three types have been too difficult to record from.

Summary

Figure 5 summarizes the model with all the anatomical data included. The heavier lines are the most significant connections functionally. What does this model do? First, it explains how different responses to different wavelengths of light can arise in the retina. Second, it explains how different responses to different retinal positions of light can arise in the retina. It does this by specifying: first, the synaptic connections of cones, horizontal cells, and bipolar cells; second, the direction of signal flow at each synapse; third, the polarities of the presynaptic and postsynaptic changes of voltage; and fourth, the relative delays of transmission via the different synapses.

What are the limitations of the model? First, it applies only to cyprinid fish, although Tom Wheeler (25) tells me that he has done the necessary electrophysiological measurements in turtles and finds that essentially the same model fits the data. Second, it applies only to voltage change signals faster than one tenth of a second. Third, it omits interactions with rods. Fourth, it omits interactions with glial cells of the retina. Fifth, it omits adaptation processes. Sixth, it omits time-dependent processes other than transmission delays.

What is the significance of the model? Notice that all the computations in this model are done by graded changes in cell membrane potentials. this is true for the cones, the horizontal cells, and the bipolar cells. It has been found that the amacrine cells of the retina, also, show only graded changes of potential. It is only at the axons of ganglion cells that regular all-or-none action potentials are observed. Thus, the retina can be said to operate as an analog computer. The retina is an extruded part of the brain, so I would expect that the rest of the brain, also, does its local computations by graded membrane potentials. The results of each local computation are then sent to other local computations networks by all-or-none potentials in the same way as the retina sends its results to the brain.

This ties in very closely with what Dr. Brown (26) has been doing, because he has developed the techniques that make it possible for researchers on the central nervous system to put their electrodes into the smaller neurons which presumably form the local computation networks. It is only with such intracellular electrodes that the graded membrane potential changes of these neurons can be detected so that we will be able to determine what kinds of computations are being done in the central nervous system.

References

1. H. G. Wagner, E. F. MacNichol, Jr. and M. L. Wolbarsht, J. Gen Physiol. 43, Suppl. 6, Part 2, 45 (1960).
2. W. B. Marks, J. Physiol. 178, 14 (1965).
3. T. Tomita, A. Kaneko, M. Murakami and E. L. Pautler, Vision Res. 7, 519 (1967).
4. A. L. Norton, J. Spekreijse, M. L. Wolbarsht and H. G. Wagner, Science 160, 1021 (1968).
5. H. Spekreijse and A. L. Norton, J. Gen. Physiol. 56, 1 (1970).
6. H. Grundfest, The Neurosciences, ed. F. O. Schmitt (Rockefeller Univ. Press, New York, 1967), p. 353.
7. D. A. Burkhardt, J. Neurophysiol. 40, 53 (1977).
8. K. I. Naka and W. A. H. Rushton, J. Physiol. 192, 437 (1967).
9. A. Kaneko, J. Physiol. 213, 95 (1971).
10. A. Kaneko and H. Hashimoto, Vision Res. 9, 37 (1969).
11. A. Kaneko, J. Physiol. 207, 623 (1970).
12. A. Kaneko, Vision Res. Suppl. 3, 17 (1971).

13. A. Kaneko, J. Physiol. 235, 133 (1973).
14. J. Toyoda, Vision Res. 13, 283 (1973).
15. R. Nelson, J. Neurophysiol. 36, 519 (1973).
16. R. F. Miller and R. F. Dacheux, J. Gen. Physiol. 67, 639 (1976).
17. W. B. Marks, Color Vision: Physiology and Experimental Psychology, eds. A. V. S. Reuck, J. Knight (Little, Brown, Boston, 1965), p. 208.
18. F. I. Harosi and E. F. MacNichol, Jr., J. Gen. Physiol. 63, 279 (1974).
19. J. Scholes, Philos. Trans. R. Soc. London Ser. B 270, 61 (1975).
20. W. K. Stell and F. I. Harosi, Vision Res. 16, 647 (1976).
21. R. E. Marc and H. G. Sperling, Science 191, 487 (1976).
22. R. E. Marc and H. G. Sperling, Vision Res. 16, 1211 (1976).
23. W. K. Stell and D. O. Lightfoot, J. Comp. Neur. 159, 473 (1975).
24. W. K. Stell, D. O. Lightfoot, T. G. Wheeler and H. F. Leeper, Science 190, 989 (1975).
25. T. G. Wheeler and K. I. Naka, Vision Res., in press (1977).
26. K. E. Brown, this symposium (1977).

Center-Surround Retinal Ganglion Cells: Receptive Field Organization with Special Reference to Light - Dark Adaptation

Christina Enroth-Cugell, M.D.
N. U. Technological Institute

Once upon a time BOB SHAPLEY and I published a paper entitled "Flux, not illumination, is what cat retinal ganglion cells really care about" (1). And, at the time, we thought that was a pretty good title. But HORACE BARLOW had some objections to it. He thought that we implied that the poor ganglion cells had a peculiar masochistic psyche caring for what turns them _off_, _i.e._, adapting flux, rather than caring for what turns them _on_, which would have been a more natural instinct.

Now, HORACE BARLOW's instincts, with regard to vision, are very sound indeed, so it is not surprising that in a sense it _was_ a bad title, although not for _his_ reasons. Let me explain what I mean.

As we all know, when steady light falls on the retina it becomes less sensitive than in complete darkness--it becomes light adapted. Ever since BRIAN CLELAND was at Northwestern in the nineteen-sixties, various co-workers and I have contended that, in the cat's concentrically organized retinal receptive fields, the center mechanism acts as a unit adding signals over a large number of receptors to produce responses to flashing lights (_i.e._, stimuli) and also adding--as a unit--adaptive effects due to steady backgrounds, to set the sensitivity of the center.

That is, we have applied WILLIAM RUSHTON's concept of summation and adaptation pools to the central response mechanism of on- and off-center cells.

Another way of expressing the idea that the center of cat ganglion cells acts as a unit is to say: provided stimulus or adaptive flux is appropriately weighted by the underlying sensitivity, it does not matter _how_ the flux is distributed within the center's summing area; only how _much_ of it there is. But from the work of EASTER (2), BURKHARDT and BERNTSON (3) and GREEN and co-workers (e.g., (4)) we have known for several years that this is not so in goldfish, frog or rat. Furthermore, there is also the finding of CLELAND, LEVICK and SANDERSON (5) that in the cat, the surrounds of brisk transient or Y-cells--and of some X-cells--contain subareas which, to an extent, adapt independently.

What I will talk about today is how the center of cat X- and Y-cells (6), behave in this respect. In particular, I want to talk about what happens to the center's light adaptation when one varies the spatial distribution of adapting flux, which has been weighted by sensitivity (i.e., effective flux; (7).

Before I go any further, let me make clear that the only kind of light adaptation that I will be concerned with is the effect upon retinal ganglion cell behavior of lights, too weak to cause any significant steady state bleaching. I will refer to this kind of adaptation with RUSHTON's term field adaptation (8).

Another thing I must do before describing any of our results is to say a few quick words about how the experiments were done. All cats were in general anesthesia--we use urethane--and the action potentials were picked up in the optic tract.

The responses that you will see in one of the figures consist of pulse density tracings, obtained by first smoothing the individual action potentials and then averaging over many stimulus cycles. In other experiments average responses were collected as peri-stimulus time histograms by a computer.

We try to keep the retinal image as sharp as possible by using the appropriate power in the contact lens--and by using an artificial pupil when not working with a Maxwellian View Stimulator.

Units were classified as X- or Y-cells with a bipartite field within which the contrast of the two halves reversed at 2.5 Hz (see ref. (9)).

The main burden of what I have to say is contained in a recent experiment, conceived and carried out by TOM HARDING in my lab. But I will begin with a much older experiment which some of you probably have had to sit through several times before. In spite of that, I will dwell on this old experiment in some detail because the interpretation of its results is very relevant to understanding the implications of TOM HARDING's findings.

The experiment I want to start with is the one in which BRIAN CLELAND and I in 1968 copied the now famous Westheimer-effect experiment. In our version, that experiment consisted of providing the middle of on-center receptive fields with a fixed stimulus while adjusting the illumination of a series of centered adapting discs, of different diameters, until--for each disc--the cell produced a certain criterion response. In other words, for each differently distributed flux we determined the total amount of flux required for always keeping the level of field adaptation constant, as evidenced by a constant output from the cell when its center mechanism was subjected to a constant input. CLELAND and I did all our experiments by listening to the cell. But since then we have obtained the same results from about 30 cells, using the objective method which you will see in the first figure which is from an (unpublished) experiment that JERRY NELSON and I did in 1972.

This is how the experiment ran. One small flashing light, i.e., the stimulus, and a series of steady adapting discs of different diameters were used. The smallest adapting disc and the stimulus had the same diameter--0.15^{o}. The receptive field was first mapped to find its most sensitive point, and then the retina was left to become thoroughly dark-adapted. Next the small flashing spot--which was green and flashing on and off continuously at 0.4 Hz--was placed in the middle of the center and the illumination adjusted so that the cell yielded a small response. This was a central response, by which I mean that the cell was driven by its central response mechanism only. From now on that central response became the criterion response for the rest of the experiment. The next step was (still in the absence of any steady adapting light) to increase the illumination of the stimulus by a factor of ten so that the cell yielded a response much larger than the criterion. From now on nothing was touched with respect to the stimulus. It remained in the same place; it remained of the same size; its color remained the same; its illumination remained the same and it kept flashing on and off at 0.4 Hz.

While this fixed stimulus continued to flash the smallest adapting disc--which like all the others was a deep red--was superimposed on the stimulus. And then the illumination of the adapting disc--not the stimulus--was adjusted so that the cell once more yielded the criterion response. (That response is shown in the left lower corner of Fig. 1). So, what we did then was to adjust the strength of the adapting light so

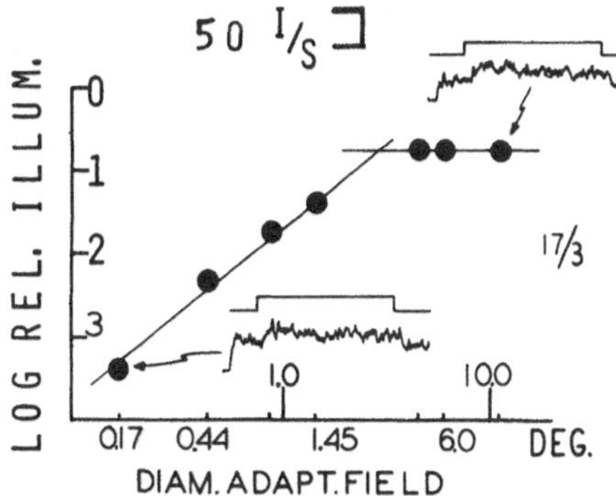

Fig. 1 Area-adaptation curve from on-center X-cell (Enroth-Cugell and Nelson, unpublished). For details, see text.

that the center mechanism was field-adapted by one log unit. This is so since a stimulus ten times stronger than the one in full dark-adaptation was now required for the cell to yield its central criterion response.

Next the 0.44° diam (Fig. 1) adapting disc was centered on the fixed stimulus; adapting illumination was adjusted until the cell once more produced its criterion response and so on for all adapting discs up to the largest one. That is, each adapting disc luminance was adjusted so that the center mechanism always remained field adapted by one log unit. The cell's response to the fixed stimulus when the largest--10° diam--disc was applied is shown in the upper right corner of Fig. 1.

On these particular coordinates a straight line of slope +2 represents a constant flux. That is, along a slope-two line the product of adapting illumination and area of the adapting disc is always the same. What is particularly noteworthy in this figure is that the data obtained with the four smallest discs can be very precisely fit with a slope-two line which means that the prerequisite for constant field-adaptation of the center mechanism is--within limits--that the adapting flux remain the same.

When the adapting light begins to fall beyond the diameter which corresponds to the intersection of the sloping line and the horizontal one drawn through the data for large adapting discs, increased area does not permit further decrease in the adapting illumination, which simply means that from here on out the additional steady light falls outside the center.

What can we conclude from this figure? Although both the small flashing stimulus and the adapting discs clearly exert some of their effect by virtue of focus error and scatter, I do not see how one can escape the conclusion that some adaptive signals spread, physiologically, over several degrees within the center. In other words, at least with respect to some stage of the setting of sensitivity, the center seems to act as a unit; and, this holds equally for X- and Y-cell.

We have also done other kinds of experiments whose results are consistent with this notion. For example, if some fraction of the field adaptation does take place in the center as a unit, then the state of adaptation produced by a steady background which extends well beyond the center's summing area, and the resulting level of adaptation, should be related; for any one background illumination, large centers should be more light-adapted than smaller centers. We tested this in experiments and found it to be true (1).

Yet, the fact remains that our early experiments did not exclude the possibility that the center mechanism also contains sub-areas which, depending upon the spatial distribution of weighted adapting flux, assume different levels of field adaptation; i.e., can be differentially field-adapted.

Our early experiments did not specifically test this but the experiment which TOM HARDING has now done (10) does explore just this possibility. Fig. 2 which is from one of his experiments is, as you can see, a classical increment-threshold curve,

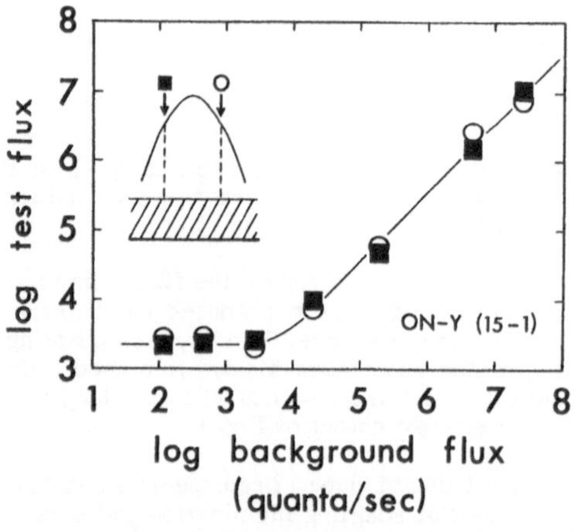

Fig. 2 Increment threshold curve for on-center Y-cell. All backgrounds are larger than the cell's central summing area. Two test stimuli applied one at a time (10).

with the exception that two test spots could be applied. And, you should also note that in this experiment the steady background was always larger than any center likely to be encountered in the cat retina. It did not, as in Fig. 1, vary in size.

The vertical axis gives the log of the stimulus flux. The horizontal axis gives the log of the adapting flux. In the previous experiment we obtained the adapting flux for the smaller discs by multiplying the area of each adapting disc by its illumination. In this case when the background is always much larger than the center, the background flux is obtained by multiplying the background illumination by the total effective summing area of the center (11).

The sketch built around a schematic representation of the center's sensitivity profile, in the upper left corner of Fig. 2, is intended to convey the following: (a) the two test spots are both 0.2^{o} in diameter. They are green while the background was a deep red, and 12° in diameter. (b) The location of the two spots was always chosen so that the cell, near threshold, would be driven only by the center mechanism. To this end the spot-separation was smaller for centers with a narrow sensitivity profile (small centers), larger for centers with a broad sensitivity profile (large centers). The spot separation varied from 1° to 2.0^{o} but for one cell it was 2.5^{o}. Furthermore, (c) within a particular center, the two spots were always at approximately equisensitive locations, and (d) from cell to cell the difference in log sensitivity between the peak and the two spot locations was always about the same, about 1 log unit.

The experiment ran as follows. First, one of the spots, say the one indicated by the square symbols, was applied alone against a very weak background, while the other spot was turned off altogether. The flashing spot was modulated at 1 Hz in a square-wave fashion. The computer then found the illumination required to elicit a small criterion response and stimulus flux was plotted versus background flux. At that same background the first spot was now turned off and the second one used in the same manner, and so on for all the different background illuminations.

There are two points I want to make with Fig. 2. First, just as one expects from any healthy, well behaved X- or Y-cell, the curve has two portions. One horizontal portion indicating that over a range of low adapting fluxes, increment sensitivity remains at its maximum. And there is a second sloping part indicating a progressive increase in increment threshold (increase in level of field-adaptation) over a range of higher backgrounds. The second, and in this particular context, more important point is that the manner in which the two types of symbol fall, shows that when background illumination is diffuse there is almost perfect correspondence between field adaptation in the two halves of the center.

But now let us see what happens if adapting flux is not uniformly distributed over the entire center. Fig. 3 shows the same experiment, carried out on the same

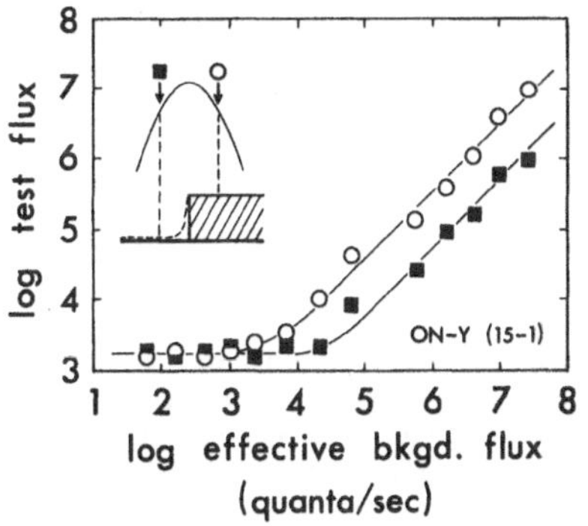

Fig. 3 Same cell, same experiment as in Fig. 2 except that background field now is a bipartite field whose dividing line passes through the center's midpoint. Broken curved line in left-upper sketch reminds reader of "light spill" from illuminated field-half onto "completely dark" field-half (10).

cell, the only difference being that now the adapting background is a 12° diam bipartite field with a vertical boundary which, as you see in the inset in the left upper corner,was positioned precisely through the middle of the receptive field center.

Theoretically the right half of the background is in complete darkness, but in reality of course the "image proper" of the edge is somewhat smudged and a fair amount of scattered light from the illuminated half falls on it; all of which I will refer to as "light spill". And I will also call this the dark half although it is not dark.

The experiment proceeded in principle exactly as in the previous slide. As you can see, first the illuminated half-field was so dim that threshold within it remained at the same value as in full dark-adaptation over about a log unit and a half of background flux. And naturally the light spilling over onto the theoretically dark half-field did not suffice either to increase threshold, so the two sets of symbol fall along the same horizontal line. As the light within the illuminated half-field is further increased, threshold does begin to rise and this happens first within the illuminated half-field; the open symbols are the first ones to deviate from the horizontal line which indicates threshold in full dark-adaptation. But then, about one log unit later, the same happens in the "dark" half-field. And once each half of the center has entered its Weber-region, each set of symbols falls along its respective Weber-line. Furthermore, the two Weber-lines are parallel and separated--in this experiment--by about 0.8 log units.

As the figure tells you this is a Y-cell and TOM HARDING did the experiment on a total of ten Y-cells. The outcome was the same for every Y-cell in the sense that there were always two parallel Weber-lines. But the vertical distance by which they were separated varied from cell to cell and ranged from 0.3 to 1.1 log units. Seven X-cells were also tested; six of them behaved in principle as the Y-cells. For five of these the vertical separation between the two Weber-lines ranged from 0.1 to 0.4 log units. For one X-cell the separation was a full log unit, and this cell had the largest center (2.0^{o}) among the X-cells. Hence, it was also tested with the largest spot separation among X-cells. For the seventh and last X-cell there was no vertical separation of the two sets of symbols--and that cell had a center which was one of the smallest X-centers.

The question is: what exactly do these observations tell us? Unfortunately there is no short or simple answer to that question--but I will tell you what I believe we can conclude with a fair degree of certainty.

To begin with, the Y-cell situation is in some respects rather straight-forward: for no Y-cell did the two sets of symbols fall neatly along the same Weber-line, although the vertical separation was as small as 0.3 log unit in one cell. This means that none of the Y-cell centers that we studied exhibited what one may call complete physiological pooling of adaptive effects. Or, expressed another way, the center did not act as one unit with regard to the entire range of its field adaptation in any of these Y-cells.

Furthermore, one can draw this conclusion irrespective of the extent to which light spills onto the "dark" half-field. For light spill could only conceal differential adaptation of sub-areas, not give a false positive.

The X-cell situation is a bit more complicated. Here it did happen--for one cell--that the two sets of symbols fell along one Weber-line while in the other six cells there was--just as in the Y-cells--a measurable difference in adaptation level between the two halves of the center, although this difference, on average, was smaller than for Y-cells.

Quite clearly, then, the difference between X- and Y-cells is not simply that one class does, the other class does not, exhibit differential adaptation within its center. But, whether the separation of the Weber-line ranged from zero to 0.4 log units for six cells of one group, from 0.3 to 1.1 log units for the ten cells of the other group, because the first group were X-cells, the latter Y-cells, is something I do not know.

It could also be that because X-cells tend to have smaller centers and were tested with smaller spot separations, imperfect imagery concealed a larger proportion of an actually existing difference in level of adaptation between the two halves of the center.

Let me conclude with two statements which I think we can safely make: 1) We studied a total of 17 center mechanisms--in the cat--with a bipartite adapting background. All but one of these centers exhibited sub-areas whose level of field adaptation depended, to some extent, upon where appropriately weighted adaptive flux fell, not only upon how much flux there was within the central summing area. 2) But, the center mechanism of cat ganglion cells also exhibits some global pooling of field adaptive effects--otherwise one could not get the results you saw in the first figure. And that result pertains equally to X and Y cells.

Supported by National Institutes of Health Grant #5 R01 EY00206 and The Rowland Foundation.

References

1. C. Enroth-Cugell and R. M. Shapley, J. Physiol. 233, 311 (1973).
2. S. S. Easter, Jr., J. Physiol. 195, 273 (1968).
3. D. A. Burkhardt and G. G. Berntson, Vision Res. 12, 1095 (1972).
4. D. G. Green, L. Tong and C. M. Cicerone, Vision Res. 17, 479 (1977).
5. B. G. Cleland, W. R. Levick and K. J. Sanderson, J. Physiol. 228, 649 (1973).
6. C. Enroth-Cugell and J. G. Robson, J. Physiol. 187, 517 (1966).
7. B. G. Cleland and C. Enroth-Cugell, J. Physiol. 198, 17 (1968).
8. W. A. H. Rushton, Proc. R. Soc. B. 162, 20 (1965).
9. S. Hochstein and R. M. Shapley, J. Physiol. 262, 237 (1976).
10. T. H. Harding, Ph.D. Thesis, Purdue University, Lafayette, Indiana, 1977.
11. B. G. Cleland and C. Enroth-Cugell, J. Physiol. 206, 73 (1970).

The Coding and Decoding of Steady State Visual Information by Patterned Pulse Trains in the Crayfish Visual System

Raymon M. Glantz
and
Howard Wood
Biology Department
Rice University
Houston, Texas 77001

A dominant thesis in sensory neurophysiology is that the "coding" of sensory information is primarily a function of the anatomical arrangements of the neurons in the sensory pathway. One implication of this concept is that sensory information is distributed on labeled lines. The information that is coded is determined by cellular connectivity. The information that is coded is determined by the receptive field and the trigger features of the cell. In recent years the thesis has been applied to visual systems with remarkable success. It has given rise to dimming units and movement detectors, color coded cells, X and Y cells and simple and complex cells to name only a few members of the growing lexicon of functional identifications.

In our determination to discover "who" a neuron is however, it is possible to overlook "what" it might be saying. The temporal features of a neuronal pulse train and its relationship to the pulse trains of parallel elements in the pathway must also constitute a significant component of the neural information code. If a neural pulse train fails to meet the requirements for successful synaptic transmission the message is lost. Furthermore, the labeled line and trigger feature concepts rely on an implicitly assumed temporal code such as mean rate or the number of impulses elicited by a stimulus.

In order to establish that a pulse train parameter (e. g., mean rate, instantaneous rate, etc.) is a component of a sensory code, the candidate parameter must meet two tests: a) the parameter must vary significantly as a function of variations in a physiological stimulus (the coding test) and b) the pulse train parameter must contribute to or modify the liklihood of successful synaptic transmission (the de-coding test) (1). If a pulse train paremeter fails to meet the coding test then the parameter says nothing about the stimulus. If it fails to meet the decoding test the parameter says nothing to the rest of the nervous system. In either case the pulse train parameter cannot serve a coding function. Furthermore, the temporal code concept and the labeled line concept can produce substantial differences in our interpretation of a neuron's coding properties. If, for instance, the synapse of a sensory neuron were to undergo significant defacilitation at high firing rates then the most effective synaptic transmission might occur at low or intermediate rates. If the sensory trigger feature is defined by an assumed mean rate code then our functional label is incorrect or at least misleading.

Studies related to the first of the two coding tests mentioned above date back to the genesis of visual neurophysiology. Visual interneurons functionally described as tonic on elements have been observed in fish, amphibians, mammals, arachnids, insects, and crustaceans. These units exhibit the common property that over a defined dynamic range of light intensities their mean rates vary as a monotonic function of the intensity of light in the excitatory receptive field. ADRIAN and MATHEWS (2) in their studies of Conger eel optic nerve observed that in addition to the stimulus dependent changes in firing rate the optic nerve fibers exhibited strong periodic modulations in discharge frequency. This periodic bursting was observed both after a step increase in illumination and in the steady state. The steady state bursting behavior typically exhibits a threshold in intensity that is considerably higher than the absolute threshold of illumination of the neuron. More recently, similar oscillations have been observed in the mammalian visual system (3, 4) and in a variety of arthropods (5, 6, 7, 8, 9, 10, 11).

It is possible that the periodic bursting behavior is a biproduct of retinal organization and that it has no particular coding function. Conversely one may conceive of any number of synaptic mechanisms such as facilitation, defacilitation, or spatial and temporal summation whereby the patterned discharge may result in a more or less effective synaptic transmission when compared to either a random sequence of interspike intervals or a continuous train of regularly spaced impulses.

Meeting the second (decoding) test is obviously a rather difficult matter in most sensory systems. The proponent of a code must demonstrate that a given pulse train parameter influences synaptic efficacy independent of correlated changes in other pulse train parameters.

One of the first documented examples of a pattern code was obtained at the crustacean neuromuscular junction. Enhanced effector responses to patterned stimuli as compared to regular trains at the same rate were obtained by WIERSMA and ADAMS (12) in a variety of crustacean neuromuscular preparations. The muscle response was facilitated up to 30 fold with patterned sequences. Furthermore, during reflex activation crustacean motoneurons commonly exhibit a strong tendency to fire in a bursty pattern with a modal interval near the optimum observed by WIERSMA (13, 14, 15).

The success with the neuromuscular junction has recently been followed by similar results with crustacean command interneurons which also exhibit enhanced effector responses or motoneuron discharges when driven by patterned trains of impulses (16, 17, 18). In addition to the crustacean studies, SEGUNDO and coworkers (19-21) have established some formal arguments and a variety of statistical techniques to demonstrate the action of patterned pulse trains on synapes in the _Aplysia_ visceral ganglion.

Based upon these results, we decided to examine the question in the crayfish visual system. The studies to be described below can be divided into three sections.

The first section deals with a review of largely published results (10, 22, 23) which document how visual stimulus parameters such as intensity, position and target velocity modify certain temporal parameters of the discharge of sustaining fibers (SF) and the correlation between SF's. The second section presents evidence for a class of descending visual interneurons labeled Tonic On cells (WOOD and GLANTZ, in preparation) which originate in the supraesophageal ganglion (i.e., the brain) and which are exicted by the same stimulus conditions which excite the SF's. In the third section it is shown that: a) The SF's transmit the visual input to the Tonic On units; b) The Tonic On cells exhibit bursts of EPSP's under conditions associated with SF bursting; and c) The Tonic On cells require temporal summation of the brief EPSP's to attain threshold. Thus these neurons selectively respond to epochs of high instantaneous frequency in the presynaptic neuron or a high correlation in firing times of the presynaptic population.

The Optic Nerve Sustaining Fibers

The basic characteristics of the crayfish SF's were originally set forth by WIERSMA and YAMAGUCHI (24, 25). The neurons are silent in the dark. Each unit has an excitatory receptive field which overlaps that of other SF's but is unique for each of the 14 members of the SF population. Within a 5° radius around the center of the excitatory field the SF's exhibit linear spatial summation (26). The regions of the retina outside the excitatory field constitute an inhibitory surround (25, 27). Upon exposing the excitatory field to a step increase in illumination the SF's exhibit a high frequency transient discharge which slowly decays to a much lower sustained rate. The frequency of the transient and steady state discharges are monotonic functions of the stimulus intensity (28). The transient discharge is frequently subdivided into a series of bursts with a period of about 250 ms. The first two bursts are generally synchronized to the stimulus and are frequently seen in the post-stimulus time histogram (26). If brief (i.e., 5 ms) shocks are delivered to the inhibitory surround the bursting can remain synchronized to the stimulus for up to five cycles (10). The interspike interval distribtuion of the transient discharge exhibits a minimum interval of 2 to 3 ms and a mode at 8-12 ms (unpublished observations).

Stimulus synchronized bursting is also observed in response to the sinusoidal modulation of a stationary point of illumination. The strength of the bursting discharge is dependent upon the stimulus intensity and the modulation frequency (10, 22). The SF's are exquisitely sensitive to moving shadows or edges. At low velocities (e. g., 2°/S) movement will produce a sustained high frequency discharge so long as the edge is contained and moving within the excitatory field. This discharge will exhibit a weak to modest burst structure. At higher velocities the bursting becomes quite pronounced (25).

In steady state conditions the discharge becomes increasingly ordered as the stimulus intensity is raised. Fig. 1, A-C is a series of interspike interval histograms from the sustained discharge of a unit at three intensities, 10X, 100X, and 500X threshold. As the intensity is increased the discharge changes from a random sequence of intervals to a regular pacing discharge with deletions. Further increases in intensity result in smaller deviations from the modes but no change in the modal intervals. The half width of the principal (modal) peak (at 65 ms in Fig. 1) decreases from 40 ms (Fig. 2B) to 25 ms (Fig. 2C) as the intensity is increased five fold and the mean rate increases from 8.1 to 11.9 impulses/s. For a given neuron the principal interval, which is generally between 50 and 200 ms, is independent of the mean firing rate over the range of 10 to 40 impulses/s. Changes in mean rate are incorporated into the rhythmic discharge by increasing the number of spikes/burst and decreasing the percentage of deletions from the regular train. In Fig. 1C brief interspike intervals associated with bursting first appear at the highest intensity.

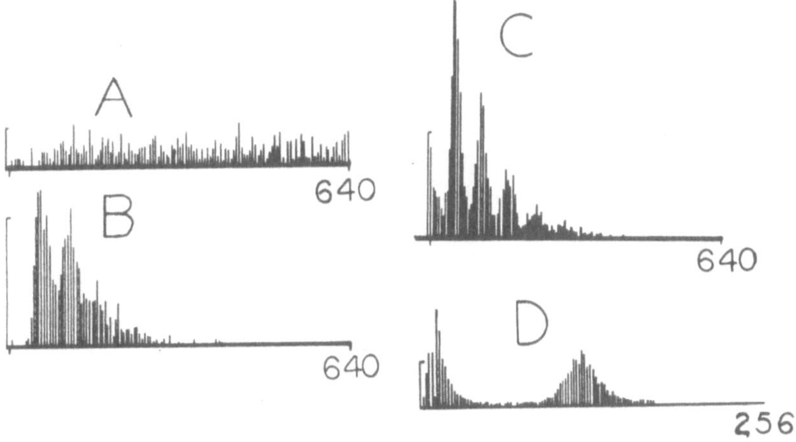

Fig. 1 A-C. Influence of light intensity on SF interspike interval distributions. Interspike interval histograms, 5 ms/bin, of SF steady-state discharge to stimulus intensity of 10X, 100X and 500X threshold respectively. Calibration is 30 events. D. Interval histogram, 2 ms/bin, of a SF exhibiting an intensity dependent steady-state bursting discharge. Calibration is 100 events (modified from GLANTZ and NUDELMAN (10)).

Furthermore, the proportion of intervals $\pm$ 15 ms from the principal mode increases from 7.2 to 10.3% as the intensity is increased.

Fig. 1D is the interval histogram of a rhythmically bursting unit. The principal interval was 122 ms and the mean rate was 10.0 Hz. The bursting behavior is invariably associated with a markedly disproportionate number of brief interspike intervals. In this case 24% of the intervals were less than16 ms.

The evidence suggests that the SF's exhibit an intensity dependent ordering of the discharge. This conclusion is further supported by autocorrelation functions of the steady discharges which exhibit intensity dependent periodic oscillations. Both the amplitude of the correlations and the number of cycles observed (up to 30) greatly exceed those of the function calculated by repeated convolution of the first order interval histogram (10, 23). This result indicates that high light intensities produce a long term ordering of the SF discharge (29). The result contrasts markedly with the maintained activity of cat retinal ganglion cells (30) for which the convolution functions precisely predict the oscillations in the empirical autocorrelation functions.

Simultaneous recordings from pairs of SF's indicate that the neurons burst in approximate synchrony. The magnitude of the crosscorrelations varies with stimulus intensity, Fig. 2. The correlations strengthen in parallel with the strength of the autocorrelation functions. Finally it was noted that the time lag to the peak of the crosscorrelation function which estimates the phase lag between the units may exhibit a stimulus dependence of up to 25% of the natural period. Phase shifts can be induced by variations in stimulus intensity and/or position. These results raise

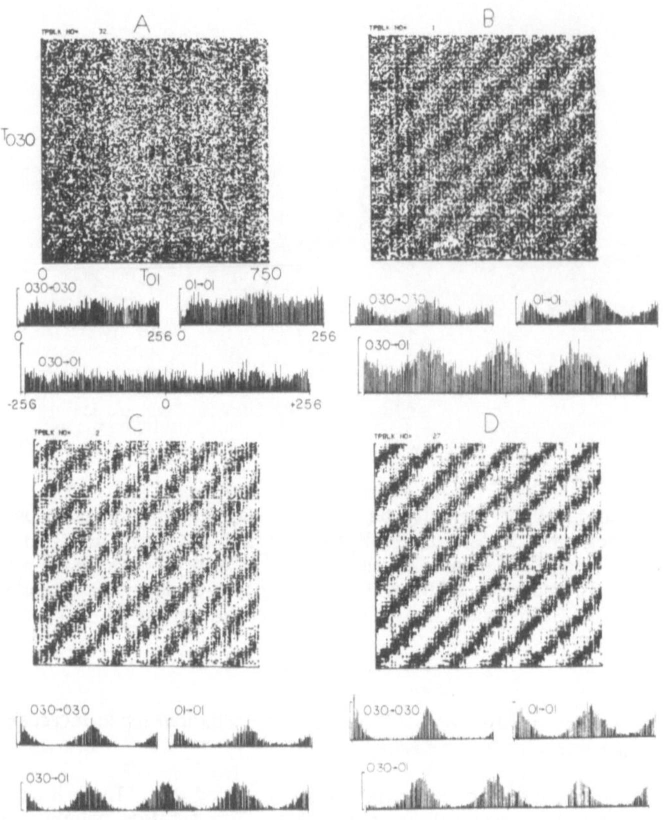

Fig. 2 Influence of light intensity on the inter-SF crosscorrelation. Joint peristimulus time histograms, auto- and crosscorrelograms of SFs 030 and 01 at four stimulus intensities with collimated light cast primarily on the 030 excitatory field, but partially overlapping the 01 field. Stimulus intensities were 0.35, 0.64, 2.28, and 4.79 mW/cm^2 in A through D, respectively. All histograms are 2 ms/bin, horizontal scales in B through D as in A; other parameters as follows: A: (030) = 87 ms, (01) = 119 ms, NE(030) = 3,090, NE(01) = 3,500, NE(030-01) = 1,500; B: (030) = 61 ms, (01) = 89 ms, NE(030) = 2,268, NE(01) = 2,100, NE(030-01) = 2,500; C: (030) - 37 ms, (01) = 111 ms, NE(030) = 507, NE(01) = 1,270, NE(030-01) = 1,600; D: (030) = 28.4 ms, (01) = 59 ms, NE(030) = 545, NE(01) = 666, NE(030-01) = 431. (Reprinted with permission (10)).

the possibility that sensory information maybe encoded by the correlation and/or time delays between members of a parallel ensemble.

To summarize, the SF's exhibit pronounced bursting during the initial transient discharge associated with step increases in illumination level. The intensity of bursting is a function of the rate of rise of the luminous intensity. In steady state conditions, as the intensity is raised, the SF discharge goes from a random sequence of intervals to a pacing discharge and finally to rhythmic bursting. As the order of the discharge increases there is an increased correlation between neighboring units. At the highest steady state intensities the SF's burst in approximate synchrony or with a tendency to assume a fixed phase relationship.

The Descending Tonic On Cells

To determine whether the rhythmic bursting behavior and/or the ordering of the SF discharge by ambient light is decoded, it is necessary to determine whether the discharge of cells postsynaptic to SF's are differently influenced by the pattern of the SF discharge.

This phase of the work commenced with a detailed analysis of visual units that arise near the terminations of the SF's in supraesophegeal ganglion (WOOD and GLANTZ, in preparation). Descending units excited by the onset of illuminaiton were first established as discrete entities by WIERSMA and MILL (31). Several hundred recordings were made from single unit and multiple unit extracellular leads both in the ganglion and in the circumesophageal connectives (the connectives link the brain with the more caudal portions of the central nervous system). The results revealed several classes of units which could be segregated by factors such as the form of the impulse response as measured by the poststimulus time histogram, the peak firing rate, the rate of habituation, movement sensivitiy, threshold to a step of illumination, etc. The tonic on cells fire at low rates, usually less than 75 impulses/s at peak, Fig. 3. Step increases in illumination result in a very weak transient discharge with a latency of about 100 ms followed by a tonic response. The response exhibits only a modest degree of habituation to repetitive stimuli (Fig. 3). The receptive fields are generally monocular and from the ipsilateral eye.

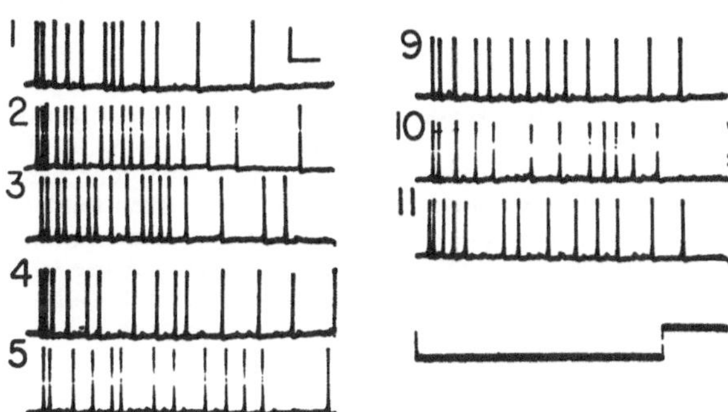

Fig. 3 Responses of a descending Tonic On cell to a rectangular pulse of illumination. Calibration is 20 mV and 20 ms. Numbers at left of each trace indicate the serial position of responses to successive stimuli. (This data and that of all subsequent figures is unpublished).

Figure 4A is a photograph of the dorsal view of a whole mount cerebral ganglion. The bar at the lower left indicates 500 mm. The Tonic On cell was injected with Procion Yellow (M-4 RAN). The brain was dehydrated and cleared in Methyl

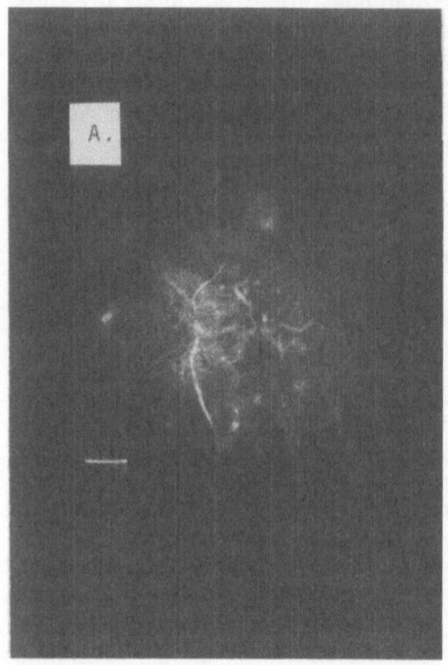

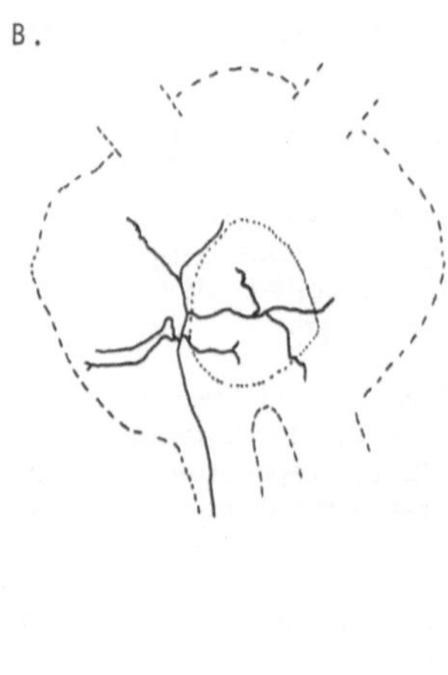

Fig. 4 Crayfish supraesophaseal ganglion, dorsal view. Calibration is 500 µ m. The Tonic On cell was injected with Procion yellow (M-4 RAN). B. Tracing of the axon and major neurites of Tonic On cell constructed from photographs at three focal planes in the ganglion. Dotted area indicates zone of most frequent SF recordings. Dashed line indicates the cerebral ganglion, circumesophageal connectives at bottom and optic peduncles at top.

Salicylate. Florescence was obtained with a xenon source and a BG-12 excitation filter. The film was exposed through a 580 nm barrier filter. The drawing at the right, Fig. 4B, was constructed from tracings of photographs at three different focal planes in the 1.5 mm thick ganglion. The tracings indicate the large axon (40 µ at maximum width) exiting in the left circumesophageal connective and several major dendritic branches. The region enclosed by the dotted line is particularly dense with the axons and terminals of optic nerve SF's as indicated by focal extracellular recordings and a large number of successful impalements of SF's in this zone.

Sustaining Fiber Activation of Tonic On Cells

The most direct means of demonstrating a functional relationship between the SF and Tonic On cell is to impale the SF and determine how selective presynaptic activation modified the activity of the presumptive postsynaptic cell. Ideally the postsynaptic recording should be done with an electrode capable of monitoring subthreshold events but at present this would require an unlikely simultaneous impalement of the two cells. Thus the experiment was performed with extracellular recordings from a Tonic On cell and an intracellular electrode in a SF.

Figure 5A (GLANTZ, previously unpublished data) illustrates the response of the two cells to a rectangular pulse of illumination. The delay from the SF light

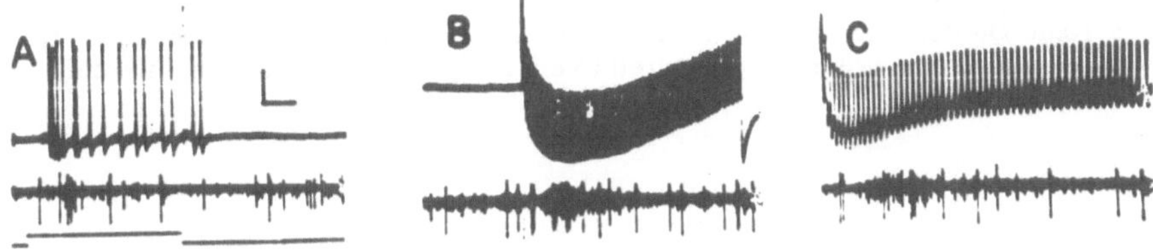

Fig. 5 Simultaneously recorded SF, top trace, and Tonic On cell, middle trace, response to a rectangular pulse of illumination. B. Tonic On cell and SF responses to a 25 nA depol of the SF. C. As in B at faster sweep. Calibrations, 5 mV for the SF, 200 μV for the Tonic On cell, 100 ms in A, 200 ms in B, 100 ms in C.

evoked response to the Tonic On response was about 30 ms. Fig. 5B illustrates the response of the two neurons to a 25 nA depolarizing (i.e., outward) current delivered to the SF. Fig. 5C illustrates the same response at slightly higher sweep speed (upper trace is now A.C. coupled) and with the onset of the horizontal trace synchronized to the onset of the depolarizing pulse. The latency from the first SF impulse to the first Tonic On cell spike was 54 ms. The shorter interresponse delay associated with the light pulse is probably a consequence of the spatial summation of light activated convergent SF inputs to the Tonic On cell. The long delay and weak response of the Tonic On cell to strong SF activation suggests that the SFTonic On cell functional connection is mediated by a poly-synaptic pathway or a synapse of very low gain. The latter hypothesis would require that EPSP's (generated by SF's) in Tonic On cells undergo substantial temporal summation to cross the spike threshold.

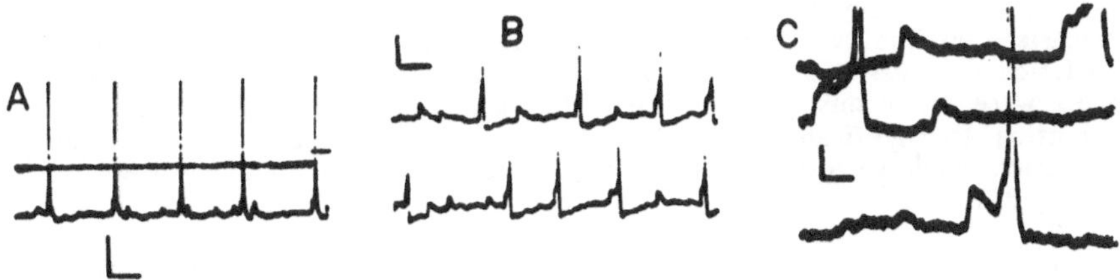

Fig. 6 A. Tonic On cell response to a pulse of illumination. B and C segments of responses similar to A at higher gain and sweep speed. Calibrations: A, 10 mV, 100 ms; B. 5 mV, 50 ms; and C. 2 mV, 10 ms.

Figure 6A illustrates an intracellular recording of a typical Tonic On cell response to a pulse of illumination. Notice that each spike is associated with a burst of EPSP's. At higher gain and faster sweep, Fig. 6B, C it can be seen that most spikes are preceded by the temporal summation of EPSP's. The spike threshold of this Tonic On cell was 4.2 mV while the mean EPSP amplitude was 3.47 mV. Thus two synaptic events are usually required to elicit a postsynaptic action potential.

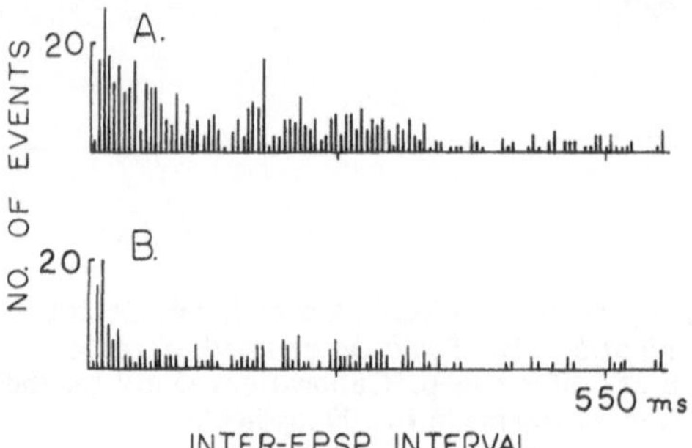

Fig. 7 Inter-EPSP interval histograms from steady-state response to continuous illumination. A histogram of all EPSP intervals. B intervals followed by a spike. A. N = 1031, B. N = 271, 5.5 ms/bin.

The inter-EPSP interval distribution was therefore examined to determine the relative frequency of EPSP's pairs that could support temporal summation. Fig. 7A displays the distribution of inter-EPSP intervals observed in steady state conditions, generated by continuous illumination. The data indicate a preponderance of intervals less than 50 ms and a modest second mode at 180 ms. The bimodal disbribution closely resembles the interspike interval distribution of weakly bursting SF's. For the sample considered here 37.1% of EPSP's resulted in spikes. Fig. 7B displays the inter-EPSP intervals (drawn from the population in Fig. 7A) in which the second EPSP of the interval pair resulted in an impulse. This is a retrospective probability distribution since one looks back in time from an impulse to determine the relative efficacy of the different classes of EPSP intervals (19, 20). A comparison of the two distributions indicates that 10.8% of the total interval population was less than 20 ms but 21.4% of the EPSP intervals that resulted in spikes were from this group. The difference suggests that postsynaptic spike generation is significantly enhanced by temporal summation. The period of summation should be reflected in the EPSP decay time constant which is about 12.3 ms, Fig. 8A.

One way to consider the contribution of temporal summation to spike generation is to determine the probability that a spike will be generated for each EPSP interval class, Fig. 8C. This is the prospective probability distribution (19). It should be noted that the prospective probability of the 5.5 to 11 ms interval class is an impressive 0.89 compared to a probability of 0.265 for very long intervals. If temporal summation is the only mechanism influencing this distribution then the prospective probability should decline with increasing intervals at a rate that reflects: a) the time course of decay of the EPSP, Fig. 8A; and b) the amplitude distribution of the EPSP's, Fig. 8B. If the synapse exhibits linear temporal summation then for any particular EPSP interval the prospective probability, P(t), should be equal to the probability that the amplitude of the second EPSP, P(V), will

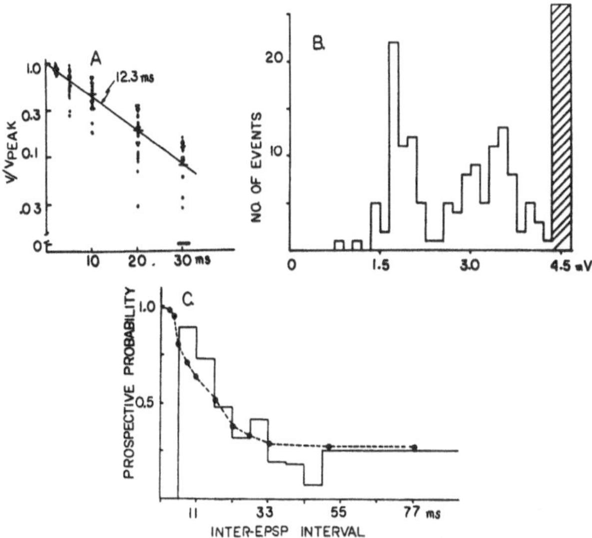

Fig. 8 A. Semilog plot of EPSP amplitude as a function of time after peak for 30 EPSPs. The arrow indicates the EPSP decay time constant. B. Amplitude histogram of EPSPs, N = 187. Hatched area at right indicates EPSPs above spike threshold. C. Observed (histogram) and predicted (closed circles, dashed lines) prospective probabilities of spike generation for EPSP intervals of 0 to 82.5 ms. Bin width is 5.5 ms. The predicted probabilities assume linear temporal summation of pairs of EPSPs.

equal or exceed the difference between the spike threshold (4.2 mV) and the amplitude of the passively decaying EPSP, f(t). Thus:

$$P(t) = P(V) \quad (- f(t)) \qquad (1)$$

Figure 8B is the amplitude histogram of the EPSP's obtained in steady state conditions. Impulse generation was not blocked. Thus the amplitude of EPSP's which elicited spikes can only be classified as suprathreshold (Fig. 8B hatched area). These EPSP's constituted 26.5% of the sample. Since it is only possible to assign a minimum value to these EPSP's (i.e., the spike threshold) the calculation of the mean EPSP amplitude may be slightly biased toward a value less than the true mean. The application of Eq. (1) to the data of Fig. 8A and B, results in an estimate of the prospective probability (indicated by the closed circles and dashed lines in Fig. 8C). The histogram indicates the prospective probabilities derived from actual spike generation by pairs of EPSP's separated by the intervals plotted on the horizontal axis. The tendency of the calculated function to slightly underestimate the observed prospective probabilities is most likely attributable to the slight bias discussed above.

The only other deviation from linear temporal summation is the tendency of the observed probabilities to be less than predicted in the 33 to 50 ms interval classes. This span encompassed 40 EPSP pairs which elicited 5 impulses for an overall prospective probability of 0.125. The probability expected from linear temporal summation of EPSPs is 0.265 more than twice the observed result. The discrepancy suggests an interval dependent dimunition in spike generating ability that is noramlly masked by temporal summation.

We hestiate applying the term defacilitation since the data do not allow us to distinguish between homosynaptic and heterosynatpic effects or between pre or postsynaptic effects. Attempts to obtain further information from our data samples were frustrated by an apparent conflict. The inter-EPSP interval distribution was consistent with a single SF input but the obviously multimodal character of the EPSP amplitude histogram is suggestive of multiple inputs. Thus we will use the conservative terminology and refer to the effect as an interval dependent depression.

To examine this depression more directly the sample of EPSP's in Fig. 8B was divided by span into two populations of approximately equal prospective probability. This was done to minimize the bias associated with the unobserveable amplitudes of the spike eliciting PSP's. The short interval group spanned the interval range of 22 to 100 ms. The long interval group consisted of all longer PSP intervals. The EPSP pairs in which the second PSP was followed by a spike were deleted. The distribution of amplitudes of the two groups is shown in Fig. 9. The mean amplitude for the longer interval group was 3.04 ± .87 mV (mean ± S.D.) and the mean amplitude of the shorter interval group was 2.28 ± .72 mV.

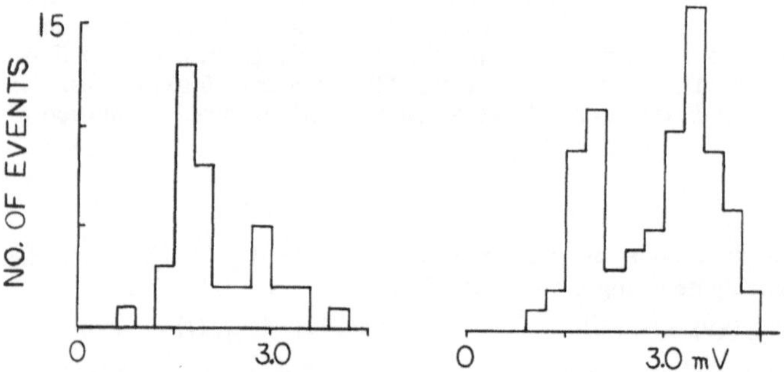

Fig. 9 Evidence for short-term depression of EPSP amplitude. The data of 8B was divided into two amplitude histograms based upon inter-EPSP interval (greater than 110 ms right and 22-110 ms, left). EPSPs eliciting spikes were deleted.

These results indicate a critical function for the SF bursting behavior. In stationary conditions under high levels of uniform illumination the mean SF discharge rate is 11.7 ± 4.6 impulses/s (mean ± S.D. for 21 SF's, GLANTZ and NUDELMAN (10)). This represents a mean interval of 85 ms. At this interval only 26% of the SF elicited EPSP's would result in spikes. The presence of high frequency bursts however raises the mean spike eliciting probability to a higher value.

The interval dependent depression has the effect of increasing the minimum instantaneous frequency required for temporal summation. Thus only those PSP's with intervals of 16 ms or less exhibit the enhanced prospective probability of spike generation associated with temporal summation (Fig. 8C). This result is particularly interesting since the steady-state bursting behavior of the SF's (Fig. 1D) results in a remarkably large percentage of interspike intervals in this class (up to 35%). The consequence of this frequency dependent enhancement can be calculated from the data of Fig. 2B and C. Increasing the ambient intensity 3.5 fold is associated with a 60% increase in SF 030 mean firing rate (from 16.7 to 27.0 Hz) as the SF bursting threshold is exceeded. The percentage of intervals less than 16 ms climbs from 2% to 35%. The Tonic On cell however would exhibit a rate increase from 4.3 to 11.0 impulses/s. Thus, under similar conditions the Tonic On cells would exhibit a 155% rate increase compared to a 60% increase for the SF. This calculation assumes that the EPSP amplitude distributions and the decay time constant would resemble those observed in Fig. 9. This is a modest bursting dependent enhancement of synaptic efficacy. At higher intensities virtually all of the SF steady-state activity is confined to bursts with interspike intervals of 2.5 to 10 ms. Such a pattern could exhibit an overall prospective probability of about 0.7 if longer term interactions did not play a role.

The same arguments could also provide a functional role for the near synchronous firing of the population of bursting SF's. The convergence ratio of SF's to Tonic On cells however is not known. The Tonic On cell receptive field generally covers the entire ipsilateral eye. The input for this field could be provided by a single identified SF, unit 01 (24) or the entire population of 14 ipsilateral SF's. Thus the role of synchronous firing, if any,cannot be determined from the data available at present.

As noted earlier the EPSP amplitude histogram of Fig. 8B suggests multiple SF inputs. The modes of the histogram occur at the amplitude intervals of 1.65 to 1.8 mV, 3.45 to 3.60 mV and presumably a third made above 4.3 mV. It is possible that each mode represents the EPSP amplitude associated with a different SF. Alternatively it is possible that the first mode at 1.72 mV is the characteristic mean EPSP amplitude for all SF inputs. If this were the case then the mode at twice this value could result from the near simultaneous firing of two or more SF's spatially summed to elicit an EPSP of twice the unitary amplitude. A third mode associated with impulse generation could be the result of 3 near simultaneous spikes in 3 SF's. Our previous findings regarding the cross correlation of simultaneously recorded SF's (Figs. 2 and 3) indicate a substantial probability that impulses in just a pair of parallel SF's may occur within 1.0 ms of one another. The likelihood of such an occurrence in the whole SF population is presumably much higher. An analysis along these lines would suggest that every postsynaptic impulse is associated with spatial and/or temporal summation of at least 3 EPSP's. The resolution of this question will require more information regarding the relationship between the presynaptic pulse train and postsynaptic potentials as well as a more detailed consideration of the SF to Tonic On cell convergence pattern.

Acknowledgements

We thank Dr. Arnold Eskin for helpful discussions during the experimental phases of these studies. The new data reported here was obtained with the support of National Science Foundation Grants No. BNS 576-80507 and No. BNS 72-020100, and an N.I.H. Eye Institute predoctoral Fellowship to H. W. No. 5 T 32 EY-0702403.

References

1. D. H. Perkel, G. L. Gerstein and G. P. Moore, Biophys. J. 7, 391 (1967).
2. E. D. Adrian and R. Mathews, J. Physiol., London 65, 273 (1928).
3. P. O. Bishop, W. R. Levick and W. V. Williams, J. Physiol., London 170, 598 (1964).

4. A. C. Sanderson,W. M. Kozak and T. W. Calvert, Biophys. J. 13, 218 (1973).
5. A. L. Adolph, J. Gen. Physiol. 62, 392 (1973).
6. E. D. Adrian, J. Physiol., London 91, 66 (1937).
7. A. D. Blest and T. S. Collett, J. Insect Physiol. 11, 1079 (1965).
8. F. Crescitelli and J. L. Jahn, J. Cellular Comp. Physiol. 19, 47 (1942).
9. R. M. Glantz, J. Neurobiol. 4, 301 (1973).
10. R. M. Glantz and H. B. Nudelman, J. Neurophysiol. 39, 1257 (1976).
11. T. Yamaguchi and T. Ohtsuka, J. Fac, Sci. Hokkaido Univ. 19, 15 (1973).
12. C. A. C. Wiersma and R. T. Adams, Physiol. Comp. Oecelo. 2, 20 (1950).
13. H. L. Guillary and D. Kennedy, J. Neurophysiol. 32, 607 (1969).
14. D. O. Smith, J. Neurophysiol. 37, 108 (1974).
15. D. Wilson and W. J. Davis, J. Exp. Biol. 43, 193 (1965).
16. H. C. Atwood and C. A. G. Wiersma, J. Exp. Biol. 46, 249 (1967).
17. H. L. Gillary and D. Kennedy, J. Neurophysiol. 32, 595 (1969).
18. R. M. Glantz, In Identified Neurons and Behavior in Arthropods, G. Hoyle (ed.), 1978. In press.
19. J. P. Segundo, G. V. Moore, L. J. Stensass and T. H. Bullock, J. Exp. Biol. 40, 643 (1963).
20. J. P. Segundo, D. H. Perkel and G. P. Moore, Kybernetik 3, 67 (1966).
21. J. P. Segundo, D. Perkel, H. Hegstad, H. Wymen and G. Moore, Kybernetik 4, 157 (1968).
22. W. H. Gordon, Doctorial Thesis, Houston, Texas: Rice University, 1975.
23. H. B. Nudelman and R. M. Glantz, Fed. Proc., 36, 2042 (1977).
24. C. A. G. Wiersma and T. Yamaguchi, J. Comp. Neurol. 128, 333 (1966).
25. C. A. G. and T. Yamaguchi, J. Exptl. Biol. 47, 409 (1967).
26. R. M. Glantz, Vision Res. 13, 1801 (1973).
27. H. Arechiga and K. Yanagisawa, Vision Res. 13, 731 (1973).
28. R. M. Glantz, J. Neurophysiol. 34, 485 (1971).
29. D. H. Perkel, G. L. Gerstein and G. P. Moore, Biophys. J. 7, 391 (1967).
30. R. W. Rodieck, J. Neurophysiol. 30, 1043 (1967).
31. C. A. G. Wiersma and P. J. Mill, J. Comp. Neurol. 125, 67 (1965).

Photic Sensitivity of Macaque Monkey and Pulvinar Neurons

M. L. J. Crawford
and
S. Espinoza
The University of Texas
Health Science Center at Houston
Graduate School of Biomedical Sciences
Houston, Texas 77025

Introduction

The pulvinar complex of the primate brain, by sheer mass alone, is the major nucleus of the thalamus having visual input. The evolution of the pulvinar appears to culminate in primates where the nucleus is largest and becomes most differentiated. The following figures will serve to illustrate the relative size and location of the divisions of the primate pulvinar. Figure 1 illustrates the locus of the three divisions of the pulvinar complex at the level of the caudal pole of the lateral geniculate nucleus (LGN) while Fig. 2 outlines the major divisions of the pulvinar at their largest extent in the posterior thalamus. Clearly these nuclear groups make up the bulk of the thalamus at this level.

More than mere size, the pulvinar complex is of interest to researchers of visual information processing because of its connections. Being retinotopically organized itself (clearly the inferior pulvinar is), the pulvinar is one synapse away from a number of other retinotopically organized structures with which it shares reciprocal connections. Figure 3 illustrates the principal extra-pulvinar connections. First, from the striate and the prestriate cortices there are projections to all divisions of the pulvinar as well as projections to the superior colliculus of the tectum. CHOW (1) and CAMPOS-ORTEGA, et. al., (2) showed that the prestriate areas project principally to the inferior pulvinar while the upper parietal cortex projected principally to the lateral division of the pulvinar. CHOW (1) also found that the anterior pole of the temporal lobe projected principally to the medial pulvinar. TROJANOWSKI and JACOBSEN (3) using combined horseradish peroxidase (HRP) and tritiated leucine have verified these pathways and shown that in the squirrel monkey the superior temporal gyrus projects to all three divisions of the pulvinar.

Another pathway (which was inadvertently omitted from this illustration) is one

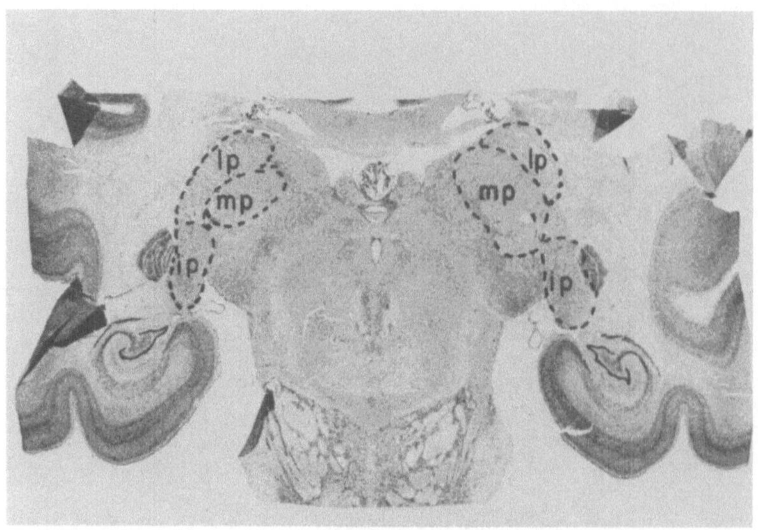

Fig. 1 Coronal representations of the divisions of the rhesus monkey pulvinar nuclei at the level of the posterior lateral geniculate nucleus. Medial pulvinar (mp); lateral pulvinar (lp); and inferior pulvinar (ip).

linking the medial pulvinar with the frontal eye fields (FEF) as described by TROJANOWSKI and JACOBSEN (4) and by BOS and BENEVENTO (5). Additionally, JONES and BURTON (6) have demonstrated a projection from the medial pulvinar to the lateral nuclear group of the amygdala.

Recent metabolic tracer studies have verified the intimate and reciprocal relationships of the striate cortex with all divisions of the pulvinar, LGN, and the superior colliculus (7, 8, 9). Layers I, V, and VI of the striate cortex project to both lateral and inferior divisions of the pulvinar. BENEVENTO and FALLON (10) produced lesions in the superior colliculus and concluded that the superficial layers of the colliculus projected to the inferior pulvinar (among several other structures including the dorsal LGN) while lesions deep within the superior colliculus produced degenerated terminals within the medial pulvinar nucleus.

At present there is only scant evidence that the pulvinar has a direct ganglion cell input from the optic tract. CAMPOS-ORTEGA, et. al., (11) have suggested such an input and have presented some tentative evidence to support the claim. Similarly, the evidence for direct connections between nuclei of the pulvinar and the adjacent LGN is not very good. TROJANOWSKI and JACOBSEN (3) injected HRP into the inferior pulvinar and noted retrograde reaction product in cells of the dorsal LGN. However, from the locus of the injection, one could not be sure that the HRP did not get to the LGN merely through broken geniculo-cortical radiation fibers.

In summary, the anatomical evidence places the pulvinar complex in a prominent position between virtually all of the known structures of the cortical mantle purporting to deal with color analysis, spatial pattern vision, visual memory,

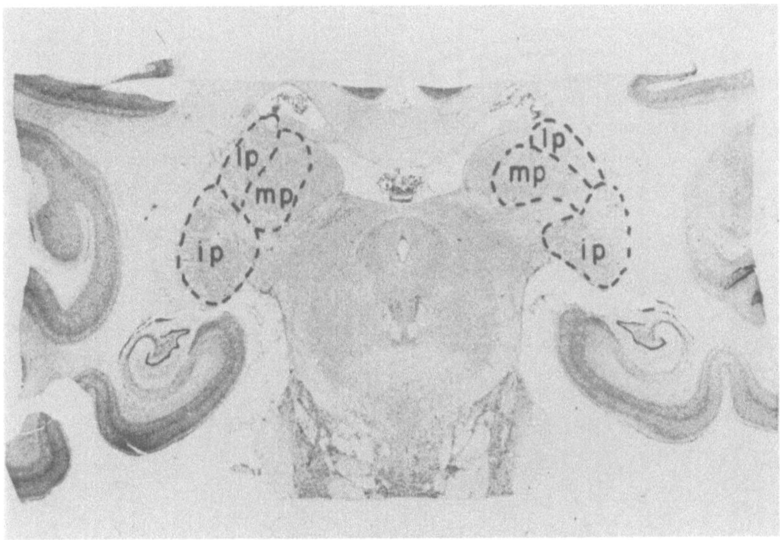

Fig. 2 As in Fig. 1 but at a more posterior section where the pulvinar nuclei reach their largest size.

and with the superior colliculus which has been suggested to be involved with the process of foveation of targets in visual space. GROSS, et. al., (12) in speculating on the function of the infero-temporal cortex suggested that the geniculo-striate system contained the information "what it is" about a visual target while the tecto-pulvinar system could supply the "where it is" information. It could be that the pulvinar plays some integrative function on these sorts of inputs.

What are some of the receptive field and stimulus feature properties of neurons of the pulvinar? ALLMAN, et. al., (13) have described in the owl monkey the receptive field properties for units of the inferior pulvinar. DAVID BENDER (personal communications) has made comparable observations in the rhesus monkey. The inferior pulvinar neurons appeared to have receptive fields somewhat larger than those described for striate cortex, or the superficial layers of the superior colliculus, but not nearly as large as those described by GROSS, et. al., (12) for the infero-temporal cortex where a 20° diameter receptive field was common. An interesting aspect of this study by GROSS, et. al., (12) was that when the inferior pulvinar was lesioned the already large receptive fields of infero-temporal cortex increased in size to exceed the size of the 60° tangent screen. The inferior pulvinar, therefore, appears to determine in part at least the size of receptive fields of the infero-temporal cortex.

Most cells of the inferior pulvinar have been described to have binocular input, some are sensitive to orientation to a slit of light, others are sensitive to the direction of movement of a target, while some respond to any flashed small target.

The role of the pulvinar in behavior and, in particular, visual behavior of the monkey has had little investigation. DENNY-BROWN and CHAMBERS (14) using an

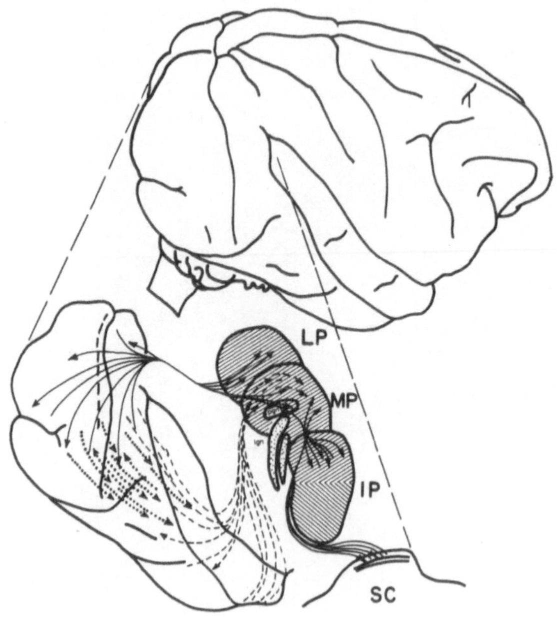

Fig. 3 Representation of the principal connections of the various divisions of the pulvinar complex of the monkey brain. Superior colliculus (sc); lateral geniculate nucleus (lgn).

assortment of observational tests on rhesus monkeys having had large cortical lesions which produced massive degeneration of the inferior pulvinar, described a behavior pattern consisting of a "staring" countenance where the monkey ignored food and test objects presented to the peripheral visual fields. With only a slight movement of the test object, however, the monkeys easily converged their eyes and fixated the target. this approach of producing large cortical lesions and suggesting a behavioral role for any subcortical structures has obvious faults in that many visual functions are simultaneously affected.

The effects of electrolytic lesions made in the pulvinar upon a more structural behavioral test faired somewhat better. Using the Wisconsin General Test Apparatus (WGTA), a multiple choice selection procedure for the monkey, CHOW (15) and MISHKIN (16) found no loss in a visual discrimination which was learned before lesions were made in the pulvinar. THOMPSON and MYERS (17), using a similar behavior test, did report a loss of retention of the preoperatively learned task where the medial and lateral pulvinar were lesioned.

More recently, CHALUPA, COYLE and LINDSLEY (18) demonstrated that, if a behavioral test was used which presented the test stimuli for only brief periods (10 ms), then indeed lesions of the inferior pulvinar hindered the learning of visual discrimination task. Monkeys having medial pulvinar and lateral pulvinar lesions learned the task as readily as did controls. These authors suggest that the inferior pulvinar is involved in visual pattern discrimination and that attentional factors controlled by or through the inferior pulvinar are absent in lesioned animals, therefore impeding or indeed preventing visual learning when the behavioral task requires extraordinary attention and where the significant visual stimuli are presented only briefly. These speculations by CHALUPA, et. al., (18) on the role of inferior pulvinar and task attention in visual learning are interesting since the inferior pulvinar communicates intimately with the superior colliculus where GOLDBERG and WURTZ (19) described changes in single unit response as a function of the visual attention of the monkey.

In summary, today very little is known of the function of the pulvinar.

Here, we report observations made on neurons of the inferior pulvinar nucleus. These observations came about when we made repeated recordings just posterior and slightly medial to the caudal termination of the lateral geniculate nucleus in the inferior and possibly in the medial pulvinar. These data were collected in the course of a study to measure the concurrent behavioral spectral sensitivity of an unanesthetized rhesus monkey and that of a single neurons of the lateral geniculate nucleus.

Methods

Figure 4 presents some of the features of the method. We set out to record from single neurons as the monkey worked on a test flash detection task. The idea was to gain behavioral control where the animal looked, hold his foveational fixation, search the visual field for the receptive field of a neuron, and then do the stimulus threshold series simultaneously for both monkey and neuron. With the electrode carrier in place, the monkey sits with his head restrained as to prevent gross head movements. The triple path Maxwellian view optical system presents a monochromatic test flash along the top; a neutral 5500 K^{o}, 3000 Troland adaptive field through the middle, and an intense chromatic adapting path along the bottom. Within the middle path, a portion of the light is diverted through a small rectangular aperture to form a movable fixation target, a 5' by 10' low contrast bar of light.

The task of the monkey is to depress and hold the lever with the right hand, whereupon the small fixation target (S1) appears somewhere within the 18° background field. The fixation target is of such low contrast that it must be foveated directly by human subjects to ascertain its horizontal or vertical orientation. By flashing the fixation target, it can be located by the monkey using peripheral retina, but must be foveated directly in order to detect changes in the orientation of the bar. It is this change in orientation (S1Δ) which the monkey must detect and report by release of the lever within 500 ms.

On half of the trials and at a variable interval preceding (S1Δ), there occurs a test flash termed S(2) which is manipulated as to size, wavelength, and intensity.

If the monkey releases the lever within 500 ms, a tone comes on signaling a required choice response on one of the two buttons. The monkey must push the right button if a flash was presented on that trial, or the left button if the flash did not occur on that trial. Correct choices are reinforced by a tone and by juice on a variable schedule. The use of these procedures allows the positioning of S(2) upon the area of the retina being seen by the electrode advanced to the pulvinar or LGN, thereby making co-extensive the receptive field for the neuron and the retinal field for the monkey's thresholds at peripheral as well as central retinal locations.

Recently, we have truncated the program in order that an increased number of

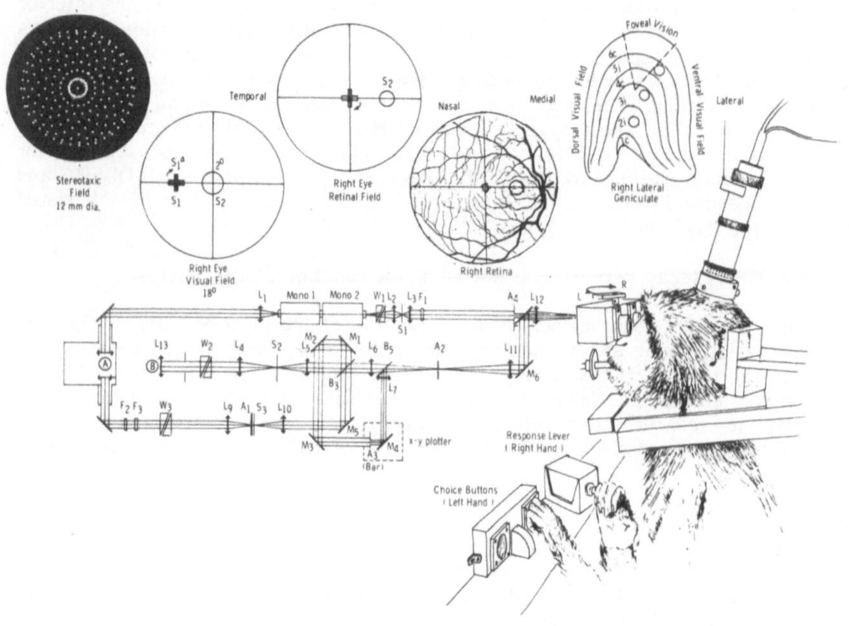

Fig. 4 Schematic representation of some of the features of the methods used to measure single neuron response and concurrent behavioral response. See text for description.

stimulus flashes could be presented to the receptive field of the neuron. The full behavioral program allowed a test flash to be delivered, on the average, only half of the time, resulting in a very low N for statistical analysis for any single neuron. The shortened version requires that the monkey only fixate and detect the change in the fixation bar. During this fixation period, the test flash is flashed repetitively upon the receptive field, thereby allowing the average unit response to be calculated from many responses to the test flash. Here, the monkey does not have to report whether or not the test flash is seen--(monitoring only the change in orientation of the fixation bar results in the same reinforcement payoff). Within the same session, after the unit is lost, or on the next session, the test flash is presented at the same retinal locus as before and the full program is used which requires that the monkey report whether or not the test flash has occurred for each trial. Preliminary comparisons fail to show any difference in the threshold value for the neuron as measured by either the truncated or the full program, i.e. "attention" does not appear to be an inportant variable in pulvinar or LGN unit thresholds.

When a single neuron is isolated, the fixation bar is slowly moved within the 18° field as to present the test flash in succession to the 4 quadrants of the field. Any change in the unit behavior to the test flash is further isolated, spectrally and spatially, by reducing the size of the test flash and adjusting the wavelengths of the test flash and background. If the unit responds to some parameter of the test flash (color, intensity, or spatial location) a higher resolution of response is sought. A small test flash of "optimal" wavelength and intensity is used to isolate the spatial field of the unit by changing the fixation point by very small increments, measuring the response to such changing retinal locations until the unit fails to respond. The wavelength of the test flash is then changed and the scan repeated to determine if there is change in spatial distribution of the "field" of the unit for different wavelengths. Having isolated the spatial and chromatic specifications of the unit, the threshold intensity required at various selected wavelengths is determined for the unit. Test flashes of decreasing intensity are presented until the unit fails to detect the flash, and where the criterion for threshold is the stimulus intensity at which the unit response does not differ from background activity. Within any intensity series for any wavelength, about 20 flashes are delivered at each intensity as to allow subsequent calculation of the average and variance of the response.

All test flash values at threshold are converted to quanta (Q) per flash, and plotted as $\log_{10}$ 1/Q to define spectral sensitivity across the visible spectrum. Single unit responses are counted for various periods before and after the test flash to form ratios of response change as a function of wave-number and intensity (in log 1/Q Quanta/Flash Units), as to be comparable with the behavioral threshold data.

Figure 5 illustrates one of two ways used to establish threshold for a neuron. In the upper part of the figure, the raw neuronal response to a 50 ms flash of light is shown. The middle trace illustrates an adjustable window used for gating and counting neuronal spikes which were detected at the level indicated by the bright enhancement near the top of each spike. The bottom trace of A indicates the occurrence of the test flash. Part B of this figure illustrates the enhanced dot display for the neuron for an intensity series. The log number of quanta contained in the test flash is indicated on the ordinate. Here multiple flashes are each represented by single traces over the face of a storage scope (only the spike occurrence being displayed). The individual traces are grouped according to the intensity of the test flash.

As the number of quanta contained in the test flash is reduced, there is a decrease in the number of spikes and a shift in the latency of the response. This latency shift was common for pulvinar unit while LGN units did not show this degree of latency shift. It is reasonably easy to see, and hear, when the unit fails to respond to the test flash. When the intensity is again increased, the response clearly returns.

A second, more laborious method for determining threshold is shown in Fig. 6. Here the average and standard deviation of the number of spikes evoked by 20 test flashes is shown for 3 wavelengths, 450, 500, and 620 nm. The same measurements taken before the test flash was presented are indicated by the horizontal lines, i.e., background activity. As the quantal energy of the test flash is reduced, the average response approaches the background level. We have arbitrarily defined unit threshold by fitting a straight line to the last three points which includes the first point where the average response is equal to, or below, the background value, and calling the associated quantum value threshold. We have used both of these methods for defining neuronal threshold and have found no difference in the threshold points selected by either method.

Results and Discussion

Figure 7 shows the receptive field maps for 14 pulvinar units recorded from stimulation of the contralateral eye. Several units had receptive fields in center fovea and all were within the central 8° of vision. These units had long latencies of response (70-80 ms to onset) and showed a shift of as much as 50 ms in latency with decreases in stimulus intensity.

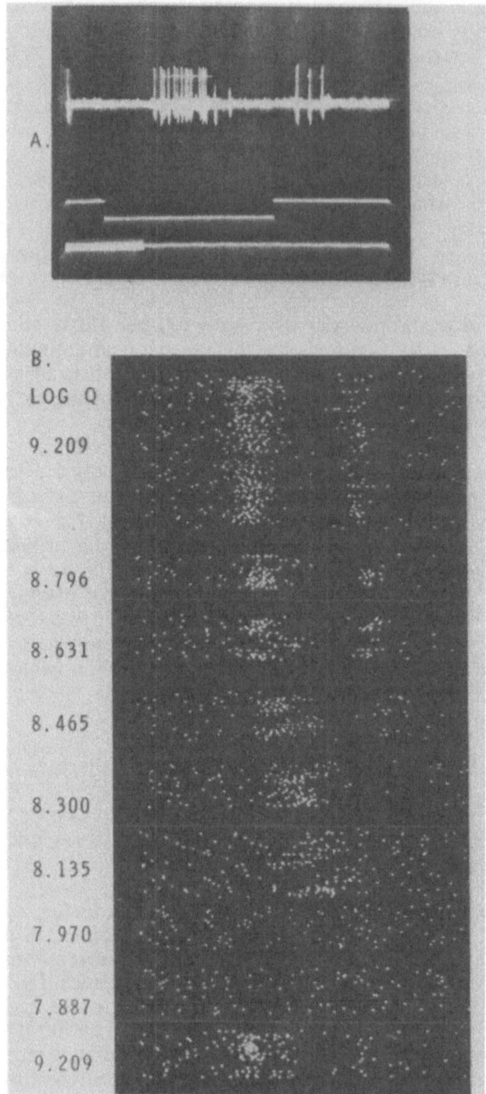

Fig. 5 Response of a neuron of pulvinar to a stimulation by 470 nm test flashes. A. Illustration of the unit response to a single flash (top trace) showing voltage discriminator level as the white dot near the peak of the response. The counting window (middle trace) and the test flash duration of 50 ms (lower trace) are also shown. 20 ms/division time scale. B. An example of a threshold intensity series ($\log_{10}$Q/flash) for the neuron of A, using a 470 nm test flash wavelength. Note the change in latency as well as a decrease in the number of spikes to threshold (7.970) where no change in neuronal activity is observed following the test flash. Note the return of the response as the intensity is again increased to (9.209).

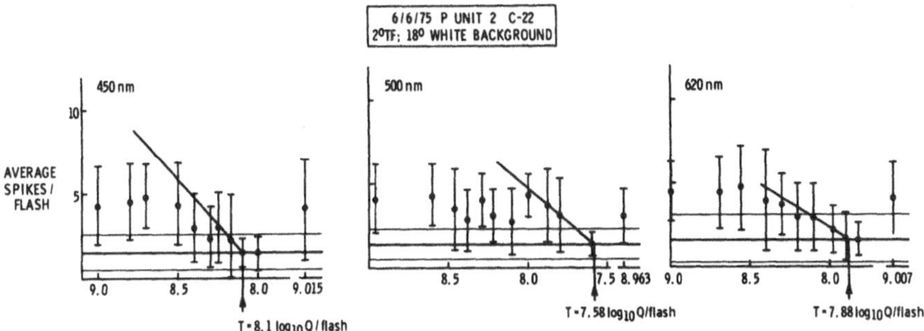

Fig. 6 The responses of a neuron of the pulvinar to intensity series of three test flash wavelengths. The average neuronal response (±1SD) is plotted as a function of the number of quanta for 450 nm, 500 nm, and 620 nm test flashes. The light horizontal line represents background neuron response.

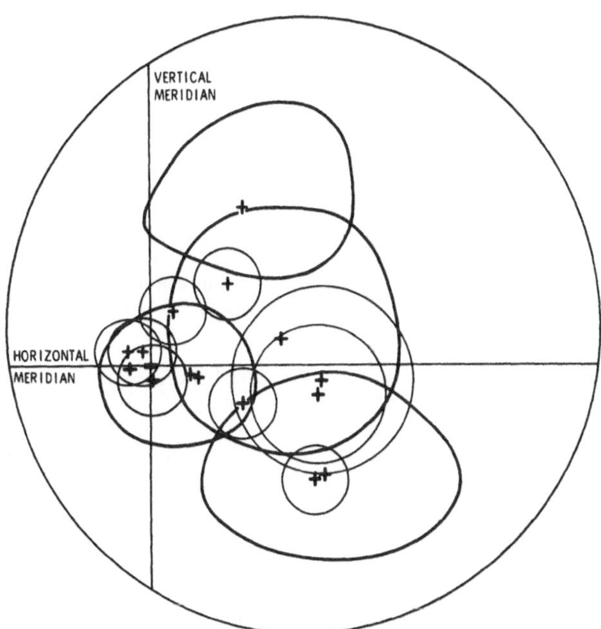

Fig. 7 Retinal location of receptive fields for 14 single units of the pulvinar. "+" indicates the field centers while the solid heavy lines indicate the boundary for the fields which were mapped in detail. The light circles represent the positions of 2°, 4°, and 6° diameter test flashes. All fields were in the contralateral hemifield and were driven by stimulation of either eye. "—" indicates the location and relative size of the fixation target. Background field: 18° diameter.

Figure 8 illustrates the change in reciprocal latency for a number of threshold intensity series across the visual spectrum. A straight line, fit by eye, to each set of data points indicates no change in latency which could be associated with the wavelength of the test flash (except for the extreme of 660 nm).

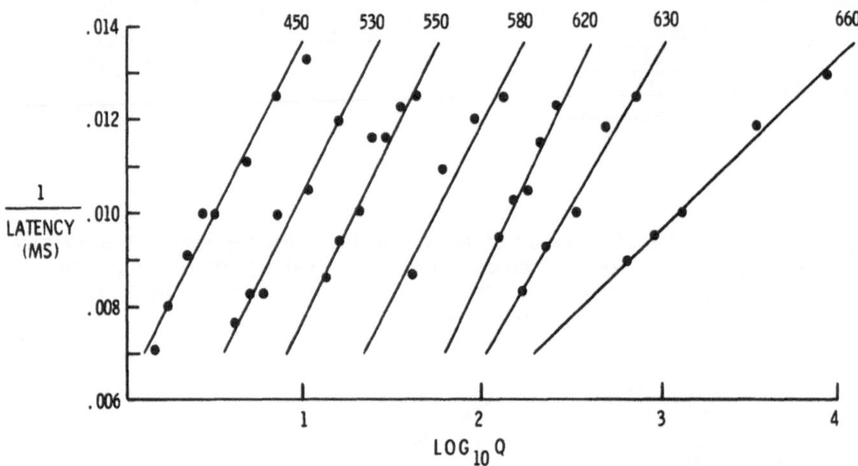

Fig. 8 The change in the latency of response of a pulvinar unit as a function of the number of quanta contained in the test flash. The intensity series for seven wavelengths are shown on a $\log_{10}$ scale, while the reciprocal latency in MS is shown on the ordinate. A line, fit by eye, shows that for all wavelengths the change in response latency is approximately linear with a change in $\log_{10}$ quanta. Except for the extreme wavelength of 660 nm, the slope of the latency change for all wavelengths is the same, suggesting comparable sensitivity over the visible spectrum. Conditions: 2° diameter test flash; 50 ms duration; 18° diameter, 3×10° Trolands (0.472 lumens/steradian) white background.

When the threshold sensitivity for these neurons is measured and compared with the behavioral threshold, the two curves are alike in absolute sensitivity and shape as shown in Fig. 9. Here, comparisons of behavioral thresholds (open circles) with neuron thresholds (filled circles) show them to be the same. Single unit and behavioral curves of A., having fewer points, show a reasonable match. The middle and lower sets of curves have been each shifted down from the top set of curves for easier inspection. Clearly, the neuron and the monkey have the same sensitivity throughout. When the neuron is at threshold, the monkey fails to report the presence of the flashed test light.

Figure 10 illustrates the change in sensitivity of the last unit of Fig. 9, when 6000 Trolands of monochromatic adapting light is added to the background. The threshold for the unit is shown as the filled point connected with the comparison behavioral threshold. Red (650 nm), the triangles - 550 nm are the squares; and 450 nm the diamonds. The solid line represents the threshold curve on the white background. In each case, the threshold for the unit varies with the behavior.

It appears that the neurons of the inferior pulvinar show a uniformity of response to flashing test lights from throughout the visible spectrum. When the unit

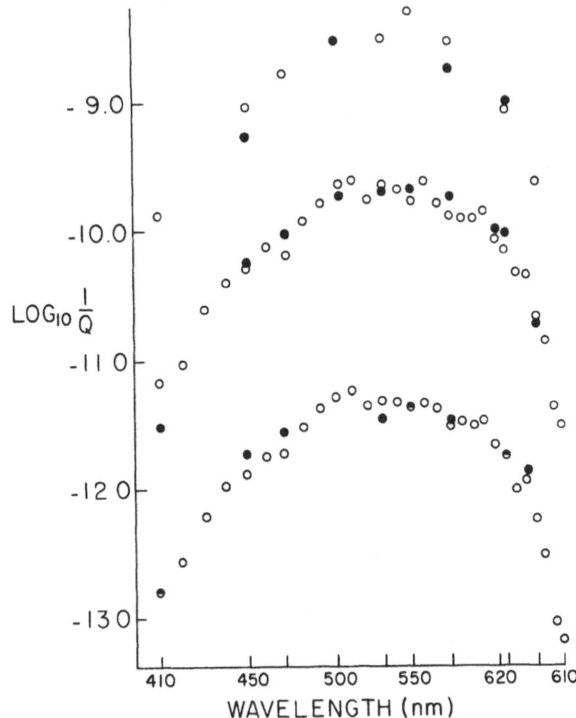

Fig. 9 Spectral sensitivity of the rhesus monkey compared with the spectral sensitivity of single neurons of the medial pulvinar. A. Open data points are threshold values for the monkey; the filled data points are comparable threshold determinations for a neuron having a receptive field at the same retinal locus where the behavioral thresholds were determined (2.8° upper retina x 1.7° nasal retina). Conditions: 2° diameter test flash; 50 ms. duration; Background, 3 x 10^3 Trolands (0.47 lumens/steradian). B. A second comparison of threshold of neuron and monkey (Stimulus locus: 3.2° upper retina on the vertical meridian). Conditions: 2° diameter test flash on a 3 x 10^3 Troland white background. This pair of curves have been displaced downward from the curves of A., for the sake of clarity. Note the similarity of the two curves indicating nearly identical sensitivity of neuron and monkey. C. Another comparison of neuron and monkey spectral sensitivities for a retinal location 2.8° upper retina x 1.7° nasal retina on a 3 x 10^3 Troland white background. Curves displaced downward from B.

fails to respond to the test flash, the monkey fails to report seeing the test flash. When adapted with monochromatic lights, the threshold of the neuron shifts as does the sensitivity of the monkey. These neurons have a spectral sensitivity comparable to other broad-band units of the geniculo-striate system. Figure 11, for comparison, shows two of our pulvinar neurons and the broad-band neurons of striate cortex described by GOURAS (20). It would appear that they are very much the same.

In summary, for any requirement for signaling the presence and location of a target in central vision, these units could clearly supply such information.

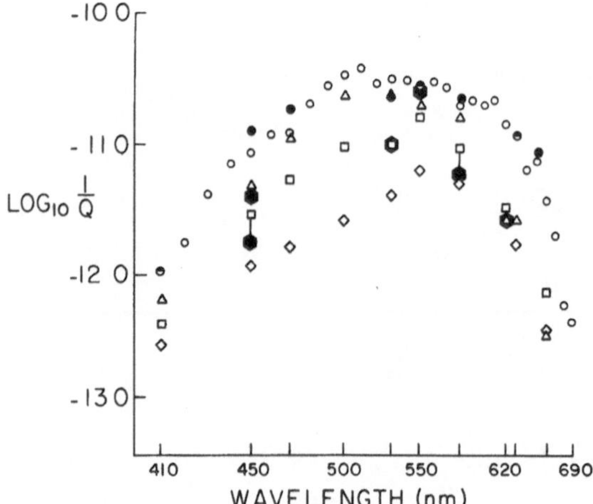

Fig. 10 Comparisons of behavioral and single neuron thresholds on intense monochromatic adaptive backgrounds. (3×10^3 Trolands white + 9×10^3 650 nm Δ ; + 9×10^3 555 nm □; + 9×10^3 450 nm ◇). Filled points are thresholds for the neuron; open points are matching behavioral points. (Vertical bars link behavioral and unit thresholds for the same conditions). Same unit and retinal locus as in C of Fig. 9. Solid curve represents behavioral threshold values on the "white" background of C, with the thresholds on the monochromatic backgrounds for both neuron and monkey being plotted on the same relative scale. Note that thresholds for the monkey and for the neuron vary together with monochromatic adaptation.

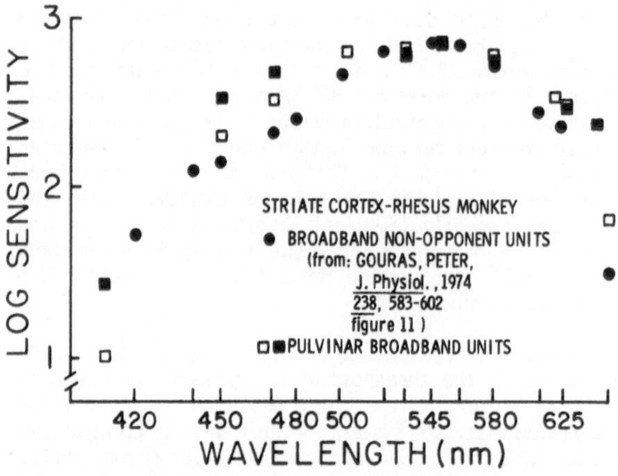

Fig. 11 Comparison of the shape of the broad-band, non-color opponent spectral sensitivity curves recorded from striate cortex with similar curves for pulvinar neurons. All curves normalized for 550 nm.

Acknowledgement

The collaboration and assistance of Dr. H. G. Sperling, Dr. Louis Meharg, Mr. Scott Marks, Mr. David Garrett, Mr. Joe McAlexander and Mr. Dick Kelly is herewith acknowledged. This work was supported by NIH Grant EY-00381.

References

1. K. L. Chow, J. Comp. Neurol. 93, 313 (1950).
2. J. A. Campos-Ortega, W. R. Hayhow and P. F. De V. Cluver, Brain Res. 22, 126 (1970).
3. J. Q. Trojanowski and S. Jacobson, Brain Res. 85, 347 (1975).
4. J. Q. Trojanowski, and S. Jacobson, Brain Res. 80, 395 (1974).
5. J. Bos and L. A. Benevento, Exp. Neurology 49, 487 (1975).
6. E. G. Jones and H. Burton, Brain Res. 104, 142 (1976).
7. M. Ogren and A. Hendrickson, Exp. Neurology 53, 780 (1976).
8. L. A. Benevento and M. Rezak, Brain Res. 96, 51 (1975).
9. J. S. Lund, R. D. Lund, A. E. Hendrickson, A. H. Bunt and A. F. Fuchs, J. Comp. Neurol. 164, 287 (1975).
10. L. A. Benevento and J. H. Fallon, J. Comp. Neurol. 160, 339 (1975).
11. J. A. Campos-Ortega, W. R. Hayhow and P. F. De V. Cluver, Brain Behav. Evol. 3, 368 (1970).
12. C. Gross, D. B. Bender and C. E. Rochamoramda. Science 166, 1303 (1969).
13. J. M. Allman, J. H. Kaas, R. H. Lane and F. M. Miezin, Brain Res. 40, 291 (1972).
14. D. Denny-Brown and R. A. Chambers, Arch. Neurol. 33, (1976 AMA).
15. K. L. Chow, Amer. Med. Asso., Arch. Neurol. Psychiat. 71, 762 (1954).
16. M. Mishkin, In Brain and Human Behavior, ed. by A. G. Karczmar and J. C. Eccles (Springer-Verlag, New York, 1972).
17. R. Thompson and R. E. Meyers, J. Comp. Physiol. Psychol. 74, 479 (1971).
18. L. M. Chalupa, R. S. Coyle and D. B. Lindsley, J. Neurophysiol. 39, 354 (1976).
19. M. E. Goldberg, and R. H. Wurtz, J. Neurophysiol. 35, 560 (1972).
20. P. Gouras, J. Physiol. 238, 583 (1974).

Questions and Answers

Q. Yager: Could you describe your stimulus condition a little more exactly? Why did you not get the 3-bump function that Harry (Sperling) usually reports? Is the background on?

A. Yes, ordinarily that is the case. I have a slide that I will use in explaining that in a moment. As you may appreciate, it takes quite a while to train an animal to perform these various behaviors; in particular, to hold still while you attempt to either record a receptive field and, in particular, long enough to get a behavioral threshold. This experiment was carried out on a monkey that was previously trained to perform this task and one that had had some previous treatment. This treatment was an experiment to attempt to change the spectral sensitivity by intense monochromatic stimulation. Harry Sperling had put 55,000 Trolands of blue light into one eye, attempting to alter the spectral sensitivity. It was subsequently determined that there was no change in the spectral sensitivity using one particular technique. The monkey was trained, so I implanted the monkey, did the recordings, and found these unusual shapes. But I found them only for one eye. If you look at Figure 12, showing data collected from the opposite eye, and make the same sort of comparison between the behavior of the monkey and a pulvinar neuron, you can see that there is a difference in the shape of the curves. You have a shape that is wider on the blue end. That is, there is a blue peak on this one that was not present in curves from the opposite eye. I think the problem here is that we made the error of using an animal

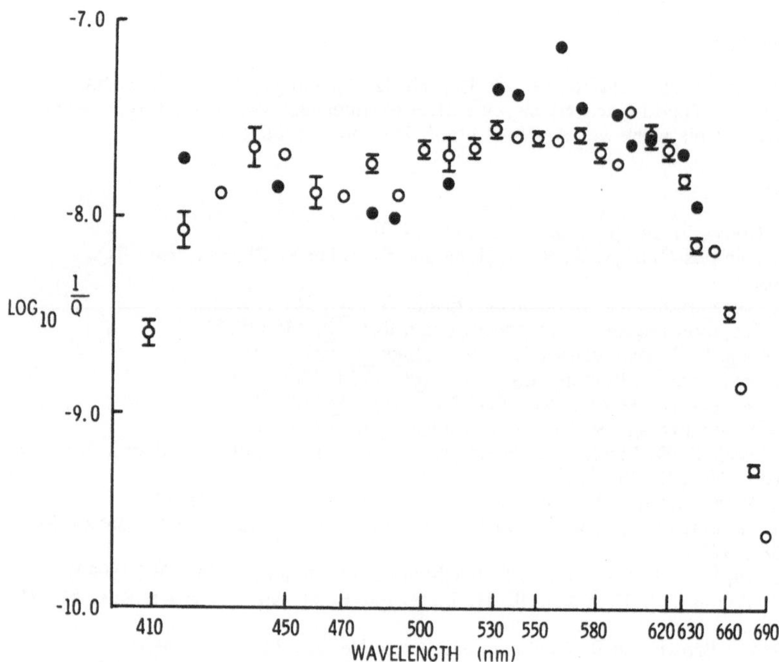

Fig. 12 The increment threshold spectral sensitivity of the rhesus monkey for a 2° test flash placed 6° into the temporal retina compared to the same threshold sensitivities for a pulvinar unit having a receptive field near this retinal locus (5° temporal x 3° upper retina). The unit threshold curve has been raised by 0.5 log units for comparison of the curve shapes. This is our worst case, where the unit thresholds are about 0.25 log units less sensitive. The mean and range for four (4) behavioral threshold determinations are shown as the open data points with bars. The filled data points are similar thresholds for the neuron. The two threshold curves are from slightly different retinal locations, taken on days about two months separate.

that had been treated, thinking that it was a normal animal. I believe that, although it was not determined at the time, the animal was blue-blinded; his sensitivity had been changed in the blue end of the spectrum. That is the only explanation that I have for the differences between the two functions.

Q. Dr. Snodderly: If you get a constant stimulus and look at the variability of the behavioral response, and the variability of the neural response, can you make any conclusions about what the fluctuations are due to? Maybe you have some fluctuating attentional factor in time, do you get a tight correlation in time?

A. Trial by trial? Well, that would be reflected in particular in the shortened paradigm, as the animal depresses the lever and fixates and we deliver from 10 to 50 flashes in a sequence. And if you inspect a train of responses to such a flash sequence, there is no suggestion that the attention waxes or wanes within that period. Now, if you go beyond that period, there are other things that might cause a change in the response, but one of the things that I have been impressed by is that we see no sign of habituation. Allman, et. al., (13) described habituation as being one of the properties of the units recorded from the inferior pulvinar of the owl monkey. However, in these cases where we have recorded from the neurons for hours, there is no suggestion that there is any habituation occurring. So attenion in this case is not a potent variable.

Q. Dr. Snodderly: What I am wondering is, if you are thinking about threshold being a fluctuating quantity, whatever it is, presumably, the monkey on some trial says, "Yes, I see it", and on the other trials says, "No, I do not". Is the neuron tightly linked to that behavioral response, or are they varying and it is a mean that you are getting?

A. That is right, you do not get that sort of response. That would be one of the obvious things you would look at first, to see if, in fact, on the "No" trials you have the same neuronal response; and that is not the case. You do not get a response when you are using the full paradigm, where the animal has to report whether the test flash does or does not occur. In fact, on the first session when we launched this program, we found a unit that responded straight forwardly whenever the test flash was presented; when it responded, the animal said, "Yes", and with an extremely high correlation; when the unit failed to respond, the monkey reported "No". In fact, you can listen to response of the unit and tell when the test flashes are presented over a wide range of stimulus intensities.

Q. Is there a level at which the neuron responds, a certain response level which perhaps never, or rarely is associated with the failure of the behavior?

A. You may recall from the graph that I presented earlier where we decreased the intensity until the neuronal response goes to the background. If the animal ordinarily responds correctly, say, 90% of the time, to a bright test flash, and it is dimmed until the monkey is responding at chance, by increasing the test flash by a quarter log unit, and the monkey will respond correctly 90% of the time. The neuronal response is equally clear. The wedge that we used to control the intensity has a unit step of about .04 log units and within two steps you can go from a unit's identifiable response to no unit response at all; so it is a very narrow range.

Q. Dr. Riggs: I notice that in one of your curves you plotted the reciprocal latency on the ordinate and log intensity on the base line. It seemed to me that in the slopes of some of the curves, particularly at the red end, there was quite a much wider range of latency than for the other wavelength.

A. Yes. That wavelength, as you recall, was 660 nm. I believe that it was the only one that deviated significantly. I did not have enough energy in the system to get very much of a curve on the other end, but throughout the middle part, they (the slopes) seemed to be quite the same, and so I concluded that I could not get very much information about the wavelength of a test light from just looking at latency.

Q. Are there spectrally opponent units of the pulvinar?

A. No. We saw none. And we were looking for them. One of the striking things about the pulvinar is the uniformity of response of the units. That is, their thresholds are the same as the behavioral threshold of the monkey, and they all have essentially the same spectral sensitivity (spectrally broadband). You also probably appreciate the number of lost units before a full curve was ever obtained in this particular experimental setup, and so we have many partial curves, but rather few complete spectral sensitivity curves. But even the partial curves are consistent with the threshold and spectral response data that we have presented here.

The Organization of the Cat Pretectum

Nancy Berman
Department of Physiology and Biochemistry
The Medical College of Pennsylvania
Philadelphia, Pennsylvania 19129

About ten years ago a number of investigators proposed that the mammalian visual system is composed of two divisions, one specialized in locating objects the other in identifying objects (1,2). According to this proposal, the geniculostriate system processes information about patterned stimuli while the superior colliculus allows the animal to orient in its visual world. This idea was based in part on an experiment done by SCHNEIDER, in which he showed that hamsters with lesions of the primary visual cortex are able to locate objects but fail pattern recognition tasks while hamsters with lesions of the superior colliculus can learn to recognize patterns but have difficulty in locating them (2). The concept of two visual systems was also based on anatomical studies which showed that the dorsal lateral geniculate nucleus and the superior colliculus are the two major targets of the retina in mammals. In this scheme other retinal targets are considered less important, partly because they are smaller and partly because there have been fewer studies of their function. More recently, however, an anatomical technique based on axoplasmic transport, the autoradiographic technique, has been developed (3, 4), and the central projections of the retina have been found to be much more widespread than previously thought. In the cat the retina projects to the dorsal lateral geniculate nucleus and to the superior colliculus, but following an injection of tritiated amino acids into an eye label can be seen in the following areas as well: the ventral lateral geniculate nucleus, the medial intralaminar nucleus: the suprachiasmatic nucleus of the hypothalamus; the dorsal, medial, and lateral nuclei of the accessory optic tract; the pulvinar, and two nuclei of the pretectal complex: the olivary pretectal nucleus and the nucleus of the optic tract (5, 6). If each of these retinal targets is considered a separate visual system, then there must be eleven visual systems. The pretectal complex receives a major retinal input, but among the way stations of the mammalian visual system, it has proved to be particularly refractory to functional interpretations. As a first step toward understanding the function(s) of the pretectal complex, I have studied its afferent and some of its efferent connections using anterograde and retrograde techniques.

The pretectal complex is a group of nuclei which lie between the superior colliculus and dorsal thalamus. KANASEKI and SPRAGUE (7) have recently divided the cat pretectal region into seven distinct nuclei on the basis of cyto- and myeloarchitectonic analyses. I have used their nomenclature in Fig. 1, which shows photomicrographs of a series of thionin stained sections at different rostro-caudal levels (A-D) through the pretectal complex.

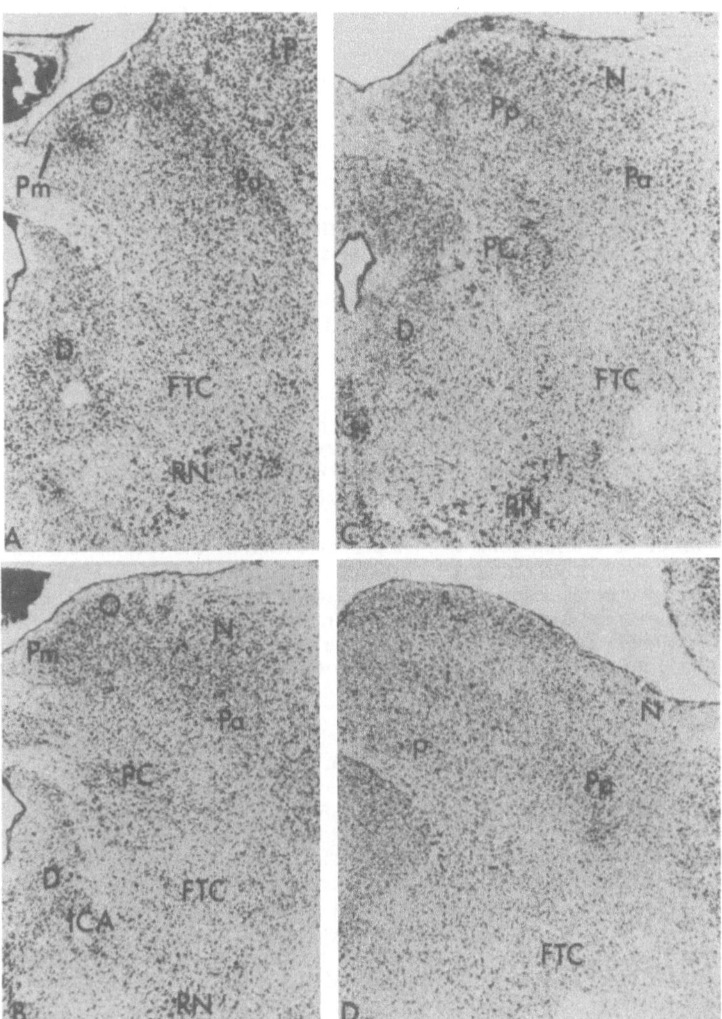

Fig. 1 Photomicrographs of thionin stained frontal sections through the pretectal complex at different rostro-caudal (A-D) levels x20.

The anterior pretectal nucleus (Pa) lies at the anterolateral border of the pretectal complex just medial to the medial medullary lamina of the thalamus and extends rostro-caudally from the level of the habenula to the most rostral part of the superior colliculus. It has been divided into two parts, nucleus pretectalis anterior, pars compacta and nucleus pretectalis anterior, pars reticularis. The compact part tends to lie dorsolateral to the reticular part and is composed of more tightly packed cells. The medial pretectal nucleus (Pm) is a small nucleus composed of tightly packed cells lying beneath the pia and dorsal to the posterior commissure in the anteromedial part of the pretectal region. The olivary pretectal nucleus (0) lies just lateral to the medial pretectal nucleus. It consists of large, deeply staining cells that are not clearly distinguishable in the cat from the nucleus of the optic tract. It is very clearly outlined, however, in experiments in which retinal fibers are labeled (5,8,9).

The nucleus of the optic tract (N) lies at the dorsolateral border of the pretectal complex along its full length and is composed of large, deeply-staining cells lying mainly among the fibers of the brachium of the superior colliculus. The posterior pretectal nucleus (Pp) lies just ventral to the olivary nucleus and nucleus of the optic tract and extends caudally into the most anterior part of the superior colliculus, where it is overlain by the superficial layers of the colliculus. More caudally, the posterior pretectal nucleus is displaced laterally by the deep layers of the colliculus and can be seen between the lateral edge of the colliculus and the medial edge of the medial geniculate body. There, it merges with the most caudal cells of the nucleus of the optic tract.

The pretectal complex is bordered medially by the nucleus of the posterior commissure, which has been considered an accessory oculomotor nucleus (10). The pretectal complex is bordered ventrally by part of the mesencephalic reticular formation termed the central tegmental field by BERMAN (11).

There is considerable disagreement on the distribution of the retinal terminals in the pretectal complex, presumably because of the difficulty of delineating the boundaries of the pretectal nuclei. Most investigators who have used silver degeneration techniques have noted degeneration in the nucleus of the optic tract and, if it was recognized as distinct nucleus, in the olivary nucleus, but they have disagreed on the extent to which the other nuclei of the pretectal complex receive retinal terminals (8, 9, 12, 13).

I have recently reinvestigated the question of the distribution of optic fibers in the cat pretectal complex using the autoradiographic technique. Drawings of frontal sections through the pretectum 200 microns apart are shown in Fig. 2, which shows the distribution of label resulting from an injection of the left eye. Contralateral to the injected eye label in the nucleus of the optic tract is confined to two or three finger-like strips, (shown in Fig. 3) elongated in the dorsolateral to ventromedial dimension. The strips that are 150-200 wide and are separated by sparsely labeled regions 400-600 wide. The label extends ventrally to the most dorsal part of the posterior pretectal nucleus but does not appear to enter this nucleus. Although the exact boundary between the two nuclei is indistinct, the labeling tends to be confined to the vicinity of the large, densely stained cells of the nucleus of the optic tract. The label in the ipsilateral nucleus of the optic tract is also in the form of oblique strips and is almost as dense as on the contralateral side. The olivary pretectal nucleus situated medial to the anterior part of the nucleus of the optic tract, is filled with label contralateral to the injected eye, while ipsilateral to the injected eye, the label in the olivary pretectal nucleus is confined to the most anterior, dorsal and lateral part of that nucleus.

One question that arises immediately is whether these strips of label from the two eyes overlap each other or interdigitate with one another, i.e., whether there is some sort of hidden lamination within this area. To answer this question I injected both eyes of a cat with tritiated leucine and proline. If the two projections are

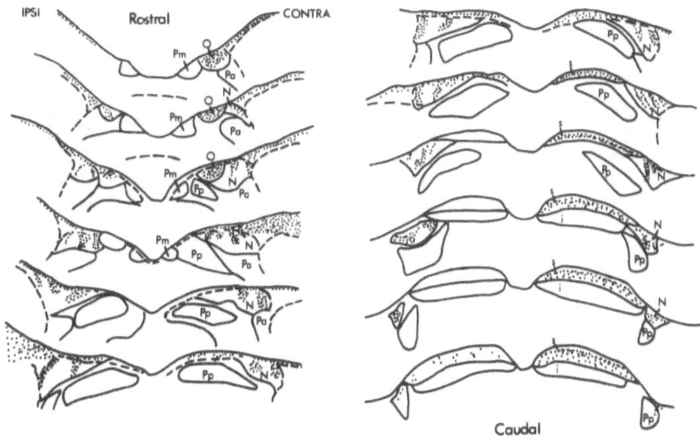

Fig. 2 Drawings of a one in ten series of 20 frontal sections through the pretectal complex showing the distribution of transported level following an eye injection.

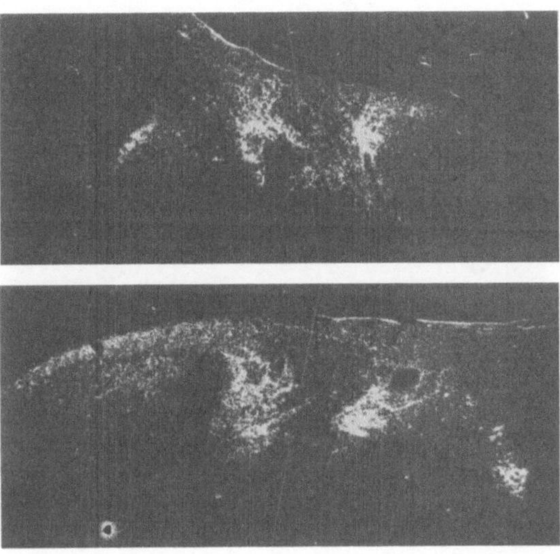

Fig. 3 Darkfield photomicrographs of frontal sections through the nucleus of the optic tract and olivary nucleus showing strips of terminal label ipsilateral (A) and contralateral (B) to an injected eye. M, medial, D, dorsal x100.

interdigitating there should be a solid band of label in the nucleus of the optic tract, whereas if they are overlapping, there should still be fingerlike strips of label. When I injected both eyes with tritiated leucine and proline, I found that there are strips of label in the nucleus of the optic tract. Grain counts on this material, comparing the number of grains in the region between the strips with the number of grains overlying one of the strips, are shown in Fig. 4. A dark circle indicates a large number of grains per square and an open circle indicates grains which are above background, but are still very low compared to the counts within the strips. There are a lot more grains in each strip than there are in between, which indicates that the projections

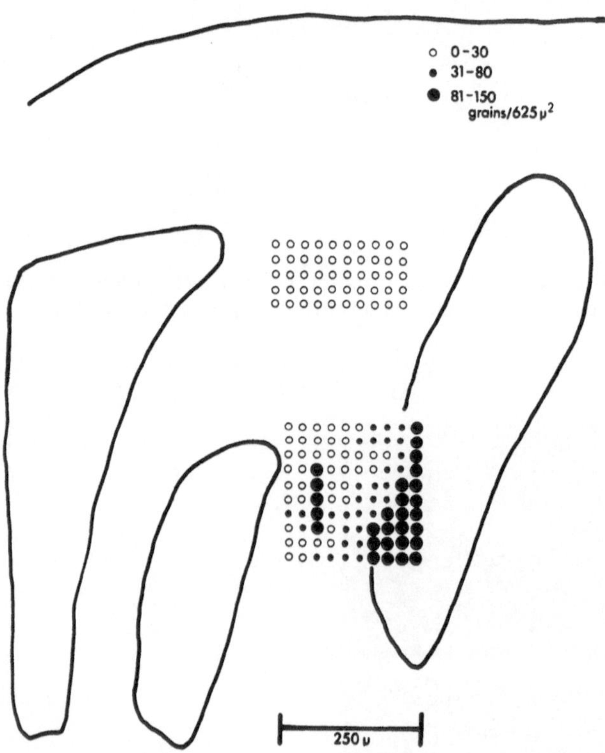

Fig. 4 Grain counts over the nucleus of the optic tract following injections of isotope into both eyes.

from the two eyes are overlapping in these strips and not interdigitating.

I made injections of isotope into each of the nuclei of the pretectal complex to determine their efferent connections. Figs. 5 and 6 show an injection confined to the nucleus of the optic tract. Labeled fibers leave the injection site, cross in the posterior commissure, and terminate into the contralateral nucleus of the posterior commissure. Another major projection of the nucleus of the optic tract is one which

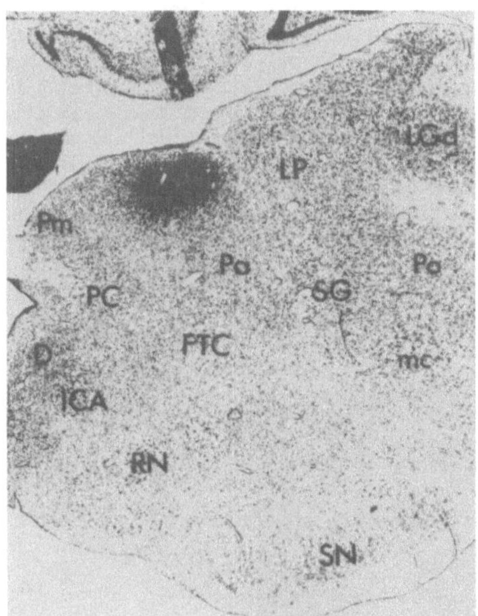

Fig. 5 Brightfield photomicrograph of an injection centered in the nucleus of the optic tract (C105) x11.4.

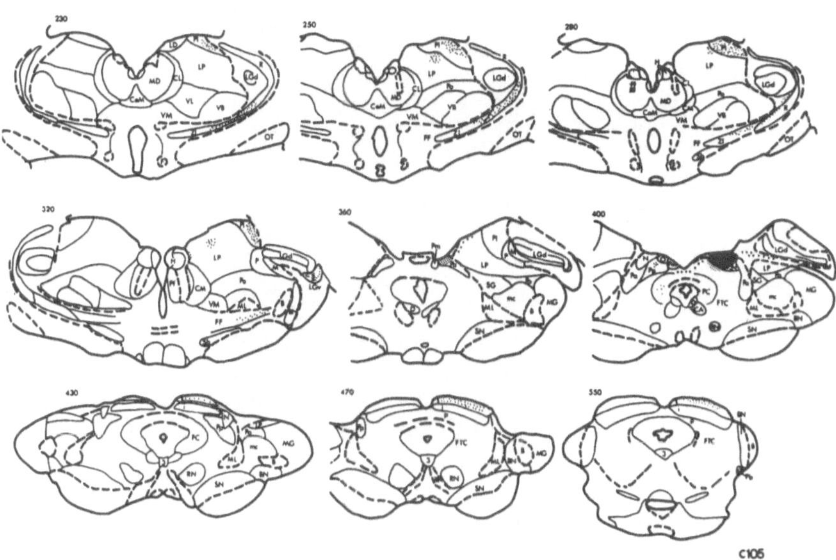

Fig. 6 Drawings showing the distribution of label in C105, in which the injection is centered in the nucleus of the optic tract.

travels laterally through the ventral ramus of the optic tract and terminates in the ventral lateral geniculate nucleus, mostly around large cells. The third major projection of the nucleus of the optic tract is to the pulvinar. In summary, the projections of the nucleus of optic tract, one crossed and two ascending, are primarily to visual areas, such as the pulvinar and the ventral lateral geniculate nucleus. Figures 7A and B show photomicrographs of the ventral lateral geniculate nucleus following an injection of the nucleus of the optic tract, bright-field and dark-field view, and the label is found both around large and small cells, but mainly around the large cells. Figures 7C and D show the ventral lateral geniculate nucleus following an injection of horseradish peroxidase into the pretectal complex. The large HRP-

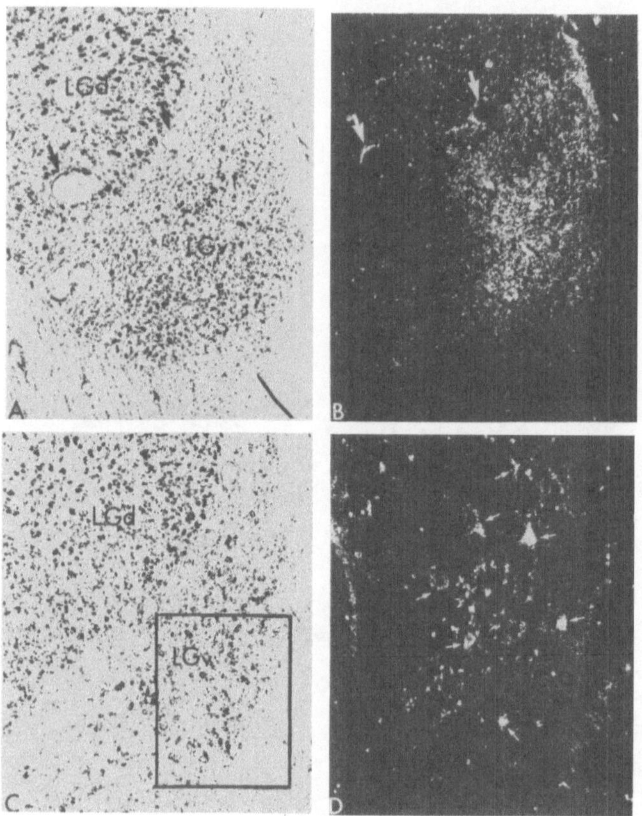

Fig. 7 Brightfield (A) and darkfield (B) photomicrographs of the ventral lateral geniculate nucleus ipsilateral to the injection site in C105, in which the injection was centered in the nucleus of the optic tract. x130.
Brightfield (C) and darkfield (D) photomicrographs of the ventral lateral geniculate nucleus ipsilateral to a large injection of horseradish peroxidase into the pretectal complex (C) x130 (D) x200.

labeled cells in the ventral lateral geniculate indicate that the pretectal-ventral lateral geniculate relation is reciprocal (see also 14,15).

Figures 8 and 9 show an injection confined for the most part to the posterior

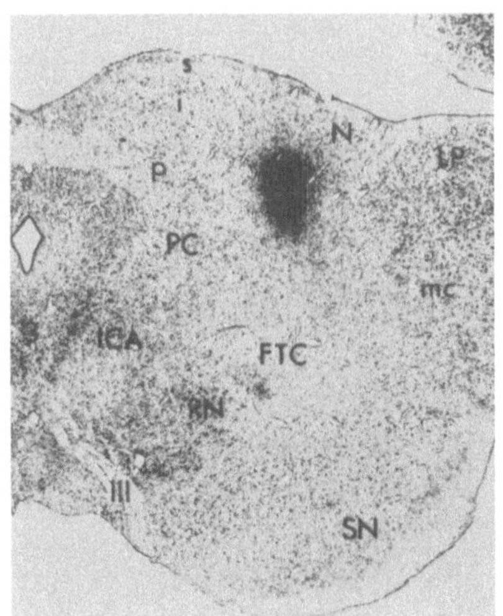

Fig. 8 Brightfield photomicrograph of the injection site in C195, in which the injection is centered in the posterior pretectal nucleus. x11.4.

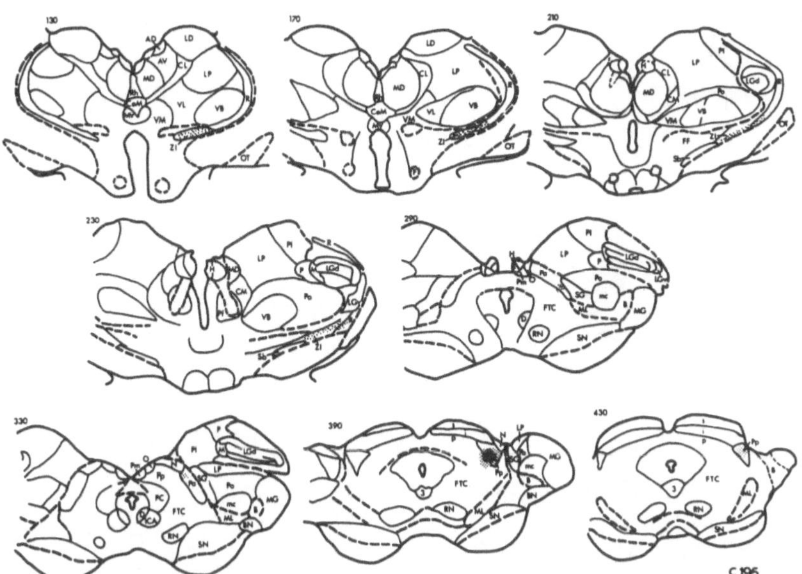

Fig. 9 Drawings showing the distribution of label in C105, in which the injection is centered in the posterior pretectal nucleus.

pretectal nucleus. In this case the only sites which were labeled by transported amino acids were the zona incerta and the adjacent part of the thalamic reticular nucleus.

Figures 10 and 11 show an injection confined to the anterior pretectal nucleus.

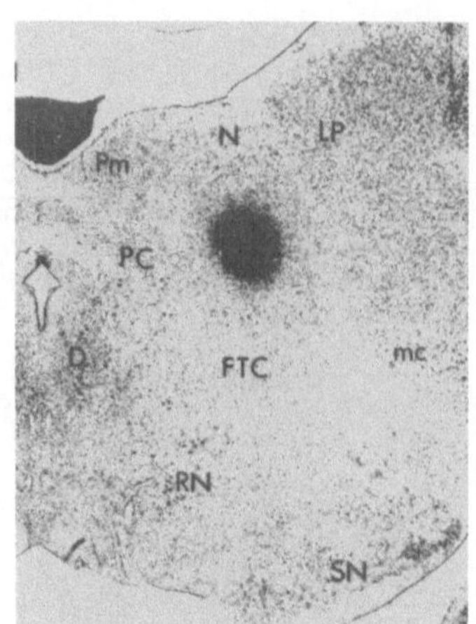

Fig. 10 Brightfield photomicrograph of an injection confined to the anterior pretectal nucleus. x11.4.

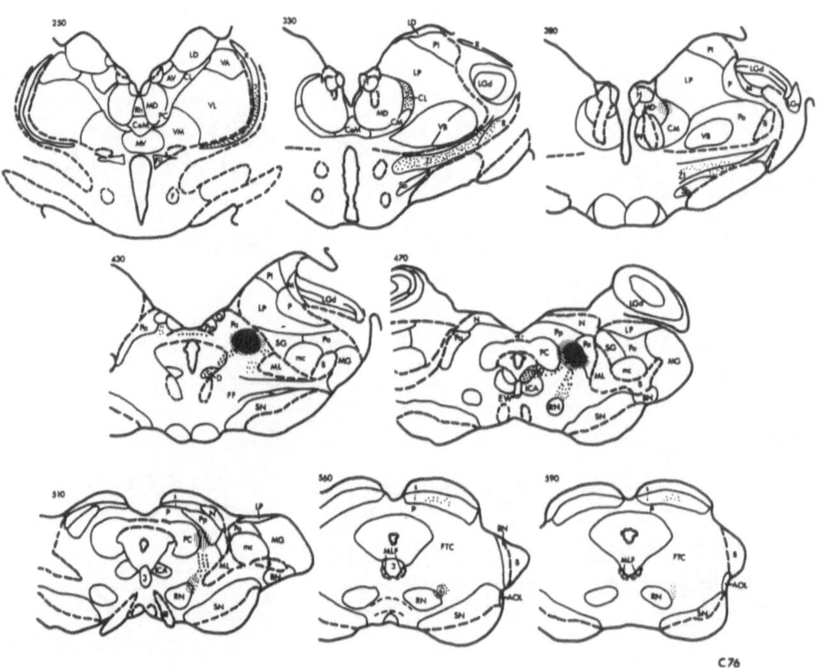

Fig. 11 An experiment in which an injection of isotope was centered in the anterior pretectal nucleus.

The anterior pretectal nucleus is a major source of the descending projections of the pretectal complex. It also projects to the thalamic reticular nucleus and zona incerta, but its major projection is to an area in the mesencephalic reticular formation which lies just dorsolateral to the red nucleus (Fig. 14). This projection sometimes encroaches on the most dorsolateral portion of the red nucleus. In this case (Fig. 10 and 11) there is a projection to the nucleus of Darkschewitsch. In cases which included only the more lateral part of the anterior pretectal nucleus and did not encroach on the nucleus of the posterior commissure that projection was absent, so I concluded that the projection to the nucleus of Darkschewitsch originates in the nucleus of the posterior commissure. But as you can imagine, it is very difficult to make that kind of distinction. Figure 12 shows the label in the reticular nucleus and

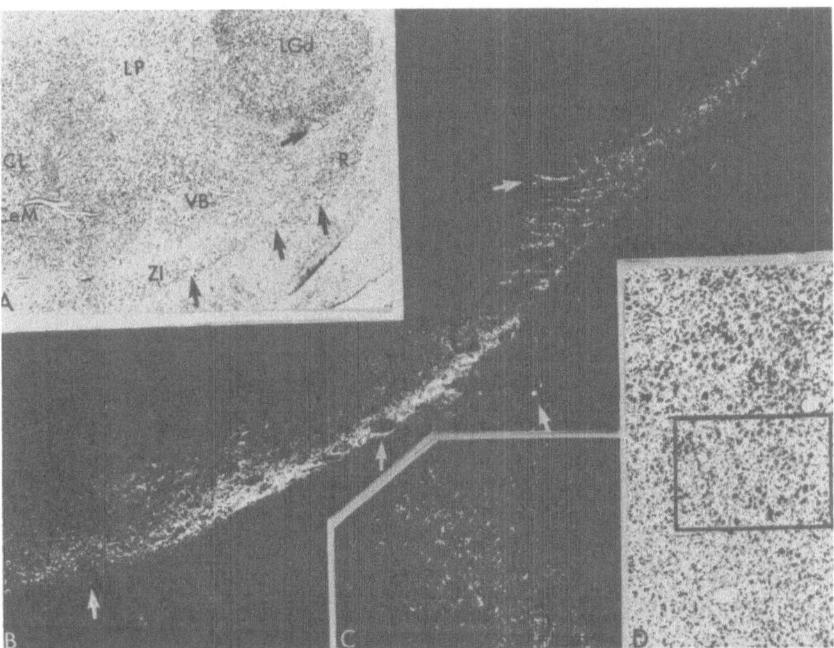

Fig. 12 Brightfield (A) and darkfield (B) photomicrographs of the reticular nucleus and zona incerta after an injection of isotope into the anterior pretectal nucleus. Arrows indicate same blood vessel. Darkfield (C) and brightfield (D) photomicrographs of the central lateral nucleus of the thalamus after an injection of isotope into the anterior pretectal nucleus. (A) x15; (B) x100; (C) x150; (D) x150.

the zona incerta following an injection of the anterior pretectal nucleus. In this material you can also see that there is a heavy projection to the central lateral nucleus of the thalamus, which is one of the intralaminar nuclei. This is a region in which there are cells related to eye movements (16) and it also receives fibers from the superior colliculus (17,18). When the injection encroaches somewhat on the dorsal part of the mesencephalic reticular formation, there is a descending projection to the tegmental reticular nucleus (not-shown), an area that projects to the cerebellum.

Figure 13 shows one of the largest injections that I made, and it has the most

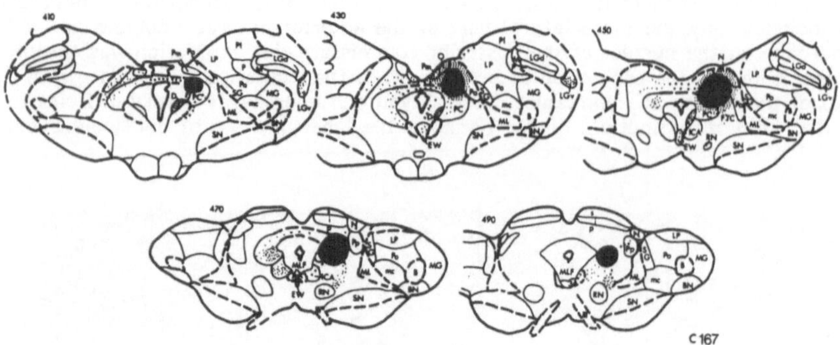

Fig. 13 Drawings of the injection site and resulting label in C167 at five levels through the pretectal complex. This is the only case in which the Edinger-Westphal nucleus was labeled; it also showed the most extensive labeling contralateral to an injection site.

extensive crossed projections, which include just about all of the nuclei of the contralateral pretectum as well as the nucleus of the posterior commissure (Fig. 14) and the interstitial nucleus of Cajal. In this case there is also a projection to the ipsilateral nucleus of Darkschewitsch (shown in Fig. 14). The nucleus of the posterior commissure, the nucleus of Darkschewitsch and the interstitial nucleus of Cajal are considered accessory oculomotor nuclei (10) and they are interconnected, with the nucleus of the posterior commissure projecting contralaterally to the intersitital nucleus of Cajal and ipsilaterally to the nucleus of Darkschewitsch.

Now if you look in any neuroanatomy textbook at the section on the pretectum you'll learn that the pretectum is the center for the consensual pupillary light reflex, (19,20) because as we all know, the pretectum receives fibers from each eye and projects to the Edinger-Westphal nucleus, which has been reported to be the origin of the parasympathetic fibers which innervate the iris. That is exactly what I expected to find but did not. Instead, the results of these studies reopen the question of the pathway for the pupillary light reflex, especially the origin of any visual input to the Edinger-Westphal nucleus.

In summary, the areas of the pretectum which receive a retinal projection are the nucleus of the optic tract and the olivary nucleus. The nucleus of the optic tract has ascending connections with major contributions to the pulvinar and the ventral lateral geniculate nucleus. The posterior pretectal nucleus is related to regions of the subthalamus, that is the zona incerta and the fields of Forel. The descending projections of the pretectal complex originate in the anterior pretectal nucleus which is strongly related to a region of the reticular formation which lies just dorsolateral to the red nucleus. It also has a major projection to the central lateral nucleus of the thalamus, a region in which there are cells related to eye movements. The nucleus of the posterior commissure projects to the contralateral nucleus of the posterior commissure, to the contralateral interstitial nucleus of Cajal and to the ipsilateral nucleus of Darkschewitsch. When an injection includes the most ventral part of the

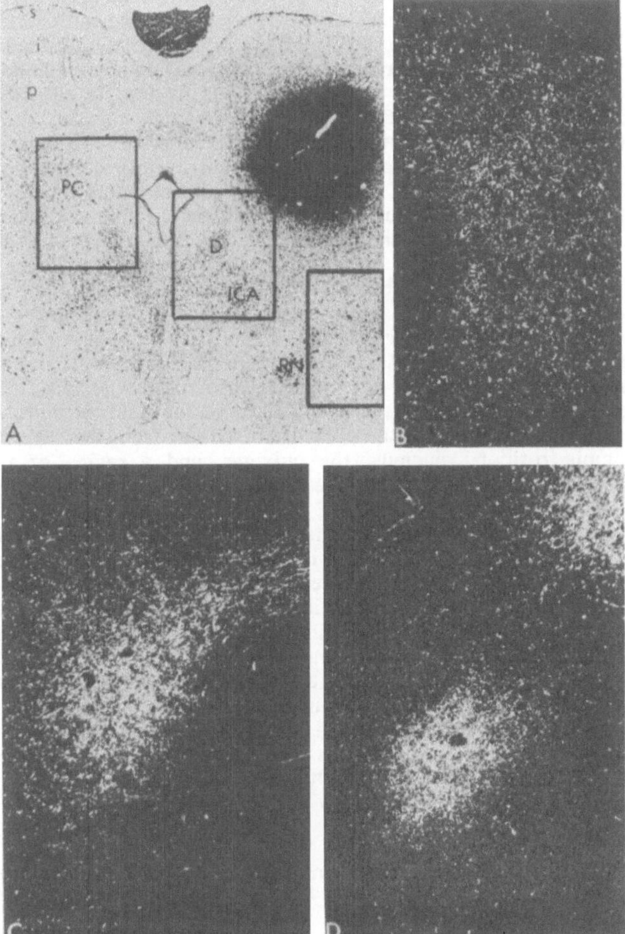

Fig. 14 (A) Brightfield photomicrograph of a section through the injection site in C167 showing regions shown in darkfield in B-D x10. (B) Darkfield photomicrograph of the region dorsolateral to the red nucleus. x100. (C) Darkfield photomicrograph of the nucleus of the posterior commissure contralateral to the injection site. x100. (D) Darkfield photomicrograph of the nucleus of Darkschewitsch ipsilateral to the injection site. x100.

anterior pretectal nucleus and the central tegmental fields, transported label is found in the tegmental reticular nucleus, which is an origin of input to the cerebellum. I could not define any of the pretectal nuclei as being a source of input to the Endinger-Westphal nucleus, and I did not see label in the inferior olive following injections into the pretectal complex.

It is interesting to compare the projections of the pretectum with those of the superior colliculus (17,18) an area we know more about. They have some similarities: The retinal input terminates in the superficial layers of the superior colliculus and in the nucleus of the optic tract, which lies dorsally in the pretectal complex. In both cases, the superficial layers of the superior colliculus and the nucleus of the optic tract have ascending connections to parts of the posterior thalamus, that is, to the lateral posterior nucleus or the pulvinar. Both the superficial layers of the superior colliculus and the nucleus of the optic tract project to the ventral lateral geniculate nucleus and in both cases there is a reciprocal connection. The intermediate layers of the superior colliculus resemble the anterior pretectal nucleus in that they both project to the central lateral nucleus of the thalamus. The deep layers of the superior colliculus, the anterior and posterior pretectal nuclei all contribute fibers to the zona incerta and fields of Forel. However, the differences outweigh the similarities. The superior colliculus has much more extensive ascending and descending projections than the pretectal complex. Some of the regions which receive fibers from the superior colliculus and not from the pretectal complex are the parabigeminal nucleus, the C layers of the lateral geniculate nucleus, the magnocellular medial geniculate nucleus, the suprageniculate nucleus and the pons. A few regions which receive connections from the pretectum but not from the superior colliculus are the thalamic reticular nucleus, the pulvinar, and a region of the reticular formation which lies just dorsolateral to the red nucleus.

We do not really know what the functions of the pretectal complex are. It may be involved in the pupillary light reflex, but if that is true, the connections may not be as direct as we originally thought. There is evidence that a lesion of the nucleus of the optic tract will abolish optokinetic nystagmus in the rabbit (21) and I think that should be followed up. In other animals there is also evidence that lesions of the pretectum may cause difficulty with the learning of pattern or diffuse light discriminations (22,23). The pretectum is related to a specific region of the reticular formation, which may be a region specialized in some aspect of visuomotor integration. More information is needed before we can decide among these possibilities. Finally, in summary, there may be as many as eleven visual systems rather than two, and the pretectum is somewhere along that list.

References

1. D. Ingle, G. E. Schneider, C. B. Trevarthen and R. Held, Psychologische Forschung 31, 42 (1967).
2. G. E. Schneider, Science 163, 895 (1969).
3. W. M. Cowan, D. I. Gottlieb, A. E. Hendrickson, J. L. Price and T. A. Woolsey, Brain Res. 37, 21 (1972).
4. R. Lasek, B. Joseph, and D. G. Whitlock, Brain Res. 8, 319 (1968).
5. N. Berman, J. Comp. Neurol. 174, 227 (1977).
6. N. Berman and E. G. Jones, Brain Res. 134, 237 (1977).
7. P, K. Blochert, R. J. Ferrier, and R. M. Cooper, Brain Res. 104, 121 (1976).
8. F. Scalia, J. Comp. Neurol. 145, 223 (1972).
9. F. Scalia, Brain Behav. Evol. 6, 237 (1972).
10. M. B. Carpenter and P. Peter, Journal fur Hirnforschung 12, 405 (1970).
11. A. L. Berman, The Brainstem of the Cat: A Cytoarchitectonic Atlas with Stereotaxic Coordinates, (Univ. of Wisconsin Press, Madison, 1968).
12. L. J. Garey and T. P. S. Powell, J. Anat. 102, 189 (1968).
13. A. M. Laties and J. M. Sprague, J. Comp. Neurol. 127, 35 (1966).
14. S. B. Edwards, A. C. Rosenquist, and L. A. Palmer, Brain Res. 72, 282 (1974).
15. L. W. Swanson, W. M. Cowan, and E. G. Jones, J. Comp. Neurol. 156, 143 (1974).
16. J. Schlag, I. Lehtinen, and M. Schlag-Ray, J. Neurophysiol. 37, 982 (1974).
17. J. Altman and M. B. Carpenter, J. Comp. Neurol. 116, 157 (1961).
18. J. Graham, J. Comp. Neurol. 173 629 (1977).

19. M. B. Carpenter and R. J. Pierson, J. Comp. Neurol. 149, 271 (1973).
20. H. W. Magoun and S. W. Ranson, Arch. Ophthal. 13, 791 (1935).
21. H. Collewijn, J. Neurobiol. 6, 3 (1975).
22. T. Kanaseki and J. M. Sprague, J. Comp. Neurol. 158, 319 (1974).
23. G. Berlucchi, J. M. Sprague, J. Levy, and A. C. DiBerardino, J. Comp. Physiol. Psychol. 78, 123 (1973).
24. A. M. Graybiel and E. Hartweig, Brain Res. 81, 543 (1974).

Acknowledgement

Supported by NSF Grant BNS 7724923 and NIH Grant EY-2088-01.

Central Mechanisms of Foveal Vision in the Monkey

Bruce M. Dow
Department of Physiology
State University of New York at Buffalo
Amherst, New York 14226

This work began about seven years ago as an examination of the orientation specificity of color cells in the foveal striate cortex (area 17) of the Rhesus monkey. At that time I joined PETER GOURAS, who was studying the color specificity of the cells in this region, using quantitative methods adapted from his earlier studies in the retina (1-3). To obtain careful measurements of orientation specificity, we had the shop build us an automated device that would generate moving visual stimuli whose length, width, orientation, velocity, direction of movement and color could be varied independently and under precise control (4).

Using this device we set out to test cell responses in anesthetized monkeys. The early experiments utilized Maxwellian view optics, with stimuli projected directly on the retina (4, 5, 6). Beginning in the summer of 1973, the experiments were moved to a larger laboratory equipped with a tangent screen. For the tangent screen experiments, retinoscopy was used to select a suitable pair of corneal contact lenses from a set of 20 pairs with base curvatures ranging from 50-60 diopters in 1/2 diopter steps. The screen could be positioned as far as 285 cm from the monkey, such that a 1 mm stimulus on the screen subtended approximately 1 minute of arc on the monkey's retina.

In all experiments the anesthetic was a mixture of nitrous oxide and oxygen (75:25). Paralysis was maintained with continuous infusions of gallamine triethiodide (5 mg/kg/hr) and d-tubocurarine (1 mg/kg/hr), and respiration was assisted with a Harvard Apparatus pump.

Our first observation was that many striate cortical cells showed specificity for line orientation (6), confirming the earlier report of HUBEL and WIESEL (7). Figure 1 illustrates responses of an orientation-specific cell recorded in an early experiment.

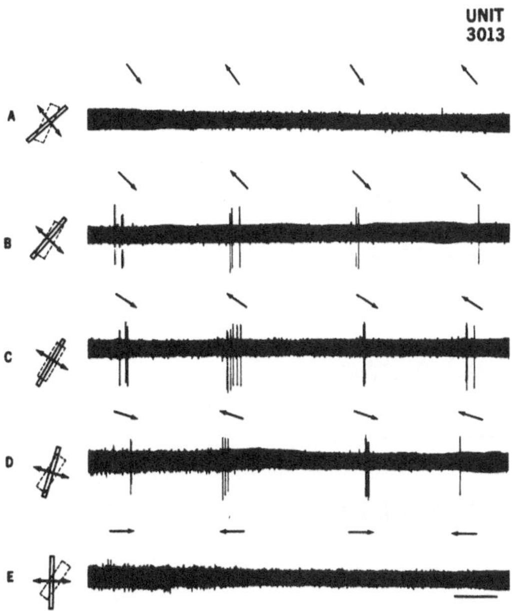

Fig. 1 Responses of a cell in foveal striate cortex to moving light bars of various orientations. Receptive field (indicated by dashed rectangles) was $1/4^{\circ}$ wide. Light bars (solid rectangles) were moved under automatic control, the directions of movement indicated by arrows. Background illumination 0.9 $\log_{10}$ cd/m^2, stimuli 3.3 $\log_{10}$ cd/m^2. Time mark 1 sec. (Reprinted, with permission, from DOW and GOURAS (5)).

Other cells, however, showed considerably less orientation specificity as illustrated in Fig. 2 (5). Here the stimulus was not moved, but was positioned carefully over the cell's receptive field and flashed on and off. Stimulus duration in this case was 250 msec. If the stimulus was made dimmer by placing neutral density filters in front of the light beam, it was possible to locate an intensity (1.2 log units of filtering, in this case) at which differential responses could be obtained to vertical and horizontal slits; but for most levels of illumination the cell was not orientation-specific.

We also used colored stimuli to test cells, and in fact, our primary goal in the early studies was to identify a population of color-specific cells. Fig. 3 shows the responses of a color-specific cell to stationary, flashing light spots. This cell responded better to a small red spot than to a small white spot. With a larger spot the preference for red over white was even more dramatic, suggesting that this cell, unlike the cell of Fig. 1, was not really very interested in line orientation.

In the course of these studies, we came upon another group of cells that were not particularly concerned about orientation either. Instead, these cells were very

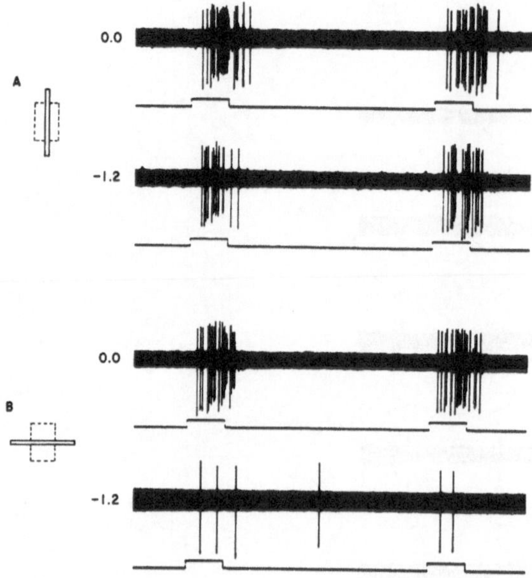

Fig. 2 Responses of a cell in foveal striate cortex to stationary orthogonal bars at different light intensities. Stimulus dimensions $1/8 \times 2^{o}$. Lower traces are photocell responses to 250 ms light pulses. Background 0.9 $\log_{10}$cd/m^2. Stimuli 3.3 $\log_{10}$cd/m^2 (0.0) and 2.1 $\log_{10}$cd/m^2 (-1.2). (Reprinted, with permission from DOW (5)).

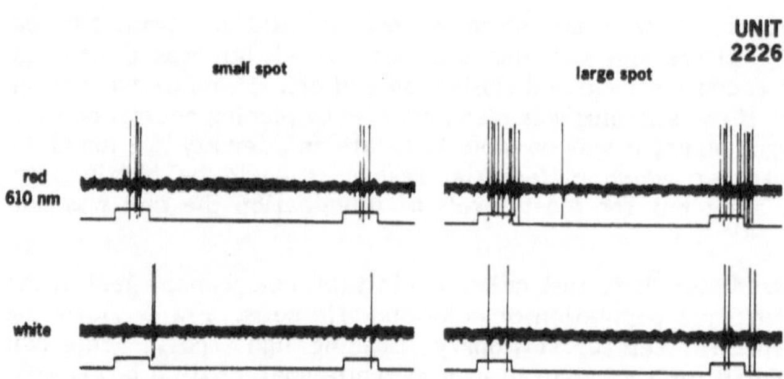

Fig. 3 Responses of a cell in foveal striate cortex to light spots of various sizes and colors. Red (610 nm) spots were produced by placing an interference filter in front of the white stimulus beam. Stimulus duration 250 msec. Small spot was $1/8 \times 1/8^{o}$, large spot 4×4°.

sensitive to direction of movement (5). Fig. 4 shows the responses of a direction-specific cell. Up-rightward movement of the stimulus can be seen to elicit vigorous responses, while down-leftward movement elicits no responses at all. The lower trace in this figure is the position signal from a linear potentiometer (4) indicating the amplitude, velocity, and repetition rate of stimulus movement.

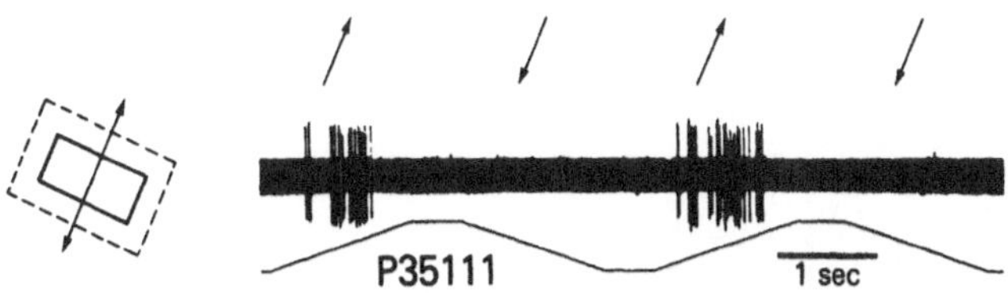

Fig. 4 Responses of a direction-selective cell in foveal prestriate cortex to an optimal stimulus moved back and forth across its receptive field. Stimulus dimensions 10' x 20'. Velocity 0.75°/sec. Lower trace is the position signal from a linear potentiometer (4) indicating amplitude, velocity and repetition rate of stimulus movement.

In some penetrations that went a little too deep, the electrode passed completely through striate cortex, across the underlying white matter, and into another cortical region located along the posterior bank of the lunate sulcus. Fig. 5 shows the geometry of this part of the brain. Fig. 5A is a photograph of the surface of a rhesus monkey's brain tipped up on end such that the occipital cortex appears as a flat, triangular-shaped sheet, bounded anteriorly by the lunate sulcus (LS) and posteriorly by the inferior occipital sulcus (IOS). Fig. 5B shows a section through the cortex at the location indicated by the dashed line in Fig. 5A. The buried cortex in the posterior bank of the lunate sulcus has been shown by ZEKI (9) to receive a dense projection from the overlying foveal striate cortex, and is designated foveal prestriate cortex, or more specifically, area 18. Fig. 5C shows a dye mark (see arrow) deposited in area 18 at the recording site of an orientation-selective cell with a foveal receptive field.

In the first few penetrations before we had the histology back, we were not even aware that the deeper cells were from area 18. In addition to orientation-selective cells, we also saw color cells and direction cells in area 18. All resembled their counterparts in area 17. In retrospect, an important distinguishing feature of cells in areas 18 is their binocular receptive fields (10, 11, 12), since many cells in area 17 are either totally or predominatly monocular (7).

Having stumbled upon area 18, and finding the same kinds of cells as in area 17, we decided to look more systematically at response properties in the two regions. We used our automated moving stimulus generator to obtain data on orientation-direction tuning from a number of cells in areas 17 and 18 in the foveal projection pathway. The data are plotted on polar coordinate graph paper, following CAMPBELL et. al., (13). Examples of polar plots are shown in Fig. 6.

The curves were obtained by having the optimal stimulus for each cell move back and forth across the cell's receptive field 5 to 10 times, and counting the spikes that occurred for an average sweep in each direction. Then the orientation and

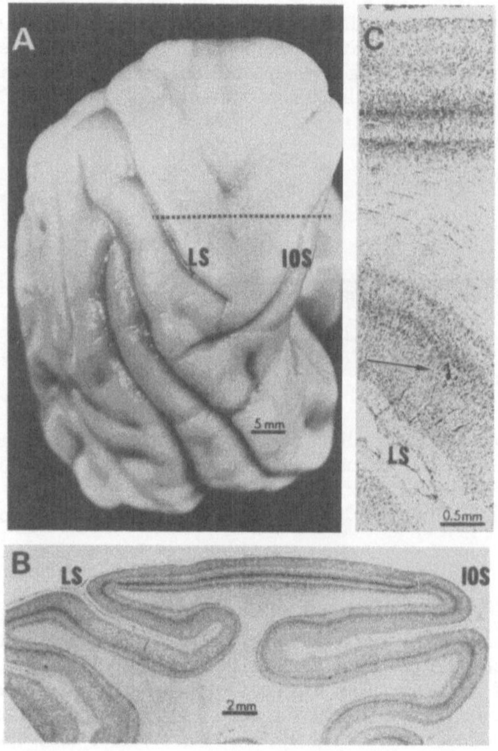

Fig. 5 Foveal striate and prestriate cortex (left hemisphere) of a rhesus monkey, as viewed on the surface (A), and in section (B and C). LS - lunate sulcus. IOS -inferior occipital sulcus. Dashed line in A indicates plane of section for B. Plane of section in C is 4 mm medial to the dashed line in A. Arrow in C points to a dye mark deposited at a recording site in area 18 by means of current injection at the tip of the recording microelectrode (8).

direction of movement were changed slightly, without changing the stimulus dimensions or velocity of movement, and the process was repeated. Most cells were tested with at least 5 or 6 different orientations, which means 10 or 12 directions.

The following conventions are used in all polar plots:

1. Each dot corresponds to a different direction of movement. The stimulus is in all cases a light bar oriented orthogonal to the direction of movement.
2. The distance of a dot from the center indicates response amplitude in mean spikes/sweep.

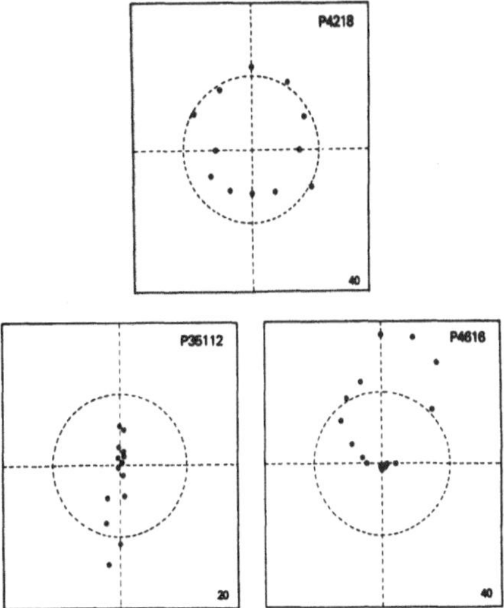

Fig. 6 Response magnitude as a function of direction of stimulus movement. Polar coordinate plots for three representative cells from foveal prestriate cortex. Stimuli were moving light bars of various orientations, direction of movement being always orthogonal to orientation. The distance of each spot from the center represents the response of the cell (in mean spikes/sweep) to a stimulus moved in that direction. Each stimulus was repeated 5-10 times, with all parameters fixed except orientation and direction of movement. The number in the lower right corner indicates the number of spikes/sweep at the dashed circle. Stimulus dimensions for cell P4218, 1.2' x 24', velocity 0.3^{o}/sec. Stimulus for cell P35112, 6' x 1°, velocity 0.5^{o}/sec. Stimulus for cell P4616, 2.4' x 24', velocity 1.5^{o}/sec.

3. The number in the lower right corner is a calibration, indicating the number of spikes/sweep at the dashed circle.

Shown here are examples of radially-symmetric (cell P4218), bilobed (cell P35112), and unilobed (cell P4616) tuning curves.

The various tuning curve shapes can be plotted as points on a scatter diagram, using angular selectivity (a measure of the fatness of the curve) as the abscissa, and directionality (a measure of how unilobed or bilobed the curve is) as the ordinate. Fig. 7 is a scatter plot based on 43 tuning curves characterized in this way. Striate and prestriate cells are represented in white and black respectively, with color sensitive cells further distinguished as stars. The points are widely scattered, but there is an apparent cluster of cells with high directionality. This cluster includes both striate and prestriate cells, only one of which is a color cell.

The highly directional cells turn out to have several additional properties in common. They respond without regard to leading edge configuration, so long as the stimulus is moving in the correct direction, and their directionality is velocity

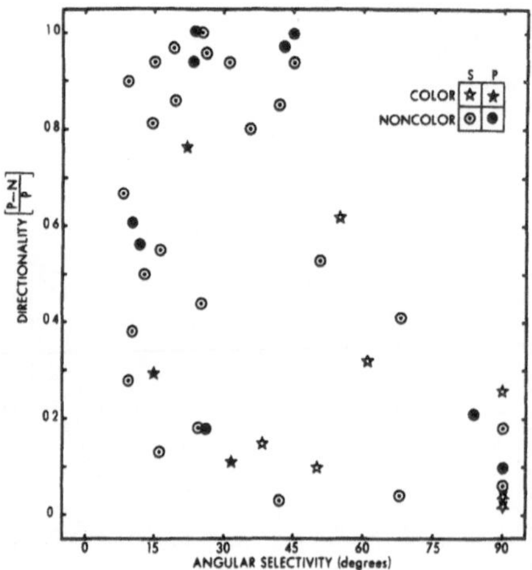

Fig. 7 Scatter diagram based on 43 polar coordinate plots obtained from cells in foveal striate (white symbols) and prestriate cortex (black symbols). Angular selectivity is the half-width at half-height of the largest lobe of a cell's tuning curve (see Ref. 3). Directionality is the preferred direction response minus the null or antipreferred direction response, divided by the preferred direction response. In each case the preferred stimulus dimensions and velocity were used for testing. Cells with color specificity (stars) were tested with stimuli of the preferred color. Monocular testing for all cells, using whichever eye gave better responses.

selective. Their directionality is maximal at velocities of about 0.5 to 1.0 deg/sec, falling off at either slower or faster velocities. Some of the cells in this group have "hypercomplex" (7) properties, responding more vigorously to shorter stimuli than to longer ones. In fact, virtually all of the hypercomplex cells we have seen in areas 17 and 18 showed a high degree of direction selectivity. DRAGER (14) has made the same observation for area 17 of the mouse.

The cells with directionality values below about 0.75 appear to be bimodally distributed, with one cluster showing narrow angular selectivity (30 degrees of less), the other showing broader angular selectivity up to a maximum of 90 degrees. The narrow angular selectivity cells turn out to have a strong preference for elongated stimuli, failing to respond until a minimum length has been reached. Compared with the highly directional cells, the narrow angle cells are quite sensitive to the leading edge configuration of a moving stimulus. Moreover, the directionality of narrow angle cells is only minimally affected by changes in stimulus velocity. These cells seem to be more sensitive to the stimulus itself, its length and orientation, than to the speed or direction of movement.

The remaining cells, with directionality less than 0.75 and angular selectivity broader than 30 degrees, are a more heterogeneous group, though the proportion of color cells is clearly higher here than in the other two groups.

The scatter plot includes only cells on which complete orientation-direction tuning curves were obtained. Given the 3 major cell groupings suggested by the scatter plot, it was possible to classify a larger population of cells whose orientation specificity had been determined using less quantitative methods. Table 1 summarizes data obtained from 268 cells (179 in area 17, 89 in area 18) of which 71 or 26%

Table 1

	Orientation Cells	Direction Cells	Other Cells	TOTAL
All Cells	106 (40%)	44 (16%)	118 (44%)	268 (100%)
Color Cells	12 (17%)	3 (4%)	56 (79%)	71 (100%)

were color cells. Only 21% of the color cells fall within either the orientation or direction groups, whereas 56% of all cells fall into one of these groups.

Furthermore, if color cells in areas 17 and 18 are considered separately, as shown in Table 2, the tendency for color cells to fall outside the orientation and direction groups is especially evident in area 18.

Table 2
Color Cells

	Orientation Cells	Direction Cells	Other Cells	TOTAL
Area 17	11 (23%)	2 (4%)	34 (73%)	47 (100%)
Area 18	1 (4%)	1 (4%)	22 (92%)	24 (100%)

Recording sites of 198 cells have been identified on the basis of extracellular dye injections from the microelectrode tip (8). The distribution of these cells among the layers of foveal striate (area 17) and prestriate cortex (area 18) is illustrated in Figs. 8 and 9. Area 17 layers are numbered according to VON BONIN (15), area 18 layers according to LORENTE DE NÓ' (16). In Fig. 8 the histograms show for each layer the proportion of cells in each group, and the dashed lines show the proportions of the 3 cell groups in the entire sample of marked cells. By scanning up and down the dashed lines one can find the layers in which a given cell type is especially common or uncommon. There is a tendency for direction cells to occur in layer 4A (just above the granule cell layer) in both areas 17 and 18. Orientation cells are most common in the upper-and lower-most layers in area 17 but not in area 18.

Figure 9 shows the laminar distribution of 52 color cells (included within the larger population in Fig. 8). Looking first at the "Total" column, color cells are rare in layers 4A and 6 of area 17 and in the middle layers of area 18, the same layers where Fig. 8 showed direction cells to be most common. This may in part relate to the very small number of direction cells which are also color cells (cf. Table 2). Orientation cells with color specificity in area 17 tend to occur in the upper and lower-most layers, where Fig. 8 showed orientation cells to be most common.

"Other" cells with color specificity are most common in layer 4B of area of area 17 (the granule cell layer) and in layers 5 and 6 of area 18. Larger numbers are needed in order to determine whether or not this reflects a real difference between the two areas with regard to color processing.

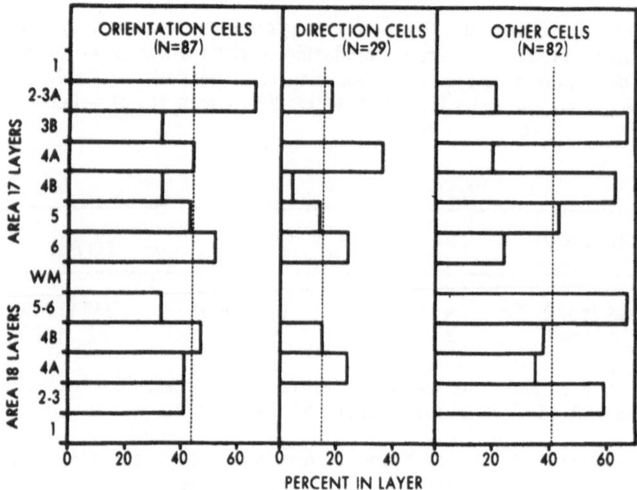

Fig. 8 Laminar distribution of 198 cells recorded in areas 17 and 18, and classified according to response properties, as in Table 1. Layers in area 17 are numbered according to VON BONIN (15). Layers in area 18 are numbered according to LORENTE DE NO's system for parieto-temporo-occipital isocortex (16, p. 279). Layer 4B is the granular layer in both cortical areas according to these numbering schemes. The dashed lines indicate the percentages of the 3 cell groups (orientation cells -44%, direction cells - 15%, other cells - 41%) in the entire sample of marked cells. The histograms indicate percentages of the three cell groups within each layer.

The main conclusion suggested by the data is that line orientation, movement direction, and color are largely processed as separate stimulus parameters by different groups of cells in the foveal portions of areas 17 and 18 in the Rhesus monkey. The area 18 results are of special interest, because they indicate that the dissociation of response specificities already reported for area 17 (5, 6) is being maintained at least as far as area 18. These results suggest that areas 17 and 18 function together as a unit, refining and separating out different kinds of visual specificity for further processing at higher levels. The generation of orientation specific responses (7) is only one of the functions of area 17; other functions include the generation of directionality and the creation of color as an independent parameter.

A lingering question remains, however, as to the differences between areas 17 and 18. There have been reports of prestriate cells, some in area 18, others located more anteriorly, that require specific binocular disparities (11, 12); striate neurons have not been reported to have such disparity requirements. Another possibility is that areas 17 and 18 are differently affected by the animal's state of attention. I am presently setting up experiments with behaving monkeys to test this latter possibility.

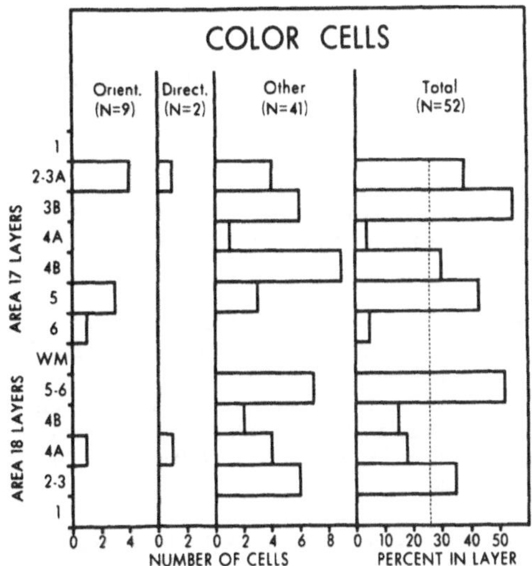

Fig. 9 Laminar distribution of 52 color cells (included within the larger population in Fig. 8). Dashed line indicates the proportion of color cells (26%) in the entire sample of marked cells. The right column of histograms indicates the percentage of color cells within each layer. The other histograms indicate the number of color cells in each of the 3 major groups of Fig. 8, and the laminar distribution of these cells.

Acknowledgements

I am grateful to Eleanor Collins for histology, Julie Lakatos for preparation of figures, and Lina Kwok for secretarial assistance.

Reference

1. P. Gouras, J. Physiol. London, 199, 533 (1968).
2. P. Gouras, Science 168, 489 (1970).
3. P. Gouras, J. Physiol. London 238, 583 (1974).
4. S. B. Leighton and B. M. Dow, Vision Res. 13, 1195 (1973).
5. B. M. Dow, J. Neurophysiol. 37, 927 (1974).
6. B. M. Dow and P. Gouras, J. Neurophysiol. 36, 79 (1973).
7. D. H. Hubel and T. N. Wiesel, J. Physiol. London 195, 215 (1968).
8. R. C. Thomas and V. J. Wilson, Nature 206, 211 (1965).
9. S. M. Zeki, Brain Res. 14, 271 (1969).
10. J. S. Baizer, D. L. Robinson and B. M. Dow, J. Neurophysiol. 40, 1024 (1977).
11. F. H. Baker, P. Grigg and G. K. Von Noorden, Brain Res. 66, 185 (1974).
12. D. H. Hubel and T. N. Wiesel, Nature 225, 41 (1970).
13. F. W. Campbell, B. G. Cleland, G. F. Cooper and C. Enroth-Cugell, J. Physiol. London 198, 237 (1968).

14. U. C. Dräger, J. Comp. Neurol. 160, 269 (1975).
15. G. Von Bonin, J. Comp. Neurol. 77, 405 (1942).
16. R. Lorente de Nó', In Physiology of the Nervous System, edited by J. F. Fulton, (Oxford Univ. Press, New York, 1944).

Discussion

Q. Dr. SPEAR: What are the other "other" cells that are not color coded? Are they just non-directional, non-oriented, and also non-color cells, like concentric cells that do not have color opponency.

A. In area 17 some of the "other" cells are broadly tuned simple or complex cells. Some are concentric cells. Some have homogeneous fields without surrounds. The homogeneous field cells can be either on (activated by light) or off (suppressed by light). Many of the same kinds of cells make up the "other" cell group in area 18 with two exceptions. First, there do not appear to be any simple cells in area 18. Second, there is a new class of cells in area 18, which JOAN BAIZER, DAVID ROBINSON and I (1) have called "border" cells.

Q. Dr. DE VALOIS: Did you break your classification down into simple and complex or X and Y?

A. I will answer the X-Y questions first, since that is easier. I have not really made any attempt to classify cells according to X and Y.

As for simple-complex, I believe I have all the information to classify cells as simple, complex, and hypercomplex, but I am not convinced that these are useful categories. In the monkey there is a class of small-field cells with separate on and off regions which I think everyone would agree are simple cells; and there is another class of large-field on/off cells, with high spontaneous activity and a preference for higher stimulus velocities, which most people would agree are complex cells. In between these two extremes, however, is a third group of cells with small receptive fields but without identifiable on and off regions. These cells respond best to moving stimuli. Some show precise orientation tuning. Others are primarily selective for direction of movement. In a previous publication I subdivided complex cells into 3 subgroups (ref. 4, Classes III, IV and V). I am now less inclined to impose artificial classifications on the cells, and prefer to let the groups fall out of data plots such as Fig. 7.

Q. Dr. RICHARDS: On the slide where you plotted directionality and angular selectivity (Fig. 7), the points were distributed very much like a hyperbolic inverse function, except for a few cells to the upper right. The striking thing is that the middle of the diagram is empty. Is it possible that the cells form a continuum, and that you have imposed cell groupings (Orientation, Direction, Other) onto the data?

A. Having devised this plot, I am now interested in adding numbers to it, to see if the center fills in. I am also interested in adding other quantative dimensions, such as color, for example, to see if cell groupings become more evident.

It may be that cells differ as to their mobility within such specificity space. Direction cells, for example, show vertical mobility as a function of velocity, and horizontal mobility as a function of stimulus length. Perhaps some cells in awake monkeys will show behavior-contingent mobility.

Q. BISHOP: A comment about hypercomplex cells. Such cells in the cat respond quite well to spots. You have to be very careful not to have the stimulus too long.

If you want to get good orientation tuning you have to use stimuli which are long enough to have orientations, but not so long as to inhibit the cell. I suspect a lot of your cells may well have had good tuning if you had picked the right stimulus lengths. The distinction between long enough and too long may be even more critical in the monkey than in the cat.

A. I did not say very much about hypercomplexity, but most of the hypercomplex cells that I see in foveal areas 17 and 18 are direction-selective. DRÄGER (6) has reported a similar phenomenon in area 17 or the mouse.

Q. Dr. BISHOP: You have to differentiate between hypercomplex I and II. Hypercomplex I cells are like simple cells, except they have end inhibition. Hypercomplex II cells are like complex cells. Hypercomplex I cells are very direction-selective, just as simple cells are. The vast majority of simple cells in the cat are direction-selective. If you are careful about the length, you find out that they are very nicely orientation-tuned.

Q. Dr. SCHILLER: I would like to say with respect to this discussion, that the evidence for hypercomplex cells as a separate category is highly open to debate. There are numerous papers which have shown a continuum between cells which are end-stopped and cells which are not. End-stopped cells can be either simple or complex. Given this evidence, I find it strange that people keep talking about this hypercomplex cell and that hypercomplex cell. There is no quantitative data to back up the idea of a separate hypercomplex group.

Q. Dr. KAAS: I would like to comment on the laminar distribution data in Figures 8 and 9, and specifically on the problem of homologizing layers from area 17 to area 18. You have obviously made a decision about this. Would you care to explain why you chose to number the layers in this way? Clearly you are not doing it the same way BRODMANN does it.

A. I find the VON BONIN system (15) more consistent with the way I have been thinking about the cortex, so I have been using VON BONIN'S system rather than BRODMANN'S for area 17. For area 18 I am using LORENTE DE NÓ'S system for parieto-temporo-occipital isocortex (13). The advantage of combining VON BONIN and LORENTE DE NO' in this manner is that the granule cell layer is defined as 4B in both areas 17 and 18,and the cell sparse layer just above the granule cell layer is defined as 4A in both areas. I wish we had some consistent terminology that we all could use.

Cortical Cells: Bar and Edge Detectors, or Spatial Frequency Filters?

Russell L. De Valois
Duane G. Albrecht
Lisa G. Thorell
Primate Vision Laboratory
Department of Psychology
University of California, Berkeley

Until the pioneering work of HUBEL and WIESEL (1, 2, 3), visual scientists could do little more than guess about the functions of cells in visual cortex and the manner in which complex patterns were analyzed. Their initial reports, however, established that cells in Area 17 of cat (and also monkey) were quite unresponsive to full-field illumination, requiring more specific patterns to elicit a response. Their receptive field (RF) maps of simple cells showed an elongated excitatory center and inhibitory flanks, or vice versa; or two adjacent excitatory and inhibitory areas. Initially HUBEL and WIESEL (2) reported that there was summation within the excitatory and inhibitory regions and an antagonism between them. As we shall see, such summation, if true, would have implications not at all foreseen at the time, and would lead one, in fact, to a quite different model of what such cells do from the one which they actually proposed.

HUBEL and WIESEL (2) then went on to describe the optimal stimulus for cortical cells as being an elongated light or dark bar of a particular width and orientation, or a sharp, correctly oriented edge between light and dark. Their descriptions of optimal stimuli in such semi-naturalistic terms (which actually implies, when one examines the question carefully, non-linear interactions quite different from the simple summation within the receptive field they described before), led others (LINDSAY and NORMAN (4); NEISSER (5)) to develop theories of pattern perception in which a complex pattern would be dissected into simple units such as bars of particular widths, and edges; these units would then be combined into more complex combinations of bars and edges. Such an approach seems intuitively reasonable and was readily accepted as being a tremendous advance over earlier notions of pattern processing.

Some years later, CAMPBELL and ROBSON (6) proposed that the visual system analyzed spatial variations in light in terms of the spatial frequency content of the pattern, rather than in terms of more naturalistic features such as edges. This notion, not at all intuitively obvious, would lead one to predict that the basic unit which visual cells should respond best to and be most selective for would be not lines and edges, but rather sinusoidal variations of the appropriate frequency in the distribution of light in space. In essence it was proposed, the visual cortex would behave in a manner similar to the cochlea in the auditory system--as a crude Fourier analyzer--but dealing with spatial rather than temporal waveforms. Considerable psychophysical evidence has in fact been reported since then supporting this view.

In our research to be summarized here, we have attempted to determine directly whether cortical cells in Area 17 of macaque monkey and cat, with RFs roughly like those described by HUBEL and WIESEL, are better described as bar and edge detectors, responsive to contrast, or as spatial frequency filters, responsive to the magnitudes of the Fourier components.

Some of this research has been reported earlier (DeVALOIS, SNODDERLY, and MORGAN, (7); DE VALOIS, VON BLANCKENSEE, READY, AND DE VALOIS, (8); DE VALOIS, ALBRECHT, and THORELL, (9)) or is being readied for more extensive publication elsewhere (VON BLANCKENSEE, READY, and DE VALOIS, (10); DE VALOIS, ALBRECHT and THORELL, (11); ALBRECHT, THORELL, and DE VALOIS, (12)).

1. Characteristics of LGN and cortical cells

In research leading up to the issue at hand, we determined the behavioral contrast sensitivity of the macaque, and examined the contrast sensitivity and orientation tuning of lateral geniculate (LGN) and cortical units. The comparison of the macaque monkey and human observers' contrast sensitivity functions showed them to be virtually identical in shape and variation with adaptation level, except that the monkey's curve was shifted to slightly lower spatial frequencies. This provides strong justification for relating macaque physiology to human psychophysics (and vice versa).

The contrast sensitivity and orientation tuning of a large population of LGN and cortical cells was then determined. The usual procedure for determining a cell's contrast sensitivity was to drift gratings of a variety of contrasts and spatial frequencies across the RF (at the optimal orientation, and at the identical, optimal temporal frequency). From these data, the contrast required for a certain criterion response was found at each spatial frequency. The cortical sample was from two loci. The majority were from an area 1° away from the fovea. We concentrated on this region to eliminate variations due to retinal locus: we wanted to examine the distribution of spatial and orientation tuning of cells all receiving from essentially the same retinal region. A second, smaller sample was obtained from a 2°-4° perifoveal area.

Geniculate cells have no orientation tuning, but show considerable individual variation in their spatial tuning. Virtually all LGN cells show a rather sharp drop in sensitivity for frequencies higher than their best frequency; the low-frequency drop in sensitivity is much gentler and sometimes almost nonexistent. They act, thus, as low-pass filters. It is important to note, however, that LGN cells found in the same probe and which have overlapping receptive fields will often show widely differing peak spatial frequencies and bandwidths. This variation in frequency tuning would of course be necessary if one were to build a system of bandpass spatial frequency filters which covered the whole visible range at a later level.

In considering the properties of striate cells, it is necessary to distinguish between two quite different cell types, corresponding to those HUBEL and WIESEL (2) termed simple and complex cells. The simple cells, one recalls, have discrete excitatory and inhibitory areas which can be mapped with spots and lines and which respond in ways predictable from such an RF map. Complex cells respond to similar stimuli, but such a stimulus can be at any of several locations within the overall RF; a conventional RF map cannot be made for a complex cell.

ENROTH-CUGELL and ROBSON (13) reported two populations of retinal ganglion cells which they termed X and Y cells. When presented with counterphase-flickering gratings at different positions with respect to their receptive fields, these two cell types responded in quite different manners. X cells showed a high degree of phase specificity--that is, they responded maximally when the grating was in some particular location with respect to the receptive field. As the grating's location was changed, the response was reduced until, at the point at which the grating was displaced from its "best" position by 90°, the cell gave no response at all, the null phase position. They point out that there must be complete linear summation within the RF to produce such a null.

The Y cells of ENROTH -CUGELL and ROBSON, on the contrary, showed no null phase; rather, the cell fired regardless of the position of the grating in the receptive field. To a counterphase flickering grating, it responded with excitation to both the light and the dark components of the grating, thus responding twice to each cycle of the flicker.

We have classified several hundred monkey and cat cortical cells as simple or complex, by the HUBEL and WIESEL criteria, and as X or Y, by the ENROTH-CUGELL and ROBSON criterion of the null phase test. A few cells are hard to classify by either set of criteria; but without exception among those that can be clearly classified, the simple cells are X cells and the complex, Y cells (THORELL, ALBRECHT, and De VALOIS (14)). It might be remarked that one can classify cells as simple (X) or complex (Y) cells much more easily and reliably from their responses to gratings than by traditional RF mapping.

It is important to emphasize that the X-Y classification should not be confused with the supposed dichotomy of cells into sustained and transient, a division which we find bears little if any correlation to the X-Y dichotomy in our cortical sample. This lack of correlation is particularly evident when one compares the temporal tuning characteristics of X cells and Y cells. Not only is the distribution of peak temporal tuning virtually identical but the proportion of cells with no low frequency attenuation (the defining property of the sustained-transient division) is also identical. In monkey, the median peak of the temporal tuning curve for X cells occurred at 3 Hertz and for Y cells at 2.75 Hertz; 24% of the X cells and 32% of the Y cells had no low frequency attenuation. The statistics for cat are quite similar.

The X and Y cells differ as described above in their responses to counterphase flickering gratings. They also differ clearly in their responses to gratings drifted across the RF. If one Fourier analyzes the post-stimulus-time histogram of a cell's response to a drifting grating, X cells will have almost all their power at the fundamental. There will be some slight DC response, by virtue of the fact that the cell is half-wave rectifying, and some upper harmonics for the same reason, but most of the power is at the fundamental. Y cells, on the other hand, respond with an almost unmodulated increase (or decrease) in mean firing rate to a drifting grating. Most of the power is thus in the DC component.

For each of a sample of 219 macaque striate cells, 145 simple (X) cells and 74 complex (Y) cells, we quantitatively determined the contrast sensitivity and orientation tuning. A similar sampling of 179 cells in cat striate has also been examined.

With respect to the orientation tuning, one point should be noted which has not been adequately made in the literature or realized by psychophysicists using physiological data in their models, that is, the presence of enormous variability, within both categories of cells, in the fineness of the orientation tuning. A fair number of monkey cortical cells have no orientation tuning at all, as HUBEL and WIESEL (3) noted. (We should note that this non-oriented population, like the rest of the cells, can be subdivided into X and Y cells or simple and complex cells, thus raising some doubt about HUBEL and WIESELS's hierarchical model.) But even among the oriented cells, the narrowness of tuning varies widely. Our most narrowly-tuned cell had a bandwidth at half amplitude of 6°; the median is about 40°; and there are numerous broadly tuned cells with bandwidths up to 100° and more. The foveal and perifoveal samples do not differ in their tuning, but X cells are slightly more narrowly tuned than Y cells. The cat cells we studied turned out to be very similar; if anything, they are slightly more narrowly tuned for orientation than monkey cells on the average, contrary to earlier reports (HUBEL and WIESEL (3)).

In examining the spatial frequency tuning of cortical cells, we find again a wide range of fineness of tuning. Some cells are very narrowly tuned, with bandwidths at half amplitude as narrow as 0.6 octaves; others are as broad as 2.5 octaves in their spatial bandwidth; and the median bandwidth is about 1.4 octaves. The X cells are slightly more narrowly tuned than Y cells, and do not differ in the two retinal areas studied. The cat cells again are somewhat more narrowly tuned on the average than monkey cells.

We find a positive correlation between the narrowness of the spatial frequency and the orientation tuning of cortical cells. There are therefore a sizeable number of cells--perhaps 25% of the total population--which are narrowly tuned spatially, with bandwidths of 1.2 octaves or less, and in orientation, with bandwidths of 40° or less. Such cells would be expected, if they show linear summation, to respond to only a small region of two-dimensional frequency space.

Another aspect of the cells' responses which bears on the question of whether such a collection of cells can act as spatial filters in analyzing the visual scene, is the range of the orientation and spatial tuning within a cortical area dealing with a given section of the retina. As is well known from the work of HUBEL and WIESEL (1, 2), cells are found, in neighboring columns, tuned to every orientation around the clock (slightly more vertical and horizontal than oblique in our sample). Within each of our foveal and perifoveal samples, we find that the cells also show a wide range of best spatial frequencies--over roughly a 4-octave range in each case. Some of the cells in the foveal sample are tuned to less than 1 cycle/deg.; the highest-tuned cell in this sample responded optimally to 16 cycles/deg. In the perifoveal sample the range was shifted to lower spatial frequencies, going from less than 0.5 cycles/deg. to 6 cycles/deg. In Fig. 1 can be seen five cells studied in a single probe through monkey cortex. The cells are tuned to a variety of spatial frequencies, two of them responding to almost non-overlapping frequency samples. It can also be seen that each cell is only responsive to a limited spatial frequency band.

These results on monkey (which are duplicated completely in cat cortex, except that the cat cells are tuned to a spatial frequency range shifted downward by about 2 octaves relative to that of monkey) are in complete agreement with the suggestions of CAMPBELL and ROBSON (6) and BLAKEMORE and CAMPBELL (15) of fairly narrow spatial-frequency channels tuned to a range of different spatial frequencies. In fact, both the narrowness of the spatial tuning (of the narrowest cells) and the spatial range covered within a given part of the visual field exceeds that which they postulated.

2. Bars versus gratings

Although these results above are entirely consistent with a spatial filtering model of visual processing, they do not directly rule out the possibility that the cells could

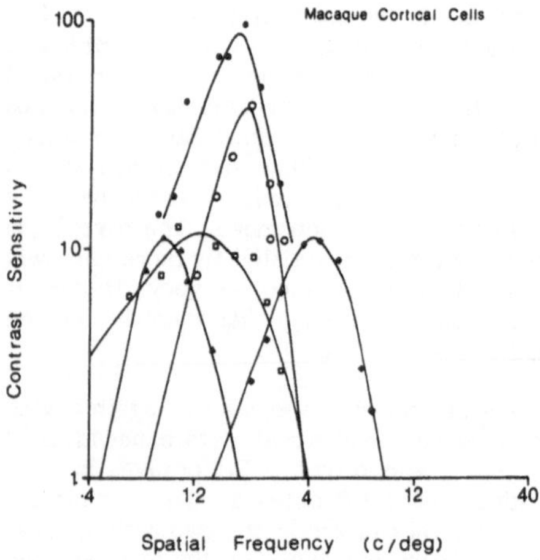

Fig. 1 Spatial frequency tuning curves of five macaque cortical cells recorded in the same electrode penetration. Contrast sensitivity (the reciprocal of the contrast required to reach a constant response criterion) is plotted as a function of spatial frequency. Symbols indicate actual data points; the curves were fitted by eye. Note the range of spatial tuning of adjacent cells.

instead be a system for analyzing the visual world into bars and/or edges, as many have concluded from the HUBEL and WIESEL statements about the optimal stimuli for the cells. As we mentioned in the introduction, and elaborate below, for that to be the case would require that the cells show quite non-linear summation properties (contrary to what HUBEL and WIESEL indicated). To choose between these alternative models a set of direct experimental tests is required. That is what we have attempted and would like to report here.

a. Selectivity

Any model which postulates that the cells in the visual pathway are analyzing the visual scene into elements along some dimension must postulate that the system is selective along that dimension. Thus any model postulating that the visual world is analyzed by cortical cells into bars of varying width and orientation must require that the cells be selective for bar width and for orientation. And correspondingly, if the scene is being analyzed into different spatial frequency bands; the system must be selective for spatial frequencies. The first question we ask, then, is: along which dimension, bar width or spatial frequency, are cortical cells in fact most selective? This we have done by examining the responses of each of a number of X and Y cells in both monkey and cat striate cortex to bars of different width and sine-wave gratings of different frequencies.

One can readily equate bar width and equivalent spatial frequency by thinking of a grating as consisting of alternate dark and light bars. The grating whose equivalent width is the same as that of a particular bar would then be one whose half-period is the same as the bar. The various bars and gratings were drifted across the RFs of the cells at the optimal orientation and rate and the responses recorded and analyzed. Bars and gratings of each width were presented at several different contrast levels, so we could compare not only the responses at a particular contrast, but could actually obtain a contrast sensitivity function for each.

The results are completely unequivocal. Without exception, in every monkey and cat cell studied, the selectivity for sine wave gratings of different spatial frequencies far exceeded that for bars of different widths. In Fig. 2 are shown four typical examples of the results. It might be noted that SCHILLER, FINLAY and

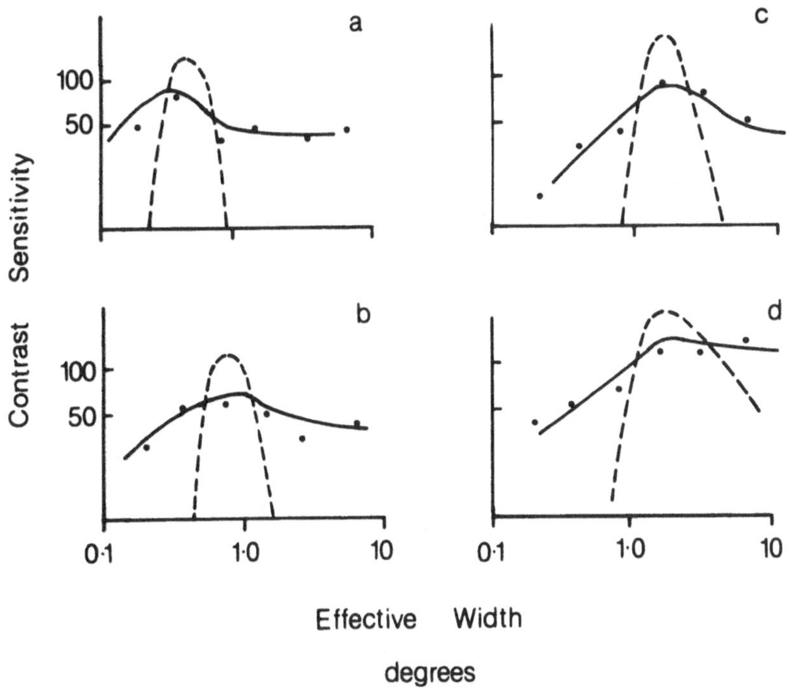

Fig. 2 Comparison of the measured bar width tuning (dots) vs the measured spatial frequency tuning (dashed lines) for two macaque monkey cortical cells (a, b) and two cat cortical cells (c, d). Contrast sensitivity is plotted as a function of the effective stimulus width. The continuous line is the predicted sensitivity to bars of different width computed from the particular cell's spatial contrast sensitivity function.

VOLMAN (16) have recently published data from monkey cortex that agrees with this also. Although they did not control for contrast and only looked at response (rather than contrast sensitivity) measures--thus making it difficult to evaluate their bandwidth figures--the direction of their results agrees completely with ours. The lack of selectivity for bar width is so great that often the response to the broadest bar does not even fall to half of the optimal width; the bar bandwidth for some cells thus approaches infinity. If one compares the bandwidth at half amplitude for bars and for gratings for all of our cells there is no overlap in the distributions.

This lack of selectivity for bar width among cortical cells makes it difficult to see how one can seriously maintain a model of visual processing based on bar width, quite independent of the fact that a superior possibility is waiting in the wings. Several ways of looking at the nature of the visual stimuli in relation to the cells' RFs, to be discussed below, may help in understanding the basis for the selectivity for sine wave gratings and lack of selectivity for bars.

b. Sensitivity or Responsivity

A second question we asked relative to these models is: which of these types of stimuli, bars or sinusoidal gratings, are optimal in driving cortical cells? To which are they more sensitive, or to which do they give larger responses? It is clearly meaningless to ask this question unless one equates the two varieties of stimuli for contrast: a cell would obviously respond more to a very bright bar on a black background than it would to a sine wave grating of threshold contrast, or respond more to a high contrast black-white grating than to a dim white bar barely visible above its background. Thus the comments of SCHILLER, FINLAY, and VOLMAN (16) on responses to bars and gratings, with no control for contrast, are uninterpretable. The problem is how to equate the contrast of bars and gratings, given that they are such different stimuli. The method we have employed is in accord with the usual definitions used by other investigators. The contrast of a bar is customarily characterized by the Weber fraction, I/I, where I is the brightness of the bar and I is the mean level. The contrast of a grating is universally defined by the Michelson formula:

$$\frac{\text{max-min}}{\text{max+min}},$$

where max and min are the peak and trough of the grating, respectively. The conventional way of characterizing the contrast of a bar and the Michelson contrast of a grating turn out to be equivalent: consider the mean luminance level of the grating to be I; then the maximum would be I + I and the minimum would be I - I. Substituting these terms in the Michelson contrast formula gives one

$$\frac{(I + \quad I) - (I - \quad I)}{(I + \quad I) + (I - \quad I)}.$$

This simplified to $\frac{2 \quad I}{2I}$, or $\frac{\quad I}{I}$, and thus a specification of contrast which is identical to that for the bar.

When we examined the responses of monkey and cat cortical cells to bars of the optimal width and gratings of the optimal spatial frequency, each equated for contrast, we found almost all cortical cells are more responsive to the grating than to the bar, and show a higher contrast sensitivity to gratings than to bars. The only exceptions among 114 cells were 4 Y cells and 2 X cells, which were about equally responsive. On the average these cells are more than twice as sensitive to gratings as to bars. Fig. 2 again shows examples of the results: it can be seen that each of the 4 cells shown is more responsive to the best frequency grating than to the optimal width bar. It is clear that the optimal stimulus is not a bar of the right width, but a grating of the correct spatial frequency.

The differences in sensitivity we found to bars and gratings are not nearly so pronounced as the differences in selectivity. But it is important to note that a spatial frequency analysis of the two types of stimuli and the contrast sensitivity of the cells would not in fact predict very large differences. The simple cells for which we did these calculations should be from 2 to 3 times as responsive to gratings as to bars, depending on the narrowness of their spatial tuning, (the narrower the spatial tuning the larger the predicted difference). This is about the range we actually found. Thus if the cells had in fact responded 10 times as much to gratings as to bars of the same contrast this would not be stronger proof for a linear spatial filtering model at all but would instead disprove it!

Both the differences in selectivity and in responsivity of cells to bars and gratings is readily understood from the spatial spectrum of the stimuli and the spatial contrast sensitivity of the cells. The fact that one can understand and indeed quantitatively predict (as we will see below) the bar versus grating data from a spatial frequency analysis of the stimulus is powerful proof both of the utility and validity of this approach.

The basic datum which one needs to know to understand these results is that a bar has a very broad spatial spectrum whereas the power in a sine wave grating is confined to a narrow spectral band (how narrow depends on the number of cycles in the grating). It can be seen in Fig. 3 that each of the bars, despite their widely

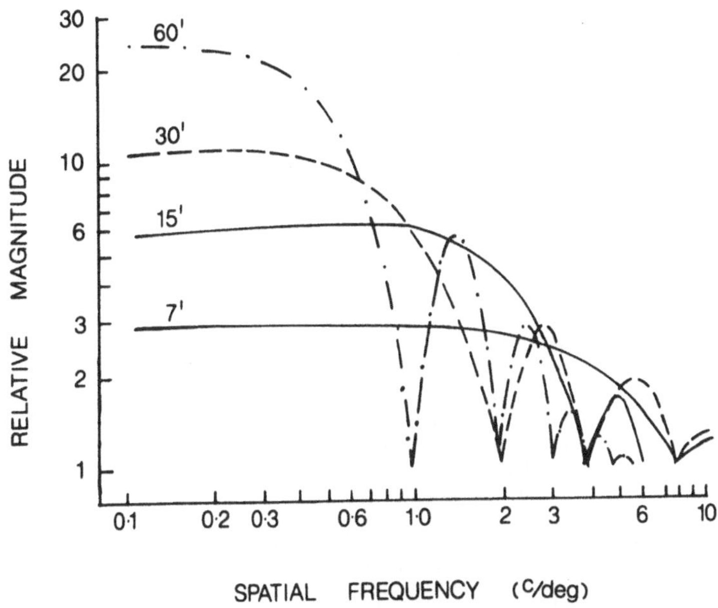

Fig. 3 Relative amplitude spectra of bars of different width for a subset of the bars actually used in the experiment. The absolute value of the Fourier transform for bar stimuli of equal contrast is plotted. Note that the bars all have very broad, overlapping spectra.

varying widths, has a broad spectrum covering most of the range to which cells are sensitive. One would therefore expect--that is, one would expect this if the cells were acting as quasi-linear spatial filters rather than as bar detectors!--that cells would quite nonselectively respond to all these bars, regardless of their optimal spatial frequency tuning. Thus all the cells shown in Fig. 1 would be expected to respond to all these bar widths (and indeed do) because each bar would have part of its spectrum within the cells' sensitivity ranges. As can be seen, each of these cells only responds, however, to a limited range of grating frequencies. That the huge difference in the cells; tuning to bars and gratings is not just qualitatively but also quantitatively predictable from the cells' contrast sensitivity functions and Fourier spectrum of the bars is shown by the fact that the lines through the bar data points in Fig. 2 are not curve fits to the data but rather the predictions from the Fourier spectrum.

Considering the Fourier spectrum of bars and gratings also enables one to understand why sinusoidal gratings rather than bars are the optimal stimuli for the cells. (We emphasize again that this interpretation assumes a spatial-filtering model, and insofar as it makes sense would thus provide support for the model.) The point to be made is that a bar, with its broad spectrum, has much of its power spread over spatial frequencies which lie outside a given cell's narrow tuning curve. The cell, in effect, is capturing and thus responding to only a fraction of the power in the bar's spectrum. A grating of even a very small number of cycles, on the other hand, has all its power in a restricted region and a cell can "capture" it all and thus respond optimally.

For some it might be easier to consider these questions not in the frequency domain, but in the space domain, in terms of cortical cell RFs. With a linear system, the Fourier transform takes one from the space to the frequency domain, and the inverse Fourier transform takes one from the frequency to the space domain, to a RF that one can think of as a spatial weighting function. If one does an inverse Fourier transform of the contrast sensitivity function of a cortical cell like those illustrated in Fig. 1, and takes into consideration the orientation tuning one ends up with a spatial weighting function that looks very similar to a simple cortical cell as described by HUBEL and WIESEL. The orientation tuning would be determined by the RF orientation and by the length/width ratio of the RF regions, the longer the RF the narrower the orientation bandwidth; the spatial frequency tuning would be determined by the width of the excitatory and inhibitory regions and the number of such regions. An excitatory center and inhibitory flanks would produce rather broad spatial tuning; additional side bands would narrow the spatial bandwidth. Another way to put the point is that if an engineer were to be asked to build a two-dimensional spatial frequency filter, he would build a device that greatly resembles a HUBEL and WIESEL simple cell, with elongated flanking excitatory and inhibitory areas and spatial summation within the device. The number of flanking areas and the degree of elongation would depend on the spatial frequency and orientation bandwidths desired.

Now, as to why a cell with such a RF should be non-selective and less than optimally responsive for bar widths, compared with gratings, consider first selectivity: a bar of a width corresponding to the RF center width would produce the best response, but the cell would fire to even very narrow bars since they would be providing net excitation to the RF center when they cross it. The cell would also respond to bars of greater-than-optimal width when they are so positioned that one edge is on the RF center. Thus the cell fires to bars of any width. But consider the responses to gratings of different spatial frequencies. The optimal spatial frequency grating produces a large response (at the correct phase) since the black and white bars of the grating then coincide with the inhibitory and excitatory areas; if one presents a grating of, say, two times this frequency, however, there will now be both white and black bars in the center; and both white and black bars in each inhibitory flank, the net result of which is no response at all. The cell will respond well to a single bar of this width but not to a grating. And at low spatial frequencies the gently changing sinusoidal grating will provide essentially uniform stimulation over the whole RF. One can thus understand in the space domain the cells' high selectivity for sine-wave gratings and vanishingly low selectivity for bars.

The larger responses cells give to gratings than to bars can best be understood from the fact that gratings simultaneously provide the appropriate stimulus to the whole RF whereas a bar does not. Consider a cell which would conventionally be described as having an excitatory center and two inhibitory flanks. A white bar on a grey background, centered on the RF, is a fine stimulus to the RF center, but not for the flanks. The flanks would best be stimulated by two appropriately separated black bars--and indeed a cell will fire as vigorously to such a stimulus as to the white bar on the RF center. This is one of many instances where confusions have been introduced by the tendency of visual scientists to concentrate on only incremental stimuli and to ignore decrements. Black objects or decrements are as numerous and as important for vision as white objects or increments, a point one might think should not have to be made to people who spend some hours every day detecting and recognizing black letters in papers and books! Nonetheless, RF maps are almost universally made only with incremental flashes, and thus this hypothetical cell we are discussing would be seen as having an excitatory center and inhibitory flanks. If decremental stimuli were used, it would have an inhibitory center and excitatory flanks! Thus it is very misleading to speak of the cell as having inhibitory flanks, or of the optimal stimulus as being a white bar in the RF center; the flanks are just as capable of firing the cell as the center, and two black bars centered on the flanks are as optimal a stimulus as a single white bar.

Now if one combines these two equally effective stimuli, the white center bar and the two black flanking bars, this three-bar combination is a considerably better stimulus than the supposedly optimal stimulus of a single white bar; it will, in fact, produce about twice as large a response. One can go one step further: since the middle of the RF center is more sensitive than the edges of the center, one would be better off to redistribute the light in the white bar to put most of it in the center and have it taper off towards the edges. So also with the black bars on the flanks where the blackest part should be in the center. What one ends up with by this rather painfully indirect procedure is the real optimal stimulus for this cell, namely 1 1/2 cycles of a sine wave grating which simultaneously and appropriately stimulates center and surround. Those (not infrequent) cells that have still further side bands would of course have an even greater preference for a grating rather than a bar.

3. Predictions from linear filter model

A final aspect of our research that we would like to summarize here is an examination of the extent to which one can predict various aspects of a cell's response just from its spatial contrast sensitivity function. One of the enormous advantages of a linear-systems analysis of the cells' responses is that insofar as the approach is valid, one can make precise quantitative predictions of a cell's response to any stimulus; this is in sharp contrast to the vague, unquantified formulation of the cells as bar or edge detectors.

a. Prediction of bar-width responses

In Fig. 2 we presented, in circles, the actual amplitude of the responses of each cell to bars of various widths. The lines through those points, however, are not curve-fits to the data, but the relative amplitudes we would predict from the observed contrast sensitivity function of the cell and the spatial frequency spectrum of the bar. As can be seen, the rather indiscriminate responses the cells give to bars of various widths is just what is predicted by the broad spatial spectrum of a bar.

b. Relation of Contrast sensitivity to RF

As we discussed above, insofar as the cells can be treated as linear spatial frequency filters, one should be able, by the Fourier transform, to go back and forth between the frequency and the space domains. That is, a cell with a certain RF shape should have a precisely predictable spatial contrast sensitivity function, and vice versa. We have examined this in a number of X cells in monkey and cat cortex, by determining the contrast sensitivity to sine wave gratings of different spatial frequencies and contrasts, then mapping the RF quantitatively by repeatedly flashing small white and black spots in each of many different locations. The question of interest is the extent to which one can predict the one from the other.

The findings are that in most cases one can make very accurate predictions of the RF from the contrast sensitivity. Broadly tuned cells turn out to have only an RF center and one or two antagonistic flanks, as predicted. Narrowly tuned cells have, as predicted, additional side bands in the RF. In Fig. 4 is the RF of a cell which could conventionally be described as having an inhibitory center and excitatory flanks (it

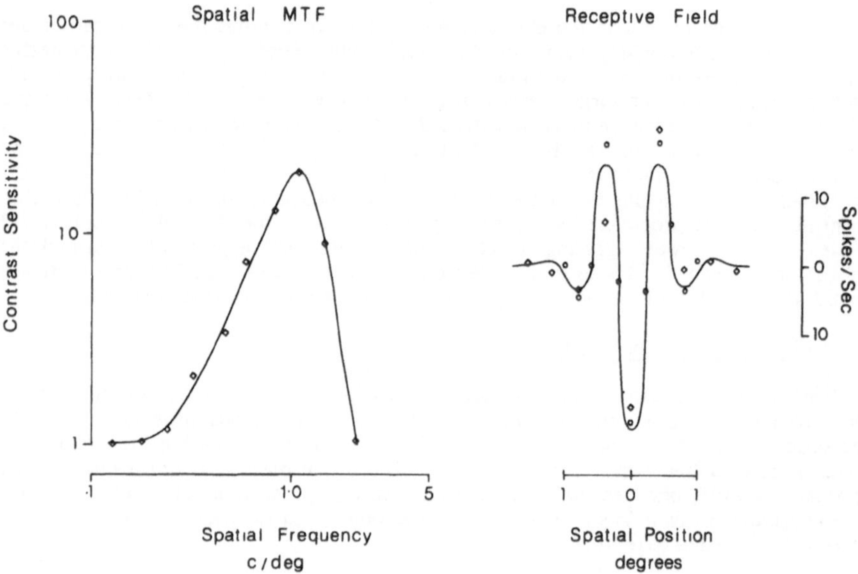

Fig. 4 Frequency and space representations of a particular X cell from monkey visual cortex. On the left is the contrast sensitivity for the cell as a function of spatial frequency (spatial MTF); the curve between the data points is fitted by eye. On the right is a receptive field response profile for the same cell. Each point is the amplitude of the response to a narrow bar in that particular position whose contrast varies from black to white; on-excitation to white is plotted upwards, on-excitation to black is plotted downwards. The RF was plotted first in 0.4° and then 0.2° steps. The continuous line is the RF profile predicted from the spatial contrast sensitivity function of the cell (using the inverse Fourier transform).

might better be described as a cell which fires to black in the center and white on the flanks). The data points are the actual responses to white and black lines in different RF loci. The solid line, which agrees quite well with the data points, as can be seen, is the RF predicted from this cell's contrast sensitivity function, by the inverse Fourier transform. The size and dimensions of the excitatory and inhibitory areas are correct. This was a fairly narrowly tuned cell, which led to the prediction of slight additional side bands, evidence for which can be seen in the data points.

c. Prediction of post-stimulus time histogram

When a white or black bar is drifted across the RF of a cell, one sees a complex temporal pattern of activity; averaged across many presentations of a given stimulus, this would constitute the post-stimulus time histogram. The shape of the PST histogram differs considerably for white and black bars, and with bar width. The question we are asking here is whether we can predict these for a cell knowing nothing but its contrast sensitivity function for sine wave gratings (and whether it has an even-or odd- symmetric receptive field). This attempt can be carried out equivalently in the frequency or the space domain. In the latter case it amounts, in effect, to predicting the precise RF shape, then determining, by linear summation, the net excitation or inhibition one should get with the bar in each of many positions across the RF.

In Fig. 5 can be seen an example of one such successful prediction for a cat X cell. The responses are predicted and observed for white and black bars of several widths. The agreement between the predicted response shape and that actually recorded is very gratifying.

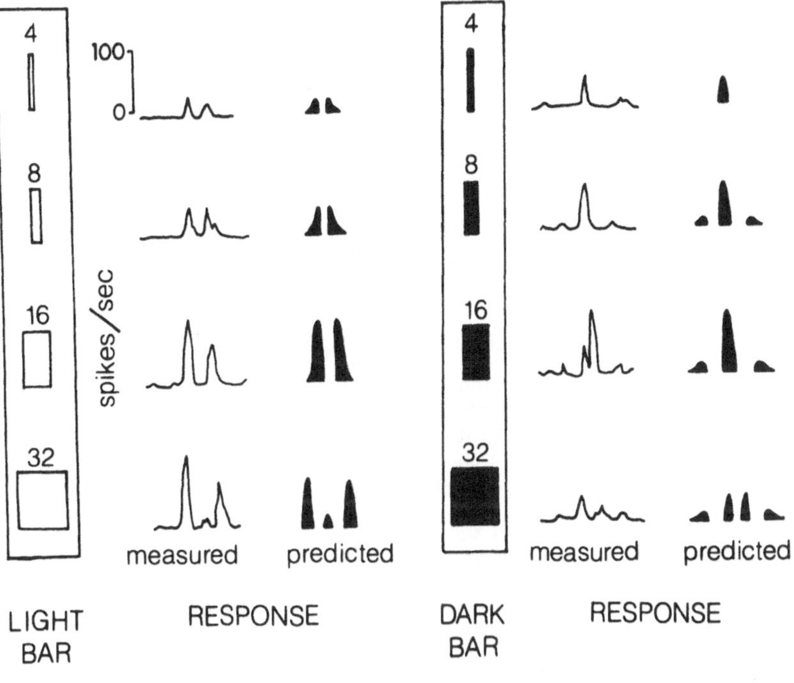

Fig. 5 Predicted and measured post-stimulus time histograms to black and white bars of different widths drifted across the RF of an X cell. The predictions were computed from the spatial-temporal contrast sensitivity function of the cell. Predictions have been rectified in accord with the half-wave rectification behavior or simple cells. The bar widths are in tenths of degrees visual angle.

In summary:

I. Cortical cells in both monkey and cat receiving from a given retinal locus show a wide range of best frequencies and orientations, and vary considerably in spatial and orientation bandwidths.

II. A sizeable proportion of these cells are quite narrowly tuned for both orientation and spatial frequency, properties that would enable them to act as two-dimensional spatial filters.

III. That they may indeed be acting as spatial filters rather than as bar or edge detectors is indicated by the fact that they are all much more narrowly tuned for sinusoidal gratings of various spatial frequencies than for bars of varying widths.

IV. Cortical cells are also more than twice as responsive to gratings of the optimal spatial frequency than to optimal width bars of the same contrast.

V. Treating the cells as linear spatial filters, we find that just from the contrast sensitivity of the cells we can predict their responses to bars of different widths, the quantitative RF shape of a cell and often the actual waveform of its responses to drifting black and white bars.

VI. We can thus conclude that the array of cortical cells related to a small section of the retina can be much more correctly, accurately, and usefully portrayed as analyzing this portion of visual space into its two-dimensional Fourier components, than as analyzing it into bar widths and orientations.

References

1. D. H. Hubel and T. N. Wiesel, J. of Physiol. 148, 574 (1959).
2. D. H. Hubel and T. N. Wiesel, J. of Physiol. 160, 106 (1962).
3. D. H. Hubel and T. N. Wiesel, J. of Physiol. 195, 215 (1968).
4. P. H. Lindsay and D. A. Norman, Human information processing (New York: Academic Press, 1972).
5. Neisser, U., Cognitive Psychology, (New York: Appleton, 1967).
6. F. W. Campbell and J. G. Robson, J. of Physiol. 197, 551 (1968).
7. R. L. De Valois, H. C. Morgan and D. M. Snodderly, Vis. Res. 14, 75 (1974).
8. R. L. De Valois, K. K. De Valois, J. Ready and H. von Blanckensee, Paper presented at the Association for Research in Vision and Ophthalmology, 1975.
9. R. L. De Valois, D. G. Albrecht and L. G. Thorell, In H. Spekreijse (Ed.) Spatial - Contrast, (Amsterdam: Monograph Series, Royal Netherlands Academy of Sciences, 1976).
10. H. von Blanckensee, J. Ready, and R. L. De Valois, Spatial tuning and linearity of macaque LGN cells. (in preparation).
11. R. L. De Valois, D. G. Albrecht and L. G. Thorell, Contrast sensitivity and linearity of cells in macaque striate cortex. (In preparation).
12. D. G. Albrecht, L. G. Thorell and R. L. DeValois, Prediction of cortical cell responses from their modulation transfer function. (In preparation).
13. C. Enroth-Cugell and J. G. Robson, J. of Physiol. 187, 517 (1966).
14. L. G. Thorell, D. G. Albrecht and R. L. De Valois, Spatial summation properties of cortical cells. (In preparation).
15. C. Blakemore and F. W. Campbell, J. of Physiol. 203, 237 (1969).
16. P. H. Schiller, B. L. Finlay and S. F. Volman, J. of Neurophysiol. 39(6), 1334 (1976).

Subdivisions and Interconnections of the Primate Visual System

Jon H. Kaas
Departments of Psychology and Anatomy
Vanderbilt University

1. Introduction

For the last ten years or so, my colleagues and I have been trying to determine the subdivisions of the visual system and the interconnections of these subdivisions in primates. Since most of our research has been on a New World monkey, the owl monkey, *Aotus trivirgatus*, this review is devoted to findings for this monkey. A more inclusive review will appear elsewhere (1).

In our research, a basic problem is how to determine those subdivisions of the visual system of functional significance. Differences in histological appearance have long been used to distinguish nuclei and cortical areas. However, architectonic distinctions are often subtle, and opinions have varied on how to histologically subdivide the brain. Thus, it is important to experimentally demonstrate the significance of perceived distinctions. A productive approach for us has been to electrophysiologically map sensory representations and to relate the findings to cortical and thalamic architecture. Microelectrode recordings followed by microlesions allow these two methods to be combined with precision. Once the architectonic features of sensory representations have been identified, studies of the connections often indicate or suggest further subdivisions of the system. Combinations of these anatomical and electrophysiological approaches have led to the following understanding of the organization and connections of the visual system in owl monkeys.

2. Visual Cortex

Some of the subdivisions of visual cortex in owl monkeys are shown in Fig. 1. Details, supporting data, and references are given in ALLMAN and KAAS (2). Two of these subdivisions, V I and V II, are recognized from traditional and textbook descriptions.

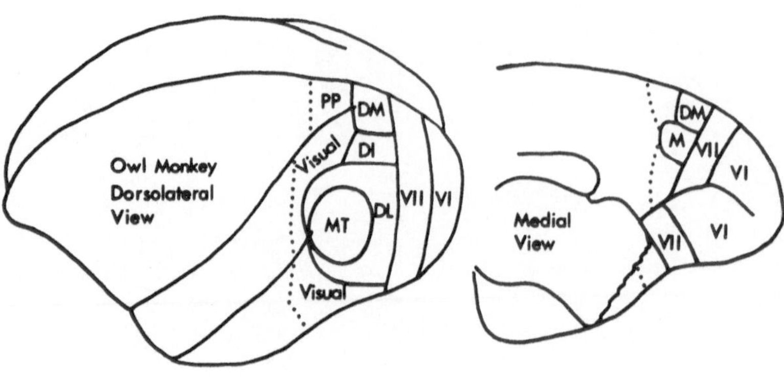

Fig. 1 Some of the visual areas of the owl monkey. The dashed line marks the rostral border of visually responsive cortex. DI, Dorsointermediate Area; DL, Dorsolateral Area; DM, Dorsomedial Area; M, Medial Area; MT, Middle Temporal Area; PP, Posterior Parietal Area; V I, Primary Visual Cortex; V II, Secondary Visual Cortex.

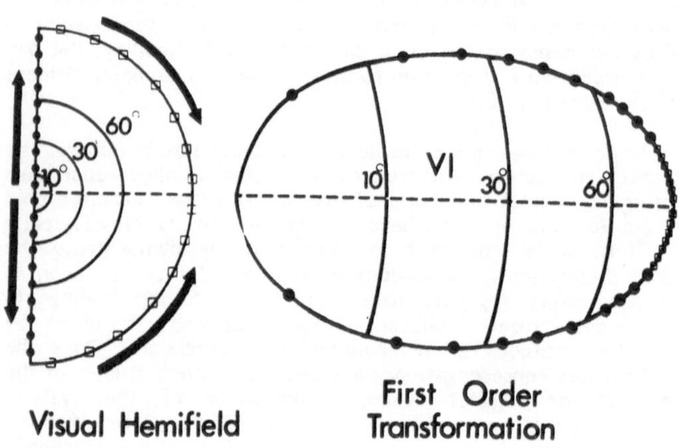

Fig. 2 A topological or First Order Transformation of the visual hemifield.

Each is a single systematic representation of the contralateral visual hemifield, and each corresponds to a single identifiable architectonic field, i.e., Areas 17 and 18, respectively. However, the portrayed organization of the cortex rostral to V II differs from traditional descriptions. Instead of a single visual representation, V III or Area 19, a number of architectonically distinct representations are found. The organization of some of these representations have been explored in detail. In other regions, further investigations are needed. Yet, it is now apparent that the traditional view of the organization of visual cortex does not apply, and that visual cortex is much more extensive and contains many more subdivisions than previously supposed. Some of the cortical visual areas, and their major connections are described below.

Striate Cortex (Area 17 or V I). Striate cortex represents the contralateral visual hemifield as a single topological transformation of a two dimensional surface (Fig. 2). The basic nature of the representation has been demonstrated in primates and other mammals by a number of methods, but only microelectrode mapping procedures have revealed the detailed organization of the area. Even in the nocturnal owl monkey, the representation is extremely distorted so that much of the tissue is devoted to central vision (3). While other cortical connections may exist, we presently have conclusive evidence for projections from striate cortex to only two other visual areas, Area 18 or V II, and the Middle Temporal Visual Area, MT.

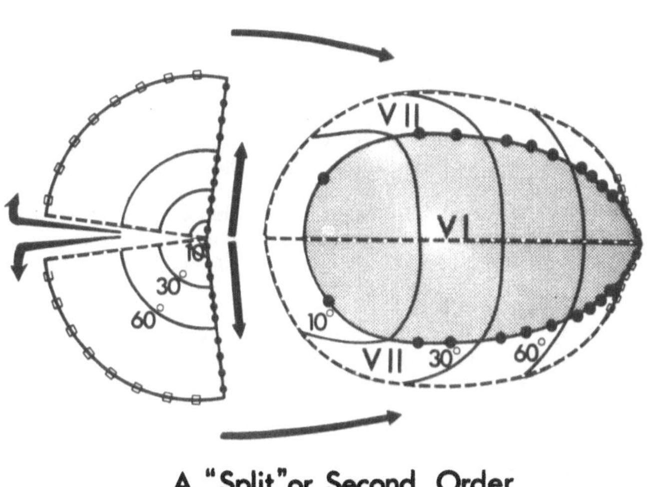

Fig. 3 A "split" or Second Order Transformation of the visual hemifield.

Area 18 (V II). The Second Visual Area borders V I along a common representation of the zero vertical meridian corresponding to the line of decussation of the retina (3). The second representation is distinctly different from the first in organization (4). V II is "split" along the horizontal midline so that the representations of the upper and lower visual quadrants are almost completely separated in the band of tissue surrounding most of V I (Fig. 3). We have referred to the topological representation such as found in V I as a First Order Transformation, and the "split" representation such as found in V II as a Second Order Transformation (5). As a consequence of the "split" second representations in V II, homotopic projections from V I originating from tissue representing the region of the horizontal meridian produce two separate foci of terminations in V II. Thus, adjoining neurons in V I sometimes project to quite distant loci in V II.

Major cortical outputs of V II are to V I, and the Dorsomedial Visual Area, DM (6). The types of interconnections between V I and V II suggest that an important function of V II is to further process information from V I, and then to feed back to modulate other outputs of V I (4).

The Middle Temporal Visual Area, MT. MT is a small topological representation of the visual hemifield that forms a reduced mirror image of V I in the temporal lobe (7). MT can be delimited with assurance by its dense myelination in brain sections stained for fibers, a characteristic useful in identifying the locations of lesions, injections, and projections in studies of connections. MT projects to a number of other visual areas (unpublished studies), including the bordering Dorsolateral Visual Area, DL; Dorsomedial Visual Area, DM; and the Posterior Parietal region. MT is the only visual area other than V II known to project back to V I.

Other Visual Areas (DL, DM, M, and PP). The details of the retinotopic organizations of three other visual areas have been determined: the Dorsolateral Area, DL (8); the Dorsomedial Area, DM (5); and the Medial Area, M (2). Each of these areas is a "split" or Second Order Transformation, a feature that allows each to have partially matched or congruent borders (5) with the outer margin of V II along the representation of the horizontal meridian. Of these three representations, the cortical connections of only DM have been reported (9). DM projects to DL, MT, and PP. A ventral focus of terminations on the medial wall of the cerebral hemisphere suggests the site of another visual area. The Posterior Parietal region, PP, responds to visual stimuli, but a map of the visual field has not been demonstrated. A major input to PP is from DM, and this may be an important source of visual information. PP, in turn, projects back to DM, as well as to M, DL, MT, and other regions of visual cortex (10).

Callosal Connections. In the above studies of cortical connections, callosal projections were also considered. There is evidence for three types of callosal connections between visual areas of the two cerebral hemispheres. One type is strictly homotopic as it appears to connect neurons in the two hemispheres with identical or nearly identical receptive fields. Homotopic callosal connections are found in the region of the V I/V II border, thus joining the two hemifields, and this is often considered to be the only type of visual interhemispheric connections. However, lesions or injections centered in MT (unpublished) and DM, and thereby in portions not representing the vertical meridian, result in degenerated or labeled terminals centered in MT or DM of the other hemisphere. These projections appear to connect an area in one hemisphere with its twin in the other hemisphere in a pattern that is mirror symmetrical. Such connections would not be homotopic since they connect parts of the right and left hemifields, but they may be called homoregional. The electrophysiological effects of these homoregional callosal connections have not been demonstrated. Other callosal connections are heteroregional. That is, callosal connections may be from one subdivision to other subdivisions of visual cortex. Thus, DM projects heteroregionally to MT and PP of the opposite hemisphere as well as homoregionally to DM. MT also has heteroregional callosal projections (unpublished).

3. Visual Thalamus

Responsiveness to visual stimuli and major connections with other parts of the visual system justify calling the lateral geniculate nucleus and the pulvinar complex the visual thalamus. The lateral geniculate nucleus, of course, receives direct input from the retina (11). Structurally, the nucleus has two important pairs of layers, the dorsal parvocellular layers and the ventral magnocellular layers (1). Electrophysiological evidence indicates that the input to the magnocellular layers is from the Y-cell type of retinal ganglion cell with transient responses to visual stimuli while the parvocellular layers are activated by X-cells with sustained responses to continuing stimuli (12). Indirect evidence suggests that Y-cells projecting to the magnocellular layers have collaterals to the superior colliculus, while X-cells projecting to the parvocellular layers do not (13). Other projections to the lateral geniculate nucleus are from cortical Areas 17, 18 and MT (14). The magnocellular and parvocellular layers can be distinguished by differences in these input patterns. While all layers receive dense projections from Area 17, inputs from Areas 18 and MT are concentrated in the magnocellular layers.

The pulvinar complex receives visual inputs from the subdivisions of visual cortex and from the superior colliculus. The complex is clearly separated by a zone of fibers into two large cellular masses that we have distinguished as the superior pulvinar and the inferior pulvinar. Architectonic subdivisions within each of these masses are not pronounced enough to justify distinction without supporting experimental evidence. However, patterns of connections indicate that both the inferior and the superior pulvinar contain separate nuclei.

Data for the inferior pulvinar have been most fully analyzed, although they have appeared in only brief reports (15, 16). Patterns of connections have been related to three nuclei of slightly different cell types and cell densities. The large "central" nucleus, IPc, constitutes over 70% of the entire inferior pulvinar. This nucleus has been electrophysiologically mapped as the "inferior pulvinar", and it contains a topological representation of the contralateral hemifield (17). A smaller medial nucleus of densely packed cells, IPm, occupies 20% or less of the inferior pulvinar. Finally, a small posterior division, IPp, forms only 10% of the inferior pulvinar. Patterns of connections suggest that both IPm and IPp also contain representations of the visual hemifield.

While IPc, IPm, and IPp can be distinguished by their architectonic characteristics, they were first identified by differences in their patterns of connections. For example, IPp receives a dense input from the superior colliculus, IPc receives a more diffuse input, while few if any axons from the superior colliculus terminate in IPm. Cortical connections also differ. IPm is strongly and reciprocally interconnected with MT, IPp projects to visual cortex in the temporal lobe rostral to MT, and IPc projects to V II and visual areas adjoining V II in the occipital lobe. Areas V I, V II, MT, DM, M and PP project to both IPm and IPc, but they do not appear to project to IPp; instead cortical inputs to IPp appear to originate in visual cortex of the rostral temporal lobe.

The superior pulvinar complex receives visual input from V I, V II, MT, DM, M, and PP. Preliminary considerations of the patterns of these inputs suggest that the superior pulvinar complex consists of at least three separate nuclei, but further statements would be premature.

4. Superior Colliculus

The superior colliculus of owl monkeys (18) and other primates topologically represents the visual hemifield, and it is not an extended representation including the binocular portion of the ipsilateral hemifield as it is in some mammals. Direct projections from the retina form a continuous terminal zone in the superficial grey throughout the contralateral colliculus, and discrete irregular patches in the superficial grey of all but the caudal ipsilateral colliculus (19). As noted above, two major ascending projections of the superior colliculus are to IPp and IPc of the inferior pulvinar. All of the visual areas that have been investigated, i.e., V I, V II, MT, DM, M, and PP, project to the superior colliculus. The V I and V II projections terminate predominately in the superficial grey and thereby overlap the terminations from the retina. Other projections tend to terminate deeper.

5. Conclusions

Our studies of the forebrains of owl monkeys indicate that the visual system contains more subdivisions, and is more complexly interconnected than previously supposed. Both extrastriate visual cortex and the pulvinar complex are divided differently than in the traditional descriptions based primarily on cytoarchitecture. Most subdivisions of the visual system form systematic representations of the contralateral visual hemifield. These representations are of two types. One type is the topologically confluent or First Order Transformation; the other type is the "split" or Second Order Transformation. Each representation is homotopically connected with other representations of the same cerebral hemisphere. Interhemispheric connections appear to be of three types that are distinguished as homotopic, homoregional, and heteroregional.

Our ongoing studies and studies by others on prosimians, other New World monkeys, and Old World monkeys demonstrate that the above conclusions are valid for primates in general. However, important differences in the total number of subdivisions, the interconnections of subdivisions, and the internal organizations of subdivisions probably occur. Until the organizations of the visual systems of primates in the major taxonomic groups are better understood, we are left with only intriguing suggestions of species differences.

The multitude of visual areas and nuclei and the complexity of the interconnections indicate that most schemes of neural processing in the primate visual system are obviously too simple. Neurons in any structure must be influenced by a multitude of inputs, and the overall processing may be more adequately described as manifold rather than serial, hierarchial, or parallel, although serial, hierarchial and parallel components certainly exist.

References

1. J. H. Kaas, In: W. P. Luckett and C. R. Noback (Eds.), Advances in Primatology: Neurobiology of Primates, (New York: Plenum, in press).
2. J. M. Allman and J. H. Kaas, Science 191, 572 (1976).
3. J. M. Allman and J. H. Kaas, Brain Research 35, 89 (1971).
4. J. M. Allman and J. H. Kaas, Brain Research, 76, 247 (1974).
5. J. M. Allman and J. H. Kaas, Brain Research 100, 473 (1975).
6. J. H. Kaas and C. S. Lin, Vision Research 16, 739 (1977).
7. J. M. Allman and J. H. Kaas, Brain Research 31, 85 (1971).
8. J. M. Allman and J. H. Kaas, Brain Research 81, 199 (1974).
9. E. Wagor, C. S. Lin and J. H. Kaas, J. Comp. Neur. 163, 227 (1975).
10. J. H. Kaas, C. S. Lin and E. Wagor, J. Comp. Neur. 171, 387 (1977).
11. J. H. Kaas, C. S. Lin and V. A. Casagrande, Brain Research 106, 371 (1976).
12. S. M. Sherman, J. R. Wilson, J. H. Kaas and S. V. Webb, Science 192, 475 (1976).
13. R. E. Weller and J. H. Kaas, Neuroscience Abstracts 3, 581 (1977).

14. C. S. Lin and J. H. Kaas, J. Comp. Neur. 173, 457 (1977).
15. C. S. Lin, Anat. Rec. 187, 637 (1977).
16. C. S. Lin and J. H. Kaas, Neuroscience Abstracts 1, 44 (1975).
17. J. M. Allman, J. H. Kaas, R. H. Lane and F. M. Miezin, Brain Research 40, 291 (1972).
18. R. H. Lane, J. M. Allman, J. H. Kaas and F.M. Miezin, Brain Research 60, 335 (1973).
19. J. K. Harting, V. A. Casagrande, J. H. Kaas and R. W. Guillery, Neuroscience Abstracts 1, 47 (1975).
20. J. H. Kaas, R. W. Guillery and J. M. Allman, Brain Behav. Evol. 6, 253 (1972).
21. C. S. Lin, E. Wagor and J. H. Kaas, Brain Research 76, 145 (1974).

Discussion

Q. DR. PETTIGREW: MT gets input from the medial pulvinar. What is the input to the medial pulvinar?

A. MT gets input from what we call the medial division of the inferior pulvinar. The superior colliculus does not appear to project to the medial inferior pulvinar, but important inputs come from visual cortex. A major input is from MT. The neurons in the medial inferior pulvinar are very responsive to visual stimuli, but they do not get a direct input from the superior colliculus. There may be other subcortical inputs, but there is no evidence for this.

Q. DR. FOX: How do your results relate to Area V III that has been reported in the rhesus monkey?

A. As you know, one of Brodmann's contentions was that Area 17 of primates is surrounded by Area 18, which in turn is surrounded by Area 19. It is now popular to conceive of these three fields as a central representation of the visual hemifield surrounded by two ring-like additional representations. I think that the evidence for New World monkeys and prosimians is quite conclusive that a single area, V III or Area 19, does not adjoin V II. Instead it is quite clear that a number of visual areas adjoin V II. It would be quite surprising to me if Old World monkeys, apes, and humans have a single area, V III, adjoining V II since you donot see this in either prosimians or New World monkeys. We need to gather more evidence on anthropoid primates.

Gross Electric Recording from the Human Retina and Cortex

Lorrin A. Riggs
Brown University

I do not read the Good Housekeeping magazine regularly, but my secretary does and she gave me a clipping the other day from a current issue.

"In a few years when you go for an eye examination, you may be hooked up to a computer. Three computerized vision tests called VER, ERG and EOG are already in use in colleges of optometry. They are more efficient than the simple eye chart and should be employed in private practices soon."

"The VER (visual evoked response) uses electrodes that are attached to the patient's head and can test the vision of young children and retarded persons by asking the brain directly, 'Are you seeing more clearly or not?' as the doctor changes lenses."

"Night blindness is now being diagnosed by the ERG (electroretinogram), which records electrical responses from the retina. And the EOG (electrooculogram) can detect nystagmus, a disorder in which the eyes constantly move."

What more do I need to say? Good Housekeeping has already given my talk for me. Actually, what I will try to do today is to spell out some of the applications of electrical recording techniques, and how they may be useful in some individual cases of experimentation and clinical problems. Now there are three people attending this meeting who could give this talk better than I. One of them is your chairman, Dr. WALTERS. I had the pleasure of looking at his laboratory yesterday and seeing the wonderful equipment that he has available, and which causes great envy in my eyes. Another is JOHN ARMINGTON, who wrote the book, The Electroretinogram, that was published by the Academic Press, in 1974. Another man on the program is D. MARTIN REGAN, who wrote the book on Evoked Potentials in Psychology, Sensory Physiology and Clinical Medicine, that was published by Chapman and Hall in 1972. I might also mention that Dr. G. A. FISHMAN wrote a little pamphlet called, The

Electroretinogram and the Electrooculogram in Retinal and Choroidal Disease. It is a manual published by the American Academy of Ophthalmology and Otolaryngology in 1975. So what will follow in the next few minutes is a sort of distillation of some of the things that are to be found in those sources, with a few figures that are mostly from our own work at Brown.

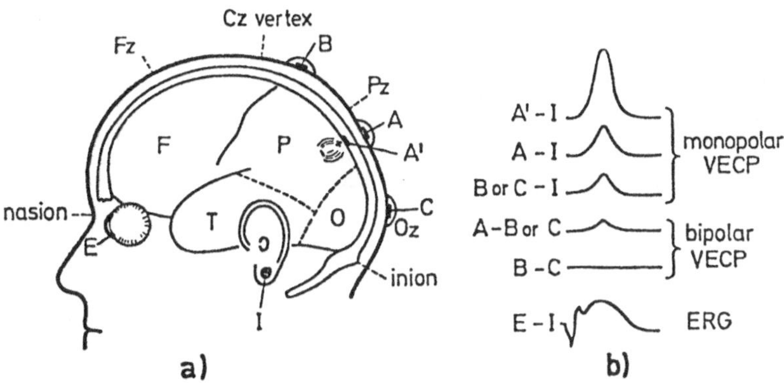

Fig. 1 Recording from electrodes on the human eye and scalp (from RIGGS and WOOTEN (1)).

Let us look at Fig. 1. This is just to remind us of the places in which we can put electrodes on the human head without causing serious discomfort, and to some extent also with animals that we wish to keep intact and to use over and over again. For retinal recording, we make use of the contact lens for the support of an electrode on the front of the eye. The other electrode can be either on the forehead or on the earlobe. The electroretinogram (ERG) is shown as the difference of potential between the electrode at E and the electrode at I. The ERG record shown in Fig. 1b has an a-wave going downward, meaning that E is negative with respect to I, followed by a positive b-wave that has two branches, a photopic early b-wave and a scotopic, later b-wave. That is a sort of prototype ERG, of course; the individual recordings vary a lot in waveform.

For measuring activity in cortical locations, we see that if a potential difference from + to - is developed here within the brain, then we can place an electrode at A' on an animal so as to pick up from the surface of the cortex the signals that arrive there by way of the optic tract. But nearly as good, fortunately, is placing skin electrodes (the standard EEG electrodes) at A for example, on the human scalp, near the site of generation of the potential. So, for example, one may record between A and I (the earlobe) and get a rough reflection of the responding that is going on somewhat distantly inside of the cortex. That is called monopolar recording of the visually evoked cortical potential (VECP). The bipolar recording, so called, is when there are two electrodes fairly near to one another on the scalp. For example, recording from A to B on the scalp, the potential difference is likely to be smaller than it is from a monopolar recording. It has certain advantages in excluding potentials coming from elsewhere on the head.

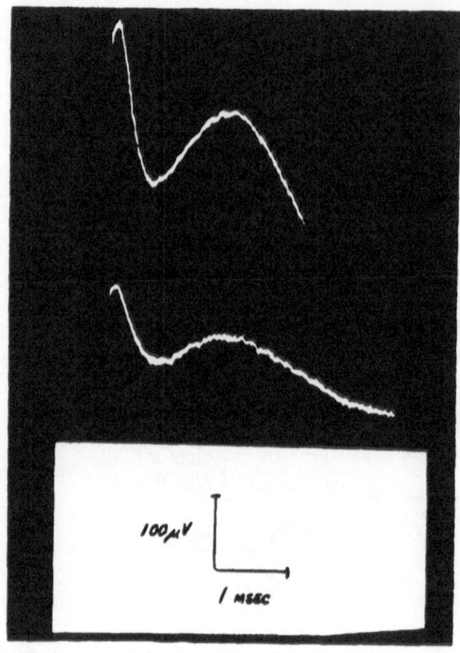

Fig. 2 Early receptor potential records from the human eye in response to intense of flashes of light. Furnished through the kindness of Dr. E. BRUCE GOLDSTEIN and Dr. ELIOT BERSON of Harvard University in 1968. Upper record is after dark adaptation and lower record after partially bleaching the retina (see GOLDSTEIN and BERSON (2)).

Now I think I will start at the very begnnning, namely at the outer segments of the receptor cells, and simply say that if we look at Fig. 2 we will see a record that is made of potentials originating there. This is the early receptor potential (ERP) described by BROWN and MURAKAMI and worked on by CONE and PAK; this particular record is by BRUCE GOLDSTEIN working with ELIOT BERSON in Boston, a record of a human ERP (2). There are various lines of evidence that show that this response originates with the photopigments in the outer segments of the receptors. For one thing, the latency is practically zero, as would be true of such a response; for another, the ERP persists even after serious damage is done to the eye as from asphyxia, anoxia, or various traumata. For another thing it is seriously knocked down by a bleach; this is predictable also from the photochemical origin of this potential. So, in short, if one wants to find out about the outer segments in a clinical case of deficiency in the photopigments, the ERP is one way to do so. It requires a very bright and very suddenly illuminated stimulus so that a strobelight of very high intensity is appropriate. It also requires that there be very good high speed registry, high fidelity if you will, in your recording system because of the very short times involved.

Now if we turn to Fig. 3, we see a record from the work of ADRIAN at the University of Cambridge way back in the 1940's. It represents a good deal of sophistication at that time on the part of ADRIAN and GRANIT and other people who were working with both human and animal ERG's. In this case we have signals arising from different wavelengths of light; and ADRIAN (3) fully realized that the photopic potentials were the early ones of these records and the scotopic potentials the later ones such as you see with the blue stimulus. It was also known at that time that the

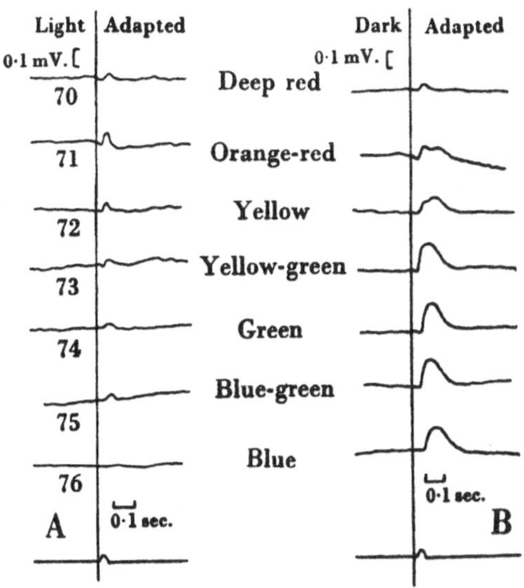

Fig. 3 The human ERG in response to various colors of light during (A) light adaptation and (B) dark adaptation as recorded by LORD ADRIAN (from ADRIAN (3)).

a-wave and the b-wave had definite properties that GRANIT summarized in some of his classic work; and I am very sorry that we will not be hearing from him at this meeting.

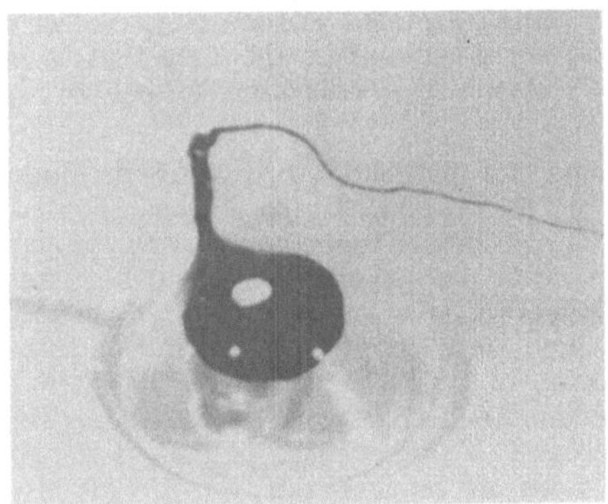

Figure 4 will show the contact lens electrode to which Dr. WALTERS referred in introducing me; it was put together in 1941, so a good many years have gone by since this happened. There is a stalk here that contains a silver disc on the inside of the contact lens. At that time I did not have a maxwellian view stimulus, so I painted an artificial pupil on the surface of the contact lens so as not to have to worry about the pupil size as one of the variables in the experiment. So it was possible, even in those days, to make measurements of the human ERG, although we were severely limited by the fact that the potentials had to be around 15 microvolts or more in order to show up on the kind of recordings that we could do at that time.

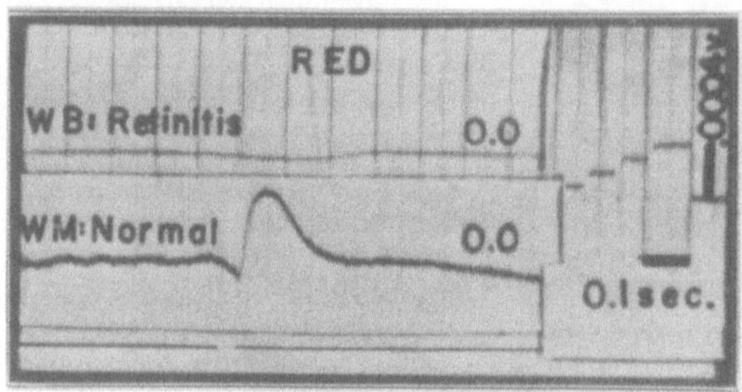

Fig. 5 Human ERG records from a patient with retinitis pigmentosa (above) and from a normal subject (below) in response to a flash of red light at the time indicated by the gap in the lowermost line (from RIGGS (5)).

Figure 5 will show comparative records in the 1950's that I got from a normal subject (below), and from a subject with retinitis pigmentosa (above). It has turned out that the ERG is a sort of test par excellence of certain hereditary degenerative diseases of the eye, especially retinitis pigmentosa. At that time KARPE and other clinical experts said that the ERG was totally extinguished by the disease. Now we know better, because we have much more sensitive means of recording; but we still know that the ERG is a valuable clinical sign, because of the fact that it is so severely attenuated even in the very early stages of the disease, before any other clinical sign is found.

I must not forget to mention that one can distinguish two different kinds of recording which we might call periodic, or "steady state" recording as opposed to single flash or "transient" recording. Up to now I have spoken only of the transient type. I think that Fig. 6 will show an example of the way in which one can do steady state recording. Here is a kind of field that we have developed in our lab to get around the troublesome problem of stray light in recording the ERG. It has been known for a long time, through the experiments of FRY, ASHER, BOYNTON and others that scattered light is a major source of the ERG as typically elicited by single flashes of light in the dark-adapted human retina. As an example of that, if you put a stimulus directly on the blind spot where there are no receptors, the ERG is just as large as if you moved the same stimulus over into the fovea or into the nearby periphery of the eye. What that means, of course, is that light scattered from the directly stimulated regions of the retina is producing a response that is mostly from the large areas of the periphery that abound in scotopic (rod) receptors. Thus if you want to work with small pattern elements or if you want to work with colors, the rod-generated signals completely obscure the signals of primary interest. To overcome this, Johnson, Schick and I (6) put together this kind of test field in which there is a counterphase alternation such that the a stripes periodically replace the b stripes and

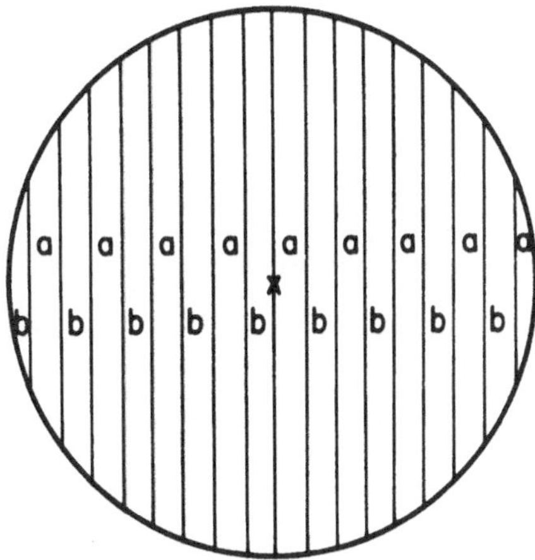

Fig. 6 Alternating stripe field in which a and b differ in luminance or wave-length, but the total area of a + b remains constant at all times. The fixation point X and the circular margin remain stationary as the pattern undergoes counter-phase alternation (from RIGGS, JOHNSON and SCHICK (6)).

vice versa. Continuous alternation of this type generates responses within the focal region of the retina. But for peripheral rods it is possible to say that the light is constant, in that the total light is constant and hence the light scattered to the periphery is constant, even though the stimulus itself is a vigorous one for the photopic or cone type of stimulation of the eye.

Now I will demonstrate for you what it is like to view the alternating stripe kind of stimulus. A rotating polaroid disc is placed in front of the projector so that the polaroid stripes on the slide are now undergoing counterphase alternation. We have some interesting perceptual experiences with this. You may think that there are black stripes and white stripes; but if you move your eye through one stripe width you realize that all the stripes are the same, they are doing the same thing. We now shift to a similar demonstraton of counterphase alternation in a checkerboard. This one is a favorite for people doing VECP work because the checkerboard turns out to be a more powerful stimulus for that than are the stripes. I think we have now had enough of that. We pity the poor subject who has to look at these things for minutes at a time in order to accumulate sufficient sizes of response. Fortunately, the alternating stripe pattern did eliminate the scattered light problem. It turned out that we could prove that the scotopic component of the ERG was negligible, and we therefore could move on to experiments on pattern vision and experiments on color contrast in which the luminance was kept constant at all times and such things as the wavelengths or the spatial frequencies were varied.

In recent years, the alternating stripe method of stimulation has increasingly been adopted for psychophysical and VECP experiments on spatial and temporal discrimination by the visual system. CAMPBELL, ROBSON and others have applied methods of Fourier analysis to the results of experiments on contrast sensitivity with sinusoidal distributions of light rather than rectangular ones. You all know, I think, the fruitfulness of that approach for the study of pattern vision. Fig. 7 will illustrate the fact that we can also apply this to wavelength experiments, and generate signals that come from a pattern of constant luminance but two different wavelengths. The slide shows VECP records obtained when these wavelengths were undergoing

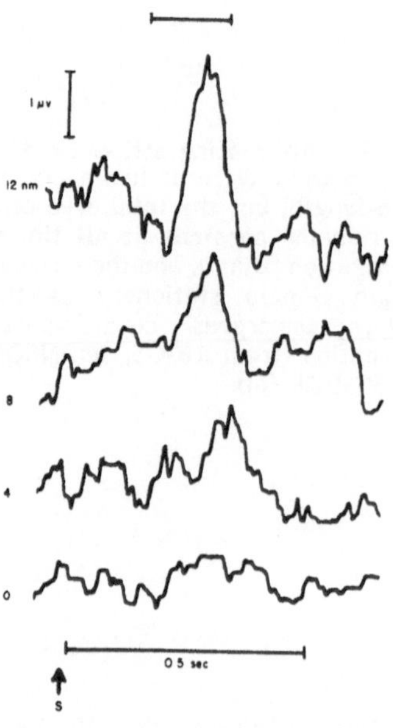

Fig. 7 The human VECP in response to transient shifts of wavelength within the stripes of a pattern. Shifts of 12, 8, 4 and 0 nm have occurred at time S. Response amplitude declines and latent time increases as the wavelength difference is reduced (from RIGGS and STERNHEIM (7)).

counterphase alternation. It shows that VECP amplitude declines as the difference in wavelength is reduced. Hence it is possible now to study human or animal wavelength discrimination functions by purely objective (VECP) techniques.

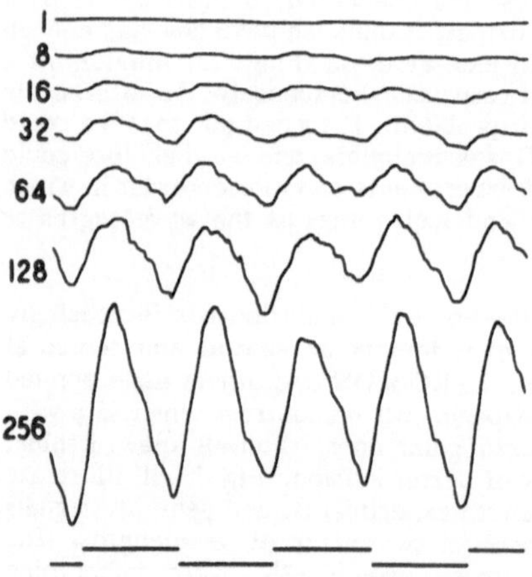

Fig. 8 Human ERG records obtained in response to counterphase alternation of red stripes with green ones of the same luminance in the pattern of Fig. 6. Several hundred cumulations are needed for achieving a satisfactory ratio of signal to noise (from RIGGS (8)).

An important question is whether ERG and VECP measurements can be made at stimulus intensities that are at or near the psychophysical threshold. The answer is that within the last 15 years the electrical tricks of signals averaging and filtering have indeed achieved this goal. Fig. 8 shows ERG signals that have been averaged over increasing numbers of cumulations during counterphase alternation in one of these electroretinogram experiments. This one happens to be the result of alternating red stimuli against blue. At the bottom of the slide we have a scale that indicates 0.4 microvolt as the size of the little displacement on the record. Thus we see that computer averaging has greatly extended the sensitivity of the electrical measures; this trick alone has taken the minimum recordable signal down from 15 μv to 0.4 μv or so.

Experiments by CAMPBELL, KULIKOWSKI, MAFFEI and others have shown that the VECP approaches zero amplitude as stimulus strength is reduced, and that zero VECP amplitude nearly coincides with the psychophysical threshold. Likewise, the ERG experiments of FINKELSTEIN, GOURAS and HOFF, and later those of HOLUB, have shown that responses can be obtained from the human eye in full dark adaptation in response to stimulus flashes in the vicinity of the absolute psychophysical threshold. This can only be achieved with extensive accumulation on the average response computer and also with ganzfeld types of stimulation. The above examples of ERG and VECP recording show that we are no longer compelled to use extraordinarily bright lights as we were in the old days with our electrical recording experiments, and we now rival the psychophysical threshold in our sensitivities.

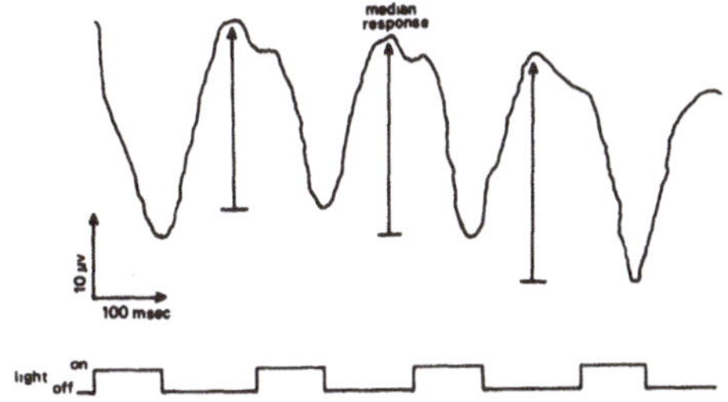

Fig. 9 ERG responses recorded with skin electrodes attached to eyeglass frames worn by a 4 year old child. Stimulus rate is 4 HZ (from DOBSON, RIGGS and SIQUELAND (9)).

Another important question is whether the ERG and VECP techniques are truly applicable, as Good Housekeeping has assured us, to young children or other difficult patients. Fig. 9 shows ERG records from a child who was wearing a pair of child's goggles with little silver plates down under the goggle where it rested against the cheek. This turns out to be quite an efficient and easy way of making an electroretinogram in a child who would not easily tolerate a contact lens. So I think I can recommend it rather highly for clinical work in which you need to have electroretinogram records but it is not easy to persuade your patient to use the contact lens electrode. This too can only be done with an averaging computer; because while the signals are relatively large (about 25 microvolts), they would not show up adequately without an averaging procedure. These records were made at

Brown University by DR. VELMA DOBSON, in a study of children about four years of age (9).

Now I think we must jump along to talk a little more about the VECP. It is fortunate that the occipital pole of the human brain is the place in which the center part of the visual field is represented. This occipital pole happens to be the most easily accessible area from the point of view of location of scalp electrodes. What this means, of course, is that the evoked potentials represent very well the stimuli that are placed at the center of vision and not too well the stimuli that are placed further out in the periphery. The center of vision is represented in the cortex by a relatively large area devoted to the small macular region of the field. So that you have a sort of natural amplifier in the brain for those potentials that are most significant and most clearly imaged within the center of the field. I should like to illustrate this point with Fig. 10, taken from a recent paper by BERKLEY and WATKINS (10). Fig. 10 illustrates a thing of optometric importance, I think, namely that the sharpness of image makes a great deal of difference in the size of VECP that

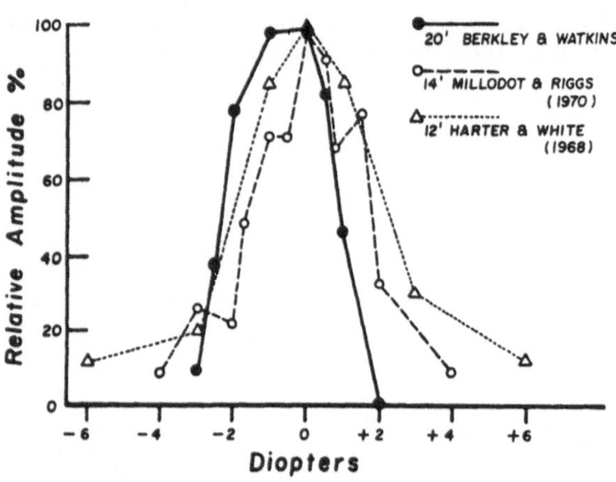

Fig. 10 Pattern resolution and refraction as measured by relative amplitude of VECP in the cat (solid line) and human (dashed line) with counterphase alternation. Human results with flashed checkerboards (dotted line) are also shown (from BERKLEY and WATKINS (10)).

one records. In this case we are concerned with stimuli that are patterned, in some cases with a checkerboard and in some cases with stripes. The various data were plotted by BERKLEY and WATKINS in this composite figure. The solid line represents their data on evoked potentials in the cat. The other lines represent the data of MILLODOT and RIGGS and those of HARTER and WHITE on human subjects. In all cases the sharpest image is at the center, where the zero diopter point is found. The other points on the abscissa represent insertion of spherical lenses in front of the eye to defocus the eye for the grating or for the stripe pattern. You will notice the rather rapid drop on either side when a plus lens or a minus lens causes the image to be less sharp on the retina. This of course gets pretty close to what the Good Housekeeping article was referring to and points out the possibility that in children at least, or in certain kinds of patients that are hard to deal with in terms of verbal responses, this may be a useful tool for refraction of the eyes.

In concluding this talk I should just point out that one must be wary, extremely wary, in one's use of VECP methods of recording. The ERG is relatively much less variable from one normal subject to another with respect to its waveform, its latency characteristics and even to some degree, the size of the potentials. The evoked potentials are notoriously undependable and variable from one subject to another, and even from one time of day to another. I think that many of us have had the experience of going home at night and swearing never to do an evoked potential experiment again. However, with sufficient care one can often use the latency of the transient evoked response or the amplitude of the periodic response to get reasonably reliable data in many cases. Perhaps it is not surprising that many different conditions of VECP recording have been used. For example, you can find periodic stimuli that alternate at the rate of 8 times per second, 14 times per second or 20, depending on which lab you visit in search of ideal frequencies. All of them will fail from time to time; and the reason they fail, I think, is that people differ so much. In working with a given subject, you must find out a favorable frequency with which to work and favorable conditions of other kinds before you will get reliable data on the evoked potential.

What I have tried to do in this talk is to outline some of the possibilities for obtaining data on vision with the ERG and VECP methods of recording. I have not said much about evaluating clinical conditions. I am sure we will hear from DR. ARDEN about that, and about the use of computers that are making the work much easier to apply in hospital experiments and clinical trials. When you visit the exquisite facilities right here in this building, you will also see how it is really done in both clinical and experimental situations.

References

1. L. A. Riggs and B. R. Wooten, In Visual Psychophysics, Volume VII/4, Handbook of Sensory Physiology, D. Jameson and L. M. Hurvich, Eds., (Springer-Verlag, Berlin, 1972).
2. E. B. Goldstein and E. Berson, Nature 22, 1272 (1969).
3. E. D. Adrian, J. Physiol. 104, 84 (1945).
4. L. A. Riggs, Amer. J. Optom. and Physiol. Optics 51, 725 (1974).
5. L. A. Riggs, Amer J. Ophthal. 38, 70 (1954).
6. L. A. Riggs, E. P. Johnson, and A. M. L. Schick, J. Opt. Soc. Amer. 56, 1621 (1966).
7. L. A. Riggs and C. E. Sternheim, J. Opt. Soc. Amer. 59, 635 (1969).
8. L. A. Riggs, J. Opt. Soc. Amer. 59, 1158 (1969).
9. V. Dobson, L. A. Riggs, and E. R. Siqueland, J. Pediatrics 85, 25 (1974).
10. M. A. Berkley and D. W. Watkins, Vis. Res. 13, 403 (1973).

Properties of Cortical Electrical Phosphenes

D. N. Rushton and G. S. Brindley
MRC Neurological Prostheses Unit, Institute of Psychiatry,
Denmark Hill, London SE5 8AF

The second patient with an inductively-linked visual prosthetic implant (F.B.) has now been tested at intervals for 5 1/2 years. Reports on this patient have considered the extent of visual cortex that gives fixed phosphenes (1); mapping methods and phosphene Braille reading (2); the degree of stability of phosphenes in the map (3); aspects of the representation of the visual field (4); and a method for generating an optimum phosphene map from a large number of observations of the relations between phosphene pairs (5).

The present paper reports observations 1) on the effect on phosphene threshold and appearance of variations of stimulus parameters; 2) of the range of phosphene brightness available and the number of discriminable steps; 3) of the colours of phosphenes; 4) of the interactions between phosphenes successively and simultaneously presented; 5) of the latency of phosphenes; and 6) of the effect of variation of stimulus parameters on pain threshold.

F. B.'s implant is activated by induction through the scalp, so the stimulating current pulses generated by the implant cannot be measured directly. They can, however, be recorded through scalp electrodes, amplified with a conventional differential amplifier, and displayed on an oscilloscope. For any given electrode, the recorded voltage pulse (a few mV in size) will be proportional to the pulse current delivered, so that proportional changes in pulse current can be measured. The scalp pulse voltage is proportional to the stimulating pulse current, but not to the voltage generated by the stimulating circuits, since the tissues behave nearly ohmically but the electrode-electrolyte junction does not. Different electrodes deliver their current in different directions in the brain, so pulse currents cannot be directly compared between different electrodes. The construction of the implant has been described in detail by DONALDSON (6).

The scalp pulses decay in a roughly exponential manner, with a time constant ranging between 1 and 2 msec for different electrodes. When long pulses are used, as in some of the threshold experiments below, they are significantly far from being rectangular. Their height is then recorded as the height of a rectangular pulse of the same duration whose area would equal that of the actual pulse.

A. Effect on Phosphene Threshold Pulse Current of Variations of Other Stimulus Parameters.

The four main variables on which the threshold pulse current for giving a phosphene may depend are the pulse length, pulse frequency, train length and pulse polarity. The polarity can not be varied for any electrode in this implant, but each of the other parameters was varied, for several phosphenes, over a wide range.

The psychophysical criterion in the experiments reported in this section was the threshold. The threshold for detecting a phosphene is likely to be influenced by the brightness and colour, and the degree of non-uniformity of brightness and colour (both in space and time) of the background against which the phosphene has to be detected. No doubt this background will vary between blind subjects, and there is no evident way of controlling it. Phosphenes are usually presented to F.B. in a regular rhythm at 1 train/sec, and this regular rhythm helps him to distinguish the phosphene from the background, and give a reliable threshold. He describes his background as being quite prominent, varying in colour and markedly non-uniform.

(i) Pulse Duration

Figure 1 gives a threshold strength-pulse duration curve, averaging results from

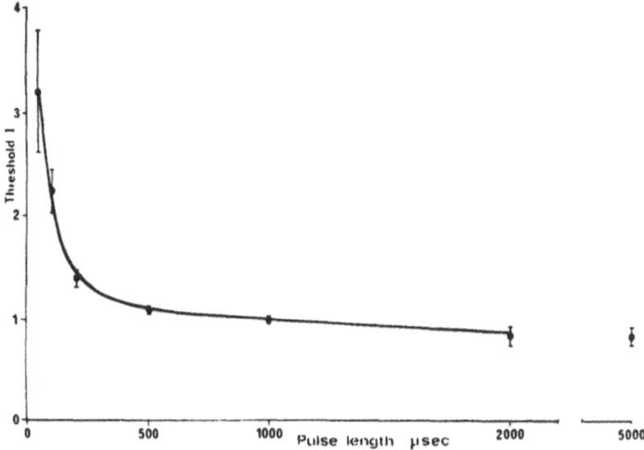

Fig. 1 Threshold pulse current (arbitrary units) is plotted against pulse duration. Experiments on five phosphenes (B2, D2, D3, H3, +) are averaged, the pulse current being scaled to 1 at pulse duration 1000 μsec. Pulse frequency 100/sec; train length 10 pulses.

experiments on five different phosphenes. Since different electrodes give different sized scalp pulses at threshold, the scalp pulse sizes were scaled so that for each electrode the mean pulse size at pulse duration 1000 μsec is 1. The threshold strength was that at which the phosphene was just visible when stimulation was given in trains each of 10 pulses, at a frequency of 100 Hz within trains, and 1 train/sec. It is seen that the chronaxie of the structures whose stimulation gives rise to phosphenes is about 200 μsec.

Since the scalp pulse voltage is proportional to the stimulating current, the charge passed during each stimulating pulse will be proportional to the product of scalp pulse height and pulse duration. Figure 2 is derived from 1 in this way.

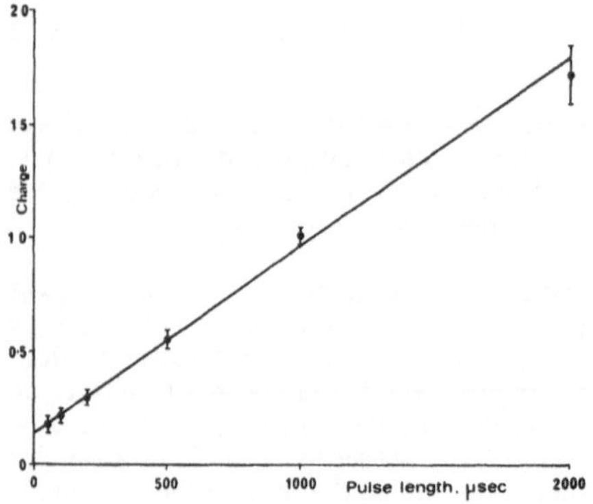

Fig. 2 Same data as Fig. 1. Charge passed (pulse duration x pulse current - arbitrary units) is plotted against pulse duration (Weiss-Hoorweg plot: Katz, 1939). Calculated regression line is shown (r = +0.993) and is extrapolated to predict phosphene threshold pulse charge for extremely short pulses. 5,000 µ sec point could not be accommodated in the Figure, but was used in calculating the regression line, and lay on it.

Charge passed (arbitrary units) per pulse at phosphene threshold is plotted against pulse duration. The relation approximates to a straight line, and has a positive intercept. The slope can be interpreted as an index of shunted current, and the intercept as the charge that would be required for phosphene threshold using an instantaneous pulse, with no time for shunting of charge. But very short pulses would imply very large pulse currents, and hence voltages, and are impracticable.

Figure 2 indicates that at 200 µsec the charge passed at threshold is about twice the intercept, while at 1000 µsec it is about six times the intercept. 100-200 sec is therefore taken as an economical pulse duration. I^2t (pulse energy) is minimal at 200 µsec, and is only 12% greater at 100 or 400 µsec.

(ii) Pulse Frequency

Figure 3 shows the variation in phosphene threshold when the frequency of pulses in a train lasting 1 sec is varied from 2 Hz to 1000 Hz. The threshold pulse current falls with increasing pulse frequency mainly between 5 and 200 Hz.

The fall in pulse current threshold with increasing pulse frequency is not very great, suggesting that a fairly low rate, not higher than 100 Hz, is desirable in order to minimize stimulation charge.

On the other hand, although this subject's phosphenes always flicker, very low pulse rates are associated with increased flicker, as discussed below, and should therefore perhaps be avoided.

(iii) Train Length

Figure 4 gives, for short bursts of stimulating pulses, the relative threshold pulse currents for different train lengths, from 1-8 pulses. The Figure shows that a train of 4 pulses gives a minimum threshold; lengthening the train prolongs the phosphene, but does not lower the threshold pulse current further.

B. Effect on Phosphene Appearance of Varying Stimulus Parameters

(i) Range of Brightness Available

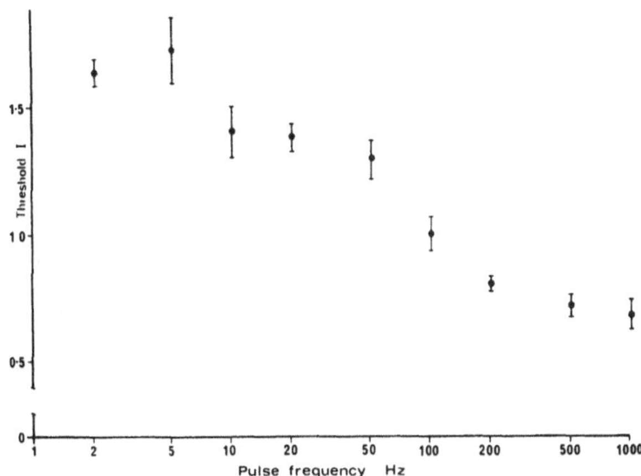

Fig. 3 The variation of threshold pulse current when 1 sec trains of pulses at different frequencies are given. Results for 3 different phosphenes have been averaged (B2, D2, H3), each experiment being scaled to mean threshold 1 at 100 Hz. Pulse length 200 μsec.

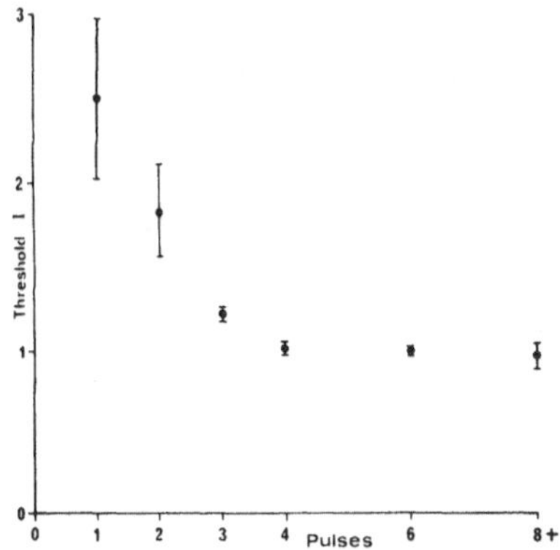

Fig. 4 Variation of phosphene threshold pulse current with train length. Average of experiments with 3 different phosphenes (B2, H3, L3), each experiment being scaled to mean pulse current threshold 1 at train length 4 pulses. Pulse length 200 μsec, pulse frequency 100 Hz, train frequency 1 Hz.

When a weak phosphene is strengthened by an increase of pulse current or pulse length, it first becomes better defined and more coloured, and then increases in brightness; further increase results in enlargement as well as further brightening. The useful range of brightness of a phosphene is limited by its threshold at one end, and the intensity at which it becomes uncomfortably bright or too large, or alternatively at which phosphene persistence occurs, at the other. Persistence of cortical phosphenes beyond the end of strong stimulation was described for their patient by BRINDLEY and LEWIN (7), and is believed to be due to a local cortical after-discharge (see also ref. 8).

A phosphene at threshold is distinguishable from the background mainly by its temporal pattern, since stimulation is intermittent, at 1 train/sec; a bright but not uncomfortable phosphene is described as "fairly bright" although a subjective scaling of the brightness range would probably be unreliable. Nevertheless, F.B. is confident that several brightnesses of a suitable phosphene (one that can with the available power of transmitter be made bright) could be distinguished. In order to test this supposition, and to attempt to estimate the number of different brightnesses that could be reliably distinguished, brightness difference thresholds were measured over the available range.

Brightness difference thresholds were measured for the phosphene K2 by the method of constant stimulus differences, using successive presentation of the phosphene at a standard and a comparison brightness. The subject was asked to state whether the comparison brightness was greater or less than the standard. "Equal" judgements were not permitted. The different standard brightnesses and differences, and the direction of the differences, were given in random order. 516 comparisons were made, and scalp pulse amplitude was measured from the oscilloscope for standard and test brightness for each comparison. The phosphene threshold was at 0.5 cm scalp pulse (2mV/cm). The strongest available stimulus using the standard apparatus (corresponding to 4.25 cm scalp pulse), gave a bright phosphene, but was not strong enough to cause persistence.

In the calculation of difference thresholds, for each reference stimulus amplitude the proportion of test stimuli of each strength that were identified as brighter is plotted against test stimulus strength. Inversely, the proportion of test stimuli of each strength identified as weaker is plotted against test stimulus strength. An example of this procedure is given in Fig. 5 for reference pulse amplitude 2 cm.

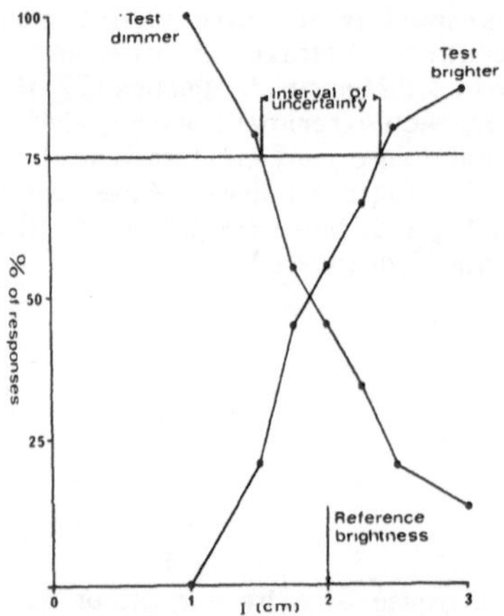

Fig. 5 Percentage of judgments "brighter" and "dimmer" for different test brightnesses when compared with a standard brightness giving a scalp pulse of 2 cm (2 mV/cm). The 75% criterion is shown, as is the interval of uncertainty for this example. The difference threshold is half of the interval of uncertainty.

The interval of uncertainty is defined as that interval within which less than 75% of judgements are correct, and the experimental points are joined by linear interpolation. The difference threshold at that reference brightness is taken as half of the interval of uncertainty so obtained.

The advantages of linear interpolation over curve-fitting procedures are that it is simple, and that it makes no assumptions about the nature of the function relating stimulus differences and differences of sensation. The disadvantage is that it does not use all of the information available in the data.

At the ends of the available range of brightness the difference thresholds were taken as the difference between the reference stimulus and the 75% correct point, obtained by linear interpolation as above, on the one side of the reference intensity where observations could be made.

The difference thresholds so obtained for each reference amplitude are plotted against reference stimulus amplitude in Fig. 6.

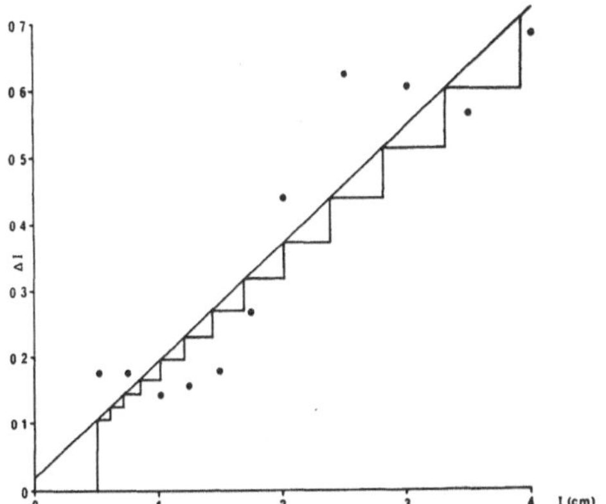

Fig. 6 Calculated difference threshold points for each reference strength between 0.5 cm (threshold) and 4 cm scalp pulses. Calculated linear regression line superimposed (slope 0.175, intercept + .021, r = 0.923). A difference threshold staircase is added, using the regression line, and starting at the phosphene threshold. Thus the length of each tread of the staircase is read from the difference threshold scale opposite it.

The points of Fig. 6 suggest that the relation between pulse current and difference threshold is not very far from linear, although many thousands of observations would probably be necessary in order to be certain of the function.

On the provisional assumption of a linear relation, a linear regression line was calculated, and is shown on Fig. 6. The difference threshold steps between absolute threshold and the brightest strength available can then be drawn, and these steps are superimposed on Fig. 6, suggesting that for K2 there at least 12 distinguishably different brightnesses.

The number of distinguishably different phosphene brightnesses is far less than a sighted observer's number of distinguishable brightnesses for an object of about the size of a phosphene ($1\text{-}2^{o}$), which may be as great as 2000 (obtained from the difference threshold curves of STEINARDT, ref. 9).

No doubt there are several reasons for this. Firstly, the pupillary and retinal light and dark adaptation mechanisms (which serve to keep the difference threshold small throughout the range of brightness) are absent. Secondly, the brightest phosphene practicable (just insufficient to cause persistence) is probably not an extremely bright object, owing to the unphysiological nature of the stimulus. Thirdly, in this subject at least, the background from which phosphenes must be distinguished near their threshold is itself rather prominent.

(ii) Clarity, Shape and Colour

Experiments to determine whether properties of a phosphene other than its brightness can be varied independently of brightness are necessarily more subjective in their nature than threshold experiments. The best that could be devised was to determine whether the subject could tell from the appearance of the phosphene which of two sets of stimulus parameters was being applied, and this only under those conditions where brightness could be matched and train durations equalized.

These conditions can be applied to variation of the pulse repetition rate in a train of fixed duration, and here it was found that variations in the pulse repetition rate can modify the appearance of a phosphene.

Thus for the phosphene B2, a train of 6 pulses at 30 Hz gave a phosphene that was more well defined in its shape and in the clarity of its margins, being distinguishable by these characteristics from the same phosphene matched for brightness and using trains at 10 Hz or 200 Hz. The appearance did not vary greatly with repetition rate, however, and 20 Hz and 50 Hz trains were indistinguishable from 30 Hz trains.

Variations of pulse length, while equalizing brightness by adjustments of pulse current, did not give reliably distinguishable phosphene appearances.

Similarly, no variation in stimulus parameters tried has succeeded, while equalizing brightness, in modifying the basic shape or the colour of phosphenes. The range of stimulus parameters tried included the range used for the phosphene threshold experiments above, and also some non-uniform trains of pulses.

Variations of train length at constant pulse frequency (100 Hz) gave a phosphene the clarity of whose outline improved up to a train length of 6-8 pulses; but phosphene duration here varied noticeably, so that comparisons for distinguishability were not appropriate.

(iii) Flicker

BRINDLEY and LEWIN (7) reported that there was for phosphenes no clear flicker fusion frequency, and probably no flicker fusion frequency at all, since their patient reported that her phosphenes always flickered whatever the pulse repetition rate. It should not be supposed from this that the phosphene appeared to flicker with each stimulus pulse, since the flicker did not vary in character between the phosphenes seen at 20, 200 and 2000 Hz. F. B.'s phosphenes, too, always flicker, regardless of pulse repetition rate. Most of this flicker is thus intrinsic to the phosphene, and independent of the stimulus frequency, as is further illustrated by the fact that a phosphene flickers when persisting after the end of strong stimulation, just as it does during stimulation. It might therefore be supposed that the "intrinsic flicker" of phosphenes is sufficient to mask any flicker that varies with pulse repetition rate.

On the other hand, stimulation with very low pulse repetition rates (5-10 Hz) gives an intermittent phosphene, and it is the loss of this intermittency which might be termed the flicker fusion frequency, and which is masked by the intrinsic phosphene flicker.

In order to determine this flicker fusion frequency, the best that could be devised was to compare the phosphene flicker using continuous stimulation with equally spaced pulses at 300 Hz in 1 sec trains with that when the pulses were arranged in groups of different lengths, pulse rate 300 Hz within a group, the groups being separated by pauses equal in length to the groups. Phosphenes from the continuous and grouped pulses were equalized for brightness, and the subject was asked to judge, from the appearance of the flicker, whether a continuous or grouped train was being used. The shortest (group + interval) length which he could distinguish from a continuous train was 30 msec, equivalent to a flicker fusion frequency of 33 Hz.

C. Colour of Phosphenes

Like the shape and position of phosphenes, their colour is to some extent constant, and like their shape it is to some extent influenced by their position in the field. Thus, horizontal bar phosphenes in the left upper field of this subject are usually yellow. Most phosphenes in the right field are greyish white, but some in the upper right field have been blue or silver. Those in the lower left field vary from yellow to grey. Phosphenes of any colour if sufficiently weakened become colourless or grey.

It is difficult to estimate the brightness or saturation of the colours, and how closely they resemble what a sighted person would describe as the colours named: but F. B. had sufficient vision to perceive colours for the first 25-30 years of his life, and is confident and consistent in his colour naming.

Table 1 gives the described colours of a selection of 21 phosphenes recorded on several occasions over a period of four and a half years.

There is no evidence that the colour of any phosphene can be reliably altered, independently of its brightness, by altering the stimulus parameters. Negative results were obtained on altering pulse length, pulse frequency and train length, and on delivering short trains with their pulses spaced at different irregular intervals. During the course of these experiments, however, the colours of a phosphene (B2) were repeatedly recorded in detail, and although yellow remained its usual and principal colour, the depth of colour varied, and its tinge or edging of other colours varied from time to time, including green-gold, silver, pink, blue-green and brick red. These changes in colour occurred apparently at random.

D. Interactions Between Phosphenes Successively and Simultaneously Presented

(i) Successive Presentation

When phosphenes are presented sequentially, so that their pulse trains occur one after the other without overlap and separated by about 100 msec or more, the phosphenes do not interact much, and probably do not interact at all. This method of presentation was used in all of the mapping observations reported previously.

(ii) Overlapping Trains of Non-Synchronous Pulses

When a pair of phosphenes is presented "simultaneously", with the trains of pulses overlapping in time, but with their pulses not necessarily synchronous, then they may interact. Typically, each becomes brighter and larger, spreading towards the other. If the phosphenes are close, the space between them may be filled. This interaction occurs whether or not the electrodes are close or the stimulation pulses synchronous, and can occur between phosphenes in opposite half fields.

Observations of these effects are necessarily descriptive, but interactions between phosphene pairs of varying separation in the field can be roughly compared, as can interactions between phosphene pairs whose pulses are synchronous be compared with those between phosphene pairs whose pulses are non-synchronous. Four examples of such observations are given below.

Table I (part 1)

Phosphene	25.4.72.	31.11.72.	12.2.73.	12.5.73.	29.1.74.	1.6.74.	19.11.74.	6.5.75.	5.11.75.	4.5.76.	9.11.76.
A1	-	red	grey	grey	white	grey	white-grey	grey	grey-white	grey-white	-
A2	-	mauve-pink	yellow	red+ yellow	yellow	yellow	yellow+ pink	grey	red+ yellow	grey	yellow
A3	-	silver	silver	grey-white	blue	blue	silver	-	-	grey	grey
C1	yellow	blue-grey	grey	grey	white	grey	blue	grey	blue	-	-
C2	yellow	grey-yellow	yellow	yellow	yellow	yellow	yellow	blue-grey+ red	grey-yellow	grey	grey-white
C3	yellow	grey	grey	grey	grey	grey	grey-white	grey-white	-	white	grey
D1	yellow	blue	blue	grey	-	-	-	-	-	-	-
D2	white	red+ yellow	yellow	yellow	grey	yellow	yellow+ pink	blue-grey+ red	grey	grey	grey
D3	yellow	grey-white	grey-yellow	white	grey	grey	yellow	grey-yellow	-	grey	grey
I1	grey-yellow	-	-	grey	-	white	blue	blue	grey	grey	grey-brown
I3	grey-yellow	grey	-	grey	grey-blue	grey-blue	grey	grey-white	-	grey	grey-blue
J3	yellow	grey-white	white	grey	grey	blue-grey	grey	grey-white	-	grey-brown	grey

Table I (part 2)

Phosphene	25.4.72.	31.11.72.	12.2.73.	12.5.73.	29.1.74.	1.6.74.	19.11.74.	6.5.75.	5.11.75.	4.5.76.	9.11.76.
K2	yellow	white	yellow	grey-white	yellow-brown	grey	grey-yellow	grey-yellow	yellow-brown	grey	yellow cream
K3	yellow	grey	grey-white	grey	grey	grey	grey	grey	-	grey	grey
L1	yellow	blue	grey	grey	grey	grey	grey-blue	grey	grey	-	-
L2	yellow	yellow	yellow	yellow	yellow	yellow	yellow	yellow	yellow	grey-yellow	yellow
L3	grey-white	grey	grey-white	grey	grey	grey	grey	grey	-	grey-yellow	grey
M1	-	blue	blue	blue	blue	-	blue	grey	grey	-	-
M2	-	yellow	yellow	brown	yellow	yellow	yellow	yellow	grey-yellow	grey-yellow	yellow
M3	-	white	grey-white	grey	grey	grey	grey	-	-	grey	grey-white
O2	yellow	yellow	yellow-white	yellow	yellow	red-yellow	grey	red-yellow	grey	yellow	-

Table 1 Colors of phosphenes. At each of 11 visits between April 1972 and November 1976 the color of each phosphene was noted. No systematic observations of color were made during the visits of February 1972, June 1972, July 1972 or September 1973. Data for a selection of 21 phosphenes are given. A blank is left where a phosphene was unobtainable. (-) between two color names indicates a color mixture. (+) between two color names indicates that different parts of the phosphene were of different colors.

E2 and D2. Synchronous pulses, fairly close phosphenes.

When given separately, they were horizontal bar phosphenes 1 1/4" x 1/4", apparent distance from the eye 18", one above the other and separated by a gap of about twice their own thickness. When put on together they gave a single larger phosphene, still horizontal in shape. At no brightness was there any gap between.

L2 and O2. Synchronous pulses, medium distance.

Horizontal bar phosphenes 1 1/2" x 1/4" one above the other, 18" from the eye, and separated by a space of about 4-5 x their own thickness, when seen separately. When put on together, they remained as separate phosphenes, but a gossamer-like infilling appeared between them. The non-adjacent edges of the phosphenes did not move, but the adjacent edges spread a little towards each other, increasing the thickness of the phosphenes by about half.

D2 and A1. Non-synchronous pulses, medium distance, opposite half fields.

Horizontal bars each about 1 1/2" x 1/2", 18" from the eye, in line side by side and separated by a gap equal to the length of one of them, when seen separately. When put on together, they appeared as one long phosphene, with no gap at any brightness setting.

M1 and I4. Non-synchronous pulses, fairly distant, opposite half fields.

Singly; M1 squarish, 3/4" across, 14" from the eye. I4 vertical oval 3/4" major axis, 10" from the eye. Separation from each other 7" when the difference in their distance was taken into account. When put on together, they both increased in size, and became rather less well defined, the enlargement (roughly a doubling of area) being towards each other. They retained their different distances from the eye.

This phenomenon was noted, but to a lesser extent (that is, over a smaller range) by the patient of BRINDLEY and LEWIN (7), and a probably similar phenomenon is implied by the descriptions of interaction quoted by DOBELLE and MLADEJOVSKY (10). Any difference in degree may be attributable to the larger size of F. B.'s phosphenes.

If a phosphene display is to be useful, it is desirable as far as possible to minimize such interaction. There is no evidence in the case of F. B. that it can be altered independently of phosphene brightness by alteration of the stimulation parameters. It is certainly lessened by a reduction of stimulus strength, but this in its turn would reduce the available range of brightness.

(iii) Trains of Pulses Synchronous at Two Electrodes

When a pair of phosphenes is presented simultaneously with synchronous pulses, the interaction described above is similar to that seen using non-synchronous overlapping trains. The two conditions cannot easily be directly compared using the same pair of phosphenes in both cases, since the design of the implant and its driver requires that two phosphenes shown together with non-synchronous pulses in overlapping trains must be in different rows of the matrix, while two phosphenes shown together with synchronous pulses must be in the same row. If these conditions are not fulfilled, then other unwanted phosphenes will be shown as well.

Another interaction occurs only when synchronous pulses are used, and that is a modification of threshold. Current that is subliminal for a phosphene at one electrode will lower the threshold current for a phosphene at a neighbouring

Table II

Phosphene	Threshold singly	Conditioning strength	Test threshold	Separation in field	Separation on array	Threshold interaction
L2 H2	1.3 2.5	0.8 -	- 1.5	9	4	+
E2 F2	2.0 3.0	1.0 -	- 2.0	1 1/2	4	+
F2 E2	2.0 2.0	1.5 -	- 0.5	1 1/2	4	+
A1 B1	6.0 2.0	3.0 -	- 1.1	1 1/4	4	+
B1 A1	2.0 6.0	1.0 -	- 3.0	1 1/4	4	+
L2 M2	0.6 0.5	0.3 -	- 0.3	1	4	+
M2 L2	0.5 0.6	0.3 -	- 0.25	1	4	+
G2 E2	0.8 1.0	0.5 -	- 0.1	1 1/4	4	+
E2 G2	1.0 0.8	0.5 -	- 0.3	1 1/4	4	+
L2 G2	0.5 0.8	0.3 -	- 0.3	2 1/2	4	+
G2 L2	0.8 0.5	0.3 -	- 0.35	2 1/2	4	+
K2 L2	3.0 1.8	2.0 -	- 1.0	9	8	+
L2 K2	1.8 3.0	1.0 -	- 2.5	9	8	+
M2 F2	1.0 2.0	0.5 -	- 2.0	2	12	
F2 M2	2.0 1.0	1.5 -	- 1.1	2	12	
L2 E2	0.5 1.0	0.3 -	- 1.0	1 1/4	12	
E2 L2	1.0 0.5	0.6 -	- 0.5	1 1/4	12	
B1 D1	3.0 5.0	1.5 -	- 5.5	1 1/2	26	
D1 B1	5.0 2.0	3.0 -	- 2.0	1 1/2	26	

Table 2 Phosphene threshold interactions using synchronous pulses at pairs of electrodes of varying separation. Electrode separations are given in mm, since the electrode array is constructed on a metric mesh. Phosphene separations are given in inches, since all such observations are made in inches. Stimulus parameters are: pulse duration 200 μsec; train length 10 pulses; pulse frequency 100 Hz; 1 train/sec.

Table III

Phosphene	Threshold singly	Conditioning strength	Test Threshold	Separation in field	Separation on array	Separation of pulses	Threshold interaction
H2	2.0	1.0	-	3/4	6	800	-
B4	0.9	-	1.0				
B4	0.9	0.7	-	3/4	6	800	-
H2	2.0	-	2.0				
A1	6.0	4.5	-	2	8	800	-
K3	4.0	-	4.0				
K3	4.0	2.5	-	2	8	800	-
A1	6.0	-	6.5				
H2	1.9	1.5	-	1 1/4	8	800	-
G4	0.5	-	0.5				
G4	0.5	0.3	-	1 1/4	8	800	-
H2	1.9	-	1.8				
H5	0.5	0.3	-	2 3/4	18	800	-
L3	3.0	-	3.5				
L3	3.0	2.0	-	2 3/4	18	800	-
H5	0.5	-	0.5				
G4	0.6	0.4	-	1 1/4	8	400	-
H2	2.2	-	2.2				
H2	2.2	1.5	-	1 1/4	8	400	-
G4	0.6	-	0.6				
G4	0.6	0.3	-	1 1/4	8	200	-
H2	2.4	-	2.3				
H2	2.4	1.5	-	1 1/4	8	200	-
G4	0.6	-	0.6				
L3	3.4	2.0	-	2 3/4	18	200	-
H5	1.5	-	1.7				
H5	1.5	1.0	-	2 3/4	18	200	-
L3	3.4	-	3.5				

Table 3 Phosphene threshold interaction experiment using overlapping trains of non-synchronous pulses, and varying the pulse separation. Pulse separations are given in µsec separating the beginning of the pulse at the first electrode from the beginning of the pulse at the second electrode. Measurements and stimulus parameters other wise as in Table 2.

electrode, but only if the pulses are synchronous, or at least overlap. It is undetectable for 200 μsec pulses such that the end of each pulse at the first electrode corresponds with the beginning of each pulse at the second electrode (Tables 2 and 3).

The threshold interaction experiments of Tables 2 and 3 were done as follows. The threshold scalp pulse size for obtaining each member of a pair of phosphenes was measured, and one member (conditioning electrode) was then set below threshold. Stimulation strength at the second (test electrode) was then slowly increased until a phosphene was just seen. Conditioning stimulation was then switched off (and the phosphene disappeared) leaving the scalp pulse from the test electrode alone to be measured. For most pairs chosen, each member was used in turn as the conditioning electrode.

Table 2 gives the results using pairs with synchronous pulses, in an experiment designed to estimate the distance over which this threshold interaction can occur. Pairs are arranged in order of increasing electrode separation in Table 2, and it is seen that threshold interaction was found for all pairs with an electrode separation of 8 mm or less, and was not found where the electrode separation was 12 mm or more. The interaction occurred regardless of the separation of the phosphenes in the field (Table 2, column 5).

Table 3 gives the results of a corresponding experiment using pairs of electrodes stimulated with overlapping trains of non-synchronous pulses. The shortest pulse timing difference that could conveniently be used was 200 μsec, so that using 200 μsec pulses, the pulse to the second electrode began at the same instant as the end of the pulse at the first electrode. No threshold interaction was found at any of these electrode pairs.

These results indicate that the threshold interactions occur as a result of the addition of currents at the site of stimulation, rather than by the convergence of activity from two separate pools of stimulated neurones, each alone of sub-threshold size. Such physiologic convergence would not be expected to be abolished by an asynchrony of synaptic activity of 200 μsec. Further, the strength of physiologic convergence might be expected to be related to the physiological 'closeness' of the two stimuli (their separation in the field) rather than their physical closeness (their separation on the electrode array) as is found to be the case.

E. Latency of Phosphenes

In order for phosphenes to be useful for the rapid transmission of information, it is desirable that they appear and disappear as rapidly as possible.

The latency of a phosphene has been measured by recording F. B's simple reaction time to the beginning of visual cortical stimulation, by requiring him to press a switch as soon as he saw the phosphene appear. The phosphene (B2) was presented for 0.1 sec at intervals varying randomly from 5-10 sec. It was arranged to be of moderate brightness, and it caused no non-visual sensations.

The distribution of reaction times was recorded on an averaging computer (Datalab Biomac 1000) by triggering the sweep at the onset of the train of stimulation pulses and adding a small voltage step to the average store through the closing of the response microswitch. The form of the record obtained is shown for one run (32 responses) of "visual" and one run of auditory reaction times in Fig. 7, a and b.

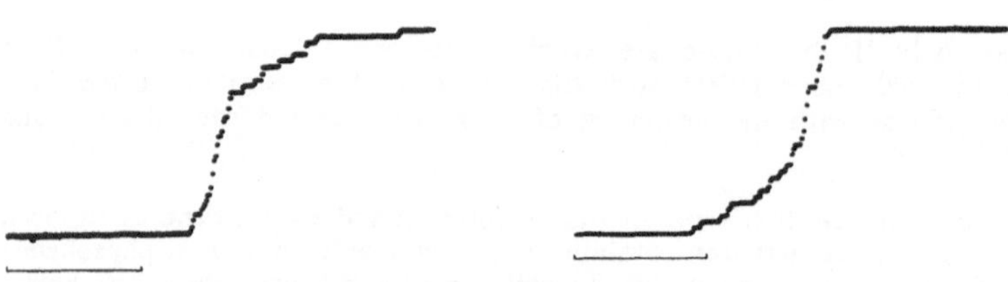

Fig. 7 Reaction times experiment. Simple R.T. to the onset of phosphene or a tone. For each trial, response switch closure makes a small step on the trace. Each run comprises 32 trials, each trial being separated by an interval varying randomly from 5-10 sec. Calibration 100 msec. a. Phosphene (B2). b. Tone (1 KHz) in earphones.

The "visual" reaction times obtained in this way were compared with the simple reaction time to the onset of a tone (1 KHz) presented for 0.1 sec through earphones, using the same response recording system.

The results are given in Table 4, where it is seen that the "visual" and auditory

Table IV

Condition	Mean R. T. (msec)	S.E.M. (msec)
Visual	176.5	8.9
Auditory	160.7	4.2
Visual	171.0	7.4
Auditory	164.9	4.2
Visual	175.4	5.9

Table 4 Reaction times to the onset of a phosphene ('visual') or a tone (auditory). Each result is the mean of 32 trials.

reaction times obtained are very similar. Auditory R. T.'s seem to be slightly shorter and less variable. In no case in Table 4 does the difference between a "visual" and auditory mean reaction time reach statistical significance (Student's t test), and nor do the pooled results for mean auditory and "visual" R. T.'s differ significantly. The impression of a difference in standard error is not statistically significant (analysis of variance).

These results are in contrast to the familiar physiological class reaction time experiment, where the visual R. T., to the onset of moderately bright light is about 20-30 msec longer than the auditory R. T., to the onset of a click or tone. This difference becomes less with increasingly bright lights, but it does not disappear, and is believed to be mainly the effect of retinal latency, and to a lesser extent the latency of conduction in the visual pathway.

"Visual" stimulation using cortical phosphenes bypasses both of these delays, and since the difference between modalities is almost abolished, the delay added by the latency of onset of a phosphene cannot be nearly as great as the delay in the retina and visual pathway.

Measurement of the rate of disappearance of phosphenes formally by reaction time has been impracticable, because this experiment would require the stimulus to be presented in a continuous train several seconds long, and of unpredictable duration. Continuous trains of this length result in fading and persistence of the phosphene. However, if a shorter train of pulses is played on a loudspeaker while being transmitted, F. B. reports that the end of the train and the disappearance of the phosphene are simultaneous.

F. Other Sensations Resulting From Electrical Stimulation

As was the case with the first patient, F. B. reports two different kinds of non-visual sensation. The first is tingling in the scalp, felt whenever several electrodes in the same row (i.e., delivering their pulses synchronously) are strongly activated. This effect is no doubt caused by current returning through the extracranial indifferent electrodes and stimulating nerve fibers in the scalp.

The second is deep pain in the head, which is felt only when certain individual electrodes are strongly activated. Those that have given deep pain are indicated by underlining in Fig. 8 a and b. The pain is lateralized according to the side of the electrode array, but does not vary reproducibly for different electrodes in the same array; it is referred to the forehead.

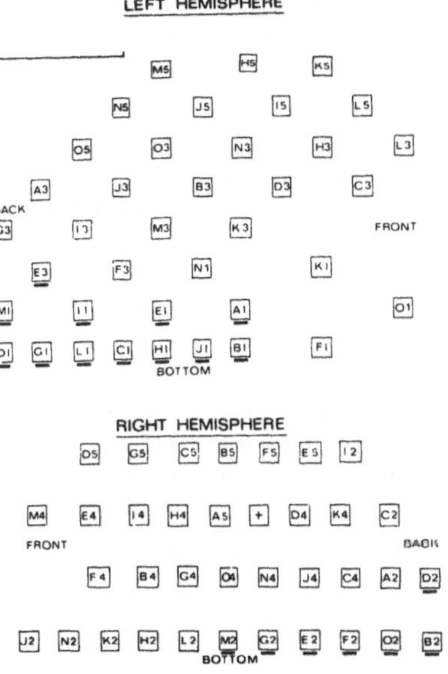

Fig. 8 Electrode arrays. Electrodes that have given deep pain in the head on strong stimulation are underlined. It is seen that they all lie in the bottom rows towards the back of the arrays, suggesting that the pain fibers concerned run in the tentorium cerebelli. Calibration 1 cm. a. Right hemisphere electrode array. b. Left hemisphere electrode array.

The distribution of the electrodes giving deep pain along the lower edge of the arrays suggests that the pain arises from stimulation of nerve fibers in the tentorium cerebelli, rather than fibers running with the calcarine artery or its branches.

When the usual stimulus parameters are used (0.2 msec pulses at 100/sec in 0.1 sec trains repeated every 1 sec) the threshold for deep pain lies well above phosphene threshold, except in the case of H1, where the thresholds were about equal.

In order to obtain as wide a range of phosphene brightness as possible without pain, it is desirable to set the stimulus parameters so that the phosphene threshold is as low as possible, and the pain threshold as high as possible. It was therefore necessary to measure the variations in threshold pulse current for pain with stimulus parameters, as was done for phosphene threshold.

Figure 9 gives the threshold-pulse length curve for deep pain, averaged from

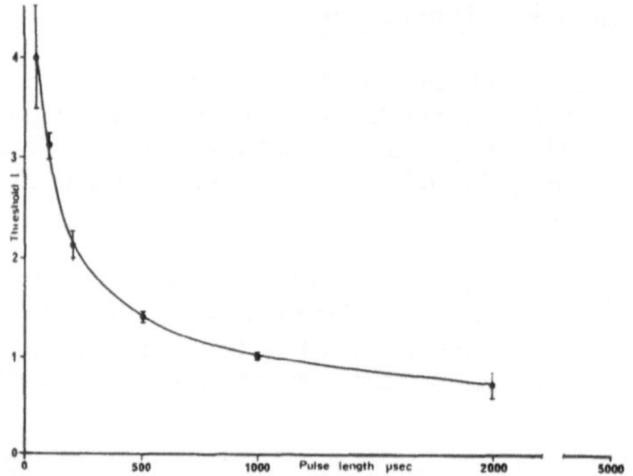

Fig. 9 Threshold pulse current (arbitrary units) is plotted against pulse duration at the onset of deep pain. Experiments on three electrodes are averaged (B1, C1, L1), the pulse current being scaled to 1 at pulse duration 1000 µsec. Pulse frequency 100/sec; train length 10 pulses.

results obtained from three pain-giving electrodes. At the longest pulse lengths used (2000 and 5000 µsec) the thresholds for phosphene and pain were typically about equal, but for shorter pulse lengths the relative threshold for pain became progressively higher. The chronaxie for pain is here about 750 µsec.

Figure 10 gives the threshold-log pulse frequency curve for pain plotted in the same way as Fig. 3, being averaged results from two different electrodes, each scaled so as to give a mean threshold of 1 at 100 Hz.

Figure 11 gives the threshold pulse current-pulse number curve for the onset of pain when short trains containing different numbers of pulses are compared, results being averaged from experiments on two electrodes each, scaled to give a mean relative threshold of 1 for long trains.

There is little fall in threshold pulse current with increasing train length, particularly beyond four pulses.

These results suggest that, in order to separate thresholds for phosphene and

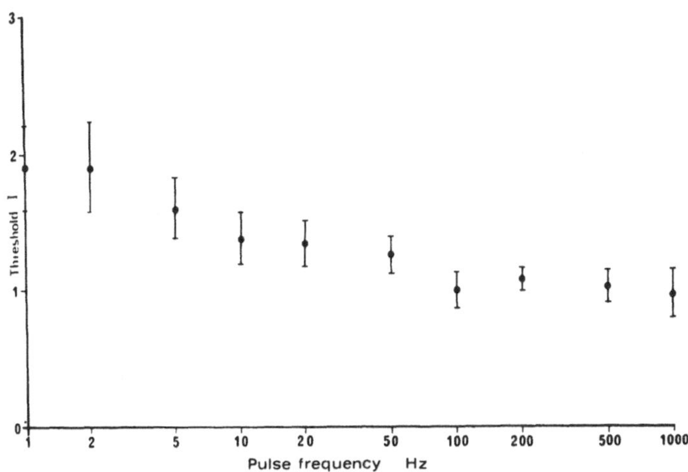

Fig. 10 Variation of threshold pulse current for the onset of deep pain when 1 sec trains of pulses at different frequencies are given. Experiments on two electrodes are averaged (C1, L1), each experiment being scaled to mean threshold pulse current 1 at 100 Hz. Pulse length 200 μsec.

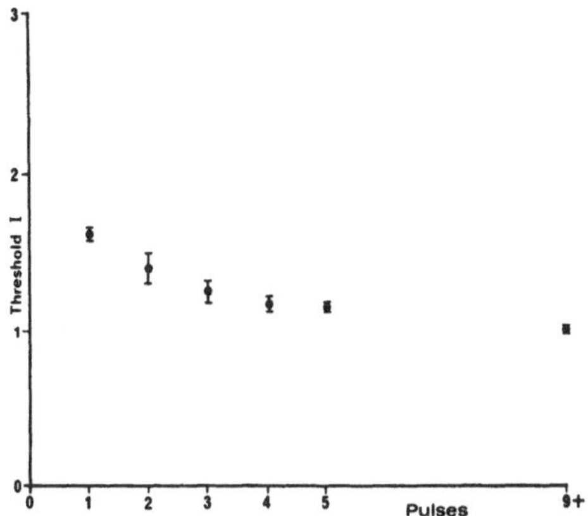

Fig. 11 Variation in threshold pulse current for the onset of pain with train length. Average of experiments with 2 different phosphenes (C1, L1) each experiment being scaled to mean pulse current threshold 1 at train length 9 + pulses. Pulse length 200 μsec, pulse frequency 100 Hz, train frequency 1 Hz.

for pain as far as possible, it is advantageous to use pulses that are short (e.g., 100-200 μsec duration), but it is not advantageous to reduce pulse frequency, and the largest threshold separation may be obtained at about 200 Hz. Similarly, the curve for train length for pain is flatter, for trains of three pulses or less, than that for phosphene, so that there is no advantage here in using trains of less than 4 pulses.

G. Anodal Stimulation

One electrode of F. B.'s implant was connected so as to deliver anodal pulses. It therefore had to be driven through its own receiver, rather than through the row-and-column matrix (6).

The phosphene evoked at this electrode resembled its neighbours in the lower left field in size, colour, approximate threshold current, and threshold variations with varying stimulus parameters. It might have been expected to be different, by analogy with the difference between the effects of anodal and cathodal stimulation of the motor cortex (11).

Unfortunately, the receiver failed after only two months, before tests on it were complete.

Anodal stimulation would need to be used in an implant employing tantalum pentoxide capacitor electrodes (12, 13).

Phosphenes from anodal and cathodal pulses could not be directly compared at the same electrode in the present implant, but the impression that they would be similar is in line with the results of DOBELLE and MLADEJOVSKY (10), some of whose patients at operation compared the appearance of phosphenes elicited by anodal and cathodal pulses (as well as biphasic pulses with either anodal or cathodal phases leading), and found no consistent difference.

POLLEN (8) has compared the neuronal responses to anodal and cathodal stimulation of the cat's primary visual cortex. Individual neurones were found to differ in the threshold and latency of their response to two types of stimulus, and in their tendency to after-discharge. There was, however, no evident consistent pattern in the direction of these differences, from one cell to another.

H. Conclusions

The parametric experiments suggest that short trains of 3-4 pulses, each of about 200 µsec, are appropriate for minimizing the energy required and nearly minimizing the charge required for eliciting a visible phosphene. Intermittent stimulation with 0.1 sec on/0.9 off has been used for many experiments, and while useful for presenting simple patterns, it would not be a rapid way of transmitting information, with an effective "frame rate" of only 1/sec.

Previous experiments (2) have suggested that such trains repeated so as to give a frame repetition rate of between 2 and 5 frames/sec will give the most rapid transfer of information. Train repetition rates higher that this not only result in reduced information transmission, but would also be likely to result in fading and persistence of phosphenes. Systematic experiments on these phenomena have not been pursued in the present subject, because the stimulus conditions that give rise to them are those that are considered likely to result in cortical damage. POLLEN (8) has shown that in the cat's visual cortex the threshold pulse current for cortical afterdischarge falls with increasing train length, apparently well beyond the train length at which phosphene threshold pulse current reaches a minimum.

References

1. G. S. Brindley, P. E. K. Donaldson, M. A. Falconer and D. N. Rushton, J. Physiol. 225, 57P (1972).
2. G. S. Brindley and D. N. Rushton, Trans. Am. Acad. Ophthalmol. Otolaryngol. 78, OP 741 (1974).

3. D. N. Rushton and G. S. Brindley. In: Physiological aspects of Clinical Neurology, ed. by F. C. Rose (Blackwell, Oxford, 1977) p. 173.
4. G. S. Brindley and D. N. Rushton. In: Functional Electrical Stimulation: applications in Neurological Prostheses, ed. by J. B. Reswick, (Dekker, New York, 1978), p. 261.
5. B. S. Everitt and D. N. Rushton. A method for plotting the optimum positions of an array of cortical phosphenes. Biometrics, In press (1978).
6. P. E. K. Donaldson, I.E.E. Proc. 120, 281 (1973).
7. G. S. Brindley and W. S. Lewin, J. Physiol. 196, 479 (1968).
8. D. A. Pollen, Brain Behav. Evol. 14, 67 (1977).
9. J. Steinardt, J. Gen. Physiol. 20, 185 (1936).
10. W. H. Dobelle and M. G. Mladejovsky, J. Physiol. 243, 553 (1974).
11. J. E. C. Hern, S. Landgren, C. G. Phillips and R. Porter, J. Physiol. 161, 91 (1962).
12. D. L. Guyton and F. T. Hambrecht, Science 181, 74 (1973).
13. P. E. K. Donaldson, Med. Biol. Eng. 13, 131 (1974).

VIII. Development of Visual System Function

Development of the Eye and Retina of Kittens

By John T. Flynn, Thomas E. Flynn, Duco I. Hamasaki,
O. Navarro, Vesna G. Sutija and Gail S. Tucker
William L. McKnight Vision Research Center
Bascom Palmer Eye Institute
Miami, Florida 33136

We studied development in the peripheral visual system of kittens and we shall present here some of the data collected on the optics, morphology, and physiology of the eye and retina during the first several postnatal months. Our studies have emphasized the critical developmental period in an effort to determine how signals being sent to higher visual centers are affected by the developmental state of the eye and retina.

We shall begin with some of the measurements which deal with optical properties of the kitten eye (see also ref. 1). Axial lengths of several eyes were measured during the first week after birth. The mean axial length at that time is 10.3 mm (Table 1, TUCKER, unreported data). Axial length increases only slightly during the next three weeks, and the kitten enters the critical developmental period with an eye having a mean axial length of 13.3 mm. This is approximately one-half that of the adult eye (21.5 mm). We find very little change in axial length during the critical developmental period (4-8 weeks of age) but from the 8th week the axial length of the eye increases rapidly.

Correlated with the short axial length, newborn kittens have a small corneal curvature. The radius of curvature of the young kitten cornea is approximately 5.0 mm, and this radius increases steadily with increasing age (Table 2, J. T. FLYNN, unreported data). At 14 weeks of age, the radius is approximately 7.0 mm, and this flattening of the cornea represents approximately a twenty diopter (20D) decrease in the refractive power of the front surface of the cornea.

The refractive error of the eye was measured by retinoscopy at different ages, and the changes in the refractive error are shown in Table 2. At the time of eye opening the kitten is extremely hypermetropic (+8.66D). As with the axial length of the eye, there is very little change in refractive error during the next three weeks. Thus, the kitten enters the critical developmental period with an eye which is still very hypermetropic (+8.34D). During the next four weeks, the refractive error decreases. However, at the end of the critical developmental period, the eye is still hypermetropic by +4.73D. Thereafter there is a steady decrease in the refractive error, but we have not determined when the adult value is attained.

These observations on the optical components of the eye show that kittens enter and pass through the critical developmental period not only with an immature but also a rapidly changing optical system. It is apparent that the size and quality of the image on the retina is different in kittens, and how this influences responses of the retina must be taken into consideration in the interpretation of physiological measurements.

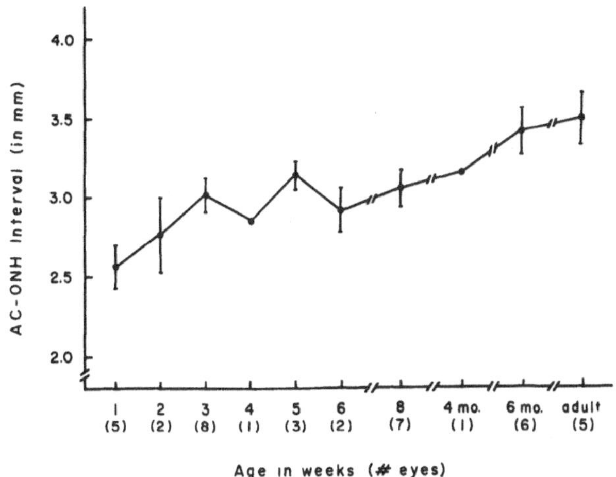

Fig. 1 The distance in mm (Y-axis) between the area centralis and optic nerve head are given for kittens of various ages (X-axis). These values are corrected for shrinkage due to fixation. The number of eyes examined are given in parentheses on the X-axis.

We have also reported on the anatomical development of the area centralis in kittens of various ages (2). The area centralis (region of highest ganglion cell density) cannot be identified in histological sections in kittens younger than 5-6 days. When it is first identified it is situated approximately 2.5 mm from the temporal edge of the optic nerve head (Fig. 1). With increasing age, there is an increase in the distance between the area centralis and the optic nerve head, and the adult position is achieved at 5-6 weeks of age. Thus the area centralis reaches the adult location early in the critical developmental period at a time when the eye is still increasing in diameter. How the distance between the area centralis and optic nerve head increases, and how the size of the eye and retina increase with increasing age is of considerable interest to us. Studies are currently in progress to examine these questions.

What is the condition of the photoreceptors during early postnatal days? About 20 years ago, ZETTERSTROM (3) reported that the b-wave of the ERG could first be recorded from 6 to 10 day old kittens. She also showed that it required about 9 to 10 weeks for the amplitude of the b-wave to become adult-like. We have repeated these experiments and have obtained similar results from the intact kitten. We also found that the stimulus intensity required to elicit a criterion amplitude b-wave was high in young (3 week) kittens, and that it required 10 to 12 weeks for the thresholds to reach adult levels. However, when the stimulus intensity required to elicit a criterion firing rate was determined for retinal ganglion cells, we found that at 3 weeks, the threshold was only 0.26 log units higher in kittens. We thus concluded that when the ERG is recorded from the intact kitten, both the size and maturation of the retina will affect the threshold and the amplitude of the response.

In order to overcome the confounding effects of size and degree of maturation, we have developed an isolated-retina preparation on which to carry out our experiments. In this preparation, the size and region of the retina to be studied can be controlled. In our study, we have followed changes in stimulus intensity thresholds and in the amplitude of the LRP (late receptor potential isolated by glutamate) in kittens of various ages. These measurements were obtained from 6 mm circular pieces of temporal retina which included the area centralis (4).

A small LRP can first be recorded at 9 days of age but the stimulus intensity must be considerably higher than that necessary for adult retinas (Fig. 2). Thereafter there is a rapid decrease in the threshold, and the thresholds are no longer significantly different from that of adults by 18-21 days of age. The maximum amplitude LRP which can be elicited from this preparation follows the same course; at 9 days, the maximum amplitude elicited is 15µ V, and by 23-26 days, the amplitude has attained adult value of about 250µ V.

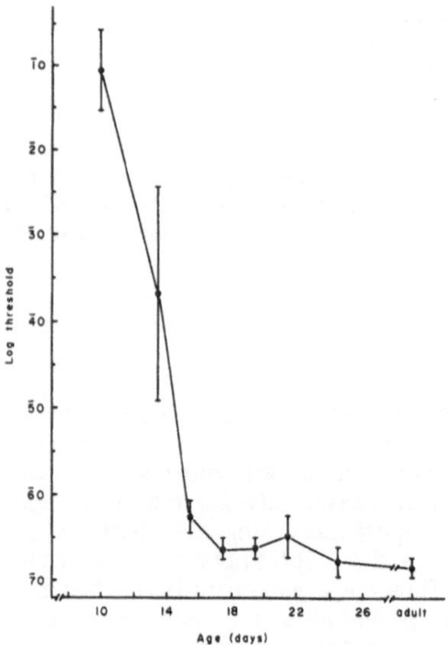

Fig. 2 Decrease in stimulus-intensity threshold with increasing age. The means and standard error of the means of the stimulus intensity required to elicit a 15µ V LRP are shown.

These retinas were fixed in glutaraldehyde immediately after recording the LRP, and the length of the outer segments and the density of the photoreceptors were determined. We have found very good correlation between the thresholds and amplitude of the LRP and outer segment length (4).

These observations show that rods are essentially adult-like by three weeks of age. Thus any differences noted in the properties of neural elements downstream from the photoreceptors cannot be attributed to the immaturity of the photoreceptors after 3 weeks of age.

We have also been following the development of physiological properties of the receptive fields of ganglion cells in kittens of various ages. At three weeks of age (5), the center-surround organization of the receptive field is present. The angular size of the receptive field center is twice as large in kittens as in adults but, when the linear size of the receptive field center is calculated and consideration is taken of its shorter axial length, the linear size of the kitten's receptive field center is approximately the same as that of adults. The threshold for eliciting a criterion firing level is 0.26 log units higher in 3 week old kittens. The maximum firing rate is significantly weaker and the intensity-response curve is flatter in kittens than in adults. The response to flickering light is also poorer in 3 week old kittens. Responses of kitten ganglion cells had a longer latency (in spite of the shorter distance between the retina and recording electrode), but no evidence was found at the retinal level for "fatigue" or "sluggish" responses when the stimulus was adequate. We conclude that the ganglion cells of 3 week old kittens have some properties which are adult-like and others which are significantly different from those of adults. From these observations it is apparent that kittens enter the critical developmental period with ganglion cells whose physiological properties are not fully developed.

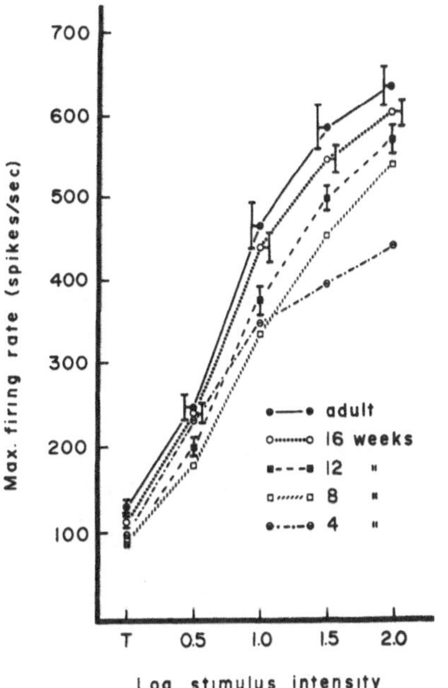

Fig. 3 Average intensity response curves for kittens and adult cats. The stimulus intensity is shown on the abscissa in log threshold units and the maximum firing rate on the ordinate. The standard errors of the means are shown for the 12 and 16-week old kittens and for the adult.

The question then arises as to when intensity-response and flicker functions attain adult values. To answer this question, kittens from two litters were tested at 4 week intervals using the same experimental and stimulus conditions. The intensity-response curves for 4, 8, 12 and 16 week old kittens and for the adult are shown in Fig. 3. The standard errors of the mean are shown for 12 and 16 week old kittens and also for the adult. At the end of the critical developmental period, the intensity-response function is still significantly different from that of the adult, and it is only at 16 weeks that the values cease to be significantly different from those of adults.

The responses of ganglion cells to stimuli of different temporal frequencies determined for the different age groups are shown in Fig. 4. At 4, 8 and 12 weeks of age, the responses are significantly weaker than those of the adult. At 16 weeks, the means do not differ significantly from adult values for lower frequencies but at frequencies higher than 10 Hz, the means are still significantly different from those of adults. At 20 weeks, the data are essentially adult-like.

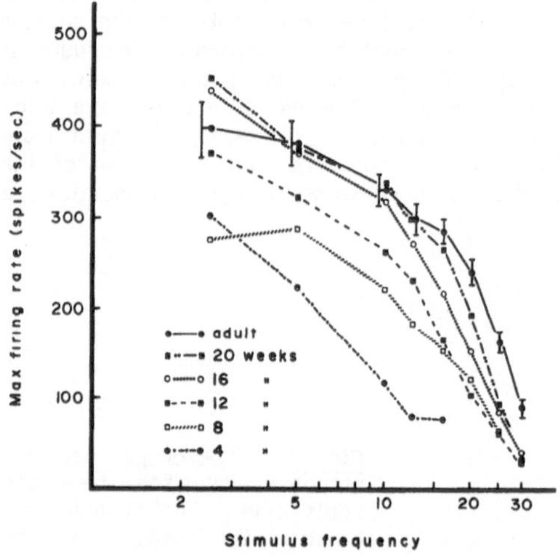

Fig. 4 Stimulus frequency-response curves for kittens and adult cats. The stimulus frequency is shown on the abscissa and the maximum firing rate on the ordinate. The standard errors of the means are shown for the adult data.

We have followed changes in the latency of the responses elicited by a stimulus 2.0 log units above threshold in these same kittens (Fig. 5). As noted for three week old kittens, the latencies were considerably longer in young kittens (4 weeks of age), and the distribution of the latencies does not become adult-like until the 12th week. These observations on the latencies correlate well with myelination of the optic nerve fibers. MOORE, KALIL and RICHARDS (6) reported that 80% of the fibers are myelinated at 4 weeks and the adult proportion is attained at 12 weeks of age.

Having demonstrated the course of development of the intensity-response and

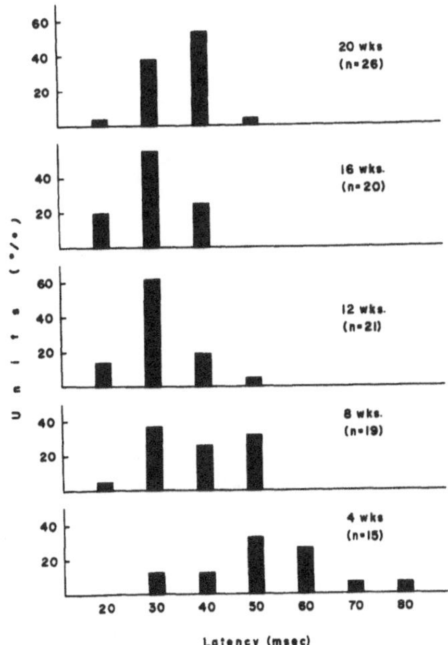

Fig. 5 Distribution of latencies for on-center units from kittens of different ages. The latency was measured for the response elicited by a stimulus intensity 2.0 log units above threshold.

flicker functions, we then asked whether visual experience was necessary for these functions to develop. To examine this question, we lid-sutured (monocularly) kittens when they were two weeks old and tested them when they were at least one year old. The experimental conditions were identical to those used for the developmental studies. The data obtained from the visually-deprived (closed) eye for the intensity-response (Fig. 6A) and the flicker (Fig. 6B) functions did not differ significantly from those obtained from the normal (open) eye of the visually deprived cats or from normally reared adult cats. The observations therefore show that visual experience is not necessary for these functions to develop.

We have also been following the development of X- and Y- retinal ganglion cells in kittens. A contrast reversal (alternating phase) stimulus was used, and responses were recorded with the stimulus placed at different positions in the receptive field. In some of the cells recorded from 3 week old kittens, the stimulus could be placed at a position in the receptive field where there was very little change in the firing rate with each contrast reversal (position 0, Fig. 7). This position was called the null position, and cells which showed a null position were classified as X-cells (7). When the stimulus was displaced slightly ($\pm$ 4.3') from the null position, each contrast reversal resulted in a large change in the firing rate.

In another group of cells from 3 week old kittens, a null position could not be

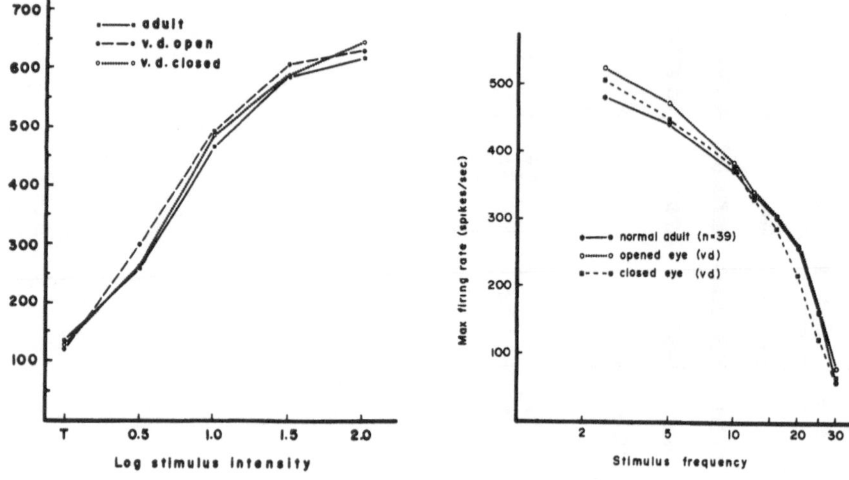

Fig. 6 A. Average intensity-response curves for units recorded from the open and closed eyes of two visually-deprived cats, and for units recorded from two normally-reared cats. The stimulus intensity in log threshold units is shown on the abscissa, and the maximum firing rate on the ordinate. B. Stimulus frequency-response curves for units recorded from the open and closed eye of the three visually-deprived cats and for the units recorded from three normally reared cats. The stimulus frequency is shown on the abscissa and the firing rate on the ordinate.

found in spite of careful movement of the stimulus across the receptive field. There was a position, however, which was symmetrically situated, where each contrast reversal elicited equal amplitude responses (position 0, Fig. 8). Cells which failed to show a null position were classified as Y-cells, and the position in the receptive field where equal responses were obtained was called the equal response position.

These responses from 3 week old kittens were qualitatively very similar to those obtained from adult cats, and it was thus easy to classify the cells into the X and Y groups. However, the majority of the units in the 3 and 4 week old kittens responded as did the unit shown in Fig. 9. A "completely null" position could not be found for this cell, and it should thus be classified as a Y-cell. However, the response at the equal response position, and also the responses elicited by the stimulus placed at different positions in the receptive field, were very different from those of adult Y-cells. In fact, the response pattern resembled more those obtained from cells in adult cats which showed a null position (8).

Because the majority of the units in young kittens showed this "mixed" type of response, it was difficult to classify them simply into X-(linear) and Y-(non-linear) cells. We thus used a method which was devised to classify the responses of adult cats (9). From the responses recorded with the stimulus placed at the null or equal response position, a ratio of the firing rates was calculated which represented the

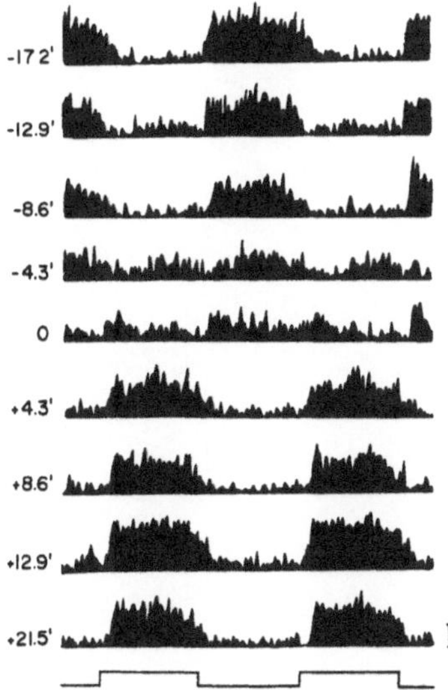

Fig. 7 Average response histograms obtained from an on-center X-cell of a 3 week old kitten. The position of the stimulus in the receptive field is shown to the left of each histogram (in minutes of arc from "0"). Calibration = 100 spikes/sec, time between contrast reversal = 500 msec.

degree of non-linearity of spatial summation. When a frequency plot of the non-linearities (null ratio) was made, the units from the adult cats fell into two groups which were shown statistically to arise from two populations (Fig. 10, adult).

The same calculation was made for all of the units obtained from kitten retinas, and the distribution of the null ratios is shown in Fig. 10. At 3 and 4 weeks, the majority of the cells show a high degree of non-linearity, and the percentage of Y-cells (70.0% at 3 weeks and 59.5% at 4 weeks) was significantly greater than that found in adult cats (38.0%). At 5-6 weeks, the percentage of Y-cells (33.8%) was not significantly different from that of the adult, but other response properties of the cells were still significantly different. The responses elicited by the contrast reversal stimulus were essentially adult-like at 12 weeks.

Summary

We have presented here some of the data we have collected on the optical, morphological and physiological condition of the eye and retina of kittens at various ages. Our results show that the photoreceptors are adult-like by the time the kitten enters the critical developmental period. However, the components of the optical system and the morphological and physiological condition of other neural elements in the retina are still immature and undergo considerable maturational changes during the course of the critical developmental period. Many of the physiological functions

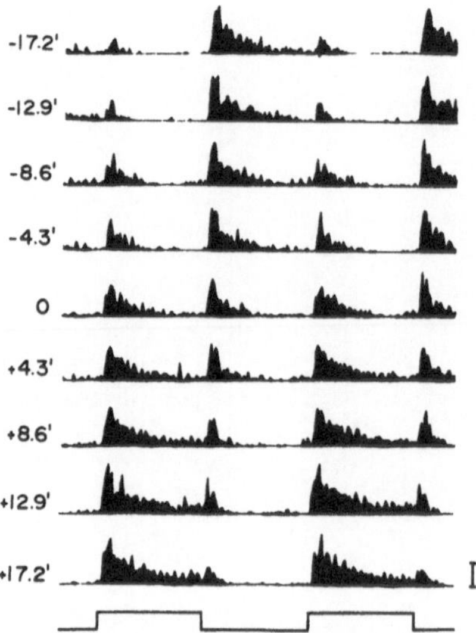

Fig. 8 Average response histograms obtained from an on-center Y-cell of a 3 week old kitten. The position of the stimulus in the receptive field is shown to the left of each histogram (in minutes of arc from "0"). Calibration = 100 spikes/sec, time between contrast reversal = 500 msec.

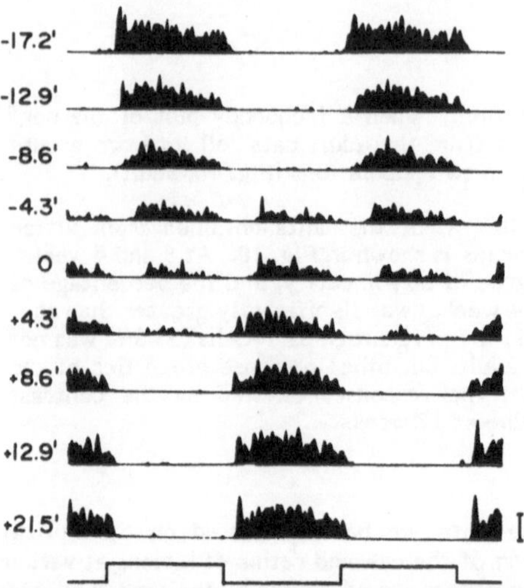

Fig. 9 Average response histograms obtained from an on-center cell of a 3 week old kitten. The position of the stimulus in the receptive field is shown to the left of each histogram (in minutes of arc form "0"). Calibration = 100 spikes/sec, time between contrast reversal = 500 msec.

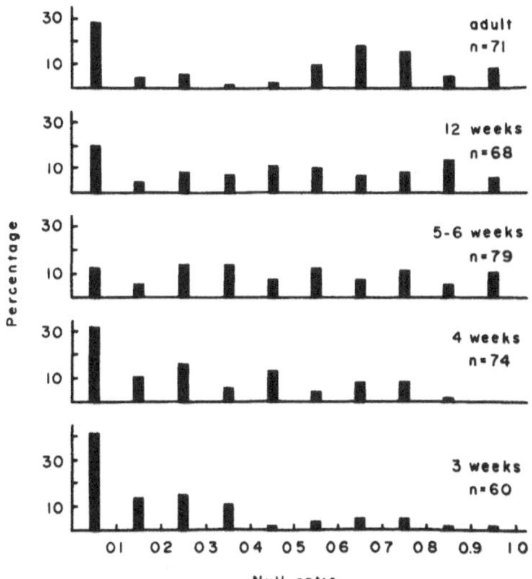

Fig. 10 Distribution of the null ratios for units obtained from kittens and adult cats.

were found to not be completely adult-like at the end of the critical developmental period.

Acknowledgements

We wish to thank Norman Bradford, Howard Cohen, Andrew Labbie, Joseph Muroff, Jose Perez and Thomas Robertson for their assistance in collecting the data reported here. We also wish to thank Barbara French for her photographic work.

These experiments were supported by PHS Grant #EY00376 from National Eye Institute to DIH, by an NIH post-doctoral fellowship #EY05021 to VGS and by a grant from the Florida Lions Eye Bank to GST.

References

1. F. Thorn, Vision Res. 16, 1145 (1976).
2. G. S. Tucker, "Light microscopic analysis of the kitten retina: Postnatal development in the area centralis" (Submitted).
3. B. Zetterstrom, The effect of light on the appearance and development of the electroretinogram. Acta Physiol. Scand. 35, 272 (1955).
4. G. S. Tucker, D. I. Hamasaki, N. Bradford and J. Muroff. "Physiological and anatomical development of the kitten retina" (In preparation).
5. D. I. Hamasaki and J. T. Flynn, Vision Res. 17, 275 (1977).
6. C. L. Moore, R. Kalil and W. Richards. J. Comp. Neurol. 165, 125 (1976).
7. C. Enroth-Cugell and J. G. Robson, J. Physiol. 187, 517 (1966).
8. D. I. Hamasaki and V. G. Sutija, "Classification of cat retinal ganglion cells into X- and Y-cells with a contrast reversal stimulus". (Submitted).
9. D. I. Hamasaki and V. G. Sutija, "Development of X- and Y-cells in kittens" (Submitted).

Postnatal Development of the Human Lateral Geniculate Nucleus

T. L. Hickey
School of Optometry/The Medical Center
University of Alabama in Birmingham
Birmingham, Alabama 35294

Introduction

During the past few years a great deal of time and effort has been spent studying the structural, functional and behavioral effects of visual deprivation. Although we are now well aware that deprivation early in life can drastically change the course of visual system development, much remains to be learned about the mechanisms that underlie these changes. In addition, very little is known about the normal development of the visual system.

A critical period of visual system development has been defined for both the cat (1,2) and monkey (3). However, we are still not certain at what point such a period occurs in the development of the human visual system. In an attempt to define the critical period for human visual system development several studies have examined clinical records of patients deprived of visual experience by one condition or another at some point during their early years of life (4,5,6). In addition, psychophysical studies using similar individuals have also been carried out (7,8,9). Much of this data suggests that the peak of the critical period in humans occurs within the first two years of life; however, the period of susceptibility may continue until the child is four or five years of age.

Critical periods of development are by no means confined to the visual system. It is generally accepted that there are periods of time during both the prenatal and postnatal development of an organism when it is particularly sensitive to outside influences. Such periods usually occur at a point in time when the organism is undergoing rapid growth (10). This raises the question of whether or not the critical period in the development of the visual system also corresponds in time to a period of rapid growth. Since most of the work on normal development has been done using the cat, it is interesting to relate the time course of normal development to the time course of the changes seen in visually deprived cats.

GAREY, FISKEN and POWELL (11) have shown that there is a period of rapid cell growth in the cat lateral geniculate nucleus (LGN) during the first 4 weeks of life. Other work suggests that growth is occurring throughout the central visual pathway during this same time period. CRAGG (12) studied the development of synapses in both the visual cortex and LGN of the cat. Synaptic development in the two structures was found to proceed at about the same rate, with synapses in the geniculate developing approximately two days earlier than those in the visual cortex. In both areas there was a rapid increase in the number of synapses between 8 and 37 days after birth. Thus, it appears that, at least in the cat, the onset of the critical period coincides in time with the end of the most rapid phase of geniculate cell growth and the period of most synapse formation in both the geniculate and visual cortex. Although not as well defined, it is also interesting to note that the end of all geniculate cell growth coincides in time reasonably well with the end of the critical period. So far such a relationship has been clearly defined only for the cat. However, one can hypothesize, with some justification, that such a relationship may hold for a number of animals, including man. If, as some of the above findings suggest, there is a relationship between the time during which the visual system is growing and the time during which it is most susceptible to outside influences, then it should be possible to define the critical period in the development of the human visual system by first defining the period during which growth is occurring. The findings presented here and elsewhere (13) show that there are two, partially overlapping, periods of postnatal cell growth in the human LGN. For cells in the parvocellular layers of the geniculate there is a period of rapid growth that ends around 6 months after birth. However, cells in the magnocellular layers continue to grow rapidly until one full year after birth and do not reach adult size until at least the end of the second year of life. Such a time course of development is remarkably similar to previous clinical and psychophysical estimates of the critical period in the human (4, 5, 6, 7, 8, 9).

Methods

Brain tissue was obtained from 53 humans ranging in age at the time of their death from newborn to 40 years. All brain tissue was collected during normal autopsy procedures at two local hospitals. Only brains showing no gross pathology or past history of neurological abnormalities were included in the present analyses. Upon receipt, all brain tissue was placed in 10% buffered formalin for at least two weeks. Blocks of tissue containing one LGN were then dehydrated in a series of alcohols, embedded in celloidin, sectioned frontally at 40μ and stained with cresyl violet. Geniculate cells showing a well defined nucleolus were drawn at a magnification of X1000 (oil) with the help of a Zeiss camera lucida. Cells were sampled from a corresponding part of the binocular segment of each geniculate (see Fig. 1). The details of this procedure have been described previously for cell measurements in the cat LGN (14). The cross-sectional area of each cell outline was later determined using a Numonics Graphics Calculator (Numonics Corporation, North Wales, Pennsylvania). One hundred cells were measured in each of the four parvocellular and two magnocellular laminae of every LGN studied. In all cases the slides from which the cells were drawn were coded so that no information was available concerning the age of the tissue.

Results

The overall change in human geniculate cell size from birth to adulthood is shown in Fig. 2. On the left side of the figure are plotted the mean cross-sectional areas for cells in each of the six geniculate laminae of fifteen newborn humans. A similar graph on the right shows the mean values for nineteen humans ranging in age from two to forty years. Comparisons between cell sizes in the latter group showed no significant increases in cell size beyond two years of age. At birth all geniculate

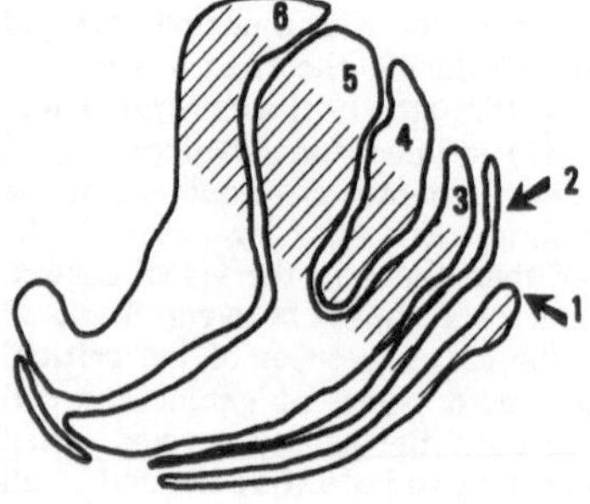

Fig. 1 Camera lucida drawing of a coronal section through the LGN of a newborn human. Following the typical system for numbering layers 1 and 2 contain large cells (magnocellular) while layers 3-6 contain smaller cells (parvocellular). Layers 1, 4 and 6 are contralaterally innervated. Layers 2, 3 and 5 receive input from the ipsilateral eye. The crosshatched region shows the area over which cells were sampled.

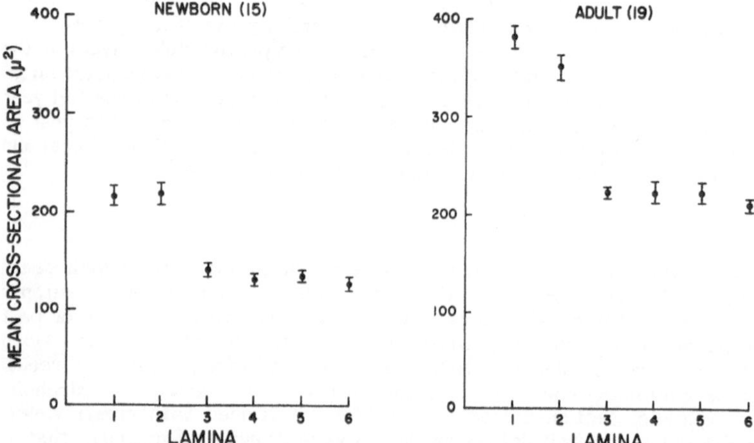

Fig. 2 Average cross-sectional areas (±S.E.) of LGN cells in fifteen newborn humans (left) and nineteen older humans (right). Each point represents the average of either fifteen or nineteen means for each of the laminae. Since all geniculate cells have completed their growth by the end of the second postnatal year the graph on the right presents data from humans that ranged in age from 2 to 40 years. The graph on the left represents data from humans ranging in age from newborn to 1 month postnatal.

cells are, on the average, 60% of their adult size (range across all laminae was 57-64%). Cells in both magnocellular layers are, on the average, equal in size. In the parvocellular layers the mean cell sizes are similar for layers 4-6. Layer 3 cells are, however, larger than those in layer 4 ($p < .001$), 5 ($p < .03$) and 6 ($p < .01$). In the adult, magnocellular layer 1 cells are considerably larger ($p < .01$) than layer 2 cells while in the parvocellular layers no differences were found between layers 3, 4 and 5. Layer 6 cells were, however, significantly smaller than those in layers 3 ($p < .01$), 4 ($p < .05$) and 5 ($p < .01$).

The results of the cell area measurements made in humans ranging in age from newborn to 40 years of age are summarized in Fig. 3 for each of the two magnocellular laminae and in Fig. 4 for each of the four parvocellular laminae. Two

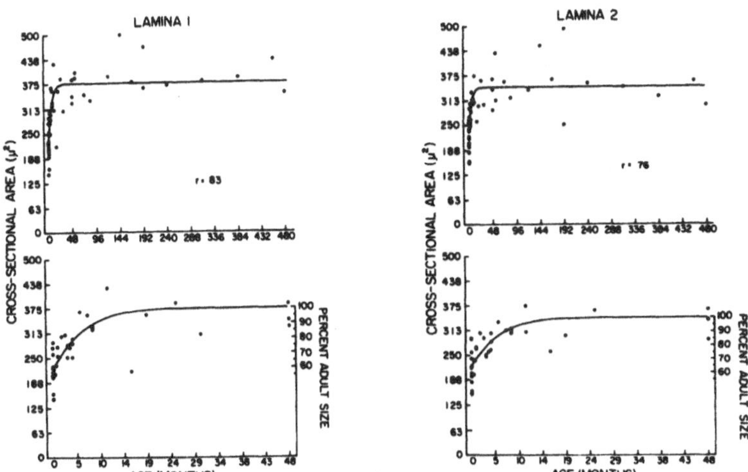

Fig. 3 Changes in cross-sectional area of human LGN cells as a function of age. In this figure one pair of graphs is shown for each of the two magnocellular laminae. The scale on the left ordinate of both graphs shows the absolute size of the cells while the ordinate on the right side of the bottom graph shows the size of the cells in terms of percent of adult size. The top graph shows all of the data collected while the bottom graph shows, in more detail, the data for only the first 48 months of life. The degree of correlation between the data points and the fitted curve is shown in the top graph of each pair. For all graphs each data point represents the mean of 100 cell measurements.

graphs are presented for each of the laminae; the top graph in each pair shows all of the data collected while the bottom graph shows, in more detail, the data covering only the first 48 months of life. In all graphs each point represents the mean of one hundred cell measurements. As in other sub-human animals (15,16,17), the variability in the average size of geniculate cells obtained from similarly aged humans is tremendous. It is important to point out, however, that although between subject variability is high the differences between mean cell sizes in the different geniculate laminae of a given human are quite predictable. Therefore, in an attempt to show an average change in cell size during development, a Gompertz growth curve has been fitted to the data using a least squares criterion. The goodness of fit of the curve, as indicated by the correlation coefficient, is shown in the upper graph in each pair.

In all cases geniculate cell growth in the human is either complete or nearing completion by the end of the second postnatal year. However, all geniculate cells do not reach their final size at the same time. The larger layer 1 and 2 cells require at least 24 months before reaching adult size (see Fig. 3). The smaller parvocellular layer cells, on the other hand, reach adult size near the end of the first year (see Fig. 4). In all cases the first six months of life is characterized by rapid geniculate

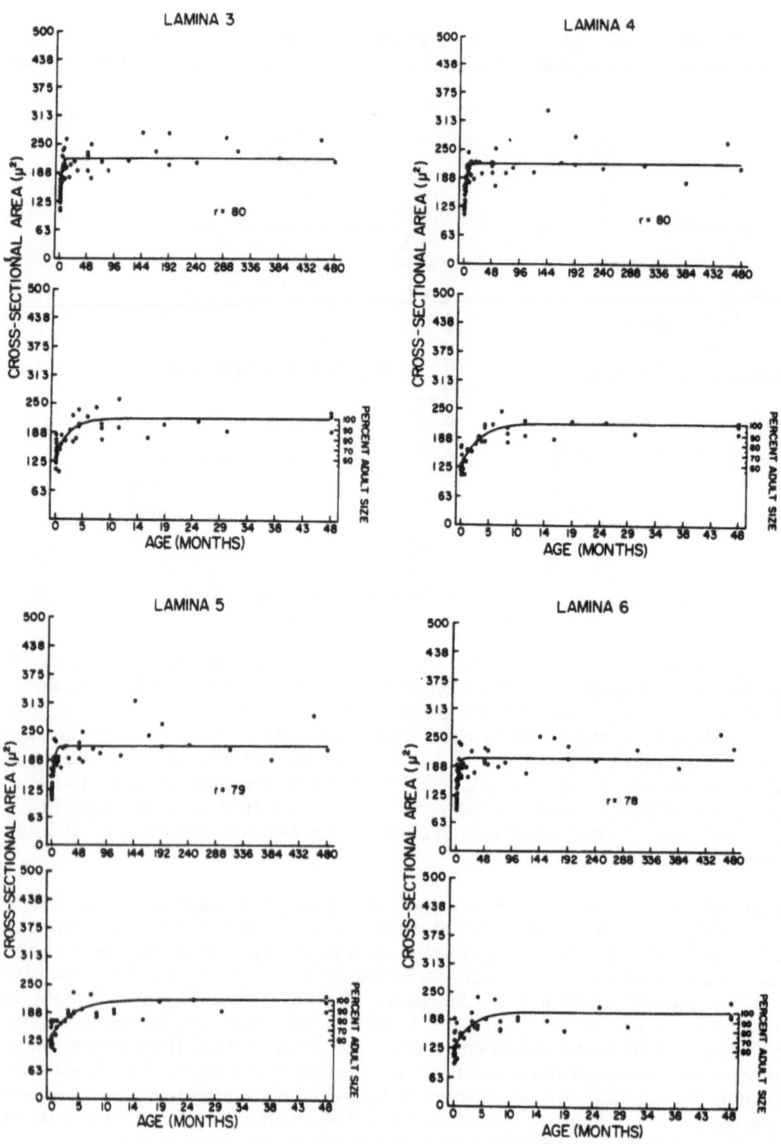

Fig. 4 Changes in cross-sectional area of human LGN cells as a function of age. In this figure one pair of graphs is shown for each of the four parvocellular laminae. All other conventions are identical to those described in Fig. 3.

cell growth. While this period of rapid growth ends around 6 months for the parvocellular layer cells it continues for as long as 12 months for the magnocellular layer cells.

The right hand ordinate in each of the bottom graphs shows the percentage of cell growth completed at a given age. When changes in cell area are viewed in terms of number of μ^2 of cross-sectional areas, all geniculate cells appear to grow at about the same rate during the first few months of life. Since the final size of the parvocellular layer cells is less than that of the magnocellular layer cells, however, this represents a greater percent increase in size for the former. For example, by six months most parvocellular layer cells are approximately 95% of their adult size. Magnocellular layer cells do not reach this point in development until near the end of the first year. Due to the variability in the data it is impossible to be sure that all geniculate cell growth is complete by a certain point in time. However, even with the variability it is clear that all measurable changes in cell size are over by the end of the second postnatal year and, even if geniculate cells continue to slowly increase in size beyond the second year, there still exists a marked difference in the time required for cells in the magnocellular and parvocellular layers to reach adult size.

Discussion

If there is a close relationship between periods of cell growth and periods of increased susceptibility then the most crucial period in the development of the human geniculo-cortical system (and most likely the visual system as a whole) would be the first 24 months of life. The present data further suggest that there could be two phases to the critical period for the human visual system. For the parvocellular layer cells the period of susceptibility would be confined to the first year. For the magnocellular layer cells, however, this period would extend into the second postnatal year. If, as suggested by some of the earlier findings, the point of peak sensitivity in the critical period corresponds in time to the end of the most rapid phase of cell growth, than parvocellular layer cells would be most susceptible near the middle of the first year while the magnocellular layer cells would not reach a similar point until near the end of the first year.

In addition to the studies presented earlier there are other reasons for believing that geniculate cell size can serve as an index of changes occurring more centrally in the visual system. Recent findings by HUBEL, WIESEL and LEVAY (18) have shown that the critical period of development in the monkey visual system corresponds to a period of time during which geniculo-cortical axons are becoming segregated into ocular dominance columns. Previous work by RAKIC (19) has shown that initially optic tract fibers from both eyes completely overlap in their projections to the lateral geniculate nucleus. The geniculo-cortical axons carrying information from each eye, in turn, overlap in layer IV of the visual cortex. Although segregation of inputs to the LGN occurs about midway through gestation, segration of geniculo-cortical inputs is not complete at birth. Therefore, according to the ideas proposed by HUBEL, WIESEL and LEVAY, competition between geniculo-cortical afferents would determine the areas (ocular dominance columns) controlled by each eye. According to their model this competition would, under normal conditions, result in the terminal arborizations of geniculo-cortical axons retracting to the borders of an appropriately sized ocular dominance column. At the geniculate level one could then hypothesize that changes in cell body size should, for the most part, be complete at the beginning of the critical period. During the critical period small changes in cell body size might be expected as the cell adjusts its metabolism to keep up with the changes in terminal arborization size occurring in the cortex. It would be interesting to know if during this "fine tuning" process some geniculate cell bodies actually decrease in size, even under normal conditions.

When one eye is placed at a disadvantage in this competition, <u>i.e.</u>, by being lid sutured, it is thought that its ability to maintain synaptic space on cortical cells is severely reduced. As a result the cortical space, and thus ocular dominance column width, controlled by the non-deprived eye is increased at the expense of the deprived eye, the degree of imbalance depending upon the amount of segregation having occurred before the time of lid suture (18). The rather drastic changes in geniculate cell body size in the laminae receiving input from the deprived eye would then, at least theoretically, simply reflect the reduced metabolic requirements of the smaller terminal arborizations. Likewise, the hypertropy of geniculate cells in the non-deprived laminae (14,20) would reflect the increased metabolic demand of the larger terminal arborizations. However, in terms of this latter point the findings of RAKIC (19) and HUBEL, WIESEL and LEVAY (18) suggest an alternative explanation. If in fact geniculate cells do, under normal conditions, decrease in size as their terminal arborizations become confined to a given ocular dominance column; then the hypertrophied cells could simply represent cells that had not reduced the size of their terminal arborizations.

The present findings also suggest a possible explanation for some of the other findings related to visual deprivation. In general, both physiological and anatomical studies (14, 17, 20, 21, 22, 23) have shown that the Y-cells (larger cells (24)) in the cat and monkey lateral geniculate nucleus suffer more from early visual deprivation than X-cells (smaller cells (24)). Since, at least in the human, the larger cells develop slower than the smaller cells, it is possible that the larger cells lag behind in the process of terminal segregation that occurs in the cortex. Thus, a deprived geniculate Y-cell would be able to maintain even fewer cortical synapses. A somewhat different, but still related, explanation is that during development X- and Y-cells compete for synaptic space on cortical cells (25). Since the smaller cells (again presumably X-cells) mature earlier they may gain an advantage in this competition that shows up dramatically when the system is visually deprived.

References

1. P. B. Dews & T. N. Wiesel, J. Physiol. 206, 437 (1970).
2. D. H. Hubel & T. N. Wiesel, J. Physiol. 206, 419 (1970).
3. M. L. J. Crawford, R. Blake, S. J. Cool & G. K. von Noorden, Brain Res. 84, 150 (1975).
4. M. C. Flom, Early Experience and Visual Information Processing in Perceptual and Reading Disorders, ed. by F. A. Young and D. B. Lindsley (Natl. Acad. Sci., Washington, 1970).
5. G. K. von Noorden, Am. J. Ophthal. 63, 238 (1967).
6. G. K. von Noorden & A. E. Maumenee, Am. J. Ophthal. 65, 220 (1968).
7. M. S. Banks, R. N. Aslin & R. D. Letson, Science 190, 675 (1975).
8. A. Hohmann & O. D. Creutzfeldt, Nature 254, 612 (1975).
9. D. E. Mitchell, R. D. Freeman, M. Millodot & G. Haegerstrom, Vis. Res. 13, 535 (1973).
10. A. N. Davison & J. Dobbing. Applied Neurochemistry, Contemporary Neurology Series: 4 & 5 (F. A. Davis Company, Philadelphia, 1968), pp. 253-316.
11. L. J. Garey, R. A. Fisken & T. P. S. Powell, Brain Res. 52, 359 (1973).
12. B. G. Cragg, J. Comp. Neur. 160, 147 (1975).
13. T. L. Hickey, Science 198, 836 (1977).
14. T. L. Hickey, D. Spear & K. E. Kratz, J. Comp. Neur. 172, 265 (1977).
15. R. W. Guillery, J. Comp. Neur 148, 417 (1973).
16. G. K. von Noorden, Invest. Ophthal. 12, 727 (1973).
17. G. K. von Noorden & P. R. Middleditch, Invest, Ophthal. 14, 674 (1975).
18. D. H. Hubel, T. N. Wiesel & S. Levay, Phil. Trans. R. Soc. Lond. B. 278, 377 (1977).
19. P. Rakic, Phil. Trans. R. Soc. Lond. B. 278, 245 (1977).
20. S. M. Sherman & J. R. Wilson, J. Comp. Neur. 161, 183 (1975).
21. S. M. Sherman, R. W. Guillery, J. H. Kaas & K. J. Sanderson, J. Comp. Neur. 158, 1 (1974).

22. S. M. Sherman, K. P. Hoffmann & J. Stone, J. Neurophysiol. 35, 532 (1972).
23. S. M. Sherman, J. R. Wilson & R. W. Guillery, Brain Res. 100, 441 (1975).
24. K. P. Hoffmann, J. Stone & S. M. Sherman, J. Neurophysiol. 35, 518 (1972).
25. S. Levay & D. Ferster, J. Comp. Neur. 172, 563 (1977).

Acknowledgements

This research was partially supported by USPHS Research Grant No. R01 EY01338 from the National Institutes of Health, National Eye Institute.

Response Properties of Retinal Ganglion Cells In Siamese Cats

Michael S. Shansky
Yuzo M. Chino
Duco I. Hamasaki
Division of Visual Science, Illinois College of Optometry
Chicago, Illinois 60616
and
Bascom Palmer Eye Institute
University of Miami, Miami, Florida

Introduction

The retina of the Siamese cat is different from common cats in that its epithelium is not pigmented (GUILLERY (1); GUILLERY and KAAS (2); KALIL, JHAVERI, and RICHARDS (3)). In addition, many of these animals exhibit a convergent squint, another trait typical of albino organisms (HUBEL and WIESEL (4); BLAKE and CRAWFORD (5); KAAS and GUILLERY (6); CHINO, SHANSKY and HAMASAKI (7,8)). There have been several studies in recent years demonstrating that the visual system of the Siamese cat is anomalous compared to common cats.

For example, GUILLERY and KAAS (2) have shown that certain optic nerve fibers, originating from a vertical strip of retina temporal to the area centralis are misrouted in the optic chiasm and thus project to the contralateral lateral geniculate nucleus. This misrouting leads to a disrupted laminar pattern in the LGN and also leads to a reversed representation of the visual field within layer A_1 of the LGN.

Other investigators (KAAS and GUILLERY (6); HUBEL and WIESEL (4); COOL and CRAWFORD (9)) have shown that as a result of this initial misrouting, certain anomalies are present within the cortex of the Siamese cat as well.

Although the retinal morphology and neurophysiology in common cats has been thoroughly investigated, similar studies have not been performed on the Siamese retina. It is worthwhile to pursue these studies since they will clarify the nature and extent of the anatomical anomalies described above.

Furthermore, it has recently been shown (BLAKE and ANTOINETTI (10); COOL and SMITH (11)) that the Siamese cat exhibits a poorer spatial resolution than do common cats. That is, the Siamese displays a narrower contrast sensitivity function, with a significantly lower overall sensitivity. These behavioral results also strongly indicate the need for neurophysiological investigations of the Siamese visual system.

We are reporting here the results of recent investigations in our laboratories on the responses of retinal ganglion cells in Siamese cats. We were particularly interested in how these responses compare to similar responses in common cats, and in addition, how the responses from the misrouted fibers within the optic tract of the Siamese compared to the responses of normally-routed fibers.

Methods

We used lacquer coated tungsten microelectrodes to isolate single optic tract fibers from 8 Siamese and 7 common cats. Surgical and recording procedures were conventional for this type of preparation. Retinal landmarks were projected onto a tangent screen using the method of FERNALD and CHASE (12).

After a unit was isolated, the edges of the receptive field center were mapped with a small spot of light obtained from an ophthalmoscope. Following this, cells were classified into X- and Y-types utilizing a contrast reversal stimulus (CHINO, SHANSKY and HAMASAKI (7)). Some cells were placed into a sustained/transient classification, according to the responses elicited by a one second stimulus. Those units which showed a mean firing rate during the last two hundred milliseconds which was approximately 100% higher than the spontaneous level were classified as sustained. The different types of classifications utilized resulted in similar relative percentages of all types, as can be seen in Table I.

Experimental Procedures

A total of 300 optic tract fibers from 8 Siamese cats were studied and compared with observations on 170 units from common cats, studied under identical experimental conditions. For the majority of the units which we recorded, the experiments included:

1) receptive field mapping
2) classification into X- or Y-type
3) intensity-response functions
4) flicker-response functions
5) responses to contrast-reversal stimuli

Results

Siamese cats do not form a homogeneous group with respect to the distribution of their interocular alignments. This is demonstrated in Fig. 1, in which the interocular

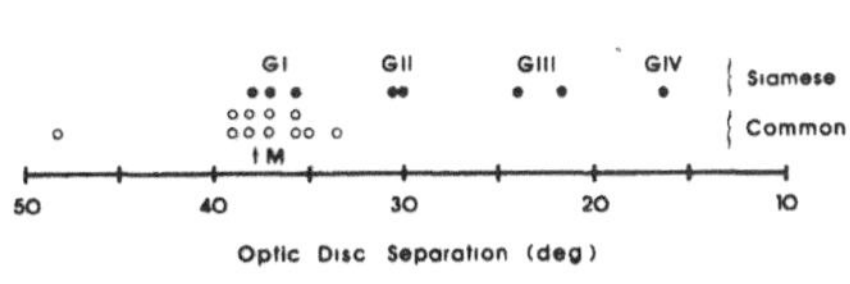

Fig. 1 Optic discs separations of eight Siamese and eleven common cats following anesthesia and paralysis. Mean separation of common cats is indicated by an arrow with M. Siamese cats were divided into four groups, depending upon severity of the eyes. GI = orthophoric cats; GII = mildly esotropic cats; GIII = strongly esotropic cats; GIV = extremely esotropic cats.

alignments of the experimental animals utilized in this study are shown. Whereas the interocular alignments (shown as the optic disc separations measured under anesthesia and paralysis) of the common cats are all clustered about a single mean and show little variance, those of the Siamese seem conveniently placed in one of four groups.

These include 3 Siamese whose interocular alignments are similar to common cats and therefore can be considered orthophoric, as well as one Siamese cat whose optic disc separation is so small as to indicate the presence of an extreme convergent misalignment. The other two groups of Siamese (GII and GIII) can be considered as mildly and strongly esotropic respectively.

The results of classifying units into X- or Y-types are summarized in Table 1.

Table I

Relative frequency of X- and Y- units in the optic tract of Siamese and common cats.

	Siamese Cats Contralateral Normally Routed Fibers	Siamese Cats Contralateral Misrouted Fibers	Siamese Cats Ipsilateral Fibers	Siamese Cats Totals	Common Cats
Sustained type*	33 (92%)	11 (79%)	0 (-)	44 (88%)	53 (62%)
Transient type*	3 (8%)	3 (21%)	0 (-)	6 (12%)	33 (38%)
Totals	36 (100%)	14 (100%)	0 (-)	50 (100%)	86 (100%)
X-types	130 (84%)	67 (87%)	17 (89%)	214 (86%)	57 (68%)
On-center	87	45	12	144	43
Off-center	43	22	5	70	14
Y-types	24 (16%)	10 (13%)	2 (11%)	36 (14%)	27 (32%)
On-center	10	5	1	16	9
Off-center	14	5	1	20	18
Totals	154 (100%)	77 (100%)	19 (100%)	250 (100%)	84 (100%)
X-sustained type	163 (86%)	78 (86%)	17 (89%)	258 (86%)	110 (65%)
Y-transient type	27 (14%)	13 (14%)	2 (11%)	42 (14%)	60 (35%)
Totals	190 (100%)	91 (100%)	19 (100%)	300 (100%)	170 (100%)

*On-center units. 17 off-center units were not classified, 5 of which were considered to be misrouted fibers.

The primary finding from these data is that there is a significantly lower percentage of Y cells in Siamese cats compared to that in common cats. For all the Siamese cats, only 14% of the units were classified as Y-type while in common cats 32% were so classified, a difference which is highly significant (corrected X^2 test, $P < .01$). A closer examination of Table I reveals that this difference is not exclusive to any one of the fiber groups or to "on" or "off" type cells, but represents lower percentages for Y-type retinal ganglion cells. This difference is even upheld when the sustained-transient classification is utilized instead of the contrast reversal stimulus.

In Figs. 2-4, the relative distributions of X- and Y-type retinal ganglion cells are shown for Siamese cats from 3 of the alignment groups described earlier. For example, the receptive fields of the Siamese cats in group GI (orthophoric Siamese)

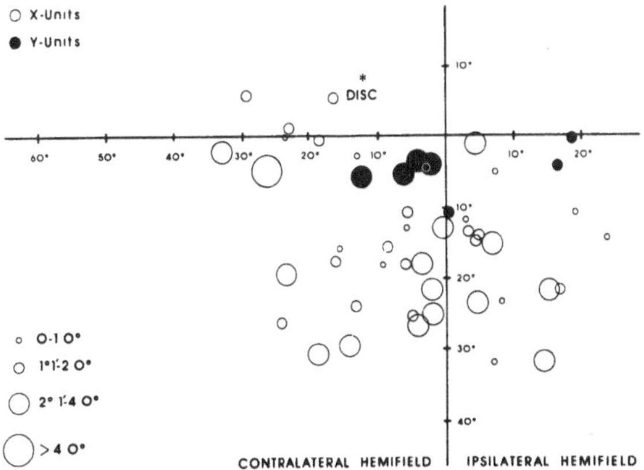

Fig. 2 Visual field locations of receptive fields of contralateral optic tract fibers in two orthophoric Siamese cats (GI). X-units are indicated by open circles; Y-units, by filled circles. The projection of the optic disc is also shown (*DISC). The horizontal and vertical meridians were estimated from the position of the optic disc.

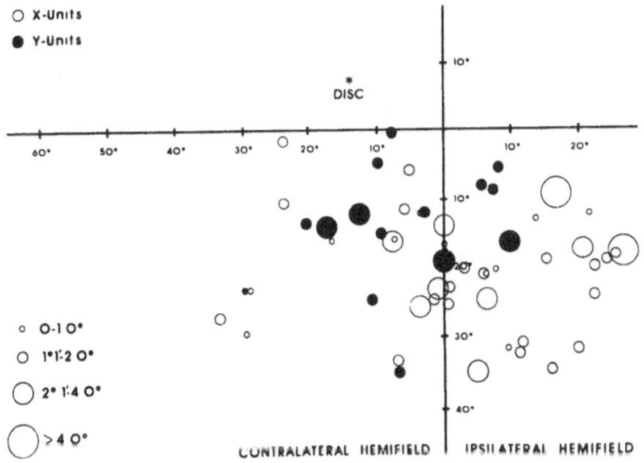

Fig. 3 Locations of receptive fields of contralateral optic tract fibers in two mildly esotropic Siamese cats (GIV). Convention as in Fig. 2.

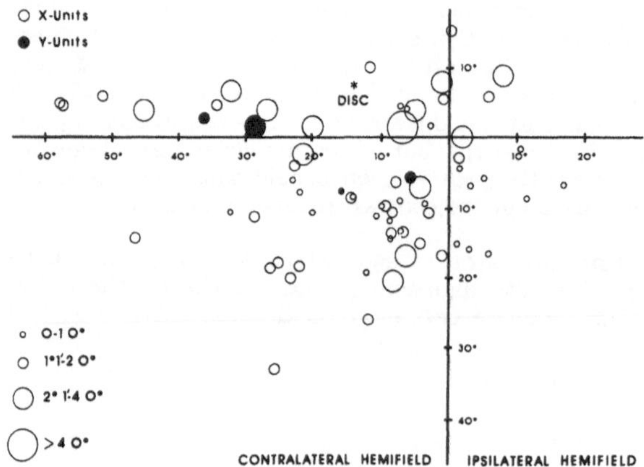

Fig. 4 Locations of receptive fields of contralateral optic tract fibers in one extremely esotropic Siamese cat (GIV). Convention as in Fig. 2.

are shown in Fig. 2. Virtually all the units recorded in these experiments had receptive fields which fell within 30 degrees of the area centralis. The receptive fields of misrouted fibers are shown in the ipsilateral hemifield, where very few are found in common cats. Even a qualitative comparison based on this figure indicates little or no difference between the receptive fields of the normally-routed and misrouted fibers within the Siamese optic tract. Fig. 3 shows a similar distribution for the Siamese cats in Group GII (mildly esotropic). The major difference between this figure and the distribution shown in Fig. 2 is the notable reduction in the number of Y-type receptive fields. This is even further dramatized in Fig. 4, where the distribution from the extremely esotropic Siamese cat is shown. Thus, there appears to be a strong relationship between the proportion of Y-type retinal ganglion cells and the interocular alignment exhibited by individual Siamese cats. This relationship is illustrated in Fig. 5, where the percentage of Y-type units encountered is plotted as a function of interocular alignment for the various experimental groups. An inverse relationship exists between the proportion of Y-type responses recorded in the optic tract and the extent of interocular misalignment exhibited by individual Siamese, as measured by optic disc separations. While the average percentage of Y-type cells in the Siamese optic tract is 14%, this percentage falls as low as 6% for the extremely esotropic Siamese cat. Furthermore, even the orthophoric Siamese cats reveal a lower proportion of Y-type responses than the common cats.

In the next series of experiments, we compared receptive field sizes between the various fiber groups. Thus, Fig. 6 indicates little or no difference in receptive field sizes between normally-routed and misrouted fibers within the Siamese optic tract. Likewise, Fig. 7 reveals little difference in sizes between X- and Y-type receptive fields for all of the fiber groups represented. Finally, in Fig. 8 it can be seen that "on" center units have significantly smaller receptive fields than "off" center units.

In the final series of experiments, we investigated responses to variations in several different stimulus parameters. Fig. 9 illustrates intensity-response functions for units in the three fiber groups which we investigated. The figure reveals significantly lower firing rates at all intensities for the Siamese as compared to the

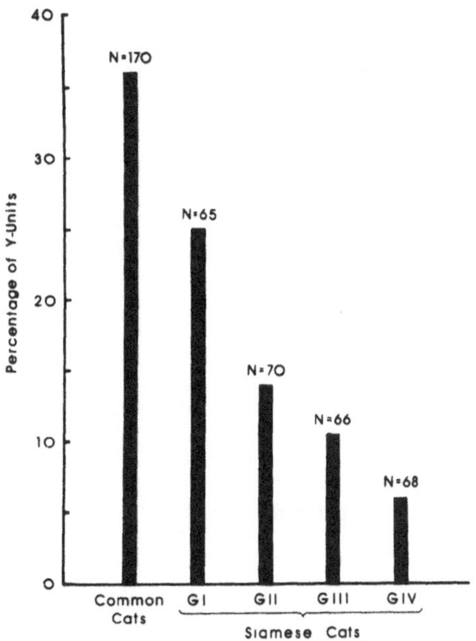

Fig. 5 Relationship between alignment of the eyes in Siamese cats and the percentage of Y-units in the optic tract. Data on common cats (N = 170) are also shown for comparison. Degrees of interocular alignment for each group of Siamese cats can be found in Fig. 1.

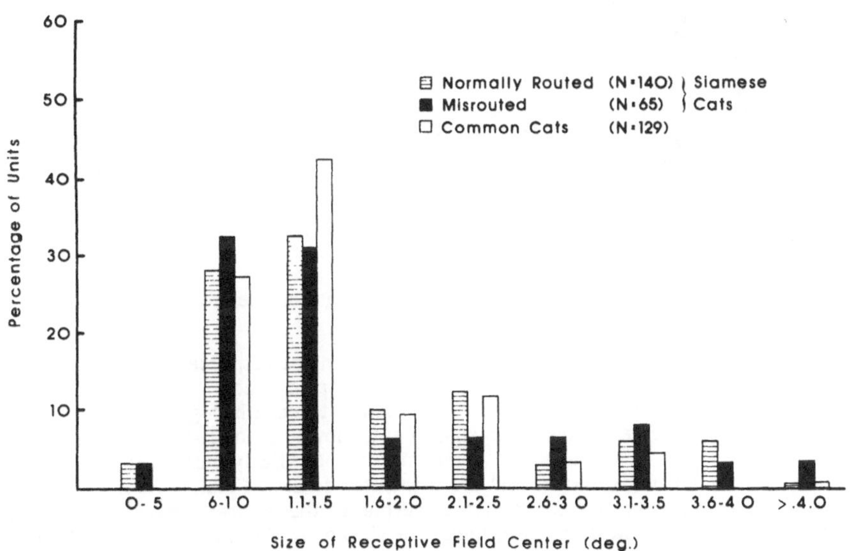

Fig. 6 Frequency histograms showing the distribution of the receptive field center sizes of 140 normally-routed and 65 misrouted optic tract fibers in Siamese cats and 129 units from common cats.

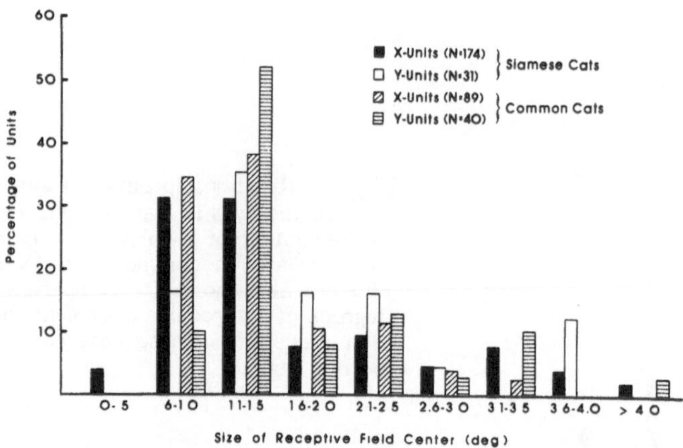

Fig. 7 RFC sizes of X- and Y- units in Siamese and common cats.

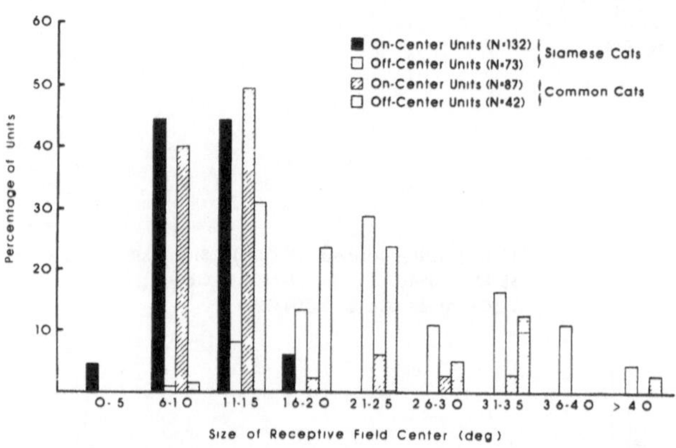

Fig. 8 RFC sizes of on-center and off-center units in Siamese and common cats.

common cat units. No difference can be seen between the responses of the normally-routed and misrouted fibers within the Siamese optic tract, however. Fig. 10 indicates flicker-response functions for the same three fiber groups (see figure legends for details). Again, there are significantly lower firing rates in the Siamese optic tract units than in those of the common cats. Similarly, no difference can be seen between the normally-routed and the misrouted fibers. Finally, we investigated the responses of Siamese and common cat neurons to the contrast-reversal stimulus. Examples of average response histograms are shown in Fig. 11 for typical X-type retinal ganglion cells. After the null position (or the equal response position for the Y-type cells) had been determined, the target was moved in one millimeter steps (4.3 min. of arc) and the response recorded for positions minus 1.0 to plus 4.0 millimeters from the "0" position. The figure indicates the poor response quality of the X-type unit in the Siamese (note the lack of an initial transient), as well as the quantitatively poorer responses (note the difference in the vertical calibration

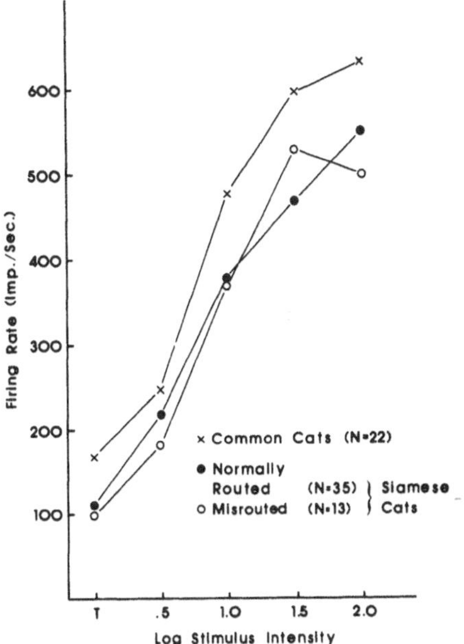

Fig. 9 Intensity-response functions. Average maximum firing rates of on-center units as a function of stimulus intensity. T: threshold intensity. A 1.2° circular stimulus was centered on receptive fields.

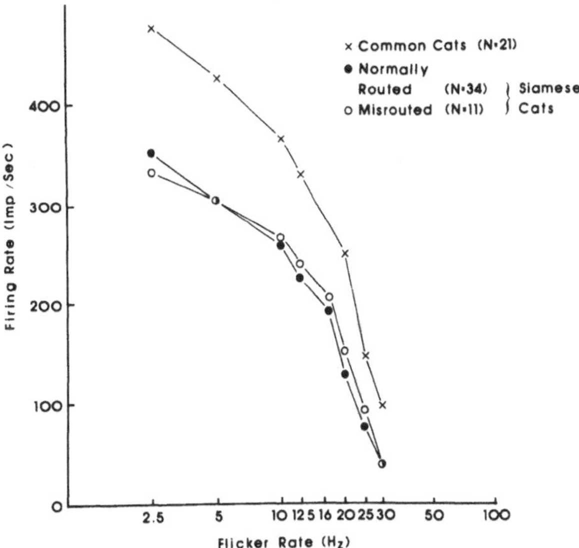

Fig. 10 Responses to flicker. Averaged maximum firing rates of on-center units as a function of flicker frequencies of a 1.2° circular stimulus centered on receptive fields. The intensity of the stimulus was set 2.0 log units above threshold. The stimulus duration was 20 msec and the interstimulus interval varied from 380 to 13.3 msec.

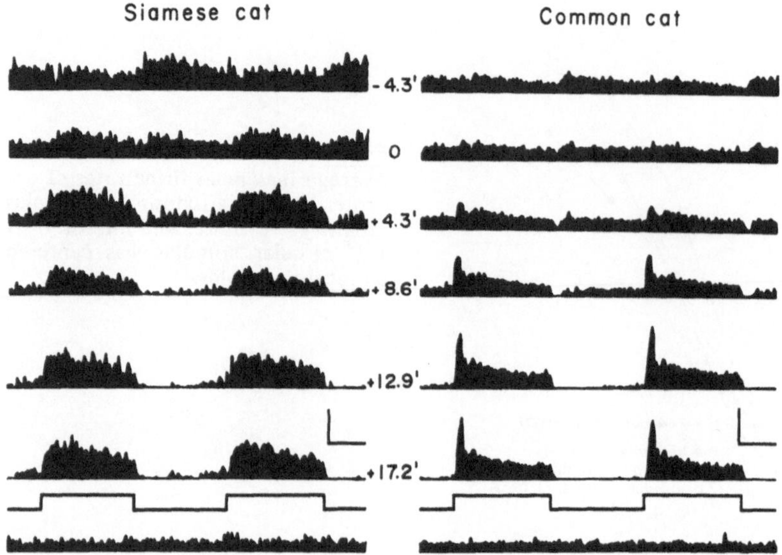

Fig. 11 Average response histograms of on-center units recorded from Siamese (left) and common (right) cat. The numbers between the histograms represent the position of the center of a contrast reversal stimulus. The null position is designated as "0". Stimulus trace indicates time course of reversals of contrast (every 500 msec). Calibration in Siamese cat, 100 spikes/sec; 0.2 sec and in common cat, 200 spikes/sec; 0.2 sec.

between the two breeds). Fig. 12 shows the mean responses from the three fiber groups for this experiment. The firing rate is plotted as a function of the stimulus

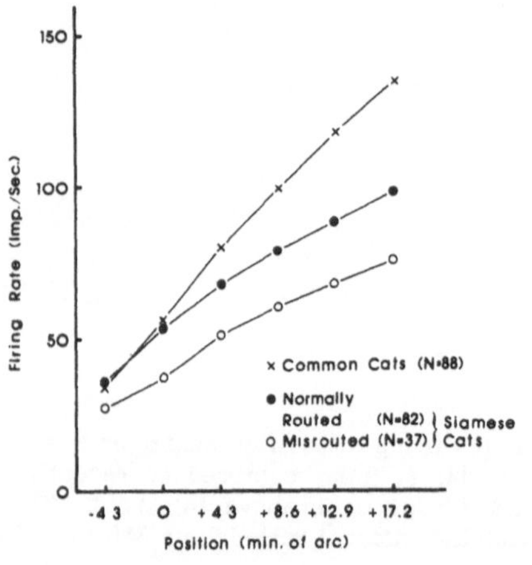

Fig. 12 Responses to contrast reversal stimuli. Averaged maximum firing rate, the mean of five 10 msec bins around the maximum, is shown on ordinate and positions of the edge on abscissa. Position 0 indicates the "null" positions of the receptive field for X-units and equal response position for Y-units.

position for all three groups. The significantly higher firing rates for common cats is quite evident at all but the two lowest positions. Furthermore, there appears to be a consistent difference between the normally-routed and misrouted fibers. That is, the misrouted fibers reveal lower firing rates at all positions of the stimulus than do the normally-routed fibers.

Discussion

The data reported here are consistent with previous anatomical, physiological and behavioral studies on Siamese cats, in that they indicate anomalies with respect to the output of the retina. That is, not only is there a reduction in the proportion of a specific cell type within the retina, but the quantitative responses of the fibers that were recorded from were notably poorer than those in common cats. Of course, there is a third sub-group of cells that was not investigated in these experiments, namely the W-cells (STONE and HOFFMAN (13); STONE and FUKUDA (14); FUKUDA and STONE (15)). These cells, with presumably small cell bodies and axons, are extremely difficult to isolate in an optic tract preparation (ENROTH-CUGELL and ROBSON (16); FUKADA (17)).

One interesting question that arises from these results is whether or not they can explain the reduced contrast sensitivity shown by recent investigators for Siamese cats (BLAKE and ANTOINETTE (10); COOL and SMITH (11)). It is not clear whether the reduced proportion of Y-type units could explain this phenomenon or whether it is simply a function of the lower firing rates exhibited by Siamese cat retinal ganglion cells. The poor responses of the Siamese optic tract fibers to changes in various stimulus parameters is of questionable origin. It may result from some photochemical anomaly in the Siamese retina or it may be a function of the light scatter known to occur in the Siamese retina because of the lack of a pigmented epithelium.

Finally, we can speculate that the consequences of a lower retinal output in Siamese may include abnormal responses in more central structures. It has been shown that monocular deprivation produces a distinct loss of Y-cells in the dorsal LGN of common cats (HOFFMAN, STONE, and SHERMAN (18)). It also results in the development of strabismus (SHERMAN (19)). To what extent these parallels with the Siamese are coincidental is not known at the present time.

The data reported here lead to some interesting questions for future investigations, including the following:

1) To what extent is the Siamese retinal anatomy different from that of the common cat, not only in terms of the ganglion cell layer but also distal to it?
2) Is it possible to mimic the effects described here by creating artificial strabismus in common kittens during the critical period of development?
3) To what extent do the anomalies revealed here lead to abnormalities in the receptive field properties of neurons in the central structures within the Siamese?

The first two of these questions are being currently investigated in our laboratory, and the third is addressed in the following paper.

Acknowledgement

This research was supported by PHS grants EYO1444 and EYOO701. We are grateful to O. Navarro for technical assistance, H. Cohen for assistance in data collection, and S. Garcia and B. Clemons for manuscript preparation.

References

1. R. W. Guillery, Brain Res. 14, 739 (1969).
2. R. W. Guillery and J. H. Kaas, J. Comp. Neurol. 143, 73 (1971).
3. R. Kalil, S. Jhaveri and W. R. Richards, Science 174, 302 (1971).
4. D. H. Hubel and T. N. Wiesel, J. Physiol. 219, 33 (1971).
5. R. Blake and M. L. J. Crawford, Brain Res. 77, 492 (1974).
6. J. H. Kaas and R. W. Guillery, Brain Res. 59, 61 (1973).
7. Y. M. Chino, M. S. Shansky and D. I. Hamasaki, Science 197, 173 (1977).
8. Y. M. Chino, M. S. Shansky and D. I. Hamasaki, Brain Res. 143, 459 (1978).
9. S. J. Cool and M. L. J. Crawford, Vision Res. 12, 1809 (1972).
10. R. Blake and D. N. Antoinette, Science 194, 109 (1976).
11. S. J. Cool and E. L. Smith, III, Am. J. Opt. and Physiol. Optics 53, 537 (1976).
12. R. Fernald and R. Chase, Vision Res. 11, 95 (1971).
13. J. Stone and K. P. Hoffman, Brain Res. 43, 610 (1972).
14. J. Stone and Y. Fukuda, J. Neurophysiol. 37, 722 (1974).
15. Y. Fukuda and J. Stone, J. Neurophysiol. 37, 749 (1974).
16. L. Enroth-Cugell and J. J. Robson, J. Physiol. 18, 517 (1966).
17. Y. Fukuda, Vision Res. 11, 209 (1971).
18. K. P. Hoffman, J. Stone and S. M. Sherman, J. Neurophysiol. 35, 518 (1972).
19. S. M. Sherman, Brain Res. 37, 187 (1972).

Discussion

Q. Dr. Bishop: You're probably aware of the specifications that have come out indicating that the situation is much more complex than just X and Y's. Y cells, in our best estimate, constitute only 3% of the total population. The bulk of the cells, of course, are X and the other ones are sluggish cells. I don't like the terms sluggish, some call them tonics, sustained, and so on. My point is that these fire very slowly and you may be muddling up your slow firing rates with different classes of cells.

A. Well, the technique that we used was from the original ENROTH-CUGELL and ROBSON criterion. Our estimates of the X and Y in common cats (65%-35%) agree very well with the brisk sustained and brisk transient breakdown shown by CLELAND and LEVICK in 1974. We're not trying to suggest what the absolute retinal percentages are.

Response Properties of Striate Neurons In Area 17 of Siamese Cats

Yuzo M. Chino
Michael S. Shansky
Wayne L. Jankowski
Division of Visual Science
Illinois College of Optometry
Chicago, Illinois 60616

Introduction

It is becoming increasingly clear that the visual system of Siamese cats is quite abnormal and provides an ideal substrate to study such important problems as acuity, stereopsis, strabismus, and the development of the mammalian visual system. The significant abnormalities are:

1. Misrouting of optic nerve fibers at the chiasm from a vertical strip of temporal retina near the area centralis (GUILLERY (1); GUILLERY & KAAS (2); KALIL, JHAVERI, & RICHARDS (3));
2. Abnormal geniculocortical projections (KAAS & GUILLERY (4); HUBEL & WIESEL (5); GUILLERY, CASAGRANDE & OBERDORFER (6));
3. Loss of binocularity in cortical neurons (HUBEL & WIESEL (5); COOL & CRAWFORD (7); KAAS & GUILLERY (4));
4. Absence of stereopsis (PACKWOOD & GORDON (8));
5. Abnormal visual fields (ELEKESSY, CAMPION & HENRY (9); GUILLERY & CASAGRANDE (10));
6. Convergent strabismus (HUBEL & WIESEL (5); BLAKE & CRAWFORD (11); KAAS & GUILLERY (4); CHINO, SHANSKY & HAMASAKI (12, 13));
7. Abnormal visual resolution (BLAKE & ANTOINETTI (14));
8. Lack of pigment in the retinal epithelium (GUILLERY (1); GUILLERY & KAAS (2); KALIL, JHAVERI & RICHARDS (3)).

In the preceding paper, we have reported anomalous responses of retinal ganglion cells in Siamese cats (SHANSKY, CHINO, & HAMASAKI (15)). In this study our objective was to investigate receptive field properties of neurons in area

17 of this organism. Such data not only are lacking in the literature, but also may have an important bearing on our findings in the retina in terms of the nature of information transfer in the cat visual system. This is, then, our initial report on the results.

Methods

Surgical procedures were conducted under halothane anesthesia (2% in 60% N_2O: 40% O_2) and the animals were subsequently switched to a mixture of nitrous oxide and oxygen (70% and 30%) for the remainder of the experiment. Body temperature was controlled and blood pressure monitored, as well as the ECG and the EEG. In addition, the animals were paralyzed with a Flaxedil/curare mixture and artificially respirated. Corneal contact lenses with 3 mm diameter artificial pupils were placed on the animals' eyes and refraction was performed to make the animals' focus conjugate with a screen placed 114 centimeters from the posterior nodal point of the eye. A small craniotomy (less than 5 mm in diameter) was performed over the post-lateral gyrus of the left hemisphere along the 17/18 border and the dura removed. The hole was covered with bone wax through which tungsten-in-glass microelectrodes were driven. Conventional AC recording procedures were employed, and the data were collected with the aid of a PDP 11/10 computer which provided average response histograms. The receptive fields were mapped both with a handheld ophthalmoscope, and with the use of a two channel projection system, employing function generators to drive mirror galvanometers in conjunction with dove prisms and X/Y positioners. All experiments were conducted against a uniform illumination of 2.7 cd/m^2 and the stimulus was approximately 1 log unit above that of the background. Following each experiment, the animals were sacrificed, the brains perfused, and cresyl violet stained sections prepared in order to confirm recording sites and the disrupted geniculate laminar pattern typical of the Siamese cat.

Results

We have investigated the receptive field properties of 73 striate neurons in 8 Siamese and 41 neurons of 4 normal cats. All of our Siamese cats exhibited the characteristics of "Mid-western" type (KAAS & GUILLERY (4)).

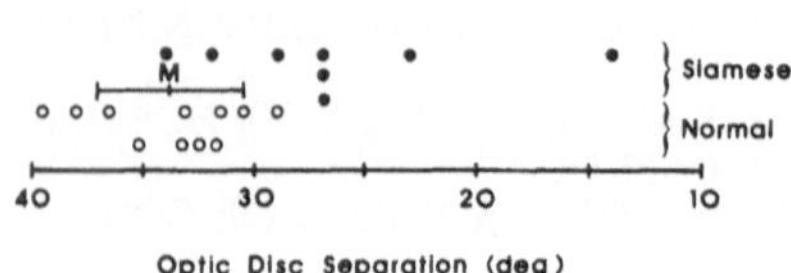

Fig. 1 Distribution of interocular alignments for Siamese (closed circles) and normal cats (open circles) utilized in these experiments. M = mean separation for normal cats; horizontal bars = ± 1 S.D. from mean.

Figure 1 shows interocular alignments of each animal, measured by recording the optic disc separations following anesthesia and paralysis (SHERMAN (16)). The optic discs were projected back and marked on a translucent screen with an ophthalmoscope (FERNALD & CHASE (17)). The discs were also used to find the position of the area centralis and the location of the receptive fields of all the units with reference to the area centralis (VAKKUR, BISHOP & KOZAK (18)). Of the 8 Siamese cats which we investigated, 2 were orthophoric (mean optic disc separation = 33°), 4 were mildly esotropic (mean optic disc separation = 27°), and 2 were esotropic (mean separation = 18.5°). The mean optic disc separation of the normal cats used for comparison in this study was 33.8° (± 3.3°) with a range of between 29° and 39.5°.

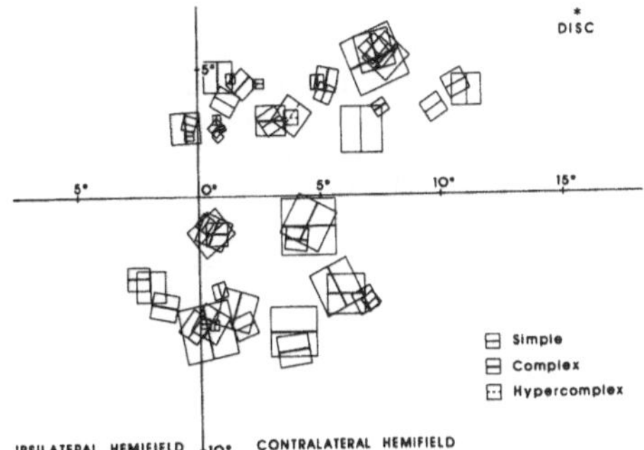

Fig. 2 Distribution of receptive fields of striate neurons in eight Siamese cats.

The location of the receptive fields is shown in Fig. 2. The type, width, and optimum orientation of each cell are also indicated in the diagram. It is important to note that over 90 percent of the cells sampled in this study had their receptive fields within 10° of the area centralis.

Binocular Properties

Figure 3 shows the ocular dominance distribution for the Siamese striate neurons and those obtained from normal cats. There are substantial differences to be seen between these distributions. For example, the Siamese distribution is predominantly contralateral and no ipsilaterally dominant units were found in the

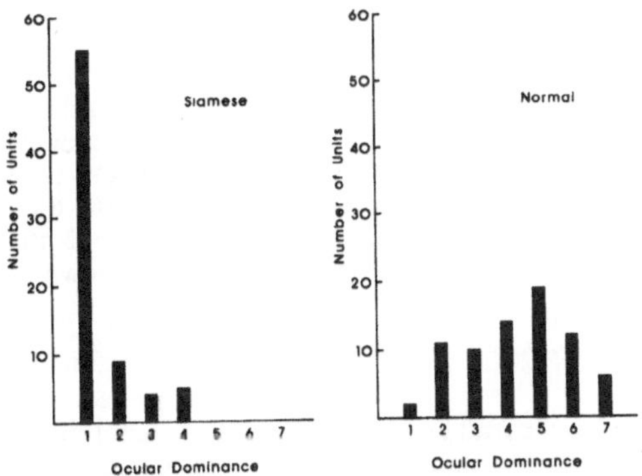

Fig. 3 Ocular dominance distributions for Siamese and normal cats (after HUBEL & WIESEL, 1962).

Siamese sample. Although 75% of the Siamese units were driven only by the contralateral eye, 25% of the Siamese cells could be driven by either eye. These units exhibited binocular occlusion, summation, or facilitation, as has been described for the cortical cells of normal cats (BARLOW, BLAKEMORE & PETTIGREW (19); NIKARA, BISHOP & PETTIGREW (20), BISHOP, HENRY & SMITH (21)).

A striking finding from this experiment can be seen in Fig. 4 where the proportion of binocularly activated units is shown as a function of interocular alignment for both normal and Siamese cats. There is an inverse relationship between the proportion of binocularly driven units and the extent of the interocular misalignment exhibited by individual Siamese cats. In the extreme cases (the esotropic Siamese), there were no binocularly activated cells encountered. On the other hand, 40% of the cells in orthophoric Siamese exhibited binocular interaction. Of these, all but one belonged to ocular dominance groups 3 and 4. Finally, in no Siamese cat was the proportion of binocularly drived cells equal to that of normal cats (approximately 80%).

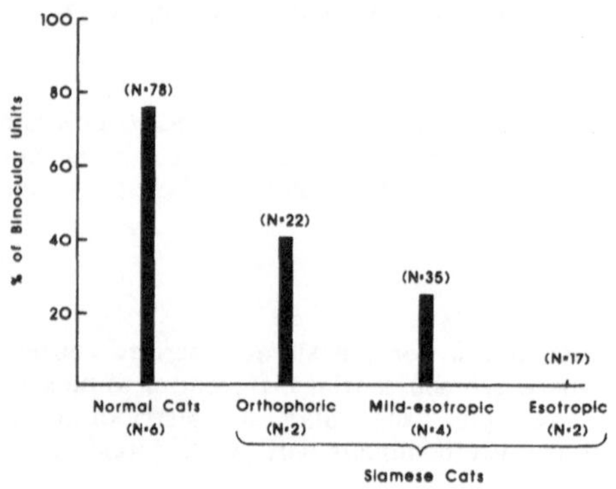

Fig. 4 Relationship between interocular alignment of normal and Siamese cats and proportion of binocularly excited striate neurons.

Monocular Properties

(a) Classification

Each cell was classified into simple, complex, or hypercomplex on the basis of criteria originally established by HUBEL & WIESEL (22), in conjunction with moving slit criteria (PETTIGREW, NIKARA & BISHOP (23); BISHOP & HENRY (24); SHERMAN, WATKINS & WILSON (25)). The relative percentages of simple (46.6%), complex (43.8%), and hypercomplex (8.2%) cells, encountered in area 17 of the Siamese cat, are comparable to those found in normal cats, and this is shown in Table 1.

(b) Sizes of Receptive Fields

In Table 1, the receptive field sizes are indicated by the field width in each type of striate neurons. Complex receptive fields are significantly larger in Siamese cats (mean width = 1.60°) than in normal cats (mean width = 1.03°). The receptive field widths of simple cells are also somewhat larger in Siamese.

TABLE 1

Cats	Cells	N(%)	Width of R.F. Range	Width of R.F. Mean (S.D.)	Spontaneous Activity Mean Spikes/sec (S.D.)
Siamese (N=8)	Simple	34(46.6)	$.4^{o}$-2.5^{o}	$1.16^{o}(.53^{o})$+	2.3 (2.3)
	Complex	32(43.8)	$.6^{o}$-4.5^{o}	$1.60^{o}(.77^{o})$*	14.2 (10.8)
	Hypercomplex	6(8.2)	$.5^{o}$-3.5^{o}	$1.86^{o}(1.1^{o})$	N.A.
	Other	1(1.4)	N.A.	N.A.	N.A.
	Total	73(100.0)	$.4^{o}$-4.5^{o}	$1.43^{o}(.72^{o})$	9.1 (10.1)**
Normal (N=4)	Simple	19(46.2)	$.5^{o}$-1.8^{o}	$.97^{o}(.36^{o})$+	1.2 (.9)
	Complex	20(48.8)	$.4^{o}$-1.7^{o}	$1.03^{o}(.40^{o})$*	9.1 (8.2)
	Hypercomplex	1(2.5)	N.A.	N.A.	N.A.
	Other	1(2.5)	N.A.	N.A.	N.A.
	Total	41(100.0)	$.4^{o}$-1.8^{o}	$1.00^{o}(.37^{o})$	5.2(7.0)**

* $P < .005$
\+ N.S.
** $P < .05$

Table 1 Type, size, and spontaneous activity of striate neurons in Siamese and normal cats.

(c) Spontaneous Discharge

Another difference concerns the rate of spontaneous discharge. Table 1 shows that the spontaneous discharge rate in striate neurons of Siamese cats is considerably higher than that of normal cats. This difference is a surprising one, since we have found that the mean spontaneous discharge of Siamese optic tract fibers (44.6 imp/sec, N = 40) is similar to the rate sampled in normal cats (46.0 imp/sec, N = 45).

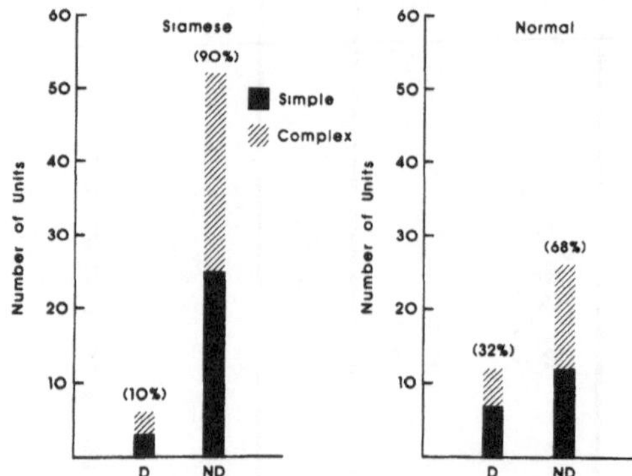

Fig. 5 Relative distributions of direction selective (D) and directionally non-selective (ND) striate neurons in Siamese and normal cats.

(d) Direction Selectivity

One of the most prominent characteristics of striate neurons in normal cats is their direction selectivity (HUBEL & WIESEL (22); PETTIGREW, NIKARA & BISHOP (23); BISHOP, COOMBS & HENRY (26); GOODWIN, HENRY & BISHOP (27); GOODWIN & HENRY (28)). In the Siamese visual cortex, however, this direction selectivity is dramatically reduced (Fig. 5). Only 10% of our sample, as opposed to 32% in normal cats, exhibited direction selectivity, and this difference is statistically significant (corrected ^{2} test, p .05). Our criterion for direction selectivity of a given neuron is that the response to movement in the null direction must be no higher than twenty percent of the response in the preferred direction (Fig. 6). Our sample is too small to justify a statistical analysis, but additional experiments indicate that the loss of directionality in Siamese visual cortex is more pronounced in complex than in simple cells.

(e) Orientation Selectivity

This receptive field property of striate neurons, systematically explored only in normal cats in recent years (HENRY, DREHER & BISHOP (29); ROSE & BLAKEMORE (30); WATKINS & BERKLEY (31)), was investigated in the Siamese striate cortex by measuring the half-width at half-height of orientation tuning curves. For this purpose, the orientation of a stimulus slit was changed by a 10° step from the optimum orientation and the average peak firing measured at each orientation. Regression lines were derived from data points to characterize tuning curves. The sharpness of orientation tuning curves was then determined by taking the half-width at the half-height.

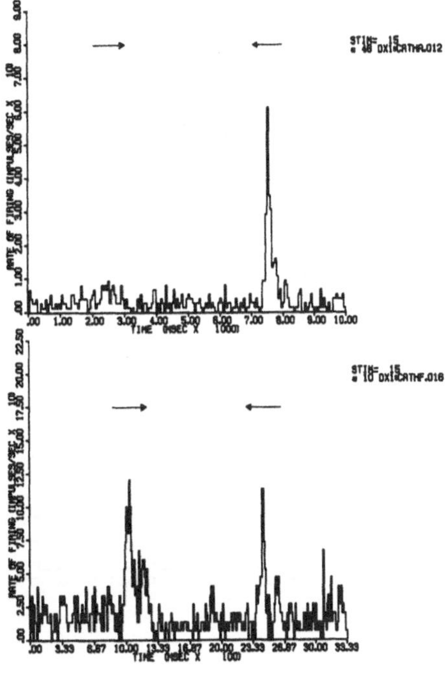

Fig. 6 Examples of average response histograms for a complex cell in Siamese (lower) and normal (upper) cats. The normal (but not the Siamese) unit is direction selective according to our criterion (see text). Calibration: 10 msec binwidth; 15 stimulus presentations. Arrows indicate direction of stimulus movement. Stimulus speed = 9°/sec.

Figure 7 shows a few examples of such tuning curves. The simple cells in Siamese cats appear to exhibit a similar half-width as that in the normal cat, and the tuning curves of simple cells in both cats are sharper than either of the complex tuning curves. On the contrary, the Siamese complex cell appears to have a much wider tuning curve than the complex neuron in the normal cat.

The above observation is confirmed by the systematic analysis of the half-width at half-height in 78 neurons (Fig. 8). The average half-width of Siamese simple cells was 21.9° ($\pm$ 8.4°) with the range between 8.5° and 36.5°. This is comparable to the sharpness of the tuning curves in normal cats, i.e., the mean half-width equalling 19.5° ($\pm$ 8.5°) and a range of between 6.8° and 38.0°. These values in the simple cells of normal cats are quite similar to those reported by other investigators (HENRY, DREHER & BISHOP (29)).

The sharpness of the tuning curves of complex cells, however, is substantially reduced in Siamese cats. The mean half-width in Siamese was 48.4° ($\pm$ 8.1°) with a range of 19.5° to 45.0° in normal cats. This difference is statistically significant (t-test, p .0001). It is clear in Fig. 7 that over half of the Siamese complex neurons sampled had wider tuning curves than any of the complex units in normal cats.

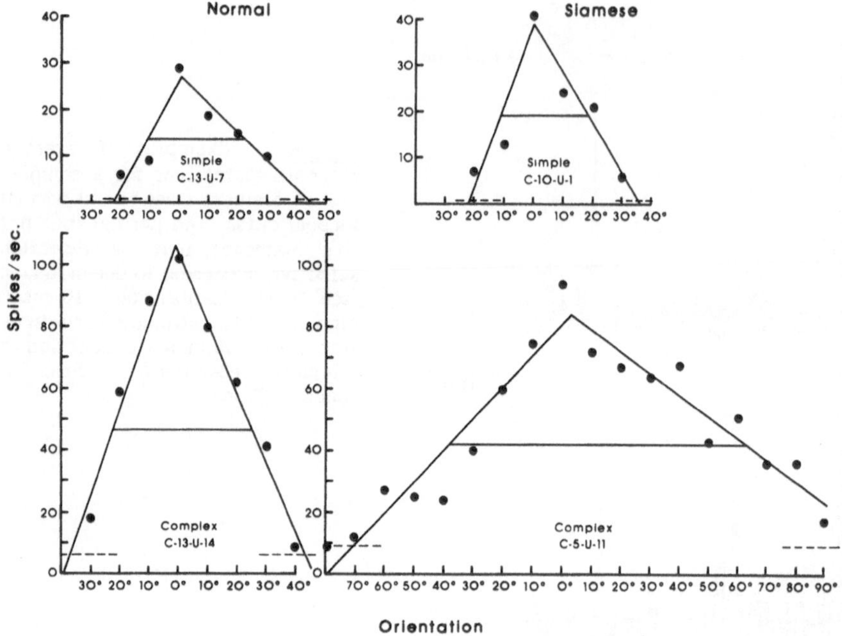

Fig. 7 Typical orientation tuning curves for normal (left side) and Siamese (right side) striate neurons.

Discussion

Binocular Properties

A great majority of neurons in the striate cortex of the normal cat can be excited by stimulation of either eye (BARLOW, BLAKEMORE & PETTIGREW (19); BISHOP, HENRY & PETTIGREW (20); PETTIGREW, NIKARA & BISHOP (23)). On the contrary, in Siamese cat striate cortex little or no binocular excitation has been reported in previous investigations (COOL & CRAWFORD (7); HUBEL & WIESEL (5): KAAS & GUILLERY (4)). Thus, it is quite surprising to find that 25% of the units sampled in this study were binocularly activated. Furthermore, we find that there is an inverse relationship between the proportion of binocularly driven units encountered and the severity of esotropia exhibited by individual Siamese cats. In orthophoric Siamese, the proportion of binocularly excited neurons was over 40%, while in extremely esotropic animals, this proportion was as low as 0%.

GUILLERY & KAAS (2) have suggested that the size of the medial normal segment in layer A1 of the dorsal LGN is critical for the development of proper interocular alignment, since esotropic Siamese cats tend to exhibit smaller medial

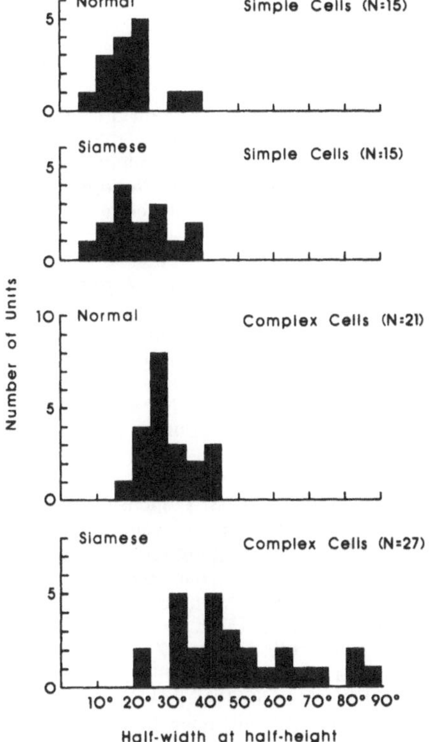

Fig. 8 Distribution of orientation tuning curves for simple and complex striate neurons in normal and Siamese cats (see text for details).

normal segments. The results of this study are consistent with their observations. Since all the cortical units studied here had their receptive fields in or around the area centralis, the binocular units must receive ipsilateral inputs from cells in the medial normal segment of layer A1.

According to some investigators (BLAKE & CRAWFORD (11), COOL & CRAWFORD (7); GUILLERY (1)), esotropia in Siamese cats is not present at birth, but develops during the several post-natal months and results from abnormalities in central mechanisms stimulated by the binocular and patterned visual input. In other words, a minimum amount of binocularly driven cells may be necessary for the development of orthophoria in Siamese cats and this minimum is far below the percentage found in normal cats. Our data are certainly consistent with such a view, although they do not directly conflict with the speculation (HUBEL & WIESEL (5)) that a loss of binocularity in Siamese striate cortex is secondary to the existence of squint. According to this view, orthophoric Siamese cats should have a higher proportion of binocular units (over 80%) than that found in this investigation (40%). Unfortunately, the former view is also obscured by the finding of SHERMAN (16) that postnatal visual cortex ablation in common kittens does not disturb the normal development of orthophoria in these animals.

Monocular Properties

The sharpness of orientation tuning curves is drastically reduced in Siamese complex cells, but not in simple cells. Thus, this abnormality must originate within the striate cortex of Siamese cats. The same argument can be applied to the differences observed in the receptive field sizes, since our studies in the optic tract show that the RFC sizes of retinal ganglion cells in Siamese cats are comparable to that in normal cats (CHINO, SHANSKY & HAMASAKI (12, 13)).

The loss of direction selectivity in Siamese striate neurons, especially in complex cells, may be attributed to a loss of Y-units in the optic tract of this animal, although the exact nature of such a relationship is not clear. There is a striking parallelism between this result and the finding by HOFFMAN & SHERMAN (33, 34) that pattern deprivation in kittens after birth leads to a loss of Y-units in the LGN (HOFFMAN, STONE & SHERMAN (35); SHERMAN, STONE & HOFFMAN (36)) and the loss of direction selectivity in superior colliculus neurons (presumably due to a similar loss in the visual cortex).

There are many other parallelisms between the abnormal responses found in the Siamese retina and cortex, and responses in the visually inexperienced animals. For example, such response properties of Siamese retinal ganglion cells as the shape of the average response histograms, intensity-response functions and flicker responses, are not only abnormal, but also strikingly resemble properties of ganglion cells in 3-week-old kittens (HAMASAKI & FLYNN (37)). In addition, it has been repeatedly shown that orientation selectivity of cells in area 17 is abolished or severely reduced in visually inexperienced kittens (BARLOW & PETTIGREW (38); BLAKEMORE & VAN SLUYTERS (39, 40); IMBERT & BUISSERET (41)). This type of parallelism must be analyzed with much caution, but will nonetheless serve to confirm Guillery's claim that this anomalous organism may prove invaluable for understanding the mechanism of normal visual development (GUILLERY & CASAGRANDE (42)).

Our future investigations in the Siamese visual system will include a systematic exploration of areas 18 and 19, along with more quantitative analyses of striate responses. A comparison of strabismic Siamese responses with those of surgically-induced esotropic cats is currently in progress in our laboratory.

Acknowledgement

This research was supported by PHS grant EY01444 and funds from the Illinois College of Optometry research committee. We thank Dr. Douglas Beemer for technical assistance and Drs. Morris Berman, Peter S. Nelson, Gary Porter, and Thomas Stelmack for clinical examination and refraction of the experimental animals. We are also grateful to Dr. Steven H. Barry for helpful suggestions concerning data analysis and to B. Clemons and S. Garcia for manuscript preparation. Finally, we thank Dr. Alfred A. Rosenbloom, Jr., for the support and encouragement which has made this work possible.

References

1. R. W. Guillery, Brain Res. 14, 739 (1969).
2. R. W. Guillery and J. H. Kaas, J. Comp. Neurol. 143, (1971).
3. R. Kalil, S. Jhaveri and W. R. Richards, Science 174, 302 (1971).
4. J. H. Kaas and R. W. Guillery, Brain Res. 59, 61 (1973).

5. D. H. Hubel and T. N. Wiesel, J. Physiol. 218, 33 (1971).
6. R. W. Guillery, V. A. Casagrande and M. D. Oberdorfer, Nature 252, 195 (1974).
7. S. J. Cool and M. L. J. Crawford, Vision Res. 12, 1809 (1972).
8. J. Packwood and B. Gordon, J. Neurophysiol. 38, 1485 (1975).
9. E. I. Elekessy, J. E. Campion and G. H. Henry, Vision Res. 13, 2533 (1973).
10. R. W. Guillery and V. A. Casagrande, J. Comp. Neurol. 174, 15 (1977).
11. R. Blake and M. L. J. Crawford, Brain Res. 77, 492 (1974).
12. Y. M. Chino, M. S. Shansky and D. I. Hamasaki, Science 197, 173 (1977).
13. Y. M. Chino, M. S. Shansky and D. I. Hamasaki, Brain Res. 143, 459 (1978).
14. R. Blake and D. N. Antoinette, Science 194, 109 (1976).
15. M. S. Shansky, Y. M. Chino and D. I. Hamasaki, This Symposium (1978).
16. S. M. Sherman, Brain Res. 37, 187 (1972).
17. R. Fernald and R. Chase, Vision Res. 11, 95 (1971).
18. G. J. Vakkur, P. O. Bishop and W. Kosak, Vision Res. 3, 289 (1963).
19. H. B. Barlow, C. Blakemore and J. D. Pettigrew, J. Physiol. 193, 327 (1967).
20. T. Nikara, P. O. Bishop and J. D. Pettigrew, Exp. Brain Res. 6, 353 (1968).
21. P. O. Bishop, G. H. Henry and C. J. Smith, J. Physiol. 216, 39 (1971).
22. D. H. Hubel and T. N. Wiesel, J. Physiol. 160, 106 (1962).
23. J. D. Pettigrew, T. Nikara and P. O. Bishop, Exp. Brain Res. 6, 373 (1968).
24. P. O. Bishop and G. H. Henry, Invest. Ophthalmol. 11, 346 (1972).
25. S. M. Sherman, D. W. Watkins and J. R. Wilson, Vision Res. 16, 919 (1976).
26. P. O. Bishop, J. S. Coombs and G. H. Henry, J. Physiol. 219, 659 (1971).
27. A. W. Goodwin, G. H. Henry and P. O. Bishop, J. Neurophysiol. 38, 1500 (1975).
28. A. W. Goodwin and G. H. Henry, J. Neurophysiol. 38, 1524 (1975).
29. G. H. Henry, R. Dreher and P. O. Bishop, J. Neurophysiol. 37, 1394 (1974).
30. D. Rose and D. Blakemore, Exp. Brain Res. 10, 1 (1974).
31. D. W. Watkins and M. A. Berkley, Exp. Brain Res. 19, 433 (1974).
32. D. E. Joshua and P. O. Bishop, Exp. Brain Res. 10, 389 (1970).
33. K. P. Hoffman and S. M. Sherman, J. Neurophysiol. 37, 1276 (1974).
34. K. P. Hoffman and S. M. Sherman, J. Neurophysiol. 38, 1049 (1975).
35. K. P. Hoffman, J. Stone and S. M. Sherman, J. Neurophysiol. 35, 518 (1972).
36. S. M. Sherman, K. P. Hoffmann and J. Stone, J. Neurophysiol. 35, 532 (1972).
37. D. I. Hamasaki and J. T. Flynn, Vision Res. 17, 275 (1977).
38. H. B. Barlow and J. D. Pettigrew, J. Physiol. 218, 98P (1971).
39. C. Blakemore and R. C. Van Sluyters, J. Physiol. 237, 195 (1974).
40. C. Blakemore and R. C. Van Sluyters, J. Physiol. 248, 663 (1975).
41. M. Imbert and P. Buisseret, Exp. Brain. Res. 22, 25 (1975).
42. R. W. Guillery and V. A. Casagrande, Cold Spring Harbour Symposia on Quant. Biol. 40, 611 (1976).
43. J. Stone, Invest. Ophthalmol. 11, 338 (1972).
44. D. H. Hubel and T. N. Wiesel, J. Neurophysiol. 28, 229 (1965).

Questions and Answers

Q. Dr. BLAKE: I would like to add one conclusion to your list of impressive findings and that is as proposed by some, STONE among others, that X cells project primarily to simple cells in the normal cat and Y to complex and your results are certainly inconsistent with that idea, since the Siamese cat cortex is looking at the severely depressed Y input. Yet you are finding normal encounter rates of complex cells.

A. Since we have shown that the encounter rate of Y-units in the Siamese retina is significantly lower than in normals, the frequency distribution of simple and complex cells in Siamese cortex should provide us with an interesting observation as to the nature of information transfer in the cat visual system. That is, the parallel processing theory (STONE (43)), which states that simple cortical cells in cats receive direct input from LGN X-cells, and complex cortical cells from Y-cells, would predict that there should be a significant reduction in the percentage of complex cells in the visual cortex of strabismic Siamese cats in comparison to normal domestic cats. On the other hand, the serial processing theory, (HUBEL & WIESEL (22,44)) would predict no difference in frequency distribution of simple and complex cells between Siamese and normal cats. Our cortical data show that there is no difference in the ratio of simple to complex cells between the two breeds, and thus appear to favor the serial processing theory.

On the other hand, we have found that the orientation selectivity and direction selectivity of complex cells in the adult Siamese cats are abnormal whereas those of simple cells are normal, when compared to normal cats. Furthermore, the sizes of the receptive fields of complex cells in Siamese striate cortex are significantly larger compared to those of normal cats. The conclusions from these data described here also include an important one with respect to the nature of information transfer in the mammalian visual system. That is, our findings of reduced encounter rates of Y-type retinal ganglion cells and abnormal properties in striate complex cells are consistent with the parallel theory of information processing. The serial theory of information processing might also be able to explain our results, but with more difficulty.

The origin of the apparent conflict between our findings on the simple/complex ratio and the data on receptive field properties in relation to the nature of information transfer is difficult to explain at the present time. One obvious explanation would be that the loss of Y-cells in the retina resulting in abnormal receptive field properties of complex cortical cells, is not severe enough to reduce the encounter rate of complex cells in the striate cortex. We are currently involved in several projects which may help us to reconcile these problems, as I mentioned in the conclusion of my talk.

Role of Binocular Interactions in Visual System Development in the Cat

Peter D. Spear
Department of Psychology
University of Wisconsin
Madison, Wisconsin 53706

Introduction

In normal cats, the majority of striate cortex cells can be excited by stimulation of both eyes (1, 2). This is illustrated in Fig. 1., which shows the ocular dominance distribution of striate cortex cells in normal cats studied in our laboratory. Numbers on the abscissa represent ocular dominance groups: Group 1 includes cells driven only by the contralateral eye, group 4 includes cells driven about equally by both eyes, and group 7 includes cells driven only by the ipsilateral eye. Groups 2, 3, 5 and 6 include binocularly driven cells with intermediate degrees of dominance by the two eyes. Most of the cells are binocularly driven in these normally reared animals (groups 2-6).

A very different picture emerges if the cats are reared with monocular pattern vision deprivation (3, 4), which can be produced by suturing the lids of one eye together prior to the time of natural eye-opening. When these animals reach maturity, the deprived eye is capable of driving only about 5% of the striate cortex cells. This is illustrated in Fig. 2. In this figure, the results for the hemispheres ipsilateral and contralateral to the deprived eye are shown separately. The experienced eye dominates nearly all of the cells and the deprived eye is able to drive few cells in both hemispheres. Furthermore, the receptive field properties of cells which respond to the deprived eye tend to be very abnormal.

Evidence from a variety of sources indicates that these effects of monocular deprivation are due in large part to a competitive interaction between inputs from the two eyes during development, and not simply from the lack of pattern vision *per se*. For example, portions of striate cortex which receive inputs from only the deprived eye, such as the region representing the monocular crescent of the visual field, are much less affected by monocular deprivation than are binocular portions of the cortex (5, 6). In addition, if both eyes are deprived instead of just one, the cortical abnormalities for either deprived eye are much less severe than following monocular deprivation (4, 7). Results such as these suggest that binocular competition plays an important role in the effects of monocular deprivation.

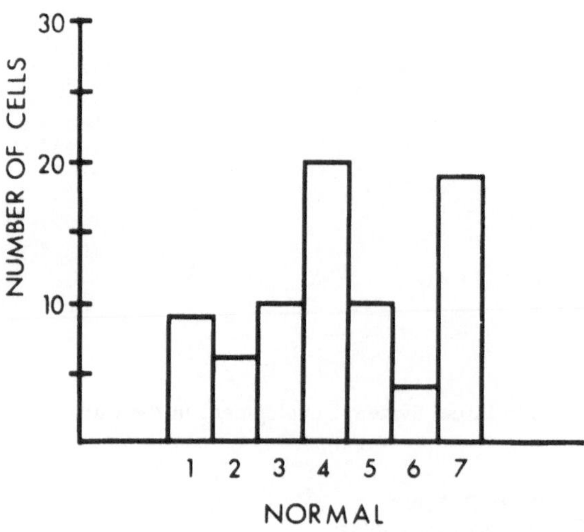

Fig. 1 Ocular dominance distribution of striate cortex cells in 4-8 month old normally reared cats. From KRATZ and SPEAR (7).

HUBEL and WIESEL (8) have shown that there is a critical period of development during which the striate cortex is susceptible to the effects of monocular deprivation. The critical period ends at about 3 months of age for cats. Monocular deprivation prior to this age produces marked changes in cortical ocular dominance, like those shown in Fig. 2, whereas monocular deprivation after this period has little or no effect. Many studies also have attempted to reverse the effects of monocular deprivation after the critical period by opening the deprived eye, or by a reverse-suture procedure in which the deprived eye is opened and the experienced eye is closed. In some cases, discrimination training also has been given through the initially deprived eye following reverse-suture. These studies have found little or no reversal of the effects of monocular deprivation after the critical period (8-12).

The series of experiments I will describe today had three general purposes. The first was to investigate the mechanisms which maintain the effects of monocular deprivation after the critical period. In particular, we were interested in the possibility that continued binocular competition is a factor. The second purpose was to learn more about the functional nature of the abnormal connections from the deprived eye following rearing with monocular deprivation. The third purpose was to investigate further the role played by binocular interactions during the course of development with monocular deprivation, and the changes in functional connections from the deprived eye which occur during development. The neurophysiological experiments I will describe were done in collaboration with KENNETH KRATZ, DOUGLAS SMITH, and ARNE LANGSETMO working in my laboratory. The morphological experiments were conducted in collaboration with TERRY HICKEY, who is at the School of Optometry at the University of Alabama.

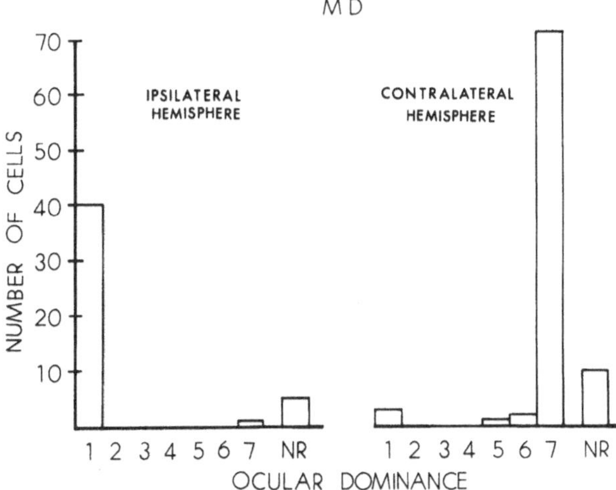

Fig. 2 Ocular dominance distribution of striate cortex cells in 4-8 month old monocularly deprived cats. NR indicates cells not responsive to visual stimulation. Ipsilateral and contralateral hemispheres are relative to the deprived eye. Data combined from KRATZ, SPEAR and SMITH (13) and SMITH, SPEAR and KRATZ (11).

Postcritical-period Reversal of the Effects of Monocular Deprivation on Striate Cortex Cells

We began by asking whether the binocular competition between inputs from the experienced eye and the deprived eye in monocularly deprived cats continues after the critical period. That is, if the influence of the experienced eye were removed altogether by enucleation, would the deprived eye be capable of some reorganization after the critical period? Further, if reorganization could be observed, we wanted to determine its time-course.

To answer these questions, we raised seventeen kittens with monocular lid-suture from the time of natural eye opening to the age of 4 or 5 months--well beyond the critical period delinated by HUBEL and WIESEL (8). The kittens were then divided into four different groups, which I will describe along with the results which are shown in Fig. 3. Each bar in Fig. 3 shows the mean percentage of striate cortex cells driven by the deprived eye for kittens in each rearing condition. The brackets show the total range for individual kittens in each condition. Typically, 30 to 35 cells were studied in each kitten.

The first group to be considered consisted of six monocularly deprived control kittens which were studied at 4-8 months of age. Their results are shown by the second bar (labeled MD) in Fig. 3. An average of 5% of the striate cortex cells could be driven by the deprived eye in these kittens, and a range of from 0% to 10% of the cells responded to the deprived eye in individual kittens. For comparison, the first bar (MD Others) in Fig. 3 shows the results of all other kittens in the literature which were studied after two months or more of monocular deprivation and which had no

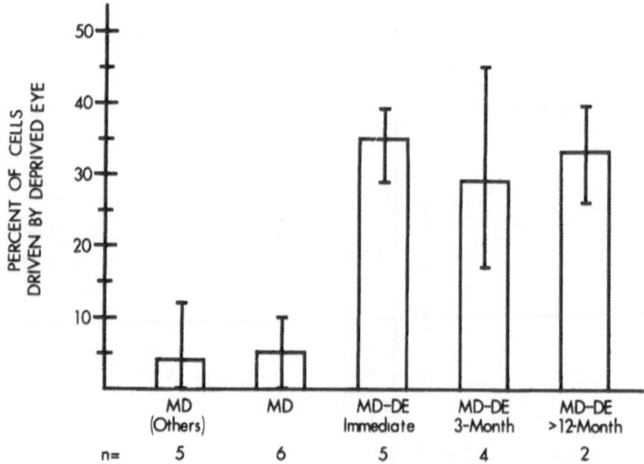

Fig. 3 Percentage of striate cortex cells which responded to the deprived eye of kittens in various rearing conditions described in the text. n indicates the number of kittens. From KRATZ, et. al. (13).

other manipulation during rearing. There were five such animals reported in the literature, all studied by WIESEL and HUBEL (3, 4). An average of 4% of the striate cortex cells responded to the deprived eye in these animals, with a range of from 0% to 12% in individual kittens. Thus, our control data agree well with those reported in the literature.

The third bar (MD-DE Immediate) in Fig. 3 shows the results of a group of monocularly deprived kittens studied immediately after removal of the experienced eye. These kittens were raised with monocular deprivation to an age of 4 (2 kittens) or 5 (3 kittens) months. They were then set up for single unit recording and several cells were sampled from striate cortex. Then, during the recording session, the experienced eye was removed under general anesthesia. Following the enucleation, recording was continued for another 24 to 30 hours. During this period, an average of 34% of the striate cortex cells responded to the deprived eye, with a range of from 29% to 39% in individual animals. Thus, acute removal of the experienced eye after the critical period produces a rapid increase in the percentage of striate cortex cells which respond to the deprived eye. A variety of control procedures indicate that this increase is not due to differences in sampling procedures, a silencing of cells connected to the experienced eye, or alterations in cortical spontaneous activity produced by the enucleation.

We were interested in knowing whether the increased responsiveness to the deprived eye would be further improved over a period of time following removal of the experienced eye. To test this, another group of animals was allowed to remain in the colony for 3 months after the experienced eye was removed. The deprived eye remained closed during this period. The results for these animals are shown by the fourth bar (MD-DE 3-Month) in Fig. 3. There was no further increase in responsiveness compared to the animals studied immediately after the experienced eye was removed. Two additional animals were studied 14 and 15 months after the experienced eye was removed. The last bar (MD-DE 12 Month) in Fig. 3 shows that there also was no further increase in the percentage of striate cortex cells which responded to the deprived eye in these animals. Thus, the rapid increase in response to the deprived eye following postcritical-period removal of the experienced eye remains unchanged over a period of more than a year.

We studied a total of 315 cells in the 11 kittens which had the experienced eye removed. Of these cells, 31% (97 cells) responded to the deprived eye, compared to 5% in the monocularly deprived control animals. We investigated the receptive field properties of the 97 responsive cells in order to determine the functional nature of the connections which were present from the deprived eye. All but 8 of the responsive cells had small, well-defined receptive fields. The receptive field size ranged from less than 1/2 degree to 7 degrees of arc, which is within the normal range. These receptive fields also followed the normal visuotopic organization of striate cortex. However, the receptive field properties were very abnormal in terms of direction and orientation selectivity. This is illustrated in Fig. 4. The figure shows

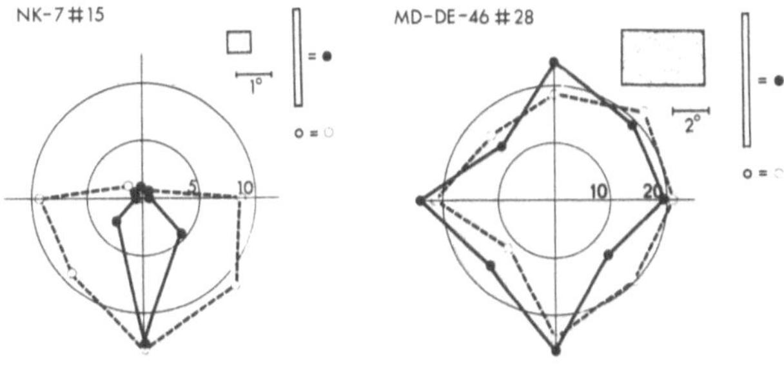

Fig. 4 Directional response range to movement of spot and slit stimuli for an orientation selective cell from a normal kitten (left) and for a non-specific cell driven by the deprived eye in a monocularly deprived kitten studied immediately after the experienced eye was removed (right). From KRATZ, et. al. (13).

polar diagrams of the responses of two striate cortex cells, one from a normal kitten (left) and one from a monocularly deprived kitten in which the experienced eye had been removed during the recording session (right). For each cell, the responses are compared for small spots and long slits moving through the receptive field in each of eight directions. Concentric circles indicate the mean number of spikes in the response for each direction. The cell recorded in a normal kitten was direction selective. It responded to downward movement but not to upward movement to both spot and slit stimuli, In addition, this cell was orientation selective because the directional tuning was much narrower for a slit than for a spot. This is one of several criteria which were used to define orientation selectivity (14, 15). In contrast, the cell which responded to the deprived eye after the experienced eye was removed showed no difference in response to spot and slit stimuli. The cell simply responded to any stimulus moving in any direction. It was neither direction selective nor orientation selective.

Among the 97 cells which responded to the deprived eye, 61% had non-specific receptive fields like that shown on the right side of Fig. 4. This is summarized in Fig. 5. Twenty-seven percent of the cells responded selectively to the direction of

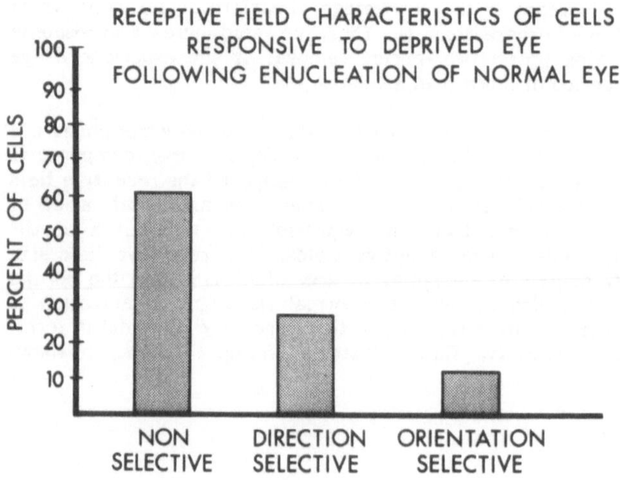

Fig. 5 Summary of receptive field characteristics of 97 cells which responded to the deprived eye following enucleation of the experienced eye in monocularly deprived cats. Data from KRATZ, et. al. (13).

stimulus movement, but were not orientation selective. Only 12% of the cells had normal appearing orientation selective receptive fields. There was no change in these properties with the length of time following removal of the experienced eye.

We also investigated the location of the responsive cells within striate cortex. Figure 6 shows reconstructions of microelectrode penetrations from three monocularly deprived animals which had the experienced eye removed. Each cell which was studied is shown by a slash mark along the penetration, and the responsive cells are indicated by a star. Two points are illustrated by this figure. First, the responsive cells were found to occur in all layers of striate cortex, with the exception of layer I which is relatively cell-free. This is best shown by cat MS-16. We do not yet have a sufficient sample to determine whether there is a higher proportion in some layers than in others. However, the combined data from all available penetrations do not yet show any marked differences. The second point illustrated by Fig. 6 is that the responsive cells occurred in clusters of several responsive cells separated by regions of non-responsive cells. This is shown best by the tangential penetration in cat MS-19. This finding suggests that the responsive cells may be related to the vertical ocular dominance columns.

It is possible to draw several conclusions from the results I have described thus far. If we assume that the results are not due to a rapid sprouting or anatomical growth of connections in the 4 to 5 month old animals, then they indicate that functional connections from the deprived eye to striate cortex remain after extended periods of monocular deprivation. However, the connections are abnormal since the receptive field properties are abnormal. In addition, the results indicate that the responsiveness of striate cortex cells to the deprived eye depends upon a continued interaction with the experienced eye well after the end of the critical period. When the experienced eye is removed, the responses to the deprived eye rapidly reappear.

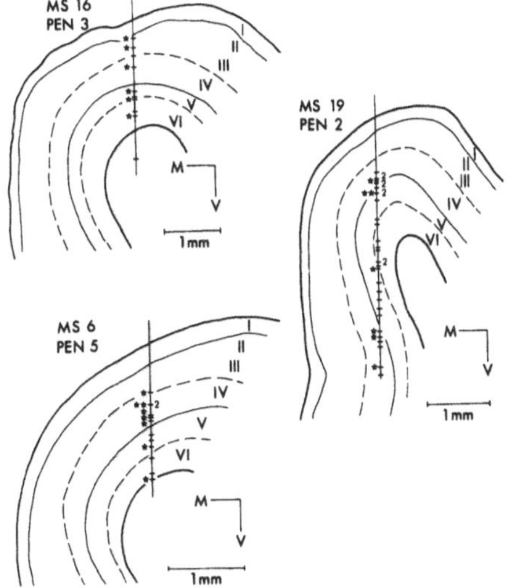

Fig. 6 Reconstructions of microelectrode penetrations through striate cortex of three monocularly deprived cats studied after the experienced eye was removed. Data from KRATZ, et. al. (13).

Presumably, this is due to release of a suppressive input from the experienced eye.

Effects of Visual Experience on Postcritical-period Reversal of Monocular Deprivation

We wondered whether removing the experienced eye in the monocularly deprived animals extends the critical period so that the connections from the deprived eye become "plastic" and can be modified by the visual environment, or whether they are simply released from suppression and cannot be further modified by visual experience. Operationally, we wanted to know whether allowing the deprived eye visual experience for some time following removal of the experienced eye would result in an improvement in the number of cells responsive to the deprived eye and/or in the receptive field properties of the responsive cells.

Figure 7 shows the results of an experiment which addressed this question. This figure shows the percentage of striate cortex cells driven by the deprived eye for animals in different rearing conditions. The bars represent the mean for each group, and the horizontal slashes represent the results from individual kittens. The first bar (MD) shows the results from 7 monocularly deprived control kittens studied between 4 and 8 months of age. The second bar (MD-DE) shows the results of 5 kittens which were monocularly deprived until they were 4 months of age, and which then had the experienced eye removed for an additional 3 months during which the deprived eye remained closed. These two groups are the same as those described in the first experiment (Fig. 3), and the results are the same. The third bar (MD-DE-0) shows the results of 5 kittens which were monocularly deprived until they were 4 months of age, and which then had the experienced eye removed for an additional 3 months during which the deprived eye was opened and allowed visual experience. The animals were kept in a large colony room in which they were free to climb on ramps and shelves

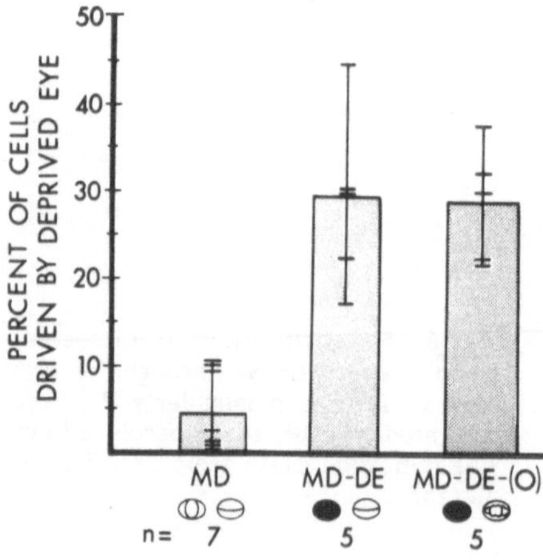

Fig. 7 Percentage of striate cortex cells which responded to the deprived eye of kittens in various rearing conditions described in the text. From SMITH, et. al. (11).

and to play with toys which were available. Their results were virtually identical to those of animals which did not have visual experience following removal of the experienced eye. There was no further increase in the percentage of cells which responded to the deprived eye.

There also was no improvement in the receptive field properties of the responsive cells. We recorded from 163 cells in the 5 cats allowed visual experience. Among these, 48 cells responded to the deprived eye (29.5%) and 77% of the responsive cells had non-specific receptive fields. They were neither direction selective nor orientation selective.

The results of this experiment indicate that enucleation of the experienced eye does not result in an extension of the period of plasticity during which connections from the deprived eye can be modified by visual experience. They again suggest that there is simply a release from suppression of connections which are already present.

Age-Related Changes in Binocular Interactions

Thus far, all of the animals I have described were 4-5 months of age at the time of removal of the experienced eye. We have seen that at this age, at least 30% of the striate cortex cells continue to have functional inputs from the deprived eye and that these connections can be revealed immediately after an acute removal of the experienced eye. We were interested in knowing whether this phenomenon changes with the age of the animal and the duration of the monocular deprivation.

There are two principal reasons for being interested in this question. First, a number of studies indicate that if a kitten receives monocular lid-suture at the time of normal eye-opening, the effects of the monocular deprivation already are present in striate cortex by 4-5 weeks of age (8, 9, 17). In cats, the age of 4-5 weeks is at the peak of the critical period (8, 9, 17). We wanted to know if the nature of the underlying binocular interactions and functional connections from the deprived eye at this age differ from those following longer periods of monocular deprivation. Second, we wanted to know if the residual connections from the deprived eye and the ability to reveal them by removing the experienced eye persisted beyond 5 months of age and into adulthood, or if they progressively declined as the animals grew older.

Figure 8 shows the results of an experiment which addressed these questions.

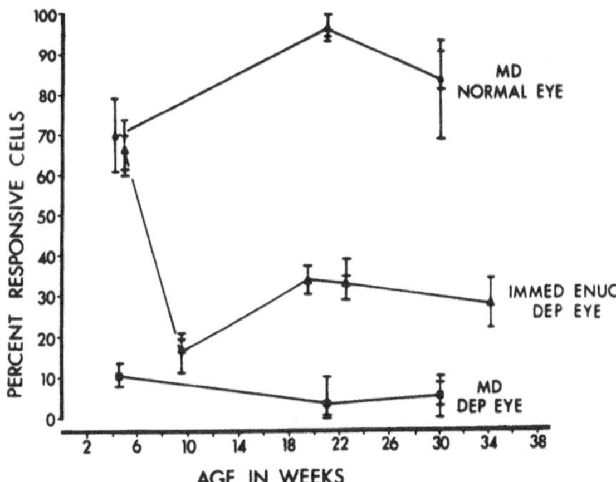

Fig. 8 Percentage of striate cortex cells which responded to the deprived eye following acute removal of the experienced eye at different ages, compared to the percentage which responded to each eye in monocularly deprived control animals. Data from SPEAR, LANGSETMO and SMITH (16).

The lower curve (MD DEP EYE) shows the percentage of striate cortex cells which could be driven by the deprived eye in monocularly deprived control animals of various ages. The results for individual animals are given by the slashes, and the average for the animals at each age are given by the filled squares. The upper curve (MD NORMAL EYE) shows the percentage of cells driven by the experienced eye in the same control animals. These two curves confirm the previous studies which found that by 4-5 weeks of age the experienced eye already dominates striate cortex, and the deprived eye is able to drive very few cells. The severity of the effects of monocular deprivation in the 4-5 week old kittens is comparable to that in the 30 week old animals.

However, the mechanisms of the effects of deprivation appear to be quite different for animals of different ages. This was demonstrated by a second series of animals. These kittens were raised with monocular lid-suture from the time of natural eye-opening to different ages, just as the monocularly deprived control animals. However, in these animals the experienced eye was removed during the recording session, and the ability of the deprived eye to drive striate cortex cells was then determined over the next 24-30 hours of recording. The middle curve (IMMED ENUC DEP EYE) in Fig. 8 shows the percentage of striate cortex cells which could be driven by the deprived eye in these animals. There were three main results. First, at 4-5 weeks of age, from 60-74% of the cells responded to the deprived eye after the experienced eye was removed. This is about the same as the percentage of cells driven by the experienced eye in control animals at this age. Thus, the percentage of striate cortex cells retaining functional connections with the deprived eye is nearly normal at 4-5 weeks of age. However, there appears to be extensive suppression of these inputs since the deprived eye drives only about 10% of the striate cortex cells in control animals with the experienced eye intact. The second finding is that by 9-10 weeks of age, near the end of the critical period, the percentage of cells responsive to the deprived eye following enucleation of the experienced eye has decreased dramatically. In fact, the percent responsive is somewhat below that seen in the 19 and 23 week old kittens, in which about 30% of the cells respond to the deprived eye immediately after removal of the experienced eye. Third, the residual functional connections with the deprived eye, and the ability to reveal them by removing the experienced eye, persists to at least 34 weeks of age (8 months). In the two animals studied at this age, 23% and 41% of the cells responded to the deprived eye after acute enucleation of the experienced eye.[1]

The receptive field properties of the responsive cells were studied to provide information about the nature of the functional connections which are present from the deprived eye at each age. These results are summarized in Fig. 9, which shows the percentage of the responsive cells which had orientation selective receptive fields. The upper curve (MD NORMAL EYE) shows that at 4-5 weeks, about 70% of the responsive cells were orientation selective through the experienced eye of monocularly deprived control kittens. This increased to about 95% in the older animals. A very different picture is seen for the cells which responded to the deprived eye after removal of the experienced eye, as shown by the lower curve (IMMED ENUC DEP EYE) in Fig. 9. Very few of these cells were orientation selective at all ages studied.

Several conclusions can be drawn from the results of this experiment. Although the outward effects of monocular deprivation appear to be the same at all ages studied, the mechanisms appear to be very different. At one month of age, the percentage of cells retaining functional connections with the deprived eye is still nearly normal, though nearly all of them appear to be suppressed by inputs from the experienced eye. In addition, while many functional connections are present at one month of age, they are very abnormal in terms of the receptive field properties they produce. Therefore, there is a dissociation during the course of development with monocular deprivation between the presence of functional connections and their functional synaptic arrangement. The connections first degrade in their spatial and temporal specificity which produce the detailed orientation selective receptive fields, and then their ability to drive the cells decreases. Nevertheless, at least 30% of the striate cortex cells retain functional connections with the deprived eye, and can be revealed by removal of the experienced eye, until 8 months of age or more.

Reversal of Dorsal Lateral Geniculate Cell Size

All of the results I have described so far have been concerned with striate cortex physiology. We were interested in knowing whether a similar reversal of the effects of monocular deprivation could occur on dorsal lateral geniculate cell size after the

[1]Since this symposium was held, we have increased the number of animals studied at each of the ages shown in Fig. 8. The additional results confirm those shown in the figure. In addition, we have now studied the effects of acute removal of the experienced eye in two cats monocularly deprived to an age of 16 and 22 months. In these two cats, 26% and 27% of the striate cortex cells responded to the deprived eye immediately after enucleating the experienced eye. Therefore, the residual connections from the deprived eye and the ability to reveal them by removing the influence of the experienced eye persists well into adulthood.

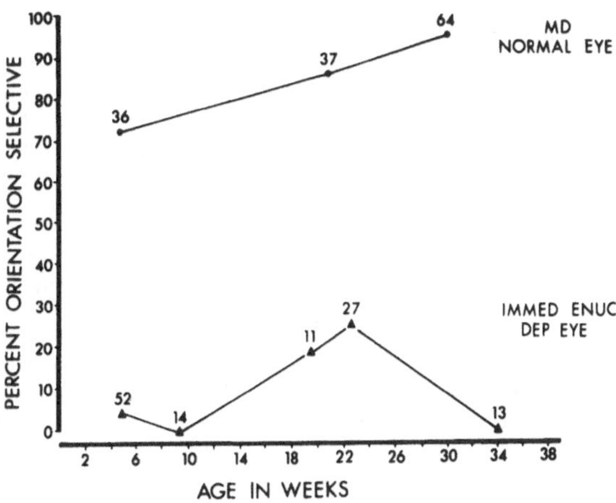

Fig. 9 Percentage of responsive cells which had orientation selective receptive fields in cats of various ages. The number above each point indicates the number of cells fully tested for orientation selectivity at each age. Data from SPEAR, et. al. (16).

critical period.

The cat's lateral geniculate is a laminated structure, and Fig. 10 is a schematic representation of the two most prominant layers. Layer A receives inputs from the contralateral eye while layer A1 below it receives inputs from the ipsilateral eye.

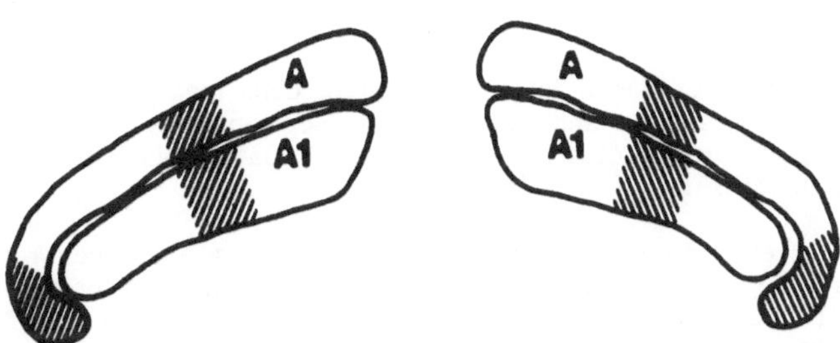

Fig. 10 Schematic representation of layers A and A1 of the two dorsal lateral geniculate nuclei of the cat.

The medial portion of the nucleus in which both layers are apposed represents the binocular segment of the nucleus. Adjacent regions of each layer, such as the two adjacent cross-hatched areas, represent corresponding locations in the visual field for each eye. The lateral portion of the nucleus consists only of layer A, and this represents the monocular crescent of the visual field seen only by the contralateral eye. This is called the monocular segment of the nucleus. Previous studies have shown that monocular deprivation produces a severe reduction in size of cells in the lateral geniculate layers which receive their inputs from the deprived eye. This reduction in cell size is much greater in the binocular segment of the nucleus than in the monocular segment, reflecting the importance of binocular competition in decreased cell size (18-21).

We investigated whether removal of the experienced eye would produce a reversal of the reduced cell-size in the layers of the lateral geniculate which receive inputs from the deprived eye. Cells were measured in the binocular segments of layers A and A1 and in the monocular segment of layer A in the regions shown by the cross-hatching in Fig. 10. One hundred cells were measured in each region of the nucleus, and mean values were determined for each region in each animal. The results of these measures for animals in different rearing conditions are shown in Fig. 11. The shaded area (NORMAL) in Fig. 11 represents the mean $\pm$ the standard error

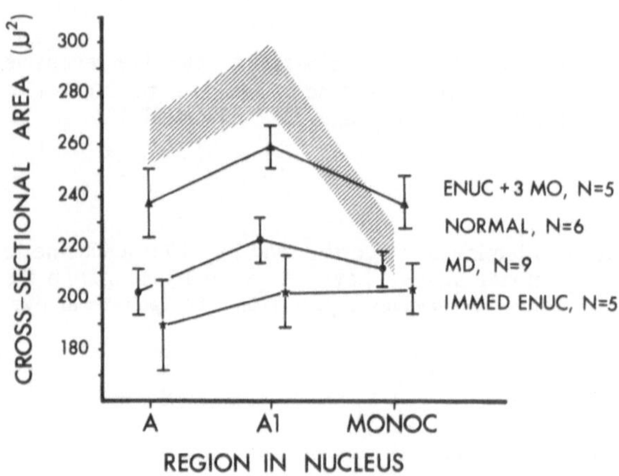

Fig. 11 Cross-sectional area of dorsal lateral geniculate cells measured in the binocular segment of layers A and A1 and the monocular segment of layer A for kittens in various rearing conditions described in the text. Data from HICKEY, SPEAR and KRATZ (21) and SPEAR and HICKEY (22).

of the values obtained for each region of the nucleus for six normal cats aged 4-8 months. Closed circles (MD) show the mean $\pm$ the standard error (brackets) for cells in the deprived layers of nine cats raised with monocular lid-suture to an age of 4-8 months. Compared to the normal animals, there was approximately a 25% decrease in cell-size in the binocular segments of both layers A and A1 following monocular deprivation. However, in the monocular segment of layer A there was only a very small decrease in cell size. These results thus replicate the previous studies of the effects of monocular deprivation on lateral geniculate cell size.

Another group of cats was raised with monocular lid-suture to an age of 4-5 months and sacrificed one day after the experienced eye was enucleated. These were the same cats which were studied in the experiments on cortical neurophysiology. The mean $\pm$ the standard error of the cell-size in the deprived layers of these animals are shown by the stars (IMMED ENUC) in Fig. 11. The cells in the deprived binocular segment of layers A and A1 were significantly smaller than normal, and they were not significantly different from the cell-size in the monocularly deprived control animals. Thus, enucleation of the experienced eye does not produce a rapid change in size of the deprived lateral geniculate cells.

Finally, a group of cats was raised with monocular lid-suture to an age of 4 months, received enucleation of the experienced eye, and then remained in the colony an additional 3 months during which the deprived eye remained closed. These also were the same animals from which cortical recordings were made. The cell-size in the deprived layers of these cats is shown by the triangles (ENUC + 3 MO) in Fig. 11. Cells in the binocular segment of layers A and A1 were significantly larger than those in both the monocularly deprived control animals and the animals studied within a day of enucleating the experienced eye. On the other hand, they were not significantly different in size from normal. Thus, postcritical-period enucleation of the experienced eye produces an increase in the size of deprived lateral geniculate cells over a period of three months following enucleation, even though the deprived eye remains closed.

These morphological results point to three conclusions. First, competition between projections from the deprived and experienced eyes also continues beyond the previously defined critical period in the case of dorsal lateral geniculate cell size. By manipulating the experienced eye alone, the cells connected to the deprived eye can increase in size, even though they are still deprived. Second, at a given point in time, there can be a dissociation in the effects observed in striate cortex physiology and dorsal lateral geniculate morphology. Acute enucleation of the experienced eye produces a reversal in striate cortex physiology but no effect on lateral geniculate morphology. Third, the effects in the two structures are nevertheless correlated. After 3 months, the deprived lateral geniculate cells did increase in size. Thus, the same manipulation which produces reversal in striate cortex physiology also produces eventual reversal in the dorsal lateral geniculate morphology, but with a longer time-course. This suggests that the effects in the lateral geniculate may be secondary to those which occur first in striate cortex.

General Conclusions

We are beginning to get a picture of the functional connections from the deprived eye and their interaction with inputs from the experienced eye during the course of development with monocular deprivation. This is represented schematically in Fig. 12. Each diagram in Fig. 12 shows the two retinae with their inputs to the separate layers of the lateral geniculate, and the binocular convergence of inputs from the lateral geniculate to striate cortex.

The first diagram depicts the system at the time of natural eye opening, which occurs at about 6-10 days of age. It is known from the work of other investigators that the striate cortex cells already have inputs from both eyes, and that many of the cells are orientation selective at this age (23, 24).

The second diagram depicts the initial effects of monocular deprivation, which is represented by a horizontal line in front of the left eye. Our results suggest that with monocular deprivation, the inputs from the deprived eye are suppressed by inputs from the experienced eye. The location of the suppressive interaction is not known

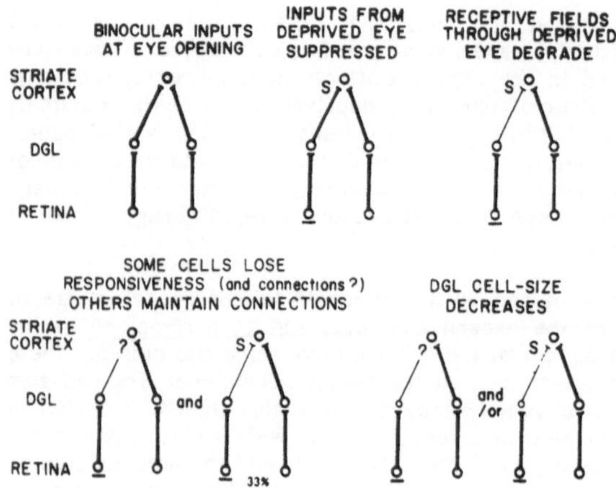

Fig. 12 Schematic representation of inputs from the deprived eye and their interaction with inputs from the experienced eye during the course of development with monocular deprivation.

with certainty, though it probably occurs in the striate cortex itself. There also is probably an interneuron involved; however, this is not shown because we do not know for sure. The suppression of inputs from the deprived eye is simply represented schematically by the letter "S" in the diagram, reflecting these uncertainties.

The third diagram depicts the system at one month of age. By this time, the suppression of inputs from the deprived eye is marked and the deprived eye can drive very few striate cortex cells in monocularly deprived kittens. However, the enucleation experiments show that many of the connections still are present at one month, since approximately the normal number of striate cortex cells can be driven by the deprived eye immediately after removal of the experienced eye. Nevertheless, the receptive fields through the deprived eye are abnormal, suggesting that the functional synaptic arrangements have degraded in their spatial and temporal specificity. This is shown in the third diagram by a thin line representing the inputs from the deprived layers of the lateral geniculate to striate cortex.

The effects of more prolonged monocular deprivation are represented by the fourth diagram (in this case, a pair of diagrams) in Fig. 12. By 2 1/2 months of age, many cells no longer respond to the deprived eye even after removal of the experienced eye. Presumably, this is due to an anatomical loss of connections from the deprived eye. However, the possibility cannot be ruled out that the connections still are present but the ability to release them from suppression has declined. The uncertainty regarding the fate of the connections to the cells which do not respond to the deprived eye after removal of the experienced eye is represented by a "?" in the left-hand member of the pair of diagrams. On the other hand, it is clear that the connections from the deprived eye persist for at least a third of the striate cortex cells. This is shown by the right-hand member of the pair of diagrams. These remaining connections from the deprived eye are present at least to 8 months of age, and possibly well into adulthood.[1] However, they are prevented from driving the cortical cells by an active binocular interaction with the experienced eye which continues well beyond the critical period of development.

The last pair of diagrams shows the decrease in cell-size in the deprived layers of the dorsal lateral geniculate. The results suggest that the reduced cell-size may be secondary to the changes in striate cortex which I have described. However, the precise relationship between the striate cortex and the lateral geniculate cell-size changes still are not fully understood. For example, we do not know the relationship between which cells change in size and which cells maintain functional, but suppressed, inputs to striate cortex. We also do not yet know the relationship between the suppressed inputs to striate cortex and the X- and Y-cell functional classes in the dorsal lateral geniculate.

In conclusion, we are beginning to understand the changes occurring during the course of development with monocular deprivation. However, many of the relationships and mechanisms still remain to be worked out.

References

1. K. Albus, Brain Res. 89, 341 (1975).
2. D. H. Hubel and T. N. Wiesel, J. Physiol. 160, 106 (1962).
3. T. N. Wiesel and D. H. Hubel, J. Neurophysiol. 26, 1003 (1963).
4. T. N. Wiesel and D. H. Hubel, J. Neurophysiol. 28, 1029 (1965).
5. S. M. Sherman, R. W. Guillery, J. H. Kaas and K. J. Sanderson, J. Comp. Neurol. 158, 1 (1974).
6. J. R. Wilson and S. M. Sherman, J. Neurophysiol. 40, 891 (1977).
7. K. E. Kratz and P. D. Spear, J. Comp. Neurol 170, 141 (1976).
8. D. H. Hubel and T. N. Wiesel, J. Physiol. 206, 419 (1970).
9. C. Blakemore and R. C. Van Sluyters, J. Physiol. 237, 195 (1974).
10. K. P. Hoffmann and M. Cynader, Phil. Trans. Roy. Soc. Lond. B. 278, 411 (1977).
11. D. C. Smith, P. D. Spear and K. E. Kratz, J. Comp. Neurol. 178, 313 (1978).
12. T. N. Wiesel and D. H. Hubel, J. Neurophysiol. 28, 1060 (1965).
13. K. E. Kratz, P. D. Spear and D. C. Smith, J. Neurophysiol. 39, 501 (1976).
14. G. H. Henry, P. O. Bishop, and B. Dreher, Vision Res. 14, 767 (1974).
15. J. D. Pettigrew, J. Physiol. 237, 49 (1974).
16. P. D. Spear, A. Langsetmo and D. C. Smith, Age-related changes in binocular interactions on striate cortex cells in monocularly deprived cats. (In preparation).
17. J. A. Movshon, J. Physiol. 261, 125 (1976).
18. R. W. Guillery, J. Comp. Neurol. 144, 117 (1972).
19. R. W. Guillery and D. J. Stelzner, J. Comp. Neurol. 139, 413 (1970).
20. T. N. Wiesel and D. H. Hubel, J. Neurophysiol. 26, 978 (1963).
21. T. L. Hickey, P. D. Spear and K. E. Kratz, J. Comp. Neurol. 172, 265 (1977).
22. P. D. Spear and T. L. Hickey, Reversal of effects of monocular deprivation on dorsal lateral geniculate cell size in the cat. (In preparation).
23. P. Buisseret and M. Imbert, J. Physiol. 225, 511 (1976).
24. D. H. Hubel and T. N. Wiesel, J. Neurophysiol. 26, 994 (1963).

Discussion

Q. Dr. Spinelli: I wonder if you have tried to block the optic nerve, let's say with anesthetic rather than enucleation.

A. No, that has not yet been tried in monocularly deprived cats.

Q. Dr. Ruskell: Were you surprised that cortex functional recovery preceded geniculate recovery?

A. To begin with I was surprised that the cortex recovered so rapidly. However, given the rapid functional recovery in cortex, it is not surprising that the changes in the lateral geniculate may be secondary if one assumes that cell-size is a reflection of the terminal arborization and/or activity of the cortical synapses that their axons support. Your question raises the whole issue of what cell size in the lateral geniculate means. It may simply be a reflection of their activity, in which case one might expect the size to follow the increased activity in cortex.

Q. Dr. Hickey: Is there any difference in the appearance of orientation selective cells through the experienced eye in a monocularly deprived animal and that in a normal animal?

A. That is a difficult question to answer experimentally. There are plenty of data to indicate that by 4-5 weeks of age many orientation selective striate cortex cells already are present in normal kittens. Therefore, any increase for the experienced eye of monocularly deprived kittens would be a matter of a relatively small percentage increase and would require many animals to demonstrate reliably. No one has yet done the experiment.

Q. Dr. Barlow: I am not quite clear what the evidence is that the effects on the lateral geniculate level are mediated by way of the cortex as you have suggested. Could it not be a matter of intrageniculate mutual inhibition between inputs for the two eyes?

A. The morphological reversal in the lateral geniculate is clearly secondary with regard to time-course. However, we do not yet know what mechanisms are involved, and the possibility that intralaminar inhibition plays a role cannot be ruled out. At this point the evidence is mainly correlational. It is possible that monocular deprivation and subsequent enucleation both produce their effects by way of entirely different mechanisms in the lateral geniculate than in striate cortex.

Animal Models for Human Visual Development

Colin Blakemore
The Physiological Laboratory
Cambridge CB2 3EG, England

Howard M. Eggers
E. S. Harkness Eye Institute
635 West 165th Street
New York, New York 10032

During the past few years there has been growing interest in disorders of development in the mammalian visual system, and, particularly, in the possibility of making correlations between experiments on animals and developmental anomalies of vision in human infants. This is an admirable strategy with some notable achievements, but it is fraught with certain inherent problems. The difficulty of making straightforward extrapolations between the results of physiological experiments in kittens and monkeys and clinical conditions is one of the themes of this paper.

The clinical condition that we wish to consider is amblyopia, usually defined as a reduction in visual acuity that cannot be corrected by optical means, but without an obvious organic cause. Its most common manifestation is the uniocular reduction in acuity associated with strabismus (especially convergent squint), anisometropia, or occlusion of one eye early in life. It is sometimes reported, however, that amblyopia can be present in both eyes, especially in cases of extreme refractive error that has gone uncorrected throughout infancy. It is also now certain that something like classical amblyopia can be restricted to a single meridian: infants that grow up with uncorrected astigmatism often have an uncorrectable deficit in acuity restricted to gratings orientated in the meridian that was originally habitually defocused (1).

Possible Animal Models of Amblyopia

Stimulus Deprivation Amblyopia

Perhaps the simplest form of amblyopia, and the one for which a physiological basis has apparently most clearly been established, is "stimulus deprivation amblyopia" (2), which is usually caused by one eye being occluded (even for a few days) within the first couple of years of life. Since the original experiments of WIESEL and HUBEL (3) it has now been clearly established that closing one eye in a kitten or a monkey, within the first few months of life, leads to the vast majority of neurones in the visual cortex losing effective input from the deprived eye. Most of them become monocularly driven by the eye that was open. Only a day or two of occlusion can cause such a change (4, 5) if it occurs at the peak of the "sensitive period" - about four weeks of age in a kitten.

Even though these consequences of monocular deprivation are perhaps the most striking example of neuronal plasticity in the mammalian visual system yet described, they are not in themselves an immediate explanation for stimulus deprivation amblyopia. Even in animals that have been continuously monocularly deprived for many months, a small but significant proportion of neurones remain excitable by the deprived eye (see KRATZ et. al, (6)), some of them being monocularly driven by that eye.

Now, although the exact neural factors that limit any sensory threshold or discrimination are far from understood, it is unlikely that simply the number of neurones responding identically to a particular stimulus is a factor of major significance in the determination of threshold. Therefore there is no reason, a priori, to believe that a monocularly deprived animal could not have virtually normal contrast sensitivity and spatial resolution, as long as those neurones that respond to stimulation of that eye (even though reduced in number) are quite normal in their individual sensitivity to contrast and ability to resolve spatial detail. The point we wish to emphasize is that it is not straightforward to argue between a developmental change in the proportion of cells responding to different stimuli and a developmental anomaly of human visual perception.

However, from published descriptions of the response properties of neurones in monocularly deprived cats, it does seem that their sensitivity, resolution and stimulus-specificity are abnormal when they are driven through the deprived eye (3). Such abnormalities go much closer to a true analogy to human stimulus deprivation amblyopia.

Anisometropic Amblyopia

The presence of a large difference in refractive error between the two eyes produces a situation that is, in some respects, a mild version of monocular occlusion. Because the accommodative state of the two eyes is usually well matched, if an anisometropic observer uses his more nearly emmetropic eye to determine his state of accommodation, the other retinal image will be habitually defocused by an amount equal to the dioptric difference in refraction between the two eyes. This blur will reduce the contrast of the retinal image in that eye and will particularly attenuate the high spatial frequency content of the visual image. Hence that eye will be subjected to degraded stimulation, perhaps only quantitatively different to the situation in a lid-sutured kitten. After all, it is clear that it is the attenuation of contrast, not the reduction in retinal illumination, which causes the cortical changes induced by monocular lid suture, since covering one eye with a translucent diffuser (virtually abolishing contrast in the retinal image) and the other with a matched neutral density filter has exactly the same effect on cortical ocular dominance as monocular lid suture (7).

Several years ago IKEDA and WRIGHT (8) pointed out that defocus would result in inadequate stimulation of those retinal ganglion cells in the area centralis that have very small receptive field centres and should therefore respond optimally to high spatial frequencies (9). In an anisometropic individual, then, the constant defocus in one eye might be expected to lead to abnormalities amongst those neurones in the visual pathway that should normally respond best to fine spatial detail in the defocused eye. Two recent studies provide complementary evidence to support this hypothesis.

First, IKEDA and TREMAIN (10) have reared kittens with accommodation and pupillary constriction continuously paralysed in one eye by the instillation of atropine. Whenever such an animal fixates near objects the image in its atropinized eye will be defocused relative to that in the other eye (assuming that the refractive state of the kitten's eye is normally emmetropic or hypermetropic). IKEDA and TREMAIN (10) recorded from the lateral geniculate nucleus (LGN) in these animals and measured the cut-off spatial frequency for each cell, by increasing the frequency of a high contrast grating, drifting across the receptive field, until the neurone just stopped responding. They found that neurones with "sustained" responses to central stimulation of the receptive field were abnormally poor in spatial resolving power in the geniculate laminae driven by the atropinized eye. Cells in the other laminae were completely normal, having cut-off spatial frequencies up to more than 5 c/deg (cycles per degree of visual angle) in the area centralis representation.

We have performed similar experiments on cortical neurones in five kittens reared with pure artificial anisometropia. The animals were kept in total darkness except for exposure in a normally lit room for an hour or two each day when they wore goggles containing a high-power negative spherical lens (-8 or -12 diopter) in front of one eye. Each animal had a total of more than 43 hours of such experience, ending at two to three months of age.

Retinoscopic examination of the kittens while they were wearing the goggles showed that they set their accommodative state according to the eye not viewing through the lens. All animals were identified by code numbers and the experimenters were unaware of which eye had been defocused. The series also contained three animals in which both eyes had been identically defocused and five completely normal kittens.

At the age of 9 to 17 weeks they were prepared for electrophysiological recording by the methods of BLAKEMORE and VAN SLUYTERS (11). Special care was taken to provide optimal optical conditions. We dilated the pupils and retracted the nictitating membranes with Homatropine sulphate and Phenylephrine HCl, and fitted contact lenses with 3mm artificial pupils. The residual refractive error was judged by ophthalmoscopy and the eyes were corrected for viewing a screen at a distance of 57 or 114 cm. The projections of the areae centrales, plotted with a reversible ophthalmoscope, were only slightly divergent, as in the normal cat, suggesting that these animals had not developed severe strabismus.

Each cortical cell was first studied by back-projecting hand-manipulated stimuli on to a tangent screen. Almost all units were orientation selective (12, 13), even when responding to stimulation of the originally defocused eye. In fact, amongst the anisometropic cats, the ocular dominance of cortical cells was substantially skewed towards the normal eye, much like the effects of extremely brief monocular occlusion (4, 5). Nevertheless, 43% of all units could be excited by stimulation of the deprived eye and 13% were monocularly driven by that eye (total number of units = 228). So, although there was a clear shift in ocular dominance, this in itself was, as in the case of monocular deprivation, an inadequate explanation for the loss of resolution experienced by human anisometropic amblyopes.

In order to examine the actual sensitivity and resolution of cortical neurones in these animals we stimulated the receptive field of each eye, in turn, with drifting gratings of sinusoidal luminance profile generated on a large display oscilloscope (28 x 25 cm) placed at a distance of 57 or 114 cm and rotated to set the grating to the optimum orientation. The stimulus was controlled by a PDP-11/10 computer which presented, in random order, 14 different spatial frequencies, all moving at the same drift rate (cycles/sec crossing any point on the screen). Each frequency was presented four times. For every stimulus, the experimenter adjusted the contrast of the grating until he judged that the cell was just detectably responding with either a modulated firing pattern or an elevated mean discharge. He pressed a button to inform the computer of this "threshold contrast" and the next grating then appeared. Finally the computer printed out all the settings, allowing the construction of a contrast sensitivity function of the type shown in Fig. 1. The ordinate plots the reciprocal of threshold contrast and the abscissa the spatial frequency of the grating. These particular data came from a cell in the cortex of a normal kitten.

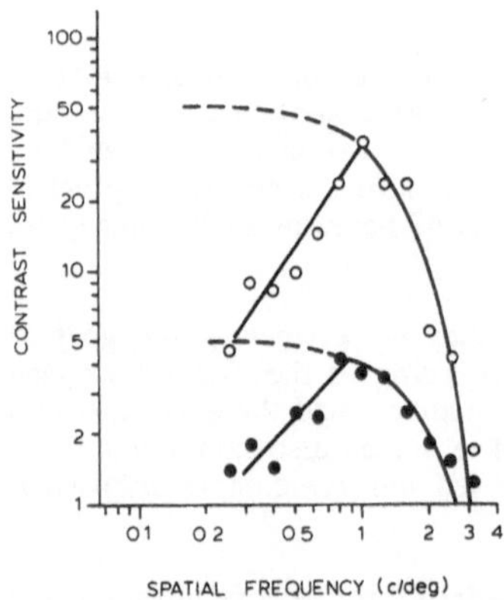

Fig. 1 Contrast sensitivity functions for a single, binocular cortical neurone from a normal kitten, determined as described in the text. Unfilled circles show data (means of four determinations) for the right eye, filled for the left eye.

Clearly the neurone has an optimum spatial frequency, with attenuation on the low- and high-frequency sides. The high-frequency attenuation is always quite well described by the exponential function:

$S = Ce^{-\alpha^2 f^2}$
where S = contrast sensitivity
f = spatial frequency
C = sensitivity constant
α = space constant

(This function was used by CAMPBELL et. al. (14) to fit the high-frequency cut-off for cortical neurones.) The low frequency attenuation was more variable in shape and we simply fitted the data points with a smooth curve.

Figure 1 shows that, in this cell (and this was true for all cells from normal cats), the properties through the two eyes differed substantially only in contrast sensitivity: the nondominant eye (in this case, the left) had lower sensitivity, although the peaks were usually very similar, as were the cut-off spatial frequencies (defined as the point at which the exponential high-frequency cut off intersects the abscissa, with contrast at 1.0 - the maximum possible).

In normal cats, cortical neurones with the highest cut-off spatial frequencies (up to 5-6 c/deg) were found within a few degrees of the area centralis representation itself, and there was a tendency for resolution to decrease in the periphery, as already described for the LGN by IKEDA and WRIGHT (15).

In the anisometropic animals the average cut-off spatial frequencies and contrast sensitivities were lower for receptive fields in the originally defocused eye, even for neurones that were monocularly dominated by that eye. Figure 2 plots the

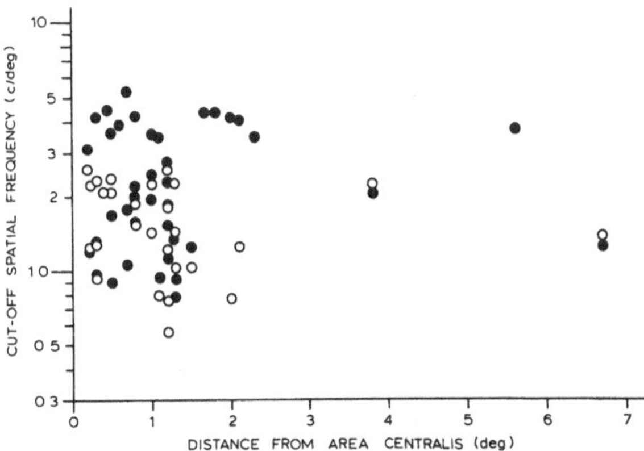

Fig. 2 Cut-off spatial frequency was determined by extrapolating the high-frequency attenuation on functions such as those in Fig. 1, for 27 receptive fields for the right eye (unfilled circles) and 39 for the left eye (filled circles). The results are plotted as a function of the distance of the center of the receptive field from the area centralis. This animal (K521) had been reared with a -12 diopter lens in front of the right eye.

cut-off spatial frequencies (derived from functions like those in Fig. 1), as a function of the retinal eccentricity of the receptive field, for an animal reared with a -12 diopter lens in front of its right eye (unfilled circles). The data for the normal, left eye (filled circles) are indistinguishable from those from completely normally-reared animals. But the cut-offs for the originally defocused eye are depressed, the maximum resolution being about an octave worse than that in the normal eye.

It has been reported that human anisometropic amblyopes have reduced contrast sensitivity, especially for high spatial frequencies (16, 17). We have confirmed this result with two anisometropic subjects, optimally refracted, viewing the display oscilloscope used in the experiments on the kittens, with the computer again presenting gratings (but stationary for these experiments) in a randomized series. Figure 3 compares the psychophysical results from one of these subjects with data from the anisometropic kitten whose cut-off frequencies are shown in Fig. 2.

To produce the diagram on the left of Fig. 3, we superimposed all the individual contrast sensitivity functions, separately for each eye. We then drew the envelope of all these functions, describing the maximum contrast sensitivity achieved by the most sensitive neurones in the sample over the whole range of spatial frequency. The envelopes (continuous lines for the left eye, interruped for the right), then, define the

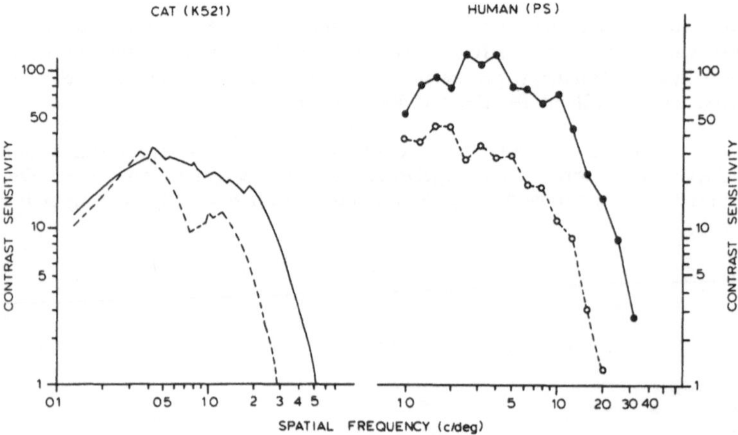

<u>Fig. 3</u> <u>Left side</u>: The 'envelope' of all individual neuronal sensitivity functions was determined as described in the text, for an animal reared with a -12 diopter lens in front of the right eye (continuous line = left eye; interrupted line = right eye). The exact level of sensitivity estimated for each cell depends on the criteria used by the experimenters to judge threshold responses. Therefore these curves cannot be taken as <u>absolute</u> determinations of neuronal sensitivity, but can legitimately be used to compare sensitivity in the two eyes. <u>Right side</u>: Psychophysically-determined contrast sensitivity functions for each eye (right eye unfilled circles; left eye filled circles) of an anisometropic subject, who was amblyopic in his right eye. Each point is the mean of four judgements of contrast sensitivity. The subject, age 39, has a refractive error of OS: +0.75 D, -1.25 D axis 170, OD +4.00, -1.75 D axis 130. His anisometropic was discovered at seven years and despite intermittent patching of the left eye from 7 - 11 years his acuity showed no maintained improvement. He has no detectable strabismus. He wore his optimal spectacle correction during the experiment. Data points for spatial frequencies below 10 c/deg were obtained with a viewing distance of 4.65 m, the remainder at 9.1 m.

maximum sensitivity seen for the eye in question over the frequency range; they are perhaps similar to overall contrast sensitivity functions, through each eye, for the whole animal, assuming that behavioural sensitivity is determined by the most sensitive neurones.

It is gratifying to see such similarity between the envelope diagrams on the left of Fig. 3 and the psychophysical contrast sensitivity functions on the right. In both cases, contrast sensitivity is depressed (as well as the cut-off frequency being lowered) and this is particularly exaggerated at high spatial frequencies.

In animals that have been reared with bilateral defocus, it is much more difficult to establish with certainty a neural deficit, because one can no longer make a comparison between the properties through the two eyes of each individual animal. One is only able to compare the overall results from experimental animals with those from normal cats, and this presents problems of obtaining exactly comparable samples. However, we and IKEDA and TREMAIN (10) have the strong impression that bilateral "amblyopia" can be produced in cats by bilateral atropinization or rearing with identical high power negative lenses in front of both eyes.

<u>Strabismic Amblyopia</u>

Deviation of one eye early in life usually leads to an impediment in binocular vision and particularly in stereopsis, but squint does not necessarily produce amblyopia. In particular, ambloypia is rare in cases of divergent squint (exotropia), where each eye is used alternately for fixation. In paralytic convergent squint (esotropia), however, the squinting eye is often profoundly amblyopic.

IKEDA and WRIGHT (15) have indeed found that LGN cells driven by the squinting eye in animals with convergent squint have the same kind of deficit in spatial resolution that is seen after unilateral atropinization. Indeed, they argue that the cause of the defect may be very similar in the two cases. If the squinting eye is never used to fixate, and therefore its accommodative state is slavishly matched to that of its more mobile fellow, its area centralis will be inevitably subjected to image defocus unless, by chance, the two eyes happen to be pointing at objects at exactly the same distance.

Exotropic kittens, however, do not have this loss of resolution in the LGN (18) because, it is argued, they tend to adopt a strategy of alternate fixation, like humans with divergent strabismus.

We have recorded from cortical neurones in three exotropic and three esotropic animals, using the same methods to measure contrast sensitivity and cut-off spatial frequency. We confirmed that esotropia, like exotropia (19) causes a wholesale breakdown in cortical binocularity. In both conditions equal numbers of cells (within sampling error) become monocularly dominated by each eye and only some 20-30% of cells are at all binocularly driven. This gross reduction in binocularity is a plausible correlate of the defective stereopsis seen in strabismic humans, but it provides no explanation, of course, for the common reduction in acuity in the squinting eye.

Using the same logical approach as that employed by IKEDA and WRIGHT (15), we have looked at cut-off spatial frequency and contrast sensitivity in these strabismic animals. For the exotropic kittens we found (as did IKEDA and TREMAIN, (18), in the LGN) that there was no detectable difference between the two eyes in cut-offs or contrast sensitivity.

We fully expected to find the situation quite different in our esotropes, in line with the findings of IKEDA and WRIGHT (15) and IKEDA and TREMAIN (18). However, in our first three animals we have again not found any dramatic difference between the performance of cells through the two eyes. Figure 4 shows "envelope" diagrams (like those of Fig. 3) for the two eyes of an esotropic animal.

We do not believe that these preliminary results contradict the observations of IKEDA and her colleagues on a large number of strabismic kittens; indeed the findings may be complementary. We think that the differences can be attributed to differences in the surgical methods used to produce the esotropia in the two studies. IKEDA and co-workers use very radical surgery, involving the complete extirpation of lateral rectus and oblique muscles, as well as removal of the connective tissue in the lateral part of the orbit and the nictitating membrane. We simply disinserted the lateral rectus, employing just the same kind of procedure used to produce an exotropia (by disinsertion of the medial rectus). Our method produced stable, large-angle squints (about 15-30 deg of deviation) but there can be no doubt that the squinting eyes were much more mobile than those of IKEDA's esotropic animals, who had a virtually total paralysis of the deviated eye.

Therefore it is, we believe, quite possible that the differences in results can be attributed to differences in ocular motility in the two experiments. This then would raise interesting questions about the origin of strabismus amblyopia, since anomalies of eye movement might play some part in its production.

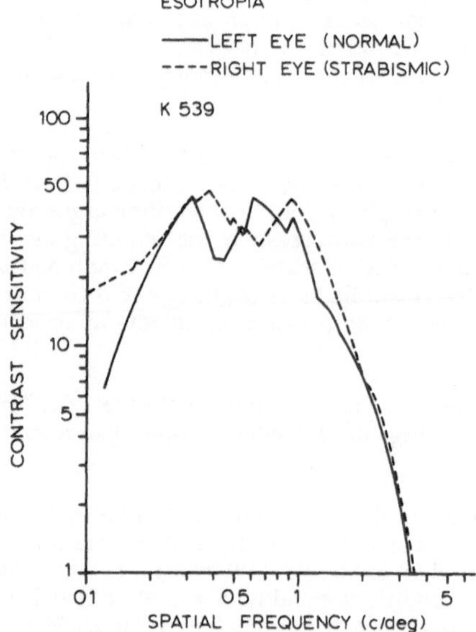

Fig. 4 'Envelope' sensitivity functions for an animal reared with artificial esotropia, induced by simple disinsertion of the lateral rectus muscle at the time of natural eye-opening.

Conclusions

The work of IKEDA and her colleagues, as well as that of MAFFEI and FIORENTINI (20) strongly suggests that the acuity loss in various sorts of amblyopia depends on defects early in the visual pathway, at the LGN or even in the eye itself. The effects we have seen at the cortex, in our anisometropic kittens, are probably simply a passive reflection of such changes at a more peripheral level. There is, however, one condition much like classical amblyopia which seems inexplicable by pre-cortical changes. The "meridional amblyopia" described by MITCHELL et, al. (1) occurs in individuals who have suffered uncorrected astigmatic defocus when young. In this case, there is an acuity deficit, and a reduction in contrast sensitivity (21, 22) but they are restricted to the meridian of original habitual defocus. Such orientationally-dependent changes in acuity are difficult to account for on the basis of changes peripheral to the visual cortex, where orientation selectivity first appears. It is therefore possible that lack of adequate stimulation can cause degradation in neuronal resolution at both the cortex and at pre-cortical levels, and that any such change can manifest itself as a reduction in acuity.

Finally, it is worth noting that although there has recently been considerable improvement in our understanding of the binocular anomalies and the simplest acuity deficits that distinguish human strabismus and amblyopia, there remain many other phenomena, such as suppression and oculomotor abnormalities, that are at present entirely unexplained.

References

1. D. E. Mitchell, R. D. Freeman, M. Millodot & G. Haegerstrom, Vision Res. 13, 535 (1973).
2. G. K. Von Noorden, Invest. Ophthal. 12, 721 (1973).
3. T. N. Wiesel & D. H. Hubel, J. Neurophysiol. 26, 1003 (1963).
4. C. R. Olson & R. D. Freeman, J. Neurophysiol. 38, 26 (1975).
5. J. A. Movshon & M. Dursteler, J. Neurophysiol. 40, 1255 (1977).
6. K. E. Kratz, P. D. Spear & D. C. Smith, J. Neurophysiol. 39, 501 (1976).
7. C. Blakemore, J. Physiol. 261, 423 (1976).
8. H. Ikeda & M. J. Wright, Vision Res. 12, 1465 (1972).
9. C. Enroth-Cugell & J. G. Robson, J. Physiol. 187, 517 (1966).
10. H. Ikeda & K. E. Tremain, J. Metabolic Ophthal. (in the press).
11. C. Blakemore & R. C. Van Sluyters, J. Physiol. 248, 663 (1975).
12. D. H. Hubel & T. N. Wiesel, J. Physiol. 160, 106 (1962).
13. D. H. Hubel & T. N. Wiesel, J. Neurophysiol. 28, 229 (1965).
14. F. W. Campbell, G. F. Cooper & C. Enroth-Cugell, J. Physiol. 203, 223 (1969).
15. H. Ikeda & M. J. Wright, Expl. Brain Res. 25, 63 (1976).
16. R. J. Gstalder & D. G. Green, J. Pediat. Ophthal. 8, 251 (1971).
17. D. M. Levi & R. S. Harwerth, Invest. Ophthal. 16, 90 (1977).
18. H. Ikeda & K. E. Tremain, J. Physiol. 269, 26P (1977).
19. D. H. Hubel & T. N. Wiesel, J. Neurophysiol. 28, 1041 (1965).
20. L. Maffei & A. Fiorentini, Nature 264, 754 (1976).
21. D. E. Mitchell & F. Wilkinson, J. Physiol. 243, 739 (1974).
22. R. D. Freeman & L. N. Thibos, J. Physiol. 247, 687 (1975).

Cortical Effects of Early Visual Experience

Helmut V. B. Hirsch
and
Audie G. Leventhal
Center for Neurobiology
The University at Albany

Cortical Effects of Early Selective Exposure

Behavioral and physiological evidence indicates that the visual system responds preferentially to horizontal and vertical patterns (see HELD's presentation at this Symposium). To determine whether this bias is a consequence of an animal's early visual experience, we have compared the effects of early selective exposure to diagonal lines with the effects of early exposure to horizontal and vertical lines (1).

Young cats, born in a laboratory colony, served as experimental subjects. Their only visual exposure was provided by masks, which they wore whenever they were in an illuminated environment; the remaining time they were housed in the dark (2). Three of these cats viewed vertical stripes with one eye and, simultaneously, horizontal stripes with the other eye. Five other animals viewed a similar pattern, but the stripes were oriented at 45° for one eye and at 135° for the other eye (Fig. 1).

To assess the cortical effects of this selective visual exposure, the response properties of single units in the visual cortex were determined (3, 4, 5, 6, 7). Tungsten microelectrodes were lowered along an oblique trajectory through the primary visual cortex (6, 8) and responses of single units were amplified and isolated. The electrode was advanced at least 75 microns between units to reduce sampling bias and to record from many different columns of orientation-sensitive cells. Once a unit had been isolated, ocular dominance was determined (3, 4), and line-shaped stimuli were presented with a hand-held projector in order to test the cell's response properties. Minimum response fields were obtained for each unit (5), and directional selectivity was assessed (9). For units showing orientation sensitivity, the preferred orientation of the cell and the range of orientations that activated the cell were determined (6).

In cats exposed only to horizontal and vertical patterns (Fig. 2), nearly all of the orientation-sensitive cells studied responded most strongly to lines oriented either horizontally ($\pm 22.5^{o}$) or vertically ($\pm 22.5^{o}$). These cells thus tended to prefer lines with the same orientation as those to which the cats had been exposed.

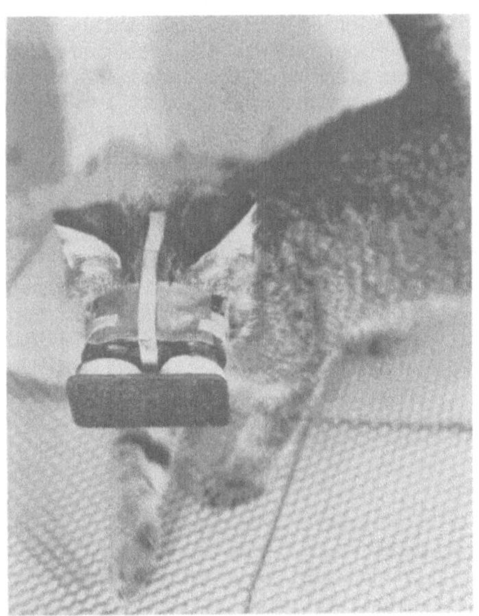

Fig. 1 Photograph of a kitten wearing one of the masks used to provide selective visual stimulation. Kittens were fitted with masks within which each eye could view a circular field, 40° to 50° in diameter, containing three parallel black lines on a white background. The lines were 1° wide, their centers separated by 6°, and their ends blurred and distorted by the lenses. At the end of an 8-hour exposure period, the animals were returned to their mother. Since the masks were put on and taken off in the darkroom, the animals' only visual experience was provided by the masks. Selective exposure was continued daily until the animals had been exposed to the patterns for approximately 100 hours. Subsequently, all of the animals remained in the darkroom until they were at least 16 weeks of age. (Reproduced with permission from HIRSCH and SPINELLI (2)).

A similar correspondence between exposure pattern and orientation selectivity was not found for cats exposed to diagonal patterns alone. In these cats, we found that 71% of the orientation-sensitive cortical cells responded preferentially to horizontal or to vertical lines, while only 29% responded best to diagonal lines ($\pm 22.5^{\circ}$) (Fig. 2) (1). In both groups of animals there were large numbers of abnormal cells which either responded indiscriminately to all of the test stimuli presented or failed to show any clear response to visual stimulation.

The cortical effects of early selective exposure to diagonal lines are thus surprisingly similar to those produced by exposure to horizontal and vertical lines. The ocular dominance distribution of those orientation-sensitive cells that responded preferentially to diagonal lines, however, did reflect the visual environment to which the animals had been exposed. Of these cells, those activated by the eye exposed to 45° stripes responded best to 45° lines, while those activated by the eye exposed to 135° lines tended to prefer that orientation.

In summary, exposing a young cat to only vertical and horizontal lines produces a dramatic modification in the distribution of the orientation preferences of its cortical neurons (2, 10). This distribution matches the animal's early experience. Exposing animals to only diagonal lines, however, does not result in as dramatic a modification of the distribution of the orientation preferences of its cortical units. In cats exposed to diagonal lines, the majority of the orientation-sensitive cells (71%) respond best to horizontal or to vertical lines; a minority (29%) respond preferentially

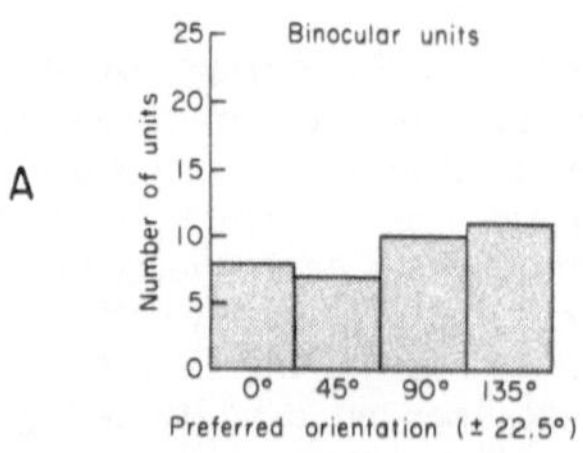

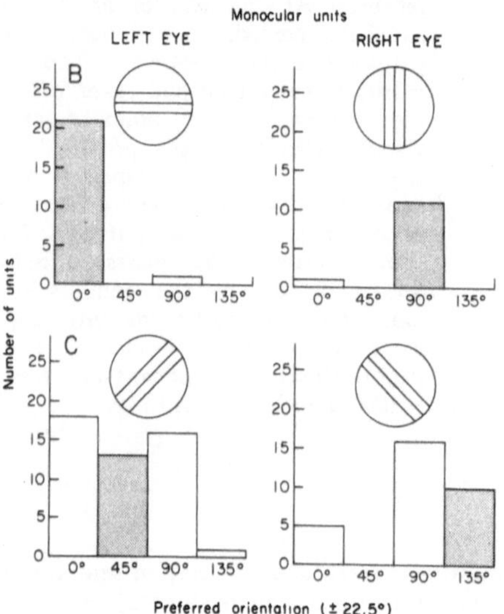

Fig. 2 Number of neurons with preferred orientations within each of four ranges. The shaded areas in each histogram indicate the orientation of the lines (± 22.5°) to which the animals were exposed during their first four months of life. The lines inscribed in each circle illustrate the patterns presented during rearing. (A) A full range of orientation preferences was found in animals raised in a normal visual environment. (B) In animals that viewed only horizontal and vertical lines, most orientation-sensitive neurons responded preferentially to horizontal or to vertical stimuli. (C) In animals that viewed only diagonal lines, most orientation-sensitive neurons responded preferentially to horizontal or to vertical lines. The rest were activated most strongly by lines with an orientation matching that of the patterns presented during rearing. The majority of the orientation-sensitive units responded to stimulation of only one eye, both in animals exposed to horizontal and vertical lines and in animals exposed to diagonal lines. For each eye of these animals, there was an absence of cells activated monocularly that preferred lines oriented at right angles to the pattern presented to that eye during rearing. (Reproduced with permission from LEVENTHAL and HIRSCH, Science, 190, pp. 902-904, Fig. 2, 28 November 1975; Copyright 1975 by the American Association for the Advancement of Science).

to diagonal lines and are activated only through the eye that was exposed to the corresponding diagonals, thereby showing a selective effect of the rearing procedure (Fig. 2).

The presence of units exhibiting preferences for horizontal or for vertical lines in cats exposed only to diagonal lines suggests to us that such neurons develop their intrinsic orientation specificity even when horizontal and vertical lines are excluded from the animal's visual environment.

To examine further the role that early visual experience plays in the development or maintenance of the response properties of cortical neurons, we compared cells in the visual cortex of normal cats with those in the visual cortex of cats reared from birth for prolonged periods without exposure to patterned visual stimuli (7).

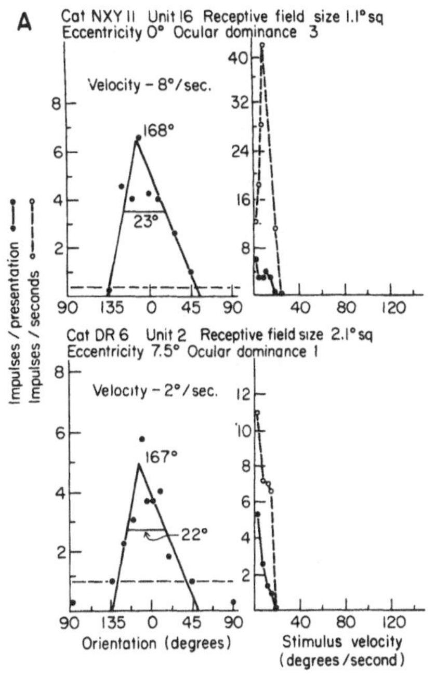
A
Cat NXY II Unit 16 Receptive field size 1.1° sq
Eccentricity 0° Ocular dominance 3
Velocity – 8°/sec.
168°
23°
Cat DR 6 Unit 2 Receptive field size 2.1° sq
Eccentricity 7.5° Ocular dominance 1
Velocity – 2°/sec.
167°
22°
Impulses / presentation
Impulses / seconds
Orientation (degrees)
Stimulus velocity
(degrees/second)

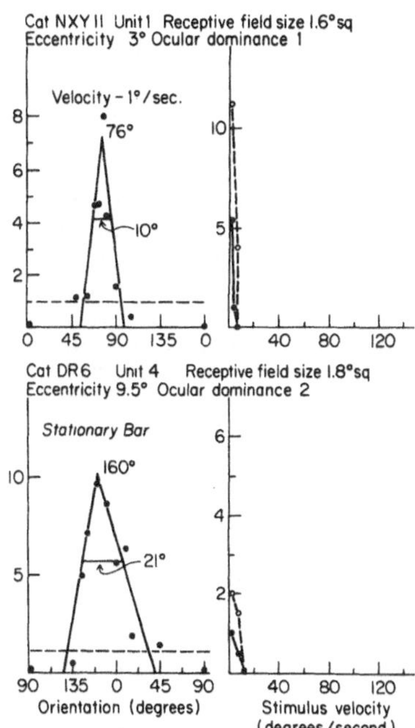
Cat NXY II Unit 1 Receptive field size 1.6° sq
Eccentricity 3° Ocular dominance 1
Velocity – 1°/sec.
76°
10°
Cat DR 6 Unit 4 Receptive field size 1.8° sq
Eccentricity 9.5° Ocular dominance 2
Stationary Bar
160°
21°
Orientation (degrees)
Stimulus velocity
(degrees/second)

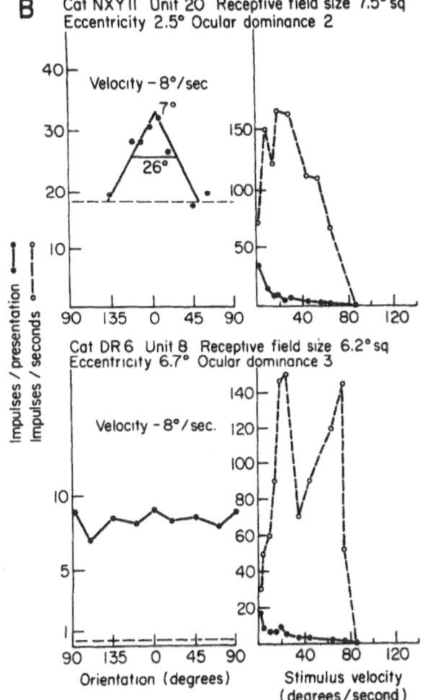
B
Cat NXY II Unit 20 Receptive field size 7.5° sq
Eccentricity 2.5° Ocular dominance 2
Velocity – 8°/sec
7°
26°
Cat DR 6 Unit 8 Receptive field size 6.2° sq
Eccentricity 6.7° Ocular dominance 3
Velocity – 8°/sec.
Impulses / presentation
Impulses / seconds
Orientation (degrees)
Stimulus velocity
(degrees/second)

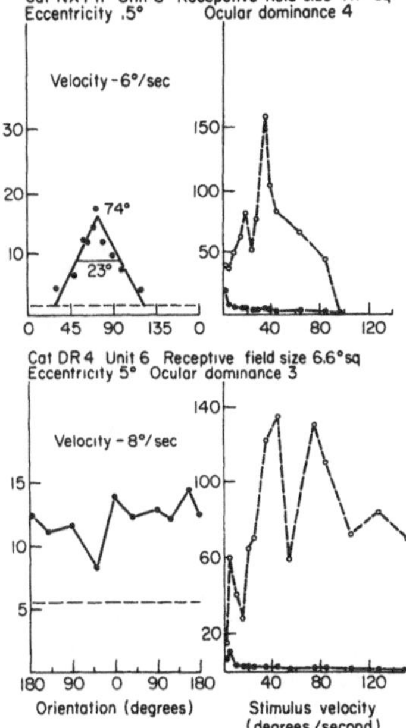
Cat NXY II Unit 8 Recepetive field size 7.7° sq
Eccentricity .5° Ocular dominance 4
Velocity – 6°/sec
74°
23°
Cat DR 4 Unit 6 Receptive field size 6.6° sq
Eccentricity 5° Ocular dominance 3
Velocity – 8°/sec
Orientation (degrees)
Stimulus velocity
(degrees/second)

Fig. 3 Orientation tuning curves and velocity tuning curves of SAS cells and of LAF cells studied in normal cats and in pattern-deprived cats. (A) (top) SAS cells: Examples of four SAS cells, the top two recorded from normal cats and the bottom two recorded from pattern-deprived cats. For each cell, an orientation tuning curve and a velocity tuning curve are presented. Data points for the orientation tuning curves have been fitted by two straight lines whose intersection determines the preferred orientation of the cell. For each orientation tuning curve, the width at half height (twice the half width) as well as the mean spontaneous discharge rate of the cell are indicated by lines parallel to the X axis. For each velocity tuning curve, the original data points (action potentials/stimulus presentation) are indicated on the lower curve (filled circles). These data points are adjusted for the time that the stimulus remained within the cell's receptive field and are displayed on the upper curve (open circles). (B) (bottom) LAF cells: Examples of four LAF cells, the top two recorded from normal cats and the bottom two recorded from pattern-deprived cats. For each cell, an orientation tuning curve and a velocity tuning curve are presented. The conventions are the same as those used in (A). (LEVENTHAL and HIRSCH, unpublished observations).

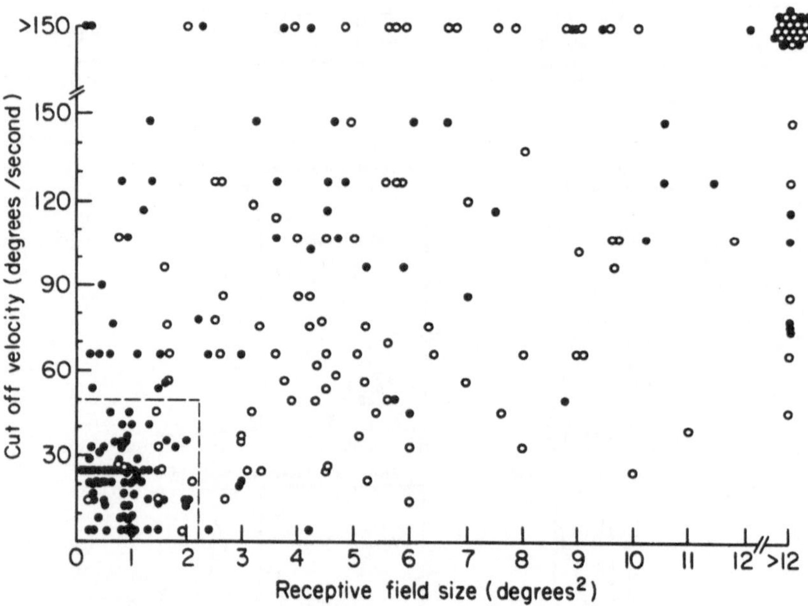

Fig. 4 Receptive field sizes and cutoff velocities (maximal stimulus velocity to which a cell responds reliably) of cortical neurons preferring either horizontal ($\pm 22.5^{\circ}$) lines or vertical ($\pm 22.5^{\circ}$) lines (solid circles) and either 45° ($\pm 22.5^{\circ}$) lines or 135° ($\pm 22.5^{\circ}$) lines (open circles). (Reproduced with permission from LEVENTHAL and HIRSCH (7)).

The Visual Cortex of Normal Adult Cats

In the visual cortex of normal cats, we have described two classes of cells: 1) cortical cells that have small receptive fields ($\leq$ 2.25 degrees^2) and that respond only to relatively slow rates of stimulus motion ($\leq$50°/sec) (Small Area--Slow or SAS); and 2) cells that have large receptive fields (>2.25 degrees^2) and/or that respond to faster rates of stimulus motion (>50°/sec) (Large Area--Fast or LAF) (7). Examples of tuning curves for both SAS and LAF cells are shown in Fig. 3. These curves illustrate both the orientation and velocity selectivity of neurons in these two classes of cells.

We have observed that a majority of the SAS cells respond best to stimuli oriented within 22.5° of either the horizontal or the vertical axis (Figs. 4 and 5). In contrast, among LAF cells a more even distribution of orientation preferences was evident (Figs. 4 and 5). Furthermore, SAS cells were found to be concentrated in cortical areas subserving central vision, while LAF cells were encountered more frequently in cortical areas subserving the peripheral visual field (Fig. 6) (7).

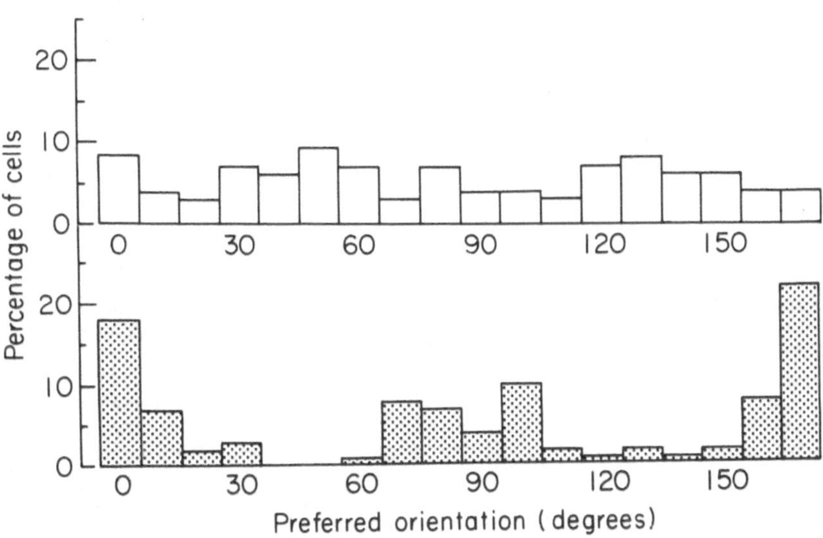

Fig. 5 Distribution of the orientation preferences of cortical neurons having both receptive fields $\leq$2.25 degrees^2 and cutoff velocities $\leq$50°/sec (shaded bars) and distribution of the orientation preferences of cells with receptive fields >2.25 degrees^2 and/or cutoff velocities >50°/sec (open bars). (Reproduced with permission from LEVENTHAL and HIRSCH (7)).

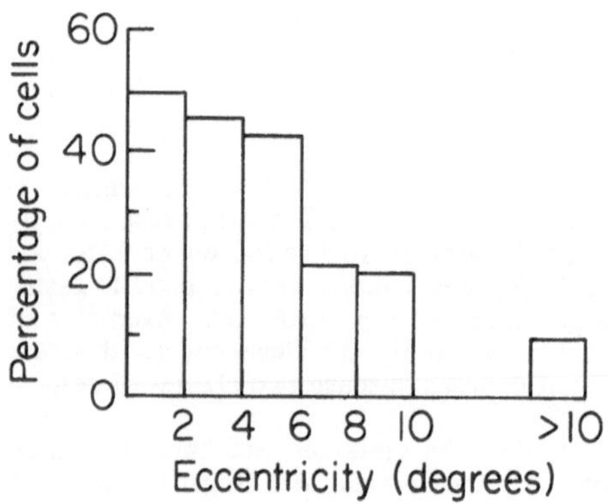

Fig. 6 Visual field distribution of neurons having cutoff velocities $\leq 50^{\circ}$/sec and receptive fields ≤ 2.25 degrees^2. (Reproduced with permission from LEVENTHAL and HIRSCH (7)).

Classification of neurons in visual cortex of normal cats as "simple" or "complex", as proposed by HUBEL and WIESEL (3, 4) fails to reveal a strong relationship between cell type and preferred orientation. In particular, in a sample of 141 cortical neurons (11), the proportion of cells responding preferentially to either horizontal ($\pm 22.5^{\circ}$) or vertical ($\pm 22.5^{\circ}$) lines was the same for cells with the characteristics of simple cells (63%) as it was for cells with the characteristics of complex cells (62%). In the same sample of cortical cells, there was a marked preponderance of cells responding preferentially to horizontal and vertical lines among SAS cells (86%), while the proportion of cells responding to horizontal or vertical among LAF cells was much lower (57%). Thus, although there are some similarities between SAS cells and "simple cells", and between LAF cells and "complex cells", the two dichotomies are not equivalent.

The Visual Cortex of Pattern-Deprived Cats

Consistent with the observations of others (6, 12, 13, 14), our investigation of the visual cortex of pattern-deprived cats (7) revealed many cortical cells that did not display the specificity of response typical of neurons studied in normal animals. In particular, over half of the neurons recorded from these animals either failed to respond to any of the visual stimuli presented or responded with no obvious preferences for the shape, orientation, or direction of movement of any stimulus presented. The remaining cells were either direction-selective (30%) or orientation-selective (15%) (6, 13).

Nearly all of the orientation-sensitive cortical cells in the visual cortex of pattern-deprived animals had properties characteristic of SAS cells: they had small receptive fields (≤ 2.25 degrees^2); they responded only to stimuli moving slowly (cutoff velocities $\leq 50^{\circ}$/sec) (Fig. 7); and they preferred either horizontal or vertical lines (Fig. 8). However, these cells were less selective for orientation than their counterparts in normal animals and nearly one fourth were entirely unselective. Finally, in contrast to normal adult cats, in which only 12% of the SAS cells were activated monocularly, in the deprived animals most cells of this type (70%), regardless of their location in the cortex, responded to stimulation of only one eye (Fig. 9) (7).

In pattern-deprived cats, most of the neurons (93%) that had response properties characteristic of LAF cells displayed little evidence of orientation selectivity, and a

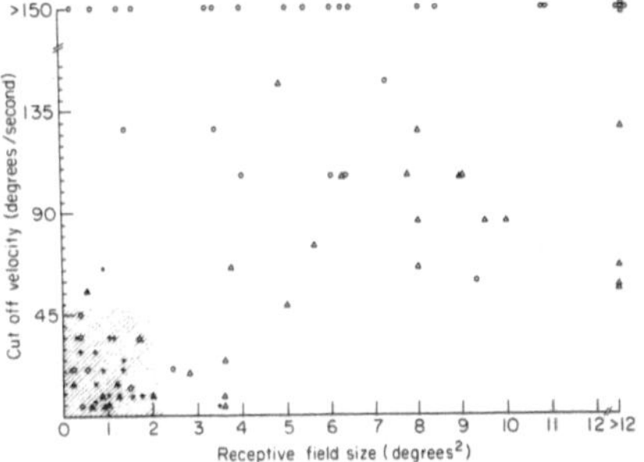

Fig. 7 Receptive field sizes and cutoff velocities of orientation-selective neurons (solid circles and triangles) and of visually-responsive cells displaying no obvious orientation preferences (open circles and triangles) in long-term pattern-deprived cats. Circles and triangles represent cells recorded from two dark-reared cats and a diffuse-reared cat, respectively. The dark-reared cats were housed in total darkness from birth until they were 8-12 months old; the diffuse-reared cat was also housed in the dark from birth to 11 months of age, except that in its second month of life it was removed from the dark daily and placed in an illuminated environment while it wore a mask in which each eye could view a blank white field. (Reproduced with permission from LEVENTHAL and HIRSCH (7)).

majority (73%) were activated binocularly. This proportion is similar to that observed in normal adult cats, in which 80% of these cells respond to stimulation of either eye (Fig. 9). However, LAF cells were encountered less frequently in pattern-deprived cats than in normal cats in all regions of cortex studied (Fig. 10) (7). Tuning curves

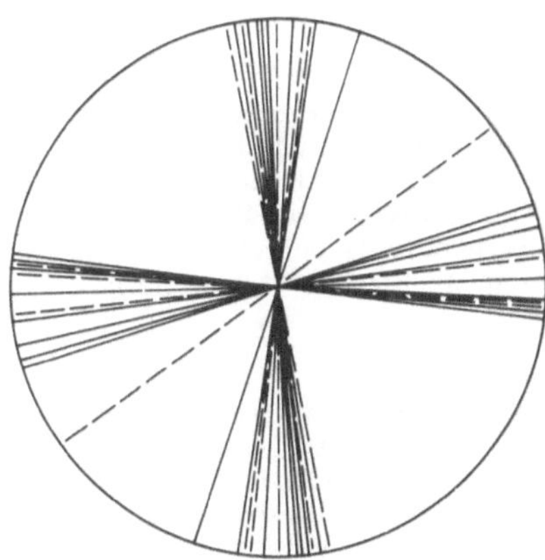

Fig. 8 Orientation-sensitive cells in the visual cortex of pattern-deprived cats. Each line represents the preferred orientation of one cortical neuron. These cells all had the characteristics of SAS cells; most responded only to stimulation of one eye. (LEVENTHAL and HIRSCH, unpublished observations).

for both cell types recorded from pattern-deprived animals are illustrated in Fig. 3.

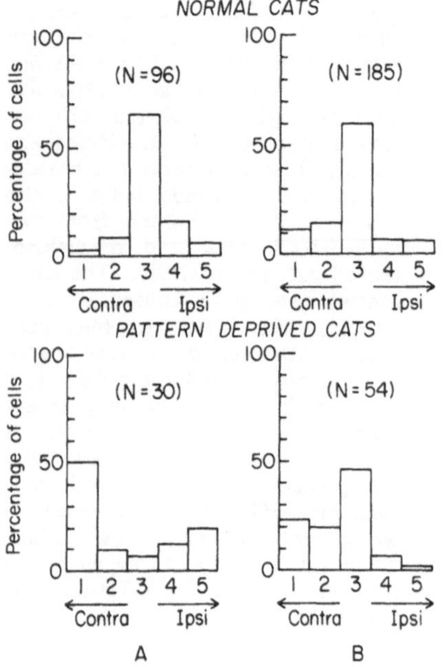

Fig. 9 Ocular dominance distributions of cells having receptive fields ≤ 2.25 degrees^2 and cutoff velocities $\leq 50^{\circ}$/sec (Column A) and of cells with larger receptive fields and/or higher cutoff velocities (Column B) in normal cats (Top) and in pattern-deprived cats (Bottom). Ocular dominance groups are defined as follows: group 1, cells activated only by the contralateral eye; group 2, cells activated by both eyes with the contralateral eye being strongly favored; group 3, cells activated by both eyes with the responses evoked by the two eyes being comparable; group 4, cells activated by both eyes with the ipsilateral eye being strongly favored; group 5, cells activated only by the ipsilateral eye. (Reproduced with permission from LEVENTHAL and HIRSCH (7)).

To summarize, SAS cells apparently do not require early exposure to patterned light for the development or maintenance of orientation sensitivity but do appear to require such experience for the development or maintenance of binocularity. On the other hand, LAF cells appear to require exposure to patterned visual stimulation for the development or maintenance of orientation selectivity but do not appear to depend upon such experience for the development or maintenance of binocularity.

Experience-Sensitive Afferents to the Cortex

The experience sensitivity of cortical neurons can be related to the afferent input which they receive from the lateral geniculate nucleus. STONE and DREHER (15) have reported that cortical neurons with small receptive fields that are selective for slow stimulus motion receive afferent inputs from X-type cells in the LGNd, whereas cells with larger receptive fields that are responsive to stimuli moving rapidly receive inputs from LGNd Y-type cells. Our results, therefore, indicate that the axons of certain X-type cells terminate on a class of cortical cells that responds best to horizontal and vertical stimuli and that is insensitive to early experience for the development of orientation sensitivity. The axons of Y-type cells, on the other hand, may terminate on neurons that require early patterned visual experience for

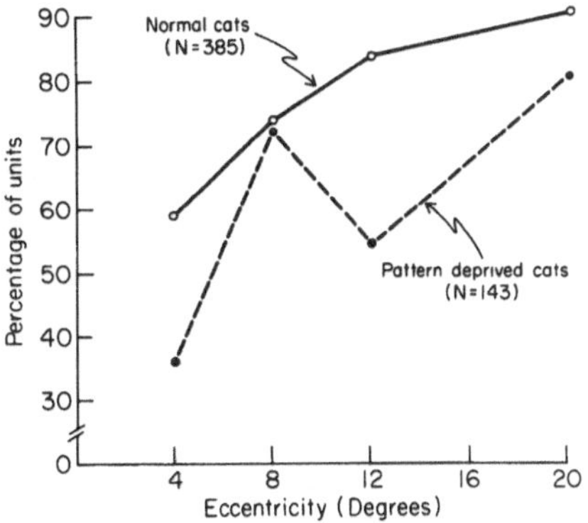

Fig. 10 Proportion of neurons having large receptive fields (> 2.25 degrees^2) and/or high cutoff velocities (> 50°/sec) that subserve different portions of the visual field. The number of units studied in normal cats (solid line) and deprived cats (dashed line) are indicated. (LEVENTHAL and HIRSCH, unpublished observations).

development or maintenance of orientation selectivity and that do not exhibit a biased distribution of orientation preferences. (A contribution of afferent inputs from LGNd W-type cells to the groups of cells we have described cannot be ruled out because the response properties of cortical cells with W-type afferents have not been determined.)

The distribution of cortical cell receptive field types is also consistent with the idea that type of afferent input is correlated with sensitivity to early experience. We have found that the relative number of SAS cells is high near the cortical projection of the area centralis and is low in more eccentric regions, while the relative frequency of LAF cells is relatively low near the projection of the area centralis and increases in portions of cortex subserving peripheral regions of the retina. These findings are consistent with the reported inhomogeneities in the distributions of X-type cells and Y-type cells in both the retina and the LGNd (16). In addition, our finding that the relative proportion of LAF cells is reduced by early binocular deprivation is compatible with the suggestion that Y-type cells in the LGNd are affected most severely by this procedure (16, 17, 18, 19, 20). Furthermore, since most SAS cells are activated monocularly in pattern-deprived cats, and possibly also in visually-deprived kittens (13, 14), their sensitivity to binocular competition still remains to be demonstrated. Our results do, however, support the hypothesis that Y-type cells in the LGNd are selectively affected by binocular competition (16, 17), because only cortical neurons receiving Y-cell afferents are binocular in the neonate. Hence, competition for synaptic space in the visual cortex is only possible among Y-type geniculate cells.

Finally, our investigation of the visual cortex following long-term pattern deprivation provides further evidence that the visual system's preferential response to horizontal and vertical patterns is determined intrinsically and reflects the response properties and distribution of certain cortical neurons, possibly a subset of those cells that receive afferent inputs from X-type cells in the LGNd.

Functional Modification of the Developing Visual System

To summarize our results: 1) In normal cats we have described a category of cells, SAS, which has small receptive fields (≤2.25 degrees^2), responds only to stimuli moving slowly ($\leq 50^{o}$/sec), and generally prefers either horizontal or vertical lines, and a category of cells, LAF, which has larger receptive fields (> 2.25 degrees^2) and/or responds to stimuli moving rapidly (> 50°/sec). We suggest that SAS cells receive input from X (and possibly W) cells in the lateral geniculate nucleus, while LAF cells primarily receive afferents from Y cells; 2) In kittens that have been reared in total darkness and are thus binocularly deprived of patterned visual stimulation, some orientation-sensitive cells are found. These neurons are mostly monocular, respond preferentially to horizontal or to vertical lines, and in general have the characteristics of SAS cells, although they tend to be somewhat less selective for orientation than SAS cells in normal cats, and, furthermore, not all the SAS cells in deprived animals are orientation sensitive; 3) The orientation-sensitive cells in the visual cortex of kittens reared with one eye viewing vertical stripes and the other eye viewing horizontal stripes are primarily monocular and respond preferentially to lines of the same orientation as those presented during development; 4) In kittens exposed to diagonal lines, however, most orientation-sensitive cells respond preferentially to horizontal or to vertical lines, and some of these can be activated through either eye. Only one third of the orientation-sensitive cells respond best to diagonal stripes, and these are activated only by the eye that viewed stimuli matching the cell's preferred orientation during the rearing.

We can account for our results using the following three factors: 1) the presence in young kittens of orientation-sensitive, monocular SAS cells, which do not depend upon sensory input for the development of orientation selectivity but do require it for normal binocularity; 2) the presence in the kitten of LAF cells, which are binocular but require sensory input to maintain or develop normal orientation selectivity; 3) the loss of binocularity that results when a LAF cell receives disparate input through the two eyes (binocular competition) (21). Thus, the only orientation-selective cells in binocularly-deprived kittens are the SAS cells, which have remained monocular in the absence of visual experience. The LAF cells remain binocular in the absence of disparate sensory input and fail to develop or maintain normal orientation selectivity.

These factors can also help to explain the cortical effects of early selective visual exposure. When an animal is exposed with one eye (for example, the right eye) to vertical stripes and with the other eye (the left eye) to horizontal stripes, then those SAS cells activated by the right eye which respond best to vertical lines (V-SAS) will be stimulated along with those SAS cells activated by the left eye which respond preferentially to horizontal lines (H-SAS). In a similar fashion, some LAF cells will respond to the horizontal or vertical stripes and develop or maintain normal orientation sensitivity; these cells become monocular because they receive unequal stimulation through the two eyes.

The situation is slightly different when cats are exposed to 45° lines with one eye and to 135° lines with the other eye. SAS cells responding preferentially to horizontal or vertical lines are stimulated minimally but equally by the two eyes. (SAS cells are broadly tuned for orientation in deprived animals.) As a result, their orientation selectivity is maintained and their binocularity is somewhat enhanced. Some LAF cells will respond to the diagonal lines and so will maintain or develop normal orientation sensitivity, while losing their binocularity as a result of the disparate stimulation of the two eyes.

In summary, using these three factors we can account rather well for the cortical effects of early selective exposure, dark-rearing and normal visual experience. In addition, our results can help explain why Y cells, which presumably supply the primary input to LAF cells, are especially dependent upon early visual experience for normal function.

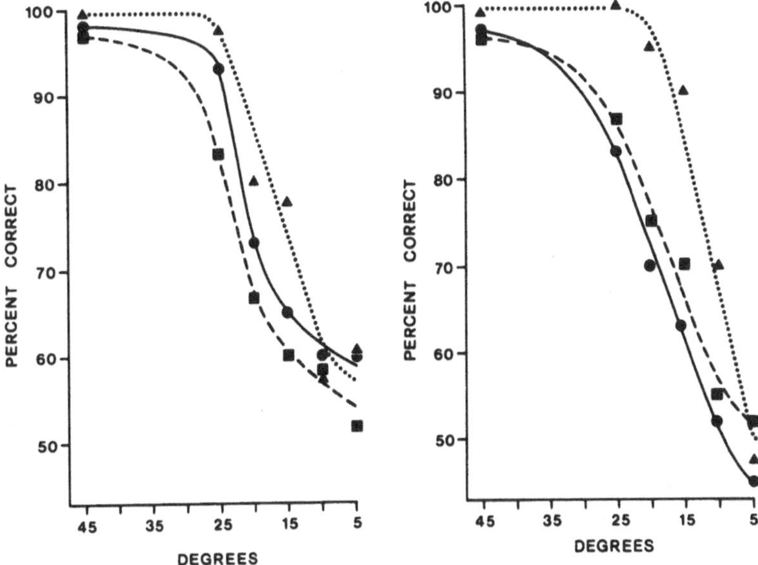

Fig. 11 Orientation Discrimination Thresholds. Mean percentage correct scores are plotted as a function of the orientation differences between the positive and negative stimuli. Performance using the eye exposed to vertical lines (VE) is represented by filled circles, performance using the eye exposed to horizontal lines (HE) by filled squares, and performance of the normally-reared control animals by filled triangles. Normal cats were tested with one eye at a time, but the data for the two eyes were then combined. Sigmoid curves were fitted by hand to the points: a solid curve for the VE data, a dashed curve for the HE data, and a dotted curve for the control animals' data. (A) Vertical Orientation Threshold (left-hand side). The positive stimulus was always vertically oriented and the negative stimuli were oriented at 45, 25, 20, 15, 10, and 5 degrees relative to the vertical axis. For all six orientation differences the experimental subjects performed better when tested with the VE than when tested with the HE. (B) Horizontal Orientation Threshold (right-hand side). The positive stimulus was always horizontally oriented and the negative stimuli were oriented at 45, 25, 20, 15, 10, and 5 degrees relative to the horizontal axis. For all but the 45 degree orientation difference the subjects performed better when tested with the HE than when tested with the VE. (Reproduced with permission from HIRSCH, (22)).

An interesting problem remains. We must explain why early selective exposure to horizontal and vertical lines produces a cortex containing a limited number of V-SAS cells activated by the eye that was exposed to horizontal lines, as well as a limited number of H-SAS cells activated by the eye that viewed vertical lines. In dark-reared cats, not all SAS cells respond in a normal fashion to visual stimulation. Exposing one eye to vertical lines and the other eye to horizontal lines thus may result in partial atrophy of those SAS cells which do not receive appropriate visual stimulation (12). Alternatively, if only some of the SAS cells are activated during an animal's early life, then the remaining SAS cells may be suppressed by a tonic functional inhibition. Such a process has been implicated to account for the cortical effects of depriving one eye of all patterned visual stimulation (23).

The Behavioral Significance of Orientation-Selective Cells?

In closing, we should perhaps question the behavioral significance of the orientation-sensitive cells found in the cat's visual cortex. As BLAKEMORE has pointed out (see BLAKEMORE'S presentation at this Symposium), although we have succeeded in altering the cortex in one way or another by varying an animal's early visual exposure, our efforts to relate this altered cortical physiology to a resulting behavioral change have met with limited success. For example, cats which were exposed with one eye to vertical stripes and with the other eye to horizontal stripes were tested on a series of orientation discrimination thresholds (22). Each cat was always tested with one eye at a time, so it was possible to compare its performance using its horizontally-exposed eye to its performance using its vertically-exposed eye. Although the experimental animals did not perform as well as normally-reared control cats, there were only small orientation-dependent changes in performance (Fig. 11). It is rather surprising that, despite what on the surface appears to be a profound difference between the cells driven by the two eyes, this behavioral task does not reveal profound differences in performance. The task of relating what we know about cortical physiology to the behavior of the intact organism remains.

Summary

To examine the role that early visual experience plays in the development of the response properties of cortical neurons, we have compared cells in the visual cortex of normal cats with those in the visual cortex of cats raised either in a controlled visual environment or without exposure to patterned visual stimulation. In all animals studied, we found neurons that responded preferentially to horizontal and vertical lines ($\pm 22.5^{\circ}$). In contrast, cells that responded preferentially to diagonal lines ($\pm 22.5^{\circ}$) were observed almost exclusively in normal control animals and in cats exposed to diagonal stripes during rearing. Furthermore, we observed that the class of neurons that has the smallest receptive fields and that responds only to relatively low rates of stimulus motion (SAS cells) is insensitive to early visual experience for development or maintenance of orientation sensitivity but appears to be sensitive to such experience for the development or maintenance of binocularity. The remaining neurons--those that have larger receptive fields and/or that respond to rapid stimulus motion (LAF cells)--appear to be sensitive to early visual experience for the development or maintenance of orientation selectivity but appear not to require such experience for the development of binocularity.

Acknowledgments

We thank D. G. Tieman and S. B. Tieman for critical comments, helpful discussions and moral support. L. M. Stern edited and prepared the manuscript. R. Loos and R. Speck prepared the figures. Support was provided by the Department of Biological Sciences at SUNYA, PHS Research Grant RO1 EY-01268 and a Fellowship from the Alfred P. Sloan Foundation.

References

1. A. G. Leventhal and H.V. B. Hirsch, Science 190, 902 (1975).
2. H. V. B. Hirsch and D. N. Spinelli, Exp. Brain Res. 13, 509 (1971).
3. D. H. Hubel and T. N. Wiesel, J. Physiol. 148, 574 (1959).
4. D. H. Hubel and T. N. Wiesel, J. Physiol. 160, 106 (1962).

5. H. B. Barlow, C. Blakemore and J. D. Pettigrew, J. Physiol. 193, 327 (1967).
6. J. D. Pettigrew, J. Physiol. 237, 49 (1974).
7. A. G. Leventhal and H. V. B. Hirsch, Proc. Nat. Acad. Sci., USA 74, 1272 (1977).
8. M. P. Stryker and H. Sherk, Science 190, 904 (1975).
9. H. B. Barlow, R. M. Hill and W. R. Levick, J.Physiol. 173, 377 (1964).
10. M. P. Stryker, H. Sherk, A. G. Leventhal and H. V. B. Hirsch, J. Neurophysiol. 41, 896 (1978).
11. A. G. Leventhal and H. V. B. Hirsch, J. Neurophysiol. 41, 948 (1978).
12. T. N. Wiesel and D. H. Hubel, J. Neurophysiol. 28, 1029 (1965).
13. C. Blakemore and R. C. Van Sluyters, J. Physiol. 248, 663 (1975).
14. P. Buisseret and M. Imbert, J. Physiol. 255, 511 (1976).
15. J. Stone and B. Dreher, J. Neurophysiol. 36, 551 (1973).
16. S. M. Sherman, K. -P. Hoffmann and J. Stone, J. Neurophysiol. 35, 532 (1972).
17. S. M. Sherman, J. R. Wilson and R. W. Guillery, Brain Res. 100, 441 (1975).
18. S. M. Sherman and J. Stone, Brain Res. 60, 224 (1973).
19. K. -P. Hoffmann and S. M. Sherman, J. Neurophysiol. 37, 1276 (1974).
20. K. -P. Hoffmann and S. M. Sherman, J. Neurophysiol. 38, 1049 (1975).
21. D. H. Hubel and T. N. Wiesel, J. Neurophysiol. 28, 1041 (1965).
22. H. V. B. Hirsch, Exp. Brain Res. 15, 405 (1972).
23. K. E. Kratz, P. D. Spear and D. C. Smith, J. Neurophysiol. 39, 501 (1976).

Neural Correlates of Visual Experience in Single Units of Cat's Visual and Somatosensory Cortex

D. N. Spinelli
Departments of Computer and Information Sciences and Psychology
University of Massachusetts

In the study of perception and pattern recognition, one is dealing with a very complex system in which some parts are, however, rather obvious. In artificial intelligence, machines built to analyze visual scenes have a preprocessor, a precessor, memory banks, and so on. A similar picture is emerging from the study of biological systems. Vision studies on the effect of early experience have provided very powerful methods and very striking results on what happens in the visual pathway. I have tried to concentrate on that part of the perceptual system that deals with memory. I will start with the first figure.

What I am showing you are some results from a very early experiment, in which the receptive fields have been mapped with a method that is objective, i.e., data is collected by computer. Naturally we also use bars and edges and other stimuli moved by hand in front of animals and listen to the loudspeaker as is done by most researchers in this field; I like to call this the subjective method. It is quick and powerful, but yields results which are difficult to compare across different experiments. In some branches of science, e.g., in engineering and physics, results are more easily compared and reproduced because experimenters have standardized their methods and use common measuring instruments. Some of the papers presented at this meeting clearly indicate that a generally accepted, objective method of mapping visual receptive fields would eliminate many controversies now raging in our field.

Another word concerning methodologies: those of you that come from medicine know that if an experimental bacteriologist is trying to find out what bacterium causes a disease, he would just run many tests on many patients until he could prove that a certain bacterium is always present when the disease is present. A pathologist, however, would not use this method at all. He knows what to look for and that gives him great power. If, in a sample, he finds a few Spirochaeta Pallida,

Fig. 1 See text for explanation.

then the patient has syphilis. He does not have to look for hundreds of thousands. He does not do a double-blind experiment.

In studying memories, in a mammalian brain, at the single cell level, using an electrode that does not have a T.V. camera at the tip, one is essentially picking blindly from a number of cells. There seems to be no way to know if one is dealing with elements which are prewired, namely the preprocessor, or elements that have been tuned or specified by experience.

SOKOLOV was once at a meeting and he suggested that only if a neural event looked very much or exactly like the experience, i.e., was isomorphic with the experience on the same dimension, we could be certain that we were looking at memory. If memories had been recorded in a non-isomorphic code then a playback could be unrecognizable. I propose to use SOKOLOV'S idea in looking at receptive fields of kittens to see if their shape resembles, in any way, the shape of the patterns the kittens viewed during development. This appraoch demands an objective, unbiased mapping method.

Figure 2 shows a receptive field mapped with a simple computerized method, which I have been using for a few years. This is a two-dimensional map in which you can actually see the "shape" of the recpetive field. For some reason, most of the time experiments concentrate on orientation sensitivity, direction sensitivity and so on. I prefer a method that also shows the shape of the receptive field because I feel that in a perceptual system that handles shapes, the shape of receptive fields is

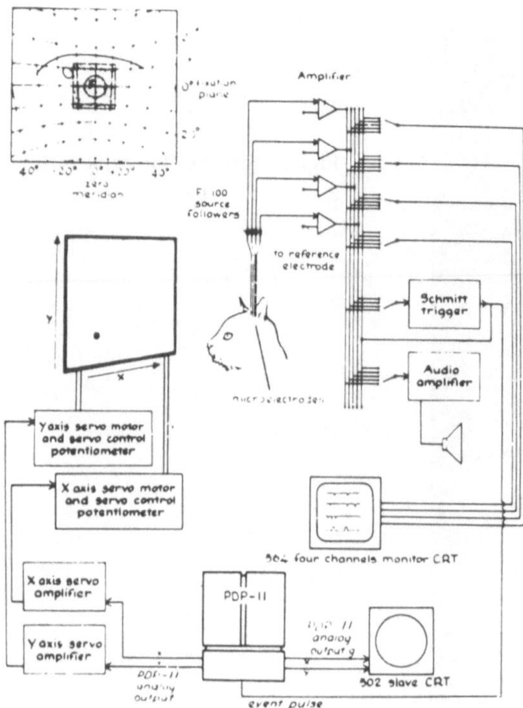

Fig. 2 See text for explanation.

important. It is what an artificial intelligence person would call the primatives in the system. Figure 2, then, shows one such two-dimensional map generated by a cell which responds only, and whose receptive field can be mapped only, through the eye that saw the three horizontal bars at the top of the picture every day during the kitten's development. The map at the bottom of Fig.2, even though it does not look like a Kodachrome picture, is definitely a recognizable image of what that animal saw through his development a few months before the recording was obtained. There are three bars in the stimulus, three in the receptive field. The bars are horizontal in the stimulus and horizontal in the map; further, the separation between bars in degrees of arc in the stimuli the kitten viewed during the experience and in the map is identical. Indeed, it would seem that the important elements of the experinece are reproduced by the cell, or more dramatically, it is now possible for the first time to "see" what another organism has seen months ago.

Let me describe very briefly the method and then review, before I come to the experiment I want to describe today, some of the work that HELMUT HIRSCH and I did in the past. The method is quite simple and in one dimension it is used all the time by people who map recpetive fields in the retina. There is a white screen and a dark spot both made of cardboard. The spot is moved under computer control back and forth and the number of spikes generated by the cell for each spot position stored. The spot is then moved up in a raster pattern so that we can, in the end,

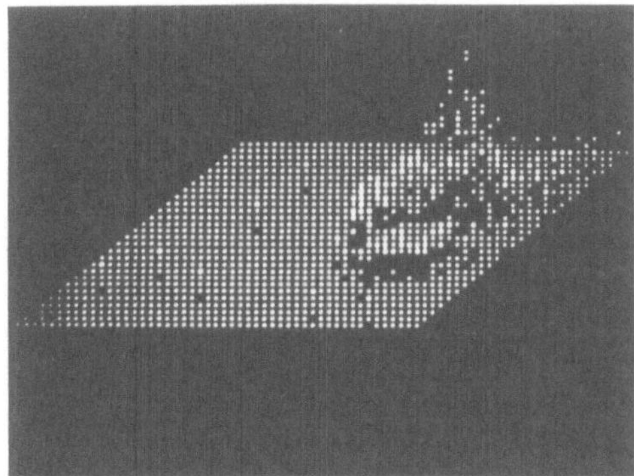

Fig. 3 See text for explanation.

obtain a two-dimensional picture. In Fig. 3 we see what the memory of the computer contains. These types of plots are used by physicists often and are called isometric displays. It is a tridimensional plot in which the X axis and the Y axis of the visual world correspond to the X axis and Y axis of the plot, while the height of the mountain represents the number of times the cell fired for the position of the spot. This is a receptive field map for a ganglion cell of the retina. We throw some data away by simply taking a cut across the mountain and now we see a concentric type of organization, excititory surround and inhibitory center. It is an off-center type of cell familiar from KUFFLER'S work on the retina, and the map looks pretty standard. With a statistical surface such as this, the cutoff plane can be moved either up or down. If it is moved up, the tip of the mountain will be shown. If moved down, the valleys, and in Fig. 4 we have an on-center cell with an inhibitory

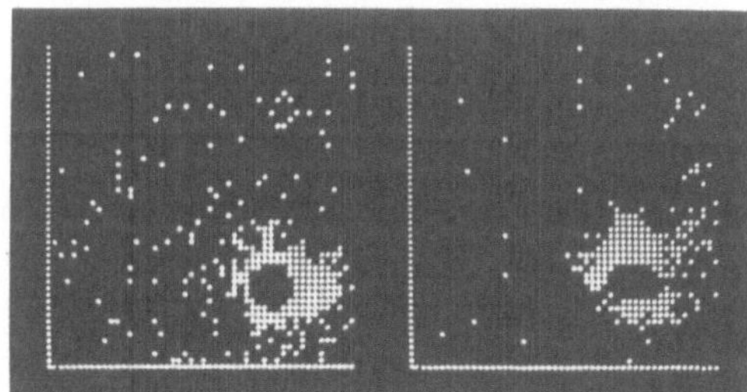

Fig. 4 See text for explanation.

surround also from the retina of the cat. Cells in visual cortex are connected to both eyes. One has to decide now what to attach to the little magnet that moves back and forth on this screen. In general, we use a bar or a dot. With a dot, as stimulus, maps are similar to those generated by the bar, but the bar is a better stimulus for elongated receptive fields. As the first example of visual cortex receptive fields I will show one which is not elongated, but circular (see Figs. 5-7), and it can be seen that this is a binocular cell, i.e., there is a receptive field for

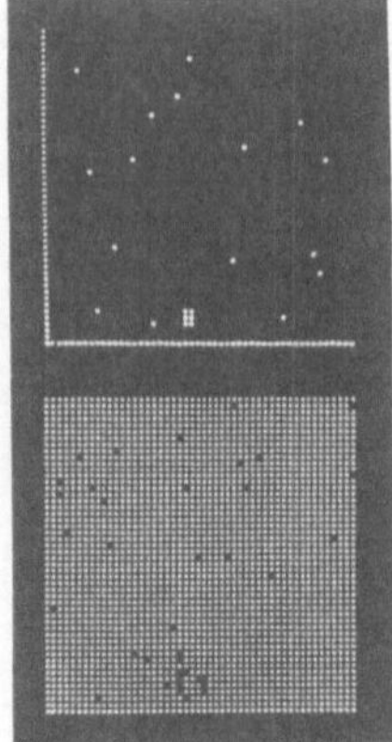

Fig. 5 See text for explanation.

each eye. This type of receptive field, in the cat, was first described by BAUMGARTEN. The visual cortex of the cat, thus, does not contain only elongated receptive fields; it contains also disc-shaped receptive fields and can see that some of them have interesting details which might be important; indeed, it is not impossible that they were derived from early experiences that the animal had. Not surprisingly, this method words quite well for what I would call HUBEL and WIESEL type cells, i.e., cells that exhibit elongated receptive fields. Figure 8 shows one with a thick excitatory bar on one side, a thin one on the other, and inhibition in the center. A bar is moved onto the receptive field and if it is parallel to it, the cell will respond vigorously. If it is orthogonal to the receptive field, there is no response. Thus, the method does work and generates predictions on what to expect with the hand method. In Fig. 9 there is a horizontal cell of that type with a thick horizontal excitatory area, and an inhibitory area in the center. Above maps are generated with a spot, below with a bar. Only the horizontal bar produces a strong field, the vertical bar nothing. We have known of the properties of these receptive

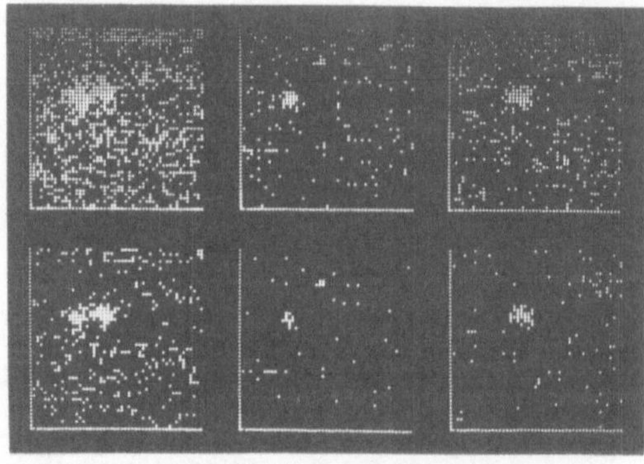

Fig. 6 See text for explanation.

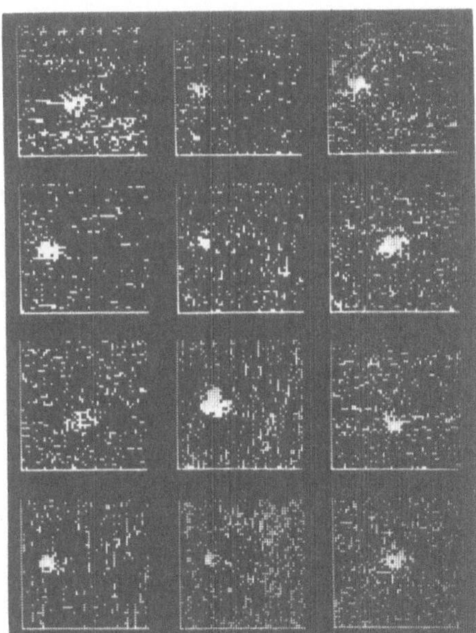

Fig. 7 See text for explanation.

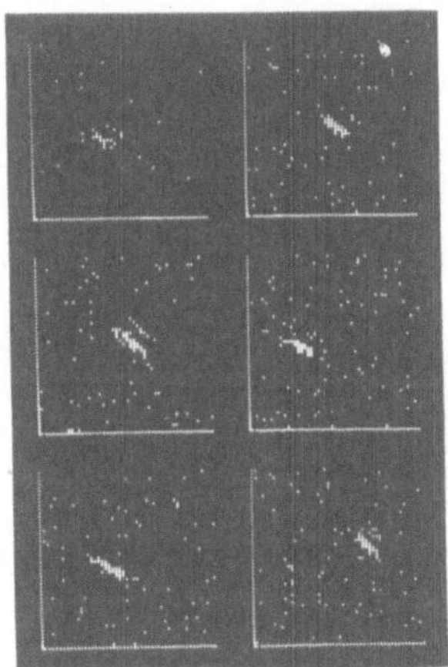

Fig. 8 See text for explanation.

field shapes which look just like the ones from HUBEL and WIESEL data, but the maps have been generated quickly and objectively. Figure 10 shows some of the work that HELMUT HIRSCH and I did at Stanford. We were trying to figure out what it is exactly that cells with elongated receptive fields do. We thought that we could take advantage of atrophy from disuse to weaken or eliminate some classes of cells for specific orientations and then study the perceptual deficits. HELMUT HIRSCH (Chapter - this volume) did show some of the results he obtained from animals that have one eye connected practically only to cells that have vertical

Fig. 9 See text for explanation.

Fig. 10 See text for explanation.

receptive fields. Comparing the performance of the two eyes shows that there is a difference, but it is a very, very small one. It is my opinion that this result compels us to at least consider that alternative hypotheses should be evaluated.

There are at least two hypotheses that one can make. And there is a third hypothesis which I will not discuss because, first of all, it is too complicated, but secondly because it is probably true, and I do not want to have it shut down before I can strengthen it enough. The two main hypotheses I wish to consider are 1) that the cat is born with all possible types of receptive fields, thus we are looking at what is left after atrophy from disuse. One knows from clinical data that patients who sit in bed for a few weeks show dramatic muscle deterioration, or hypertrophy from use, the tennis arm being a well-known example of it. 2) There is a structure at birth upon which is plastic and whose functional properties will be shaped by experience. The study of that prewired structure is extremely important because it is the boot-strap that allows the system to "become" what it will be later on. These two hypotheses are stated in a black and white fashion, but clearly both mechanisms must be operative in different ratios in different animal species.

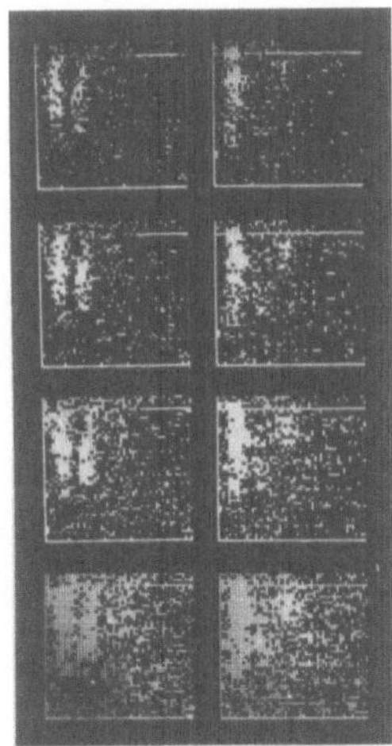

Fig. 11 See text for explanation.

Today I want to concentrate on the second hypothesis, i.e., that receptive field properties are engendered by experience. Let us begin with some old data: (Fig. 11) in the left column we have cells that respond only to the vertically exposed eye, and we can see that the receptive fields have the shape of vertical bars. If bars are moved by hand on the receptive field, the cell will, of course, fire vigorously if the bar has the same orientation of the recpetive field and not if it is orthogonal to it. The horizontally exposed eye has very little, if any, effect on the cells. Similarly cells with horizontal receptive fields can be mapped only through the horizontally exposed eye. Some of these receptive fields were rather unusual, in that their shape mimicked the three bar stimulus to an astonishing degree. Here is a map of the receptive field from a cell that was connected only to the eye that had seen vertical bars. One notices, first of all, more than one bar. Secondly, the bars are gigantic. Remember, the map represents 25 degrees by 25 degrees of visual field. In the figure different levels of activity are displayed, thus at the top we are farthest from background level, namely we are looking at regions that gave the strongest responses. Here we see two bars and measuring their separation we find that it is exactly six degrees of arc from center to center, just like in the stimulus. If we display also those regions where responses were closer to background, a faint third

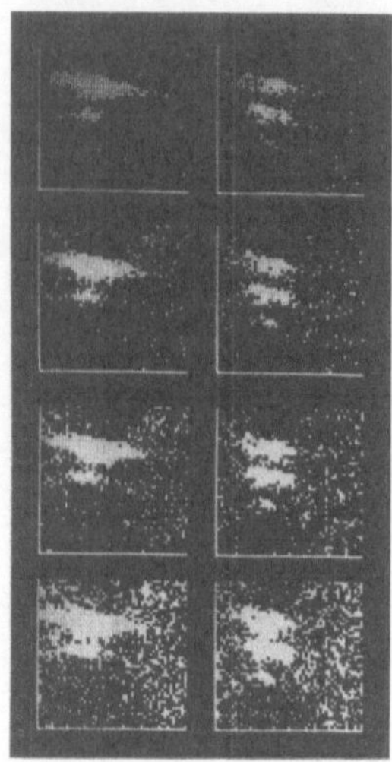

Fig. 12 See text for explanation.

bar becomes visible. Figure 12 illustrates a different cell, and again we have several views of the map in the statistically surface to levels much higher than background; now there is no questions that a third bar is visible also with the same separation of 6 degrees of arc from the next bar. Remember that when he viewed the bars during development, the cat was free to move his eyes behind the mask. These cells are already beyond, I think, the mechanism for translation constancy. Receptive fields such as these pose the greatest difficulty for the atrophy from disuse hypothesis. It is difficult to imagine that the cat was born with them and that atrophy has destroyed everything else and this is what is left over.

Can we design an experiment unexplainable by the atrophy hypothesis? Or can we demonstrate that one is dealing with a record of experience? There is no question that experience is recorded. The fact that we can learn, that animals can learn, tells us that. We just do not know how it expresses itself in the brain. If one is dealing with memory, with records of experience, what does one expect? After damage in the preprocessor of a system, the behavioral deficits are immediate, clear-cut, straight forward, nonrecoverable. A person missing one of the color mechanisms, has highly predictable, incurable, disturbances in the perception of colors. Similarly for pattern vision if the preprocessor consists only of elongated receptive fields that analyzed the visual world in terms of orientaton and all except one were liminated, pattern vision should be devastated.

We know from HELMUT HIRSCH'S work that it is just not so. It is not even mildly devastated, just barely changed, and I will contend that many of the changes we see really have to do with learning disabilities rather than damage in the preprocessor. On the other hand, let us say we are dealing with records of experience. Just after the restricted experience, the only types of receptive fields to be found mirror the experience and no others because that is all he has seen. If new experiences are allowed they will be recorded. The animal, much older now, might be a little disabled, just as I am a little disabled in speaking English. I learned it too late in my life and I do not speak it perfectly, however, I have no damage in my English detectors; in fact I never had any.

The atrophy hypothesis, then predicts devasatating behavioral deficits, the memory hypothesis mild ones or none. In the first case we expect to find that physiological properties of nerve cells have been drastically changed and permanently so. In the second case we expect to find that if we give the animal a new experience, new types of receptive fields will appear. From where? From the pool of uncommitted cells. We find 50% of cells do not have sharply tuned receptive fields. We call those cells uncommitted. They have not been tuned.

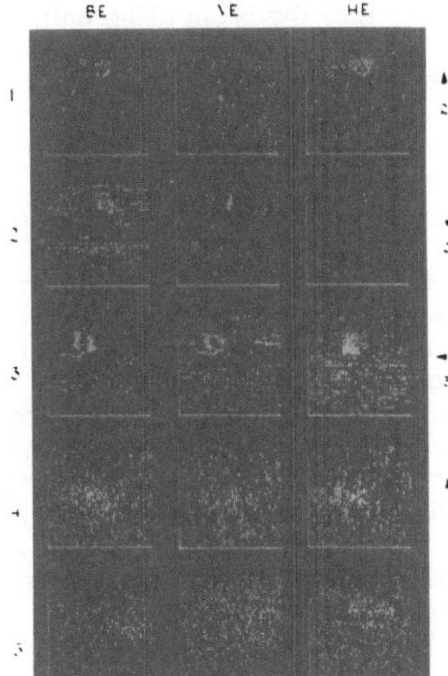

Fig. 13 See text for explanation.

In Fig. 13 we see a receptive field which is disc-shaped and binocular; we found it in one of these animals after they had been recovered from the first experiment

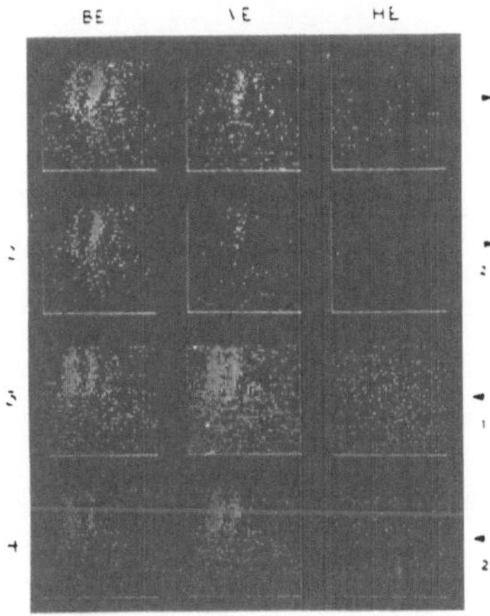

Fig. 14 See text for explanation.

and allowed normal, binocular visual experiences in the animal facility for about a year and a half. Thus, there are now binocular cells with disc-shaped receptive fields. New experineces, new receptive fields. Further, the percentage of uncommitted cells has decreased. However, we were lucky enough to find a few three bar cells, even after a year and a half (Fig. 14). The receptive field contains three vertical bars and it can only be mapped from the eye that saw those patterns, not from the other one, i.e., it seems that once cells have been committed they cannot be changed.

Well, what does one do at this point? I think one wants to do an experiment which addresses itself specifically to search for records of experience, memory traces, if you will. LASHLEY had tried to do that many years ago and had been unsuccessful, possibly because he had put an experience into the brain of an animal that had many other experiences of his own; it would be as if one inserted a record into a huge computer center and then wanted to find where it is: impossible. On the other hand if one fills the memory banks with just one number, then it is easier to find it. We designed the following experiment: we kept a number of kittens in the dark so that we could control their visual experiences completely. Vision is the sensory system in which one has full control over the experiences that the animal has because it cannot create light. Similar control in other sensory modalities is impossible. Once a day we trained our kittens to a simple task. Figure 15 shows one of our volunteers. On his head is a felt mask with two openings where the eyes are. During training, the kitten is restrained in a sling, and only one eye at a time is left opened. On each forearm we have two little electrodes so that we can give him a mild shock. The shock has to be carefully adjusted so as not to frighten or anger the kittens, otherwise, they will not work at all. In front of the animal we just have two cardboard boxes with stimuli cut into them so that we can turn on a little light inside one or the other. The box on the right will display vertical bars, the one on the left horizontal bars. The experiment is performed in the dark. The right eye is trained to the right stimulus and whenever the stimulus goes on, the kitten has to lift the ipsilateral leg or else receive a little shock on his forearm. After ten trials we shift to the other eye, show him vertical bars and he has to pull up the left leg or be shocked.

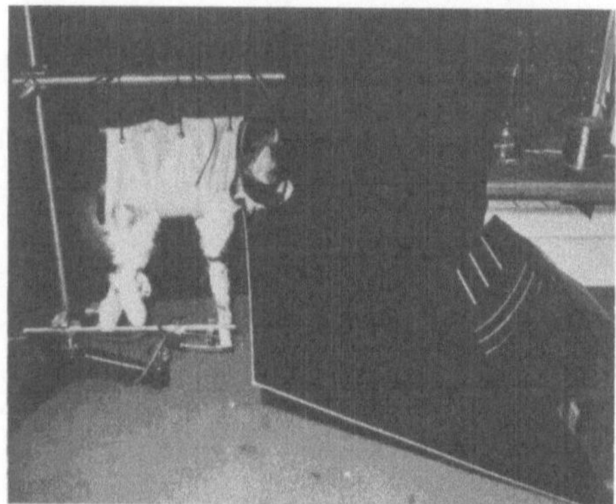

Fig. 15 See text for explanation.

This is an experiment that from a psychological point of view leaves something to be desired because the animal does not have to know the shape of the stimulus to solve the problem. All he has to know is that if the stimulus is coming to the right eye, then the right leg has to be lifted; if to the left eye, the left leg. However, from a physiological standpoint, it has the advantage that the effects of the experience are detectable not only in the shape of the receptive fields or in the orientatoin sensitivity of cells but also in the monocularity of cells. We trained our kittens every day, and then put them back in the dark; they learned the problem very rapidly, it is a simple test, and they had nothing else to do.

What do we expect to find from an experiment like this? We wanted to study two things. One, we wanted to see what happens in visual cortex. We know from the previous experiment with HELMUT HIRSCH that if kittens view vertical lines with one eye and horizontal lines with the other during development we should find monocular cells in visual cortex with receptive fields that resemble the experience for the eye they respond to. And indeed that is what we found. But there was another area of the brain that I had a great interest in and that is somatosensory cortex. This area receives inputs from the body, but a great number of cells are polysensory, i.e., many cells will respond to visual stimuli, auditory stimuli, and so on. In THOMPSON'S laboratory, it has been shown that some cells in this area will respond to visual stimuli of specific orientation. It could just be that these responses are really engendered by previous experiences of the animal. If this is so, then in somatosensory cortex we should find that if a cell responds to the left forearm area and to light stimuli, it should only respond to light stimuli coming through the left eye that was trained with that leg and to the orientation of the bars used in training that eye. When we did this experiment we wanted to attach the shock electrodes to the foot pads because that's the most sensitive part of the leg. But they just would not stay on. And I was not about to insert needles in the foot

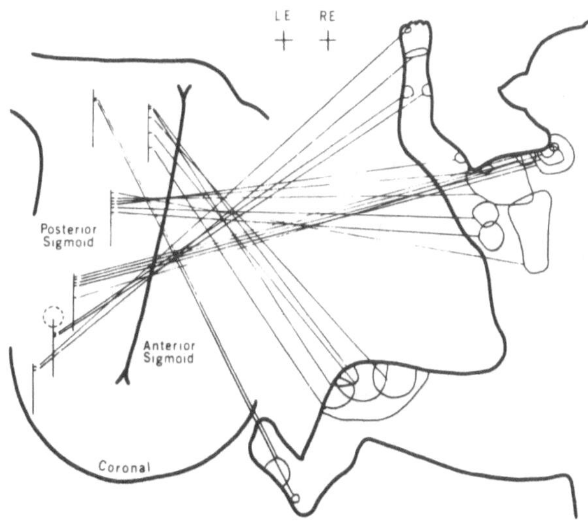

Fig. 16 See text for explanation.

pads of the kittens. The only place where the little rubber band with the stimulating electrode would stay on was the forearm. In a way that was fortunate as I will explain later. Figure 16 shows a number of penetrations in the posterior sigmoid gyrus. Everytime we found a cell in a certain position on the cortex we explored the surface of the body with light touch, little brushes, pressure, etc., to find where the receptive field was located. Naturally, we find that if the cells are medial then they respond to the hind paw and if they are lateral, i.e., on the border of the curvature of the gyrus then they respond to the front paw, the rest of the body being in between. That is a normal kitten in Figure 16. Notice how small the area dedicated to the forearm is, which is normal because the size of each area depends on the sensitivity of the body area. We also found two cells that would respond to visual stimuli; they were both binocular and one of them responded better to vertical lines while the other responded better to horizontal lines (see top of Fig. 16). The cat does not have a very detailed homuculus like primates have. There is a large, relatively speaking, of course, area for the foot pad of the foreleg, one for the hindleg, one for the face beyond the sulcus, and so on. But the forearm is very poorly represented, about half a millimeter square. In Figure 17 we have results from a trained kitten. Here, by just dropping the microelectrode in the middle of the posterior sigmoid gyrus one finds cells that respond to the forearm. The representatoin of the forearm, it appears, is very much larger than normal. Further, most of the cells in here respond to visual stimuli, and they only respond to visual stimuli that have the same orientatoin and are coming from the same eye that was used during the training (see top of Fig. 17 under LE).

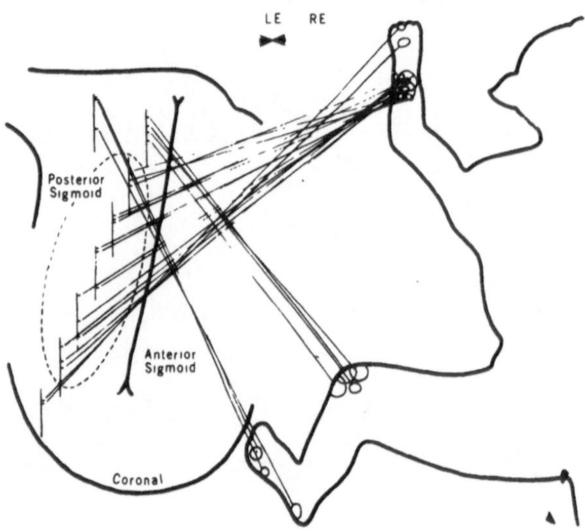

Fig. 17 See text for explanation.

Since then, we have repeated variations of this experiment in a number of cats. We have found that just stimulating the skin will enlarge the cortical representation of it, these cells however, will not have any specific visual responses. This phenomenon does not require that the animals be kept in the dark or be sensory deprived in any way. We have tried to use the foot pads. In this case the area that represents the foot pad changes only very slightly.

It appears that there is an interaction between the system which is prewired at birth, namely, the projection system and the number of cortical cells it will occupy. The projection fibers to cortex for the foot pads are richer. So that is an area that is already prepared at birth for collecting a lot of information. One might be able to change that by teaching the kitten to play the piano, which would be quite an unusual and rich experience for a kitten or by severely depriving it of experiences. But, not by teaching the kitten a simple visuo-motor task. However, the forearm is ready to accumulate only a very limited number of experiences. If the situation is important and requires more processing or storage, then more cells must be allocated to those fibers. In our first experiments, when we showed an animal a lot of vertical lines in one eye and a lot of horizontal lines to the other eye, over a long, long period of time, we ended up with a lot of cells that respond to vertical bars and horizontal bars. Here we take an animal and train him to this particular situation. We give many little electrical stimuli to his forearm every time he is wrong; many cells are dedicated to that. That must be how the experience is stored, somehow. In a way the enlargement of this area, though dramatic, is not really what we were looking for. We were not trying to make the area grow bigger, but just trying to have cells that would exhibit responses which were similar to the experience that the animal had. The experiment shows, however, that the number of cells dedicated to this function is not fixed but negotiable, and that the cortical representation of the homunculus is really determined by an interaction between projection systems, which would represent the phylogenetic memory of that particular species, and ontogenetic memories which the animal has acquired on its own. If this is the case we would expect to find that people with different motor skills have different cortical representations. Hopefully, I will find a friend who plays tennis very well and who is willing to be an experimental subject! In the meantime we have developed a system that enables us to stain nerve cells after recording from them. It is a simple system which uses tungsten microelectrodes. We deposit a little silver by our cells which acts as a seed for a silver stain which then enables us to see the cells recorded from. We are now in the process of trying to find out how those cells connect to one another to elucidate the neural networks they belong to.

Let me conclude by saying that even though the procedures I have described produce massive changes in the visual pathway, binocularly coherent information is much less effective. For example, STRYKER was unable to replicate BLAKEMORE'S results. I have tried that experiment myself with goggles and I can not replicate it. Other puzzling results have been produced by MAFFEI and also by ALLAN HEIN (see this volume). Thus, I do not expect that you will accept the memory hypothesis uncritically. Some of these results, I am convinced, do have to do with memory. However, even if this is not the case, the methodology described is very powerful in guiding brain development. The implications for human children raised in suboptimal environments are staggering, because it appears that brain modifications induced by the environment during development are irreversible.

Questions and Answers

Q. Dr. Spear: I have been hesitating to make this comment, but I guess I feel that it needs to be made, and that is that, well, all of us are obviously interested and fascinated by the plasticity that we can demonstrate in the developing animals. I really think we need to be careful about whether or not we consider this to be memory. Certainly, I think there is a lot of evidence that adult organisms, monkeys, for example, learn faster than neonatal monkeys, and I certainly like to think that I am not deteriorating yet in my ability to learn new information. Yet, all the evidence is that these physiological changes cannot be controlled specifically by the environement after a particular critical period and so I think that we are studying something which while very interesting, is very different from memory.

A. I must agree with some of your comments, but I will disagree with some others. There is no queston that some of these changes can be produced only during a critical period. Probably the usefulness of the critical period is that the animal can overimpose on his phyloenetic system, which prepares him for the class of problems that his species has encountered through the ages, another system which is the ontogenetic one, that will give him greater capabilities to cope with his environment; then that system freezes, if you will. It is true that he will go on and will keep on learning new things, but his learning predispositoin has been altered by what he learned during development. I certainly feel that adult humans can learn a lot of new things more easily than noenates provided these new experiences fit in the context of what they already know. The learning of new things, for example, a language that has no reference to languages already known, is slower in adults. I do agree with you, however, that the teasing apart of what is momory and what is development is a major quest at this time.

References

No list of references supplied by author.

Meridional Amblyopia

R. D. Freeman
School of Optometry
University of California
Berkeley, California 94720

I am going to discuss some work that began with a project on which DONALD MITCHELL and I worked while we were both graduate students at Berkeley. We were interested in what is now often referred to as the "oblique effect" (1), which pertains to the observation that visual resolution is slightly but consistently reduced when test objects are displayed obliquely as compared to horizontal and vertical orientations. Now we discovered two strange subjects who, in addition to a pronounced oblique effect, had markedly unequal acuities for vertical and horizontal gratings. We later conducted additional tests to try to find out why the data for these two subjects, Mitchell was one and I was the other, were so unusual. I shall describe some of the studies that followed.

To begin with, data on the variation of visual resolution with orientation are shown in Fig. 1. Mean acuities for six normal subjects viewing square wave or sinusoidal gratings are given. These results, obtained as control data in early studies that MITCHELL and I conducted with MICHEL MILLODOT and GUNILLA HAEGERSTROM (2, 3), agree with other experiments showing that acuities are approximately equal for horizontal and vertical orientations but are reduced, in this case from 2 to 5 cycles/degree, when gratings are oblique. These subjects are either emmetropic or have spherical refractive errors with negligible amounts of astigmatism. But, as we shall see, the situation can be quite different when subjects are highly astigmatic.

Astigmatism and perception

Astigmatism is illustrated in Fig. 2. A pattern of radiating lines, shown in Fig 2(A), is defocused with a cylindrical lens placed in front of the camera. The result, 2(B), is that vertical lines appear unaffected, while horizontal lines are substantially defocused. All other lines are differentially defocused in proportion to their angular

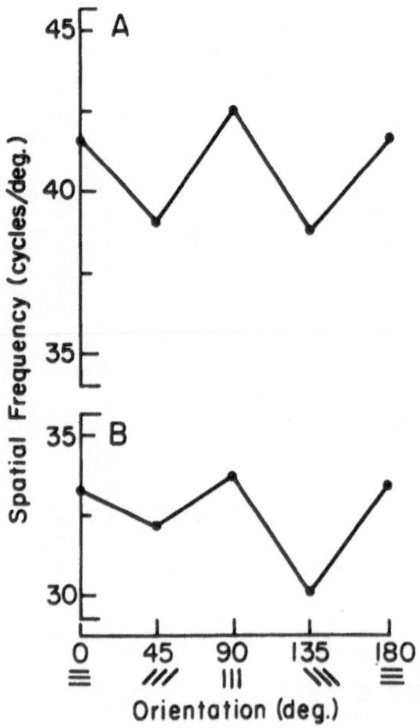

Fig. 1 Mean spatial frequency cut-off values are shown for 3 subjects tested with high contrast square wave gratings (top) and 3 subjects for whom the target was lower contrast sinusoidal gratings (bottom). Data from (3).

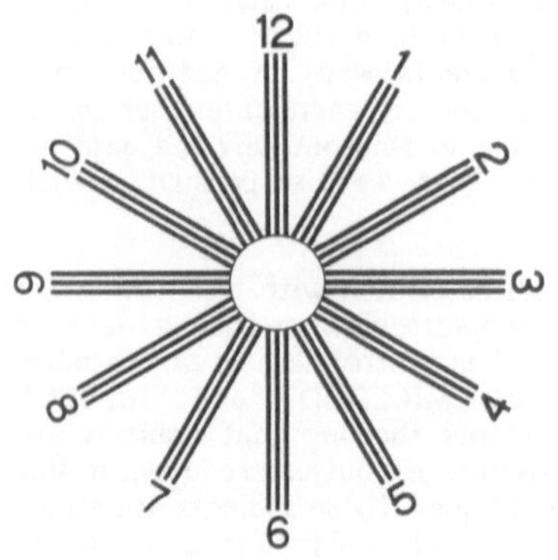

Fig. A

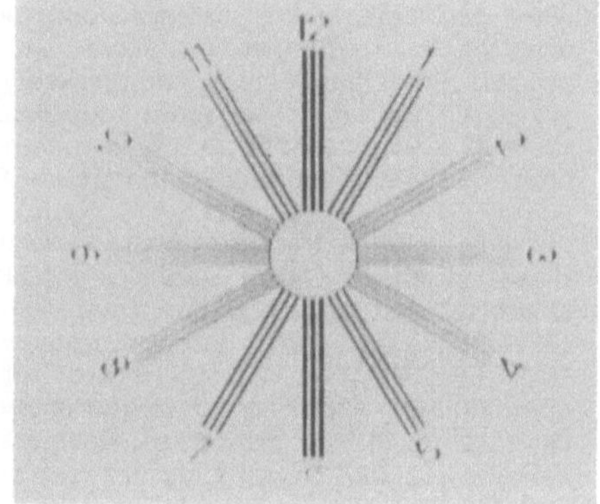

Fig. B

Fig. 2 A wheel-spoke-like patterns of radiating lines (A) is selectively defocused with a cylindrical lens resulting in the pattern on the right (B). The vertical lines remain unaffected but all other lines are blurred in varying degrees.

distance from the vertical meridian. Astigmatism, then, selectively affects features of certain orientations in a given pattern. Ocular astigmatism is illustrated in Fig. 3. In any ametropic eye, the dioptric components and linear dimensions are mismatched such that the posterior focal plane is not located at the retina. The ametropia of an astigmatic eye is meridionally anisomorphic and a point image of a point object (the Gaussian idealization) can never be formed. In Fig. 3(A), an emmetropic condition is shown, i.e., an object at optical infinity is imaged at the retina. As in any optical system, a diminished object distance is accompanied by an increased image distance. But by means of accommodation, the eye can increase its refractive power to maintain focus at the retina (Fig. 3(B)).

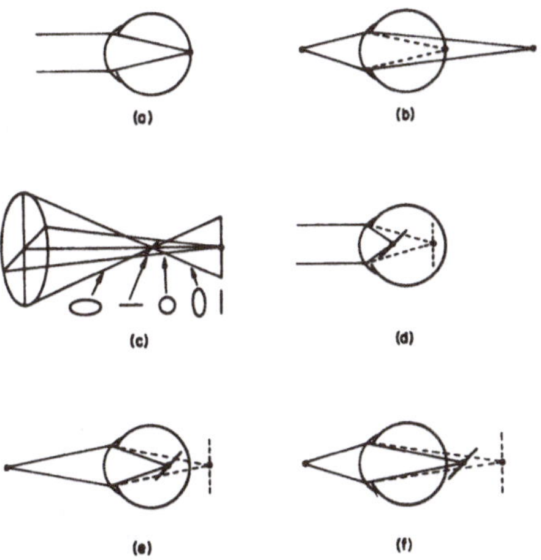

Fig. 3 Illustration of image formation by a nonastigmatic (a, b) and an astigmatic (d, e, f) eye. The astigmatic image bundle, represented in (c), shows that two mutually perpendicular focal lines are formed. Their separation is determined by the difference in focal power between the vertical and horizontal meridians.

In astigmatism, the refractive power varies in different meridians with maximum and minimum powers in mutually orthogonal meridians. As a result, the idealized focal point image becomes two focal line images, usually oriented close to vertical and horizontal. The interfocal region is called the interval of Sturm after the mathematician who first described it. Fig. 3(C) illustrates the optical condition and the lines, ellipses, and circle just beneath represent the head-on views of an image screen placed in planes perpendicular to the page at the positions indicated. The circle is historically known as the "circle of least confusion" since it represents the best image symmetry between the two image lines. In Fig. 3(D), myopic ocular astigmatism is indicated for an object at optical infinity. As the object is brought in toward the static eye, the focal lines move posterior so that they straddle the retina (Fig. 3(E)) or are both behind it (Fig. 3(F)). Once again, accommodation can move the focal lines anterior so that one, or the other, or some intermediate plane is positioned to coincide with the retina. In the unaccommodated astigmatic eye, when the focal lines reside anterior or posterior to the retina, the condition is myopic or hyperopic astigmatism, respectively. When the focal lines straddle the retina, the condition is mixed astigmatism.

Fig. 4 Photographs of the El Greco painting, St. Peter and St. Paul (c. 1592) by O. Ahlstrom with a -1 O.D. cylindrical lens at axis 15 degrees (right) and with no lens (left). From (4).

Consider how ocular astigmatism might be related to visual perception. An interesting example is the attempt to link distortions in paintings by some artists to astigmatic defects (4). Figure 4 shows a rendition of Saint Peter and Saint Paul, by the Greek artist, El Greco, who became famous in Spain. The original work, shown on the left, has been photographed again (on the right) through a cylindrical lens which has the effect of reducing the excessive elongation. A major difficulty with the idea that faulty proportions in artistic creations might result from astigmatism is that a perceived distortion will be compensated for in rendering a scene, resulting in an undistorted reproduction of external space. But it is conceivable, if astigmatism permanently modified the brain, that distortions in representation could occur. In anycase, Fig. 4 illustrates that with a simple refractive procedure, great works of art may be "corrected"!

Astigmatism and visual acuity

Returning to the question of visual resolution in astigmatic subjects, data obtained using the same stimuli and conditions as used for the data of Fig. 1, are shown in Fig. 5. Cut-off spatial frequencies are given for gratings oriented horizontally, and obliquely, for subjects whose uncorrected astigmatism would result in defocus for vertical, 5(A), or horizontal 5(C) stimuli. We have tested a large number of subjects and find data similar to those of Fig. 5 in only some. Invariably, these subjects are highly astigmatic, but tests have been conducted with ophthalmic lenses that fully correct the optical conditions. Still, it might be argued that residual optical defects are present in these subjects, and we used an additional method to counter this possibility.

The technique is based on the classic double-slit experiment THOMAS YOUNG used to demonstrate the wave nature of light, as illustrated in Fig. 6. The waves interfere constructively or destructively and create an interference pattern with a sinusoidal intensity profile. LE GRAND (5), ingeniously applied this method to the

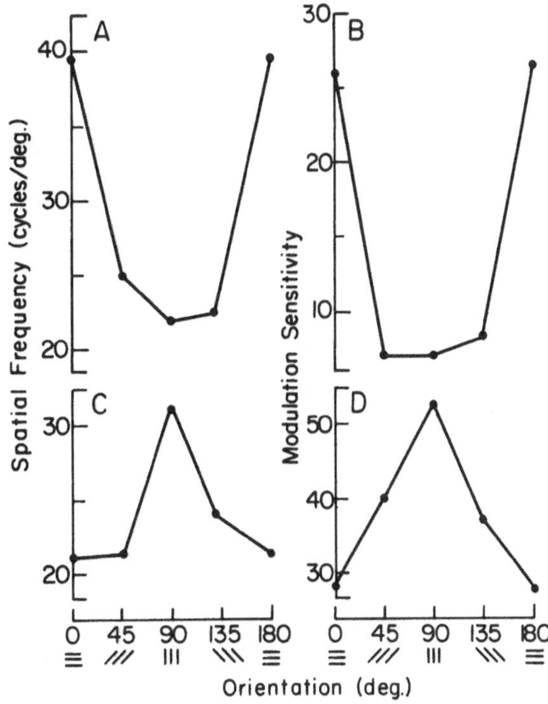

Fig. 5 Spatial frequency cut-off values using sinusoidal gratings are given for meridional amblyopes whose defective meridians are vertical (A) and horizontal (C). Modulation sensitivities for laser-created interference patterns of 16 cycles per degree are also shown for the same subjects (B and D). Data from (2).

eye and, with the advent of lasers, it became practical to test modulation sensitivity for interference patterns created within the eye, so that optical considerations are largely by-passed (see ref. (6) for data obtained using this method, and GREEN, this volume, for a detailed description of its principles and use).

Results obtained using this technique, for the subjects of Figs. 5(A) and (C), are illustrated in Figs. 5(B) and (D). It is clear that the same profiles are found for sensitivity changes with orientation. The subject for whom vertical stimuli are defocused in the uncorrected state has a deficit for vertical gratings in cut-off spatial frequency and in modulation sensitivity for interference patterns. The same applies, mutatis mutandis, for the other subject whose defect is for horizontal gratings. Since an optical explanation for the effect is ruled out by the interference fringe data, the acuity reduction must be neurally caused. This condition, called astigmatic amblyopia by the person who first described it (7) and meridional amblyopia by us, is of particular interest because it presents an opportunity to investigate a visual defect that appears to be orientation specific so that control data can be obtained in the same eye.

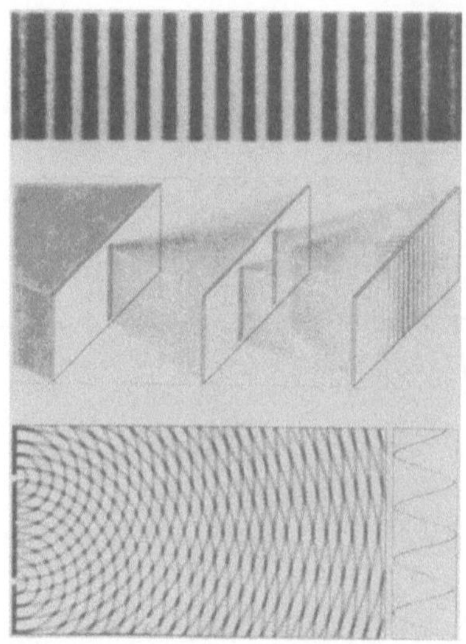

Fig. 6 Interference fringes (top) are shown from an experiment like Thomas Young's using a mercury-arc light. The formation of the fringe pattern is illustrated (middle). Two sets of wave fronts are used (bottom) to represent how the fringes are formed. The wave intensity pattern, resulting from in-phase and out-of-phase interactions, is also shown (bottom, right). From (21).

Astigmatism and accommodation

There are two major assumptions that one makes in attributing the neural phenomenon, meridional amblyopia, to the optical one, ocular astigmatism. First, one must assume that the condition was present during an important developmental period of plasticity of the visual system. Second, it must be assumed that an astigmatic eye accommodates preferentially to the meridian for which acuity is highest. The first assumption is, of course, not verifiable since we cannot know for a given subject how long the condition has been present even though clinical evidence suggests that high astigmatism is hereditary and congenital (8). We also do not know the exact period of neural sensitivity for this condition.

On the other hand, the second assumption, of accommodative bias, may be tested. I used a classical physiological optics technique to determine steady-state accommodation for targets containing features of different orientations (9). The method, called stigmatoscopy, utilizes a haploscope-optometer as illustrated in Fig. 7. One eye observes a target at a given distance, while the other, occluded from the target, views a point source (stigma) of light via a mirror. The subject's task is to view the target while adjusting the light source, seen superimposed on the target, so that it is also in focus. If one knows the optometer lens' focal length and the distance from the eyes to the target and to the light source, the accommodative state of the eye can be computed.

Results for spherically myopic or hyperopic eyes are shown in Figs. 8(A) and (B). The subtle differences involving factors such as "accommodative lag" are not of interest for this study, and it is enough to note that with correcting lenses, the accommodative responses are approximately matched to the stimuli. Without lenses, the responses are determined by the magnitudes of the refractive errors. Before presenting data for astigmatic eyes, it is useful to see how an emmetropic eye responds when made artificially astigmatic. Figures 9(B) and (C) show results for the conditions indicated by the dispositions of the focal lines. The findings indicate that the accommodative effort is aimed toward clear imagery. Contours parallel to the focal line posterior to the retina are preferentially focused upon even

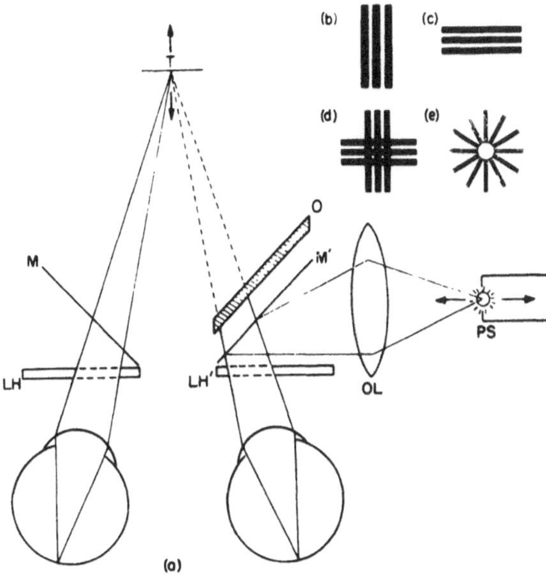

Fig. 7 Schematic representation of a haploscope optometer apparatus used to measure steady-state accommodation (a). The left eye observes one of the targets shown in the inset (b-e) and the accommodative state of the right eye is determined simultaneously. From (9).

if, in the unaccommodated state, the anterior focal line is somewhat closer to the retina. Thus, for these conditions, the textbook notion that the accommodative mechanism adjusts for the "circle of least confusion" is incorrect. Of course, this is simulated astigmatism and only the steady-state condition is measured.

What is the situation in actual astigmatism? Figure 10 contains data for myopic, mixed, and hyperopic astigmats. In each case, without lens corrections, a clear bias is shown for contours of a given orientation. This preference can be marked as, for example, with my uncorrected hyperopic left eye (Fig. 10(C)), I always accommodate selectively for horizontal features. Even with appropriate corrective lenses, the myopic and mixed astigmats show a small bias for a given orientation. Figure 10 also contains visual resolution data for the same subjects which show that accommodative orientation preferences are matched to those of acuity performance. In each case, the orientation for which acuity is high tends to control accommodation. These data are consistent with the linking assumption between accommodation and meridional amblyopia.

Astigmatism and the visual cortex

Having established that astigmatic amblyopia is a neural defect, its physiological basis is of prime interest. But one is limited using only psychophysical methods in this search, so JACK PETTIGREW and I decided to see if kittens reared with simulated astigmatism developed abnormalities that could be related to meridional

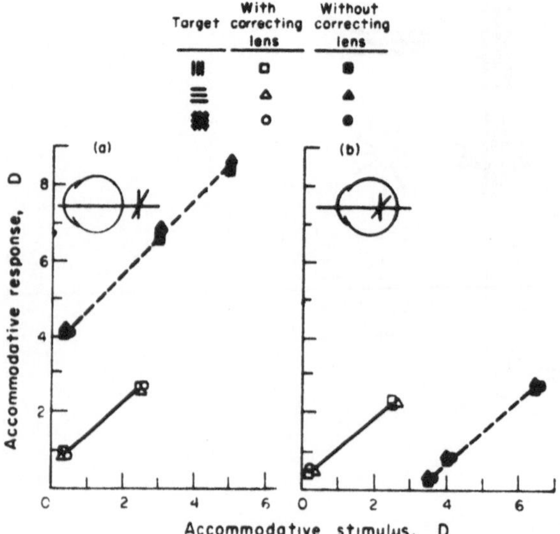

Fig. 8 The accommodative response for a hyperopic (a) and a myopic (b) eye. Measurements at different distances (equivalent to the stimulus values in diopters) are for the targets indicated. Schematic eyes are shown with optical axes to represent the disposition of the focal lines for each refractive condition. In each case, the mean of five measurements is given for target viewing with (open symbols, solid line) and without (filled symbols, dashed line) lenses used to correct refractive errors. From (9).

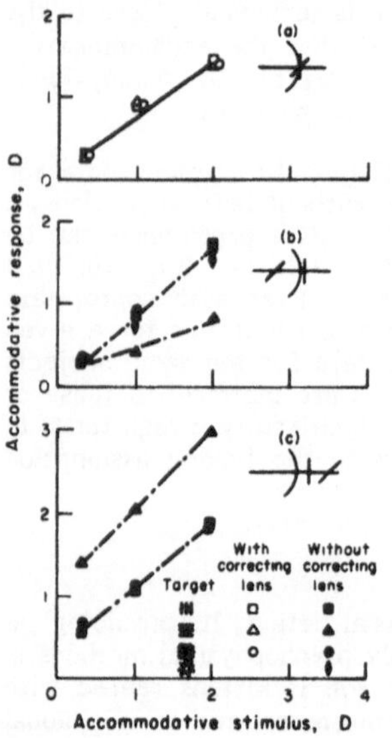

Fig. 9 In (a) the accommodative responses to the targets indicated are given for an emmetropic eye. The eye is made artifically astigmatic in (b) as depicted in the retina-focal line diagram. In (c) a different condition of astigmatism is simulated and accommodative responses are given for the targets shown. From (9).

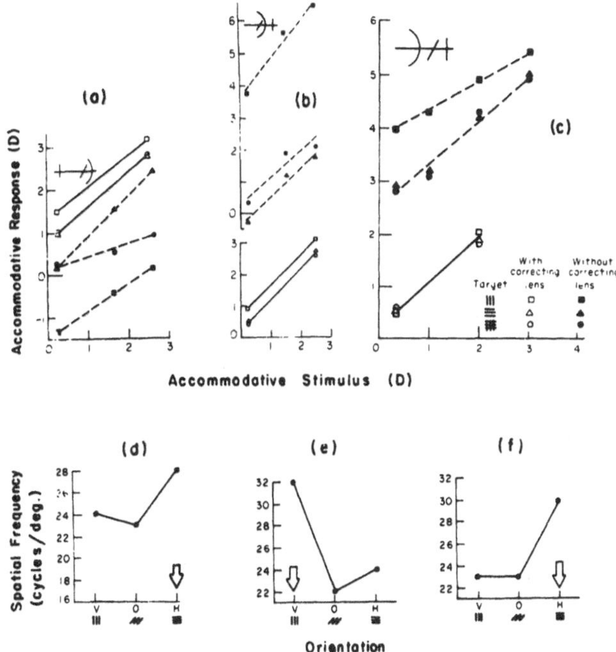

Fig. 10 Accommodative responses to the targets indicated are given for myopic (a), mixed (b), and hyperopic (c) astigmats. Data are included for the uncorrected condition (open symbols, solid lines) and with corrective ophthalmic lenses (solid symbols, dashed lines). For the subjects of (a), (b), and (c), cut-off spatial frequencies for sinusoidal gratings are also given--(d), (e), and (f), respectively. Arrows indicate the orientation that controls accommodation. V, O, and H denote vertical, oblique, and horizontal, respectively. Data from ref. (9).

amblyopia (10). We chose to investigate the most likely site of the defect, visual cortex, and recorded extra-cellular responses from individual neurons in cats whose only visual experience was obtained through goggles containing ophthalmic cylindrical lenses. Powers of +6D and +12D were used, but it must be remembered that the defocusing effect is relatively less in a small kitten eye than in a human. Kittens were reared in a completely dark room until they were 4 weeks old. During the next 25 days, they were exposed for 3 hours daily while goggles were worn. Except for that time, they remained in the dark. Photographs of kittens, during a rearing session, are shown in Fig. 11. At the end of the exposure period, animals were prepared for electrophysiological examination using standard methods. Single cells were isolated and receptive fields were plotted with particular attention to orientation preference.

The rearing conditions are such that for each animal, a small range of orientations is in focus. If there is a selective effect in the population of orientation specific neurons, one might expect a preponderance of cells whose optimal orientations fall near that of the focused meridian. Figures 12(A), (B), and (C)

Fig. 11 Photographs of kittens wearing goggles with cylindrical lenses were taken while the animals were exposed on a large Lucite platform.

contain plots of the numbers of cells with preferred orientations at different angular distances from the focused meridian. These results are for the first group of kittens, whose rearing bias was moderate. Within each 10 degree bin, the most cells are found with preferred orientations from 0 to 10 degrees of that which had been in focus during rearing in the cats of Figs. 12(B) and (C) but not (A). The effect is more pronounced in the cats of Fig. 13(A) whose rearing had been more severely biased (higher power lenses). For comparison, results for a normal control cat are shown in Fig. 13(B).

We examined qualitatively other parameters of the receptive field data including stimulus velocity, ranges over which orientation specific cells respond, directional selectivity, ocular dominance, and response vigor. In general, we found

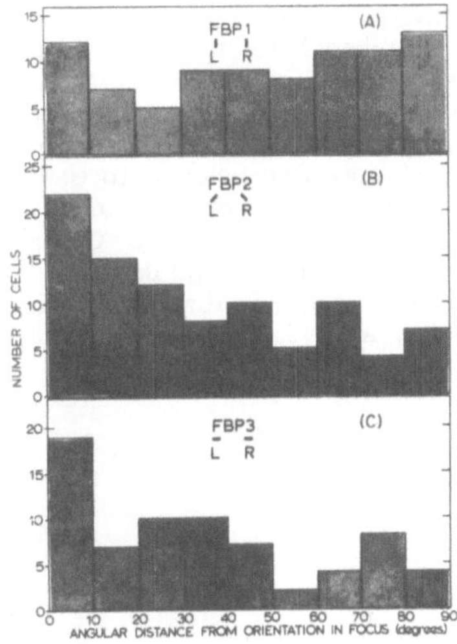

Fig. 12 Histograms show the numbers of cells in 10 degree bins with preferred orientations at different angular distances from the focused meridian. The orientations of the focused meridians for these cats, FBP 1, 2, and 3, are indicated by the lines above each left (L) and right (R) eye.

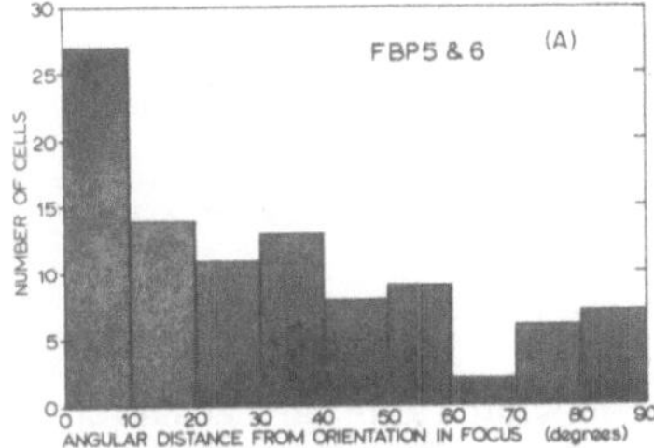

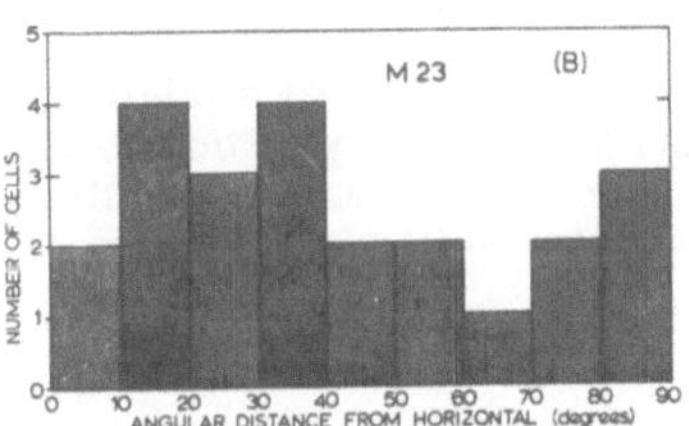

Fig. 13 Histograms as in Fig. 12. Data for FBP 5 and 6, reared as FBP 2 and 3, respectively, but with stronger defocusing lenses, are combined in (A). Results for a control cat are given in (B).

that cells of different preferred orientations could not be distinguished on the basis of response strength, velocity, or directional selectivity. A full range of tuning widths was found for orientations near the plane that had been in focus during rearing, but there was a tendency for a lack of tightly tuned cells with preferred orientations at a considerable angular distance from the focused meridian. Ocular dominance data showed that most cells were monocular. That is to be expected in the case of the animals reared with oblique cylindrical axes perpendicular to each other. In the other cats, the breakdown of binocularity was probably caused by prismatic effects of the lenses since exact centering could not be achieved. Other types of analysis were carried out but differences between focused and defocused rearing orientations were not prominent. An example is shown in Fig. 14, which includes data on receptive field area and dimensions as a function of angular distance from the orientation in focus during rearing.

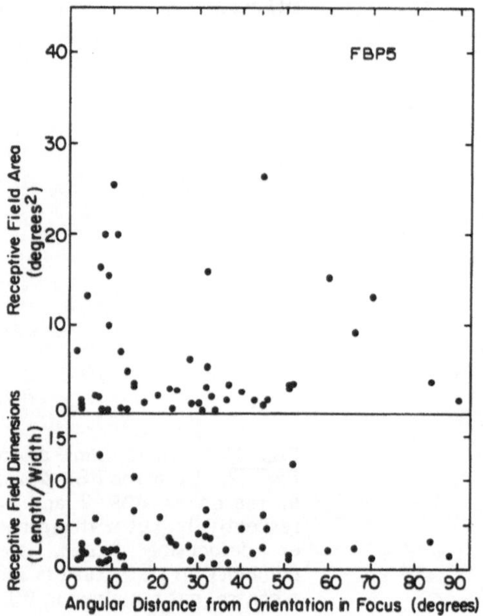

Fig. 14 Receptive field areas and dimensions (length to width ratios) are given for a cat, reared as FBP 2, who showed the orientation preference bias indicated in Fig. 13(A).

In summary, the data on astigmatically reared kittens show that a relatively subtle bias in early visual exposure can have an effect at the primary cortical level. The changes could reflect either passive degeneration or active neural reconnection or both, but our methods do not allow us to distinguish between these possibilities.

Spatial contrast sensitivity

Returning to the psychophysical phenomenon of human meridional amblyopia, it is

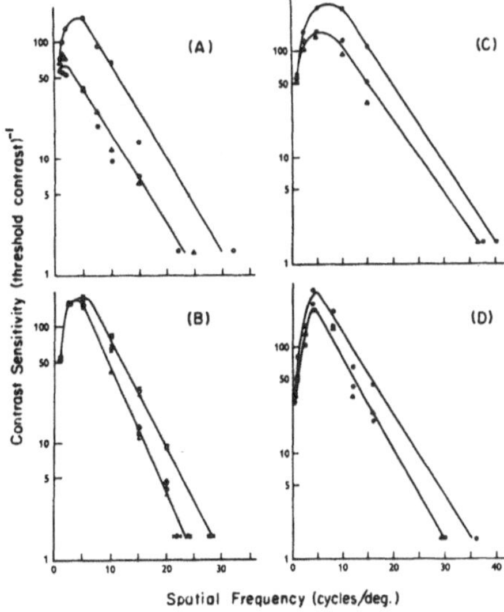

Fig. 15 Astigmatic subjects with meridional amblyopia show prominent orientation differences in cut-off spatial frequencies. Contrast sensitivities also depend on orientation and for these subjects sensitivity for horizontal (A and B) or vertical (C and D) gratings is reduced in addition to the typical oblique reduction. Without optical correction, horizontal (A and B) or vertical (C and D) gratings are defocused with these subjects, but measurements have been made using optimal ophthalmic lenses. Data for vertical (open circles), horizontal (filled circles), and oblique (open triangles) gratings are presented. Bars represent $\pm$ 1 S.E. for 5 or more measurements. Data from (13).

important to know what visual functions are affected in addition to high contrast visual acuity. To begin with, spatial frequency analysis is of most obvious interest and I initiated a study of contrast sensitivity functions in meridional amblyopes (11) which was pursued along with an evoked potential investigation with LARRY THIBOS (12-14). For the contrast sensitivity measurements, a standard apparatus was constructed to allow presentaton of sinusoidally modulated gratings at variable spatial frequencies. Typical results are shown in Fig. 15. In Fig. 15(A) and (B), contrast sensitivity functions are plotted for astigmatic subjects who, when optically uncorrected, are defocused for horizontal features. The tests were conducted with full ophthalmic lens corrections but a deficit remains. For the subject of Fig. 15(A), sensitivity over nearly the entire range of spatial freuqencies is greatly reduced, whereas for Fig. 15(B) only frequencies above about 6 cycles/degree are affected. Data in Fig. 15(C) and (D), for subjects with a vertical deficit, also show a substantial reduction throughout most of the frequency range tested. Similar results have been obtained by MITCHELL and WILKINSON (15). Figure 16 illustrates this effect more clearly by use of log-log coordinates for data obtained from a subject with an unusual oblique axis astigmatism. Lowest and highest sensitivities are found at oblique axes which correspond to the most and least defocused meridians, respectively, in the optically uncorrected state.

These contrast sensitivity data show a certain amount of inter-subject variability which must reflect factors such as age at onset of the astigmatism, duration of the condition, and, of course, magnitude of the refractive error. One can obtain an estimate for a diffraction-limited optical system, of the theoretical effect of different amounts of defocus. With suitable conversion factors, it is then possible to see if there is an approximate match between sensitivity and the magnitude of the astigmatic defect. An example of this type of comparison is shown in Fig. 17 which includes theoretical defocus curves with experimental data for a

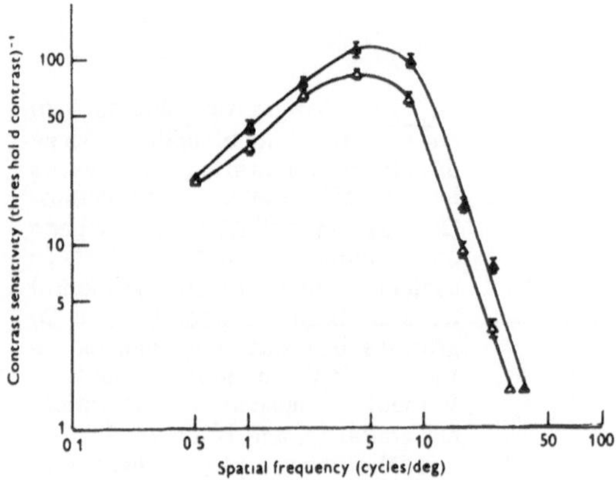

Fig. 16 Data for a meridional amblyope whose principal astigmatic axes are at oblique orientations are plotted on log-log coordinates to illustrate the orientation differences at low frequencies. Focused and defocused meridian data are given by closed and open triangles, respectively. Data from (14).

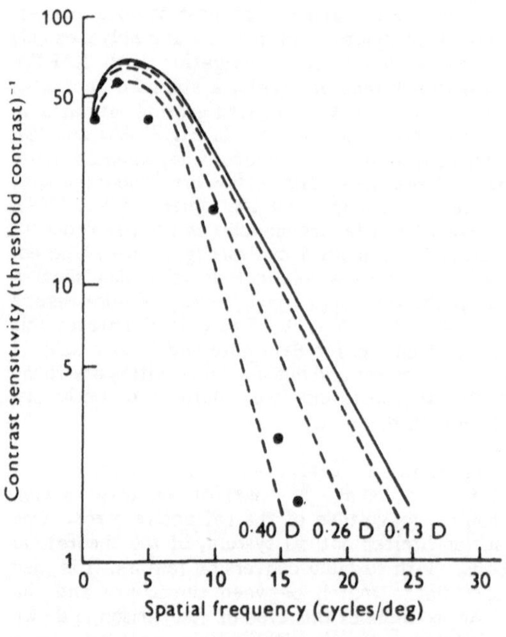

Fig. 17 Theoretical effects of defocus on a contrast sensitivity function are shown here. The solid curve represents a sensitivity function for a normal meridian of a meridional amblyope. Assuming diffraction-limited optics for this meridian, several values of defocus (in diopters) would depress normal sensitivity as indicated by the dashed curves. Data points (filled circles) are contrast sensitivity values for the deficit meridian of a meridional amblyope. Data from (13).

meridional amblyope. The result shows that the sensitivity loss is equivalent to a small amount of defocus, but it must be remembered that this estimate is based on the diffraction-limited case.

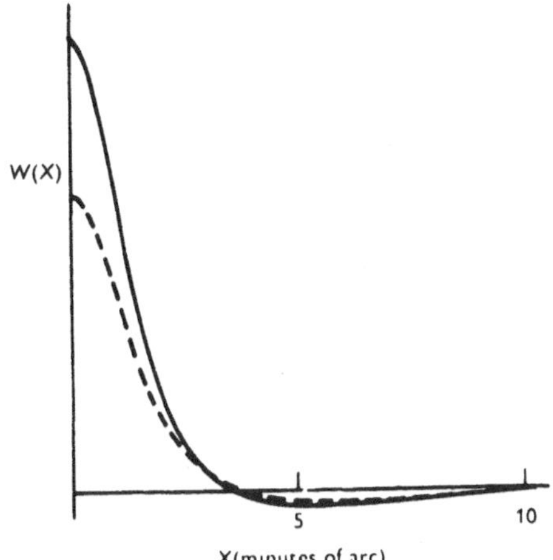

Fig. 18 Spatial weighting function of a meridional amblyope. The curves are computer-plotted inverse Fourier transforms of the contrast sensitivity functions shown in Fig. 15(A). Continuous and dashed curves represent orientations of maximum and minimum sensitivities, respectively. Data from (13).

Another type of theoretical consideration of these data that may be useful is to examine them in the spatial domain, as done previously in classic treatments (16, 17). An example of this, a spatial weighting function, is shown in Fig. 18 which is a computer-plotted inverse Fourier transform of the contrast sensitivity function given in Fig. 15(D). The main difference between the normal and deficit meridians is the decreased response for the latter. One can carry this treatment further by assuming anisotropic filtering at a peripheral stage. A continuous two-dimensional contrast sensitivity function may be inferred by interpolation of the one-dimensional functions for normal and deficit meridians. Once again, using the data of Fig. 15(D), the result of this procedure is shown in Fig. 19. The inverse Fourier transform of this function, shown in Fig. 20, may be interpreted as a two-dimensional receptive field of a single anisotropic neuron responsible for processing gratings of any orientation.

Visual evoked response

The visual evoked response in humans has been studied for a variety of purposes, but its main utility is to help bridge the gap between psychophysical and electrophysiological results. Although this connection remains somewhat tenuous, the evoked response allows an objective measure of subjective phenomena such as meridional amblyopia (12, 18). To obtain an objective analogue to the spatial frequency sensitivities discussed in the previous section, we used the same visual stimulus equipment as before to generate sinusoidal gratings, but a larger field size was used to get a relatively strong response. The gratings were temporally modulated sinusoidally at a frequency of 9 or 12 Hz. Results for the subject mentioned in the last section (Fig. 16), whose principal astigmatic axes are uniquely at oblique orientations, are shown in Fig. 21. The psychophysically determined contrast sensitivity function, included for comparison, is quite similar to the evoked potential data. This degree of concordance was not found in all subjects, but we

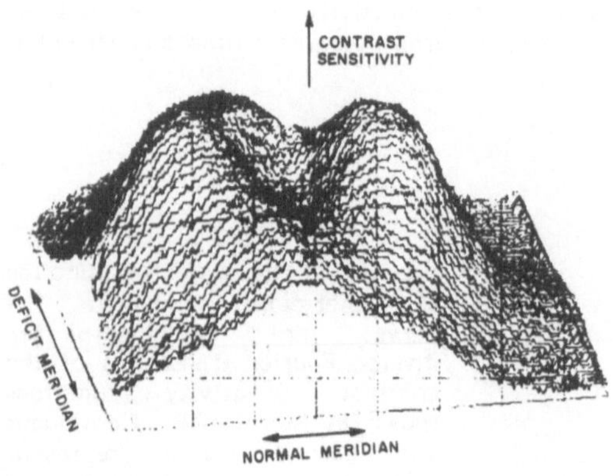

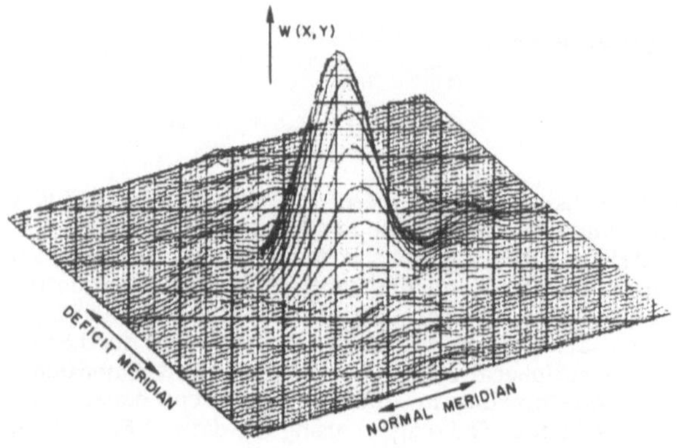

Figs. 19 and 20 An interpolation has been made by eye from the data of three meridians for the meridional amblyope of Fig. 18. It represents, as shown in Fig. 19, a qualitative form for a continuous two-dimensional contrast sensitivity function. Iso-sensitivity contours, constructed by eye, were entered into the computer which produced the isometric display of Fig. 19. The two independent variables (X and Y axes) are spatial frequencies in the normal and deficit meridians and the dependent variable (Z axis) is contrast sensitivity. The inverse Fourier transform of this function, shown in Fig. 20, may be interpreted as a two-dimensional receptive field of a single, anisotropic neuron responsible for processing gratings of any orientation. The two independent variables here (X and Y axes) are spatial dimensions in the normal and deficit meridians and the dependent variable (Z axis) is sensitivity, W (X,Y). The distortion of this receptive field is readily apparent in comparison with a normal concentric pattern.

may conclude that meridional amblyopia can be demonstrated with evoked potentials. Therefore, the physiological locus or loci of the defect must be in structures and pathways that are peripheral to or at the site of evoked potential generation, presumably visual cortex.

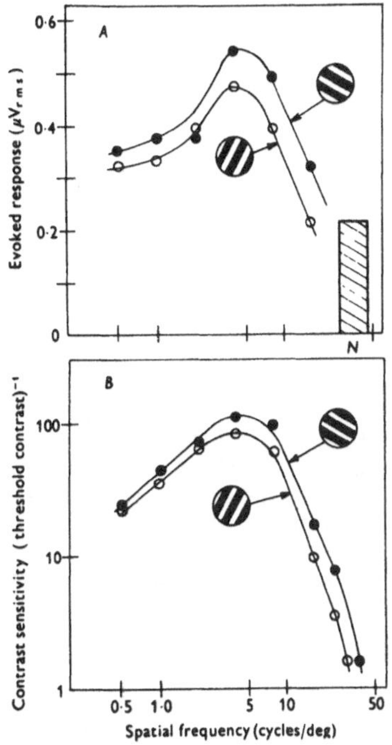

Fig. 21 Spatial frequency response functions are shown for a meridional amblyope who has oblique astigmatism. Evoked responses are given in (A) and psychophysical contrast sensitivity data are shown in (B). Filled circles represent the normal meridian and open circles are the results for the orthogonal orientation. From (14).

Increment sensitivity and spatial interactions

In the search to find the functional basis of meridional amblyopia, it is necessary to identify how fundamental visual capacities are altered. One of the basic functions, increment sensitivity, is usually determined with annular patterns. BARRY BEYERSTEIN and I used bar-shaped targets of various lengths, widths, and orientations to see if anisotropies, similar to those found for acuity tests, were exhibited by meridional amblyopes (19). We also tested increment sensitivity using a small flashing light patch with and without the presence of a second light patch at various distances from the first, along different meridians (20).

The first studies were conducted using a standard Maxwellian system, as schematically shown in Fig. 22. Test bars of various widths, lengths, and orientations were seen against the background, as illustrated, and increment sensitivity was determined by rotation of a circular wedge. Typical results for thin bars of various lengths are shown in Fig. 23 for a control subject and a meridional amblyope. In both cases, sensitivity increases with bar length, as expected, but no systematic differences with orientation are found for the control. On the other hand, marked differences are found for the astigmat, with depressed sensitivity for the horizontal orientation. This subject also exhibits poor acuity for horizontal gratings. The deficit is only expressed for bars of 7 minutes of arc or longer. For this subject and most others, reduction of the illuminance level, from 70 trolands, as used for this test, to 7 trolands, greatly reduces or eliminates the anisotropy.

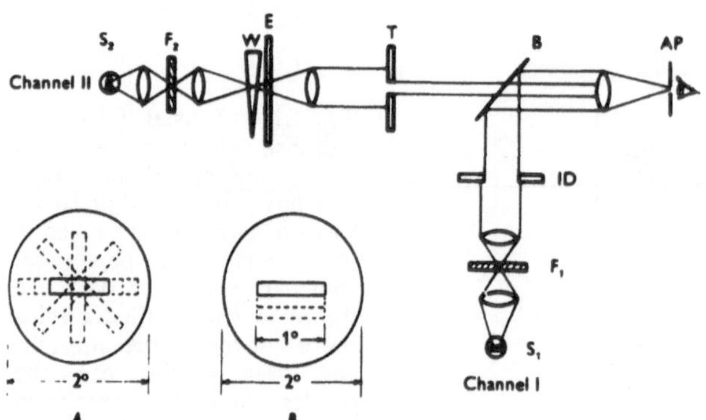

Fig. 22 The apparatus is a standard two-channel Maxwellian optical system. Channel I provides a background field from a compact filament tungsten lamp (S_1) which is collimated after passage through neutral density filters (F_1) and limited to 2° by an iris diaphragm (ID). Channel II contains target disks (T), a wedge (W), and an episcotister (E) in addition to a source (S_2) and neutral density filters (F_2). The channels are combined through a beam splitter (B) and imaged through an artificial pupil (AP) in the plane of the subject's entrance pupil. The inset depicts the subject's view of the stimulus. In (A), bar orientation and length are the variables and in (B) bar width is varied. From (19).

Turning now to the other dimension of the stimulus, bar width, sensitivity must increase as the test bar broadens, up to some level. But results for normal and astigmatic subjects might be quite dissimilar, reflecting different summation properties. Typical results are shown in Fig. 24. Sensitivity for the control subject reaches an asymptotic level quite quickly at a width of about 4 minutes of arc. Results are markedly different for the other subject, a meridional amblyope for whom data have been presented in Fig. 23 and other figures. There are clear orientational differences and summation occurs for all widths tested when test bars are oblique or horizontal. Recall that this subject has reduced contrast sensitivity for horizontal gratings. From these bar-target experiments, we may conclude that increment sensitivity mechanisms are affected in meridional amblyopia.

To do the second set of increment sensitivity experients, we modified the apparatus shown in Fig. 22 as schematically illustrated in Fig. 25. We could then determine sensitivity for a test patch with and without a satellite stimulus. Typical summary results are shown in Fig. 26. Mean values are given for normal control subjects (filled symbols) and for meridional amblyopes, who have low acuity for horizontal gratings (open symbols). Mean increment sensitivities for the test spot alone serve as baseline data (horizontal dashed lines and arrows), and increased and decreased sensitivities are indicated in log units as positive and negative values, respectively. Increment sensitivities for the test patch alone are set at zero for the purpose of this graph.

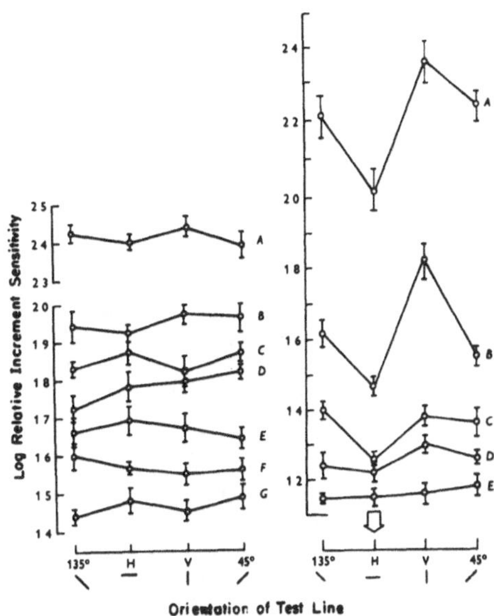

Fig. 23 When the test line is shortened or lengthened, increment sensitivities are altered accordingly. The data shown on the left are results for lines 1.5' wide with lengths of: 3' (G), 4.5' (F), 6' (E), 7.5' (D), 9' (C), 11.5' (B), and 60' (A). The subject is normal and a moderate background level is used (70 td). Line length is varied as on the left, only in this case, the subject is a meridional amblyope. Target width is 1.5' and lengths are: 4.8' (E), 5.7' (D), 6.8' (C), 12' (B), and 60' (A) (data on the right). Orientation differences are prominent in (A) and (B), diminished in (C) and absent in (E). The arrow indicates the deficit meridian. Data from (19).

Note first, that for the normal subjects there are small differences between vertical, horizontal, and oblique orientations of the satellite light patch. Sensitivity of the test spot is reduced most, over the broadest distance, when the satellite patch is positioned in the oblique meridian. The least threshold lowering effect is found for vertical satellite positions and sensitivity is apparently increased when the satellite is removed beyond about 9 minutes of arc from the test spot. Given the oblique acuity effect, one could suggest that there is functional significance of these orientational differences in spatial interactions, but the vertical-horizontal differences are problematic, since acuities for targets in these meridians are considered to be equal.

On the other hand, orientational differences for the meridional amblyopes are more easily connected to other data. The effect of vertically positioned satellites is opposite to the normal group. Sensitivity is reduced more, over a broader region, than for other meridians. In fact, all transition zones are broader than for the

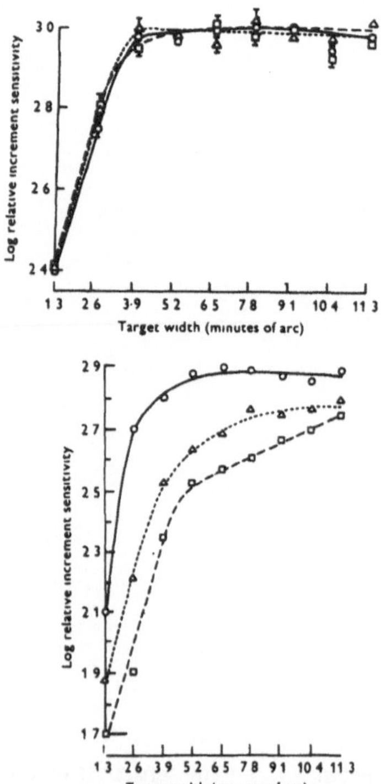

Fig. 24 Increment sensitivity data for a normal subject (top) and a meridional amblyope who has reduced acuity for horizontal gratings (bottom) are given for a stimulus bar 1° long that varies in width as indicated. Background level is 70 td. Vertical (open circles, continuous line), horizontal (open squares, dashed line), and oblique (open triangles, dotted line) orientations are included. Curves are fitted by eye through the data. Data from (19).

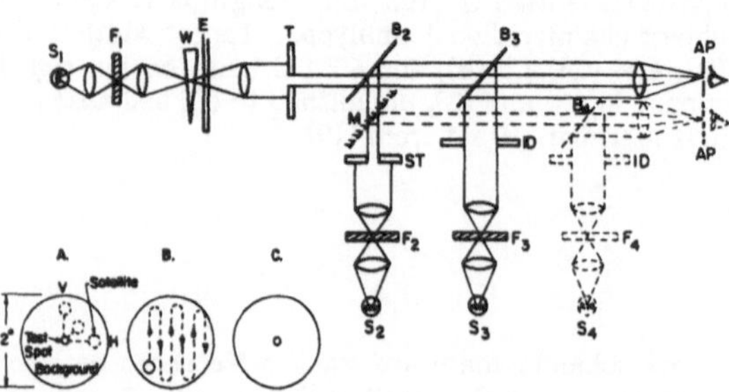

Fig. 25 A modified form of the Maxwellian system shown in Fig. 22 is illustrated here. Light from tungsten filament bulbs, S_1-S_4, attenuated by neutral density filters, F_1-F_4, is combined by beam splitters, B_2-B_4, and imaged in the plane of the artificial pupil, AP. Exposure is controlled by an epicotister, E, and incremental luminance of the 1.5' dia test patch formed at T is adjusted with the circular neutral density wedge, W. The background is limited to 2° by an iris diaphragm, ID, and the satellite patch, subtending 3.5', is formed by ST. A fourth channel (dashed lines) is added for the dichoptic experiment. The subject's view is illustred in inset (A) which shows the meridians along which the satellite may be positioned (vertical (V), oblique (O), and horizontal (H). Insets (B) and (C) depict the dichoptic arrangement whereby the satellite (B) is shown to one eye and the test spot (C) to the other. From (20).

normal group. In addition to the interaction differences, mean increment sensitivity for the test spot alone is considerably lower than that for the normal group. Therefore, these eyes are sub-normal other than just for linear targets imaged along the deficit meridian. This is illustrated more dramatically, for the interesting case of a subject with monocular astigmatism, in Fig. 27. The other eye is emmetropic. The astigmatic eye is of the same type as shown in Fig. 26, i.e., acuity is reduced for horizontal gratings. The satellite effects are quite severe, and increment sensitivity for the test spot alone is also markedly reduced compared to the normal eye. For this subject and others, we tried in several ways to obtain dichoptic effects by presenting a satellite to one eye and the test spot to the other. We could never get an effect, i.e., there was no dichoptic interaction. This result and others lead to the conclusion that the phenomenon of meridional amblyopia can be accounted for by monocular systems. This, of course, does not rule out visual cortex, but it means that peripheral stages of visual processing could be involved.

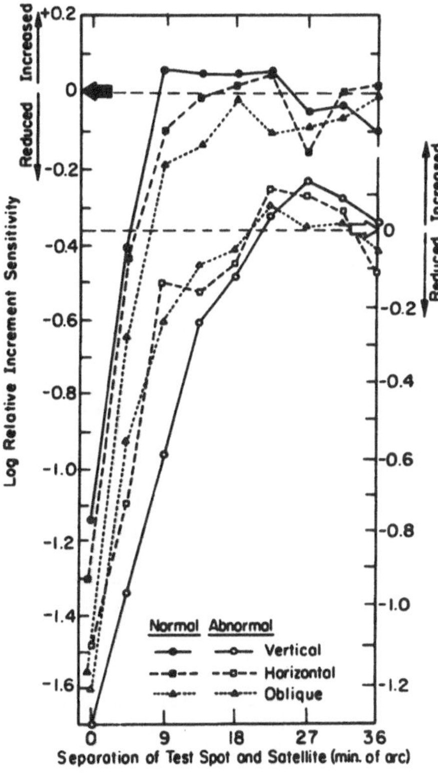

Fig. 26 Summary results are shown for increment sensitivity measurements with a satellite light patch positioned at varying distances along vertical, horizontal, or oblique meridians. Mean values for four normal eyes (solid symbols) are compared to those for three meridional amblyopes (unfilled symbols) who have low acuity for horizontal gratings. The satellite raises or lowers threshold, as shown, and ordinates give values above and below that for the test patch alone. Therefore, separation of the horizontal dashed lines and arrows represents the difference in mean increment sensitivity for the test patch alone, between normal and meridionally amblyopic subjects. From (20).

Finally, it is possible to construct perceptual "receptive fields" that represent physiological models for the types of spatial interaction results discussed above. Although, admittedly, these are somewhat fanciful, they may be heuristically helpful. The fields, shown in Fig. 28, assume "on" centers for which effective adaptation levels and associated increment thresholds are raised by the satellite light patch.

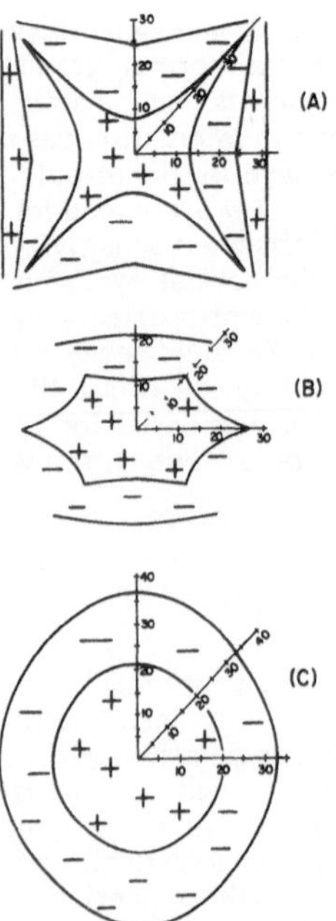

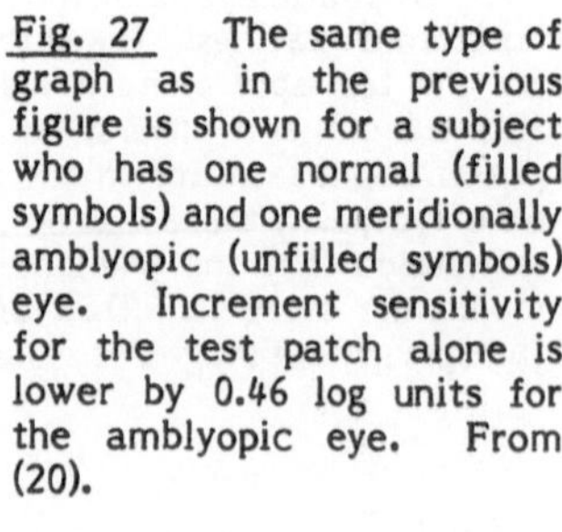

Fig. 27 The same type of graph as in the previous figure is shown for a subject who has one normal (filled symbols) and one meridionally amblyopic (unfilled symbols) eye. Increment sensitivity for the test patch alone is lower by 0.46 log units for the amblyopic eye. From (20).

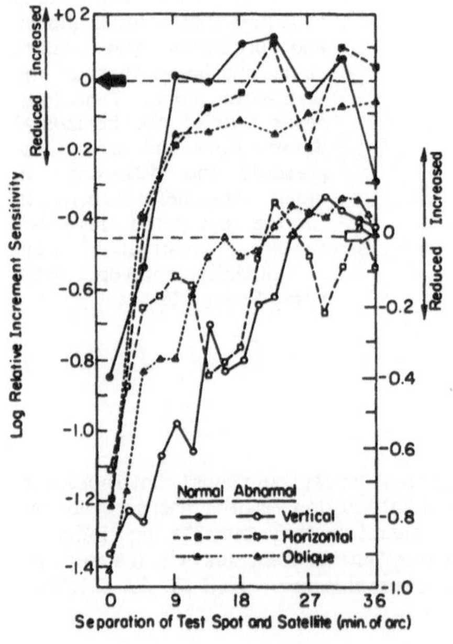

Fig. 28 Perceptual "receptive fields" are depicted in (A) for the normal subjects shown in Fig. 26. Fields for meridinal amblyopes of the type shown in Fig. 26 are illustrated in (C) and those with an orthogonal deficit meridian, i.e., low acuity for vertical gratings, are represented in (B). The model assumes "on" centers for which adaptation levels and associated increment thresholds are raised by the satellite. So, areas with plus signs represent reduced sensitivities. Axes indicate distances (min arc) at which transitions occur between reduced and increased sensitivities. For presentation, the quarter field data used to construct there fields are extrapolated to form symmetrical full fields. From (20).

Conclusions

A neural condition, meridional amblyopia, may be confidently attributed to faulty development caused by marked ocular astigmatism. An analogous condition may be simulated in astigmatically reared kittens who show physiological effects of the selective visual experience at the level of visual cortex. For meridional amblyopes, the orientation for which acuity is highest also tends to control accommodation. Tests of contrast sensitivity show that meridional amblyopia is not simply a reduction of the high-frequency cut-off, since a broad range of frequencies are affected. This effect may also be demonstrated by use of visual evoked potentials. Increment sensitivity for a bar-shaped target is affected in meridional amblyopia, but no oblique effect is found in normal subjects. Spatial interaction between a test spot and a satellite light patch differs with orientation of the satellite, and suggests possible mechanisms that underlie meridional amblyopia. Several experiments show that monocular pathways can account for the condition.

Acknowledgement

The work described here was supported by research grant EY01175 from the U.S. Public Health Service, N.I.H. The author is supported by Research Career Development Award EY 00092 from the U.S. Public Health Service, N.I.H.

References

1. S. Appelle, Pyschol. Bull. 78, 266 (1972).
2. R. D. Freeman, D. E. Mitchell and M. Millodot, Science, 175, 1384 (1972).
3. D. E. Mitchell, R. D. Freeman, M. Millodot and G. Haegerstrom, Vision Res., 13, 535 (1973).
4. P. Trevor-Roper, The World Through Blunted Sight (Thames and Hudson, London, 1970).
5. G. Le Grand, Qe Reunion de L'Institute d'Optique (Paris, 1937).
6. F. W. Campbell and D. G. Green, J. Physiol. 181, 576 (1965).
7. G. Martin, Bull. Soc. Francaise Ophthalmol. 8, 217 (1890).
8. W. S. Duke-Elder, The Practice of Refraction (Mosby, St. Louis, 1969).
9. R. D. Freeman, Vision Res. 15, 483 (1975).
10. R. D. Freeman and J. D. Pettigrew, Nature, Lond. 246, 359 (1973).
11. R. D. Freeman, Invest. Ophthalmol. 14, 78 (1975).
12. R. D. Freeman and L. N. Thibos, Science, 180, 876 (1973).
13. R. D. Freeman and L. N. Thibos, J. Physiol. 247, 687 (1975).
14. R. D. Freeman and L. N. Thibos, J. Physiol. 247, 711 (1975).
15. D. E. Mitchell and F. Wilkinson, J. Physiol 243, 739 (1974).
16. F. Ratliff, Mach Bands: Quantitative studies in neural networks in the retina (Holden-Day, San Francisco, 1965).
17. R. W. Rodieck and J. Stone, J. Neurophysiol. 28, 833 (1965).
18. A. Fiorentini and L. Maffei, Vision Res. 13, 1781 (1973).
19. B. L. Beyerstein and R. D. Freeman, J. Physiol. 260, 497 (1976).
20. B. L. Beyerstein and R. D. Freeman, Vision Res. 17, 1029 (1977).
21. G. Feinberg, Sci. Amer. 219, 50 (1968).

Development of Visual Acuity in Normal and Astigmatic Infants

Richard Held
Department of Psychology
Massachusetts Institute of Technology
Cambridge, Massachusetts 02139

Renewed interest in the visual acuity of infants and, especially, its meridional variation was triggered by reports of the stripe-reared kitten experiments (1, 2). The important result was the claim that the distribution of orientation sensitive single units, in the visual cortices of these kittens, had been biased by rearing in the striped environments. For example, if the kitten had been reared under conditions allowing it to see only vertical stripes, the incidence of units sensitive to vertical was high, relative to that of the normally reared animal, and there were very few if any units responsive to horizontal edges. Just the opposite occurred if the kitten had been reared with exposure only to horizontal stripes. In the wake of that claim, many visual scientists began to wonder about its potential relevance to human vision. Two human parallels were suggested and this report will deal with each in turn.

The first alleged parallel was the idea that the oblique effect--the small reduction in acuity that adult human observers show for oblique-edged targets as compared to verticals or horizontals--was the result of living in an urban environment. It was supposed that in such an environment, people were exposed more to vertical and horizontal than to oblique edges and that habitual condition would constitute a somewhat attenuated parallel to the kitten rearing experiments. Consequently, exposure to the urban environment would sharpen up the vertical and horizontal detection system more than the oblique one. On the other hand, if one came from an environment containing more obliques, the effect might either be greatly reduced or even reversed. And, in fact, ANNIS and FROST (3), who measured acuity in the Cree Indians of northern Canada, who apparently live in environments of this kind, seemed to find just such a reduced oblique effect.

How can we directly test the notion that the oblique effect results from biased

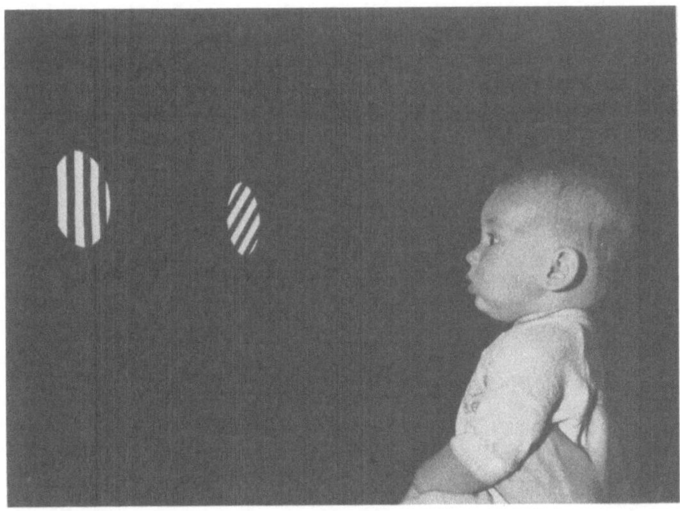

Fig. 1 Eight month old infant viewing grating stimuli in the oblique versus main axis choice.

exposure? The logic of our approach is as follows. If it is true that the viewer has to be exposed to the environment to develop the biasing effect, then prior to such exposure there should be no such bias. In order to test the effect of prior exposure one has to go back as early as one can in the history of the perceiver. And that, of course, took us to infants. How does one test the acuity of infants? There are a number of methods but here we shall be concerned only with the preference procedure previously used by STIRNIMANN (4), FANTZ, ORDY and UDELF (5), TELLER, et. al., (6), and others. We call this procedure "looking at looking" because the observer/experimenter looks at the infant's eyes. Two stimuli are shown to the infant. They are rear projected onto two translucent windows which the infant views binocularly at a distance of 50 cm in an otherwise dark room (see Fig. 1). The observer does not see the stimuli but does watch the infant's eyes and must make a decision, within a few seconds, as to which of the two stimuli the infant prefers to look at. Unlike the pair shown in Fig. 1, the typical stimuli that are used to measure acuity are a grating of some orientation paired with a homogeneous field of equivalent space-averaged luminance. We know from many previous studies that the infant prefers to look at the more figured of the two targets. Hence, he will prefer to look at the grating as long as he can see it. Initially, a coarse grating is presented since it should be above threshold. Then in subsequent trials, the spatial frequency of the grating (the density of the bars) will be increased until the preference of the infant drops to 50%. The data from that procedure yield a psychometric function from which one can define a preference threshold. We have used the 75% preference and the 58% preference and the resulting curves of Fig. 2 show the result. Quite a number of investigators have used this procedure and the results of Fig. 2 are in close agreement with those of other laboratories using this technique. Acuity rises monotonically from about 2.0 cy/deg, which is an acuity of 20/300 in the Snellen notation, at ten weeks up to at least 12.0 cy/deg at about one year, a Snellen acuity of 20/50.

Using different techniques, several investigators have made different claims about the growth of acuity (8, 9). The evoked potential findings are said to show a much more rapid growth of acuity, reaching the adult level by four to six months of age. The discrepancy has been claimed to be a result of a criterial difference. But such a difference could not explain the continued increase of acuity after four to six months of age (7).

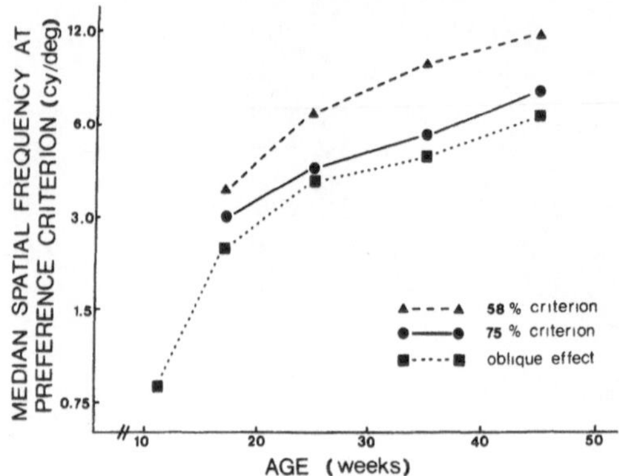

Fig. 2 Circles and triangles show median spatial frequencies for four age groups at which infants preferred a vertical grating over a homogeneous field at the 75% and 58% criteria, respectively. Squares show median spatial frequencies for each of five age groups within which infants preferred either a vertical or a horizontal grating over an oblique grating at one spatial frequency at least 67% of the time. Mean ages ranged from 11 to 45 weeks.

Returning to our original question--does the oblique effect occur in infants--- we measured acuity for oblique as well as vertical gratings on the same infants. Figure 3 (from ref. 7) shows the results. There is no difference at the earliest age level but the difference increases during the second semester of life. This result agrees with that of TELLER, et. al., (6) who failed to find an oblique effect in infants up to six months of age.

Suspecting that we should be able to demonstrate the oblique effect in younger infants, we decided to use a different technique. Instead of pairing a grating with a homogeneous field of equivalent luminance, we thought we might get greater sensitivity by directly pairing a grating which is either horizontal or vertical with an oblique grating (see Fig. 1). The two gratings were matched in spatial frequency, space-averaged luminance, and contrast. The only difference between them was the orientation of the gratings--one oblique and one either horizontal or vertical. A series of paired stimuli, each presented for a duration of 10 sec, varied only in their spatial frequency, below and above what we presumed was the threshold. We asked the question, will the infant show any preference which would indicate the presence of this difference in acuity between oblique and either vertical or horizontal? Figure 4 shows the result obtained on one infant tested for 60 trials on the paired gratings. It was one of the clearest results we obtained but it makes the point I want to make. It shows that when both vertical and oblique gratings are at high frequencies, hence below the threshold of acuity, there is no preference because the

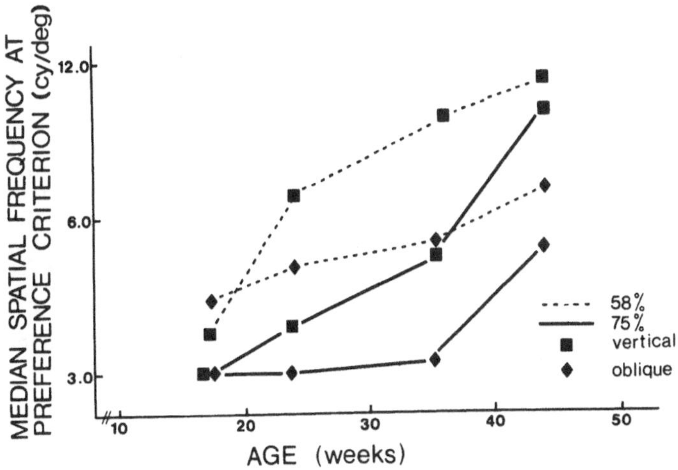

Fig. 3 Median spatial frequency at 75% and 58% criteria for preferring either a vertical or an oblique grating to a homogeneous field. Mean ages ranged from 17 to 44 weeks of age.

infant sees no bars. When the gratings are at low frequency, both gratings are above threshold and the infant has no preference. But, when these two gratings are at or near the acuity threshold (see Fig. 2), then the small difference in visibility between the two gratings results in a preference for the vertical and horizontal over the oblique gratings. The vertical or the horizontal grating is slightly more visible than the oblique grating.

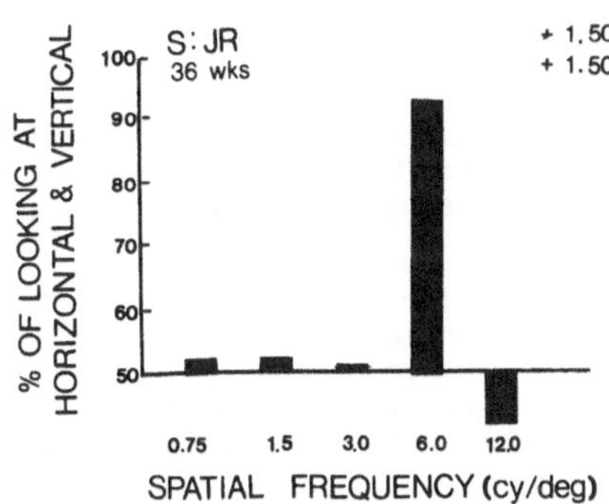

Fig. 4 Preferences of a single infant for oblique vs. main axes gratings. Refraction given in upper right corner.

When we tested a series of infants with this procedure, seven or eight in each age group, and pooled their data, we obtained the results shown in Fig. 5 (from ref. 10). This evidence shows that the oblique effect is demonstrable by a few weeks of age and argues that it develops endogenously. Whatever may happen due to environmental influence, the oblique effect is already present in some form in the developing nervous system. A second point is that the preference maxima move out

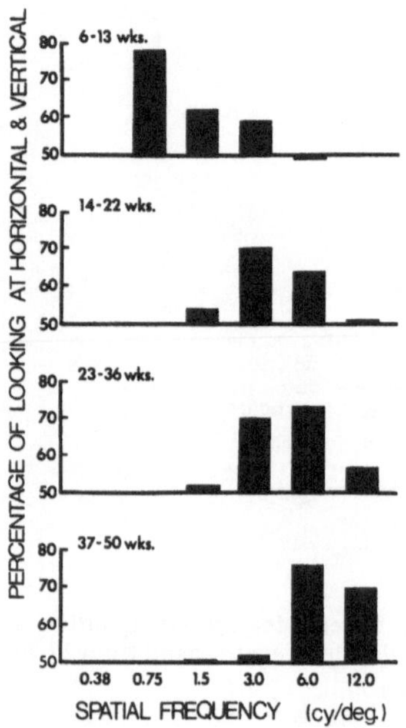

Fig. 5 Mean percentage of vertical and horizontal gratings looked at for the four age groups. Data for spatial frequencies of 0.75, 1.5, 3.0, and 6.0 cy deg are given for the youngest group and for spatial frequencies of 1.5, 3.0, 6.0, and 12.0 cy/deg for the three older groups.

along the spatial frequency or acuity dimension with age in a consistent way. The maximum preferences appear to occur at spatial frequencies at or near the acuity threshold. In an experiment by GWIAZDA et. al., (7), a series of infants were tested with the technique just described and the frequency of maximum preference was plotted together with the acuity thresholds previously discussed (see Fig. 2). The curve of oblique effect parallels that of the acuity thresholds and lies about one-quarter octave below that taken at the 75% preference threshold. These data confirm our hunch that the maximum preference for main axes gratings occurs at or near the acuity threshold.

We have concluded that the oblique effect is not a parallel to the stripe-reared kitten experiments. At least one portion of the effect seems to be there from the earliest time we have tested and, while exposure may have some influence, the simplest interpretation is that the oblique effect results from endogenous maturation of the visual system.

The second parallel to the stripe-reared kitten result that has been proposed is the irreversible acuity losses that are suffered by some strongly astigmatic adults despite careful refractive correction, and even despite creating retinal images by bypassing the lens system of the eye (11). The astigmat tends to have one sharply focused orientation for edges with increased blurring as the angular deviation from that orientation increases up to 90°. The optically uncorrectable acuity loss is called meridional amblyopia and it tends to be greatest for edges along the axis of greatest habitual blurring despite current refractive correction. Unless one believes that such acuity losses in the visual system are somehow genetically linked to the presence of cylindrical power in the ocular system, these findings are a strong argument for environmental determination.

If it is exposure to the habitual blurring that produces the loss of acuity in chronic astigmats, we would like to know when in the life of the infant the visual

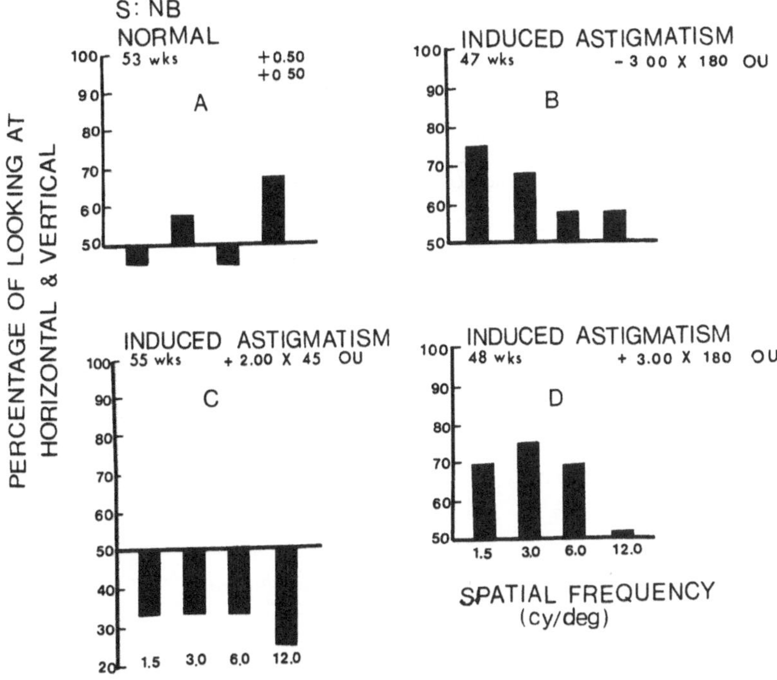

Fig. 6 The effects of induced astigmatism (amount shown in upper right corner) on the preference for oblique vs. main axes gratings.

system is susceptible to the loss. In other words, what is the timing of the modifiable period? This is a question not only of theoretical interest, but obviously of great importance for clinical practice. The procedure for defining the critical period entails finding a group of astigmatic infants as soon after birth as possible and tracing both their refraction and their acuity with and without optical correction as they grow older. When one finds that their vision can no longer be fully corrected optically, one has detected the onset of meridional amblyopia. We have been fortunate in having as a collaborator a skilled refractionist capable of retinoscoping and measuring refraction of infants beginning at birth and without the use of cycloplegics (12). Using the preference procedure with an oblique grating gave us the following kind of results. Figure 6A shows the preference graph of a visually normal infant of 53 weeks showing an oblique effect by the preference for horizontal and vertical at 12.0 cy/deg. The same infant, when we put either a plus or a minus 3.0 diopter cylinder at axis 180 over his eyes, showed the profiles of Fig. 6B and 6D, respectively. What happens is that the preference which is usually only evident at a high frequency (12.0 cy/deg in this case) extends down into lower frequencies when the cylinders are present. The same infant looking through a +2.0 cylinder at an oblique axis inverts the preference and yields an anti-oblique effect (Fig. 6C). These results can be explained as the consequence of the optics of the cylindrical lens plus the accommodative facility that the infant has. How does this work? In brief, target edges that are either parallel or orthogonal to the axis of a spherocylinder may be brought to sharp focus on the retina by a sphere of appropriate power. Edges of all other orientations are inevitably blurred to the extent that they deviate in orientation, up to a maximum of 45° from the principal axes (13). Accommodation can provide the spherical power needed provided that it is within the range available in the infant. The results of testing a real infant astigmat that we have pulled out of a large sample is shown in Fig. 7. The eye diagram is for a target at 2.0 diopters distance (at 50 cm). The more conventional diagram would be for a distant target point and can be derived from that of the figure by making the line foci 1.25 diopters more myopic (12).

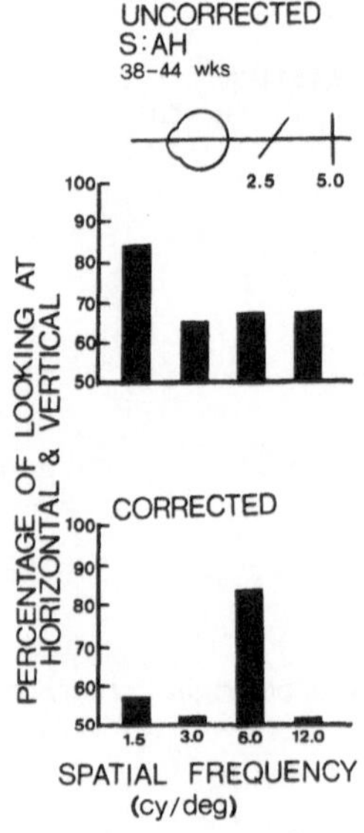

Fig. 7 Preferences of natural astigmat uncorrected and corrected.

When we tested this uncorrected astigmat we found the same results as in optically induced astigmatism (see Fig. 6B). Horizontals and verticals were preferred at all the spatial frequencies tested. What happens if we correct these infants by adding the necessary cylindrical power? The result is shown in the lower graph of Fig. 7. The curve is not significantly different from that of a non-astigmatic infant. In other words, as far as we can tell, this 2.5 diopter astigmatic infant shows no sign of having developed meridional amblyopia. A further question can be raised. Is the preference different for horizontal vs. oblique as compared to vertical vs. oblique? When we break down our data in that way, we find no difference. This result reinforces our view that accommodation sharpens either of the two astigmatic foci, depending upon the orientation of the target viewed. Of course, this result is not always the case since in severe myopic and hyperopic astigmatism the range of accommodation may be insufficient. In such cases we do find a preference in direct comparisons of horizontal and vertical gratings. But these, too, do not survive optical correction and we have, so far, detected no meridional amblyopia in our sample of infants before one year of age.

In the course of refracting by retinoscopy and tracing acuity in hundreds of infants, we discovered to our surprise that at about six months of age, 50% of the infants showed astigmatism of one or more diopters, correlated with the usual preference profile of the astigmat (Figs. 6 and 7). In the normal adult population, that figure is probably between 5-10%. At first, thinking we had simply found an aberrant sample, we were pleased to have found so many to study. Then we began to realize that our population of astigmatic infants was rapidly declining as the infants grew older. What was happening to them? They were still being tested but they were no longer as strongly or at all astigmatic. By one year, the majority had lost or reduced their astigmatism (14,15). Initially, we were disappointed to lose the bulk of our sample but then we realized that we had a natural experiment on the development of amblyopia during the first year since we had infants who had had astigmatism at one age and then no longer had it when they grew older. The question was, given this natural correction, or emmetropization, do these infants show the development of any meridional amblyopia? The answer in brief is that, as in the case of the corrected astigmats, we have not detected amblyopia. Within the limits of sampling--naturally corrected and corrected by optics--it does not occur to a significant extent prior to one year of age.

Returning to our original question of the presumed human parallels to the stripe-reared experiment, the oblique effect, said to result from exposure to the environment, does not appear to be a very good parallel. We are convinced, on grounds other than our experiments (11), that the meridional amblyopia resulting from astigmatism is a parallel, but to date we have found no evidence up to a year of age for the development of meridional amblyopia. Therefore, we have to conclude that this form of amblyopia must develop after one year of age.

Acknowledgement

This research was funded in part by research grants from: The National Eye Institute (NIH # 5-R01-EY01191), the National Institute for Neurological and Communicative Diseases and Stroke (NIH # 5-P01-NS-12336-03), and the Spencer Foundation

References

1. C. Blakemore and G. F. Cooper, Nature 228, 477 (1970).
2. H. V. B. Hirsch and D. N. Spinelli, Science 168, 869 (1970).
3. R. C. Annis and B. Frost, Science 182, 729 (1973).
4. F. Stirnimann, Ann. Paedia. 163, 1 (1944).
5. R. L. Fantz, J. M. Ordy and M. S. Udelf, J. Comp. Physio. Psych. 55, 907 (1962).
6. D. Y. Teller, R. Morse, R. Borton and D. Regal, Vision Res. 14, 1433 (1974).
7. J. Gwiazda, S. Brill, I. Mohindra and R. Held, Vision Res. (in press).
8. E. Marg, D. N. Freeman, P. Peltzman and P. J. Goldstein, Invest. Ophthal. 15, 150 (1976).
9. S. Sokol, Vision Res. 18, 33 (1978).
10. S. C. Leehey, A. Moskowitz-Cook, S. Brill and R. Held, Science 190, 900 (1975).
11. D. E. Mitchell, R. D. Freeman, M. Millodot and G. Haegerstrom, Vision Res. 13, 535 (1973).
12. I. Mohindra, J. Am. Opt. Assoc. 48, 518 (1977).
13. D. D. Michaels, Visual Optics and Refraction: A Clinical Approach (C. V. Mosby Company, St. Louis, 1975).
14. R. Held, I. Mohindra, J. Gwiazda and S. Brill, A.R.V.O., Sarasota, Florida, April 25-29, 1977.
15. I. Mohindra, R. Held, J. Gwiazda and S. Brill, A.R.V.O., Sarasota, Florida, April 30-May 5, 1978.

Eye Motility and Visual Motor Development

Alan Hein
Department of Psychology
Massachusetts Institute of Technology
Cambridge, Massachusetts

Exposure of a kitten in light when changes in visual stimulation are systematically related to self-produced movement supports development of visually guided behavior (1-3). When opportunity for such feedback is not provided, stimulation with patterned light is not sufficient for acquisition of these capacities. If certain sources of feedback from movement are made available to the kitten, while others are not, we observe independent acquisition of several coordinations. For example, if a kitten locomotes in light only while wearing an opaque collar that prevents view of limb and torso, it will develop the ability to locomote so as to descend to the shallow side of a visual cliff, and to avoid objects in an obstacle course (4). However, it cannot guide its limbs to contact a small object. Similarly, if one forelimb has been viewed, but not the other, visually guided reaching is restricted to the limb that had previously been viewed (5). Additional evidence for a precise relation between opportunity for feedback and development of specific visual motor coordinations has been provided in a number of studies in cat (5, 6) and in monkey (7, 8). We suggested that these findings do not imply that opportunities for motor visual feedback alone determine acquisition (9). In particular there may be a prescribed sequence in which certain visual motor capacities may develop. Recently we examined the possibility that development of the ability to locomote in visual space might be prerequisite to development of visually directed reaching (10).

When the capacity for visually guided locomotion is displayed, utilization of a body-centered representation of visual space is implied. The limbs, like all other objects, may be given a visual position by the coordinates of this map. Visual feedback from self-produced movements of a forelimb appears important for development of visually guided reaching with that forelimb (5, 11, 12). If the information which supports this development derives from a body centered representation of visual space, it may not be possible for accurate reaching to visual targets to develop in an animal that cannot perform visually guided locomotion.

The experiments which have implicated visual feedback from limb movement in acquisition of guided reaching have permitted kittens to view the surrounding during the same exposure periods which provided opportunity to view the limb. Visual stimulation from stationary objects correlated with self-produced locomotory movement supports the development of visually guided locomotion (1, 6). To test the hypothesis that the development of a body-centered map of visual space is prerequisite to acquisition of guided reaching, opportunity to view the surround must be segregated from opportunity for visual feedback from limb movements and an appropriate sequence of exposure conditions arranged.

Visual feedback form the moving forelimb, without visual stimulation from any other source, was provided in our experiment by permitting the kitten to locomote in a totally dark enclosure with a spot of chemiluminescent material applied to one forepaw. This exposure condition was imposed both before and after a period during which the conditions essential for development of visually guided locomotion were provided. Animals in a control group were not provided with exposure supporting the acquisition of visually guided locomotion but were, during the comparable period of time, given exposure in patterned light. Visually guided reaching was tested following each of the three exposure periods.

The experimental method was as follows: each of the kittens was kept in total darkness from birth until 5-12 weeks of age. At that time, a strip of surgical tape was wrapped around one forepaw and a small spot of chemiluminescent material applied to the tape. Then the animal locomoted in a totally dark, non-reflective enclosure for three hours daily. After three hours, the kitten was returned to total darkness. This procedure was repeated for 10 days, with the same forelimb bearing the luminous spot each day. Then the animals were brought into light and tested for visually guided reaching to the prongs of an interrupted surface (4). In this test the kitten is permitted to reach with only one forelimb at a time, so that visual guidance of each limb may be assessed separately. Normally reared kittens tested with this procedure strike a prong on 95% of all trials. With neither paw did the performance of these kittens approximate that of the normal animals. They contacted a prong on only 44% of trials, a chance level of performance.

Next, the group of kittens was subdivided. Half of the kittens spent three hours each day locomoting in a normally illuminated room while wearing light-weight collars which prevented view of the limbs. The other kittens spent three hours each day in the same room in holders which minimized movements of torso and limbs and did not permit locomotion. This procedure was repeated for 10 days with return to total darkenss between daily exposure sessions. Then two tests of visually guided locomotion were performed, results of which were in agreement. When placed in an obstacle course consisting of wooden blocks arranged to provide narrow channels of passage (6), the kittens that had locomoted while in light threaded their way among the blocks, avoiding contact with the obstacles. The kittens restrained in holders while in light repeatedly collided with the blocks and startled upon contact with the obstacles. The other test examined behavior on a visual cliff (13), with a criterion of 7 out of 8 descents to the shallow side. All of the animals that had locomoted in light met this criterion but none of the kittens that had spent the period of exposure in holders did. Both tests revealed that the animals given opportunity to locomote in light developed visually guided locomotion while the others did not. At this point both groups of kittens were re-tested on the interrupted surface and all performed at chance level.

Now the third exposure period was begun. All kittens were again given opportunity to locomote in a totally dark enclosure with visual feedback from forelimb movement provided by a spot of chemiluminescent material, just as in the initial exposure period. This time, however, the chemiluminescent was applied to the forelimb contralateral to that to which it had first been applied. During the next 10 days, each animal received visual feedback from movements of one forelimb in a totally dark surround, for three hours each day, with return to total darkness between exposure sessions. Subsequently, the kittens were brought into light and re-tested for visually guided reaching to the interrupted surface. The two groups of animals now clearly differed in the capacity for guided reaching with the limb that had borne a luminous spot during the last exposure period. Those animals that had previously displayed visually guided locomotion, by negotiating the obstacle course and meeting the criterion for descent to the shallow side of the visual cliff, struck a prong on 33 of 36 extensions of that limb. The kittens that had not developed guided locomotion struck a prong on 18 of 36 limb extensions, a chance level of performance. With the other forelimb, the forelimb that had borne a luminous spot only during the initial exposure period, accurate reaching was absent. Animals from both groups performed at chance level when reaching with that limb.

We concluded that visually guided reaching was acquired by animals provided with visual feedback from forelimb movements, after, but not before, the development of visually guided locomotion. Kittens displayed accurate reaching only when using the limb from which they had received movement-produced visual feedback subsequent to acquiring a body-centered map of visual space. The limb from which comparable feedback had been provided prior to development of visually guided locomotion was not visually guided. The results of this experiment identified a constraint on the sequence in which two components of visual motor coordination, guided reaching and guided locomotion, may be acquired. Recently, we examined the possibility that development of all visually-coordinated movement is constrained with relation to one component, movements of the eye. We hypothesized that these movements provide an innate link between the visual and motor system essential to utilization of motor-visual feedback.

Fixating movements of the eyes provide a basis for the mapping of position of the eye in its orbit onto visual space. These movements are displayed by dark-reared kittens when they are first presented with a high-contrast visual target, and may be used to establish a correspondence between each point in visual space and the retinal locus to which it will project, given a particular posture of the eye. Eye movements can elicit, either on an innate bases or after appropriate exposure, orienting movements of the head toward visual targets. When this eye-head link has been established, exposure with visual feedback from locomotion permits the construction of a body-centered representation of visual space. This representation supports locomotion with respect to visual objects. Subsequently, when a limb has been localized within this map, visual feedback from limb movements permits acquisition of visually-guided reaching (9).

This hypothesized sequence suggested that, in the absence of eye movements, opportunity to locomote in light would not support development of visually-guided locomotion. To investigate this possibility we have employed two surgical techniques to immobilize the eye of a kitten. In each case cranial nerves III, IV and VI, which innervate the ocular musculature, are sectioned on one side. One procedure developed in collaboration with F. VITAL-DURAND is a modification of a technique which BERLUCCI, MUNSON and RISSOLATTI (14) used in the adult cat. A ventral approach through occipital bone between the acoustic bullae is made to nerve VI and it is cut. Nerves III and IV are approached laterally in the <u>fossa cranialis media</u>, ventral to the temporal lobe. In a second technique, developed in collaboration with

WALTER SALINGER, the nerves are approached through the roof of the mouth. The soft palate is sectioned and the overlying bone removed with a rose burr to reveal the optic chiasm together with portions of one side of the optic nerve and optic tract. Lateral to these structures, cranial nerves III, IV, V and VI appear just caudal to the orbit. Careful dissection permits isolation of nerve V and section of III, IV and VI.

The first technique was used to produce monocular paralysis in five dark-reared kittens 6-8 weeks of age. While the kitten was anesthetized, the eye contralateral to the side of cranial nerve section was suture shut. The animals recovered from surgery in light and were kept in a normal laboratory environment thereafter. Observations during the next 12 weeks confirmed the absence of eye movements. Eye movements could not be elicited in response to moving high-contrast targets and attempts to elicit post-rotatory nystagmus also failed.

The kittens were tested repeatedly for the capacity to make visually-guided movements under control of the immobilized eye. Visually-guided locomotion was tested in an obstacle course; visually-guided reaching was tested in a bridge box (6). The kittens remained unable to perform these behaviors, repeatedly colliding with objects and failing to step across a gap onto a pier when it was not straight ahead. During the next week (13 weeks after surgery) eye movements began to appear, indicating reinnervation of the eye muscles. One week after the eye movements were first observed visually-guided behaviors were re-examined. At this time the kittens traversed the obstacle course rapidly without collision and stepped accurately on to the pier of the bridge box.

These observations support the suggestion that visually initiated eye movements are important for the development of visually-guided behavior. Moreover, the late acquisition of visual-motor coordination (when the animals were 20-24 weeks old) is evidence that this system remains labile well after the critical period for influencing characteristics of single units in visual cortex.

The technique in which cranial nerves III, IV and VI are sectioned just caudal to the orbit results in permanent immobilization of the eye. This technique was used with dark-reared kittens in which the contralateral eye was sutured shut at the time of cranial nerve section. The animals were kept in light continuously following surgery. Visually-guided behaviors were absent one month after surgery and had not been acquired when the animals were retested one year later. The lack of visually-guided behaviors in these animals is consistent with the failure of eye movements to re-appear.

Immobilizing the eye by cranial nerve section produced emmetropia and mydriasis. These changes in the optical properties of the eye degrade the retinal image for nearby objects. Decreased acuity might disrupt acquisition of visually-guided behavior regardless of the effects of the absence of eye movement. We examined the contribution of reduced visual acuity to visual-motor development in a new group of dark-reared kittens. One eye was sutured shut. The animals locomoted in light with the contralateral eye exposed in a state of mydriasis and emmetropia induced by the topical application of atropine. Ophthalmic atropine was applied 20 minutes before exposure in light. While paralyzing accommodative and pupillary reflexes, atropine does not interfere with eye movements. When tested after one week, these kittens were able to use the exposed eye to traverse an obstacle course and to mediate guided reaching in a bridge box. We concluded that the absence of eye movement rather than degredation of the retinal image had been the effective deprivation in preventing the acquisition of visually-guided behaviors.

These results support our hypohesis that development of visually coordinated behavior is constrained with relation to one component, movements of the eye. Eye movements provide an essential link between visual and motor systems. In their absence, visually coordinated behavior fails to develop.

Acknowledgement

This research was supported by the Spencer Foundation, the National Institues of Health Research Grant NS09279, awarded by the National Institutes of Neurological Diseases and Stroke, and the National Aeronautics and Space Administration Grant NGR 22-009-798.

References

1. A. Hein, in Minnesota Symposia on Child Development, Vol. 6, ed. by A. Pick (University of Minnesota Press, Minneapolis, 1972) p. 53.
2. A. Hein & R. Held, in Biological Prototypes and Synthetic Systems, ed. by E. E. Bernard and M. R. Kare (Plenum Press, New York, 1962) p. 71.
3. R. Held & A. Hein, J. Comp. Physiol. Psychol. 56, 872 (1963).
4. A. Hein & R. Held, Science 158, 390 (1967).
5. A. Hein & R. M. Diamond, J. Comp. Physiol. Psychol. 76, 219 (1971).
6. A. Hein & R. M. Diamond, J. Comp. Physiol. Psychol. 76, 31 (1971).
7. J. A. Bauer & R. Held, J. Exp. Psychol.: Animal Behav. Processes 1, 298 (1975).
8. R. Held & J. Bauer, Science 155, 718 (1967).
9. A. Hein & R. M. Diamond, J. Comp. Physiol. Psychol 81, 394 (1972).
10. A. Hein, Brain Res. 71, 259 (1974).
11. A. Hein, E. Gower & R. Diamond, J. Comp. Physiol. Psychol. 73, 188 (1970).
12. A. Hein, R. Held & E. Gower, J. Comp. Physiol. Psychol. 73, 181 (1970).
13. R. D. Walk & E. J. Gibson, Psychol. Monogr. 75, 15 (1961).
14. G. Berlucci, J. B. Munson & G. Rissolatti, Electrocenceph. Clin. Neurophysiol. 21, 504 (1966).

Psychophysical Functions in Fish with Respecified Retinotectal Connections

Dean Yager
State College of Optometry
State University of New York
New York, New York 10010

Introduction

In most submammalian species, the primary visual projection area within the brain is the optic tecum, with each eye projecting to the opposite tectum. It has been shown through both anatomical and electrophysiological studies that there is a continuous projection of the retinal surface onto the contralateral optic tectum, and there is no ipsilateral tectal projection, in the goldfish. This means that the axons of ganglion cells that are adjacent in the retina project to adjacent areas in the optic tectum. The posterior visual field projects to the anterior retina, and then projects to the posterior part of the optic tectum on the opposite side of the brain, and *vice versa.* The apparent simplicity of this mapping makes it possible to investigate the time course, and factors that influence the formation and reformation of retinotectal connections.

Unlike most mammalian species, fish have a remarkable ability to restore the retinotectal map after damage at various levels of the visual system. More than 30 years ago ROGER SPERRY (1, 2) showed that after sectioning the optic nerve, retinal ganglion cell axons regenerate, and an orderly re-establishment of the normal topographic retinotectal projection occurs. This result has led to a great number of experiments on reformation of these connections, with the implicit assumption that if the mechanisms responsible for the orderly remapping could be understood, then we might have some understanding of how these connections are formed during embryological development. I will not discuss the various hypotheses that have been proposed to account for how the central connections of the optic nerves are formed, or to review the anatomical, physiological, and biochemical evidence for or against the hypotheses. What I will discuss is one of the crucial questions in most of the regeneration work, a question to which there was no good answer until very recently. And that question is, "Have new functional synapses been formed after regeneration?" Electrophysiological mapping studies do not provide evidence that the new maps are functional, since they depend on presynaptic recordings from the terminal arborizations of the ganglion cell axons. It is obvious that in order for the regenerating visual system to be useful as a model for embryological development, the regeneration must result in synaptic connections that can transmit visual information. The demonstraton that the regenerated map is functional may be accomplished in at least two ways. The first is by recording visual responses from postsynaptic tectal neurons, or, secondly, by eliciting behavioral responses from the fish that depend on visual input that must cross the presumed synapses.

In this paper I will summarize the results of several experiments, from a number of different laboratories, that were carried out to assess the visual capacities of goldfish that had undergone a variety of lesions and regenerations in the visual system. In addition to demonstrating functional capacities of the regenerated visual system, some of these studies also provide information about the normal function of certain brain structures by demonstrating the absence or impairment of these functions following lesions.

We have known for many years that the regeneration of the optic nerve, which SPERRY first described, results in the retention of a visual discrimination that was established before surgery. ARORA and SPERRY(3) reported in 1963 that the Oscar, a teleost fish, could be trained to discriminate between a variety of colored stimuli, and if the training was done before sectioning the optic nerve, regeneration of the nerve resulted in reinstatement of the discrimination with no further training required. This result shows that some very specific connections have been reformed in the tectum, although their method did not allow them to test for subtle deficits in color discrimination. Another experiment from SPERRY'S laboratory, by MARJORIE SCOTT (4), demonstrated recovery of function after a different kind of damage to the visual system; this result will be discussed below.

The "Compression" Phenomenon

GAZE AND SHARMA (5) showed that if the caudal half of the optic tectum is removed, in a goldfish, and the contralateral optic nerve is also cut, then when the nerve regenerates and grows back into the front half of the optic tectum, the entire visual field is compressed onto the remaining rostral half of the tectum. This means, of course, that although there still is a point-for-point retinotectal map, there are now different corresponding points between the retina and the tectum. In an early psychophysical experiment on this type of tectal lesion, SCHWASSMANN and KRAG (6) showed that immediately following lesions in the tectum, goldfish have a behavioral scotoma in the appropriate part of the visual field. SCHWASSMANN and KRAG terminated their experiment at this point, and they did not allow for recovery to occur. SCOTT (4) showed that if the scotoma is mapped repeatedly as the compression of the retinotectal map takes place, by determining which parts of the visual field can mediate a simple color discrimination, the behaviorally blind field contracts as the formation of connections proceeds. Thus, the altered connections retain enough specificity to keep color information intact, even though each ganglion cell in the retina is now projecting to a different point in the optic tectum.

Psychophysical Experiments

In our lab, we are in the midst of a series of experiments in which we are looking at a number of other kinds of psychophysical functions in goldfish that have undergone a variety of different lesions and respecification of connections. Before proceeding with these studies, though, the visual capacities or completely tectumless fish must be investigated, because there are a large number of retinal projection sites in goldfish in addition to the tectum. Only after these controls have been run can we assess the function of new retinotectal connections by comparison of visual function in subjects that have the new connections with normal and tectumless subjects.

We used a conditioned respiration technique that was developed by DAVID NORTHMORE (7); this is illustrated in Fig. 1. The fish was held in front of a screen on which stimuli could be projected, and a thermistor, operated in its heating mode, was suspended in front of the fish's mouth. Respiration by the fish cooled the thermistor, and the temperature changes were electronically converted to pulses, which enabled us to count breaths. A brief electric shock delivered through two electrodes beside the fish's body caused a deceleration of breathing rate. Through pairing a visual stimulus with the shock, according to a Pavlovian, or classical, conditioning paradigm, the presentation of the visual stimulus alone produced the slow-down of breathing. An estimate of visual threshold was obtained by computing the ratio of breathing rate during stimulus trials, which were 10 sec in duration, to the rate during blank trials, and plotting this suppression ratio as a function of stimulus radiance.

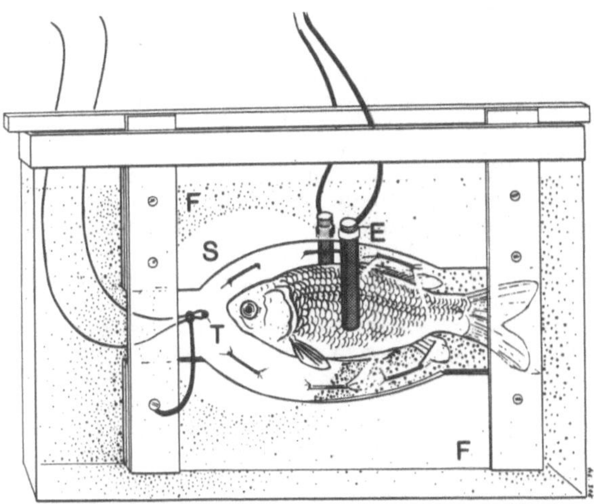

Fig. 1 Conditioning apparatus. The fish is restrained between two sheets of transparent plastic clamped between Plexiglass formers (F). Respiration is detected by a thermistor, T, and shock is administered through carbon electrodes, E. The conditioned stimulus consists of a stimulus patch, S, projected onto a translucent screen 33 cm from the eye.

Experiment 1: Time Course of Recovery

Four normal fish were conditioned to a disc of light, 5° in diameter, with a luminance of 95 mL superimposed on a background of 1.04 mL. The disc was presented in 200 msec flashes at a rate of 1.3 flashes/sec, for 10 sec. After training, the average breathing rate during stimulus presentation was 0.87 of the rate during blank trials. Immediately following bilateral tectal removal, the suppression ratio was not reliably different from 1.00 (no response). This surgery left intact the direct retino-thalamic and pretectal projections, which comprise less than 10% of the total population of optic nerve axons; the psychophysical experiment failed to show that these normal non-tectal projections could mediate the detection of light.

Conditioned responses began to appear about three weeks following surgery, which is approximately the time required for regeneration of optic nerve fibers following a cut at such a proximal location (8). This result suggests that when the severed optic axons regenerate, their heavy projections onto the pretectal and thalamic visual areas (9) are able to mediate a simple detection of change in illumination.

Experiment 2: Comparison of Thresholds of Normal and Tectumless Fish

Psychometric functions were obtained from a group of normal fish and for the tectumless group between 4 and 5 months following surgery. Stimulus conditions were the same as for the previous experiment. The normal group was found to be approximately 1.0 log unit more sensitive than the group with only non-tectal projections.

Experiment 3: Object Perception

Several tests demonstrated reduced or absent object perception in tectumless fish for as long as 15 months following surgery. (a) If a small (5mm) piece of charcoal affixed to the end of a glass rod was swept back and forth in the home tank of normal fish, pursuit and attack movements generally occurred; such responses were never observed with tectumless fish. (b) When placed in a white styrofoam tank (60 cm x 60 cm x 12 cm), normal fish explored freely and rapidly, spending a good proportion of their time moving through the middle of the tank. When tectumless fish were placed in this tank, they moved more slowly, and exploratory movements were almost entirely restricted to swimming along a wall until they bumped into the corner where they turned and swam along the next wall to the corner, and so on. (c) In such a tank, a 5 cm black disc affixed to a glass rod was swept across the fishes' path. Normal fish turned and darted away before the disc approached to within 10 cm; tectumless fish did not reliably avoid. When they did, a turning movement was not executed until the disc was within about 4 cm of them. The movement was slower, and the distance moved was much smaller. (d) When tectumless fish were pursued with a net in their home tanks, no avoidance movements were ever made, but an exaggerated startle response occurred when they were touched with the net. (e) When food pellets were dropped into the home tank, normal fish accurately struck at pellets at the top, bottom, or while they were sinking through the water. Tectumless fish showed a very different feeding pattern. A few seconds after pellets were dropped into the tank, these fish began rapid striking movments to the top or bottom, which eventually brought the fish to the area of the food pellets, usually within 10 sec. Thus, feeding movements appear to be directed by chemical rather than visual cues.

Experiment 4: Areal Summation

Thresholds for two different stimulus diameters were measured in tectumless and normal fish; ablations had been made 6 months to one year prior to training. The circular stimulus, having an angular subtense of either 5.4^{o} or 25.8^{o}, was presented continuously for 10 sec superimposed on a background of 1.0 mL. Psychometric functions for the two groups and two conditions are shown in Fig. 2. Linear summation of luminous flux (Ricco's law) from the circular stimuli would produce a difference in log threshold luminance of 1.36 for these two diameters. As can be seen from the figure, the tectumless subjects had a threshold difference of about 1.30 log units; they acted as almost perfect integrators of luminous flux with summation up to at least 25.8° stimulus diameter. The normal subjects had a log threshold difference of about 1.00.

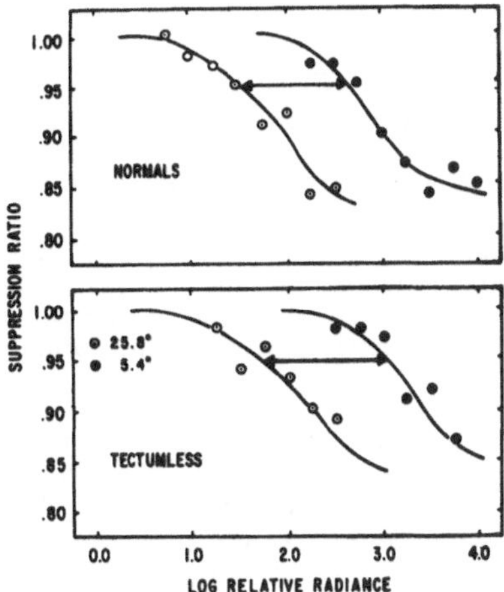

Fig. 2 Suppression ratio as a function of stimulus luminance for normal and tectumless subjects, for two different stimulus diameters. Four animals in each group. Each point is the mean of two sessions per subject, 50 trials per session.

Discussion of Experiments on Tectumless Fish

These results show that simple detection of a change in illumination may be mediated by structures other than the optic tectum; nevertheless, the tectum normally is involved in such a discrimination, since an intact subject is approximatley ten times as sensitive as the tectumless. In many experiments, in which the role of certain structures in making a particular discrimination or detection is assessed by noting the presence or absence of the behavior following extirpation of the structure, little attention is given to the possibility that the threshold for a response may be raised by the lesion but that necessary structures are still present. For example, if, in testing for the ability of a tectumless fish to detect lgiht, we had merely used a moderate stimulus luminance that elicited strong responses from the normal subjects, we might have concluded that the tectumless subject is completely blind. Whenever possible, psychometric functions and thresholds should be obtained, rather than obtaining responses at only one stimulus level.

This criticism may, of course, be made of our observations on object perception (experiment 3) since they were made only at the level of room illumination. Experiment 4 provides more definitive evidence that tectumless fish have a deficiency, and possibly a complete lack, of ability to respond to spatial aspects of a stimulus. The tectumless fish behaved like an integrating sphere up to a very large stimulus diameter: thresholds in terms of total luminous flux were the same for 5.4° and 25.8°, whereas the normal fish integrated over a smaller diameter on the retina. The tectumless animal appears to have a spatial summation function similar to that of monkeys with complete striate cortex removal (10), whereas spatial integration in normals is probably restricted by the angular extent of receptive fields in the retina and tectum (11-13). Experiments on shape discrimination and visual acuity are needed to elucidate further the role of the tectum in spatial vision; some data that provide some partial answers to the visual acuity question are reported in the next section of this paper.

Experiment 5: Visual Acuity with a Compressed Retinotectal Map

There are several laboratories that are conducting psychophysical studies on goldfish with compressed maps and, taken together, we are beginning to get a better idea of their visual capacities. I reported above that SCOTT demonstrated that the new connections can be used to mediate a color discrimination. DAVID NORTHMORE has measured visual thresholds in compressed-map goldfish and has shown that they have normal detection sensitivity (personal communication). He also investigated summation area in these animals and found that they also have about the same critical diameter as normals: that is, their spatial integraton seems to be normal, even though only half of their tectal cells are present.

A compressed projection has also been shown to mediate simple pattern discriminations. Three different groups have reported on this problem. HODOS and YOLEN (14) have used a two-choice operant conditioning technique to show that animals with compressed retinotectal maps can acquire discrimination between coarse horizontal and vertical gratings almost as rapidly as normal. CAMPBELL, STIRLING and INGLE (personal communication) have reached the same conclusion with a more complicated form discrimination. And finally, we have adapted the conditioned-respiration procedure to demonstrate discrimination between 5° horizontal and vertical black bars on a bright background. In this procedure, we delivered the conditioning shock only following presentation of the horizontal bar and presented the vertical bar on non-shocked trails. We then computed the ratio of breath rate during horizontal presentations to the rate during vertical presentations. A clear departure from a suppression ratio of 1.0 was obtained within about 150 training trails in this crude horizontal-vertical discrimination. Acquisition was as fast for a compressed eye-tectum pair as it was for the normal.

These results, i. e., the lack of a deficit in the acquisition of a simple form discrimination, would lead many people who do work with lesions and visual deficits to conclude that the compressed eye-tectum system is essentially normal. Often, this kind of trials-to-criterion measure is taken as the only index of a deficit, and much potentially good lesion work has been wasted because more subtle psychophysical functions have not been investigated.

The three form discrimination studies I have just described were all followed up by obtaining measures of visual acuity. Figure 3 shows the discrimination suppression ratios in my study for a series of bars of different angular subtenses. The subject was used as its own control, since there was one eye-tectum pair that was normal, and there was one compressed eye-tectum pair in the same fish. As can be seen from the data, when the stimulus was seen by the normal system, this subject continued to show good discrimination down to 16 minutes of arc, and his limit of detection had not yet been reached. The compressed system produced normal good discrimination for the 5° stimulus, so there was no deficit in his ability to be conditioned to the visual stimuli; however, discrimination for smaller stimulus sizes declined markedly. HODOS and YOLEN (15), and CAMPBELL, et. al., obtained qualitatively similar results in their studies of visual acuity with compressed retinotectal maps.

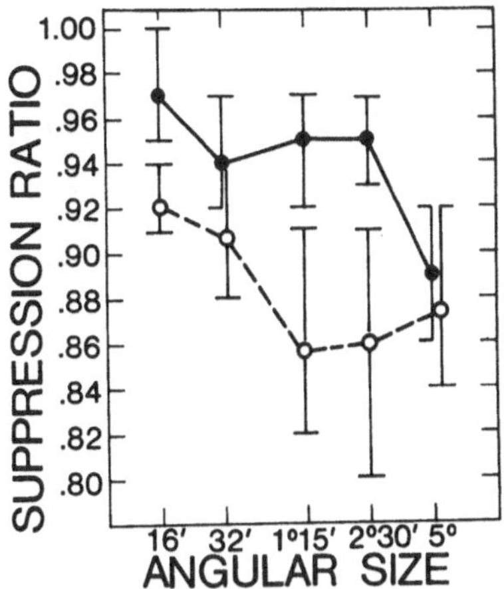

Fig. 3 Suppression ratio as a function of width of a single black line. Closed symbols: compressed eye-tectum system. Open symbols: normal eye-tectum system. Vertical bars: inter-quartile range of ten measurements per point, 50 trials per measurement.

It would be premature to speculate in detail about what these results imply for the function of the tectum in determining the limits of visual acuity. But let me suggest, as an example, the kind of psychophysical analysis that might be done in the future: since with caudal ablation, the compression of the retinotectal map occurs only in the anterior-posterior axis, it might be interesting to test whether acuity loss occurs for vertical gratings, but not for horizontal ones. There are a number of other kinds of psychophysical functions that are at present being investigated, in several laboratories, on goldfish with various kinds of "rewiring" of their visual systems. Within a relatively short time, we will have a fairly complete description of the visual behavior of this kind of subject.

Acknowledgement

All of the experiments of my own that are reported here were carried out in collaboration with Sansar Sharma of New York Medical College, and he is responsible for all the surgical, anatomical, and physiological aspects of the studies. The research reported here was supported by Grants 01426 from NIH-NEI and GB-43506 from NSF to Professor Sharma. Some of the results reported here have been published previously (16, 17).

References

1. R. W. Sperry, Physiol. Zool. 21, 351 (1948).
2. R. W. Sperry, In Handbood of Experimental Psychology, ed. by S. S. Stevens (Wiley, New York, 1951).
3. H. L. Arora and R. W. Sperry, Devel. Biol. 7, 234 (1963).
4. M. Y. Scott, Exp. Neurol. 54, 579 (1977).
5. R. M. Gaze and S. C. Sharma, Exp. Brain Res. 10, 171 (1970).
6. H. O. Schwassmann and M. H. Krag, Vis. Res. 10, 29 (1970).
7. D. P. M. Northmore and D. Yager, In Vision in Fishes, ed. by M. A. Ali (Plenum, New York, 1975) p. 689.

8. R. M. Gaze, The Formation of Nerve Connections (Academic Press, New York, 1970).
9. D. Yager and S. C. Sharma, Neurosc. Abstr. 2, 841 (1976).
10. R. W. Doty, In Handbook of Sensory Physiology, Vol. VII/3, Part B, ed. by R. Jung (Springer, Berlin, 1973) p. 483.
11. S. S. Easter, J. Physiol. (Lond.) 195, 253 (1968).
12. H. O. Schwassmann and L. Kruger, J. Comp. Neurol. 124, 113 (1965).
13. H. G. Wagner, E. F. MacNichol, Jr. and M. L. Wolbarsht, J. Opt. Soc. Amer. 53, 66 (1963).
14. N. M. Yolen and W. Hodos, Brain Behav. Evol. 13, 451 (1976).
15. W. Hodos and N. M. Yolen, Brain Behav. Evol. 13, 468 (1976).
16. D. Yager and S. C. Sharma, Nature 256, 490 (1975).
17. D. Yager, S. C. Sharma, and B. Grover, Brain Res. 137, 267 (1977).

The Cheshire Study:

Change in Incidence of Myopia Following Program Intervention

John W. Streff
Southern College of Optometry

Introduction

The evolution of treatment programs used by clinical disciplines are fundamentally determined by the accepted basic hypotheses relating to the etiology of a condition. In vision care this can readily be seen as it relates to the etiology of myopia. When considering the etiology of myopia, one is faced with the relative contribution of heredity and environment in the development of the condition.

If a clinician accepts myopia to be primarily a physiological variable determined by heredity, clinical regimens will mainly be concerned with compensation and possible attempts to control the progression of myopia. Even the attempts to control progression might be questioned within the confines of the hypothesis. On the other hand, if one accepts the regimens that there is a significant adaptive process related to environmental variables in the development of myopia, then regimens of care might be evolved to prevent and control the development.

In teaching institutions and in the care programs being offered to the public today the emphasis, either implicit or explicit, is predominantly derived from the hypothesis of an heredity etiology of myopia. The principal clinical approach is to prescribe optical compensation for myopia. Relatively little effort is made to attempt to modify the expected progression of the condition. Little attention has been given to prevent the development of school age myopia by modifying environmental influences. One reason for this has been the lack of statistical research evidence to support the significance of the relative contribution of adaptation and environment in the etiology of myopia. This lack of evidence is cited as an important factor to help account for the position of the clinical eye and vision care professions.

It has been difficult to obtain statistical evidence because most previous studies using environmental variables have attempted to deal with clinical populations. The present study was done with public school populations, a suggestion offered by

SORSBY (1) when he stated, "It is likely that we will solve the problems of myopia without intensive study in its breeding ground - the schools". The purpose of this paper is to present evidence of a significant positive change in the incidence of school age myopia after an intervention program to effect changes within the classroom environment.

Background

Today most educators and clinicians do not take an "either-or posture" as related to the development of school age myopia. They agree that both heredity and environment play a role. The important issue is the relationship, the relative role of heredity and the relative role of adaptation and the environment. The environmental role has recently gained some theoretical acceptance. But this acceptance has neither significantly changed clinical regimens of care nor educational programs in professional schools and colleges.

The role of environment in the etiology of myopia was suggested as early as 1813 when JAMES WARE (2) noted that officers of the Queen's Guard frequently were nearsighted, but that the foot solders were rarely myopic. He attributed this difference to the officers perusal of books and other forms of studying. His conclusion was that reading was the probable cause of myopia. COHN (3) was more specific in defining possible variables. He listed the causes as poor reading, poor illumination, and poor printing in books. In general, this explanation has come to be known as the 'use-abuse theory'.

The use-abuse theory to explain myopia has been attributed to different functions. PASCAL (4) believed convergence was involved, especially where exophoria exists. LEVINSOHN (5) and RUSMUSSEN (6) offered a theory of 'gravitational pull' related to the lowered head posture in reading. The most popular concept over the years is related to the accommodative function for near work. ERISMAN (7), SATO (8), AMANO (9), BALDWIN (10), and YOUNG (11, 12) all support the concept that accommodation is related to the development of school myopia.

Statistical studies supporting the environmental hypothesis have dealt with patterns of increase in myopia or attempts to control myopia with cycloplegia or bifocals. The major study showing the increase in incidence of myopia was performed by YOUNG *et al.* (13) using an Eskimo sample of Barrow, Alaska. YOUNG (14, 15) also demonstrated a significant development of myopia in monkeys by restricting visual space and varying illumination levels. Previous to the Cheshire Study, no other study has demonstrated a significant relative reduction of the incidence of myopia today by varying environmental factors.

ROSONES (16) presented a study which showed the psychological correlates to myopia. Using the Rorschach Test, she supported the work of KELLY (17), demonstrating that myopes manifest a characteristic personality pattern expressing increased anxiety.

The hereditary hypothesis of today is founded on the work of STEIGER (18) who concluded that myopia is a biological varient. STENSTROM (19) concluded that a major factor was abnormal axial length of the eye. SORSBY, *et al.* (20, 21) gave support to the biological variant concept extending the considerations to the relationship of optical variables. HIRSCH (22) supports this concept and summarized by stating that myopia has not yet been demonstrated to be preventable.

The evidence is quite conclusive that there are biological changes related to myopia. There are often axial length changes of the eye as well as other optical changes when an individual manifests myopia. This would seem obvious by the definition of myopia (manifestation of reduced distance acuity, against motion in the retinoscopy measurement, and a positive response in acuity to concave lenses); the eye is different in its optical measurement. To accept the fact that biological changes occur neither supports nor contradicts an etiological hypothesis. Hereditary factors or adaptation, due to environmental conditions, are capable of producing biological changes. While there remains a variance of opinion related to etiology, there are some areas of general agreement. The following list describes some of the areas of agreement pertaining to the development of myopia.

1. Myopia shows an increase in incidence and amount as a child progress from grade 2 through high school (23).
2. Recent studies support the concept that today's incidence of myopia is higher than it was earlier in the century, or even 20 years ago (23, 24).
3. Significant amounts of myopia occur at an earlier age (grade) tody, than in the past (24).
4. Girls manifest school age myopia earlier than boys, although this eventually equalizes or reverses (25).
5. The difference related to sex, girls manifesting myopia earlier than boys, has been consistent in all other studies. This pattern was evident in studies showing increases from 7 to 9 years of age as well as studies showing peaking from 11 to 14 years of age (12, 24, 25, 26).

Methodology

The purpose of the study is to evaluate the effects of a school intervention program on the development of school age myopia. The school program is broad in scope with changes in the physical, social, emotional and teaching environments.

This project will test the null hypothesis, which is, environmental factors have no significant effect on the incidence of school age myopia.

To test this hypothesis, a study was done to compare the incidence of myopia within four sample groups of sixth grade children. The experimental sample was made up of all the sixth grade students in 1974 who were always enrolled in the Cheshire Public Schools. The experimental sample also included full time Cheshire sixth grade students from one school building in 1975. All students in the experimental group were participants in the environmental changes during all their school years.

Three different control samples of sixth grade students were used for comparison to reduce the possibility of chance. One control sample was made up of sixth grade students in the Cheshire Public Schools before the start of the intervention program. This data was collected in 1970 and 1971. The second control group was a sixth grade sample in a neighboring school system in 1974. A third control sample was added after most of the data had been collected. This sample is composed of students who transferred into the Cheshire Public Schools after first grade.

For some of the analysis procedures, the three different control samples are treated as one sample after testing for statistical significance. No significant statistical difference, using the t-test, was found between the control samples. This makes it possible to compare the full treatment sample with all other samples.

The Intervention Program

The intervention program consisted of homogenous grouping according to developmental levels, classroom equipment, furniture and lighting changes which were limited to one or two rooms in each of the five school buildings; and a program to train teachers having them spend one half hour a day doing motor-sensory developmental problem solving activities. Relatively few of the tested children were exposed to the classroom equipment, furniture and lighting changes. All teachers were encouraged to rearrange their classrooms to orient them more to the childrens' physiological needs and the expanded program activities.

Grouping Changes

The homogenous class grouping according to developmental age began in the early 1960's. Each child was tested using the modified Gesell test and grouped according to behavior age score. Evaluations by teachers and reading achievement scores were utilized in the formula for class make-up.

Under six years of age, developmental behavior scores are in half year divisions. To make it possible for children to progress relative to their behavior age, additional grade steps were introduced (4 1/2 years and 5 1/2 years). This positive procedure effectively delays less mature children early in school life lowering the risks of a possible failure at a later grade. The net result of this action was that the average sixth grader in the experimental sample was six months older with a lower standard deviation of age difference.

The Physical Environment

Students from Yale University, School of Architecture and Design, under the direction of Professor Felix Drewry, became involved in the program in 1967. After spending time in school buildings and teacher workshops, the students recommended and constructed physical environmental changes. These varied from changing room lighting and designing of classroom equipment, to more dramatic physical changes. While the architectural contributions were significant, a relatively small number of the children in the test sample were exposed to the physical changes. Because of the timing of the architectural program and its focus on the early grades, later classes (not in the selection sample) had more benefit from these changes.

Activities Program

The motor-sensory development and problem solving activities program started in February, 1966 with a pilot program in the second grade. Following that experience, the program shifted to the pre-first grade level and then progressed one grade level each of the following years. The program is described by AMES, GILLESPIE and STREFF (27) and was called a perceptual program.

Except during the 18 one-half day workshops with each classroom teacher, the activities were performed in the classrooms by teachers. Each teacher was asked to perform a 1/2 hour of daily activities, oriented to motor learning and motor-sensory development. The activities were structured as problem solving tasks. I had responsibility for the activity program design, and translation of optometric vision training concepts and procedures into educational activities. The program was designed to help children develop motor-sensory skills and recognize the use of these skills as tools for problem solving. Emphasis was placed on the role of feedback using information differences and similarities. The overall objective was directed toward helping children and teachers trust themselves and their abilities, the process of decision making related to experience, and to the achievement of results that worked for children. An attempt was made to avoid exercise type activities, unless they were a game or related to a task. However, some of the procedures did resemble tasks used in optometric offices.

An attempt was made to relate tasks to abstract codes, symbol representation and processes of communication. Sharing of perceptions was encouraged, as was cooperation by working in small group activities. No lenses, prisms, or stereoscopic instrumentation was used. I did spend time with individual teachers and occasionally with individual children. As the year progressed teachers were encouraged to use the concepts related to the problem-solving approach while teaching the traditional curricular material.

My success with teachers was varied. In general, about 1/3 of the group became enthusiastic and greatly changed their teaching styles and approach. Another 1/3 did follow the activity schedule part of the time and selected components of the program to incorporate within their classroom. The remaining 1/3 attended the workshops, tried some of the activities, but generally did not feel they could give up their planned teaching time for the program. As might be expected, teachers in grades kindergarten through two, were usually more enthusiastic and found it easier to incorporate the program activities into their daily schedules.

Test Population

The test population for this study is made up of 822 sixth grade children, 384 girls and 438 boys. Except for the control groups from a neighboring school system, all the test population was from the public school system in Cheshire, Connecticut.

Cheshire's students come from families which are mainly an upper middle class socio-economic population. A small number of inner-city children were bussed into the school system. In 1974, the mean I.Q. of the sixth grade population was 107. The mean achievement composite test scores (Iowa Test) in various school buildings ranged from 85 to 95 percentile on national scales in 1974.

Pre-treatment Group

The pre-treatment group is composed of all the sixth grade students in one of the five school buildings in the Cheshire system. This data was collected in 1970 and 1971 with children who were ahead of the intervention program. The pre-treatment sample consists of 121 children, 61 girls and 60 boys. This group is labeled as comparison sample #2 in the analysis.

Treatment Group

The treatment group is composed of all the sixth grade students in the Cheshire Public School system in 1974. Additionally, all the sixth graders from the largest building were added in 1975. The purpose of adding the 1975 sample was that all of the 1974 data was gathered by the author. The 1975 data was gathered by an examiner who had no connection with the project and was used as a control for possible examiner bias. Additionally, it was used to confirm the consistency of results for more than the one year.

The total treatment sample size was 538 children, 237 girls and 301 boys. The 1974 sample was 401 children, 175 girls and 226 boys. The 1975 sample was 137 children, 62 girls and 75 boys.

After most of the data had been collected, the treatment group was divided into two categories. More than 1/3 of the Cheshire sixth grade population had moved into the community and transferred into the school system after first grade. This group did not have the benefit of the program in their early school years. The sample for "transfer-in students" consists of 192 children, 81 girls and 111 boys. This sample is labeled as comparison sample #1.

The experimental sample was composed of all the sixth graders (1974 and 1975) who had been full time Cheshire students or came into the system before the end of the first grade. This sample was made up of 346 students consists of 156 girls and 190 boys.

The Cheshire population in both 1974 and 1975 contains more boys than girls. This was a pattern within the school system itself, and cannot be attributed to any aspect of selection by sampling or slowing down of boys more than girls early in school years. The pattern was relatively consistent for the two grades below those tested.

Neighboring School

The comparison population from a neighboring school consisted of the entire population of one school building. A previous study, McWEENEY (1971) comparing the two schools systems, showed that these systems were not significantly different in I.Q. scores or socio-economic status (father's occupation status) using the scale of the U.S. Bureau of Census. The population consisted of 163 children, 86 girls and 77 boys. The data was collected in 1974. The classrooms of this school building were about equally divided between traditional and cooperative teaching in more open space environments. This sample is labeled as comparison sample #3.

As mentioned earlier, for part of the analysis procedures, comparison samples #1, #2 and #3 are combined as one group labeled as the combined control sample.

Data Collection

The data for the study was collected from February 1st to April 30th in each respective year. The pre-treatment data was collected as part of other studies, but the procedures, were similar. No history of previous school experience was collected for the pre-treatment sample. Retinoscopy was performed by two optometry residents in 1970 and 1971.

The examination routines were done in the school buildings. The equipment used in this study consisted of a stand and phoropter, an acuity chart projected for twenty feet, and a retinoscope. The data were collected by an assistant and an optometrist. I was the optometrist for all data collected in 1974.

The sequence of tests, after history by the assistant, was unaided Snellen acuity for each eye and both eyes, Snellen acuity with glasses (if worn) for each eye and both eyes, static retinoscopy (non-cycloplegic) at twenty feet for each eye, acuity through static retinoscopy. The history consisted of name, date of birth, school history, ocular history if the child had ever worn glasses, and when indicated, date of first glasses worn.

In 1975, the examiner collected the history as well as the refractive data. The plan was to use two examiners collecting the same data independently. After about 1/3 of the data was collected, one examiner became seriously ill. This necessitated using a single examiner's data. Because of the delay, a small number of cases were seen in May of 1975.

Results

Comparison of Examiners

The first analysis was to compare the results of the examiners in 1974 (J.W.S.) and 1975 (D.T.) to determine if examiner bias existed.

In January, 1975, D.T. and I each examined ten fifth grade Cheshire students for the exclusive purpose of comparing our retinoscopy measurement results. The data shows D.T.'s retinoscopy measurements were consistently more minus than mine. The mean difference was -0.40 diopters with a standard deviation of 0.156 diopters, indicating the difference was relatively consistent. Table I presents these results expressed in the mean spherical equivalence of the two eyes.

TABLE I

Retinoscopy Measurement Comparisons
For Two Examiners

Subject	JWS	DT	Difference
1	+0.62 D	+0.37 D	-0.25 D
2	+0.50 D	Plano	-0.50 D
3	+0.50 D	-0.25 D	-0.25 D
4	+0.25 D	Plano	-0.25 D
5	-0.50 D	-1.00 D	-0.50 D
6	+0.25	-0.25 D	-0.50 D
7	Plano	-0.25 D	-0.25 D
8	+0.37 D	Plano	-0.37 D
9	+0.25 D	-0.12 D	-0.37 D
10	+0.75 D	+0.50 D	-0.25 D

Measurements are binocular spherical equivilant.
Mean of differences = -0.40 D, Standard Deviation = .156 D.

Table II presents the comparison of the two examiners results for the Cheshire

TABLE II

Comparison of Two Examiners

	Examiner: JWS Cheshire, 1974			Examiner: DT Cheshire, 1975			Examiner: DT Corrected by 0.37 D		
	Full time	Transfer in	Total	Full time	Transfer in	Total	Full time	Transfer in	Total
N	254	147	401	92	45	137	92	45	137
Myopic	9.4	16.3	12.0	12.0	22.2	15.3	8.7	17.8	11.7
Non-myopic	90.6	83.7	88.0	88.0	77.8	84.7	91.3	82.2	88.3
Total Population	100.0	100.0	100.0	100.0	100.0	100.0	100.0	100.0	100.0
Comparison									
X^2	--	--	--	.47	.82	1.03	.05	.05	.008
P	--	--	--	.50	.63	.31	.83	.81	.92

sixth grade samples in 1974 and 1975. The table presents the percentage of myopic and non-myopic students divided into full time students and transfer-in students. For this comparison, myopia was defined as more than -0.50 diopters, using the mean of the spherical equivalents for the two eyes. The table also presents D.T.'s data corrected by +0.37 diopter (nearest clinically measurable increment of mean retinoscopy difference from Table I). Chi square analysis, performed on both the uncorrected and corrected data for comparison, indicates that there was not statistical difference between the 1974 and 1975 results. When the retinoscopy correction was considered, the results were very similar. The full time students who were myopic in 1974 consisted of 9.4% of the sample. In 1975, they were 8.7% of the sample. The transfer-in myopic samples were 16.3% and 17.8% of their groups for the respective years.

Comparison of Experimental and Comparison Samples

Because the 1974 and 1975 data was so similar, they were combined for the analysis of the experimental and comparison samples. The corrective factor (+0.37 diopters), which was used for the analysis in comparing examiners, was not used in comparing the experimental and combined control groups. The 1971 pre-treatment sample (60 subjects) was corrected for examiner retinoscopy difference (+0.50 diopters), arrived at in a similar manner as the 1975 difference. The only rationale for using this approach (correcting one control comparison but not the experimental) was to be as conservative as possible.

Refractive Distribution

Percentage distribution curves of refractive measures are presented in Table III, and plotted in Fig. #1. The mode of both distributions is the same. The median is +0.25 diopters respectively. The differences in curves are mainly in the tails, and the areas from plano to -0.50 diopters. The mean of the experimental sample is $\bar{X}$ = +.064 diopters (standard deviaiton = 1.20) and of the combined control sample is $\bar{X}$ = -0.17 diopters (standard deviation is 1.31). Throughout the entire myopic tail of the distribution, the control sample shows a higher percentage of myopia. The two samples are very similar (0.58% and 0.63% respectively) at the greater than -6.00 diopter category. This category is made up of children who had developed pre-school myopia.

To better illustrate the relative changes, the data from the myopic tail of the distribution was grouped into three categories; -0.51 to -1.50 diopters, -1.51 to -3.00 diopters, and -3.01 diopters and greater. Table IV presents this data.

Figure 2 presents graphs showing the percentage of myopia for the experimental and combined control sample. This presentation clearly shows that the girls in the combination control sample manifested a higher incidence of myopia than the boys, 24.12% vs. 14.92%. In the experimental sample, the reverse is true. The percentage of myopic girls was 8.97% against 11.05% for the boys. The presentation also illustrates that the difference in incidence of myopia, between the experimental sample and combined control sample, was greater for girls than for boys. The girls in the control sample manifested 2.69 times more myopia than the experimental girls. The same relative comparison shows the control boys manifesting only 1.35 times more myopia than the experimental sample for boys.

Analysis

The Cochran Test (28) indicated that it was inappropriate to use an analysis of variance method. The reasons can be seen by examining the refractive distribution curves (Fig. 1). This figure shows that statistically the modes were identical for the

TABLE III

Refractive Measurements*
Frequency and Percentage Distribution Table

Refractive Grouping	Experimental						Three Control Combined					
	Girls		Boys		Total		Girls		Boys		Total	
	N	%	N	%	N	%	N	%	N	%	N	%
greater than +1.50	4	2.56	7	3.68	11	3.18	4	1.75	7	2.82	11	2.31
+1.50 to +1.01	3	1.92	0	0.00	3	0.87	2	0.88	1	0.40	3	0.63
+1.00 to +0.76	3	1.92	1	0.53	4	1.16	2	0.88	4	1.61	6	1.26
+0.75 to +0.51	11	7.05	8	4.21	19	5.49	22	9.65	15	6.05	37	7.77
+0.50 to +0.26	36	23.08	43	22.63	79	22.83	47	20.61	54	21.77	101	21.22
+0.25 to +0.01	39	25.00	57	30.00	96	27.75	54	23.68	77	31.05	131	27.52
Pl	30	19.23	23	12.11	53	15.32	17	7.46	25	10.08	42	8.28
0.01 to -0.25	13	8.33	23	12.11	36	10.40	20	8.77	14	5.65	42	8.28
-0.26 to -0.50	3	1.92	7	3.68	10	2.89	5	2.19	14	5.65	19	3.99
-0.51 to -0.75	0	0.00	3	1.58	3	0.87	8	3.51	3	1.21	11	2.31
-0.76 to -1.00	4	2.56	2	1.05	6	1.73	4	1.75	9	3.62	13	2.93
-1.01 to -1.50	4	2.56	6	3.16	10	2.89	11	4.82	4	1.61	15	3.15
-1.51 to -2.00	1	0.64	1	0.53	2	0.58	12	5.26	5	2.02	17	3.57
-2.01 to -2.50	1	0.64	2	1.05	3	0.87	4	1.75	3	1.21	7	1.47
-2.51 to -3.00	1	0.64	0	0.00	1	0.29	4	1.75	3	1.21	7	1.47
-3.01 to -6.00	2	1.28	6	3.16	8	2.31	12	5.26	7	2.82	19	3.99
greater than -6.00	1	0.64	1	0.53	2	0.58	0	0.00	3	1.21	3	0.63
Totals	156	100%	190	100%	346	100%	228	100%	248	100%	476	100%

*Spherical equivalent, binocular, as measured by static retinoscopy.

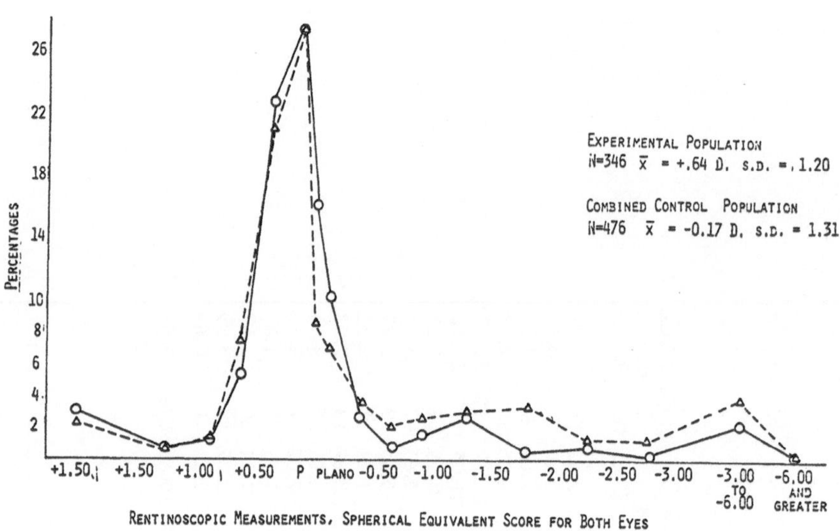

Fig. 1 Percentage distribution of retinoscopic refraction.

TABLE IV

Frequency and Percentage Distribution Table of Myopic Incidence in boys and girls comparing experimental sample and combined control sample.

Myopic Categories	Experimental Sample						Combined Control Sample					
	Girls N = 156		Boys N = 190		Total N = 346		Girls N = 228		Boys N = 248		Total N = 476	
	N	%	N	%	N	%	N	%	N	%	N	%
from -0.51 to -1.50 D.	8	5.13	11	5.79	19	5.49	23	10.09	16	6.45	39	8.19
from -1.51 to -3.00 D.	3	1.92	3	1.58	6	1.73	20	8.77	11	4.44	31	6.51
More than -3.00	3	1.92	7	3.68	10	2.89	12	5.26	10	4.03	22	4.62
Totals	14	8.97	21	11.05	35	10.26	55	24.12	37	14.92	92	19.33

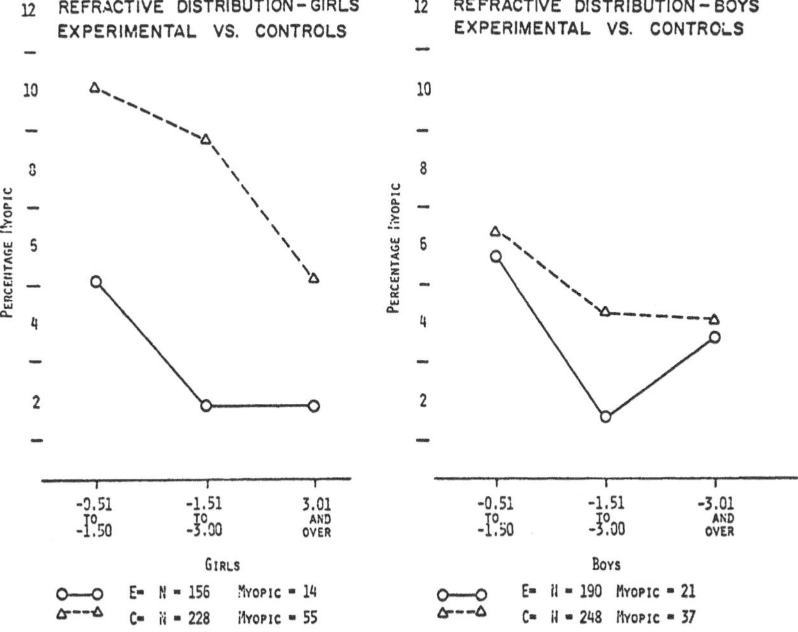

Fig. 2 Refractive distributions for girls and boys.

experimental and combined comparison samples, but the tails of the distributions were different.

Chi-square analysis, comparing the experimental samples to the three comparison samples is presented in Table V. For this analysis only, myopia was defined as the more nearsighted eye being greater than -0.50 diopters (spherical equivalent). The results show the reduction in the incidence of myopia was mainly contributed by the girls, at a significance level of .01, when comparing the experimental sample with each of the comparison sample. There is no statistically significant change for boys in any of the three samples.

When the full time students are combined with the transfer-in students (full treatment and partial treatment) and compared to the pre-treatment sample the significance is at the .01 level for girls as well as the total sample. This same comparison related with the neighboring school population is at the .05 level of confidence.

When the three comparison control samples are combined and compared with the experimental group, the significance is at the .01 level for girls, at the .05 level for boys, and at the .01 level for the total sample.

One of the more obvious conditions which might account for the differences could be age. Table VI represents a summary of mean ages in months for the

TABLE V

Percentage of Sixth-Grade Children Who Scope Nearsighted* in Experimental and Three Comparison Samples

Variable	Experimental sample: Cheshire 1974+1975, in perceptual program since 1st grade			Comparison sample #1 Cheshire 1974+1975, transfer into school system after 1st grade			Comparison sample #2 Cheshire 1970+1971, before start of perceptual program			Comparison sample #3 Neighboring school, 1974 no perceptual program		
	Girls	Boys	Total	Girls	Boys	Total	Girls	Boys	Total	Girls	Boys	Total
N	156	190	346	81	111	192	61	60	121	86	77	163
Nearsighted	9.0	12.1	10.7	22.2	14.4	17.7	29.5	18.3	24.0	23.3	16.9	20.2
Not nearsighted	91.0	87.9	89.3	77.8	85.6	82.3	70.5	81.7	76.0	76.7	83.1	79.8
Total	100.0	100.0	100.0	100.0	100.0	100.0	100.0	100.0	100.0	100.0	100.0	100.0
	Compared with experimental sample											
X^2**	--	--	--	6.92	.16	4.71	13.64	1.02	11.94	8.22	.70	7.74
P	--	--	--	.01	--	.05	.01	--	.01	--	--	.01
	Compared with total Cheshire 1974+1975 sample (Experimental + Comparison #1 samples pooled)											
X^2**	--	--	--	--	--	--	7.79	.80	8.08	3.75	.50	4.38
P	--	--	--	--	--	--	.01	--	.01	(.06)	--	.05

*More than -0.50 diopters in the more nearsighted eye.

**X^2 with Yates' correction for continuity.

Note -- For comparison of all four samples together, for girls X^2 = 16.49, p .01; for boys X^2 = 1.95, p .05; for total sample X^2 = 13.35, p .01.

TABLE VI

Age Comparison, Experimental vs. Controls

	Experimental Group Cheshire, since gr. 1			Control Group All others combined		
	N	Age in months	Standard Deviation	N	Age in months	Standard Deviation
Girls	156	144.3	5.4	228	139.3	7.4
Boys	190	148.1	6.8	248	141.4	7.5
Total	346	146.4	6.5	476	140.4	7.5

experimental and control samples as well as standard deviations. A breakdown of boys' and girls' ages is also presented. For the purposes of calculating ages, a March 1 date was used for each respective year. (Data were collected from Feb. 1 to April 1.)

In all categories, the experimental sample was older. The girls were on the average, 5 months older, while the boys were 6.7 months older. The total experimental groups was older showing a smaller age spread than the control group. One of the intervention programs (homogenous grouping by developmental ages), delayed children early in school.

If the development of myopia was primarily a growth function, the experimental sample should have manifested a higher incidence because their average age was 146.4 months versus 140.4 months for the combined control sample. The results show the experimental sample, although they averaged 6 months older, exhibit a statistically significant lower incidence of myopia.

Test Anxiety Study

After the data had been gathered and tabulated in 1974, I attempted to obtain additional data relating to test anxiety. The reason for attempting this was to see if information relative to this area might help explain some of the results.

It was difficult to rationalize the results in terms of accommodation demands, for the experimental sample demonstrated a high level of academic success. One could present an argument that the experimental subjects were allowed more freedom to vary their activity, thereby not sustaining accommodation at near for the usual long periods of time. However, this did not explain the sex related differences which were revealed.

Because it was near the end of the school term, we were forced to be selective in the sample we could survey in order to complete the data collection in the remaining time. Three test populations were selected. These were: Highland School in Cheshire, the school building with the lowest incidence of myopia in the system (N=136), Darcey School in Cheshire, the school with the highest incidence of myopia in the system (N=43), and about 1/3 of the students in the neighboring school system (N=57).

The SARASON TEST-ANXIETY SCALE (29) was administered by a research assistant. This test is a group administered test consisting of thirty questions. The children respond by circling either 'yes' or 'no' on their answer sheet. The test was administered within the child's respective classroom. The administration followed SARASON's directions and introductory remarks were given verbatum (page 308). A score of "1" is given for each yes answer and "0" for each no answer. Accordingly, a higher score is indicative of more anxiety related to testing and test situations.

Anxiety Results

Table VII presents the results of the mean scores for the three populations divided

TABLE VII

Mean Anxiety Scores

Group	Neighbor School (23% myopes)		Darcey (19% myopes)		Highland (10% myopes)		All (15% myopes)	
	N	Mean	N	Mean	N	Mean	N	Mean
Myopes								
Girls	6	15.0	3	14.3	5	13.4	14	14.3
Boys	7	11.7	5	13.0	9	14.0	21	13.0
Both	13	13.2	8	13.5	14	13.8	35	13.5
Non-myopes								
Girls	22	15.6	14	16.7	53	11.6	89	13.4
Boys	22	13.1	21	13.8	69	8.8	112	10.6
Both	44	14.4	35	15.0	122	10.0	201	11.8

Analysis of Variance

	df	Mean Square	F
Covariate (age	1	14.20	.47
Main effects			
School	2	400.06	13.12**
Visual status	1	32.06	1.05
Sex	1	368.74	12.09**
Interactions			
School x visual status	1	93.95	3.08*
School x sex	1	1.03	.03
Visual status x sex	1	18.32	.60
School x visual status x sex	2	13.51	.44
Residual	223	30.50	
Total	235	35.48	

[a]Children who scoped more than -0.5 diopters in the more nearsighted eyes were classified as myopes, those who did not as non-myopes.

*F significant beyond .05 level.

**F significant beyond .01 level.

into categories of myopes and non-myopes. Using an analysis of variance procedure the main effects were the school and sex (p = .01). Girls had significantly higher test anxiety scores. The schools where the incidence of myopia was the greatest (Darcey and neighbor school) had higher anxiety scores. The only interaction which was significant was school related to visual status (p = .05).

Figure 3 presents a graph of the anxiety scores for myopes and non-myopes in

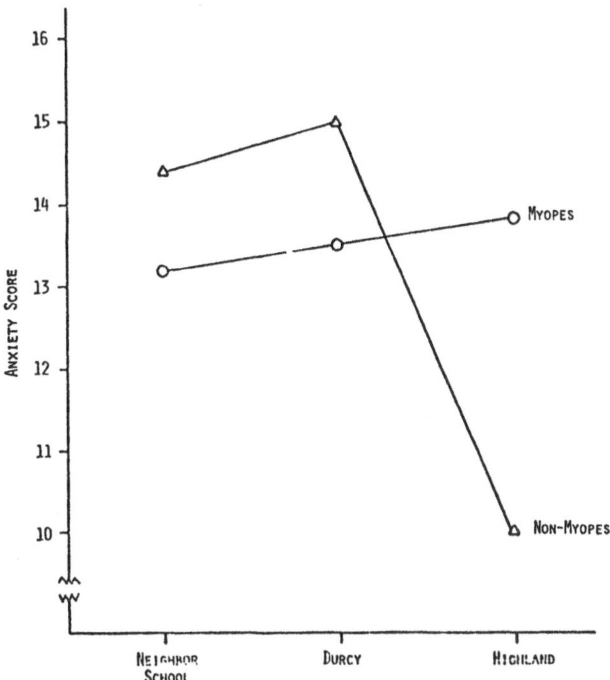

Fig. 3 Interaction of school and visual status.

the three sample school buildings. In all three samples the myopic children showed similar test anxiety scores, ranging from 13.2 to 13.8. The non-myopic subjects in the buildings with a high incidence of myopia (neighbor school and Darcey) scored higher in test anxiety than the building with the lower incidence of myopia (Highland). The non-myopic subjects in Highland, where the incidence of myopia was lower, had a lower mean anxiety score than the myopic population (10.0 versus 13.8).

Discussion

The null hypothesis, environmental changes have no significant effect on the incidence of school age myopia, was rejected by the data from this study. The results show a significant difference in the incidence of myopia between the experimental sample after intervention as compared to the control samples.

The significant difference was contributed mainly by the females in the sample population. In all the control comparison samples, girls consistently showed a greater incidence of myopia than the boys. In the experimental sample, the reverse was true; the girls showed a lower incidence of myopia than the boys.

While no attempt was made to attribute specific aspects of the program to the changes, it is reasonable to assume that the activities and teachers training program were a contributing factor. The architectural changes, physical and lighting environmental modifications, were not initiated early enough to affect a large number of the experimental sample. The analysis performed showed no relationship of myopia to age. This does not rule out the effect of the developmental placement program especially in view of the anxiety findings. One of the effects of placement may be a reduction of test anxiety.

The differences in the results of the experimental sample, (full time students in Cheshire from first grade), and the comparison sample #1 (transfer-in students) suggests the importance of the early school years in affecting changes in children. It also suggests that a further analysis should be made of transfer-in students according to grade when entering the program. The transfer-in sample did show the lowest incidence of myopia of the three control comparison samples indicating some effect of partial treatment.

Why the girls responded differently than the boys is still a puzzling question. Possibly the anxiety data does shed some light on this. Girls do show a higher mean test anxiety score. To explain the results it would be necessary to assume that when children develop anxiety related to tests or doing well in school, they are more likely to make a myopic adaptation to help them in that task. The relative task success should result in a lowering of anxiety.

The intervention program did lower the test anxiety with many children, but not with all children. To the degree this was successful, children did not 'need' to make the adaptation into myopia. One suggestion to test this assumption could be by longitudinal study comparing pre- and post-myopia test anxiety results.

The children from Cheshire who received the results of the intervention program were tested for behavior variables at earlier grade levels. This information was reported by WALKER and STREFF (30) and showed that the boys made more significant behavioral changes than the girls. Boys showed statistically significant changes in self-control, self-concept and class room attitude. Boys demonstrated behavior changes while girls demonstrated visual changes in response to the intervention program.

The results from this study present statistical evidence against the single role of herdity in the etiology of myopia. If myopia were simply a biological variant related to heredity and growth, the experimental population should have had the highest incidence of myopia because the subjects were 6 months older. It would be very difficult to explain the resultant reduced incidence of myopia found unless one assumed that the intervention program successfully retarded the growth and development of children. Even this unlikelyhood would not explain the incidence reversal related to sex differences. Previous studies have demonstrated a pattern of girls developing myopia earlier than boys.

Another aspect which deserves consideration is the level of vision care children received. I am quite sure that when the school system and teachers became involved in this program, they became more aware of the significance of good vision. Cheshire did refer children for vision care for reasons other than reduced acuity. Thus, it may be argued that the experimental population received better vision care. If vision care was a significant factor in reducing the incidence of myopia, this supports the role of adaption and environment in the etiology of myopia.

One positive conclusion that can be made is that this study supported SORSBY's (1) recommendation, "To understand the etiology of myopia, we need to study its breeding grounds - the schools".

References

1. A. Sorsby, Brit. J. Ophth. 16, 217 (1932).
2. J. Ware, Phil. Trans. Roy. Soc. London 103, 31 (1813).
3. H. Cohn, Lupzig, H. Cohn. (1867).
4. J. I. Pascal, Studies in Visual Optics (C.V. Mosby Co., St. Louis, 1952), Chap. 20.
5. E. Levinsohn, Arch. Ophth. 54, 434 (1925).
6. O.D. Rasmussen, Thesis on the Cause of Myopia, (Tonbridge, Egn., Tonbridge Free Press, 1951).
7. F. Erissmann, Albrecht von Graef's Arch., fur Ophth. 17, 1 (1871).
8. T. Sato, The Causes and Prevention of Acquired Myopia, (Kanchara Shuppa Co., Tokyo, 1957).
9. K. Amano, Acta Soc. Oph. Jap. 62 (1958).
10. W. R. Baldwin, 1st Int. Cont. on Myopia, Myopia Research Foundation, N.Y. (1964).
11. F. A. Young, Am. J. Optom. and Arch. Amer. Acad. Optom. 42, 438 (1965).
12. F. A. Young, Contacto. 15, 36 (1971).
13. F. A. Young, et. al., Am. J. Optom. and Arch. Am. Acad. Optom. 46, 676 (1969).
14. F. A. Young, Am. J. Ophth. 52, 799 (1961).
15. F. A. Young, Am. J. Optom. and Arch. Am. Acad. Optom. 39, 60 (1962).
16. M. B. Rosones, J. of Proj., Tech. and Pers. Assoc. 31, 31 (1967).
17. C. R. Kelley, The psychological factors in myopia, Unpublished Ph.D. dissertation, New York School for Research, New York (1953).
18. A. Steiger, Die Entstchung der Sparischen Refraktionen des menschlichen Auges, (S. Karger, Berlin, 1913).
19. S. Stenstrom, Am. J. Optom. and Arch. Am. Acad. of Optom. 25, 218 (1948).
20. A. Sorsby, et. al., Med. Res. Council. Spec. Rep. Service 293 (1957).
21. A. Sorsby, et. al., Med. Res. Council, No. 301, Her Majesty Stationary Offices, London (1961).
22. M. J. Hirsch, Amer. J. Optom. and Arch. Amer. Acad. Optom. 42, 327 (1965).
23. F. A. Young, et. al., Am. J. Optom. and Arch. Am. Acad. Optom. 31, 111 (1954).
24. H. M. Coleman, JAOA 41, 341 (1970).
25. M. J. Hirsch, Amer. J. Optom. and Arch. Amer. Acad. Optom. 30, 135 (1953).
26. H. Hendrickson, JAOA 21, 428 (1950).
27. Ames, Gillespie and Streff, Stop School Failure, (Harper and Row, N.Y., 1972), Chap. XII.
28. W. G. Cochran, Biometrika 37, 256 (1950).
29. S. Sarason, et. al., Anxiety in Elementary School Children, (John Wiley and Sons, N.Y., 1960).
30. Walker and Streff, Gen. Psych. Monograph 87, 253 (1973).

Visual Deprivation Studies and the Therapy of Strabismus, Amblyopia and Learning Disorders

Merrill J. Allen
School of Optometry
Indiana University
Bloomington, Indiana 47401

Introduction

The current expansion of research in the neurology of vision is both gratifying and welcomed. It is not possible to review very much of this research for this paper, but certain things have become clear to me. There are oculomotor control centers in the occipital cortex, the frontal cortex, the 8th nerve centers, the superior colliculus, the cerebellum and no doubt several other areas, suggesting that the individual has several ways to gain access to motor control of his extraocular muscles. There are visual fibers to the superior colliculus, the pretectal region, the pineal body, the hypophysis, the occiptal cortex and apparently diffusely distributed over some other parts of the cerebral cortex.

Laboratory animals have been shown to develop what appears to be analogous to amblyopia and strabismus in humans upon interruption of the sensory inputs during early life. This can happen to one or both eyes depending on the experimental set up. Interruption of the motor circuit by surgery has shown that strabismus can also be induced and that amblyopia sometimes developes as well. In both sensory and motor insult to the test animals, binocular sensory cooperation is reduced or becomes nonexistent.

The human rarely suffers the degree of insult to the sensory or motor system imposed on the laboratory animal, and the treatability of humans with amblyopia and strabismus appears to be greater than similar defects induced in animals. The usual therapy applied to damaged animals is to cover the better eye just as we patch the better eye in humans with amblyopia. The success rate of this technique among humans in only about 50% with the younger patients being more responsive. In fact, an often repeated clinical statement is that anyone older than eight years cannot be treated if they have amblyopia. Another clinic "rule" is that corrected acuities poorer than 20/200 are probably organic, and, therefore, cannot be treated.

Fortunately, treatements that are more aggressive than the occluder are available, and the eight year rule and the 20/200 rule appear to be myths. For example, Fig. 1 shows the recovery curve for a nine-year-old female having an initial corrected acuity of 20/2600 who was treated with flash therapy (see Fig. 2C) applied in the macular region of the amblyopic eye. Similar recoveries have been observed for patients ranging into their 30's and with acuity of light perception only. Older patients have not been available for treatment. My oldest patient for the successful treatment of lifetime strabismus was 58 years of age.

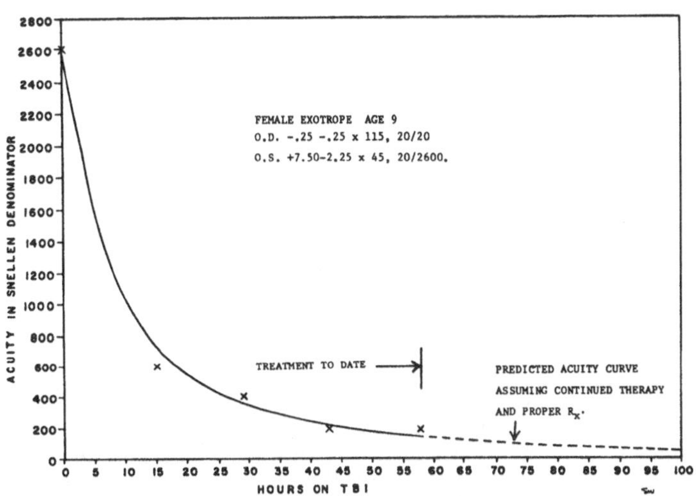

Fig. 1 The visual acuity recovery curve of a 9 year-old girl treated in the Indiana University Optometry Clinic.

The shape of this visual acuity recovery curve seems to be typical based on clinical experience and may be used for prediction purposes. All patients do not respond at the same rate, either due to treatment differences, to depth of the anomaly or to age, but the curve shape seems to be similar for all. Thus if it took 45 hours to attain 20/200, 20/100 will take approximately 90 hours, 20/50 approximately 180 hours and 20/25 approximately 360 hours of treatment. At the rate of one hour per day, 20/25 would thus be expected in one year starting at 20/2600 in this nine year old's case, assuming the same intensity of therapy continued throughout.

The use of several procedures or changes in procedure or the amount of effort could produce irregularities in the recovery curve and may result in a series of plateaus. Such other largely empirical procedures include use of the major amblyoscope, pleoptics, afterimage transfer, Haidinger's brush, Maxwell's Spot, cheiroscopic drawings and many others. In amblyopia and strabismus, the retina, optic nerve and lateral geniculate functions apparently are hardly distinguishable from normal while binocularly responding cortical cells are fewer. I deduce that the neural activity is essentially normal except in the striate cortex and that in the final analysis, the deficits are mostly due to the competetive advantage of the good eye in driving cortical cells compared to the deprived (or turned) eye. The application of special treatments such as in Fig. 2 to overcome the unilateral dominance has speeded the response to therapy in humans and, to my knowledge, has not been therapeutically applied to animals.[1]

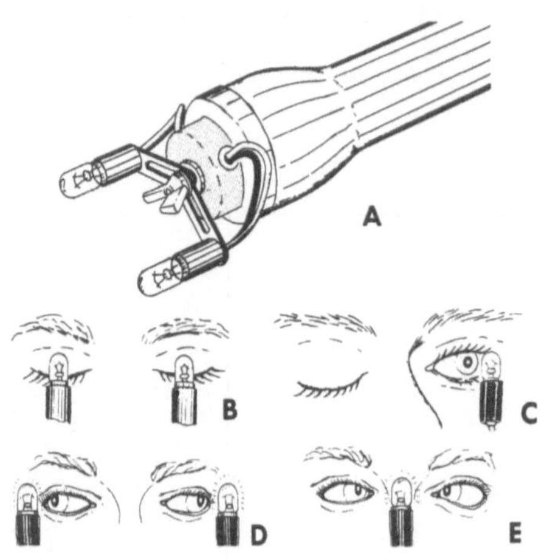

Fig. 2 a) The TBI (Translid Binocular Interactor) flash training device developed at Indiana University School of Optometry. b) Treatment with light bulbs resting on the closed eyelids. c) Treatment with strong foveal stimulation of the weaker eye. d) Binocular treatment to force normal binocular function in an exotrope. e) Binocular treatment to force normal binocular function in an esotrope. Note that corneal light reflections can be used to approximately align the bright filaments on the lines of sight of each eye.

To overcome the functional inability of an eye to drive the cortex, it must repeatedly send strong signals toward the cortex while the better eye is kept in darkness. The better eye may have to be kept in partial darkness (be attenuated) for many months. Strong stimulation of the amblyopic or suppressed eye requires high levels of illumination, rapid fluctuations in light intensity, concentration of stimuli in foveal and macular regions and the use of targets of high contrast and complex patterns as acuity improves.

[1]Unintentionally researchers have found that frequencies of stimulation in the 4 to 12 cycle per second range have caused functional changes in a few minutes. The tissues are reprogramming even while measurements are being made.

Overcoming a binocular vision deficit requires binocular, simultaneous stimulation of the two retinae when the eyes are exactly in binocular alignment, using objective criteria at first or a repeating pattern of vertical lines. Amblyopia therapy alone should also work to restore the function to a considerable number of binocular cortical cells.

These procedures are, in my experience, the ones most successful in visual therapy of strabismic and amblyopic defects. The human treatment problem now seems to be optimization of stimulus parameters and the elimination of those older techniques which are slower or are less cost effective. One suggestion from Fig. 1 is that the amount of stimulus activity is important. A stimulus event in a treatment program as used on this nine-year-old girl is a flash of light. The more flashes one can crowd into a second, the more rapid should be the recovery of function, i.e., 10 flashes per second should give at least 10 times the treatment events of one flash per second. No one knows whether 20 flashes per second would be twice as good as 10 or even if it would be worse. The BARTLEY enhancement phemonemon suggests that 20 would not be better, see Fig. 3 (1).

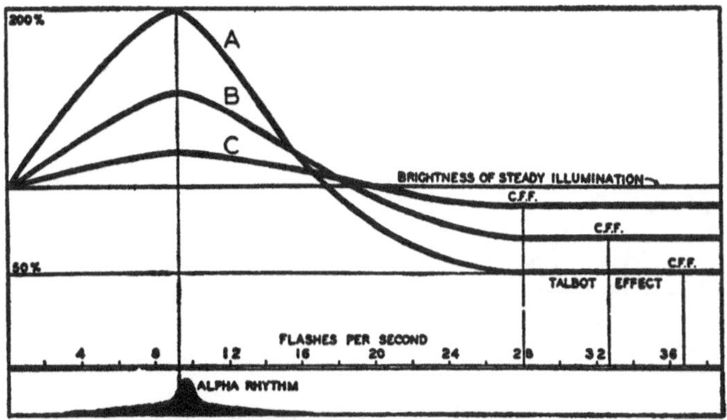

Fig. 3 The brightness of a flickering light relative to a steady light at different flicker frequencies. Curve A refers to a light-dark ratio of 1:1; curve B to a ratio of 7:2; and curve C to a ratio of 8:1. Note that the brightness of a light flickering at the rate of about 10 flashes per second is substantially brighter subjectively than a steady light of the same intensity. (From S. H. BARTLEY (1).

A simple line drawing target would not give the stimulus strength of a complex target or the complex real world in which we all learned to see. To attempt to restore visual function without using hand and body movement, auditory and vestibular inputs, etc. is reducing the strength of the stimulation and may thus be slowing recovery. To fail to provide optimum nutrition and health during therapy also is likely to reduce the responsiveness of the neural networks to treatment.

To the extent that the binocular visual performance of a learning disordered child is often similar to that of an amblyopic eye, the use of amblyopia type treatments seem appropriate to treating the visual problems in learning disorder. Many of these children behave as if vision were not able to dirve the cortex in a normal manner thus they show poor oculomotor control, poor accommodative control, poor eye-hand coordination, poor space perception and poor symbol decyphering ability. These children may have been visually deprived at an earlier age (see Fig. 5).

Figure 4 shows the refractive error distribution of a population of newborns compared to a population of children in the first grade. The newborn high refractive errors have all disappeared by grade one, but their presence for only a few weeks at birth can have an adverse effect on the development of functional visual pathways as judged by animal experiments (2-4).

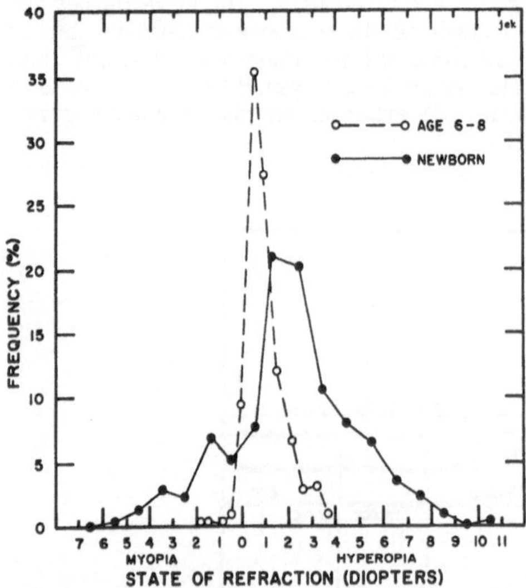

Fig. 4 Comparison of frequency distribution of refractive state at birth, data of COOK and GLASSCOCK (2), with that at age 6-8, data of KEMPF, COLLINS and JARMAN (3); Calculations and drawings by HIRSCH (4).

Figure 5 shows a schema for understanding the role that vision may play in the etiology of learning disorder (5). Keep in mind that the high ametropias at birth usually will be gone by grade one and that this schema can reflect both the adverse effects of high ametropia at birth and problems of lesser ametropias normally encountered in grade school. Compounding these visual problems can be nutritional problems, toxicity (lead, cadmium, etc.) problems, adverse home or institutional environment problems, health problems, etc. The correction of or treatment for the visual problems is no less important than the other factors. In fact, the visual problems may prevent the child making appropriate adaptations to his other difficulties and vice versa.

Appendix:

Methods to treat animals with experimental amblyopia and strabismus are as follows (based on human experience).

1) a. Occlude the better eye. Be certain the refraction of the amblyopic eye is within 1/2 diopter of proper focus on the visual detail

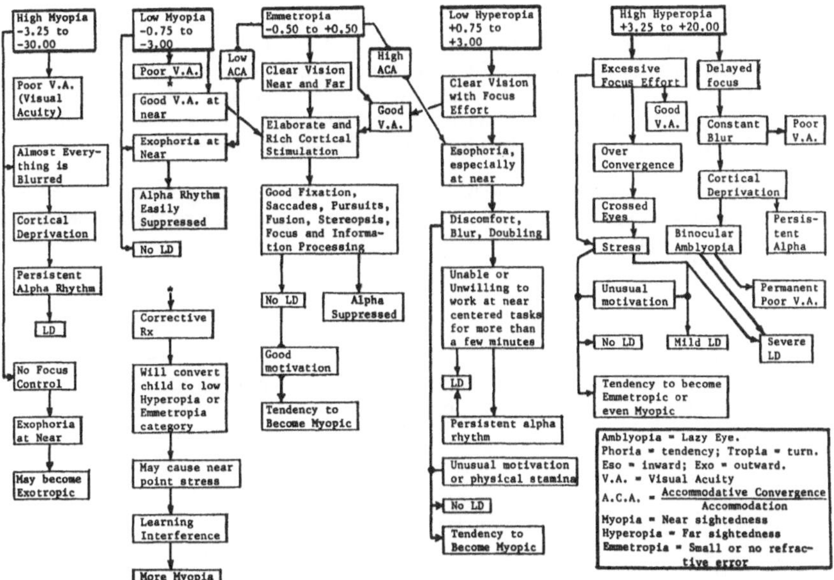

Fig. 5 The Interaction Between Vision, Environment, Ametropia and Learning Disorder. ALLEN (5).

provided.

b. Provide simple patterns made of black lines on a white background encompassing all sizes, shapes, orientations of pattern and occupying most of the field of view.

c. Provide strobe flash illumination at 7 to 9 cycles per second for three or four minutes followed by steady illumination for a like period. Repeat this training sequence for at least one hour of flash per day.

2) Consecutively, if no strabismus exists, remove the occluder from the good eye but replace it with a filter (attenuator) transmitting between 10 and 20% visible light, leaving the amblyopic eye uncovered and unfiltered. The amblyopic eye must now be checked to see that its refraction is the same as the good eye because the good eye will be in control of the amblyopic eye's focus and fixation.

3) Binocular flash therapy should not be used on animals or humans not in alignment except when repeating patterns of constant angular size cover the walls of the treatment chamber. Flashing binocularly will set a strabismus into anomalous retinal correspondence if complex patterns are used. Flashing while aligned (or with repeating patterns) will build binocularly responsive cortical cells.

4) If the amblyopia fails to respond to steps 1-3 above, provide multiple flashing bright spots of light in the field of view of the amblyopic eye or

treat with a single bright light as a #49 bulb (the #49 bulb does not get hot, hence corneal burns are impossible) at about 2 cm from the open amblyopic eye's cornea (Fig. 2C) while the good eye is kept covered. The flash rate should be 7-9 cycles per second for four to five minutes with rest periods of a few minutes for an hour or more of flashing per day. Continue step 2 as an ongoing passive therapeutic measure.

5. To ensure overcoming amblyopic and strabismic symptoms never allow binocular experience unless the eyes are aligned (or repeating patterns are viewed) and the good eye is attenuated (10-20 percent transmission filter).

Summary

The author has used the results of animal studies to develop a rationale for the treatment of human visual problems. Recommendations are made as to how to treat the deficits created in animals by sensory deprivation and by induced strabismus experiments.

References

1. S. H. Bartley, Psychol. Rev. 46, 344 (1939).
2. R. C. Cook and R. E.Glasscock, Am. J. Ophth. 34, 1407 (1951).
3. A. Kempf, S. D. Collins and B. L. Jarman, Public Health Bulletin No. 182 (Government printing Office, Washington, 1928).
4. J. Hirsch, Vision of Children (Chilton Book Co., Philadelphia, 1969) p. 149.
5. M. J. Allen, J. Learning Disabilities, 10, 411 (1977).

Optometry as Remediation and as Education

S. Howard Bartley
Psychology Department
Memphis State University
Memphis, Tennessee

I have been enchanted with the kind of things I have been hearing in these reports during these last few days, the wonderful details and discoveries and so on. What I have to say is of a little different nature. I am interested in our thinking framework and our action framework for dealing with matters of vision for example. In order that I may get said all I want to say in this short time and say it the way I want it, I am going to read to you what I have to say. What I have to say is addressed primarily to optometrists, and thus to the remediation and education of visual performance. In this matter, two problems emerge. What is to be done and whose job it is to do what. This recognizes that it is not only optometrists that may be involved. Since the concerns we are going to discuss have to do with humans as subjects, or as patients, it is fundamental that we all understand the principles operating in the human organism as a person. We need to understand what personness this is and how it is involved in what often seems to be less than personness.

I think I can make clearer what I shall have to say if I pause a moment to review some generalizations. They should throw light on all that will be dealt with more specifically. These generalizations are implicitly, but inescapably involved in the thinking that underlies scientific procedure in all disciplines and professions.

Two contrasting ways of reasoning exist. One we speak of as atomism. Not simply the notion of atomic particles but the notion that the road to disclosing causation lies in the almost exclusive concentration on gathering isolated bits of information, supposing that these in themselves will surely go together to form the overall understanding needed. It is a view that fails to make explicit the whole of which the items are a part. Thus it fails to begin by identifying the whole. It is a view that holds that items have intrinsic and fixed properties of their own. The other view is that no single datum is understandable until it is placed in some larger context until it has been first discussed or supposed. Further, that this context or whole is not a passive structure but has inherent in it the dynamics that gives the parts their roles. This is organicism.

Over the centuries we have, for the most part, been adherents to atomism. Whole professions have been built on that foundation. In medicine, it has been the isolation of specific, separate diseases and the treatment of diseases as apart from the patient. In optometry, it was the fitting of glasses as the exclusive task.

Here and there, various deviations from this have emerged. In psychology, configurationism, or the Gestalt theories were an abortive move away from atomism. Some healers and theorizers have had at least transient insights and have momentarily indulged in organismic thinking. But for the most part, they did not recognize what they were thinking and doing was an example of the broad, underlying principle of organisms I just stated. Hence, they naturally tended to mix this in curious ways with a traditional and prevalent atomism. This has precluded the full success that they and those around them might have expected.

Some medical men, unlike their associates became interested in constitutional medicine. Some optometrists have sought to understand the patient as a person. Such men have supposed that peripheral mechanisms are devices the organism uses in characterizing its behavior. Furthermore, some optometrists have come to recognize that all patients need not utilize similar peripheral inputs fully alike.

When cause is attributed to what is labeled as the person, or personality, one appears to be talking about something that can not be made concrete or objective. This makes those who carefully plan and carry on experimentation to obtain valid data, either uneasy or impatient. Hence, two styles of outlook are perpetuated both in the professions and in the scientific disciplines. How to reconcile this polarity is a problem. That is part of our problem at the moment.

Can those who reason organismically from the patient as a person down to behavior detail somehow conceptualize ways of putting their ideas to experimental tests and can those who devote themselves to studying precise detail get together? The main hope for this, as I see it, lies in something further back. A general view of the nature of Nature.

I happen to have been taught something that has served me well over the years in this connection. It was this: that there is only one causal system, the energistic system exemplified in nuclear physics and chemistry. This means that basically, when causation is to be explained, it must be in terms of this system. The same doctrine recognized that there is another category, namely, the emergent phenomena. While these phenomena are as diverse as our experiences have shown us, they in themselves are not causal. One author gives us a good example of an emergent phenomenon. A lap emerges when we sit down, and disappears when we stand up. A lap does not cause anything. Most of the matters that we are concerned with from day to day are emergent phenomena. If we adhere strictly to this, we, like B. F. SKINNER, will say that emotions and other experiences, for example, do not cause anything.

The Nature of Disciplines

In addition to this, the failure to understand atomism and its shortcomings and an unsophisticated notion of the nature of Nature, the prevalent view of the natural disciplines in professions has gotten in the way of scientific progress.

There are two major views regarding the characteristics of a discipline. One sees it as an area of information which has emerged by tradition. This area is an information bank. It is as if once it gets a label, then most of the informational items that traditionally are in the bank, are owned by it. If some of the information is used by other disciplines, it is by way of borrowing.

The other view, is, instead of being an ownership view, a functional view. It regards items of information not as having intrinsic properties, but in terms of their relevance. The issue is whether or not they contribute to the solution of the problem cluster that identifies the discipline in question. If they do, they are part of that discipline when they play that role. The same data may just as truly contribute to the solution of the problems of other disciplines. All data must be seen in light of the roles they play. No single discipline owns anything but its own identity and purpose. The prime concern for a discipline is to be able to state its purpose and state what gives it identity. It does not seem as though currently disciplines are viewed this way. Atomism is still in vogue.

The Nature of Professions

In essence what I want to say is that professions must at all times be as explicit as possible about their objectives and that these objectives determine what can be included in their practices. It is true that professions contain and support membership organizations which have constitutions and bylaws. It might seem that these constitutions already do what I am suggesting, namely, that they state objectives and purposes. But somehow it does not seem clear to me that the members of some professions do represent anything like a unified body. The diversity is sometimes self-defeating. At best, it is confusing.

What each profession sets out to do involves areas that do overlap with other adjacent and even formally more distant professions. There is a cluster in this respect. As far as optometry is concerned, the cluster includes ophthalmology, and public school education. Among the scientific disciplines, there is primarily psychology. Out of all the sciences, psychology stands out because it pretends to deal with the organism as a person. Psychology, in its major organization, the American Psychological Association, explicity claims psychology to be a science, a profession, a human welfare movement. And this badly confuses matters. On the other hand, physiology, for example, is considered by everybody as exclusively, a basic science. What is optometry's task and how can it perform it in the current context? And if the current context is not what optometry ideally envisages it to be, how can needed changes be brought about? Optometry, like several other professions has arisen as a form of remediation. It stems from the belief that there was a human need for the intelligent optical examination of patients and the prescribing of corrective glasses.

In some quarters, the profession has, over the years, developed greatly in insights regarding the relation of physiological optics and effective visual perception. It is not certain that all groups in and outside of optometry are fully agreed upon what optometry's task is in this respect. A large factor that differentiates patient from healer and one healer from another is simply the question of what is involved in seeing. What does seeing consist of? Right here is where the patient and some of the practitioners are likely to think that dealing with a local condition achieves the remote end and takes care of the person. In some quarters, particularly in optometry, we hear the statement that seeing is a skill. This would seem to imply that the organism as a person is involved and that troubles with seeing are not all localized in the eye.

We all know that children come in for considerable concern relative to vision. This is occasioned by their failure to achieve well in school. A variety of reasons are now assumed possible. But malfunctioning vision is one that receives a great deal of attention. Often the eyes are taken to be the culprits and the child is taken either to an optometrist or an ophthalmologist. Seldom is there an intelligent reason for which is chosen. Sometimes a report given by the examiner is that "There is nothing wrong with the child's eyes." This is a surprise when the child is a poor reader and makes curious mistakes in dealing with a printed page.

Somehow the word perception crops up. When it does, in most quarters, it is a rather empty term. Here, as in many other cases, words substitute for ideas. People follow words around. Once it is thought that perception is the process to look to for the child's inability to read, psychology is turned to, if any discipline is consulted.

The common views of perception all stem from the fact that all of us have been trained as dualists, people who believe in something corporeal and in something in a totally different category, which we call mind. Not only can body processes be thought of as belonging to a causal system, but mind is supposed to be able to cause things, too. And, this belief has provided most people with a logically unsatisfactory basis for handling the problem of cause-and-effect. It shows up in the conceptualization of the organism and it's internal mechanisms and it shows up in the false ideas about the relation of the organism to its surrounds; and, thus, it shows up in our ideas about perception.

Seeing is a perceptual act. That is, it involves the immediate use of data from the environment. Reading like other perceptual activities involves other things as well. It involves understanding that the individual has gained prior to the specific activity of reading. It involves personal orientation toward the task or performance. This personal orientation has certain facets of its own, namely, the idea of what reading is all about and the idea of whether it is worthwhile and interesting or not. Since emotion is the substitution reaction emerging when immediate overt avoidance cannot occur; or when the individual's own behavior is not successful in his own estimation, emotion is involved.

Everything that is to be described in consciousness, such as seeing, hearing, and the rest, everything that is described as either feeling, wishing, or knowing, is an emergent phenomenon, and emerges from body process. So, for explanation, ideally, it would be necessary to understand body processes. We have not gotten very far in doing this, as yet, but I would say that there are some nice things coming along in what I have been hearing in the last few days.

Visual identification of objects is not dependent on a single orienation of the things seen. If we apply this to letters of the alphabet, we should not be startled by the fact that a lower case "b" and "d" are equivalent for that child. They are the same thing in different positions. This is true also, of "Z" and "N". The one position is simply a rotation from that of the other. "3" and "E" are also equivalent in shape, but not in position. Something further can be said about symbols. Recall letters of the alphabet. We tacitly assume that letters and words, visually, are to the child what they are to us as adults. This is definitely not the case, and it is difficult for us to understand and accept that this may be so. NELSON and BARTLEY, in 1962, found in studying children's ability to reproduce such materials as circles, squares and diamonds, that the young child is primarily oriented to the visual appreciation of area, hence, tends to see shape rather than contours. Edges become salient only with age. Letters of the alphabet have little area. That is, little area is enclosed by the edges of the letters. Thus, they are primarily material without area, simply with contours. The shapes that would be there if letters possessed areas are largely absent. These authors studied children from those in their third to those in their eleventh year and compared them with each other and adults. The studies showed that the appreciation of flat materials changes, in keeping with the experience the child has with it. The child begins by seeing area and shape; he does not appreciate contour, but only increasingly becomes aware of it. The seeing of area becomes more marked again in the eleventh year when the area features of tracing are brought to attention in school. This is the year that measurement in arithmetic is being taught to children and the problems have to do with area.

I assume that you will be interested in hearing about the experimentation from which the findings and conclusions just mentioned were obtained. The materials were a set of test items, the standard or reference item being a line drawing of a diamond. The other items, five in number, were items to be compared with it. The first of these was a black wire formed into an outline of a diamond, the same size and shape as the markings on the paper, constituting a standard. The second comparison item was a white cardboard diamond cutout, the same size and shape as the standard line drawing. The third was another cardboard identical to the previous one, but having a definite black edge drawn around it. The fourth was a card in which a diamond shaped hole was cut. The hole was the same size and shape as were the previous items. The fifth and last comparison item was another card with an identical hole but with the hole having a black line around it. In order to preclude other surfaces in view having an incidental effect on the test items, they were all laid on a glass plate supported some distance from the table which was covered with black cloth. All of the test items were arranged randomly in view, and the subject was to select from them one that most resembled the standard. After the subject made the selection, that item was removed and the same question was asked regarding which one now most resembled the standard. When this second selection was made, the item was removed by the experimenter. This procedure was repeated till all comparisons were made.

If I have made my description of the test items clear enough, you will have realized that these several comparison items differ from each other in whether they were areas, contours, or in the fourth and fifth cases, whether the concrete area involved was outside the diamond itself. The comparisons that children were called upon to make gave clues to whether it was area of contour that was seen. While common sense notions attribute the inability of a child to reproduce drawings to the immaturity of the motor system, the study ably showed that the differences in the way the material was perceived accounted for the manner in which the material was reproduced in children's drawings. This, and what I have already said about orientation of the letters seemed to go together to begin to describe some of the perceptual features involved in the ability to read.

ROSEN, PORITSKY and SOTSKY, in 1971, have reported a study, which, in its way, bears out first, what I have already said about children's aversions to reading, and second, what has been discovered about areas versus contours in children's perception. These authors substituted Chinese characters for English words. These they gave to a number of very poor readers, and in a short time, had them reading sentences shown to them and composing sentences of their own. To the children, the endeavor did not seem like reading and this bears on what I said earlier about aversions. I said that one of the possible virtues of certain tests and training procedures might be their seeming dissimilarity to reading. Secondly, the Chinese symbols stood for words, whereas our written words are combinations of letters standing for phonemes which apparently are not recognized as representing components of the child's own, or other people's, speech. The Chinese symbol study was somewhat analogous to the current work of NELSON at the University of Alberta in which he is utilizing areas and textures as symbols instead of letters, which are characters composed mainly of contours. The disclosures are not necessarily meant to advocate doing away with our way of symbolizing spoken words, but rather to help clarify the nature of visual perception and how we may be inadvertently making reading more difficult than some other scheme of symbols would entail.

If more were understood about the human organism as a person, exceptional insights regarding patients would become the rule. And if, in general, the present secrets were to become the understanding of those in the public school teaching, many of the visual problem cases would never develop. Optometry has the job of setting its house in order and from this, educating those around it, including the public as to what it understands vision to be and how it goes about dealing with visual problems. The fact that starts the ball rolling is the finding that the child can't see or can't perform as expected. When the trouble is inferred to be visual, it is referred to an optometrist or to an ophthalmologist. There, refractionists, of course, administer routine tests and determine a number of features which in general have to do with physiological optics, features that are attributed to the stimulus of the eye and how they work together. All this has to do with image formation. As varied as these tests are, they do not include all that the refractionist needs to find out about the patient.

What I am saying is that they do not encompass the problem of how the patient uses his optical input. Some of the most enlightening cases that have come to my attention, if I am to believe what I am told about them, have been the cases in which vision was not what it ought to be, even though the patients had 20/20 visual acuity, and tested satisfactorily in other areas. This was taken to mean that the patient did not have a visual problem. But some optometrists, supposing the patients had visual problems, went further and attempted to see how the patients used their visual resources. In so doing, they discovered that the trouble of utilization of resources lay in that area.

Here is a subtle area of remediation in patient education. It is an area in which optometrists and others do not see alike for they do not understand alike, partly because they have not given their attention alike to such matters. It would seem that those who look with disfavor and perhaps disdain on dealing with what I call personalistic factors and visual inadeqaucy should apply themselves to appropriate experimental investigation of them. I look to schools of optometry to be the hotbeds of experimental research in the area of what is involved in adequate visual performance, just as we have expected universities to be the realms in which non-applied scientific investigation is centered.

Since the overall behavior of a pateint or a child presents a picture of general maladjustment, socially, and from the child's standpoint, emotionally, the specific basis of the trouble is generally masked. The tasks that the child is called upon to perform in school involve vision. In reading, they are most critically and artifically visual. When inability to read occurs, it is attributed to all sorts of causes, including brain damage. Some children are referred to neurologists, some to psychologists, some to ophthalmologists and some to optometrists. Since reading involves a critical use of the eyes, some are sent to the eye specialists first. When they are sent to these two professions, it is supposed that certain routine tests will disclose the presence or absence of ocular anomalies. That is all that is expected. Some children are sent back with a report that "There is nothing the matter with the child's eyes." This is supposed to resolve the question of a visual basis for the trouble; but generally, it does not.

Unless a profession is prepared to deal with individual cases as individual cases, it is not prepared to do what it should do. It is not prepared for its task. Even some public school educators are beginning to point out that standardized tests do not disclose the behavioral basis for performance deficits. Yet, these tests are used. And it has been the general practice to use the same training procedures for many different children manifesting a variety of educational handicaps. OWEN, BRAGGIO and ELLEN, in 1976, point out that, "The educator must specify the relationship between the nature of the task and the occurrence of correct and incorrect responding." They say that it is only when this specific relationship is examined and then understood that ways for remediation can be discovered.

The same principle applies to optometry, which is called upon to deal with some of the very same children. It can not be supposed that every educational problem, is, at the bottom, a visual problem. Vision is in the picture, somehow, of course, when the child can not read. It may enter in secondarily, or it may be primary. HAMILL states that "If reading comprehension and visual perceptual ability are unrelated to any educationally significant degree, one would expect to be unsuccessful in improving reading through training visual perception. Yet many such programs are begun in schools for the expressed or implied purpose of enhancing reading comprehension in some fashion." HAMILL disclaims mass training procedures that do not improve reading ability. Into this argument he injects what he calls visual perceptual ability. As he uses it, he appears to imply that it is an identifiable, intrinsic something, like a tool to be used in reading, if possessed, but which may be absent in some children. I cannot assume such a specific tool, such an individual unit. If what HAMILL calls visual perceptual ability is not a pin-pointable characteristic of the human organism, naturally it cannot be used to account for the failures he points out of formal mass training devices to enhance reading comprehension.

Throughout, we should be concerned with whether there are broad, common disabling factors and what they are, if they exist, and, on the other hand, whether there are specific factors pertaining to individual cases that are crucial. You see, it is some of these individual cases that I do not have time to illustrate, so that cuts down what I am trying to say. If we happen to be doing the wrong thinking, many of our efforts are redundant and are irrelevant. I am saying that if what gets children into trouble is not what is commonly supposed, then using remedial means based on such a wrong supposition should not be expected to get the children out of their trouble. Behind it all is the failure to recongize the need for dealing with individual cases. I would suppose that neither the advocates of mass visual training nor their critics have quite hit the nail on the head. We cannot expect physicians to make the same diagnosis to every patient that comes into their offices and prescribe the same medicines or other treatment to all. Just so, we yet cannot expect routine corrections for reading disability.

IX. Vision Health Care Delivery

Tasks and Goals of a Clinically Oriented Vision Research Laboratory: Its Role in Research and its Application to Serve Patients

Edgar Auerbach, M.D.
Vision Research Laboratory, Hadassah University Hospital
Jerusalem, Israel

About 18 years ago I was asked to set up a clinically oriented vision research laboratory, which I did under rather pioneering conditions. Looking back, I am naturally aware of things I neglected and of those I would now do differently. Also, during this period of time new findings influenced or changed our attitude and new concepts matured. This resulted in a gradual development of ideas and of methods: on the one hand, examination of patients for diagnosis and prognosis, and on the other, research based mainly on problems and data provided by our patients.

When invited to this symposium, I considered it appropriate to talk about the ideas and the philosophy behind establishing such a laboratory in view of the special occasion. I proposed to speak about the tasks and goals of a laboratory in which research is carried out along with very specialized clinical examinations and to give suitable examples in order to show the central position of research, namely its clinical application and how dysfunction may lead back to research and eventually to basic findings.

Over the years various approaches were tried regarding both the attention given to the patients and the testing methods used. As to the former, although a laboratory, a personal and positive physician-patient relationship developed and proved highly satisfying to both. As to methods, they are always in flux and every effort is made to keep them in line with current understanding and progress; several trends developed which eventually were accepted after modifications; others were abandoned altogether.

The general character of a newly established laboratory and the emphasis given in its development depend, at least at the beginning, largely on the main interests of the person who initiates it or on an obvious, predominant problem which may be present. If the laboratory is clinically oriented as well, and one for which no standard type yet exists, it depends on how he understands the function of its service. We were interested mainly in the events which lead to vision, and so we concentrated on retinal function and its interaction with the visual cortex. But equally well, more emphasis in the clinical service could have been given, for example, to the cerebral functions connected to attention or other subjective phenomena, or to the motor events such as those in the external ocular muscles.

The establishment of a vision research laboratory in a clinically oriented enviornment requires a kind of dual loyalty, one toward the patient and another toward research. It also requires that the latter ultimately will serve the former.

To make the background clear to myself, I devised a diagram. To my satisfaction, when it was finished it became obvious that it suits not only a vision research laboratory but could be applied to any laboratory in a similar environment.

Figure 1 illustrates the activity of a university laboratory as opposed to that of

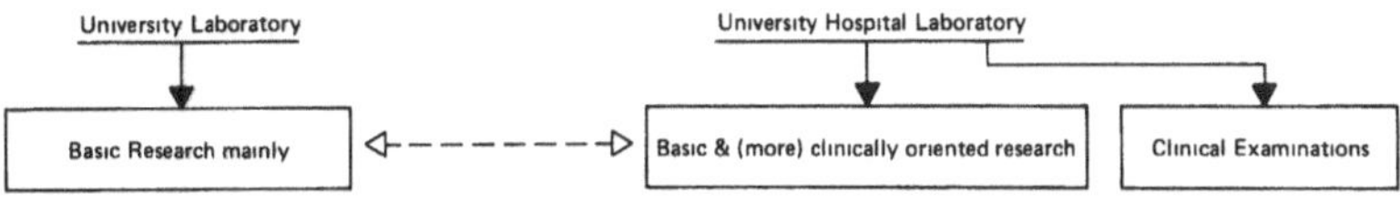

Fig. 1 Diagram: university laboratory - university hospital laboratory.

a university hospital laboratory. While in the former mainly basic research has been performed, the latter carries out basic and applied, clinically oriented research as well as clinical examinations.

However, a strict division between so-called pure research and applied research is no longer tenable. The barrier between them is gradually torn down and barely discernible. It becomes increasingly clear that the normal condition can be understood not only by working with the normal organism or living material but also by searching for the cause of disease, disorder, dysfunction, or missing function. Life includes death as well as sickness. Nature performs experiments with its organic material which we are not often able to devise and only rarely to duplicate. Nature enables us then to uncover indirectly the physiological mechanism and the normal function by the study of deviations from the normal.

This requires a few remarks on normality, a problem which certainly cannot be exhaustively treated here. In short, the very concept of normality does not coincide with that of constancy. It would be all too simple to define it by the absence of dysfunction and to consider normality a stationary condition of perfect adaptation and adaptability to environmental changes within a certain physiological range. Normality is subject to the laws of Nature just as life is. It is, therefore, also determined and characterized by the ageing of the organism which in turn leads to gradual changes in the adaptability of the organism toward the environmental influences. Only if,due to internal and/or external causes, this adaptability fails, the entire system breaks down. To make a differentiation between normal and abnormal function possible, the subtle and continual change of normality during developing and ageing has to be accounted for in our standards of physiological functions. Strictly speaking a standard curve, e.g. of dark adaptation, is a standard only for a certain age group.

On a different level, but again demonstrating the approximation of the two lines of work, during the last 10-20 years or so, researchers from the exact sciences as well as from biology were becoming increasingly interested in the pathological organism, in the direct study of the phenomenon "disease". This interest became furthered and often even prompted by the great technological achievements of our age and the search for their application. So, presently the physician works at equal level together with members of other scientific disciplines, and their cooperation, way of thinking, and educational background are indispensible to him.

In Fig. 2 I compare more specifically the function of a university hospital

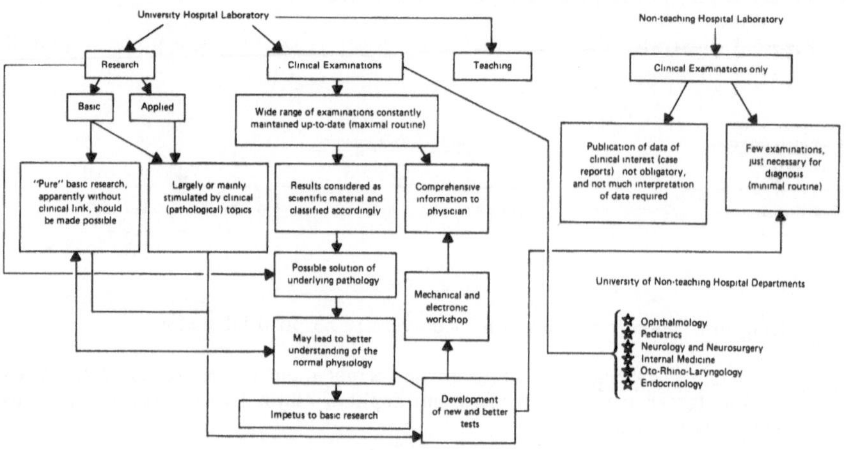

Fig. 2 Diagram: university hospital laboratory - non-teaching hospital.

laboratory, such as a vision research laboratory, with that of a laboratory in a non-teaching hospital. I am aware, however, that things are not always exactly as they are outlined here; exceptions and transitions are found and often necessitated by local conditions.

While the former is characterized by three main activities, namely research, teaching and clinical examination, in the latter generally only clinical examinations are carried out, adequate to provide just a diagnosis for the physician, often on a "yes-no" basis. In short, it is a minimal routine.

By contrast, the clinical examinations in a university hospital laboratory should be designed for a maximal routine; it should contain a wide range of examinations in keeping with scientific progress and using up-to-date apparatus constantly realigned with current knowledge. More specifically, a maximal routine certainly does not imply a rigid adherence to once-recognized principles but should constantly be improved; on the other hand, it means, in our laboratory, a systematic examination of the visual system beginning from the pigment epithelium up to the striate cortex and often to the conscious response. I will come to that later on in more detail. However, this routine has the disadvantage that the number of patients examined per day is considerably smaller, which is inefficient from the standpoint of the physician and of the medical service rendered. This shortcoming is only partially compensated by the advantage of comprehensive information.

A maximal routine is partially justified by the fact that the results from the examinations simultaneously represent scientific material. Pathological entities, and all data available, therefore, should be classified according to fixed criteria so that they are within reach for clinical information and for research. To facilitate this and to avoid the loss of irretrievable, valuable information accumulated over years, the data should be fed into a general computer from the very beginning, if at all possible.

However, this does not yet satisfy the dual loyalty toward patient and research. Full justification of a maximal routine is achieved only when clinicians may work and travel outside the confines of the laboratory using mobile units built for electrophysiological and psychophysical examinations. These permit rapid clarification of a diagnostic problem on a yes-no basis and performance of a short preliminary examination; it does not exclude the laboratory when more accurate and comprehensive information is requested. The mobile units should be used in villages and remote areas. For example, Israel, among other countries, has provided medical help to developing nations in Asia and Africa. Here, at a far-away place where an operation is about to be performed, such as extraction of a cataract, on-the-spot diagnosis is important as to whether the eye is still functioning or the visual pathways are still conducting information to the brain; it may well save unnecessary interventions. These mobile units also find use in clinical departments such as the eye department, neurology, neurosurgery, intensive care, emergency, pediatrics, etc. Moreover, direct contact between the laboratory and the clinical departments in the form of joint meetings and discussions should be maintained. The project with mobile units was not carried out by our Vision Research Laboratory.

Results both from minimal and maximal routine examinations in the two kinds of laboratory do not necessarily go to the physician alone. The clinically oriented associate, an M.D. or a Ph.D., or a well-trained technician, should be encouraged to report interesting findings encountered during examination. If no research facilities are available, the time until publication of his findings reaches the attention of the researcher will be greatly extended. However, if he works in a university hospital laboratory or is affiliated with one, a great shortcut will be achieved since all procedures take place in the same laboratory. This implies that members of a non-teaching hospital should maintain a close relationship to a teaching hospital, as is indeed the case in larger centers.

In this way, a phenomenon not understood during clinical examination feeds directly into applied and/or basic research and becomes a research project (diagram, Fig. 2). Another advantage, if all this is carried out in the same laboratory, is that solutions found can be checked without delay in its clinical section. If the problem is solved, the result leads to better understanding of the underlying pathology and, by including it in future routine examinations, more complete information reaches the physician. The research which is carried out with clinical material, moreover, may not only reveal the underlying pathology but may also lead to a better understanding of the normal physiological mechanism. And this in turn will complete the circle by giving impetus to basic research, as indicated in the diagram. With this, the problem moves into the foreground of interest while the human being, the patient, shifts into the background for a while. Together with this goes the realization that standstill means slipping backward. It follows that it is imperative for a laboratory of this kind, which provides direct service to patients, to make every effort to keep abreast of increasing knowledge as well as to make its own contribution.

This activity leads also into a technical direction for it is necessary to be able to develop and to build apparatus in the laboratory for a particular task or project for which no ready-made design, no precedent is available, and in order to save time. Therefore, electronic and mechanical workshops are useful, even necessary for research, for maintenance and for increasing the efficiency in the subdepartment of clinical examination.

We now have more patients than we are able to handle. Despite the rationale to justify the maximal routine, this statement reflects a certain discontent since patients may have to wait for months, except in urgent cases. We are then not quite able to serve our patients fully satisfactorily. On the other hand, a great demand for a service is simultaneously a sign of recognition. Therefore, to put this difficult problem in perspective, an increasing scale of the demand for a service is a criterion as to whether the work is good and considered necessary. Moreover, it is almost a rule that space and equipment sooner or later become inadequate and have to be enlarged and exchanged.

One comment I wish to make about "pure" basic research which does not appear to have a clinical link. It is only very rare that a research in the end will not lead to the human being and eventually serve humanity. Man is intensely interested in man; even research which seems motivated solely by intellectual curiosity indeed does satisfy his urge for knowledge, all the same in the end scientific discoveries will find their application. This has been aptly discussed by J. G. CROWTHER in his book, The Social Relations of Science (1).

I will now go into some detail and discuss examinations of the visual system (2-5) according to a maximal routine. Of the battery of examinations in our laboratory, the most frequently used are color vision tests, the electro-oculogram (EOG), the electroretinogram (ERG) and the visual evoked potential (VEP). The ERG, which reflects retinal function, is a very useful test if one bears in mind that it becomes more meaningful in conjunction with the VEP, and that it reflects, among other things, the process of dark adaptation. When specific information on the visual pigment is necessary, we employ the early receptor potential (ERP) as the last of the battery of tests because of the strong bleaching effect by the stimulating light.

As to dark adaptation, the psychophysical method is one of the best research tools and provides a world of information on the visual system and its functions. These measurements, however, are used now only occasionally in patients for the clarification of ambiguous electrophysiological examinations. They are used extensively for research. Moreover, the fact that psychophysical dark adpatation measurements require active participation of the subject, disqualifies many patients. As JOHNSON and RIGGS (6) have shown in 1951, dark adaptation is a function of the retina. Therefore, the ERG reflects this process and especially its duplex nature, because it shows systematic changes of cone (photopic) and rod (scotopic) components while it recovers from a preliminary light adaptation. We have substantiated this as well (Fig. 3) by the fact that the curve showing the increase in ERG amplitudes as a function of time in the dark tallies essentially with the psychophysical curve of dark adaptation (7).

Since dark adaptation is one of the fundamental functions of the visual system, we examine it routinely in patients and use the ERG elicited by white stimuli as the method of choice. Figure 4 is self-explanatory, showing schematically the elements on which we base our evaluation of clinical ERG recordings (7). I prefer this method for the maximal routine to recordings with blue and red stimuli (8), although the latter is a good method since it separates the rod and cone components from each other, which is an advantage. It is also useful for a fast diagnosis. However, the other method gives more information on the dynamic aspect of dark adaptation and on the interaction of the photopic and scotopic mechanisms.

Although proposed and discussed for many years, an international standardization of the clinical ERG concerning methodology, apparatus and analysis could never be accomplished, perhaps because basic principles are still not fully comprehended roughly 30 years after the pioneering work of GRANIT (9, 10). There are still surprising findings. I mention, e.g. those of MILLER and DOWLING (11)

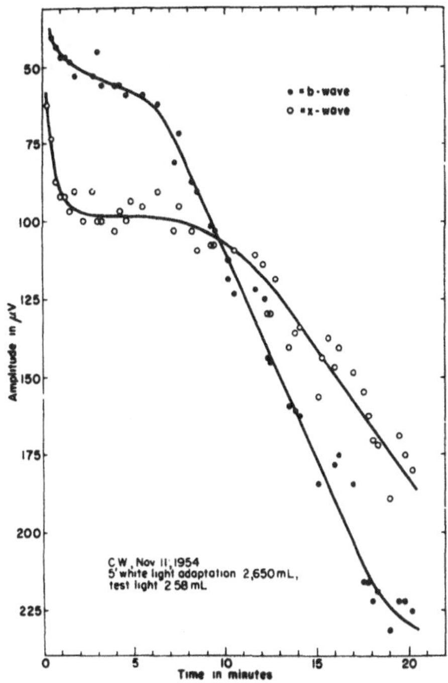

Fig. 3 Graphic representation of an experiment in a normal subject showing the relation of the amplitudes of the photopic (X) and scotopic (b) ERG components during dark adaptation (7).

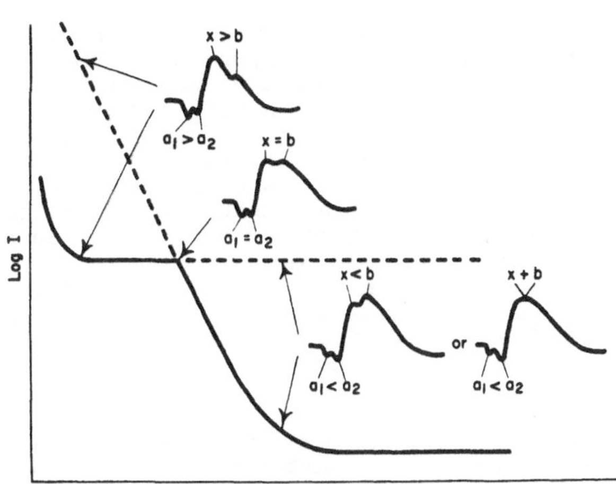

Fig. 4 Schematic presentation of dark adaptation curve with corresponding diagrammatically represented ERGs showing relationships of photopic and scotopic negative and positive components (a_1, a_2, X and b) during different phases of dark adaptation (7).

which involve the glial Müller cells in the generation of the ERG positive potential and, if applicable to the mammalian ERG, necessitate changes in its interpretation. Of course, in hindsight these data should not be entirely surprising because glial cells contribute to the EEG as well.

However, because an international standard is lacking, it is indispensable to establish normal standards in each laboratory for the test conditions used, such as the growth of ERG amplitudes either as a function of stimulus intensity (Fig. 5) or as a function of time in the dark (i.e. of dark adaptation). The latter is better illustrated by the responses obtained in an experiment (Fig. 6) than by a graph because here it is the response pattern which counts. Both the characteristic way in which the overlapping ERG potentials increase in amplitude and the shift of the positive wave to the right during recovery in the dark have become clear diagnostic criteria (2, 3). This shift, which is produced by the phenomenon that the positive scotopic component gradually seems to surpass in amplitude the faster photopic component within the first eight minutes or so of the recovery, was called the Transition Phenomenon. It may point to a suppression of photopic activity by the scotopic. Its diagnostic significance became clear to us only when we found pathological cases where it regularly occurs after more than 8 minutes of the recovery (Fig. 7). The latter finding points to congential nyctalopia (night blindness) and to a special type for that matter since the patient ultimately reaches normal threshold, as reflected by the height of the negative and positive amplitudes at full recovery and substantiated by a psychophysical dark adaptation test (2).

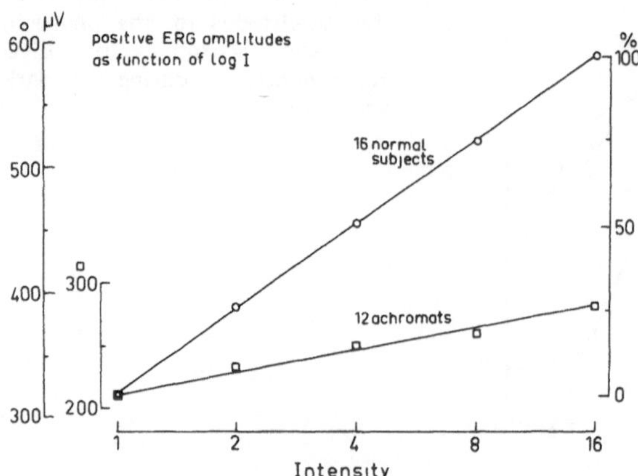

Fig. 5 Graph of ERG amplitude of positive wave as function log stimulus intensity in normals and achromats tested over a 1-2 log range (12).

Here, obviously standard values of the normal range of the negative and positive amplitudes at the steady state of dark adaptation are needed in order to be able to assess the ERG recovery during dark adaptation (Fig. 8). Only then are we able to estimate the so-called late receptor potential (negative wave) and the retinal function reflected by the positive complex. I will not go into further detail, but the criteria of the ERG test are quite reliable and serve us well. They serve also in cases in which the ocular media are not transparent so that neither fundoscopy nor examination of visual acuity is possible.

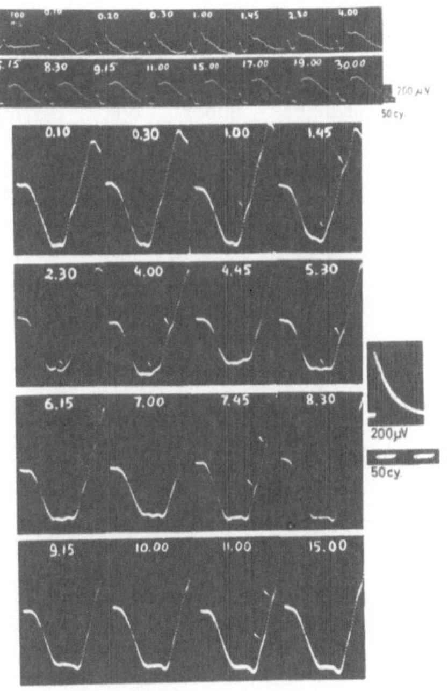

Fig. 6 The normal recovery of the ERG following a preliminary light adaptation. The Transition Phenomenon occurs here after 4 min in the dark. Two top rows: complete responses; below: enlarged negative wave and ascending branch of positive wave; numbers above each recording: min • sec in the dark. The arrows denote the oscillatory potentials and serve simultaneously to show the shift of the positive wave to the right together with the extension of the negative wave (2).

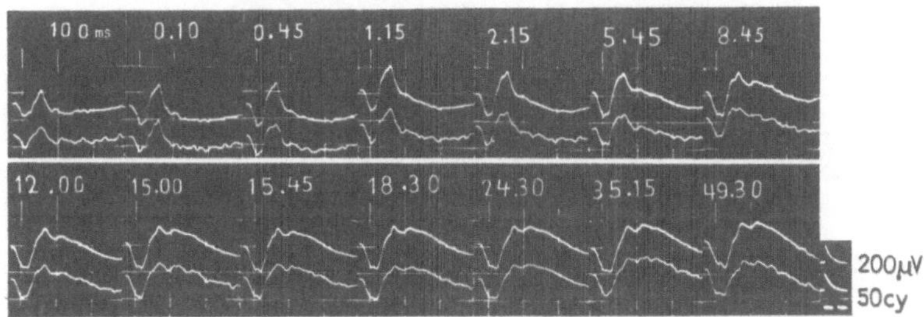

Fig. 7 The delayed ERG recovery of a congenital nyctalope: the Transition Phenomenon occurs here only after about 18 min in the dark and the amplitudes achieve belatedly normal values. Also characteristic for this type of night blindness is the dip between the photopic and scotopic positive components. Numbers above each recording: min • sec in the dark (2).

The two extremes in which there is no Transition Phenomenon found are total color blindness (achromatopsia) (12) (Fig. 9) and two other forms of congenital nyctalopia (13): the Riggs type (14) (Fig. 10) and the Schubert-Bornschein type (15) (Fig. 11). In the former, the scotopic component never recovers completely, while in the latter the positive scotopic and photopic components are either almost completely absent or very small, leaving a normal negative wave. This is clearly

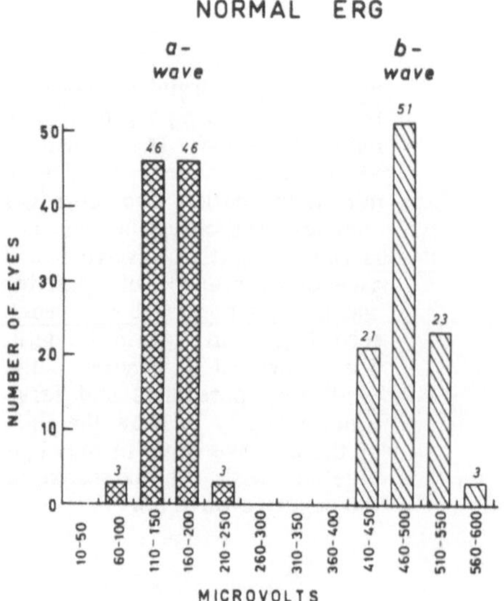

Fig. 8 Histogram of the normal ranges under our testing conditions of ERG amplitudes of negative and positive waves at the steady state of dark adaptation (16).

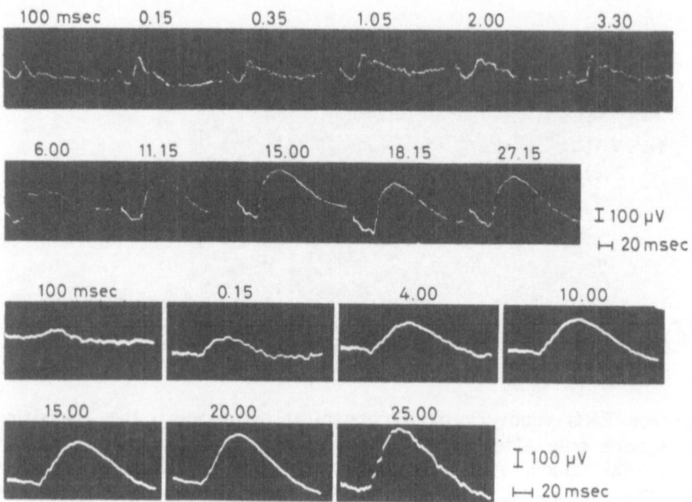

Fig. 9 Two top rows: normal ERG recovery; two bottom rows: ERG recovery in an achromat. Note here in contrast to the normal ERG the long extension of the shallow negative wave pointing at the missing photopic component and the long latency of the positive wave characteristic for the scotopic response. Numbers above each recording: min · sec in the dark (3).

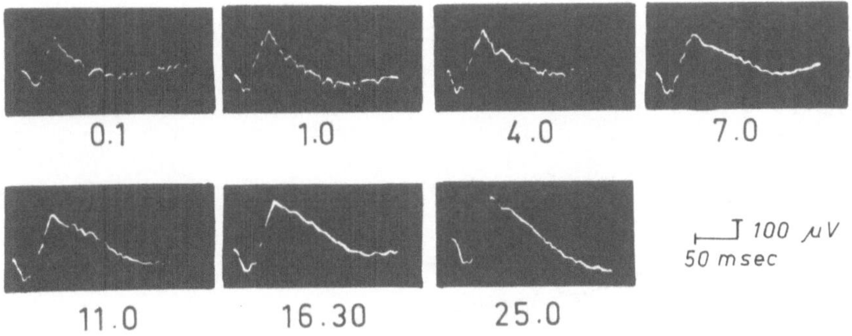

Fig. 10 The recovery of the ERG of a congenital nyctalope of the Riggs type (14). Note that the scotopic positive component appears as a hump at the descending branch of the positive wave and never exhibits the Transition Phenomenon. Numbers below each recording: min • sec in the dark (13).

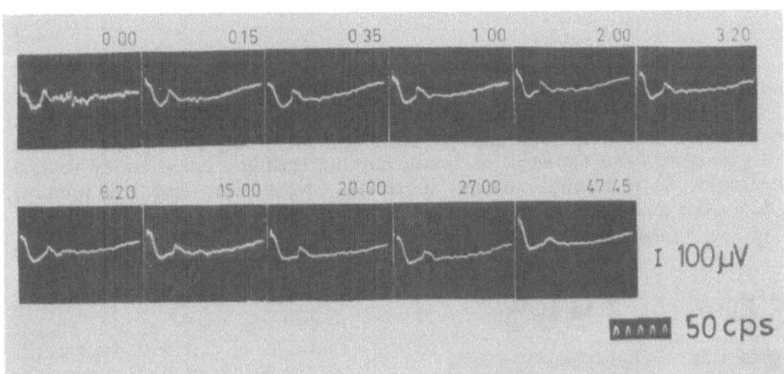

Fig. 11 Congenital nyctalope of the Schubert-Bornschein type (13, 15). Note the normal negative wave and the missing (a) or very small positive components (b). ERG recording and psychophysical dark adaptation wave of the same subject (points omitted in the psychophysical curve) (13).

demonstrable psychophysically (Fig. 12). Both types of nyctalopia are regarded as stationary. In this context, the finding that in some nyctalopic patients the condition was progressive, seems to me very important (13). Because of the serious connotation to RP, we now re-examine nyctalopic patients several times over the years as a rule. Compare Fig. 13 with Fig. 14, the latter showing an extinct response at the end.

However, the exclusive use of the ERG is valuable only in a restricted number of cases. Since vision in primates is mainly a cortical function of the brain, an

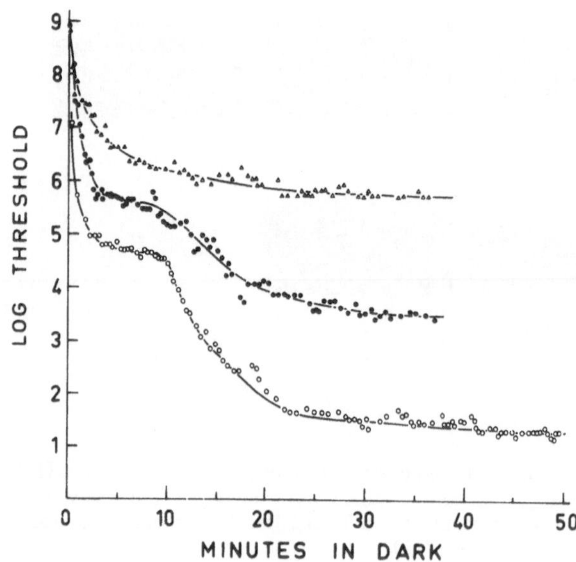

Fig. 12 Psychophysical dark adaptation curves of: 1) congenital nyctalope, Schubert-Bornschein type (top); note that there is still some scotopic function since the plateau is reached only after about 25 min in the dark. The normally occurring break between the photopic and scotopic phases is immeasurably small; 2) congenital nyctalope, Riggs type; here appears the break but the scotopic phase never reaches absolute threshold; 3) a normal subject (bottom). Note that also the photopic mechanism functions at higher threshold in the nyctalope (13).

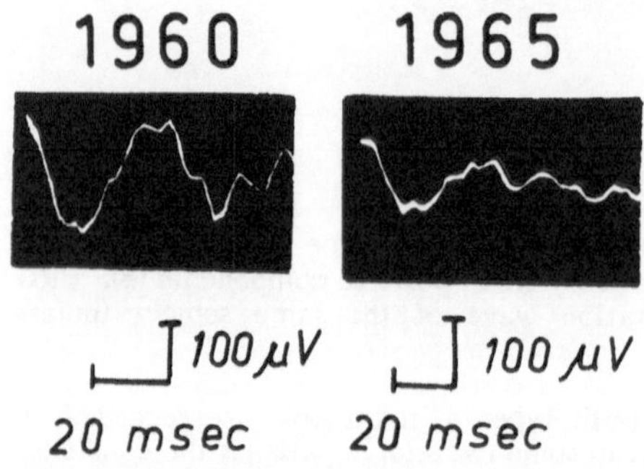

Fig. 13 ERGs in progressive nyctalopia found in a small percentage of patients (apparently of the Schubert-Bornschein type) recorded from the same subject under identical conditions in 1960 and 1965. We consider this a sign of danger (13).

examination which concentrates only on the retina would often be incomplete, e.g. in cerebral traumata of the visual system, and in many cases almost meaningless. A normally recovering ERG during dark adaptation can be obtained in the absence of afferent impulse propagation, the amplitudes of which may range from subnormal to supernormal (16) (Fig. 15). This may be due, e.g., to space-occupying lesions in the chiasma or other regions of the visual system (17), to retrobulbar neuritis, e.g. in

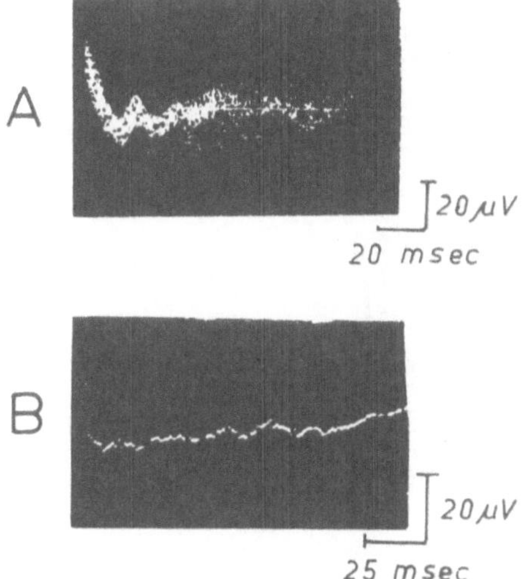

Fig. 14 ERGs in progressive "nyctalopia" recorded under identical conditions of the patient in 1960(A) and 1964(B). In contrast with Fig. 13, here we deal with a clear case of advanced retinitis pigmentosa, although in A the pattern is typical for Schubert-Bornschein nyctalopia. A: 70 recordings were superimposed on one frame of the film and display still a very small response (no CAT available in 1960). B: 70 responses were averaged by CAT showing that 4 years later the response was extinct.

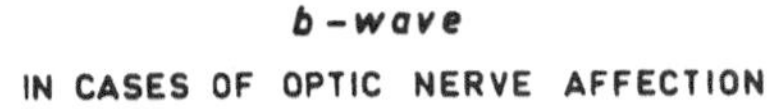

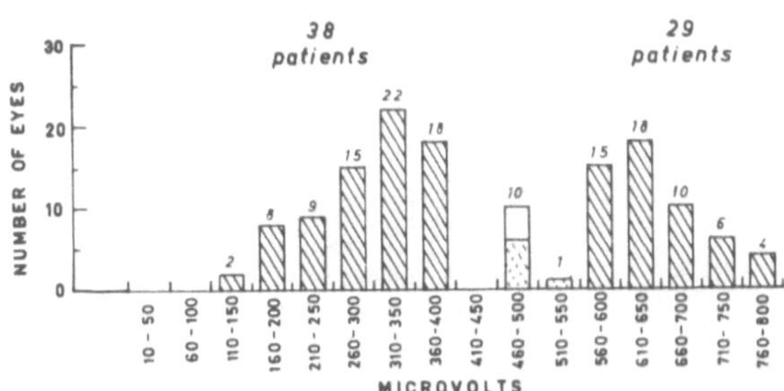

Fig. 15 Histogram of the distribution of amplitudes of the ERG b-wave in affection of the optic nerve (including multiple sclerosis.) Compare with the normal distribution of the b-wave in Fig. 8. The responses are either subnormal (up to 400 μV) or enhanced (above 560 μV). Two patients had normal ERGs (empty bars) and seven patients had a normal ERG in one eye (dotted bars) (16).

multiple sclerosis (16, 18, 19), or due to optic nerve atrophy or lesion (16) or transection (16, 20). Therefore the VEP is most important: it demonstrates the extent to which information, if any, reaches the striate cortex, and it indicates its velocity (normal latency of the VEP 15-20 msec as opposed to 2-3 msec in the ERG) (Fig. 16). Its importance is enhanced by the fact that the fovea is largely responsible for it; this is due to the great extension with which the fovea projects to the striate cortex (21). For example in RP at an advanced stage at which there is only foveal vision left, the ERG is extinct or very small, but a sizeable VEP is still recordable.

The VEP conveys then insight also in the condition of the retina and into what

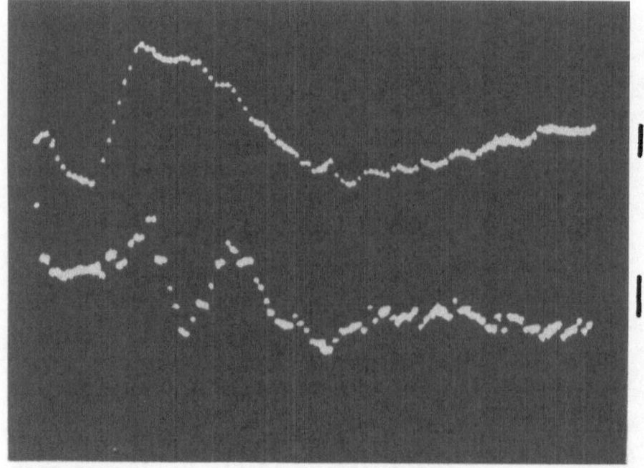

Fig. 16 Normal ERG (top): average of 10 responses showing the negative late receptor potential with a latency of about 2 msec; normal VEP: average of 100 responses at a rate of 1/sec showing the presynaptic initial positivity with a latency of about 20 msec. Calibration: 25 msec, 100 µV (top); 10 µV (bottom). We represent the VEP like the ERG with the positivity upward.

happens to the afferent impulses, such as in cortical blindness where the presynaptic initial positivity with normal latency is recordable and nothing else or, for example, in head injuries (20) such as the one shown in Fig. 17; here a fracture of the right frontal bone resulted in blindness in the right eye which could be diagnosed by the

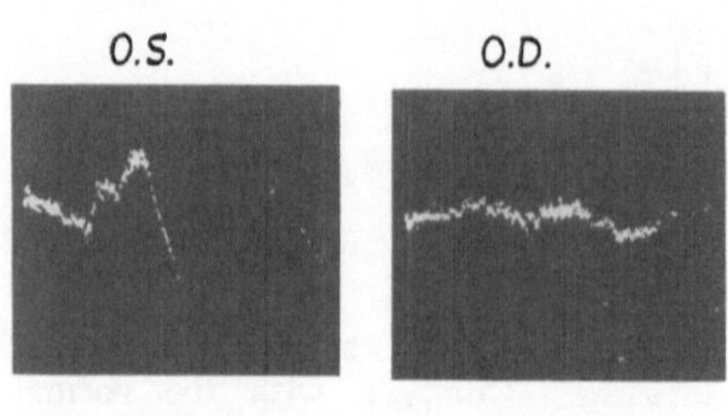

Fig. 17 Averaged VEP recorded by CAT from stimulation to each eye separately. The patient was stuporous. Note the practically absent VEP to stimulation of the right eye (OD) which was later found to be blind. One month later optic nerve atrophy became visible fundoscopically. The ERG was normal. Calibration: 10 µV, 25 msec (20).

VEP while the patient was still stuporous. There are other cases which show different degrees of optic nerve lesions up to transection (Fig. 18a,b,c). The VEP should be recorded following monocular and binocular stimulations. In the latter case the processing of the information may be shown by the very fact that the VEP following binocular stimulation is generally larger in amplitudes than any one of the VEPs following the two monocular stimulations.

Flashes are the most unsophisticated stimuli for such complex responses as the VEP. In spite of this, the flash-elicited VEP recorded by an averager is still a good method to clearly demonstrate a defect in the visual pathways, even a subclinical one

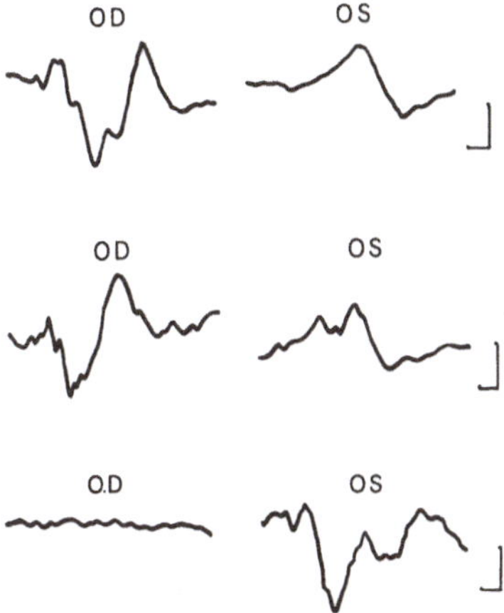

Fig. 18 Traces from averaged VEPs recorded by CAT. A, B: left (OS) partial optic nerve lesion showing in lengthened latency and small amplitudes; C: right (OD) optic nerve transection, as in Fig. 17, no impulses reach the striate cortex. Calibration: 5 μ V, 25 msec (41).

as in multiple sclerosis (18, 19, 19A) as well as in the visual cortex. They all show changes both in the latent period for the afferent impulses and of its amplitudes (Fig. 19), an information especially necessary in cases of multiple sclerosis and retrobulbar neuritis.

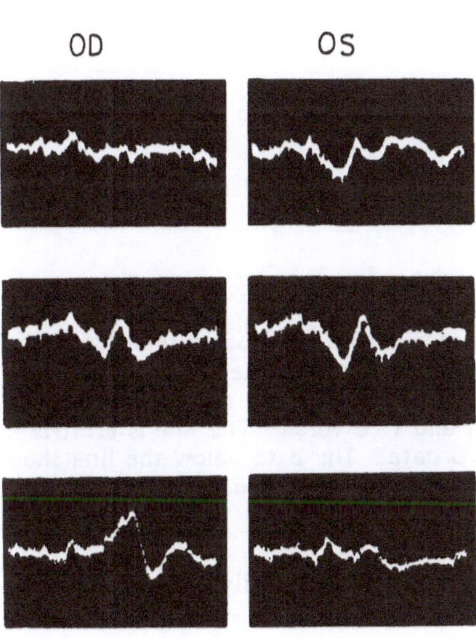

Fig. 19 Averaged VEPs in a patient suffering from multiple sclerosis. The diagnosis was confirmed when, following the retrobulbar neuritis, pyramidal signs appeared. Upper row: no visual symptoms in the left eye (OS) although the VEP is subnormal and the latency lengthened to stimulation of this eye; the VEP is by far more subnormal to stimulation of the right eye (OD) where a central scotoma was found. Medium row: the right eye has recovered after 3 weeks and the visual acuity is 20/20 in both eyes. However, there is still evidence of a subclinical conduction defect because of the abnormal VEP. Lower row: six months later the patient suffered from left retrobulbar neuritis; this is seen by the now very subnormal VEP to stimulation of OS. The VEP to stimulation of the right eye has improved but is abnormal and the latency is lengthened although visual acuity is normal. Calibration 5μV, 25 msec (18).

Another, more sophisticated physiological method for studying basic visual functions, such as contrast vision, is the pattern VEP by means of stimuli which are constantly in motion, e.g. an alternating checkerboard pattern (22, 23). However, this examination can be carried out only in a relatively small number of patients because it depends very much on the patient's cooperation and his ability to focus and attend to the everchanging checkerboard. In fact, even with the flash VEP it is advantageous if the patient can focus on the light source. For this purpose, all examinations should be carried out with corrected refraction.

The test with which we generally begin our series of examinations is the EOG (24-27), a test which does not exclude the ERG. There are some cases of a near-normal EOG and a very defective ERG. For instance several forms of congenital night blindness as well as a number of retinitis pigmentosa (RP) cases, but certainly not all, exhibit an easily measurable EOG and an extinct or very small ERG recordable only by means of the CAT or a similar averager. These cases of which I possess tentative evidence (Fig. 20) seem to contradict the assumption that the

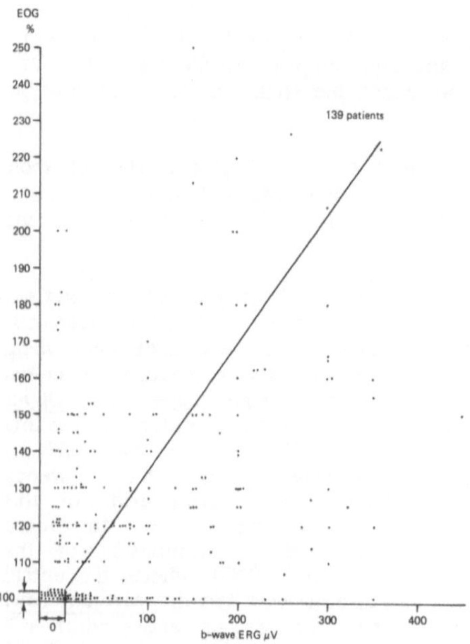

Fig. 20 Tentative result from 139 retinitis pigmentosa (RP) patients showing the distribution of the EOG values versus the ERG b-wave. While there are quite a number of data of extinct EOG and ERG, there are others with the ERG extinct or very small but still displaying a sizeable EOG and vice versa. The line is arbitrary, indicating an approximate fit of EOG and ERG data. The data below the line show the ERG too large for the small EOG. The data above the line show the EOG too good for the small ERG.

pigment epithelium where the EOG is mainly generated is primarily responsible for all cases of RP. Other cases where both the ERG and the EOG are extinct clearly show the disease to be very advanced. More about RP later on.

There are certain cases in which the pathology demands the verification whether visual pigments are present in the receptor cells. For this purpose there are two methods available: one is reflection densitometry (28-30) and the other the electrophysiological method of the ERP (31, 31A) the early receptor potential which measures at the molecular level. We ourselves use the ERP in such cases but we are neither fully satisfied with the results yet nor convinced of the clinical significance and practicality of the method. Therefore, we are about to introduce densitometry.

We regularly test color vision, using part of the Farnsworth battery, which also provides the possibility to distinguish tentatively between a congential and acquired color defect. However, when the patient fails these tests, the information on the photopic mechanism and foveal vision becomes complete only if large colored surfaces with a subtense larger than the macular area are examined as well. This makes it possible in certain cases to distinguish, for example, between macular degeneration and achromatopsia, both of which display the "scotopic line" with the Farnsworth D-15 test. If there is peripheral color vision, it is more likely that we deal with macular degeneration. However, this diagnosis has to be substantiated at least by combined ERG and VEP tests.

As mentioned, we had to exclude largely psychophysical adaptation and sensitivity measurements from the routine examinations since they demand too much cooperation and are, therefore, by far not reliable enough in the majority of cases. It is essential, however, to realize that the electrophysiological examinations often do not show the complete picture. This is reflected in our psychophysical study of achromatopsia (32). It is a rare dysfunction mainly identified by means of color vision tests and the ERG, and interesting results were obtained even by the pattern VEP which cannot yet be satisfactorily interpreted (33). We examined 39 achromats, and of these five were examined psychophysically. Their spectral sensitivity measured against background illumination displayed one maximum around 515 nm as compared to the three maxima of the normal subject (Fig. 21). The measurements of dark adaptation in different retinal locations, to our surprise, displayed up to three

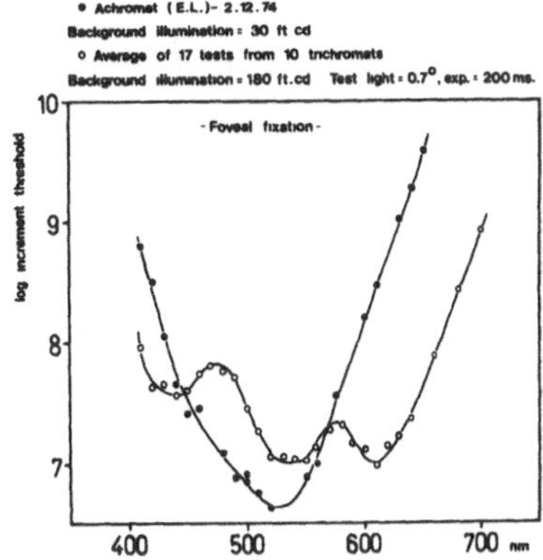

Fig. 21 Psychophysical measurement of spectral sensitivity against background illumination of 60 ft⁻cd in a complete congenital achromat showing maximal sensitivity of 525 nm as compared to our 3-peaked standard curve.

plateaus in those achromats (Fig. 22) during the normally photopic phase of dark adaptation in a circumscribed retinal area; this finding was the result of trial and

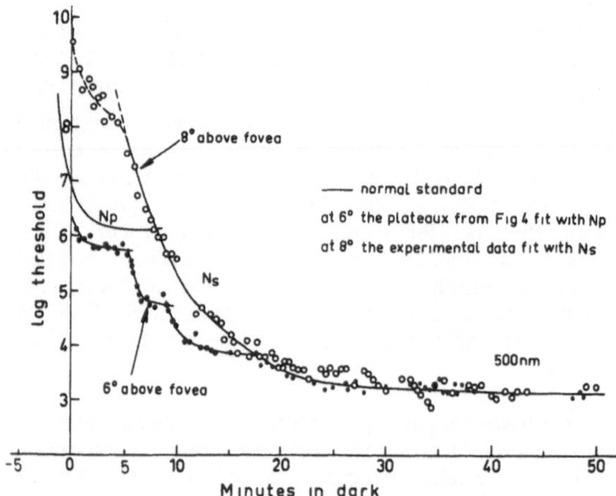

Fig. 22. Psychophysical curve of dark adaptation after a 3 min white preadaptation of 3.10^3 ft-cd in a congenital achromat. At 6° above fovea several plateaus were obtained partially during the normally photopic phase. At 8° above fovea, the scotopic branch could be followed over a range of about 5 log units. Note that the normal standard curve had to be shifted to the left for best fit because the achromat being photophobic did not get the same degree of preadaptation as the normal. Np: photopic branch of standard curve, Ns: its scotopic branch.

error (32). Outside this area, in one of the achromats it was possible to follow the scotopic curve above the normal photopic plateau (Fig. 22) which was shown before by RUSHTON (28). These plateaus were noticed at first in one patient only as small irregularities which were considered artifacts; they were identified years later in a different retinal locus as plateaus after we found plateaus also in three other achromats (Fig. 23). The spectral sensitivity curves, determined from each of these plateaus plus the scotopic one, had similar maxima between 500 and 520 nm (Fig. 24).

We have hypothesized that some small retinal areas contain cone clusters or remnants of cones which show the fast photopic kinetics but scotopic sensitivity. These cones, and for that matter possibly three different kinds of cones, may well be filled with the wrong pigment--rhodopsin. (32).

A typical example of an electrophysiological finding which is not understood and has thus become a research topic is the following. In 75% of 130 otherwise healthy young persons with intraocular foreign bodies (IOFB), no matter whether metallic, plastic, glass or wood, we noticed that the ERG recovery during dark adaptation displayed the typical pattern and course of Riggs nyctalopia. It was interesting that several of these patients complained, unasked, about nyctalopia since they had been injured. In order not to be misled by a technical artifact, we examined dark adaptation psychophysically in seven of these patients with normal visual acuity (Fig. 25). Six patients showed a significant delay in dark adaptation. Three of these

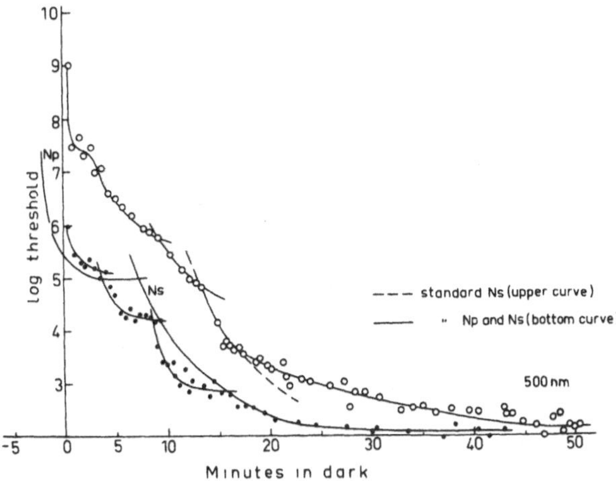

Fig. 23 Psychophysical dark adaptation in another achromat. At 6° above fovea the curve showed several irregularities (open rings). About a year later the other curve (dots) was obtained with the patient fixing at 6° above and 15° to the nasal side.

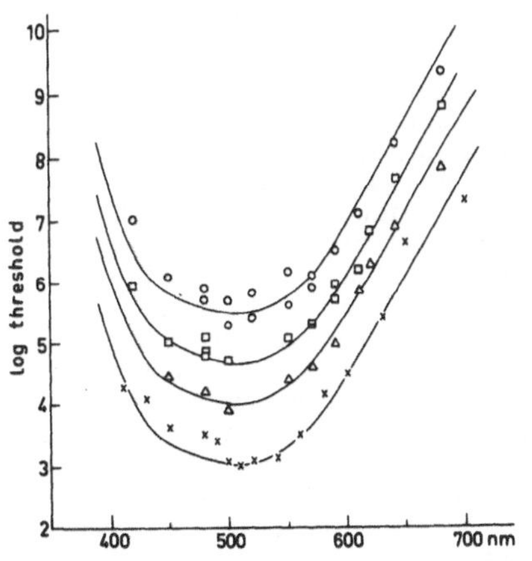

Fig. 24 Log threshold as a function of wavelength obtained from individual dark adaptation curves. In this achromat 3 plateaus were measured whose threshold was above the final plateau at full dark adaptation. The 4 continuous curves are the 1951 CIE scotopic function.

never reached the absolute threshold and the curve from only one patient tallied with our normal standard curve.

I mentioned retinitis pigmentosa several times. It is a subject which occupies us very much and which brings me to the topic of patient care and rehabilitation, especially necessary in hereditary diseases such as RP. In connection with institutions for the blind, we make it our task to counsel, to warn and to direct the patients into the right channels where they receive maximal help.

Years ago the extinction of the ERG was considered for a short period of time

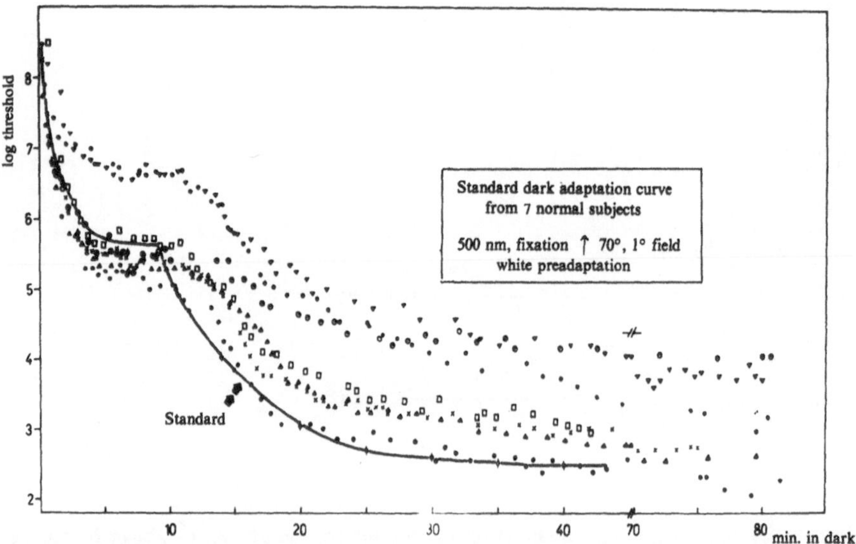

Fig. 25 Psychophysical dark adaptation measurements of 7 subjects of ages 18-23, each having an intraocular foreign body in one eye but still good visual acuity. The curve is our normal standard. Note that the measurements of only one subject tally with the threshold of the standard. Fixation 6° above fovea.

to be a "symptom" of RP. This is wrong, a fact which can be shown in responses to single stimuli and, very strikingly, even in very advanced cases by means of an averager. In some cases, the responses may display the Schubert-Bornschein pattern before they become extinct, as shown earlier in Fig. 13 and in Fig. 26, which latter could have easily been missed in responses to single stimuli (Fig. 27) which appear extinct. This makes our above mentioned finding of progressive congenital nyctalopia even more important. Due to technological progress, all stages of retinal responses can now be shown in RP from almost normal ERGs down to virtual extinction. (This is equally well achieved with cerebral responses.) We have examined a very large number of RP patients--more than 700 (see our Review on RP) (34)--and many of them were followed over years. In some we also have records for several members of the same family belonging to several generations. The accumulation of so much material dealing with this tragic disease demands intensive research, which we carry out both with our patients and by using an animal model for RP, the RCS rat, which exhibits an inherited retinal dystrophy appearing 18 days after birth. It is the best model available at present, although, as mentioned before, I have evidence (Fig. 20) to doubt that it fits with all the hereditary modes of the human disease.

With these animals we conduct a behavioral and electrophysiological (ERG) study in an attempt to quantify and qualify the effect of various environmental variables which may affect the progress of RP. We were able to confirm that light precipitates the anatomical degeneration which is recorded electrophysiologically (35). The more intense the light exposure, the faster the ERG becomes extinct.

Correspondingly, MARSHA KAITZ in our laboratory works with RCS rats reared in darkness from birth and obtained ERG amplitudes nearly four times those recorded

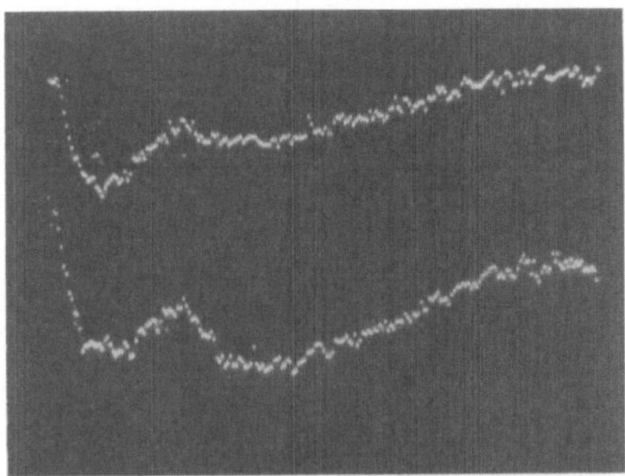

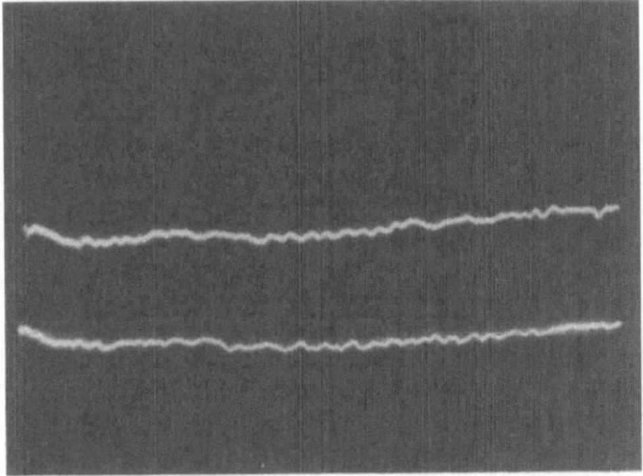

Figs. 26 & 27 ERG in an advanced case of retinitis pigmentosa measured at the steady state of dark adaptation. The conventional recording (Fig. 27) with single stimuli displayed responses which appeared extinct in both eyes (upper and lower trace). That this is not so was shown in the averaged responses (Fig. 26) to 20 stroboscopic stimuli which showed a typical nyctalopia pattern of the Schubert-Bornschein type in each eye of very small amplitudes (note: the normal range of the positive wave is between 420 and 550 µV and of the negative wave between 120 and 200 µV). Calibration: Fig. 26, 10 µV, 25 msec; Fig. 27, 100 µV, 20 msec.

in littermates reared in ordinary lighting conditions (Fig. 28). She also showed that rats reared in light and later transferred to darkness from the 15th to the 30th days of life show ERGs in adulthood equal to animals reared in darkness from birth.

Another clinical subject became a research project several years ago--functional amblyopia, a disorder which we saw in a large number of patients. In humans, electrophysiological (36, 37) and psychophysical experiments (38) and in cats microelectrode experiments (39, 40) were performed. In the latter we studied, among other things, artificial strabismus and light deprivation. It also led to studies of plasticity of the brain.

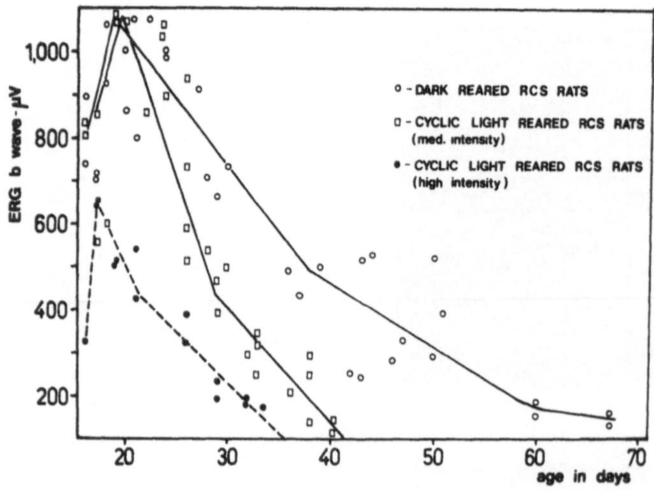

Fig. 28 The progressive deterioration of the ERG (b-wave) of RCS rats as a function of age and rearing conditions. Rats reared in complete darkness show a substantial savings over cyclic light-reared rats who in turn show larger b-wave amplitudes throughout life compared to rats reared in higher intensity cyclic light conditions.

This paper attempts to describe the scope and tasks of a clinically oriented vision research laboratory in the framework of a university hospital. It is essentially based on my own experience. The conditions under which research and examinations of the visual system in patients are compatible with each other are critically dealt with as well as the disadvantages and advantages if the dual loyalty toward patients and research is adhered to. Examples were given as to how the rich clinical material may be put to good use for the patient and for research. The point is especially stressed that the normal, physiological condition can be elucidated by searching for the causes of abnormal conditions, so that by the study of disease and dysfunction not only the underlying pathology is better understood.

References

1. J. G. Crowther, The Social Relations of Science (Macmillan & Co. Ltd., 1942).
2. E. Auerbach, Doc. Ophthal. 22, I (1967).
3. E. Auerbach, ISCERG Symposium, Ghent, 1966, (Karger, 1968, p. 162).
4. E. Auerbach, in: Textbook of the Fundus of the Eye, ed. by A. J. Ballantyne and I. C. Michaelson, (E. and S. Livingstone, London, p. 22, 1970, 2nd ed.).
5. E. Auerbach, in: ibid, p. 512.
6. E. P. Johnson & L. A. Riggs, J. Exp. Psychol. 41, 139 (1951).
7. E. Auerbach & H. Burian, Am. J. Ophthal. 40 Pt. II, 42 (1955).
8. P. Gouras, Invest. Ophthal. 9, 557 (1970).
9. R. Granit, Sensory Mechanisms of the Retina, (Oxford University Press, 1947).
10. R. Granit, in: The Eye, ed. by H. Davson, Vol. 2, (Academic Press, New York/London, 1962, p. 727).
11. R. F. Miller & J. E. Dowling, J. Neurophysiol. 33, 323 (1970).
12. E. Auerbach & S. Merin, Doc. Ophthal. 37, 79 (1974).

13. E. Auerbach, V. Godel & H. Rowe, Invest. Ophthal. 8, 332 (1969).
14. L. A. Riggs, Am. J. Ophthal. 38 Pt. II, 70 (1954).
15. G. Schubert & H. Bornschein, Ophthalmologica 123, 396 (1952).
16. M. Feinsod, H. Rowe & E. Auerbach, Doc. Ophthal. 29, 169 (1971).
17. M. Feinsod & E. Auerbach, Ophthalmologica 163, 360 (1971).
18. M. Feinsod, O. Abramsky & E. Auerbach, J. Neurol. Sci. 20, 161 (1973).
19. M. Feinsod & W. F. Hoyt, J. Neurol. Neurosurg. & Psychiatry 38, (1975).
19A M. Feinsod, W. F. Hoyt, W. B. Wilson & J. P. Spire, Arch. Ophthal. 94, 237 (1976).
20. M. Feinsod & E. Auerbach, European Neurology 9, 56 (1973).
21. R. G. Devoe, H. Ripps & H. G. Vaughan, Jr., Vision Res. 8, 135 (1968).
22. H. Spekreijse, L. H. van der Tweel & T. Zuidema, Vision Res. 13, 1577 (1973).
23. D. Regan, Evoked Potentials in Psychology, Sensory Physiology and Clinical Medicine, (Chapman & Hall Ltd., London, 1972).
24. P. Gouras, in: The Clinical Value of Electroretinography. ISCERG Symp. Ghent 1966. (Karger, Basel/New York, 1968, p. 66).
25. G. B. Arden, A. Barrada & J. H. Kelsey, Brit. J. Ophthal. 46, 449 (1962).
26. G. B. Arden & J. H. Kelsey, J. Physiol. 161, 189 (1962).
27. G. B. Arden & J. H. Kelsey, J. Physiol. 161, 205 (1962).
28. W. A. H. Rushton, J. Physiol. 156, 193 (1961).
29. W. A. H. Rushton, T. W. Campbell, W. A. Hagins & G. S. Brindley, Optica Acta 1, 182 (1955).
30. R. A. Weale, J. Physiol. 122, 322 (1953).
31. K. T. Brown, Vision Res. 8, 633 (1968).
31A. R. A. Cone & W. L. Pak, in: Handbook of Sensory Physiology, Vol. I, Chapt. 12, (Springer-Verlag, New York, 1971, p. 345).
32. E. Auerbach & B. Kripke, Doc. Ophthal. 37, 119 (1974).
33. H. van der Tweel & E. Auerbach, Doc. Ophthal. Proc. Series 11, 105 (1977). XIIIth ISCERG Symp. Israel, 1975.
34. S. Merin & E. Auerbach, Survey Ophthal. 20, 303 (1976).
35. J. E. Dowling & R. L. Sidman, J. Cell. Biol. 14, 73 (1962).
36. I. Nawratski, E. Auerbach & H. Rowe, Am. J. Ophthal. 161, 430 (1966).
37. U. Yinon & E. Auerbach, Invest. Ophthal. 13, 538 (1974).
38. E. Auerbach & R. Tsvilich, Doc. Ophthal. Proc. Series 11, 27 (1977).
39. U. Yinon, F. Jakobovitz & E. Auerbach, Invest. Ophthal. 13, 538 (1974).
40. U. Yinon, C. Shaw & E. Auerbach, Advances in Exp. Biol. & Med. 15, 41 (1975).
41. M. Feinsod, in: Head Injuries, ed. by R. L. McLawrin, (Grune & Stratton, Inc., 1976, p. 95).

Trends in Higher Education and Health: Opportunities for Optometry

Florence Kavaler, M.D., M.P.H.
Staten Island, New York

The word "crisis" is not new to either the academic world or the health field. The dictionary defines it as a "stage in a sequence of events at which the trend of all future events, especially for better or worse, is determined; turning point". Synonyms include exigency, strait, pinch. (Random House Dictionary of the English Language). The word when we use it in our fields does not seem to have that clear connotation of an acute turning point. Rather we experience concurrent and sequential chronic crises. My contention is that the identification and full exploration of these crises, with an intension of prescribing alleviation through responsive societal efforts is a positive force for change and is a valid reflection of our national attitude now and predictor of direction and movement for the future.

The role of the university in society has undergone much introspection and re-evaluation in recent years. As the American technocracy becomes more complex, the demand escalates for new and more appropriate responses from institutions concerned with preparing individuals to meet society's constantly changing needs.

The philosophic basis for educational responsibility in shaping the affairs of nations extends back at least to PLATO, culminating in NEWMAN'S monumental articulation of academic prerogatives and moral obligations, The Idea of a University. But until recently there had been no material updating of the educational processes underlying these prerogatives and obligations; and as a result, the development of American Higher education has lagged while other aspects of our cultural life have been advancing at a rapid, even frenzied, pace.

The restructuring of national priorities following World War II produced significant changes in the traditional curricular and structural apparatus. Academic elitism and idealism have gradually given way to more pragmatic conceptions of educational goals.

We have witnessed the active transformation of the institution of higher education from societal observer and sometime ex cathedra advisor to active participant in the life and values of the community.

Like every other area of our inflationary society, higher education is entering a period of increased financial stringency, and severe cutbacks have been experienced at all education levels.

The decline in availability of monies reflects the redirection in federal educational spending away from facility construction and institutional support and toward aid to individual students. Graduate education has been particularly hard hit; grants to departments and specialized centers have been slashed, and the rate of increase in federally sponsored research has slowed.

In its report of August, 1970, the PRESIDENT'S TASK FORCE ON HIGHER EDUCATION recommended as an immediate priority support for the education of health care professionals. As a result, the health fields have suffered somewhat less from the federal budget squeeze than other disciplines.

Most often educational institutions have been depicted as irrelevant and anachronistic, and in fact university interests have on more than one occassion been shown to operate counter to prevailing societal norms. Educators have consistently deplored, but accepted with benign resignation, this propensity of the university and the society it serves to work against each other.

However, the professional health related service orientation of higher education in disciplines such as optometry operates at a distinct advantage here, for the field is able to capitalize immediately on its stated goals by involving the community in the educational, clinical and research process. The direct relationship of health students and faculty to society via fieldwork, consultation, and research in community problems can, if properly exploited, mitigate to a great extent the prevailing suspicion and antagonism of the layman.

Financial pressures have brought into relief the internal power struggle endemic to higher educaton. The balance of power among administration, faculty and students is rapidly changing. A survey of college and university presidents conducted by the Carnegie Commission in 1968-69 revealed an overwhelming recognition of increased student and faculty control as the single most significant change of the last decade.

As if in a deliberate attempt to offset this revolution in internal governance, there has arisen a counter trend toward the concentration of power within a nucleus of professional administrators promoting operational efficiency. Faculty, the "deliverers" of instruction, traditionally propose, but it is the administrator who disposes; and these two factions often perceive educational priorities quite differently.

The ultimate resolution of this potentially creative tension is of particular importance to the health field, for the continued expansion of the health science complex at multiple campus sites under varying auspices depends upon the existence of a viable credible leadership. The model here might be among the University of Texas, Baylor and the University of Houston. Mutual trust, reinforced by an effective communications network between administrative and instructional personnel, must be developed and nurtured if the complex is to coordinate its activities successfully.

The enrollment explosion of the 1960's produced a doubling of undergraduates within a ten-year period (from 923,069 in 1960 to an estimated 1,885,000 in 1971), and a quadrupling of graduate enrollments between 1950 and 1970. Already the slow down and in some cases the reversal of this trend is becoming apparent. Undergraduate enrollment is expected to have peaked temporarily, and an actual decline is anticipated during the 1980's.

There has been an unusually rapid rise in enrollment for public institutions of higher education. This shift to mass public higher education is due to a variety of socioeconomic pressures which need not be elaborated here. Suffice it to say that as the prevailing enrollment pattern has shifted from private to public institutions, governmental funding and the distribution of public financial support has followed.

In recent years the federal government has directed its considerable educational resources to the development of graduate programs in health. This outpouring of federal support, in concert with the need for health personnel and the job market's demonstrated success in placing graduates from health programs, has stimulated institutional support. As a direct consequence of this chain of encouragement, enrollments in educational programs for the health professions have increased disproportionately to those of other disciplines at virtually every level.

Changes in the job market have had enormous implications for higher education. During the past decade the baccalaureate degree has declined sharply both in prestige and earning power, while the availability of graduate and professional credentials has accelearted disproportionately to actual need, witness the current glut and oversupply of Ph.D.'s in the arts and humanities and other fields.

This situation implies several caveats for higher education generally, and for graduate and professional education in particular. It is clear that the current wholesale injection of "nonessential" personnel into the job market is in great measure due to the failure of academic institutions to take into consideration the effect of their actions on the economy. Responsibility for providing a curriculum and setting professional standards relevant to the actual needs of society (specifically, of the job market) devolves squarely upon the university.

The rapid rate of proliferation of new knowledge has widened the gap in its ability to be applied to people at the time and place where necessary. It is in response to this profound manpower crisis that the direction of resource allocation was made to create this new academic environment for learning and research in the field of optometry.

Sociologists point out to us, many times over that health is not a primary concern to most of us - certainly not amongst those with most health needs and least access to it. Health ranks fourth after food, shelter and job, all of which have an impact on health.

The current health care scene is another exhibit of constant chronic crisis-most of which is characterized as "financial" rather than depicted as chaotic, fragmented, uncoordinated, unpredictable and unresponsive to peoples needs. The primary focus appears to be on illness and its treatment with government increasing its involvement in financing, regulating and assuring or appraising quality of services.

There has not been a national health policy - an operable public or social policy about health, nor any part of accountability in the system, nor professional health leadership acknowledged by all. We have bits and pieces of legislation to help the medically poor, or people with certain diseases or disabilities, some help for the elderly, some manpower legislation and attempts to control environmental hazards and keep the work force safe. However, more curative wonder drugs and surgical and technological feats for certain illnesses and disabilities have not daunted the non-traditional methodologists who prescribe and proclaim (transactional analysis, yoga, massage, naturopathy, acupuncture, folk medicines). Changing patterns of illness,with more chronic disease and disability, and more mental illness and emotional disturbances, have not created professional enthusiasts, or altered our perceptions of long-term care institutions as warehouses and storage bins for people rather than "health maintaining environments." The health problems of depressed inner-city and rural residents and the meager concerted societal efforts to alleviate poverty seem to ensure that the disadvantaged will remain that way.

Perhaps this litany frightens you - or turns you off, as they say. However, it is important for us all to recognize and acknowledge that which is happening. We cannot afford to be a sideliner, a watcher, a critical observer but non-participator.

There are trends that are emerging that need your support and more importantly your participation.

All sources of predictions indicate that the cost of health care will continue to rise with the federal government having an increasing burden of the total dollars. The prospects of a national health insurance coverage for various groups of people will probably emerge along with the growth of the private insurance sector. While this can occur, as did with Medicare, as solely a fiscal response, unless there are alterations in the distribution of manpower and the delivery of services there will be little effect on the people and minimal impact on cost.

The increasing governmental stress on local planning - via the Health Systems Agencies - will aid in new facility construction and new services but may have no voice in the containment of cost potential of closing excess institutional beds or location or relocaiton of manpower and services.

The public regulatory mechanisms of quality review of care and practitioners services - via the PSRO mechanisms - is an acknowledgement of the right to independent review but the "1984 Big Brother is watching you" concerns of professionals are at a counterpoint to professional choice of isolation and solo uncontrolled professional activities.

The emerging solutions to financing health care, planning new services to meet needs and quality assurance mechanisms imply professional equality of participation in the processes, and a private public interface which should be approached aggressively as an opportunity for enhancing the profession rather than being viewed as professional sefl-interest. The critical analysis of professional activities is a fertile field of exploration, investigation and research. For example, in the optometric field (1) expansion of early detection rather than emphasizing treatment, (2) new modalities of delivering care in group practice and interdisciplinary settings, (3) participation in health insurance prepayment plans, (4) active interest in quality review and methods of assessment.

The explosion in information and technology of the past two decades has forced professionals to reassess their functional effectiveness. As a professional response the demand for continuing professional education is growing, particularly in the health fields. In addition to the manifest problem of updating professional knowledge in areas whose information base is constantly expanding, new conceptions of professional roles and functions are emerging. In every discipline the professional is confronted with new questions of ethics, personal privacy, escalating specialization, and interdisciplinary approaches to problem-solving.

Perhaps because of the obvious need for health professionals to keep abreast of new developments in their respective fields, continuing education has probably been more thoroughly accepted and more highly regarded by the health professions than by any other vocational field.

Institutions for professional study are beginning to intensify their outreach programs in an effort to tackle the problems of maintaining effective health manpower and filling new consumer demands for knowledge.

Several of the health professionals already require some form of continuing education as a prerequisite for licensure and recertification; and as more states move twoard mandatory licensure and re-licensure in the health sciences it is likely that the importance of continuing education for other professions will increase as well.

There has been a dramatic reversal of attitude toward continuing education by university faculty. With the popularization of continuing education programs by major governmental and professional groups, and the gradual recruitment of distinguished faculty members into special seminar, conference and short-term programs, the prestige of continuing education with the academic community has increased considerably and has gradually been recognized by the higher education establishment as a valid professional interest.

Continuing education can and should be articulated with other elements of professional education. Moreover, it is vital that the concept of lifelong professional learning to be integrated into the total structure of the basic professional preparation.

Although health professionals are highly skilled and equipped with modern scientific knowledge and a rational basis for practice, the armamentarium against illness and disability is not always met with enthusiasm by the increasingly aware public.

There are several socio political trends which continue to mitigate against the professional accomplishment of professional. Vis-a-vis the changing roles of males and females, the high mobility and restlessness of the population, the widening gap in income and distribution of wealth, the increase in leisure time and declining commitment to work, all enhance the feelings of depersonalization, isolation, independence and apparent aloofness to professional encounters. Self damaging behavior (i.e., alcohol, drugs, suicide, speed driving) is increasing and there is societal neglect of the adolescent maturation process, the aged, and the severely physically or mentally handicapped, and a lack of acknowledgement of the basic need of a family structure. Health cannot be achieved without the active understanding and participation by individuals in their own promotion and maintenance of a healthy status.

The simple dissemination of faculty information has not succeeded in bringing about the desired behavioral changes necessary (i.e., screening in glaucoma with mass numbers of people asking for testing). Similarly, massive smoking cessation, etc. What people do is not based solely on knowledge but on their attitudes towards health, health services and the health professions. The apathetic response to scientific data and facts may be related not only to the sociodemographic variables (such as age, sex, ethnic identification, occupation) but also to the psychosocial milieu engendered by factors such as alienation and stress. A sense of personal control may be necessary to make knowledge motivationally relevant.

Our professional education process in the health disciplines should heighten the awareness of the intricacies of that special relationship - the doctor patient interaction - so that effective behavior is manifested on both sides. An aggressive exposure to the social and behavioral sciences during the academic phase of professional development will enable health professionals to maximize each interaction not only with individuals but also with identified social groups.

Higher education must reflect the growth of the American Health industry and the need for people devoted to human services. One must acknowledge the responsibility for training practitioners as well as researchers and the emphasis on utility and breadth of study.

Interdisciplinary activity is becoming more common and more consistently sought by students and institutions. Closer and more meaningful ties are being encouraged among the members of the academic power triangle (students-faculty-administration); and attempts are being made to define and place in perspective the precise relationship between institutions of higher education and the society within which they function.

Perhaps at this dedication symposium of this new facility we have tangible evidence that innovation and change are possible and that in essence, the values of the community are gradually becoming those of the university; and in educating professionals to enter a real rather than idealized society, higher education is at least conspicuously beginning to recognize the true depth and breadth of its task.

The Future of Public and Community Health in Optometry

Alden N. Haffner
State College of Optometry
State University of New York

The rubric, public and community health, implies a variety of policy concerns for the profession and for optometric education. Issues such as manpower, health economics of eye care, health services delivery models, interdisciplinary integration, health services research, eye care consumerism and quality care maintenance and review are among those that impact upon optometry's structure. The subject areas that relate to these and other issues are of concern for the development of curricula in the schools and colleges of optometry. This paper will explore possible developments in the profession during the period of the next generation (to the end of the century). The forces likely to play a role in shaping the profession will be discussed in the context of each of the topical areas mentioned above. That these areas have implications for curriculum development will be evaluated.

Though there was a scattering of individual interests in the subject of public and community health in the 1930's and 1940's, very little was done on an "organization level" until the late 1950's with the creation of the Committee on Social and Health Care Trends of the American Optometric Association. Originally conceived to view the narrow issue of third party payment mechanisms, its focus quickly broadened to consider the impact of public policy forces, particularly the growing role of the Federal Government in health matters. 1962 is the Year that the A.O.A. (not without some considerable internal opposition) changed the name of the committee to that of the committee on Public Health and Optometric Care. The archivists and historians may quibble but, in this author's opinion, the official introduction of the term public health marked a significant and irrevocable organizational shift of serious concern toward the subject. It was quickly followed the next year by the establishment in the American Academy of Optometry of the Section on Public Health and Occupational Vision. This author, along with a small but zealously dedicated cadre of "public healthniks", takes pride in those significant steps.

That public and community health had finally been raised to a level of visibility permitted several national conferences to be held under A.O.A. auspices. These pioneer meetings helped significantly to raise both the academic conscience and the profession's responsibility level. By the mid-1960's, all of the schools and colleges of optometry had installed, to some extent, at least a survey course in public health. Several institutions additionally pursued special topic areas of particular public health concern to the profession. In 1964, one institution developed a "second level" course in public health which this author was privileged to teach for seven years.

The 1960's not only witnessed developed organizational relationships between optometry and public health, but a more lasting and critical development began. Young and talented graduates of the schools and colleges of optometry, men and women, were encouraged to pursue formal graduate studies in the nation's schools of public health and in the schools of public administration. While the number has not reached 100, this corps of dedicated professionals has fostered, in a competent and professional way, the interpretation of optometry and optometric care in a public health context--and both disciplines have significantly benefited. That this development should continue to be encouraged, fostered and significantly expanded needs little elaboration. But what does need a comment is that this professional group, a very special cadre in optometry, now must also turn its professional energies toward the bureaucracies of governments, at all levels. This is necessary in order to project, in a formal way, the rationale of the profession of optometry into the decision making and policy formation and administration of government bureaus and agencies concerned with health affairs. Optometry must be "on the inside" and the public health optometrist is the most capable person available to carry out this mission. It is a role critically needed to be fulfilled and it has enormous growth and professional potential for service in the public sector.

Higher Education for Public Health is the 1976 report of the Milbank Memorial Fund Commission whose members carefully studied the subject for the period of three years. Chapter 11 of the report discusses the subject area related to the various health professions schools. The following two paragraphs pertain to schools of optometry.

> "A national study of optometric education in 1973 recommended that 'the curriculum of schools of optometry should give greater emphasis to . . . public and community . . .' among the other areas needing strengthening (Havighurst, 1973). The Commission surveyed all 12 schools and colleges of optometry, and concluded that there is need for more instruction in public health and community medicine. Most of the full-time faculty for these subjects have only master's level degrees from schools of public health.
>
> Little knowledge of public health is expected from a student when he completed his academic education at a school of optometry. Public and community health is buried in the 'Social, Legal, Ethical, Economic and Professional Aspects of Optometry' section of the examinations given by the National Board of Examiners in Optometry. More and more optometrists are now active in administrative planning and design of vision care programs in official health agencies."
>
> The chapter concludes:
>
> "Recommendation
>
> 33. The curriculum in schools of the health professions such as nursing, dentistry, veterinary medicine, pharmacy, optometry, and podiatry should be strengthened, with improved and extended instruction in the elements of the knowledge base of public health relevant to the practice of their respective professions. This should include instruction in the measurement sciences of public health, environmental health concerns, the organization, delivery and

evaluation of health services, and field experience in public health programs."

With a major education report of such recency, it is entirely propitious for this subject to be explored.

It is often said that education looks ahead. Indeed, if today's students are tomorrow's leaders, then they must be educated in order to be able to cope with the complexities and issues likely to be encountered. In the absence of a crystal ball, some elements of the optometric and health care conditions of the next generation, to the year 2000, can be postulated. They are:

1) The planning process in all phases of health care including optometry will permanently be installed as a fixture of public policy determination at all levels of government and community involvement;
2) The "controls" for decision making processes at all levels of public policy determinations will substantially be directed and decided by the consumers of health care with only that input needed as the expertise of the providers;
3) Health care in the public sector will be the dominant force and health care in the private sector will have a progressively decreasing base;
4) The community health enterprise will show progressive efforts to evolve into a public utility;
5) National health entitlement through a system of tax supported mandatory federal insurance will have evolved into a comprehensive entity;
6) Group and clinical entities, free standing and hospital affiliated, of single disciplines and of interdisciplinary character, will displace solo private offices as a growing and forceful movement of health services delivery models;
7) Women in optometry will constitute 50% or more of the classes in the schools and colleges of this profession and of the other primary care disciplines;
8) Quality standards and quality review will be accepted and ongoing mechanisms for "control" of quality assurance accountability in all disciplines including optometry will become institutionalized;
9) Institutional forms of public information dissemination will become more pervasive in all aspects of health care replacing advertising in the commercial sense;
10) Diagnostic drugs will be utilized in routines throughout all of the states as part of the authorities of optometry and the use of therapeutic drugs for the treatment of diseases of the eye will be fairly though not totally disseminated;
11) There will be a significant reduction in the numbers of ophthalmologists relative to optometry and optometric manpower numbers will remain somewhat static at current numbers with significant relative increases in numbers and utilization of ancillary personnel;
12) Optometric care and services will be more cost controlled and cost effective along with other forms of health care services;
13) Variance in costs of ophthalmic materials and dispensing will be more cost controlled in the public sector;
14) A major federal initiative will have been undertaken to change substantially the present methods of licensure and professional regulation in order to introduce the concept to make institutions more responsible for licensure;
15) Geographic distribution of health manpower will become exponential of the factor of governmental support of increasingly rapid rises in tuition which will be waived and/or more substantially tax supported;
16) There will be an open movement to limit control and even employ directly-by government of health personnel;
17) Great emphasis will be placed on prevention, health education and healthy life styles; and
18) The concept of licensure maintenance through a system of evaluated continuing competence will become institutionalized.

In order to relate the above enumerated policy concerns in health care and in optometry, a series of five generalized core, areas are, herewith, proposed as constituting appropriate groupings for tracts or courses in public and community health education in an optometric curriculum.These five subject area groupings respond to and confront the problems cited by the Milbank Commission and the recommendations made in its report.

Conceptualization of each of the five core areas of study is suggested as follows:

I. Generalized Introduction to Public and Community Health

A basic course content in health policy should provide an historical overview of public health and publicly financed programs including Medicare and Medicaid. Consideration of determinants of demand for health services should, among others, include socio-economic and demographic characteristics. An analysis and overview of the scope and cost of health services and the extent and potential growth of health insurance should be explored. Analysis of social factors in the delivery of health care services and the contribution of medical care and optometric care should be evaluated.

This basic course content should be constructed:

1. To provide students with insights into health maintenance, concepts of illness, disease and disability so that, as health professionals, optometrists can understand and appropriately deal with health related problems;
2. To introduce students in optometry to the social and political structure of the pluralism of United States health care systems and to those agencies and institutions responsible for the delivery of health care services;
3. To clarify the interrelationships, roles, and functions of those involved in the delivery of health services and to explore the methods by which they relate to changing social, economic and political forces;
4. To recognize the critical importance of social and behavioral factors as determinants of health and illness and the place of optometry and vision in this schema;
5. To recognize the role of public and/or private resources and of growing governmental involvement in the financing and regulation of health services and of health policy;
6. To define the relationships among hospitals, neighborhood health centers, federal/state and municipal agencies, and voluntary and proprietary institutions so that as primary care input health professionals the O.D. may act as patient advocates referring the patient to appropriate sources of services;
7. To recognize the inequitable distribution and lack of quality of care provisions in health services and the consequent impact of health services on health status; and
8. To teach principles of comprehensive health care and to demonstrate a health care system that emphasized the team approach and health maintenance through interdisciplinary integration without subversion of the peer and primacy principles of the profession of optometry.

II. Public Social and Economic Policy in Health Care

This course content should present an analysis of social, economic and political influences on the pluralism of the American health care system. Comparisons should be made of various health care organizations and finance mechanisms. Consideration of the social and economic effects of prepaid group practice, health maintenance organization, national health insurance and the reciprocal impact on federal/state policies are key areas to be explored. Topics must include neighborhood health centers, transitional regional medical programs and comprehensive health planning, health manpower issues, health care financing systems, methods of reimbursement, public intervention through financing and public imperatives of cost escalation and cost effectivity.

This basic course content should be constructed:

1. To recognize organizational problems and potential alternatives to the current delivery of health services when viewed as social and economic policy issues;
2. To develop an awareness of federal/state health legislative initiatives and developments and newer governmental policy as they affect optometry;
3. To identify new directions in federal/state health financing and potential policy implications of federal health insurance proposals;
4. To provide students with an understanding of the growing complexities of social and economic policies pertaining to health resources and health manpower; and
5. To recognize the advantages and disadvantages of different health delivery and financing models including fee-for-service, pre-payment, indemnification, group practice arrangements, medical foundations, health maintenance organizations, neighborhood health centers, and tax supported mandatory social insurance.

III. Biostatistical Methods, Epidemiology and the Epidemiology of Vision Problems

The principles of biostatistics and quantitative methodology should be related to the concepts of epidemiology and demography. Students should be exposed to the application of epidemiological principles and methods in optometry in relevant exercises. An overview of the present knowledge of the distribution of vision and health problems in various populations should be explored with due emphasis on sociological determinants, research design analysis with related epidemological implications, both descriptive and quantitative.

The content of this course should be constructed:

1. To enable the practical applications of basic principles of epidemiology to problems of vision in the practice of optometry;
2. To emphasize as a methodologically oriented course the strategies of data collection and analysis;
3. To enable students to develop a more realistic practitioner understanding of complex interactions among social, psychological, economic, environmental and biological factors;
4. To develop an understanding of the concept of multiple etiology and to recognize the importance of the "web of causation" in presenting, alleviating and eradicating health disorders including the affectations and disabilities of vision;
5. To provide students with methodological tools to design, initiate and evaluate epidemological studies including those related to vision and optometric care; and
6. To prepare students to do empirical research, formulate testable hypotheses and to set up study and control groups in vision care investigations.

IV. Public Health Methodology and Optometric Community Health Practice

This key course content area should provide understanding of the application of community health principles and methods to optometry. The policy, organization and implementation of optometric participation in multidisciplinary community health delivery systems should be reviewed. Participation in community health planning, consumer health education, screening programs, and the administration of optometric programs, clinics and outreach projects need presentation with viable and realistic examples.

Course content should be constructed:

1. To develop a positive attitude toward comprehensive patient care and to the health team approach with an intensive relatedness to optometric care;
2. To create a flexible and adaptable attitude toward different forms of current health delivery and possible future forms of delivery of primary, secondary and tertiary care and their implications for optometry;
3. To introduce students to the administration of optometric programs and community health organizations and to relate them to current forms of optometric care delivery;
4. To describe optometry's present and potential roles in the delivery of a total public health program and optometry's function in an interdisciplinary health delivery system;
5. To recognize the role and function of consumerism and community involvement and direction in health care planning and the delivery of health care services; and
6. To orient students to the evaluation of health services and to the mechanisms used to insure quality of care with particular emphasis on optometric care and services.

V. Optometry in a Social Context as an Applied Life Science

This course should provide seminar interaction and should be concerned with those issues, concepts, ideas, and social values which relate to the construction, development and performance of the optometrist in relation to his occupational and life systems. The course should examine the special social, legal, ethical, professional and scientific obligations of the optometrist as a licensed professional.

The content of this course should be viewed:

1. To examine optometry as a licensed discipline and whose practitioners have leadership roles in health care and as social change-agents in the community;
2. To evaluate the impact of the optometrist as a leader in the community and one who changes the behavior, performance, potential and well-being of the people with whom he/she comes in contact;
3. To examine the special social, legal, ethical, professional and scientific obligations of the optometrist as a licensed professional which marks him different from other persons by virtue of the special responsibilities and authorities which are inherent in licensure; and
4. To provide the student with an understanding of the many vantage points, other than that of the optometric discipline, when looking at human problems which are encountered and they should be viewed from standpoints of those in social work, higher education, administration, public health, law, behavioral science and others.

Social change is inevitable, constant, flowing, made by mankind and directed by it. Nowhere are the social forces more in evidence than those vividly relating to health care. Optometry is changing with dramatic intensity and so is the health care milieu of which the profession is an integral part. Conceptually, these are two areas in a state of social change, one within the framework of the other. Their interactions are complex and intricate. Perhaps the most pervasive view of the role and status of health care in the changing American society and in its value system stems from the vantage point of public and community health. To deny optometric education, and thereby the future of optometry, the opportunity to be at that "cutting edge" would be to deny an essential and critical ingredient for the profession's growth, development, evolvement and maturity. More serious, would be the potential loss to human visual welfare. With this essential understanding, optometric education bears a substantial and inevitable responsibility to the future of the profession.